5th International Seminar on Aerospace Science and Technology (ISAST 2017)

Development of Aeronautics and Space Technology to Support Maritime Applications

Journal of Physics: Conference Series Volume 1005

Medan, Indonesia
27 - 29 September 2017

Editors:

Soleh Fajar Junjunan
Afrido Prayogi
Sridana Windya

ISBN: 978-1-5108-6286-9
ISSN: 1742-6588

Printed from e-media with permission by:

Curran Associates, Inc.
57 Morehouse Lane
Red Hook, NY 12571

Some format issues inherent in the e-media version may also appear in this print version.

Printed by Curran Associates, Inc. (2018)

For permission requests, please contact the Institute of Physics at the address below.

Institute of Physics
Dirac House, Temple Back
Bristol BS1 6BE UK

Phone: 44 1 17 929 7481
Fax: 44 1 17 920 0979

techtracking@iop.org

Additional copies of this publication are available from:

Curran Associates, Inc.
57 Morehouse Lane
Red Hook, NY 12571 USA
Phone: 845-758-0400
Fax: 845-758-2633
Email: curran@proceedings.com
Web: www.proceedings.com

Table of contents

Volume 1005

5th International Seminar on Aerospace Science and Technology

27–29 September 2017, Medan, Indonesia

Accepted papers received: 9 April 2018
Published online: 4 May 2018

Preface

Papers

IOP Conf. Series: Journal of Physics: Conf. Series **1005** (2018) 011001 doi:10.1088/1742-6596/1005/1/011001

Preface

International Seminar on Aerospace Science and Technology (ISAST) is yearly seminar sponsored by Indonesian National Aeronautics and Space Agency (LAPAN), to publish the results of research and development the aerospace science and technology conducted mainly in Asian region. The seminar has been held since 2013, therefore in 2017 is its fifth.

The theme of this year seminar is Development of Aeronautics and Space Technology to Support Maritime Applications. The topics covers aerodynamics, astrodynamics, propulsions, light structures, propellant, energetic materials, avionics, flight dynamics, guidance & control, rocket technology, satellite technology, aircraft technology, flight sensor, and UAV applications. For such topics, seven distinguished scientists become the keynote speakers for the two-days seminar, i.e. Prof. Dr. Erna Sri Adiningsih from LAPAN representing Remote Sensing Satellite technology, Prof. Dr-Ing. Nicolas Gascoin from France representing rocket propulsion technology, Prof. Yoshifumi Inatani from JAXA, Japan, representing interplanetary space mission, Dr. Rais Zain from Institut Teknologi Bandung(ITB) representing aircraft design, Prof. Qin Shiyin from Beihang University, China, representing Guidance and Control technology for UAV, Dr. I Gusti Agung Aditya Jaya from Indonesia representing aircraft maintenance technology, and Dr. Himsar Ambarita from North Sumatera University (USU) representing computational science in fluid mechanics.

From 79 papers submitted, 50 papers were selected to be presented in the seminar held in Medan, North Sumatra, Indonesia, on 27-29 September 2017. The seminar participants from Indonesia came from ITB, USU, Universitas Gajah Mada, LAPAN, Agency for Technology Assessment and Application (BPPT), and Indonesian Aerospace (PTDI). International participants came from Hongkong, Universiti Teknologi MARA and Universiti Sains Malaysia, Malaysia, Hongkong University of Science and Technology, and University of Surrey, UK.

Bogor, October 2017

ISAST Scientific Committee

IOP Conf. Series: Journal of Physics: Conf. Series **1005** (2018) 011002 doi:10.1088/1742-6596/1005/1/011002

Committees

Patron : Head of Indonesian National Institute of Aeronautics and Space (LAPAN)

Advisers : Deputy of Aeronautics and Space Technology, LAPAN
Director of Aeronautic Technology Center, LAPAN
Director of Rocket Technology Center, LAPAN

Steering Committee

Prof. Dr. Wahyu Widada.
Prof. Dr. Heri Budi Wibowo
Lilis Mariani
Agus Bayu Utama

Scientific Committee

Dr. Robertus Heru Triharjanto	LAPAN, Indonesia
Dr. Bagus H. Jihad	LAPAN, Indonesia
Dr. Mabe Siahaan	LAPAN, Indonesia
Dr. Kendra Hartaya	LAPAN, Indonesia
Prof. Dr-Ing. Nicolas Gascoin	Institut National des Sciences Appliquees, France
Prof. Dr. Josaphat Tetuko Sri Sumantyo	Chiba University, Japan
Prof. Dr. Aizawa	Shibaura Institute of Technology, Japan
Prof. Dr. Qin Shi Yin	Beihang University, China
Prof. Dr. Edy Hermawan	LAPAN, Indonesia
Prof. Dr. Rini Akmeliawati	International Islamic University Malaysia
Dr. Winardi Sani	UTHM, Malaysia
Prof. Dr. Heri Budi Wibowo	LAPAN, Indonesia
Prof. Dr. Wahyu Widada	LAPAN, Indonesia
Dr. Ridanto Eko Poetro	ITB, Indonesia
Dr. Arif Nurhakim	LAPAN, Indonesia
Dr. Akhmad Farid Widodo	BPPT, Indonesia
Dr. Firman Hartono	ITB, Indonesia
Dr. Arnaldo Marulitua Sinaga	IT Del, Indonesia
Dr. Himsar Ambarita	North Sumatera University, Indonesia
Agus Bayu Utama	LAPAN, Indonesia
Herma Yudhi Irwanto	LAPAN, Indonesia
Atik Bintoro	LAPAN, Indonesia
Albert Sagala	IT Del, Indonesia

Executive Committee

Chairman : Dr. Effendi Dodi Arisandi
Secretariat : Nugroho Agung Prasetyo

Anita Pinalia	Surianawanti
Imas Tri Setyadewi	Ida
Kandi Rahardiyanti	Iqbal Sasmi M.
Hilda Panca Setiawati	Dian Noviwiyanah
Edi Sutanta	

Editor : Soleh Fajar Junjunan
Afrido Prayogi
Sridana Windya

IOP Conf. Series: Journal of Physics: Conf. Series **1005** (2018) 011003 doi:10.1088/1742-6596/1005/1/011003

Participants

No	Name	Institutions
1	Achmad Pradivta Nugraha	ITB
2	Adi Wirawan	LAPAN
3	Aditya Prayoga	ITB
4	Adityo Suksmono	BPPT
5	Afni Restasari	LAPAN
6	Afni Restasari, S.Si	LAPAN
7	Afrido Prayogi	LAPAN
8	Agastya Bagaskara	ITB
9	Agit Zulharman	Universitas Komputer Indonesia
10	Agus Bayu Utama	LAPAN
11	Agus Mulyana	Universitas Komputer Indonesia
12	Ahmad Fariz	ITB
13	Alfonso Gamero Navarro	Spanish Innovation Agency (CDTI)
14	Ali Mukhsin	ITB
15	Andrey Stephan Siahaan	Muroran Institute of Technology, Hokkaido Japan
16	Angga Dwi Saputra	BPPT
17	Anggi Dewita	Indonesia Bureau of Meteorology Climatology and Geophysic
18	Anggi Dewita	Indonesia Bureau of Meteorology Climatology and Geophysic
19	Annisa Syafira Gusnop	USU
20	Aprian Rizki Fauzi	LAPAN
21	Aswan Tajuddin	Universitas Suryadarma
22	Athir Muhammad Fakhar	USU
23	Awang Rahmadi Nuranto	ITB
24	Bagus Bambang Riswandi	ITB
25	Bambang K. Hadi	ITB
26	Bruno Gaignard	Consultant - Satellite & Launcher
27	Budi Sampurno	BPPT
28	Buldan Muslim	LAPAN
29	Cahya Amalinadhi Putra	ITB
30	Cahya Amalinadhi Putra	ITB
31	Cahya Amalinadhi Putra	ITB
32	Chandra Nurohman	ITB
33	Dede Andhika Purnamasari	LAPAN
34	Dede Andhika Purnamasari	LAPAN
35	Derajat	PT. DI

IOP Conf. Series: Journal of Physics: Conf. Series **1005** (2018) 011003 doi:10.1088/1742-6596/1005/1/011003

36	Dewi Kesuma NST, S.S., M.Hum	UMSU
37	Dian Noviwiyanah	LAPAN
38	Dimas Bahtera Eskayudha	BPPT
39	Dony Hidayat	LAPAN
40	Dony Hidayat, ST	LAPAN
41	Dr. Himsar Ambarita	USU - Indonesia
42	Dr. Ir. Rais Zain, M.Eng	IAEC - Indonesia
43	Dr. Muhammad Qorib, M.A.	UMSU
44	Dr. Ony Arifianto	ITB
45	Dr. Taufiq Mulyanto	ITB
46	E.M. Alkoy	Beihang Univ.
47	Edi Sutanta	LAPAN
48	Effendi Dodi Arisandi	LAPAN
49	Eko susilo	Institute for Marine Research and Observation
50	Ema Amalia	ITB
51	Emerson P Sinulingga	USU
52	Faber Yosua O Silitonga	ITB
53	Fahmi	USU
54	Ghifari Adam Faza	ITB
55	Ginaldi Ari Nugroho	Universitas Komputer Indonesia
56	Gunawan Setyo Prabowo	LAPAN
57	Gunta Akhiri	PT Aering Indonesia
58	Hari Muhammad	ITB
59	Heri Budi Wibowo	LAPAN
60	Herma Yudhi Irwanto	LAPAN
61	Herma Yudhi Irwanto, M.Eng	LAPAN
62	Hilda Panca Setiawati	LAPAN
63	Hindawan Hariowibowo	PT. DI
64	Husnul Khotimah	USU
65	I Gusti Agung Aditya Jaya, Ph.D	GMF - Indonesia
66	Ida	LAPAN
67	Imas Tri Setyadewi	LAPAN
68	Indra Hartanto Tambunan, Ph.D.	IT Del
69	Indra Sarito Tobing	IT-DEL
70	Iqbal Sasmi Ma'arif	LAPAN
71	Ir. Agus Budi Djatmiko, MT	LAPAN
72	Ir. Ediwan, MT	LAPAN
73	Kandi Rahardiyanti	LAPAN
74	Katia Mayang Mahasti	BPPT
75	Kei-ichi Hirako,	Keio University, Japan
76	Kendra Hartaya	LAPAN
77	Ken-ichi Anma,	Mitsubishi Heavy Industries, Ltd, Japan.
78	Kevin Setiawan	ITB

79	Khodijah Kholish Rumayshah	ITB
80	Kosim Abdurrohman	LAPAN
81	Kosim Abdurrohman, ST	LAPAN
82	Kuncoro	LAPAN
83	Liana	LAPAN
84	Lilik Supriyatun	LAPAN
85	Liu Feng Gouw	Beihang Univ.
86	Luqman Fathurrohim	ITB
87	M. Fajar	LAPAN
88	M. Fajar, ST	LAPAN
89	Mabe Siahaan	LAPAN
90	Maria Carine PAD	Indonesia Bureau of Meteorology Climatology and Geophysic
91	Marshell Antonius	ITB
92	Masahiro Koyama ,	The University of Tokyo, Japan.
93	Mitra Khasyofi	PT. DI
94	Mochamad Agoes Moelyadi	ITB
95	Mochammad Agoes Moelyadi	ITB
96	Mochammad Agoes Moelyadi	ITB
97	Mochammad Agoes Moelyadi	ITB
98	Mohamad Arif Andira	ITB
99	Mohamad Arif Andira	ITB
100	Mohamad Ivan Aji Saputro	BPPT
101	Muhamad Luthfi Imam Nurhakim, M.T	ITB
102	Muhammad Azwan Nasirrudin	Universiti Sains Malaysia
103	Muhammad Farraz Al Farizi, ST, MT	ITB
104	Muhammad Fazlur Rahman	ITB
105	Muhammad Fikri Zulkarnain	ITB
106	Muhammad Iqra Harahap	USU
107	Muhammad Luqmanul Hakim	ITB
108	Muhammad Rafii	UMSU
109	Muhammad Taufiq	BPPT
110	Mujtahid	LAPAN
111	Nina Kartika	BPPT
112	Nugroho Agung P	LAPAN
113	Nur Azizah Dewi Kurniasari	ITB
114	Nurhayyan Halim Rosid	ITB
115	Oktina Nugraheni	Inspektur
116	Prof. Dr-Ing. Nicolas Gascoin	Kedutaan Perancis
117	Prof. Eddy Hermawan	LAPAN
118	Prof. Erna Sri Adiningsih	LAPAN

119	Prof. Igif G. Prihanto	LAPAN
120	Prof. Muchlisin Arief	LAPAN
121	Prof. Qin Shiyin	Beihang University - China
122	Prof. Yoshifumi Inatani	JAXA - Jepang
123	Purwadi	BPPT
124	Putra Adnan Fadilah	ITB
125	Putra F. Adnan	ITB
126	R. Wibawa Purabaya	BPPT
127	Rahmat Alfi Duhri	ITB
128	Rahmat Hadiyatul Akbar	ITB
129	Ratih Julistina	ITB
130	Regia Dwisanti	ITB
131	Rhea P. Liem	Hongkong University of Science and Technology
132	Rianto Adhy Sasongko	ITB
133	Rianto Adhy Sasongko	ITB
134	Rianto Adhy Sasongko	ITB
135	Ricardo Situmeang, M.A.	IT Del
136	Ridanto Eko Poetro	ITB
137	Rika Andiarti	LAPAN
138	Rika Suwana Budi	Beihang Univ.
139	Rika Suwana Budi	Beihang Univ.
140	Rika Suwana Budi	LAPAN
141	Rini Ekayati, S.S., M.Hum	UMSU
142	Rini Hutahaean	IT-DEL
143	Rio Rheza Harris	ITB
144	Rizka novia dwi asmi lubis	UMSU
145	Rizki Hanintyo	Institute for Marine Research and Observation
146	Robertus Heru Triharjanto	LAPAN
147	Romie oktovianus Bura	ITB
148	Rommy Hartono	LAPAN
149	Ronald Gunawan Putra, S.T.	LAPAN
150	Roy Deddy Lumban Tobing, S.T., M.ICT.	IT Del
151	Ruth Natalie Napitupulu	IT-DEL
152	Saad Al Ahmed	Ajwa Aero
153	Sayuti Syamsuar	BPPT
154	Seiko Shirasaka,	Keio University, Japan
155	Shin-ichi Nakasuka,	The University of Tokyo, Japan.
156	Shinobu Nakamura,	Japan Science and Technology Agency, Japan.
157	Soleh Fajar Junjunan	LAPAN
158	Sri Sukarni	Inspektur
159	Sridana Windya	LAPAN
160	Stanislaus Billy Novrando	ITB

161	Starida Moranova	ITB
162	Subagyo	BPPT
163	Sulistyo Atmadi	IAEC
164	Suprapto	LAPAN
165	Surianawanti	LAPAN
166	Syafruddin Hasan	USU
167	Takeshi Tohara	Japan Science and Technology Agency, Japan.
168	Teungku Erry Nuradi	Gubernur Sumut
169	Toto Indriyanto	ITB
170	Tricory indahsari	UMSU
171	Triyono Widi Sasongko	BPPT
172	Wahyu Widada	LAPAN
173	Wigati	LAPAN
174	Wiji Hastuti	LAPAN
175	Wisnu	ITB
176	Xiao Wei Shu	Beihang Univ.
177	Yanto Daryanto	BPPT
178	Yulis Amanah Tri Rohmahwati	ITB
179	Zakaria	LAPAN

Acknowledgements

International Seminar on Aerospace Science and Technology (ISAST) 2017 was supported by Indonesian National Aeronautics and Space Agency (LAPAN).

IOP Conf. Series: Journal of Physics: Conf. Series **1005** (2018) 011005 doi:10.1088/1742-6596/1005/1/011005

Conference Photograph

Background
The importance of remote sensing technology for natural resources
and environmental monitoring in Indonesia.
The availability of satellite-based remote sensing data in Indonesia,
provided by LAPAN, as the mandate of Indonesian Space Acts
No.21/2013 on Space Activities.
Lake Toba as a unique & complex ecosystem ~ the largest lake in
Indonesia, comprising land and water ecosystems with various
types/uses ~ experiencing environmental problems &
pressures.
gal basis by Government Regulation ~ PP no. 81/2014
anning of Lake Toba & Its Surrounding ~ continuous
d monitoring.
e supports of remote sensing technology from operational
ellites and future satellites for Lake Toba natural resources and
vironmental monitoring are necessary.
vious studies on the use of remote sensing for natural resource
ervation ~ as the basis for further assessment.

LAPAN
International Semi
eme
velopment of Aeronautics and Sp
aritime Application

nternational Seminar on Aeros
ment of Aer
Technology
rt Maritime

Th
De
to

International Seminar
pace Technology
Hotel Aryaduta

REGISTRATION
IMPORTANT DATES
FREE

5th ISAST 2017
International Seminar on Aerospace and Technology

IOP Conf. Series: Journal of Physics: Conf. Series **1005** (2018) 011006 doi:10.1088/1742-6596/1005/1/011006

Peer review statement

All papers published in this volume of *Journal of Physics: Conference Series* have been peer reviewed through processes administered by the proceedings Editors. Reviews were conducted by expert referees to the professional and scientific standards expected of a proceedings journal published by IOP Publishing.

IOP Conf. Series: Journal of Physics: Conf. Series **1005** (2018) 012001 doi:10.1088/1742-6596/1005/1/012001

Experimental Tensile Strength Analysis of Woven-Glass/Epoxy Composite Plates with Central Circular Hole

Bambang K. Hadi* and Bima K. Rofa

Faculty of Mechanical and Aerospace Engineering,
INSTITUT TEKNOLOGI BANDUNG,
INDONESIA

*bkhadi@ae.itb.ac.id , bambang.hadi60@gmail.com

Abstract. The use of composite materials in aerospace engineering, as well as in maritime structure has increased significantly during the recent years. The extensive use of composite materials in industrial applications should make composite structural engineers and scientists more aware of the advantage and disadvantage of this material and provide them with necessary data and certification process. One of the problems in composite structures is the existence of hole. Hole can not be avoided in actual structures, since it may be the necessity of providing access for maintenance or due to impact damage. The presence of hole will weaken the structures. Therefore, in this paper, the effect of hole on the strength of glass-woven/epoxy composite will be discussed. Extensive tests have been carried out to study the effect of hole-diameter on the tensile strengths of these specimens. The results showed that the bigger the hole-diameter compared to the width of the specimens has weakened the structures further, as expected. Further study should be carried in the future to model it with the finite element and theoretical analysis precisely.

1. Introduction

The use of composite materials in aircraft structures dates back in 1970s, mostly for military aircrafts. During those years, boron-epoxy, carbon-epoxy were developed to meet the needs for military aircrafts. The application of these materials began with secondary structures in 1980s. Since then, the use of composite materials has been flourished to primary structures, such as wing and recently fuselage. The latest Boeing aircraft, B-787 Dreamliner, uses 50% of the total structural weight in composite materials [1]. Their design was based on Boeing's long experience on using composite materials since B-737 programs. On the other hand, Airbus A-350 XWB uses 52% of the total weight with composite materials [2]. In the marine structures, composite materials was used various applications, such as small crafts, fishing boats, naval vessels, underwater vehicles and others [3]. The continuous development in composite materials in both aerospace and marine structures gives opportunities and challenges to engineers and scientists working with this material [4] and [5]. One of the challenges is in the field of airworthiness process. Extensive tests should be carried out in order to fulfil the airworthiness requirements.

The presence of hole in actual structures usually cannot be avoided, in order to give access for maintenance or due to impact damage. The hole will weaken the structures, since it gives high concentration factor in the adjacent to the hole. Therefore, extensive tests should be carried out to study this phenomenon.

Kim et al [6] investigated failure strength of CFRP composite structure with open hole under tensile loading. The method developed was in a good agreement with the failure criterion within an error rate of 10%. Tino and de Aquino [7] studied the fracture characteristics of notched glass fiber reinforced plastic under tensile load. The effect of stress concentration on the failure mode was investigated. Meanwhile, Soutis and Fleck [8] investigated the failure of carbon fibre T800/924C composite plate with hole under compression load. Both experimental and theoretical analyses were done. They concluded that the agreement between the theory and experiment data was acceptable.

Arslan, et al [9] studied the effect of circular hole on the failure of composite plate. Tsai-Hill failure criterion was used, together with first-ply-failure assumption. Finally, Toubal, et al [10] studied stress concentration in a circular hole in composite plate experimentally, using a non-contact measurement method namely electronic speckle pattern interferometer (ESPI). The data was then compared to the theoretical solution of Lekhnitskii, as well as finite element study. The results are mixed, depending on the lay-up configuration of the composite plate.

In this paper, experimental study on the tensile strength of woven-glass/epoxy composite plate with circular hole was presented. The effect of lay-up configuration on the tensile strength was studied, namely $(0^o/90^o)_s$, $(\pm45^o)_s$ and $(0^o/90^o/\pm45^o)_s$.

2. Specimen Preparation and Testing Method

2.1 Woven-glass/epoxy Specimens

The specimens were made using woven-glass type WR200, and the epoxy was Bisphenol A-epichlorohydrin with hardener type Versamid 140. The fibre volume fraction was 50%. The manufacturing method was a standard hand lay-up technique. The lay-up configuration was $(0^o/90^o)_s$, $(\pm45^o)_s$ and $(0^o/90^o/\pm45^o)_s$. The specimens were then cured in room temperature for 24 hours, and cut with geometry of 200 mm x 25 mm. A circular hole is manufactured by drilling at the centre of the specimens. Different hole-diameters are: 3 mm; 6 mm; 10 mm; 12 mm and 16 mm. Fig.1 shows the specimens.

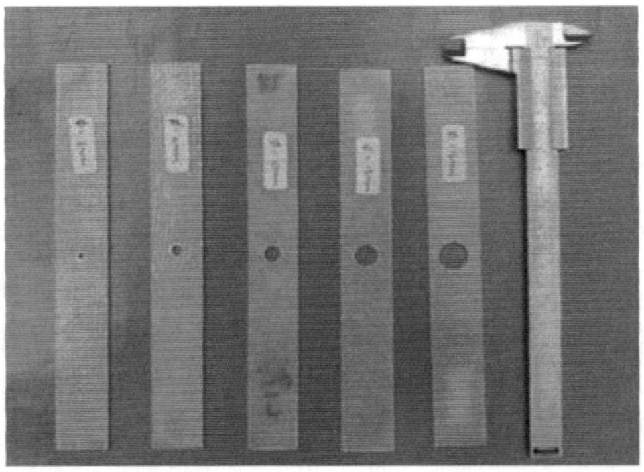

Figure 1: Woven-glass/epoxy specimens with central circular hole.

2.2 Testing Method

All testing has been done using universal testing machine Tensilon RTF-1310 with crosshead speed of 2.0 mm/second. Fig.2 shows tensile testing arrangement.

Figure 2: Testing arrangement

3. Data Analysis

Two-parameters Weibull distribution was used in the data analysis [11]. The Weibull distribution is given by Eq (1):

$$F(x;b,c) = 1 - exp\left(-\left(\frac{x}{b}\right)^{c}\right), b \geq 0, c \geq 0 \quad (1)$$

where *b* and *c* are two-parameters, which can be calculated from the experimental data. Taking a double logarithm of Eq. 1, we get a linear equation in the form of $Y = m X + r$ and is given in the form of Eq. 2:

$$\ln\left[\ln\left(\frac{1}{1-F(x;b,c)}\right)\right] = c\ln(x) - c\ln(b) \quad (2)$$

The value of $F(x;b,c)$ is unknown. A good estimator of $F(x;b,c)$ is the median rank of $x(i)$ in the form of Eq. 3 [12]:

$$F\left(x_{(i)};b,c\right) = \frac{i-0.3}{(n+0.4)} \quad (3)$$

where n is the number of data observation and $x_{(i)}$ denotes the i_{th} smallest observations ($i = 1$ corresponding to the smallest data observation and $i = n$ corresponds to the largest). By arranging the data and use linear regression in Eq.3, the parameters b and c can be calculated, and hence the probability and reliability of the materials can be computed using Eq.1. In this paper, reliability of 50% was taken as the specimen strength.

4. Results and Discussion

4.1 Tensile Test Results

The results are given in Fig.3, for the hole diameter of 3 mm, for different lay-up configuration. Results for different hole diameter are also available. Fig.3 shows that most specimens failed in elastic mode, except specimens configurations of $(\pm 45^{o})_{s}$, which failed in elastic-plastic mode; where plastic failure was the dominant one. It is a well-known fact that specimens having $(\pm 45^{o})_{s}$ lay-up configuration has matrix-dominated properties. Therefore, it is expected that these specimens will produce large plasticity mode.

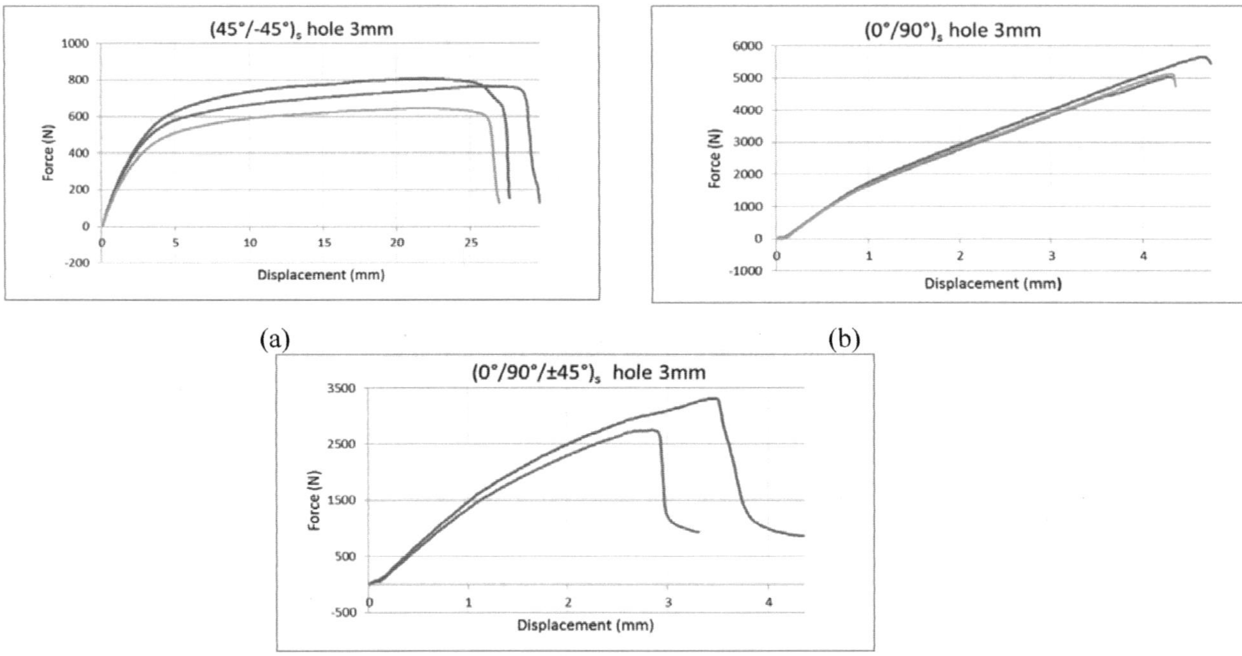

Figure 3: Load-displacement curves for specimens with 3 mm hole-diameter. (a) (0°/90°)$_s$; (b) (±45°)$_s$; (c) (0°/90°/±45°)$_s$.

Specimens having (0°/90°)$_s$ as shown in Fig 3(a) produces a kink in their load-displacement curves. These indicate the occurrence of first-ply and last-ply failure within the specimens. When the (0°/90°)$_s$ specimens are given tensile loads, the 90° ply will fail first, which is called first-ply failure; leaving the 0° ply to resist tensile loading until it all fails, that is called last-ply failure.

The diagram of maximum tensile loads for all lay-up configuration and hole-diameter is given in Fig.4. It shows that lay-up configuration of (0°/90°)$_s$ have the highest tensile strength, while the (±45°)$_s$ are the weakest. The (±45°)$_s$ lay-up is the strongest in shear loads, but the weakest in tensile loads. The strength of (0°/90°/±45°)$_s$ lay-up configuration is between the other two configurations. Fig. 4 also shows that the bigger hole-diameter will significantly reduce the plate strength. The reduction of tensile strength is more prominent in (0°/90°)$_s$ compared to other lay-up configuration.

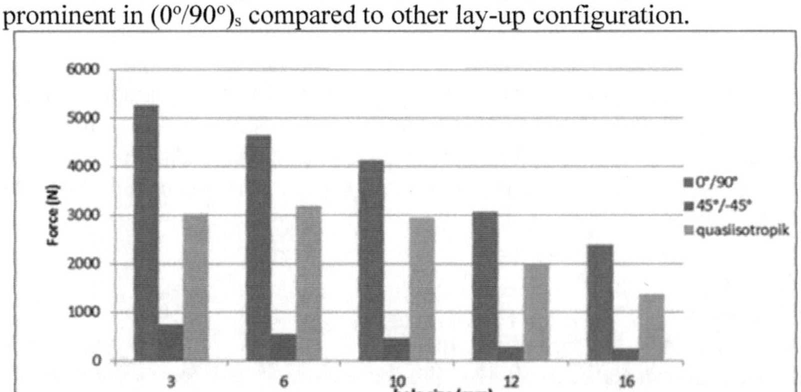

Figure 4: Tensile strength diagram for different lay-up configuration and hole-diameter

5. Failure Modes

Failure modes for different hole-diameter and lay-up configuration were given in Fig.5.

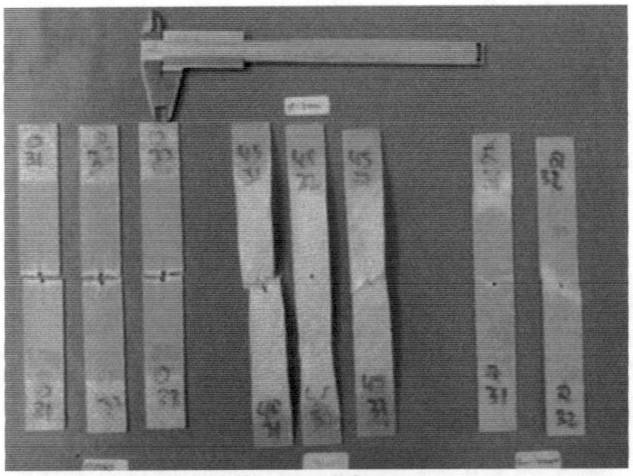

(a) Hole-diameter 3 mm.

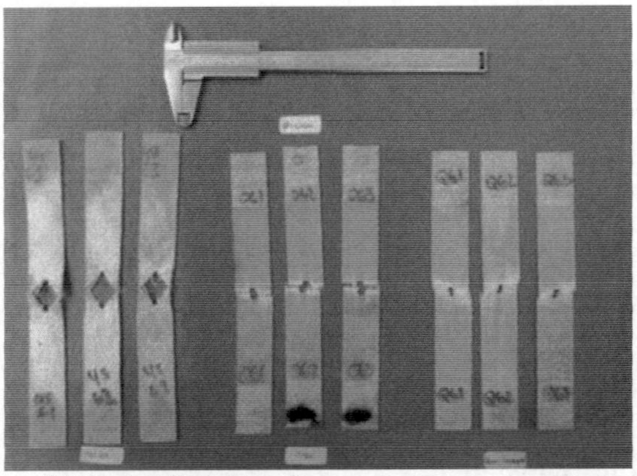

(b) Hole-diameter 6 mm

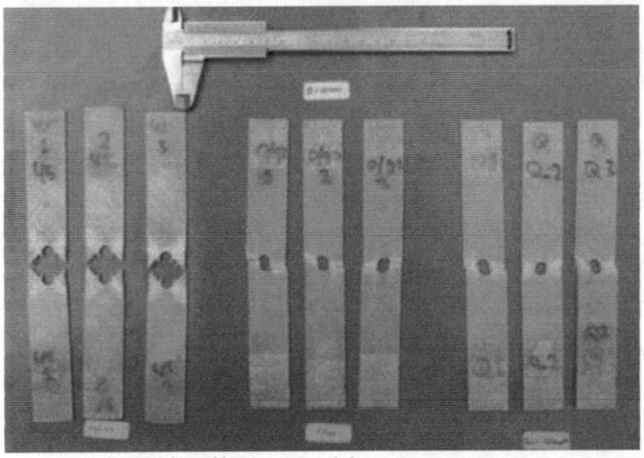

(c) Hole-diameter 10 mm

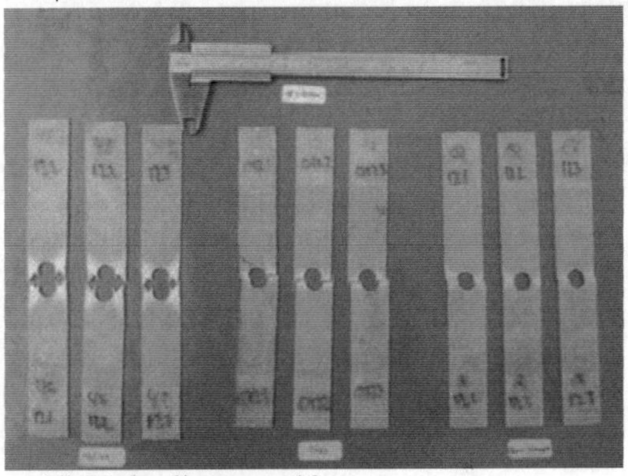

(d) Hole-diameter 12 mm

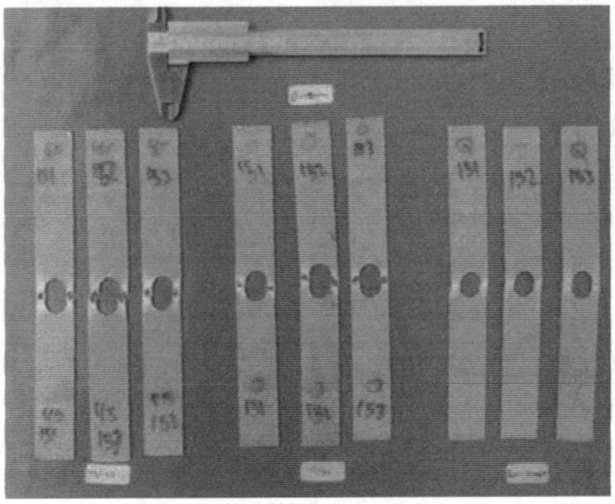

(e) Hole-diameter 16 mm

Figure 5: Failure modes of uniaxially loaded specimens for different lay-up configurations and hole-diameters.

Fig.5 shows that specimens having $(\pm 45^{o})_s$ failed in 45° direction; while the other lay-up configurations failed in 0°. For smaller hole-diameter, however, the results are mixed. The failure directions are not fixed in 0°, but they diverted for 45°. The failure direction may coincide with the maximum stresses in the adjacent of the hole. Therefore, for $(\pm 45°)_s$ lay-up configuration, the maximum stresses may not occur in 0° direction. This should be investigated further using finite element or theoretical analysis. The investigation should determine exactly the direction of the failure modes

6. Conclusion

In conclusion, it shows that layup configuration of $(0°/90°)_s$ gives the highest strength, while the $(\pm 45°)_s$ gives the weakest strength. The lay-up configuration of $(0°/90°/\pm 45°)_s$ is between the other lay-up configurations. The $(\pm 45°)_s$ specimens failed in 45° failure direction, while the other two lay-up failed in 0° direction. This coincides with the maximum stresses occurred in the adjacent to the hole. Therefore in structural application where hole is necessary, it is recommended to use more $(0°/90°)_s$ lay-up configuration. This is the first step in predicting the strength of composite plate with central circular hole. Further study will be carried out regarding to modeling of this plate using finite element method or theoretical analysis.

Acknowledgement

The research was partly supported by the Ministry of Research, Technology and Higher Education, Republic of Indonesia through INSINAS Research Grant 2017. The authors wish to thank the Ministry for the generous funding.

References

[1] A.J. Fawcett and G.D. Oakes. "*Boeing Composite Airframe Damage Tolerance and Service Experience*", Chicago Damage Tolerance Workshop, July 19-21, 2006.

[2] G. Hellard. "*Composite in Airbus: A Long Story of Innovations and Experiences*", Global Investor Forum, Paris, 2008.

[3] C.S. Smith. *Design of Marine Structures in Composite Materials*, Elsevier Applied Science, 1990.

[4] C.E. Harris. "Opportunities for Next Generation Aircraft Enabled by Revolutionary Materials", *AIAA SDM Conference*, Denver, CO, April 4-7, 2011.

[5] F. Smith. "*The Use of Composites in Aerospace: Past, Present and Future Challenges*", Avalon Consultancy Services, 2013.

[6] S-Y. Kim, J-M. Koo, C-S. Seok,"Failure Strength of CFRP Composite Structure with Open Hole under Tensile Loading", *International Journal of Advances in Science and Technology*, pp. 110-114, ISSN 2348-5426, 2011.

[7] S.R.L. Tino, E.M.F. de Aquino. "Fracture Characteristics and Anisotropy in Notched Glass Fiber Reinforced Plastics", *Materials Research*, Vol.17, No.6, pp. 1610-1619, 2014.

[8] C. Soutis, N.A. Fleck. "Static Compression Failure of Carbon Fibre T800/924C Composite Plate with a Single Hole", *Journal of Composite Materials*, Vol. 24, pp.536-558, 2009.

[9] H.M.. Arslan, M.Y. Kaltakci, H.R. Yerli, H.R., "Effect of Circular Holes on Cross-ply Laminated Composite Plates", *The Arabian Journal for Science and Engineering*, Vol. 34, No. 2B, pp. 301-315, 2009.

[10] L. Toubal, M. Karama, and B. Lorrain, "Stress concentration in a circular hole in composite plate", *Composite Structures*, Vol. 68, pp. 31-36, 2005.

[11] A. Weibull, "*Statistical Theory of the Strength of Materials*", Royal Swedish Institute for Engineering Research, 1939.

[12] M.H. Dirikolu, A. Aktas and B. Birgoren. "Statistical Analysis of Fracture Strength of Composite Materials Using Weibull Distribution", *Turkish J. Eng. Env. Sci.*, Vol. 26, pp.45-48, 2002.

Fatigue life estimation on coke drum due to cycle optimization

Andrey Stephan Siahaan[1], Himsar Ambarita[2,*], Hideki Kawai[1], and Masashi Daimaruya[1]
[1]Department of Mechanical Systems of Engineering, Muroran Institute of Technology 27-1 Mizumoto-cho, Muroran 050-8585, Japan
[2]Sustainable Energy Research Centre, Faculty of Engineering, University of Sumatera Utara, Jl. Almamater Kampus USU Medan 20155, Indonesia

E-mail: himsar@usu.ac.id

Abstract. In the last decade, due to the increasing demand of petroleum product, the necessity for converting the heavy oil are increasing. Thus, demand for installing coke drum in whole world will be increase. The coke drum undergoes the cyclic high temperature and suddenly cooling but in fact is not designed to withstand that kind of cycle, thus the operational life of coke drum is much shorter in comparison to other equipment in oil refinery. Various factors determine in order to improve reliability and minimize the down time, and it is found that the cycle optimization due to cycle, temperature, and pressure have an important role. From this research it is found that the fatigue life of the short cycle is decrease by a half compare to the normal cycle. It also found that in the preheating stage, the stress peak is far exceed the yield strength of coke drum material and fall into plastic deformation. This is happened because of the temperature leap in the preheating stage that cause thermal shock in the upper part of the skirt of the coke drum.

Keywords: Coke Drum, Fatigue Life Estimation, Preheating Stage, Cutting Stage

1. Introduction

Coke drum are vessels critical to the operation and profitability of today's oil refinery and tar sand facilities. These vessel operate in batch cycle at relatively high temperature; approximate 430°C to 500°C, and about once a day, they are filled with vapor and hot oil and finally quenched with water. Coke drum designed according to ASME Section VIII division II pressure vessel with little or no account for the cyclic operation fatigue service in which they operate [1].

In general temperature history can be divide into four stages. First is preheating stage from 0 to 440 minutes, second is filling stage from 440 to 2000 minutes, third is cooling stage from 2000 to 2400 minutes, and the rest is cutting stage. Typical operational temperatures for one cycle are shown in fig. 1. In the figure, temperature of point T1 and T8 are only shown. As a note, temperature of the shell portion of the coke drum are represented by points T5 and T8.

In the preheating stage to avoid harmful thermal shock, the coke drum is preheated. Thus, all temperature increase as the time increasing. Since the coke drum is heated from inside, increase of shell portion is bigger than the skirt [2]. As the result, temperatures of points T1 to T4 are relatively bigger than T5 to T8. In other words, the shell is hotter than the skirt. However, or a while, at the beginning of preheating stage, normally the lower part of the skirt is hotter than the upper one. This is because of remaining heat from previous cycle.

After the preheating stage is finish, in the filling stage, the hot feed material is injected into coke drum. Since the temperature of coke drum is lower than the hot feed, at the beginning of filling stage, coke drum temperatures increase rapidly. After that temperature become flats or quite constants on filling stage. As a note, heat is transferred from coke drum to surrounding through the wall and through the skirt to the base. Since so, the shell of the coke drum is hotter than the skirt. As the matter of fact, during the filling stage the coke drum falls into steady state condition.

In the cooling stage, quenched water is injected and cools the coke drum from inside. Decrease of temperatures with time of the shell is more rapid than the skirt. As a result, temperature of points T1 to T4 are now relatively lower than T5 to T7. In other words, the shell portion is colder than the skirt. However, there is an exception at point T8, which its temperature is lower than the shell portion. This is because point T8 contact with base which has surrounding temperature.

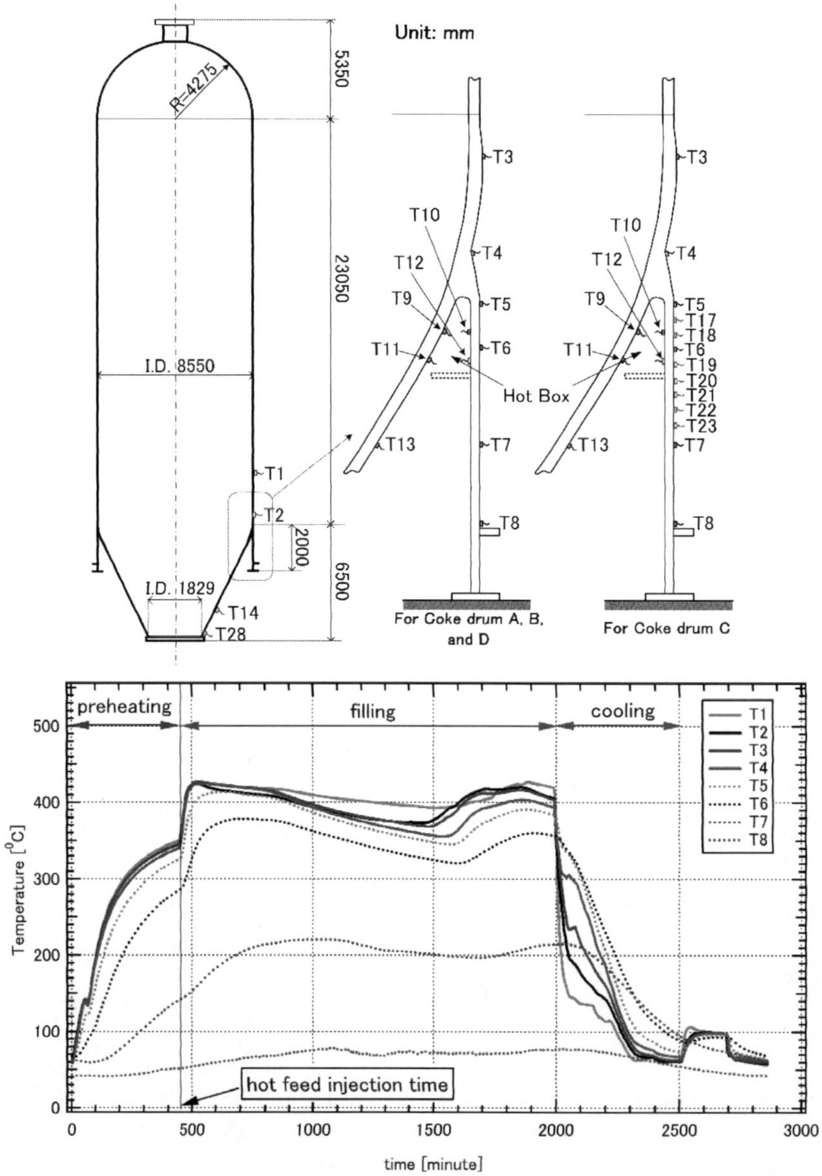

Fig. 1 Coke Drum and Typical Coke Drum Cycle

In the past ten years, the frequency and severity of these cycles dramatically increased while the cycle times steadily decreased from 12 to 24 hours and more recently to as low 10 hours cycles. As a result, coke drum with few or no previous problems began to crack through wall in the shell and/or skirt region. In addition, many new coke drums are cracking through wall in as few as four or five years because they are not designed for the more severe fatigue service in which they currently operated.

This paper focused on the coke drum operating cycle behavior. Some statement says that longer cycle will effect on longer fatigue life of coke drum, but it also occurred that the operational stages also plays an important role to prolong coke drum's fatigue life. Thus, this paper attempt to investigate some selected cycle of coke drum to understand the correlation between cycle period and the fatigue life of coke drum. The next attempt is to compare the cycle which can be categorized as normal cycle and another cycle which is assumed as a bad operation cycle due to very short cycle period.

2. Method

For this coke drum 28 thermocouples and 6 strain gauges are installed in the skirt attachment area as shown in fig. 2. Three pairs of the strain gauges are installed at points C4, C5, and C6 respectively. Each pair is aimed to measure strain in axial and hoop directions. More refined thermocouples were attached at the upper part of the skirt. The objective is to capture a more realistic temperature distribution.

Temperature boundaries at two measured points are assumed to be linier. Temperature variation in radial direction is assumed to be zero. Since the analyses is focused at the skirt area and it is convinced that temperature variations above point C1 do not strongly affect the skirt area. Thus, temperature of the coke drum above point C1 is assumed to be the same as temperature of point C1.

Coke drum material is considering by activating gravitational force in –z directions. Here the coke drum material density of 7850 kg/m^3 is imposed. The weight of the coke and quenched water are considered. At the end of the cooling stage, the coke drum is filled by the coke and quenched water with a level which is assumed be 80% of its capacity. Here the height of the coke and water is assumed to be 25.6 m. this load is imposed as hydrostatic pressure in the inner surface. And the density of combination of coke and water is assumed to be 1481 kg/m^3.

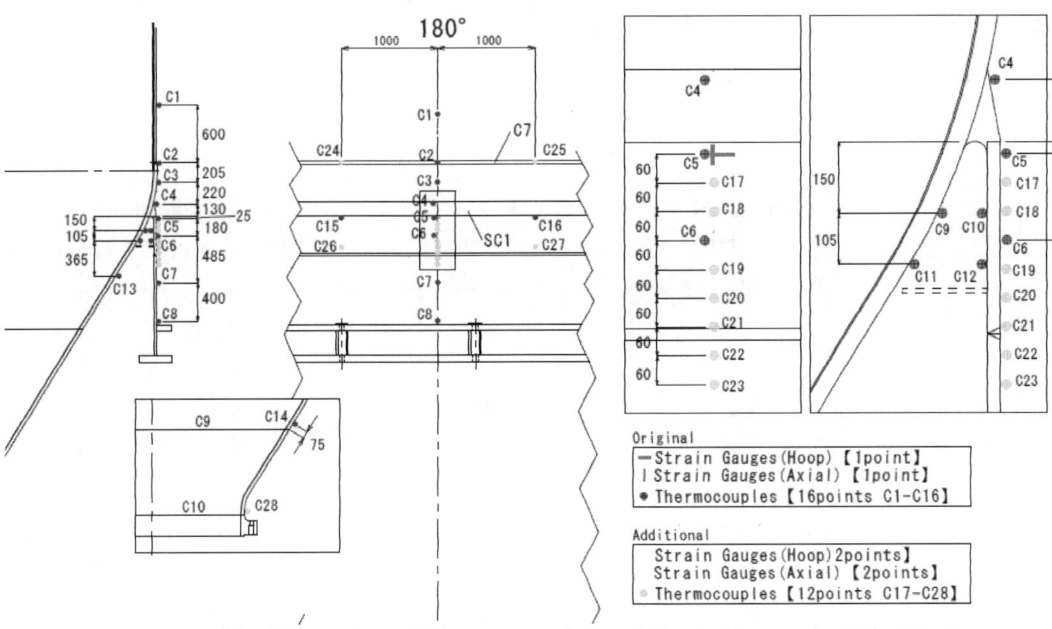

Fig.2 Location of Thermocouples and Strain Gauges in Coke Drum

Numerical analyses were performed by using ANSYS code [3]. The coke drum was analyzed by three dimensional model, which is developed based on fabricated drawing of the coke drum. The model and constrains are shown in fig. 3.

IOP Conf. Series: Journal of Physics: Conf. Series **1005** (2018) 012002 doi:10.1088/1742-6596/1005/1/012002

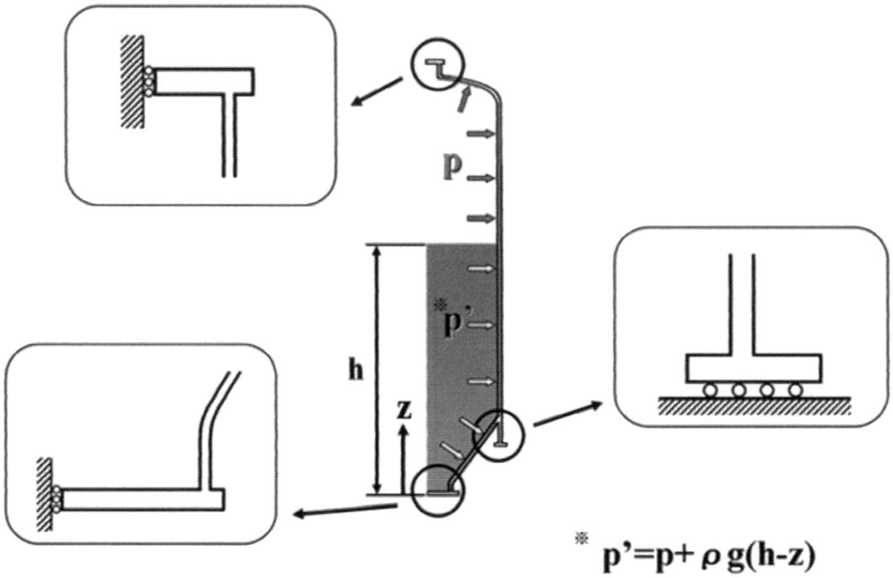

Fig. 3 Model of Structural Boundary

The coke drum and the skirt are made of low alloy steel SA-387 Gr. 11 CL. 2 (1.25Cr-0.5Mo-Si). The mechanical properties are obtained from ASME boiler and pressure vessel codes II section D. Young modulus (E), thermal conductivity (γ), and coefficient of thermal expansion (α) are temperature dependent and shown in table 1. Poisson's ratio is constant at 0.3.

The model for numerical analyses is half segment of the coke drum. There are two mesh model were tested in this simulation which is 25675 and 41351 nodes. The strain result is almost the same but for the better result the model with 41351 nodes was selected for analyses.

Table. 1 Temperature dependent material properties

Temperature (°C)	Properties		
	Young modulus (GPa)	Thermal Conductivity (W/mK)	Coeff.of thermal exp (/°C)
0	205.5	41.05	-
25	204	40.9	0.0000116
100	200	40.6	0.0000121
150	197	40.4	0.0000124
200	193	40.1	0.0000128
250	190	39.5	0.0000131
300	186	38.7	0.0000132
350	183	37.8	0.0000135
400	179	36.8	0.0000139
450	174	35.8	0.0000142
500	169	34.8	0.0000143
550	164	33.9	0.0000146
600	157	32.8	0.0000148

IOP Conf. Series: Journal of Physics: Conf. Series **1005** (2018) 012002 doi:10.1088/1742-6596/1005/1/012002

3. Result and Discussion

As for numerical analyses validation, strain date which gathered from strain gauges is compared to the numerical result. The selected cycle have total time of 2771 minutes. Similar with the temperature plot history we can also divided the strain result to preheating, filling, cooling, and switching stage.

The comparison between the numerical result and measured data at point C5 is shown in fig. 4. The figure shows that, even though there are some differences between those results, the results of present numerical solution show the same trend and they do agree.

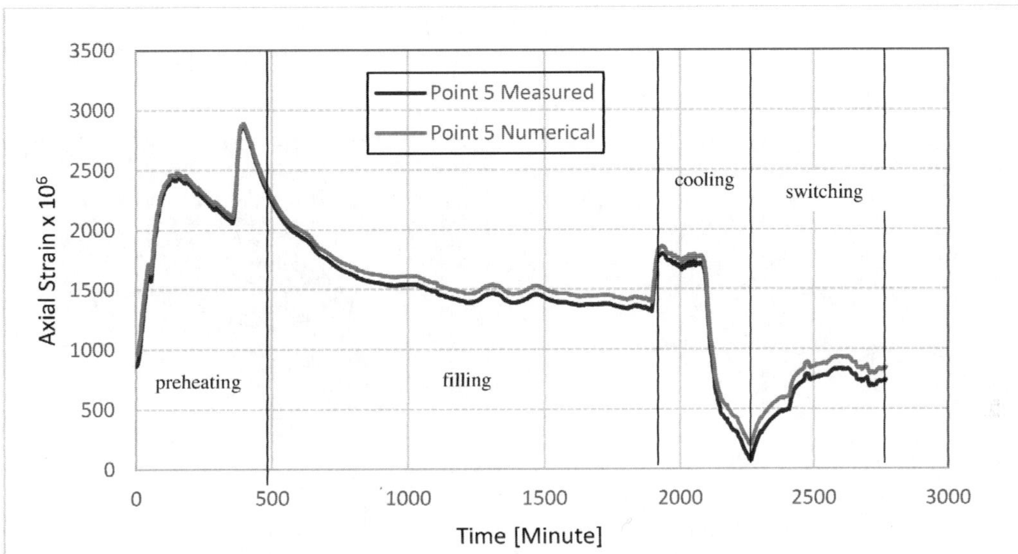

Fig. 4 Comparison of Numerical and Measured axial strains

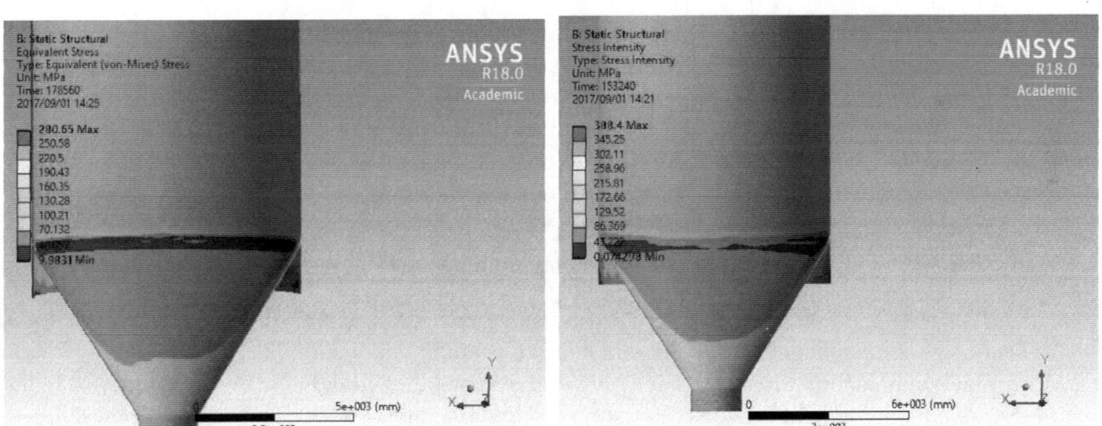

(a) (b)

Fig. 5(a) Contour of stress intensity for normal cycle and (b) short cycle

After validating result between numerical and measured data, fatigue life estimation of some selected cycle will be performed. First is the cycle that can be categorized as the normal cycle with cycle period of 2976 minutes or we can call it as the normal cycle. Another cycle is the cycle that can be categorized as a bad operation cycle with cycle period of 2554 minutes, or we can call it as a short cycle.

In order to examine whether the skirt area falls into plastic deformation or not, contour of stress intensity from aforementioned selected cycles is shown in fig. 5. From the contour of stress intensity we found that with the normal cycle maximum stress is 292 MPa, fig. 5(a), and for the short cycle maximum stress is 388 MPa, fig. 5(b). The highest stress intensity occurred in preheating stage, both for two selected cycles. From the numerical result it is found that, for the normal cycle stress intensity have a peak at minute 595. Than can be categorized as the switching stage from preheating to filling stage. The peak stress intensity of the short cycle is occurred at 320 minutes in preheating stage.

According to material data properties, the yield strength of SA-387 Gr. 2 CL.11 is 310 MPa. From this fact we can concluded that for the normal cycle, the stress peak is still tolerable because the peak stress is below the limit of yield strength of material. For the short cycle, stress peak is far exceed the yield strength of coke drum material. In this kind of cycle, the upper part of the skirt suffers from the most severe thermal stress and they fall into plastic deformation in the preheating stages. The main conclusion here is that for the selected short cycle can be categorized as a bad operation, and should be avoided to extend the useful life of coke drum.

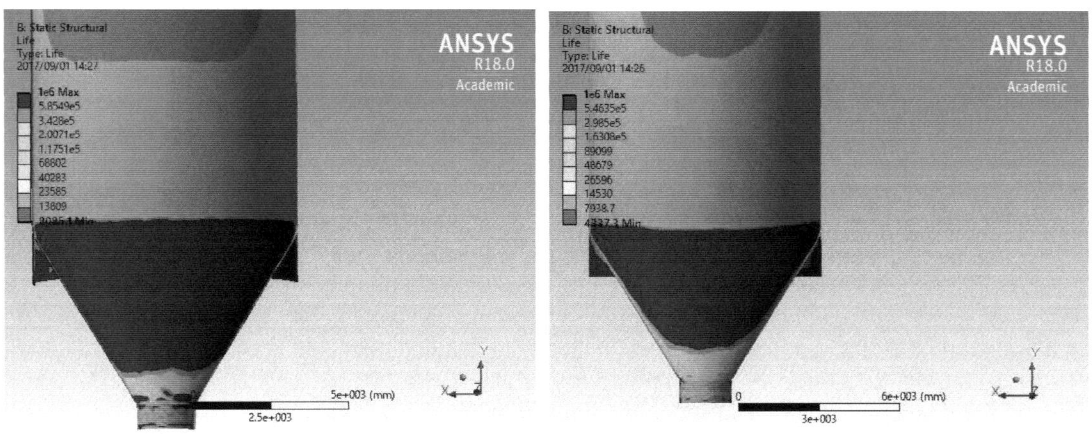

(a) (b)

Fig. 6(a) Contour of Fatigue Life for Normal Cycle and (b) Short Cycle

From the contour of fatigue life in fig. 6(a) above we can see that the fatigue life estimation of coke drum for normal cycle is 8085 cycles and for the short cycle coke drum, the fatigue life estimation is decrease to 4333 cycles, almost the half compare to the normal cycle, shown in fig. 6(b). In the normal cycle, preheating stage with moderate rise in temperature gradient, generate a moderate thermal stress in the shell to skirt junction. However, for the short cycle the short cycle period produce a higher temperature gradient in the preheating stage. Effect of the cycle period will be discussed in the next paragraphs.

As the aforementioned before, in the preheating stage temperature rise first in the shell portion of the coke drum. From the coke drum picture, we can see that the connection between the shell portions to the skirt of the drum is only the small part of the junction or we call it the shell to skirt junction. Thus, in the preheating stage the heat transfer from the shell part to the skirt part is merely conduct only by the conduction heat transfer of the junction. This is causing a high temperature difference between the shell and the skirt part which generate high thermal stress [4].

Actually some research have been implemented to try to reduce the high temperature difference between the shell and the skirt portion of the drum. One of the research focused on the attachment of hot box in the shell to skirt junction area to enhance heat transfer due to the generation heat transfer. Nevertheless, the result show that the natural convection have only a small part to enhance the hat transfer in the shell to skirt junction area or only around 10% for the total heat transfer [5]. Thus new method to enhance the heat transfer in shell to skirt junction need to be developed.

For further understanding about the mechanism of thermal stress in the coke drum, temperature plots for two selected cycles which were taken from point C5 is shown in fig. 7. From the temperature plot we found that the average temperature gradient in the beginning of preheating stage is 1.3°C/minute for the normal cycle, and 1.2°C/minute for the short cycle. Thus, we can concluded that in the beginning of preheating stage, there are no significant differences between these two cycles. However, between preheating stages to the filling stage for the short cycle there occurred a temperature leap at minute 319 to minute 320, from 310°C to 410°C. At this time, temperature inflation is around 100°C. This fact also show that the stress peak occurred in the same time of the occurrence of the temperature leap.

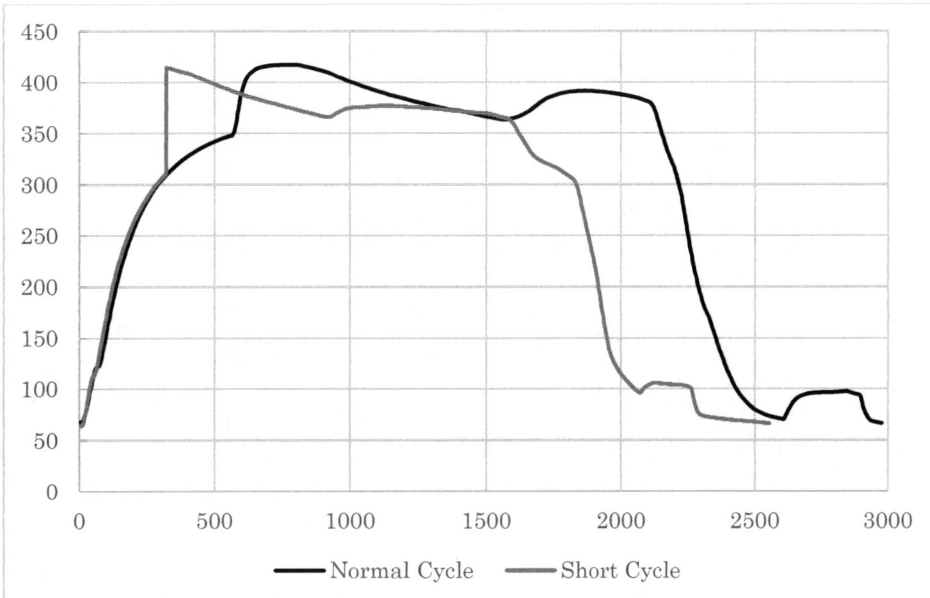

Fig. 7 Temperature plot for Normal Cycle and Short Cycle

The sudden change in temperature can caused a sudden change in stress rate due to thermal stress, called thermal shock. In thermal shock the rate application of the stress is very rapid. In some ways, the thermal shock also affected material to become brittle material by rapid application of stress and therefore may not be able to withstand a thermal-shock stress which they could readily absorb if it were slowly applied [6]. Even though the coke drum material SA-387 Gr. 11 CL.2 is categorized as a ductile material, but under repetitive application of thermal stress, the coke drum material can fall into brittle due to plastic deformation, work hardening and accelerate the thermal stress fatigue of material.

The cycle behavior have the important role to determine the fatigue life of the coke drum. As the aforementioned before, the sudden temperature increasing due to the bad maintaining during the cycle period can reduce the life span of the coke drum by a half, compare to the normal cycle. Thus the optimization of the cycle behavior have an important role to prolong the fatigue life of coke drum and need to be implemented.

4. Conclusion

In this research, numerical analyses to estimate fatigue life of coke drum have been implemented. It was found that the numerical analyses and measured data have a good agreement when it compare the strain result.

Two kind of cycles have been observed, which are categorized as normal cycle and short cycle. Two kind of cycles have been observed, which are categorized as normal cycle and short cycle. From the normal cycle it was found that the fatigue life of coke drum is 8085 cycle and the fatigue life of the short cycle is 4333 cycle, according to numerical analyses.

There is a significant fatigue life deflation for the short cycle compare to the normal cycle, which assumed because of the high stress peak in that occurred in the preheating stage of the short cycle that exceed the yield strength of coke drum material. This happened due to the leap of temperature in the preheating stage that caused a thermal shock and turn the upper part of the skirt area of the short cycle into plastic deformation and accelerate the thermal stress fatigue. For the aim of longer life span of coke drum material it can be concluded that the cycle optimization and behavior is have an important role to

IOP Conf. Series: Journal of Physics: Conf. Series **1005** (2018) 012002 doi:10.1088/1742-6596/1005/1/012002

control the cycle in moderate rise of thermal stress and avoid the thermal shock.

References

[1] ASME 2007 ASME Boiler & Pressure Vessel Code, Section II, Part D, Properties-Material (ASME)

[2] Oka A, Ambarita H, Daimaruya M, Fujiki H 2011 Study on the effects of switching temperature on the thermal fatigue life of the shell to skirt junction of coke drum *ASME Pressure Vessel Technology* 133

[3] *ANSYS Fatigue Module* 2006 ANSYS, Inc

[4] Zihui Xia, Fengju, Pierre Du Plessis 2010 Heat Transfer and Stress Analyses of Coke Drum for Complete Operating Cycle *ASME Journal of Pressure Vessel Technology 132*

[5] S Andrey, Ambarita H, Daimaruya M, K Hideki Transient Natural Convection and Conducting Heat Transfers on Hot Box of a Coke Drum in Pre-heating Stage IOP Conf. Seried: Journal of Physics. Series 801

[6] S. S Manson 1966 *Thermal Stress and Low-Cycle Fatigue* (McGraw-Hill) p 3

IOP Conf. Series: Journal of Physics: Conf. Series **1005** (2018) 012003 doi:10.1088/1742-6596/1005/1/012003

Spring-back of Thick Uni-Directional Carbon Fibre Reinforced Composite Laminate for Aircraft Structure Application

M Zakaria[1,4], Y Aminanda[1*], S A Rashidi[2] and M A Mat Sah[3]

[1] International Islamic University Malaysia, IIUM, Selangor, Malaysia
[2] Aerospace Malaysia Innovation Centre, AMIC, Kuala Lumpur, Malaysia
[3] CTRM Test Laboratory, Melaka, Malaysia
[4] CTRM AC, Melaka, Malaysia

*corresponding author: yulfian@iium.edu.my

Abstract. The springback phenomena of CFRP after curing process through autoclave manufacturing method results on the out of tolerance for its utilisation in aerospace industry. This paper relates to the measurements of springback for Uni-directional flat laminate as a first steps to the springback study for the real aircraft composite laminate structures. A flat laminate with dimension of 300 mm x 300 mm, 400 mm x 400 mm and 500mm x 500 mm with different number of ply; 20, 24 and 28 are manufactured. The choice of dimension and number of lay-up corresponds to the dimension and lay-up of rib structure. After process, the springbacks are measured using 3D scanner (optical-based three-dimensional) with an accuracy of 42 micrometers to obtain an accurate measurement. The analysis of the effect of dimension and number of ply to the magnitude of springback are presented within the range of specimen studied in this work.

1. Introduction

The increasing use of fiber-reinforced polymers in aerostructures as well as other industrial applications necessitates that their manufacturing quality and performance be continuously improved to allow rapid product development. In applications such as aerospace where precision is paramount, warpage of composite parts should be minimized to meet tight tolerance specification. Parts that warped beyond tolerance would demand excessive pre-stressing during assemblies to allow mating parts to get fastened. This pre-stressing would reduce the strength margin of the composite structures, risking premature failure.

Structures that undergo elevated temperature curing would always present post-cure warpage due to thermal expansion or contraction. This warpage is also known as springback. The thermal expansion mismatch between tools and laminates causes residual stresses to build up within the laminates [1-11]. Nonetheless, a practical solution such as predictive numerical model is crucial in quantifying the springback deformation [12 -14].

A previous study [13-14] has been carried out which managed to develop a predictive numerical model to predict the warpage or springback deformation of flat thin laminates. Prediction accuracy was excellent for some configurations but mediocre for some others. However, the validation of the model was only limited to thin specimens (four plies) since no empirical data was available for thicker ones. The actual composite parts are five to six times thicker, which may exhibit a totally different behavior that the existing model may not be able to predict. This work is aimed at extending the applicability of that model by providing validation baseline of thick laminates through empirical database.

2. Specimen fabrication process

The material used in this work is a typical one for aircraft structure application which a uni-directional (UD) carbon fibre pre-impregnated in epoxy resin (AIMS05-27-002). This material underwent the following manufacturing processes; cold storage, kitting, layup, curing, demould, trimming, non-destructive test (NDT), painting, assembly, and packaging.

Springback can be observed after demould process and therefore for this study the specimens underwent processes only until demould. In terms of inspection, only visual inspection was carried out, observing attributes such as surface appearance and colour. NDT, which can defect internal defects was not performed due to cost constraint. Furthermore, these specimens were cured together with production parts in the same autoclave using the same curing recipe so that any non-conformance would be reported through the production parts.

The composite specimens were made up of multiple plies laid up in a specific sequence and orientation [15]. These plies were cut to their shapes from the raw material roll using automated ply cutters. As such, the machines required input data in the form of nesting programs. Using a CAD software together with a nesting software, the nesting programs were created in such a way that will maximize the use of the materials. After being cut, the plies were sorted and kitted according to each specimen to be made.

The mould tools would determine the final shape and surface finish of the specimens. One flat mould tool with a dimension of 1.2 m x 1.2 m was used for this work as shown in Figure 1. The mould has a surface roughness, R_a of 1.0 to 1.6 μm as required for industrial application.

Figure 1. Mould tool used for the layup of the specimens

After all plies had been laid up, final bagging was performed using standard bagging procedure. The curing recipe followed the production standard, with double dwell. Due to confidentiality issue, the exact recipe is not shared in this text.

Using this manufacturing methods, a certain number of square flat UD laminates were fabricated (as shown in Table 1) for the springback study.

Table 1. Number of square flat UD laminates fabricated for springback study

No. of Ply	--------- 20 ---------			--------- 24 ---------			--------- 28 ---------			Total
Length	300	400	500	300	400	500	300	400	500	
Number of Panel	3	2	3	3	4	3	3	3	3	27

3. Springback measurement method

The springback deformation or warpage of the specimens was quantified through the measurement of surface deviation on the tool side the mould tool from its nominal contours. To get the most accurate reading as possible, a non-contact optical-based three-dimensional measurement was employed. A non-contact measurement was chosen to avoid any possible flexing of the specimens due to contact force from any contact-based measurement device. Thus, the most accurate reading is guaranteed. Plus, the speed at which a typical laser scanning device acquires measurement data is many times that of conventional contact-based probing [16].

This study requires a measurement accuracy of at least 150 micrometers (equivalent to one-third of the tolerance of a typical aerospace composite parts). The type of measurement device used in this case was the latter, a triangular-based 3D laser line scanner mounted on a six degrees of freedom portable measuring arm, as shown on the left side of Figure 2.

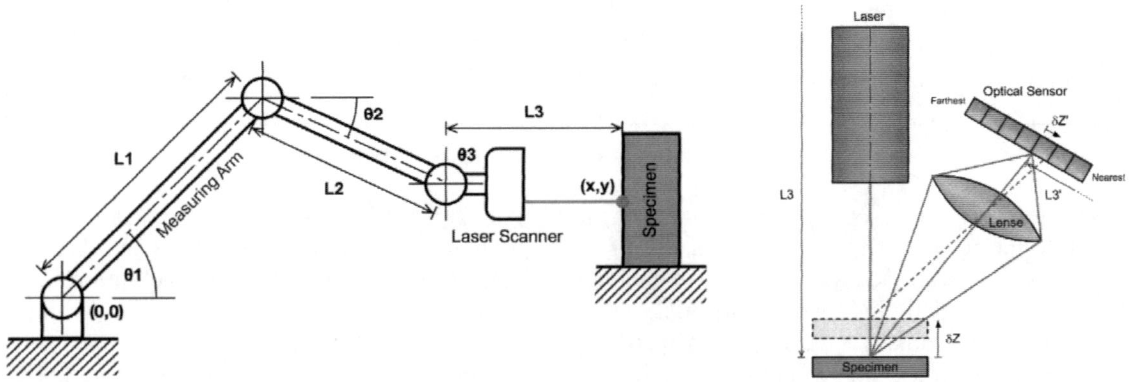

Figure 2. A two-dimensional representation of the measuring arm with the laser scanner (left) and the measurement principle of triangulation method (right)

There are two measurement principles used in this setup, one for the measuring arm and the other for the laser scanner. The measuring arm uses rotation-to-translation coordinate transformation based on input from rotary encoders built in each of its joint. For example, to get the coordinate of a two-dimensional measurement (x,y), the following transformation equation is employed;

$$x = L_1 \cos\theta_1 + L_2 \sin\theta_2 + L_3 \cos\theta_3 , \qquad y = L_1 \sin\theta_1 + L_2 \sin\theta_2 + L_3 \sin\theta_3$$

L1 and L2 are known from the physical dimension of the arm while L3 is acquired from the distance measurement of the laser scanner based on triangulation method as shown on the right side of Figure 2 [17]. The term triangulation stemmed from the fact that the three components (laser source, specimen, and sensor) form a triangular. When the laser beam strikes the surface of the specimen at a distance, some components of the reflected light ray would fall onto a spot along the optical sensor. A calibration is thus required to accurately map the laser spot position along the sensor with its physical position counterpart. Furthermore, when the distance of the specimen relative to the scanner is changed (δZ), a corresponding change in the position of the reflected spot along the sensor (δZ') can be measured. This change also needs to be calibrated accordingly to reflect the actual physical change of the distance, taking into account non-linearity effect of the triangulation. There will therefore be a range limit at which the scanner will be able to register its distance from the specimen, depending on the length of the sensor.

The basic specification of the measuring equipment used for this study is shown in Table 2 [16][18].

Table 2. Basic specification of the measuring arm (left) and the 3D laser scanner (right)

Parameter	Specification
Measuring Range (m)	2.0
Point Repeatability (mm)	0.030
Volumetric Accuracy (mm)	0.042
Arm Weight (kg)	7.4

Parameter	Specification
Accuracy (mm)	0.040
Sampling Rate (Hz)	45,000
Line Width at mid-field (mm)	80
Stand-off Distance (mm)	135 ± 45
Minimum Point Spacing (mm)	0.08

The result of the measurement from the laser scanner are clouds of point scattered throughout the three-dimensional space. An average number of measurement points for a specimen is 1.5 million. Considering that the actual number of measurement points to be included in the database is less than 20, the scanning capability in terms of resolution is somewhat an overkill. Nevertheless, more is always better when it comes to data especially when resources are abundant. More data means that more information could be extracted from the measurement results alone without the need to refer back to the physical specimens, which is a real time saver.

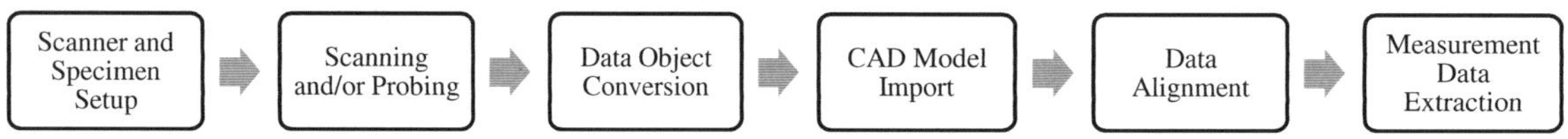

Figure 3. General workflow of the measurement process

The general workflow of measurement is shown in Figure 3. Scanning involved sweeping the scanner normal to the surface of specimens at a stand-off distance of about 150mm at a rate of about 50mm per second. Both tool and bag sides of each specimen were scanned in order to obtain as much information as possible, especially in regards to its thickness. Inspecting their thickness acts as a validation tool to ensure that all specimens were fabricated according to their respective specification.

After a complete points cloud was acquired, some trimming was done to remove unwanted noise from the model. Then the points cloud was converted to polygonal mesh. Polygonal mesh was used in this study since it can inherently define the normal direction of any point on a surface to be used in analyses such as thickness and surface deviation colour plots. Later, the CAD model of the mould tool would be imported into the project file to be used for data alignment, which is explained in the following paragraphs.

From the polygonal mesh of the specimens, nine extraction points were defined. The standard extraction pattern is depicted in Figure 4.

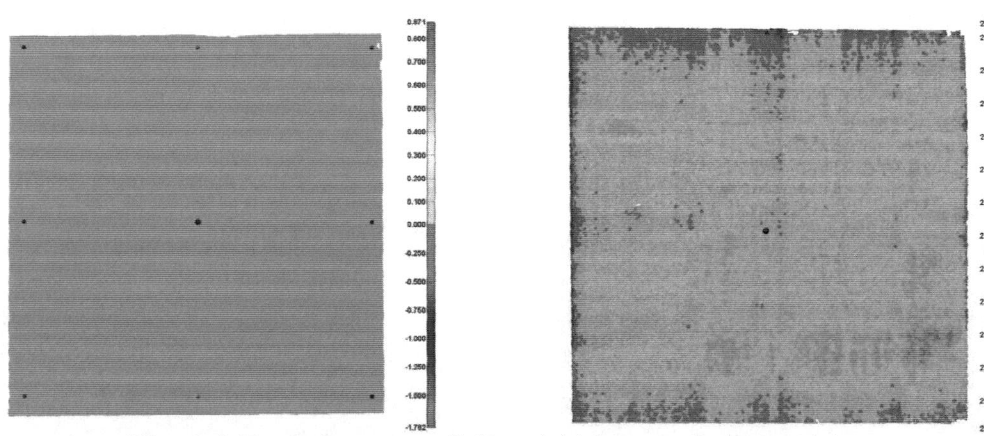

Figure 4. Extraction pattern (left) and thickness colour plot (right)

Each extraction point, known as comparison point, as its name implies, compares the three-dimensional location of the measurement point against the nominal surface of the CAD model of the mould tool. However, a valid comparison can only be done after a proper alignment be made between the data object (polygonal mesh) and the reference object (CAD model). Alignment for the specimen was accomplished through a best-fitting of the data object with the layup surface of the mould tool.

The in-plane position of each polygonal mesh with respect to the CAD model was approximated to be at its centre. This assumption was made since there is no record of the exact in-plane position of each specimen during the layup process. Furthermore, a thorough inspection of the mould tool itself revealed that the tool has satisfied the fabrication tolerance of 0.15mm throughout its layup surface. So any in-plane alignment position of the specimen should give similar output.

The extracted points were exported as spreadsheet workbook to be compiled into a database as shown in **Table 3**.

Table 3. Average of warpage for each specimen configuration

No. of Ply	----------- 20 -----------			----------- 24 -----------			----------- 28 -----------		
Length	300	400	500	300	400	500	300	400	500
Average of Warpage (mm)	0.033	0.379	0.955	0.135	0.296	0.361	0.106	0.187	0.284

4. Springback data analysis

In the dataset of this study, two parameters, size and thickness, are considered as the factors for the warpage or springback effects. The warpage was derived by calculating the difference between the highest and lowest measurement point along fibre direction. Graphs of the out-of-plane warpage versus the specimen size and thickness are depicted in Figure 5.

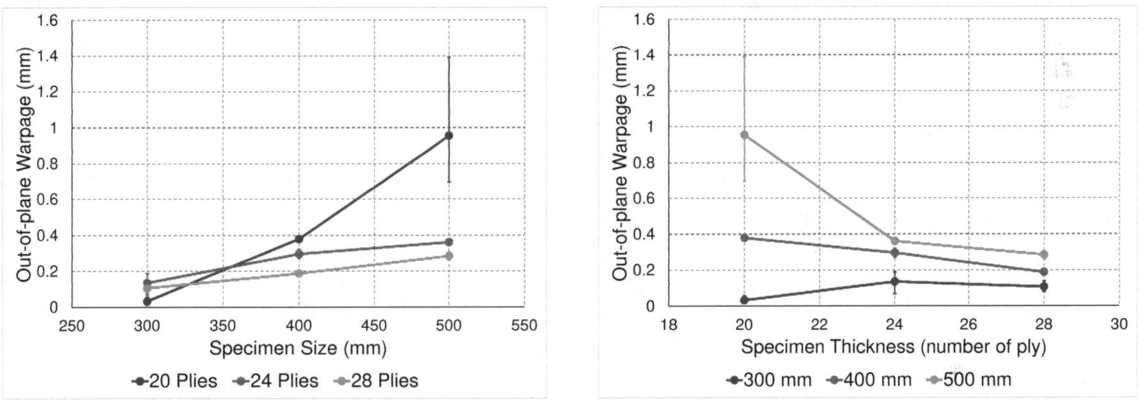

Figure 5. Graphs of the effect of specimen size (left) and thickness (right) on the warpage

Each data point in both graphs above has been averaged among three measurement data points. The error bars indicate the range between maximum and minimum values for respective data points. From the graphs, all data series are showing similar trends which increasing with size and decreasing with thickness, except for specimen thickness of 20 plies at sizes 300 mm. This can also be seen from their rather large error bars. The reason behind this outlying data is still under investigation. Nevertheless, these outliers have been excluded from the regression analysis. The mathematical models with respect to both parameters were derived through the best-fitting of normalized measurement data. Plots of the normalized warpage as a function of both parameters are illustrated in Figure 6.

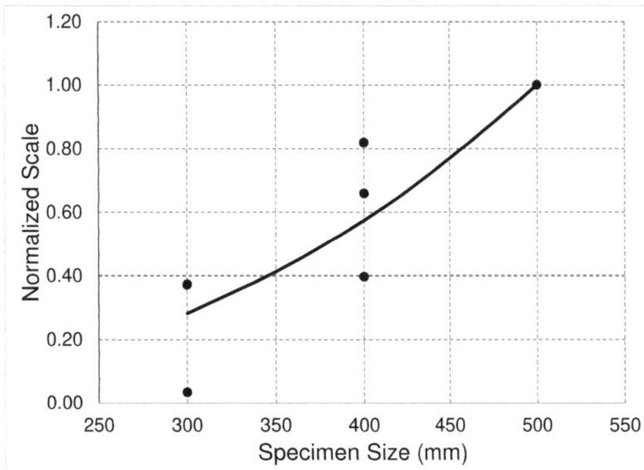

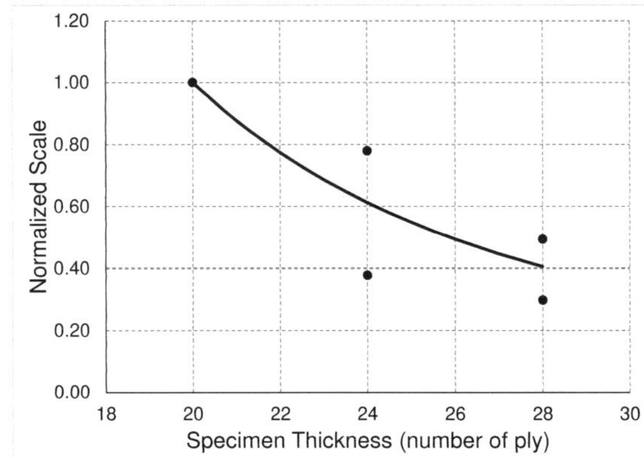

Figure 6. Plots of normalized out-of-plane warpage as a function of specimen size (left) and specimen thickness (right)

The result of the regression analysis yielded two mathematical equations, one for each factor, as follows;

$$W(L,t) = \left(\frac{L}{500}\right)^{2.5} \times \left(\frac{20}{t}\right)^{2.7} \times 0.955$$

Where W, L, and t are warpage, specimen length, and thickness respectively. Using these equations, the warpage of a specimen of different size and thickness can be predicted.

5. Conclusion and future works

The springback measurement and analysis for thick flat UD laminates have been carried-out. The normalized warpage trend takes the form of power function which conforms with thin laminates one with a difference on its magnitude. The warpage measurements in this work completes the range of number of UD layers for ribs aircraft structure application. A follow-up from this finding is the development of predictive numerical models in order to postulate mechanisms that govern the spring-back behavior. Ultimately, this numerical model will be utilized in the design of mold tools in order to compensate for the effect of spring-back and reduce manufacturing cost.

Acknowledgement
This work has been supported by MITI (Ministry of International Trade and Industry) of Malaysia through AMIC (Aerospace Malaysia Innovation Centre). The authors would like to thank AMIC for providing facilities and financial supports for this research project.

References

[1] Olivier PA. A note upon the development of residual curing strains in carbon/epoxy laminates. Study by thermomechanical analysis. Compos Part A 2006; 37: 602-16.
[2] Wisnom MR, Gigliotti M, Ersoy N, Campbell M, Potter KD. Mechanisms generating residual stresses and distortion during manufacture of polymer–matrix composite structures. Compos Part A, 2006; 37: 522–529.
[3] Twigg G, Poursartip A, Fernlund G. An experimental method for quantifying tool–part shear interaction during composites processing. Compos Sci and Technol 2003; 63: 1985–2002.
[4] Twigg G, Poursartip A, Fernlund G. Tool-laminate interaction in composites processing. Part I: experimental investigation and analytical model. Compos Part A 2003; 35: 121-33.
[5] Darrow DA, Smith LV. Isolating components of processing induced warpage in laminated composites. J Compos Mater 2002; 36: 2407-19.

[6] Tarsha-Kurdi KE, and Olivier P. Thermoviscoelastic analysis of residual curing stresses and the influence of autoclave pressure on these stresses in carbon/epoxy laminates. Compos Sci and Technol 2002; 62: 559-565.
[7] Marc Waris M, Liotier PJ, and Drapier S. Effect of the mold on the residual strain field monitored with optical fibers sensors in resin transfer molding processes. J Compos Mater 2014; 48:2589-2601.
[8] Hahn HT. Effects of residual stresses in polymer matrix composites. J Astro Sci. 1984; 32: 253–67.
[9] Kappel E, Stefaniak D, Hühne C. Process distortions in prepreg manufacturing – An experimental studyon CFRP L-profiles. Compos Struct 2013; 106: 615–625.
[10] Twigg G, Poursartip A, Fernlund G. Tool-laminate interaction in composites processing. Part II: numerical modeling. Compos Part A 2003; 35: 135-41
[11] Griffith J. Finite element based prediction of process-induced deformation of autoclaved composite structures using 2D process analysis and 3D structural analysis. Compos Struct 2003; 62: 223–234.
[12] Erik Kappel E, Stefaniak D, Fernlund G. Predicting process-induced distortions in composite manufacturing – A pheno-numerical simulation strategy. Compos Struct 2015; 120: 98–106.
[13] Mezeix L, Nasir MNM., Aminanda Y, Rivai A, Ali KM. Parameter study of tool-laminate interface through simulation for composite manufacturing using autoclave process. Appl Mech and Mater 2014; 606: 113-17.
[14] L. Mezeix, A.Seman, M. N. M. Nasir, Y. Aminanda, A. Rivai, B.Castanie, P.Olivier and K. M. Ali. Spring-back Simulation of Unidirectional Carbon epoxy flat laminate composite manufactured through autoclave process. Composite Structure vol 124, 2015, page 196 – 205
[15] Airbus Operation GmbH. RIB-MAIN 05. *Drawing Number V57591144-02, Issue B00, 2011.*
[16] Hexagon Metrology. HP-L-8.9 LASER SCANNER. *Product Brochure, No. 839991, Wetzlar 2015.*
[17] Zaharia, V D; Holonec, R; Crisan, S; Muresan, C; jr., R Munteanu. Three-Dimensional Scanning and Graphic Processing System. *Mediamira Science Publisher, Volume 55, Issue 1-2, Cluj-Napoca, 2014.*
[18] Hexagon Metrology. Romer Absolute Arm. *Factsheet No. 797356, Wetzlar 2015.*

IOP Conf. Series: Journal of Physics: Conf. Series **1005** (2018) 012004 doi:10.1088/1742-6596/1005/1/012004

Effect of Manufacturing Method to Tensile Properties of Hybrid Composite Reinforced by Natural (Agel Leaf Fiber) and Glass Fibers

A Nugroho[1)]*, K Abdurohman[2)], Kusmono[1)], H Hestiawan[1)] and Jamasri[1)]

[1]) Department of Mechanical and Industrial Engineering, Faculty of Engineering, Gadjah Mada University,Jln. Grafika No. 2 Yogyakarta 55281, Indonesia
[2]) Aeronautics Technology Center - LAPAN, Indonesia
*afidnugroho83@gmail.go.id

Abstract. This paper described the effect of different type of manufacturing method to tensile properties of hybrid composite woven agel leaf fiber and glass fiber as an alternative of LSU structure material. The research was done by using 3 ply of woven agel leaf fiber (ALF) and 3 ply of glass fiber (wr200) while the matrix was using unsaturated polyester. Composite manufacturing method used hand lay-up and vacuum bagging. Tensile test conducted with Tensilon universal testing machine, specimen shape and size according to standard size ASTM D 638. Based on tensile test result showed that the tensile strength of agel leaf fiber composite with unsaturated polyester matrix is 54.5 MPa by hand lay-up and 84.6 MPa with vacuum bagging method. From result of tensile test, hybrid fiber agel composite and glass fiber with unsaturated polyester matrix have potential as LSU structure.

1. Introduction

Lembaga Penerbangan dan Antariksa Nasional (LAPAN) was developing unmanned aircraft called Lapan Surveillance UAV (Unmanned Aerial Vehicle - LSU. There were some kind of UAV developing by LAPAN such as LSU-01, LSU-02, LSU-03, and LSU-05. They have some missions such as air picture or mapping of certain region.

The main materials used for LSU-02, LSU-03, and LSU-05 are polymer matrix composit using glass or carbon fiber and epoxy or polyester resin. The fiber is synthetic fiber type like glass, carbon, and aramid that widely used in composite because of its high stiffness, strength, and light weight. The disadvantages of shyntetic fiber are weakness of biodegradability, high cost in manufacturing, recycle, energy consumsion, engine abrasion, and healthy hazard. It changes interest of synthetic to natural fiber.

The advantages of natural fiber are environmental friendly, low density, recycle ability, no toxic, low price, low energy consumsion, high specific mechanical properties, good non-abrassive and heat resistance [1]; [2]; [3]. The disadvantages of natural fiber are high humidity, water resistance properties, and not uniform size variation [4].

Development of natural fiber is widely used because of limited metal material source in the world [5]. Despite couldn't fully change synthetic fiber, the used of natural fiber is a wise step in environmental issue. One of Indonesian natural fiber is agel leaf fiber from gebang tree (Corypha gebanga). Its wide leaves are often used for traditional roof, and the fibers are used for producing traditional strings, bags, caps, chairs, and other handicraft.

IOP Conf. Series: Journal of Physics: Conf. Series **1005** (2018) 012004 doi:10.1088/1742-6596/1005/1/012004

Figure 1. Gebang tree

Mechanical properties of composite are influnced by several factors such as fiber type, matrix type, fiber to resin ratio, geometry, and fiber orientation [6]. There are 3 types of matrix; thermoplastic, thermoset, and elastomer. Widely used matrix is thermoset such as epoxy, polyster, and phenol resin [7]. The development of composite materials improving their performance limits based on the reinforcement of two or more fibres (synthetic fibre with another synthetic fibre or synthetic fibre with natural fibre or synthetic fibre with metallic fibres) in a single polymeric matrix, which leads to the advanced material system called hybrid composites with a great diversity of material properties, is still in its infancy [8].

The properties of a hybrid composite mainly depend upon the fiber content, length of individual fibres, orientation, extent of intermingling of fibres, fiber to matrix bonding and arrangement of both the fibres. The strength of the hybrid composite is also dependent on the failure strain of individual fibres. Maximum hybrid results are obtained when the fibres are highly strain compatible [9]. Fiber woven fabrics are probably the most commonly used textiles in structural applications [10]. This research use woven fabric glass fiber and hybrid by agel leaf natural fiber.

The objectives of this research is to know the influence of manufacturing method; hand layup and vacuum bagging in the manufacturing process of hybrid composite agel leaf fiber (ALF)/WR 200 glass fiber/unsaturated polyster in relationship to tensile properties.

2. Method

Manufacturing of hybrid composite woven ALF/WR 200 glass fiber/unsaturated polyster are used hand layup and vacuum bagging method. Braiding process of ALF are using simple machine. Braiding process are did in handicraft center of natural fiber in Sentolo, Kulonprogo, Yogyakarta. WR200 fiber glass and unsaturated polyester (UPRs) 157 BQTN with hardener MEKPO (metil etil keton peroksida) 1% are supplied by PT. Justus Kimia Raya Jakarta.

Fig. 2 showed the stacking sequence of hybrid composite consist of 6 layers of fiber (3 layers ALF and 3 layers WR200 glass fiber) where the laminate are ALF-FG-ALF-FG-ALF-FG. Hand layup used mould that refer to tensile test standard of ASTM D638 (see fig. 2). It used post processing by press. Vacuum bagging used release film and breather fabric to reserve resin during vacuum process. Vacuum bagging equipment are showed in fig. 3. Composite panel 100x170x2 mm is cutted to tensile test specimen refer to ASTM D638.

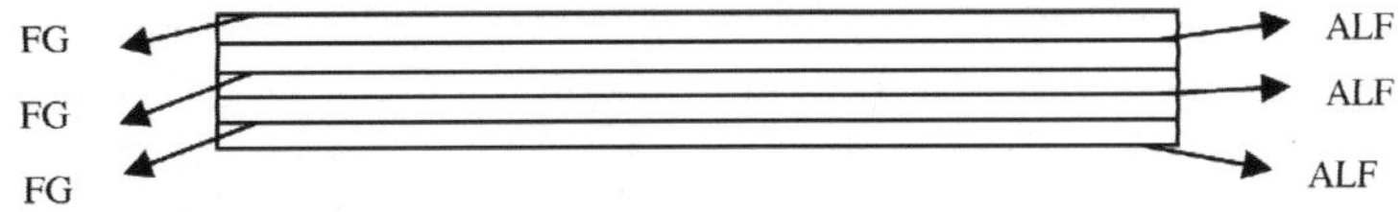

Figure 2. Stacking sequence of composite

Figure 3. Hand lay-up mould

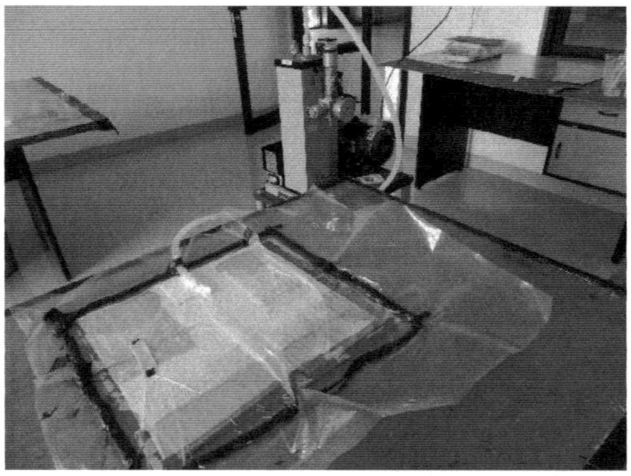

Figure 4. Vacuum bagging equipment

3. Results and Discussions

The result of manufacturing process showed the physic differences between hand layup and vacuum bagging, specially in composite thickness. With the same layer of fibers there are different thickness of composite wich are 3 mm for hand layup and 2 mm for vacuum bagging. Vacuum bagging utilized vacuum pressure during the process, so it pressed the laminates and partly resin come out from the laminate. It caused the vacuum bagging thickness lower than hand lay up. Otherwise, it described that the matrix fraction of vacuum bagging lower than hand layup (fiber fraction higher) because there are resin reduction during the process caused by vacuum pressure.

The tensile strength of both hand layup and vacuum bagging are presented in Fig. 5. It showed the ultimate strength for hand layup composite is 54.5 MPa and vacuum bagging composite 84.6 MPa. It showed that composite tensile strength using vacuum bagging 36% higher than hand layup composite. It was caused by higher fiber fraction in vacuum bagging as a result of vacuum process. K. Jarukumjorn and N. Suppakarn 2009 [11] have reported the properties of hybrid composite glass and sisal natural fiber–polypropylene. The best tensile strength for glass-sisal composite is 31.59 MPa. It means that tensile strength of glass-ALF-UPRs better than glass-sisal-PP composite. Other research by V.P. Arthanarieswaran et. al. 2014 [12] have reported the properties of banana fiber reinforced epoxy composite with glass fiber hybridization. It used the laminate variation of glass (G)/banana (B)/glass (G) and glass/banana/glass/banana/glass. The results showed the tensile strength are 46 MPa for G/B/G and 88 MPa for G/B/G/B/G. It indicated that the tensile strength of glass-ALF-UPRs higher than G/B/G-epoxy but lower than G/B/G/B/G-epoxy. This comparative tensile strength value could be suggestion to the next research of ALF hybrid composite with the upper and lower skin laminate using glass fiber to be symmetric balanced. It possible that the ALF skin of laminate become the first failure of the composite because the mechanical properties of ALF is lower than glass fiber.

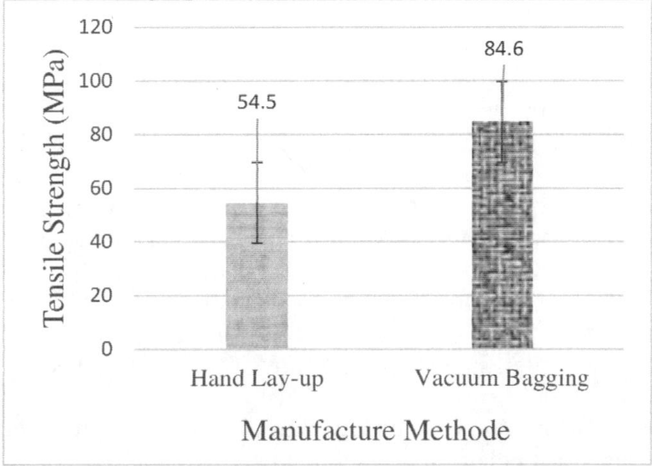

Figure 5. Ratio of manufacturing method to tensile strength

Fig. 6 presented the elastic modulus value for both vacuum bagging and hand layup composite. Elastic modulus value for hand layup composite is 3494.8 MPa and vacuum bagging composite 3046.1 MPa. It showed that vacuum bagging elastic modulus 13% lower than hand layup. Other research reported that elastic modulus is 2430 MPa for sisal-glass-PP [11] and 1940 MPa for banana-glass-epoxy composite [12]. It indicate that elastic modulus value of ALF-glass-UPRs composite higher than other glass hybridization composite for sisal and banana fiber.

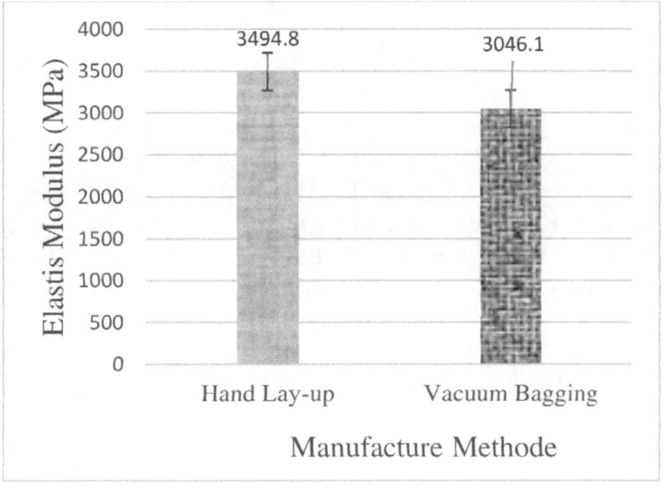

Figure 6. Ratio of manufacturing method to elastic modulus

4. Conclusions

Conclusion of this paper are vacuum bagging strength has 36% higher than hand layup but 13% lower in elastic modulus. It tensile strength and elastic modulus are higher compare to other glass hybridization for sisal and banana composite.

Acknowledgement
The authors would like to thank director of Aeronautics Technology Center (Pustekbang) LAPAN Mr. Gunawan Setyo Prabowo and head of progfas department Mr. Agus Aribowo for providing research support facilities.

References

[1] A. K. Mohanty, M. Misra, and L. T. Drzal, "Sustainable Bio-Composites from Renewable Resources : Opportunities and Challenges in the Green Materials World," vol. 10, no. April, pp. 19–26, 2002.

[2] V. S. Sreenivasan, D. Ravindran, V. Manikandan, and R. Narayanasamy, "Mechanical properties of randomly oriented short Sansevieria cylindrica fibre / polyester composites," *Mater. Des.*, vol. 32, no. 4, pp. 2444–2455, 2011.

[3] C. W. Nguong, S. N. B. Lee, and D. Sujan, "A Review on Natural Fibre Reinforced Polymer Composites," vol. 7, no. 1, pp. 33–40, 2013.

[4] C. Baillie, *Green Composites: Polymer Composites and the Environment*. CRC Press, Taylor & Francis Group, 2005.

[5] S. V Joshi, L. T. Drzal, A. K. Mohanty, and S. Arora, "Are natural fiber composites environmentally superior to glass fiber reinforced composites ?," *Compos. Part A*, vol. 35, pp. 371–376, 2004.

[6] W. D. Callister, *Materials Science and Engineering An Introduction*, SEVENTHE. John Wiley & Sons, Inc. New York, 2007.

[7] L. Y. Mwaikambo and M. P. Ansell, "Chemical modification of hemp , sisal , jute , and kapok fibers by alkalization Chemical Modification of Hemp , Sisal , Jute , and Kapok Fibers by Alkalization," *J. Appl. Polym. Sci.*, vol. 84, no. June, pp. 2222–2234, 2002.

[8] R.T. Durai Prabhakaran, Tom L. Andersen, C.M. Markussen, Bo Madsen, Hans Lilholt, "Tensile and Compression Properties of Hybrid Composites – A Comparative Study", The 19th International Conference on Composite Materials, pp. 1029-1035,2013.

[9] Ali I. Al-Mosawi, Mohammad H.Al-Maamori, Zaynab A.Wetwet, "Mechanical Properties of Composite Material Reinforcing by Natural-Synthetic Fibers", Academic Research International, vol. 3, no. 3, November, pp. 108-112, 2012.

[10] Mohd Aidy Faizal, Yeo Kiam Beng & Mohd Noh Dalimin, "Tensile Property of Hand Lay-UpPlain-Weave Woven E-Glass/Polyester Composite: Curing Pressure And Ply Arrangement Effect", Borneo Science, The Journal of Science & Technology, pp: 27-34, ISSN: 2231-9085.

[11] Jarukumjorn, K., & Suppakarn, N., "Effect of Glass Fiber Hybridization on Properties of Sisal Fiber-Polypropylene Composite", Composite: Part B, pp: 623-627, April 2009.

[12] Arthanarieswaran, V.P., Kumaravel, A. and Kathirselvam, "Evaluation of Mechanical Properties of Banana and Sisal Fiber Reinforced Epoxy Composite: Influence of Glass Fiber Hybridization", Materials & Design, pp. 194-202, August 2014.

Analysis of Aerodynamic Load of LSU-03 (LAPAN Surveillance UAV-03) Propeller

Awang Rahmadi Nuranto[1)]Ahmad Jamaludin Fitroh[2)]and Hendri Syamsudin[3)]

[1)] Student, Faculty of Mechanical and Aerospace Engineering, InstitutTeknologi Bandung, Indonesia
Aeronautics Technology Center, National Institute of Aeronautics and Space (LAPAN), Indonesia
[2)] Rocket Technology Center, National Institute of Aeronautics and Space (LAPAN), Indonesia
[3)] Faculty of Mechanical and Aerospace Engineering, InstitutTeknologi Bandung, Indonesia

Abstract. The existing propeller of the LSU-03 aircraft is made of wood. To improve structural strength and obtain better mechanical properties, the propeller will be redesigned usingcomposite materials. It is necessary to simulate and analyze the design load. This research paper explainsthe simulation and analysis of aerodynamic load prior to structural design phase of composite propeller. Aerodynamic load calculations are performed using both the Blade Element Theory(BET) and the Computational Fluid Dynamic (CFD)simulation. The result of both methods show a close agreement, the different thrust forces is only 1.2 and 4.1% for two type mesh. Thus the distribution of aerodynamic loads along the surface of the propeller blades of the 3-D CFD simulation results are considered valid and ready to design the composite structure. TheCFD results is directly imported to the structure model using the Direct Import CFD / One-Way Fluid Structure Interaction (FSI) method. Design load of propeller is chosen at the flight condition at speed of 20 m/s at 7000 rpm.

1. Introduction

Propeller is component of several types of aircraft to be local components for aircraft produced in Indonesia.

Various types of aircraft ranging from passenger aircraft and Unmanned Aerial Vehicle (UAV), turbopropeller engine to piston requires a propeller as a force-changer of rotational motion into a thrust force by utilizing the aerodynamic forces acting on the propeller. The design of the propeller is essential to improve performance with the adjustable propeller and engine that will produce matching so that high efficiency will be obtained. Well designed propellers can generally achieve 80% efficiency in their best operating areas. [1]

One of the UAV prototypes produced by LAPAN is LSU-03. That UAV uses pusher type of propeller. The existing propeller is made of wood. One of the weaknesses of wood is not withstand the age and the unfavorable mechanical properties. To improve structural strength and better mechanical properties, the propeller will then be made using composite materials. Before designing the composite material, aerodynamic load distribution is calculated along the surface of the propeller blades.

IOP Conf. Series: Journal of Physics: Conf. Series **1005** (2018) 012005 doi:10.1088/1742-6596/1005/1/012005

This research aims to obtain aerodynamic load distribution along the surface of the propeller blades. The aerodynamic load distribution is expected to be used as input in designing the composite structure of the propeller blades.

The aircraft propeller that will be simulated and analyzed for their structural strength are the LSU-03 (LAPAN Surveillance UAV - 03) propeller. The aircraft is designed as a high-performance unmanned aircraft for mapping and surveying in small to medium areas. This aircraft can fly for 4 hours with a load of 5 kg with a cruise speed of 80 km / h and a maximum speed of 110 km/h with a flying height cruise 1000-2000 m. The power used is a piston engine with a capacity of 60 cc and power output of 6 hp @ 8000 rpm using fuel octane 92. Engine power is connected through the shaft to the propeller 2 blades with the size 24 x 10 pusher type. [2]

Figure 1. LSU-03Aircraft[2]

2. Methodology

In this research, the method to calculate analytical aerodynamic load is using the blade element theory. The Blade Element Theory (BET) is a relatively simple model to predict the performance of a certain propeller geometry. In BET the propeller blade is subdivided in small elements and the two-dimensional flow around each element is analyzed individually. [3]

The BET is assumed that the behavior of an element is not affected by adjacent elements of the same blade, the airfoil characteristic is adopted for the element, and the effective velocity of the element by air is the resultant axial velocity, rotation speed and velocity imposed. [4]

The theory of aircraft propellers, following the original development of finite wing theory, has nearly always proceeded as a lifting line analysis. That is, blade elements may be considered to act as two-dimensional foils upon which the forces are the same as would be found in a uniform two-dimensional flow with the same velocity and direction as occurs locally at the blade element. This approach to the design of blade elements is continued in the present study. [5]

Data on blade elements theory are airfoil geometry of each segment, chord length, twist angle, lift coefficient, drag coefficient and rotary speed propeller.

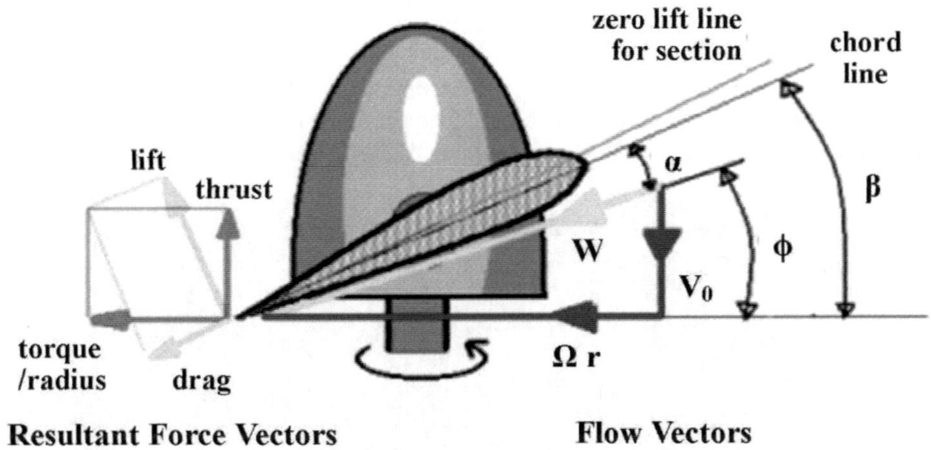

Figure 2. Classic blade element theory [4]

The formula of blade element theory:

$$l = \frac{1}{2}\rho W^2 c C_l \quad (1)$$

$$d = \frac{1}{2}\rho W^2 c C_d \quad (2)$$

$$C_l, C_d = f(\alpha) \quad (3)$$

$$W = \sqrt{V_0^2 + (\Omega r)^2} \quad (4)$$

$$T = B(l cos\phi - d sin\phi) \quad (5)$$

where l is lift, d is drag, W is resultan velocity, ρ is air density, c is chord, C_l is lift coefficient, $C_d$is coefficient drag, Vo is axial velocity, Ωr is tangential velocity, B is the number of propeller blades, ϕ is angle of velocity and T is thrust.

The propeller is rotated at a speed of 7000 RPM with axial velocity (flight speed) of 20 m/s, this condition is equal to the condition at the time of take-off aircraft which the propeller receives the maximum load. The propeller having two blades has a hub diameter of 0.052 m and a tip diameter of 0.6 m. ρ is 1,225 kg/m3.

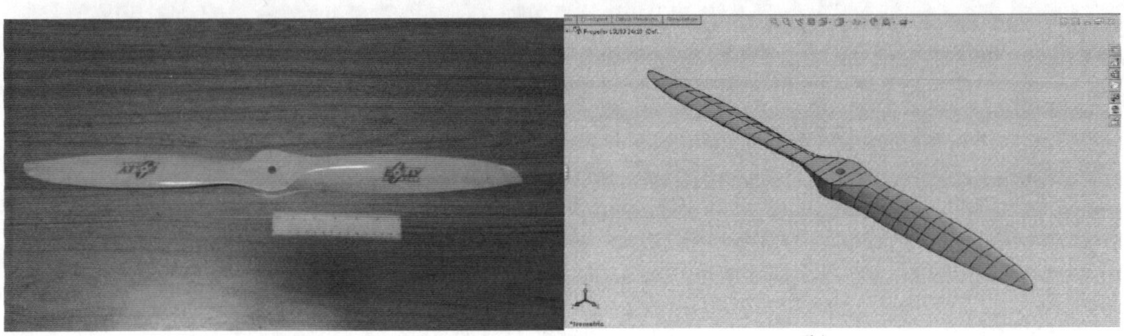

(a) (b)

Figure 3. Original propeller (a), 3-D CAD propeller model (b)

The propeller model LSU-03 by modeling the specification of an existing propeller using 3D scan and converted in Computer Aided Design(CAD) with Solidworks parametric for easy change in geometry.

Aerodynamic load calculations were performed using CFD 3-D simulation method with Ansys software with CFD CFX which is included in this software. Prior to entering the simulation stage, a 3-D CAD plane model was made using Solidworks software as an input geometry model for simulation.

The CFD simulation reliability depends on many aspects, including computational domain sizing, mesh size and quality as well chosen solution type (steady state or transient) and physical formulation (type of equation solved, additional wall formulations and corrections) [6].

Figure 4.3-D CAD propeller model has been converted to Ansys

The simulation stages to obtain aerodynamic load on the propeller are as follows: first make a 3-D CAD model of the propeller. In Figure 4show of CAD model geometry from solidworks to ansys to create fluid flow domains and their boundaries.

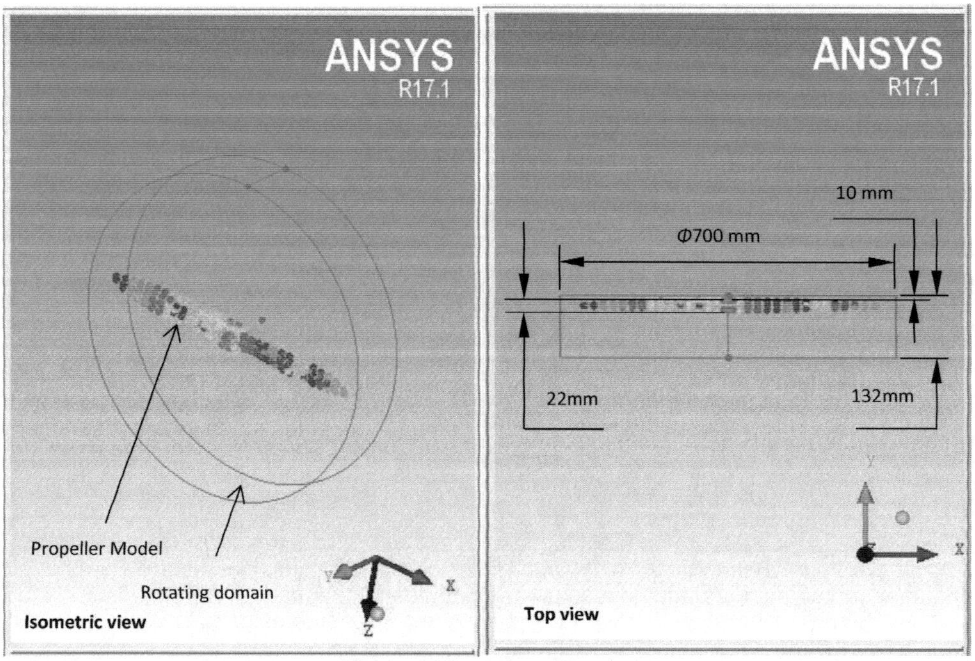

Figure 5. Rotating domain

By using the ICEM software (part of Ansys) 3-D geometry model is imported for the creation of rotating domain with space interfaces for cylindrical rotation space of 0.7 mdiameter with a distance of 10 mm from the surface of the front propeller and 100 mm from the surface of the rear propeller (see figure 5).For the cylindrical field (static domain) has a dimension of 10 m diameter with a cylinder length of 10 m and the distance from the front side of the static space to the front of the rotation space of 1.5 m (see figure 6).

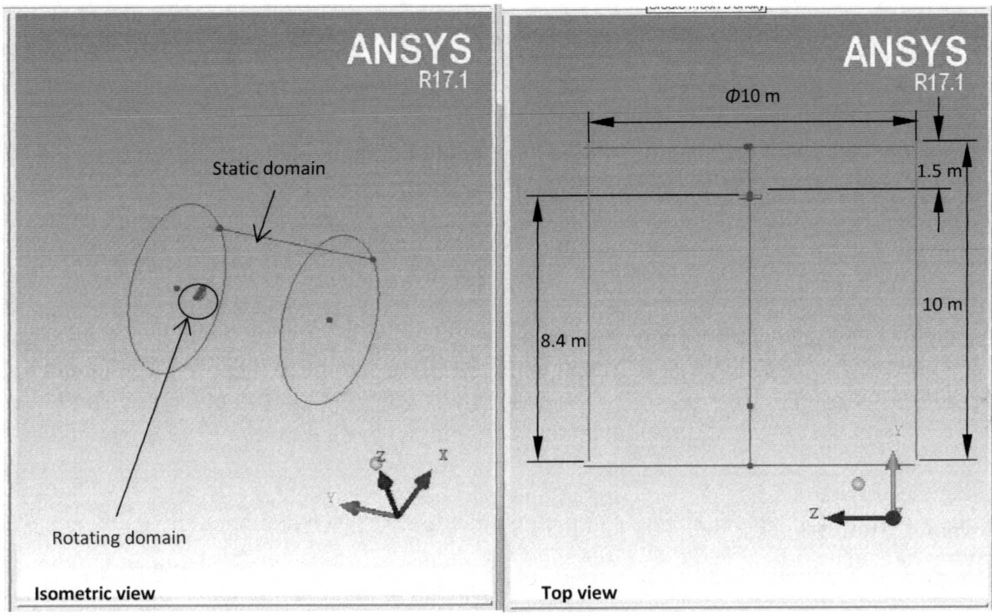

Figure 6. Static domain

After that created the input, outputto define fluid flow limitsfor each surface. Do the meshing process, with tetra/mixed mesh type and robust mesh (octree) mesh method, this meshing applied to rotation domain and static domain. The rotating mesh composition has a size of 1.25 for the propeller, 14 for the front (input) and rear (output) and 7 for the tube. And static mesh made with size 14 for the front (input), rear (output) and tube. In static mesh arranged in more detail by using the ratio at 1.4 so the denser of mesh more dense in the middle.

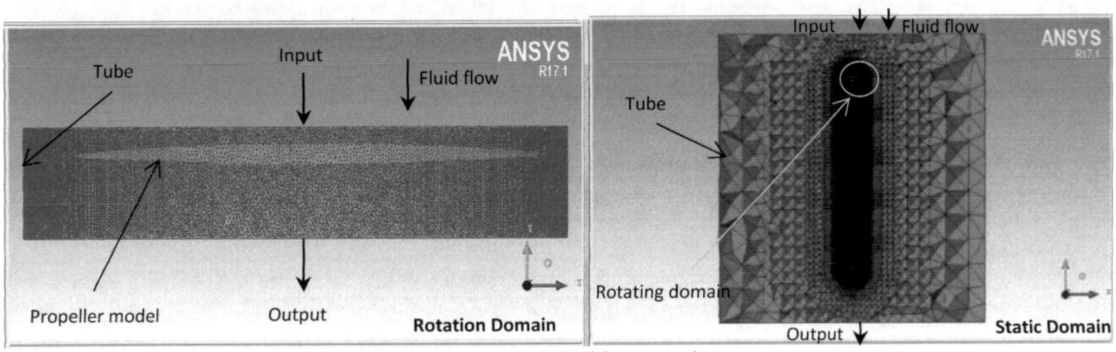

Figure 7. Meshing results

After meshing, open CFX-Pre is still part of Ansys software to enter the boundary conditions of fluid flow.

The rotation domainare set for Basic Setting and Fluid Model as follows: The rotation domain of Basic Setting are for material: *"Air Ideal Gas"*, domain option: *"Rotating"*, angular velocity: *"7000 [rev min^{-1}]"*androtation axil: *"Global Y"*. AndFluid Models arefor heat transfer option:*"Total Energy"*andturbulence option:*"Shear Stress Transport"*.

On the static domain consist of opening, inlet and outlet sideare set as follows:

1. Openingside (tube)setting, set the Basic Setting on Boundary Type: "*Opening*".The Boundary Details on Mass and Momentum Option: "*Cart Vel Components*"at V = *20[m s^{-1}]* and Opening Temp.:at "*25°C*".
2. Inletside setting, set the Basic Setting onBoundary Type is *"Inlet"*.The Boundary Details on Flow Regime Option:*"Subsonic"*, Mass and Momentum Option: *"Normal Speed"*, Normal Speed: *20 [m s^{-1}];*andStatic Frame Total Temp.:*"25°C"*.
3. Outlet side setting,as follows set the Basic Setting onBoundary Type is *"Outlet"*. The Boundary Details on Flow Regime Option: *"Subsonic"*, Mass and Momentum Option: *"Static Pressure"* andRelative Pressure:*" 0 (Pa)"*.

The simulation is carried out at the condition of the flight speed of 20 km/h and the rotational speed of 7,000 rpm. The CFD 3-D simulation results will be validated using Blade Element Theory.

From the propeller geometry, a 3-D rotational aerodynamic model (CFD CFX) is applied to obtain the pressure / load distribution on the propeller. The CFD result of the pressure distribution is compared with the analytical calculation result using the Blade Element Theory (BET) method. After CFD simulation, for the next research use the Direct Import CFD (One-Way Fluid Structure Interaction) method then mapping to match the CFD model with Finite Element Method (FEM)in Ansys software.

3. Result and Analysis

a) Analytical Result

The data of $C_l$and C_d based on the angle of attack (α) were obtained using CFD 2-D by incorporating the airfoil geometry of the propeller and its condition,as shown in Table 1 and Figure 8 below.

Table 1. Data of liftcoefficient anddrag coefficient

Angle of Attact *α*	***Lift*Coefficient *Cl***	***Drag*Coefficient *Cd***	***L/D***
-10	-0.344	0.023	-15.088
-9	-0.279	0.021	-13.037
-8	-0.214	0.020	-10.700
-7	-0.142	0.019	-7.358
-6	-0.070	0.019	-3.763
-5	0.006	0.018	0.302
-4	0.081	0.018	4.481
-3	0.153	0.018	8.453
-2	0.221	0.046	4.846
-1	0.291	0.050	5.879
0	0.374	0.051	7.305
1	0.432	0.020	21.818
2	0.497	0.021	24.010
3	0.559	0.022	25.760
4	0.617	0.023	26.826
5	0.671	0.025	27.388

6	0.723	0.026	27.490
7	0.772	0.029	27.088
8	0.817	0.031	26.355
9	0.817	0.031	26.355
10	0.817	0.031	26.355

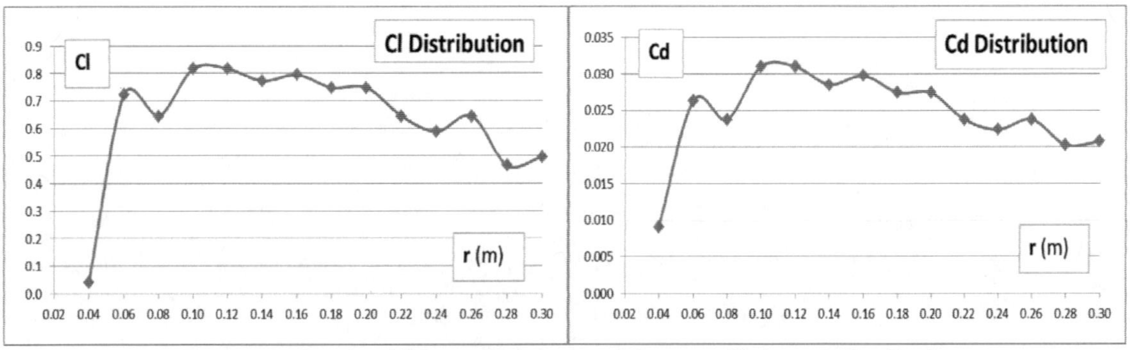

Figure 8. Graph of lift coefficient$C_l$and drag coefficient $C_d$distribution

Figure 8 shows that the value of$C_l$along the blade ranges from 0.5 to 0.8. The value ofC_l at the root is get near to zero because its section is not airfoil. $C_l$value tend to decrease from the position of root to tip because the angle of attack which also decreases. Similarly, with the value of *Cd* because the value of $C_l$and *Cd* is a function of angle ofattack.

The value of lift and drag are obtained from equations (1) and (2). Chord and twist angle distributions are presented in Table 2 as below.

Table 2. Chord and twist angledistributions

Segment	**Radius, r (m)**	***Chord*, c (m)**	**Twist Angle, β (deg)**
1	0.04	0.0413	29.36
2	0.06	0.0414	30.40
3	0.08	0.0440	23.34
4	0.10	0.0455	23.18
5	0.12	0.0469	21.20
6	0.14	0.0472	18.20
7	0.16	0.0470	17.00
8	0.18	0.0456	15.00
9	0.20	0.0435	14.30
10	0.22	0.0403	11.35
11	0.24	0.0360	10.00
12	0.26	0.0310	10.70
13	0.28	0.0236	7.00
14	0.30	0.0080	8.00

Table 2 shows the chord distribution tends to decreasefrom root to tip. It is intended that the aerodynamic load at the tip of the blade is not too large. The table also shows that the twist angle distribution also decreases from root to tip. It is also intended that the aerodynamic load at the tip of the blade is not too large. The twist angle at the root is made larger to increase the strength of the structure in that area.

The speed distribution is calculated using equation (4). The results are presented in Figure 9 as below.

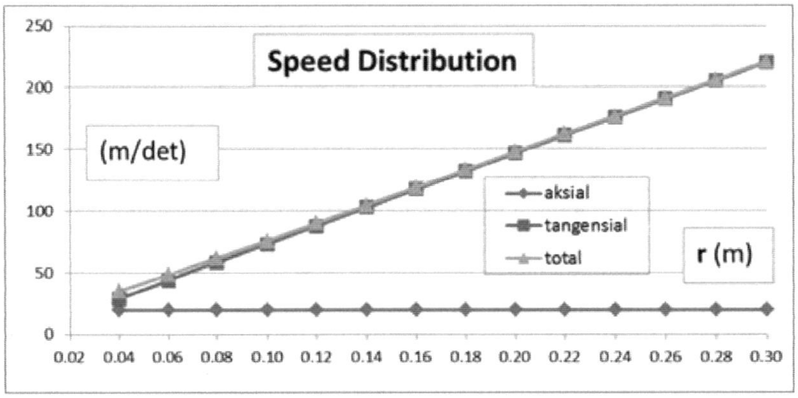

Figure 9. Graph of speed distribution

The axial velocity distribution in Figure9has constant value. Because of the assumption used is not difference in angle between the axial position of the propeller with the flight direction of the plane. The tangential velocity has enlarged to the tipand the value r increasingly large. By summing up the constantly axial velocity with the enlarged of tangential velocity at the tip of the blade, the enlarged of total velocity is obtained.

After all the necessary parameters have been fullfiled, we get the value of lift and drag in each segment of the blade. The distribution of lift and drags is shown in Figure 10.

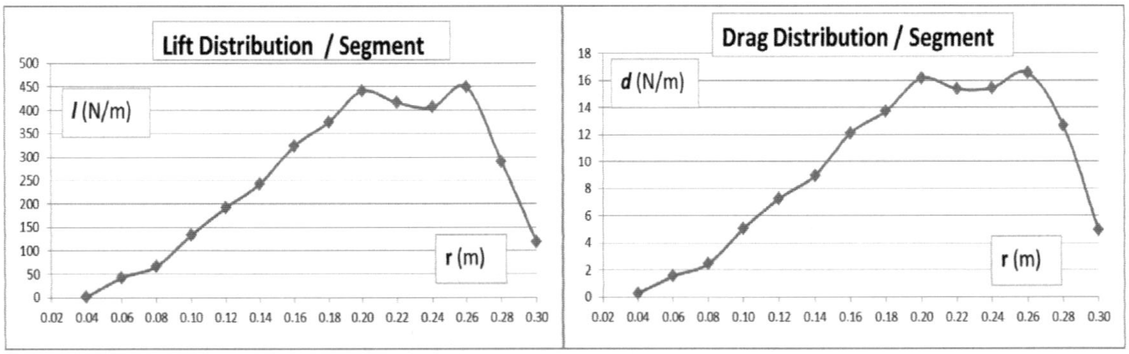

Figure 10. Graph of lift and drag distribution

Figure 10 shows that the value of lift and drags is getting increaseat the tip of the blade. That caused the value of the total speed is getting bigger in that area.

The amount of thrust in each segment is calculated by equation (5). The distribution of the calculation results is presented in Figure 11 as below.

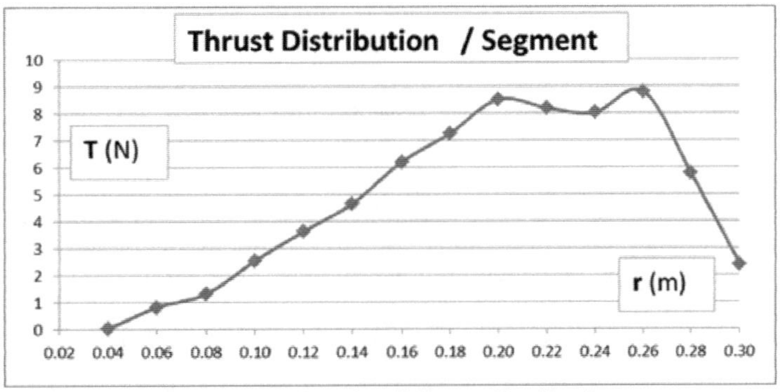

Figure 11. Graph of thrust distribution

Figure 11 above shows that the thrust is getting larger in the tip area of the blade. That caused the lift is getting bigger in that area.

The calculation results using analytical aerodynamic load calculation obtained a thrust for each blade of 68 N and the total thrust force of the propeller is 136 N.

b) CFD Result

After all is done, then the next step is done running to get the simulation results. Two simulations were performed, for the second simulation, the addition of mesh on the surface of the propeller with the prism mesh so that the boundary layer on the surface of the propeller can be analyzed in more detail.

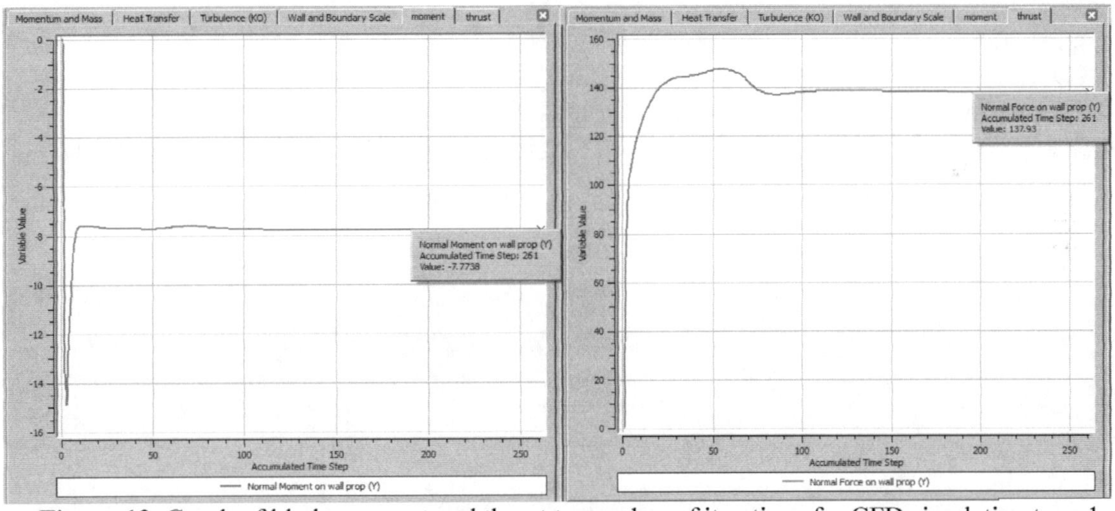

Figure 12. Graph of blade moment and thrust to number of iterations for CFD simulation type 1 (without prism mesh)

From figure 12 and 13show graph the moment and thrust on the propeller has not changed even though the iteration process is reproduced. The addition of iteration will not have much effect on the results obtained. For CFD simulation type 1 and 2, the moment and thrust obtained not too far.

IOP Conf. Series: Journal of Physics: Conf. Series **1005** (2018) 012005 doi:10.1088/1742-6596/1005/1/012005

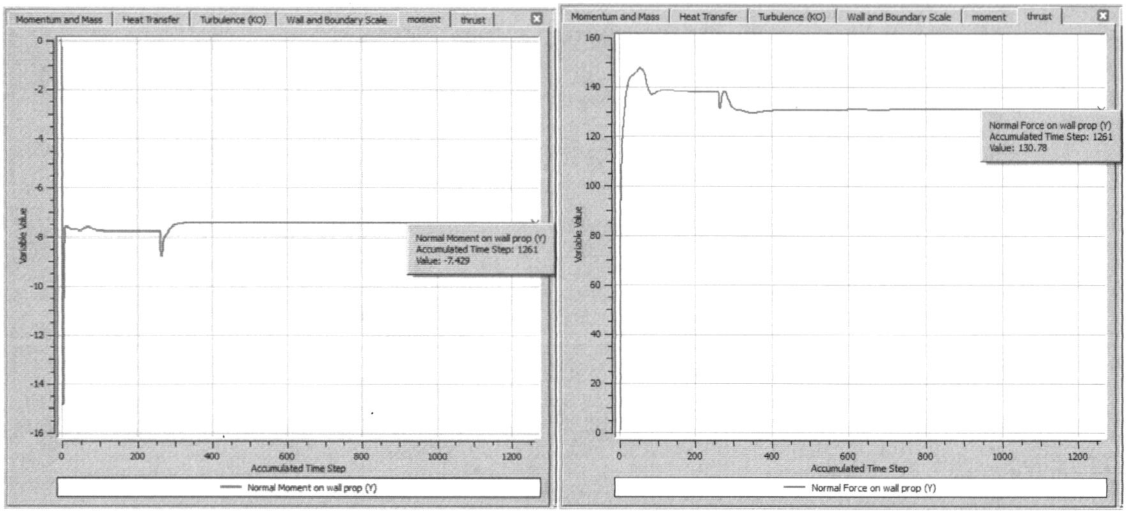

Figure 13. Graph of blade moment and thrust to number of iterations for CFD simulation type 2 (with prism mesh)

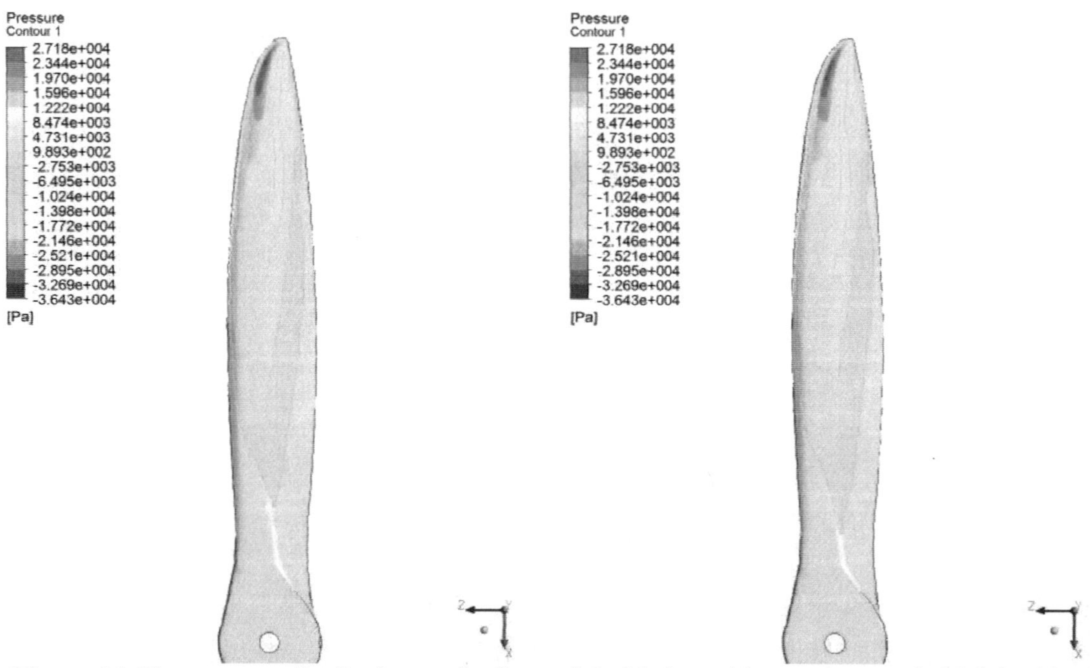

Figure 14. The pressure distribution on the front of the blades, without prism mesh (right) withprism mesh (left)

From the figures 14 and 15 it appears that the result of the pressure distribution is not very different between the simulations using prism mesh and without prism mesh.

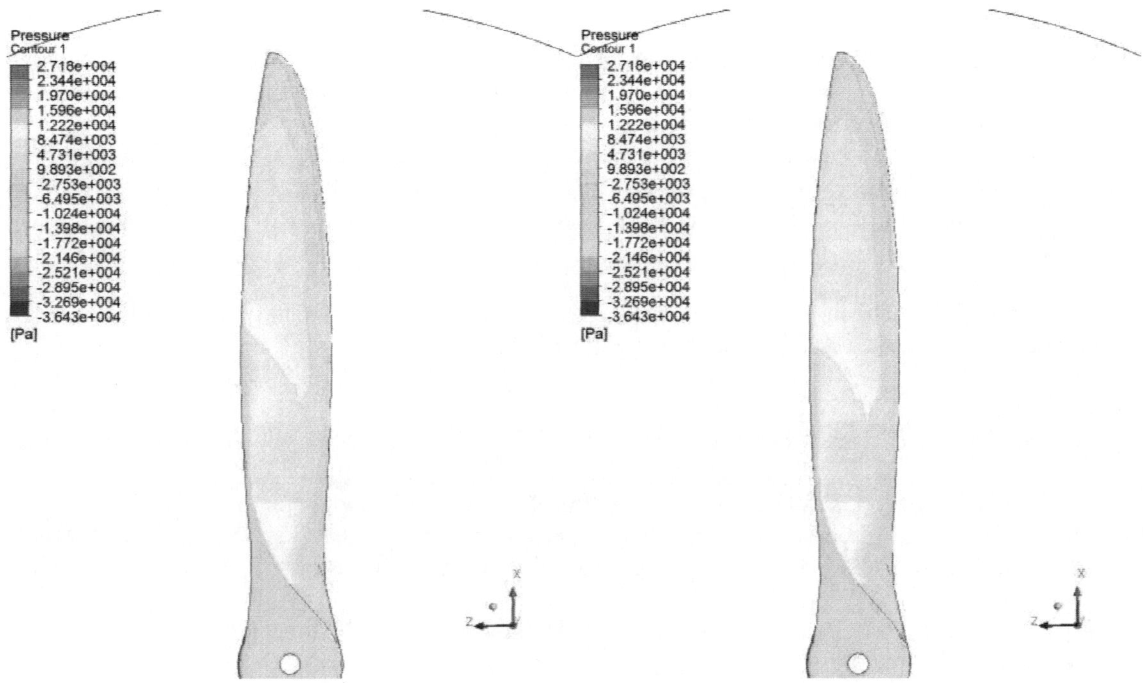

Figure 15. The pressure distribution on the rear of the blades, without prism mesh (right) withprism mesh (left)

Table 3. Result of CFD simulation

	*CFD*simulation(Type 1)	*CFD*simulation (Type 2) (with prism mesh)
Thrust (Newton)	137.7	130.5
Power (Watt)	5844	5601
Effisiensi (%)	47.1	46.6

Obtained by analytic and numerical calculation of aerodynamic load as follows:

Table 4. Comparison of analytic and numerical calculations

	Analytic	CFD simulation (Type 1)	CFD simulation (Type 2) (with prism mesh)
Thrust (Newton)	136	137.7	130.5
Percentage of Analytic vs CFD (%)		1.2	4.1
Percentage of CFDType 1 vs Type 2 (%)		5.2	5.2

From other paper references obtained, CFD analysis the thrust and torque for our propeller are 1007N and 311.19Nm which are 22.3% and 21.3% less than the theoretical calculated values. [7]The percentage of analytic and CFD calculation in this research is smaller than reference data so the comparison is acceptable.

IOP Conf. Series: Journal of Physics: Conf. Series **1005** (2018) 012005 doi:10.1088/1742-6596/1005/1/012005

4. Conclusions

From the analytic and numerical simulation results of aerodynamic load of propeller, the conclusion are: blade element theory useful for quick calculation at conseptual design stage. It gives lift and drag coefficient at varios position of the propeller.

Numerical calculation based on prismatic mesh will provide more accurate pressure distribution along the propeller span. Two mesh type gives 1.2 % differents for type 1 and 4.1% for type 2 compared to blade element theory result.Both approcech could be use for structural design stage.

Acknowledgment

I would like to thank Faculty of Mechanical Aerospace Engineering ITB and National Institute of Aeronautics and Space (LAPAN) for supporting this research.

Reference

[1] allstar**.** Propeller Aircraft Performance and The Bootstrap Approach. [Online] Allstar Network, January 18, 2011. http://www.allstar.fiu.edu/aero/BA-Background.htm.

[2] LSU-03 Datasheet, LSU-03 UAV Platform, Pusat Teknologi Penerbangan, LAPAN, 2015.

[3] Mario Heene, Aerodynamic Propeller Model for Load Analysis, Master of Science Thesis,Royal Institute of Technology School of Engineering Sciences, Stockholm Sweden 2012.

[4] F. Hartono, Metode Perancangan dan Analisis Low Reynolds Number Propeller, Aerospace Indonesia Meeting 2005.

[5] Q.R. Wald. The aerodynamics of propellers. *Progress in Aerospace Sciences*, 2006.

[6] MateuszStajuda, MaciejKarczewski, DamianObidowski, KrzysztofJozwik, Development of a CFD model for propeller simulation, Lodz University of TechnologyInstitute of Turbomachinery, Vol. 20, No. 4 (2016) 579–593

[7] Arif Ahmed Sohel, Md., Abdus Shamir Talukder & Dr. Mohammad Arif Hasan Mamun, Constant Pitch Propeller Design for Low Subsonic Airplane, Global Journal of Researches in Engineering: A Mechanical and Mechanics Engineering, Volume 1 4 Issue 6 Version 1.0 Year 2014

Three Point Bending of Top-Hat Stiffened Chopped Short Fibre Ramie/HDPE Thermoplastic Composite Beam

Bambang K. Hadi and Yogie S. Nuril

Faculty of Mechanical and Aerospace Engineering,
Institut Teknologi Bandung,
Jl. Ganesha 10, Bandung 40132, INDONESIA
Email: bkhadi@ae.itb.ac.id, bambang.hadi60@gmail.com

Abstract. The use of natural fibre and thermoplastic matrices in composite materials increased significantly during the last decade especially in the automotive industries. Ramie is one of these potential natural fibres. In this paper, a three point bending of top-hat beam made of ramie/HDPE (High-Density-Polyethylene) composites was performed. Top-hat stiffened structures were common structures found in the aerospace industries. Nevertheless, these structures are beginning to be applied in automotive structures in the forms of chassis and bumpers. The ramie/HDPE composite was manufactured using hot-press technique. The temperature was set to be 135°C and the pressure was 6 bars. Chopped short ramie fibre was used, due to good drape ability characteristics. The experiments showed that the beams produced a large non-linearity. Linear Finite Element Analysis was carried out to be compared with the experimental data. The differences are reasonable.

1. Introduction

The use of natural fibre composites have increased drastically during the last decades, especially for automotive industries in European countries. European countries and Japan has released stringent guidelines for environmental effects for their automotive products. European Union legislation stated that by 2006, 80% of a vehicle must be reused or recycled and by 2015 the percentage is increased to 85%. Japan requires that 88% of a vehicle should be recovered in 2005, rising to 95% by 2015 [1]. The new legislation forces automotive industries to evaluate the environmental impact of a vehicle's entire lifecycle. The cost and energy to produce of natural fibres are also considerably less than their glass or carbon fibres counterparts. Natural fibres cost US$200 – 1000/ton and use 4 GJ/ton to produce, compared to glass fibres (cost US$1200 – 1800/ton and energy to produce 30GJ/ton) or carbon fibres (cost US$12500/ton and energy to produce 130 GJ/ton) [2]. Kenaf, hemp, banana, sisal and also wood fibres are commonly used in natural fibre composites [3] as well as date palm fibres [4] and jute fibres [5].

Both thermosetting and thermoplastics have been used as matrices. However, thermoplastics matrices are gaining more attention due to the fact that these materials could be recycled more easily. Several thermoplastic matrices have been used, such as Polyethylene (PE) and Polypropylene (PP). The study of Ramie Fibber Reinforced Polypropylene (RF-PP) and their mechanical properties have been studied extensively [6, 7]. In this paper, Ramie Fibber Reinforced High Density Polyethylene (RF-HDPE) was used and studied. Chopped short fibre arrangement was used, due to better drape ability characteristics, compared to unidirectional or woven fabric layups. Therefore, it will be easier to be manufactured into form such as top-hat beams. The tensile strength characteristic of chopped short ramie fibre and HDPE matrix has been studied previously [8].

The use of natural fibre composites in the automotive industries usually confined into secondary structures, such as door panels, instrument panel support or interior structures [3]. However, the use of glass fibres and carbon fibres in the main structures of aerospace, rail and automotive vehicles had been developed and studied. Glass fibre reinforced plastics (GFRP) composite has been used as a structural bogie component for urban subway trains [9] and a double hat shaped tube made of carbon fibre epoxy has been studied for car vehicle body frame [10].

In the aerospace industries, the use of top-hat structures was fully established in the design of a composite wing and fuselage. For thermoplastic matrix composite, top-hat structures have been implemented in fuselage floor structures, due to ease of manufacture. Therefore, the three-point-bending of top-hat structures are important in the development of fuselage floor beam structures.

In this paper, a top-hat structural beam made of chopped short RF-HDPE composite was manufactured, tested and analysed to study its characteristic. A three point bending test was carried out to study their stiffness and energy absorption capacity, and the finite element analysis was performed to be compared with the experimental data. The structure has a potential to be applied in light-weight automotive and rail components.

2. Specimen Manufacturing and Testing

The top-hat structures were manufactured using chopped RF-HDPE composites. The HDPE in the form of powder was mixed with chopped ramie fibres and placed into a hot-press machine. The temperature was set to be 135°C with the pressure of 6 bars for 1 hour. The fibre volume ratio was set to be 50% with the rest is HDPE powder. The resulted RF-HDPE composite plate was cooled down and cut into square plate with dimensions of 120 x 120 mm. The manufacturing detail can be found in [8].

In order to manufacture a top-hat beam structure, a RF-HDPE composite plate was heated in an oven until it was soft and quickly pressed to form a hat beam. The beam was then bolted with the lower plate to form a top-hat beam structure. The beam was then tested in a three-point bending test. Fig.1a shows the beam structures, while Fig.1b shows the three-point bending test.

Figure 1. RF-HDPE beam structures (*left*) and Three-point bending test (*right*)

3. Finite Element Analysis

A Linear Finite Element Analysis was performed. ANSYS software package was used extensively. Figure 2 shows the specimen dimension and the finite element model. A three dimensional brick element was used thoroughly. Three steel bolts were inserted to join the hat structure with the lower plate. The loading was given in the upper part of the hat. The mechanical characteristics of the RF-HDPE material are: Young Modulus was 489 MPa, shear modulus was 183 MPa and the Poisson's ratio was 0.3. The mesh size was 3 mm with the skewness ratio of 0.4.

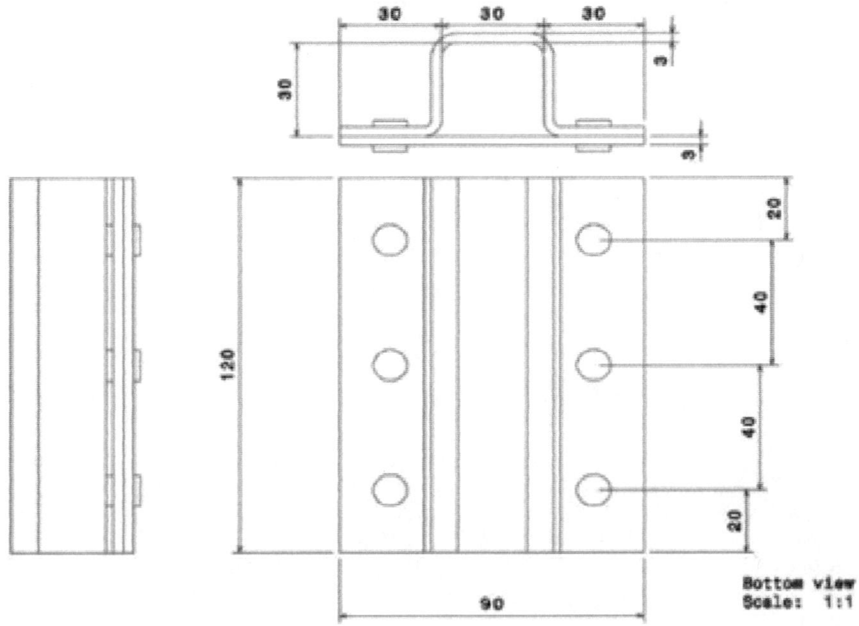

(*upper*)

(*lower*)

Figure 2. Beam dimension in mm (*upper*) and the finite element model (*lower*)

4. Results and Analysis

The experimental result in the form of load-displacement curves can be found in Figure 3 for the three specimens, while the data summary is given in Table 1. Figure 3 shows that RF-HDPE beam structures experienced high plasticity regions before failed. This is due to the behaviour of HDPE thermoplastics which has high plasticity region. This region is important in absorbing impact energy and thus increasing crashworthiness characteristics. The beam is able to absorb in the order of 37 Joule.

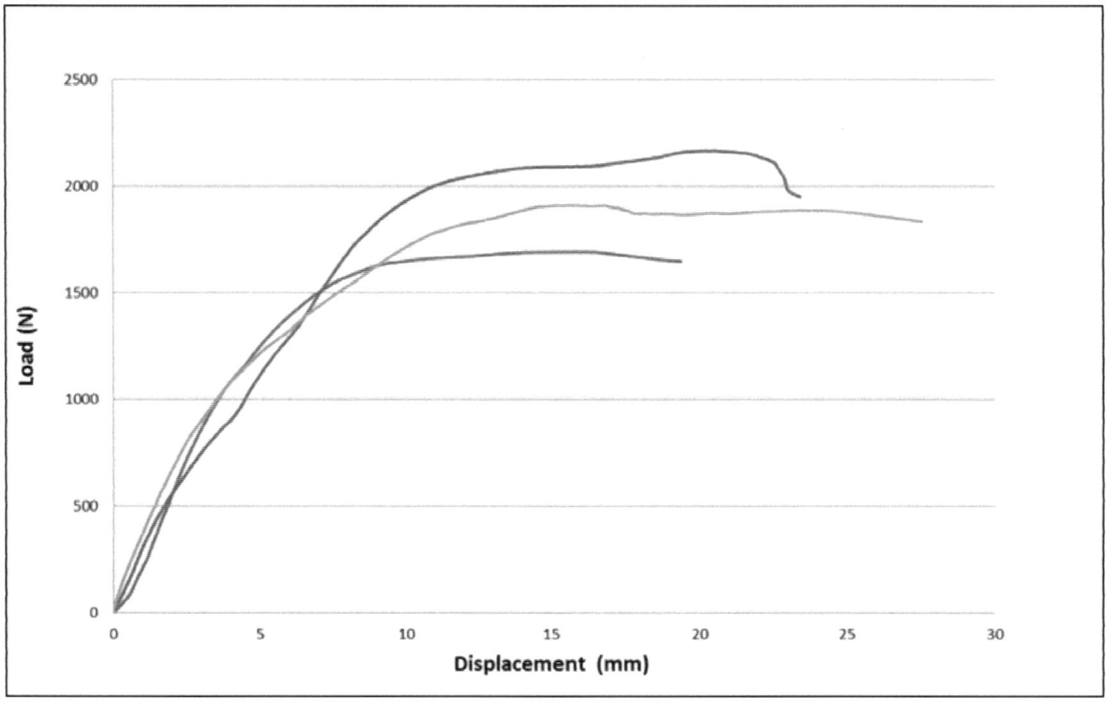

Figure 3. Experimental load-displacement curves for the three specimens

Table 1. Summary of Experimental Data

Specimen No.	Maximum Load (N)	Maximum Displacement (mm)	Absorbed Energy (Joule)
1	1690	19	27
2	2164	23	39
3	1991	29	44
Mean	1948	24	37

In this work, a linear finite element analysis was performed and will be compared with the experimental data. Since linear finite element was used, the analysis was limited to the yield point. Figure 4 shows the finite element results. It shows the loading condition and the displacement results.

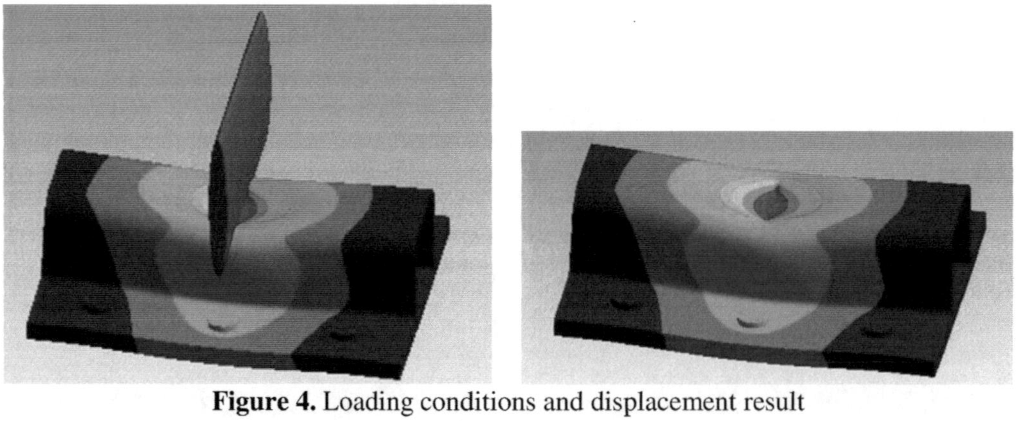

Figure 4. Loading conditions and displacement result

The maximum displacement was in the middle of the upper plate, as expected. The loading was modelled using actual plate with contact condition. The loading plate was made of steel. Since the loading was modelled using a contact problem, the maximum displacement was in the centre of the upper plate. It should also be noted that the displacement on the lower plate was considerably lower than the maximum displacement.

Table 2 shows the comparison between the experimental data and the finite element analysis in the yield region. Table 2 shows that the finite element results are reasonable compared to the experimental data.

Table 2. Comparison between experimental data and finite element results

Specimen No.	Experiment Data		Finite Element Results	
	Load (N)	Deformation (mm)	Load (N)	Deformation (mm)
1	750	2.2	750	1.84
2	700	2.1	700	1.71
3	450	2	450	1.12
Mean	633	2.1	633	1.56

The discrepancies between the finite element results and the experimental data are due to the fact that the beams are highly non-linear, due to the characteristics of the HDPE thermoplastic matrices. Figure 5 shows the failure characteristics of the beam due to three-point bending. It shows that high plasticity happened in the upper compressive plate of the beam. Therefore, it was obvious that the experimental data shows greater displacement compared with the finite element results even in the yield region.

Figure 5. High non-linear behaviour of the beam.

5. Conclusion

The paper presents a three-point bending of chopped short ramie fibre and high-density polyethylene (RF-HDPE) thermoplastic composite beam structures. Experiments were performed and the resulted data were compared with a linear finite element analysis using ANSYS. The beam structures showed high non-linear behaviour due to the high plasticity characteristics of the HDPE thermoplastic matrices. Linear finite element analysis compared satisfactory with the experimental data in the area before yield occur. The energy absorption capacity was calculated from the experimental data. From the analysis, it shows that the RF-HDPE beam structure has the potential to be applied in the energy absorbing structures, in the aerospace, rail and automotive vehicle structures.

6. References

[1] Holbery J, Houston D 2006 *JOM J. Minerals, Metals and Material Society* **58**, issue 11, 80-86.
[2] Huda MS, Drzal LT, Ray D, Mohanty AK, Mishra M, 2008 *Properties and Performance of Natural Fiber Composites*, Woodhead Publishing, 221-268.

[3] Norhidayah MH, Hambali A, Mohd Yuhazri Y, Zolkarnain M, Taufik, Saifuddin HY 2014 *Appl Mechanics and Materials* **564**, 3-7. http://doi:10.4028/www.scientific.net/AMM.564.3
[4] Al-Oqla FM, Sapuan SM 2013 *J. Cleaner Production*, 1-8, http://dx.doi.org/10.1016/j.jclpro.203.10.050
[5] Alves C, Ferrao PMC, Silva AJ, Reis G, Freitas M, Rodrigues LB, Alves DE 2010 *J. Cleaner Production* **18**, 313-327.
[6] Li-Ping H, Yong T, Lu-Lin W 2008 *Advanced Materials Research* **41-42**, 313-316 http://doi:10.4028/www.scientific.net/AMR.41-42.313
[7] Sun Y, Peng D, Li M 2011 *Advanced Materials Research* **194-196**, 1839-1844 http://doi:10.4028/www.scientific.net/AMR.194-196.1839
[8] Banowati L, Hadi BK, Suratman R 2016 *International Conference on Computational Modelling, Simulation and Applied Mathematics,* DEStech Publication Inc, Pennsylvania, USA, pp. 135-139.
[9] KIM JS, Yoon HJ 2011 *Procedia Engineering* **10**, 2375-2380.
[10] Liu Q, Xing H, Ju Y, Ou Z, Li Q 2014 *Composite Structures* **117**, 1-11.

Acknowledgments
The research was partly supported by the Ministry of Research, Technology and Higher Education, Republic of Indonesia through INSINAS Research Grant 2017. The authors wish to thank the Ministry for the generous funding.

IOP Conf. Series: Journal of Physics: Conf. Series **1005** (2018) 012007 doi:10.1088/1742-6596/1005/1/012007

Comparison Virtual Landing Gear Drop Test for Commuter Aircraft Utilize MSC ADAMS And Solidworks Motion Analysis

Dony Hidayat[1,2]*, Jos Istiyanto[2] and Danardono Agus Sumarsono[2]

[1] Aeronautics Technology Center, National Institute of Aeronautics and Space (LAPAN), Indonesia
[2] Department of Mechanical, Faculty of Engineering, Universitas Indonesia, Indonesia

*dony.hidayat@lapan.go.id

Abstract. Loads at main landing gear while touchdown impact is function of aircraft weight and ground reaction load factor. In regulation states ground reaction load factor at v_{sink} = 3.05 m/s is below 3. Contact/impact force from simulation using MSC ADAMS is 94680 N, while using Solidworks Motion Analysis is 97691 N. The difference between MSC ADAMS and Solidworks Motion Analysis is 3.08%. The ground reaction load factor in MSC ADAMS is 2.78 while in Solidworks Motion Analysis is 2.87.

1. Introduction

The design process of 19 passengers commuter airplane for flight in remote areas has reached the testing stage of its components. The requirements for this test refer to CASR (Civil Aviation Safety Regulation) Part 23 issued by the Ministry of Transportation.

Loads that occur in the main landing gear when touchdown impact is a function of the aircraft weight multiplied by the ground reaction load factor. In CASR Part 23.473 (Ground Load Conditions and Assumptions) stated that the value of ground reaction load factor is between 2 s / d 2.67 and should be proved by LGDT [1].

Computer Aided Engineering (CAE) developed rapidly in recent decades, one of which is the Multi-Body Simulation (MBS). Many MBS programs capable of automatic generation and integration of the differential equations of motion have been developed, namely DAP, MSC ADAMS, COMPAMM, and SIMPACK [2]. MBS provides very fast tools for modeling and simulating an assembly consisting of several complex components. This is one of the advantages of MBS rather than finite element analysis (FEA) that tends to partially separate components on an assembly that moves and interacts with each other [3].

Romeo et al [4] has developed a methodology for simulating landing gear drop tests using MSC ADAMS and comparing simulation with experimental testing. Krason and Malachowski [5] introduced a methodology for evaluating several landing gear models and numerical research on a complete landing gear system. Fu et al [6] modeled a LGDT using the SimMechanics module approach on Matlab/Simulink.

Sundry studies have been conducted to create a simulated LGDT using a MBS approach. However, comparison of performance (result of simulation, ease of set up simulation and speed at simulation) some MBS software in the same case is very interesting to be examined, therefore conducted a study on the comparison of MSC ADAM and Solidworks motion analysis

2. Method

Experimental test of landing gear drop test was conducted based on regulation from Ministry of Transportation which is contained in Civil Aviation Safety Regulation (CASR) Part 23 [1]. The LGDT testing prerequisites are in CASR 23.473, CASR 23723, CASR 23725, CASR 23726 and CASR 23.727 [7].

Initial condition of simulations using MSC ADAMS and Solidworks Motion Analysis comply to DRO (Design Requirement and Obejctive) condition of the 19 passenger commuter aircaraft. Simulation conditions (vLGDT) are shown in Table 1 below

Tabel 1. Simulation condition of main landing gear [7]

No.	Conditions	Load (kg)	v_{sink} (m/s)	High estimation (mm)
1	DRO	3470	3.05	470

Simulation input for contact is stiffness (k), force exponent (e), damping (c_{max}) and penetration depth (d). Whilst simulation input for rubber damper only stiffness coefficient and damping coefficient. Time step simulation for this case is 0.001 seconds.

3. Results and Discussions

To compare MSC ADAMS and Solidworks Motion Analysis , accordingly conduct simulated with 3D rigid model approach.

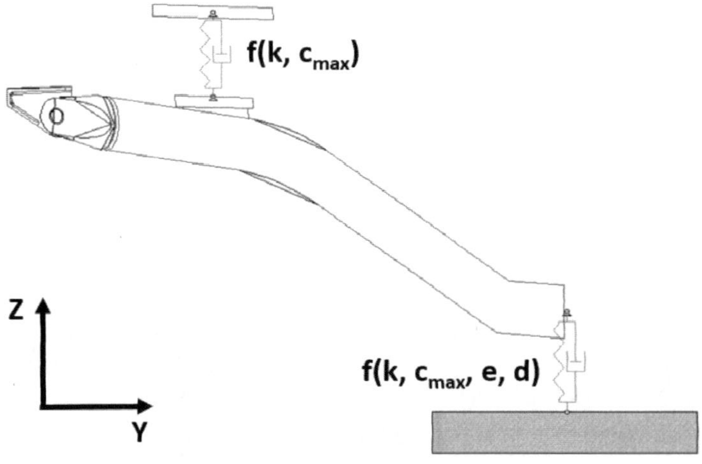

Figure 1. 3D rigid model apporoach

Rubber damper on the main landing gear is a function of stiffness (k) and damping (cmax), whereas the tire is a function of stiffness (k), damping (cmax), force exponent (e) and penetration depth (d) as shown in Fig. 1 above. 3D models built using CAD software afterward exported to MSC ADAMS and Solidworks Motion Analysis as shown in Figure 2.

Figure 2. VLGDT uses MSC ADAMS and Solidworks Motion Analysis with a 3D approach

Rubber damper stiffness (k) varied from 1500 - 12000 N/mm, whereas the damping (c_{max}) is 1% of the stiffness value [8]. For tire stiffness parameters use data from Dunlop tire vendor [9], force exponent (e) for rubber ≈ 1.1 and penetration depth (d) = 1.0e-4 m [10, 11].

Comparison of simulation results using MSC ADAMS and Solidworks Motion Analysis with rubber damper stiffness 2000 N/mm and 65 psi tire pressure shown in Figure 3 below. Vsink simulation results from MSC ADAMS of 3.03 m/s, while Solidworks Motion Analysis of 3.05 m/s.

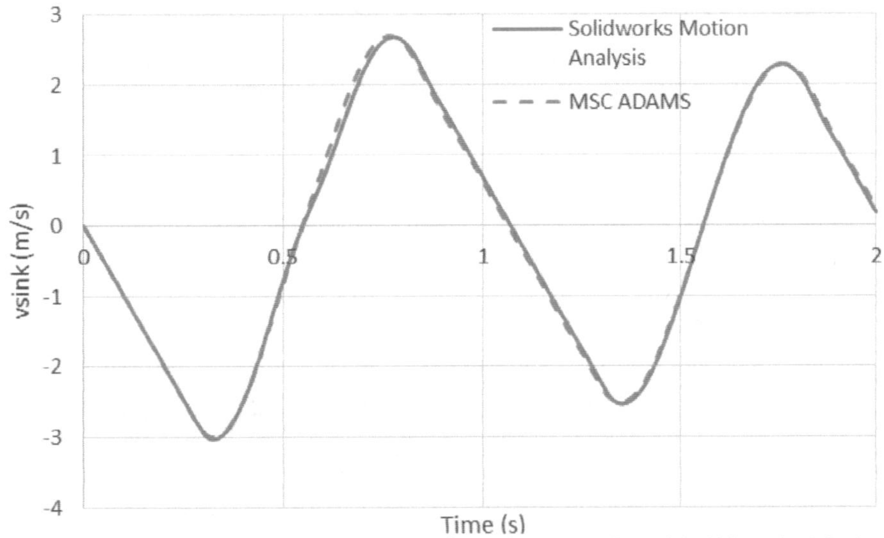

Figure 3. Vsink comparison simulation results of MSC ADAMS and Solidworks Motion Analysis

From the simulation that has been done using MSC ADAMS and Solidworks Motion Analysis software, to get the value of ground reaction load factor below 3 then the stiffness of rubber damper ranged from 1900 N/mm to 2100 N/mm and tire pressure between 60 psi to 65 Psi as shown in Figure 4 and Figure 5 below. From the figure also obtained that the stiffness of rubber damper has non linear function to ground reaction load factor. Significant changes occurred in the stiffness value of rubber damper 2000 N/mm to 3000 N/mm. Stiffness function above 3000 N/mm tends to linear, but the value of ground reaction load factor is above 4. Virtual landing gear drop test of Partenavia AP.68 TP-300 [4] with fix landing gear configuration obtained ground reaction load factor 3.87 while experimental result of 3.9. This configuration only uses leaf spring and tire as damper. While on 19 passengers commuter airplane were added rubber block as a damper.

The ground reaction factor value at the pressure of the 55 psi tire is generally below the 60 psi and 65 psi pressure, but on the rubber damper stiffness of about 2000 N/mm, the ground reaction load factor value is not below 3 such as 60 psi and 65 psi tire pressures. The pressure of a 55 psi tire is not recommended for this commuter aircraft.

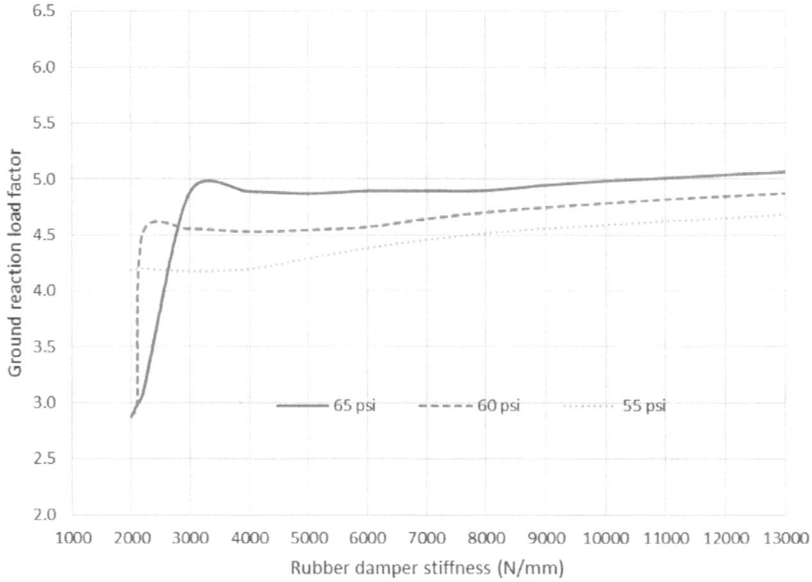

Figure 4. Effect of rubber damper stiffness on ground reaction load factor with tire pressure variation using Solidworks Motion Analysis software

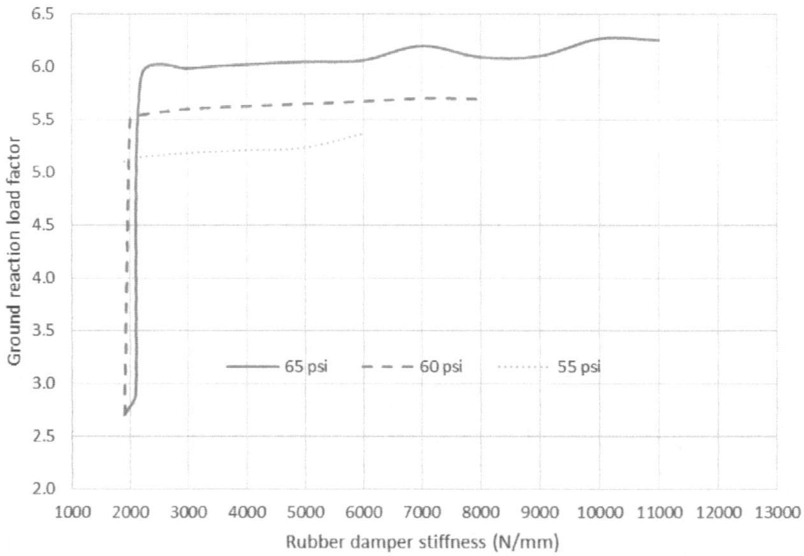

Figure 5. Effect of rubber damper stiffness on ground reaction load factor with tire pressure variation using MSC ADAMS

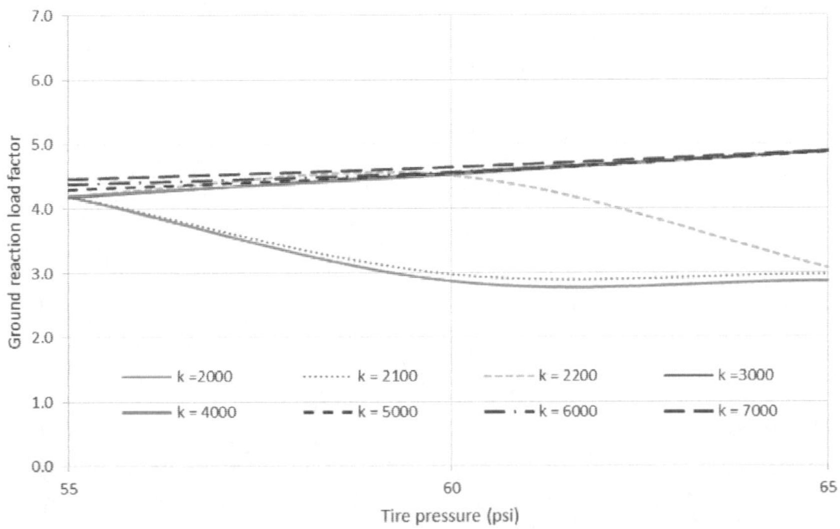

Figure 6. Effect of tire pressure on ground reaction load factor by rubber damper stiffness variation using Solidworks Motion Analysis software

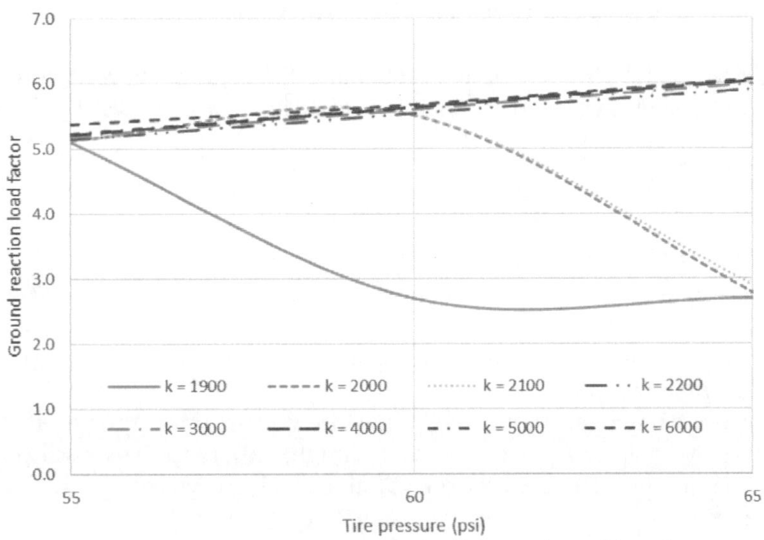

Figure 7. Effect of tire pressure on ground reaction load factor by rubber damper stiffness variation using MSC ADAMS software

Contact/impact force simulation results using MSC ADAMS software of 94680 N, while using Solidworks Motion Analysis of 97691 N as shown in Figure 8 below. The difference in simulation results between the two software is 3.08%. The ground reaction load factor of MSC ADAMS software is 2.78 while in Solidworks Motion Analysis is 2.87.

In the simulation results there is a difference in contact/impact force value between the two software. There are several factors that cause these differences, including: the difference between the assembly process of both software and the placement of the spring model. In MSC ADAMS, component assembly and spring modeling processes are still conservative like most early 2000s software, while Solidworks Motion Analysis is more detailed with accurate constraints such as modern CAD software.

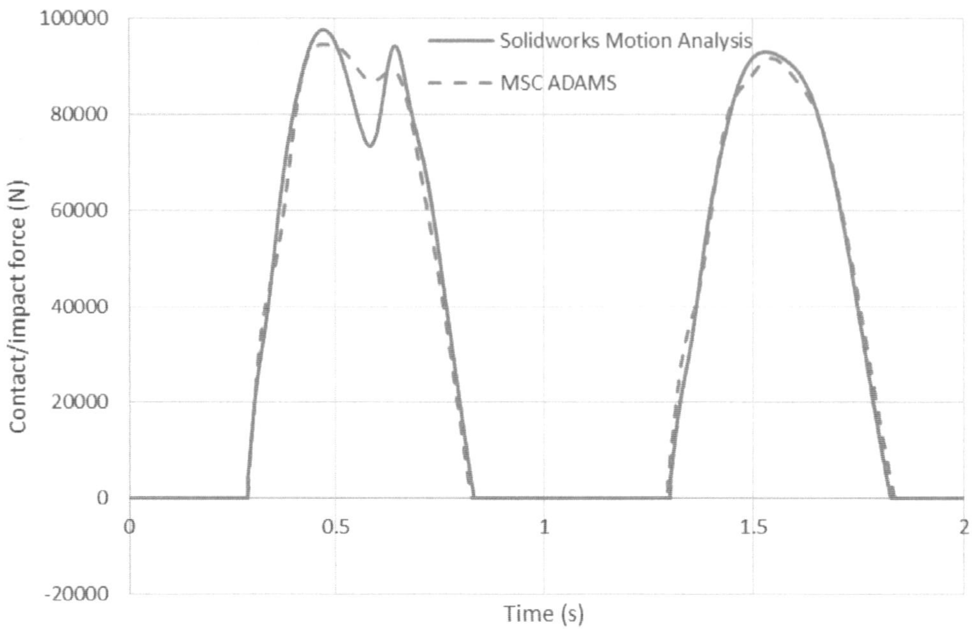

Figure 8. Comparison of contact /impact force MSC ADAMS and Solidworks Motion Analysis

In Figure 8, there is also a force drop phenomenon at 0.6 second, the contact/impact force drops suddenly and then rises again in seconds to 0.7. This phenomenon is caused by the load that occurs towards the y-axis (Fig 1) is not able to be retained by the wheel.This phenomenon is caused due to slip on the wheel toward the y-axis.

4. Conclusions or Concluding Remarks

Virtual Landing Gear Dropt Test (vLGDT) was successfully created using MSC ADAMS and Solidworks Motion Analysis software. To obtain the ground reaction load factor value below 3 then the stiffness of rubber damper ranges from 1900 N/mm up to 2100 N/mm. Tires use a stiffness of 590 N/mm or identical to a 65 psi tire pressure.

Simulation using 3D approach with MSC ADAMS, contact/impact force obtained of 73650 N while with Solidworks Motion Analysis of 74468 N with the difference between two software 1.1%. From the ease in the set up simulation, Solidworks Motion Analysis is easier to use than MSC ADAMS while from the speed at simulation, both software is relatively the same.

Acknowledgement

The authors would like to thank Mr. Gunawan Setyo Prabowo, Mr. Agus Aribowo and Mr Atik Bintoro on guidance for the preparation of this paper.

This content of paper is entirely author’s responsibility.

IOP Conf. Series: Journal of Physics: Conf. Series **1005** (2018) 012007 doi:10.1088/1742-6596/1005/1/012007

References

[1] Kemenhub, *Civil Aviation Safety Regulations (CASR) Part 23 Amd. 1*, 2001.

[2] Machado, M., et al., *Compliant contact force models in multibody dynamics: Evolution of the Hertz contact theory.* Mechanism and Machine Theory, 2012. **53**: p. 99-121.

[3] Khude, N.N., *Efficient simulation of flexible body systems with frictional contact/impact*, 2014, Citeseer.

[4] Leo, R.D., et al., *Drop Test Simulation for An Aircraft Landing Gear Via Multi-Body Approach.* Archive of Mechanical Engineering, 2014. **61**(2): p. 287-304.

[5] Krason, W. and J. Malachowski, *Multibody rigid models and 3D FE models in numerical analysis of transport aircraft main landing gear.* Bulletin of the Polish Academy of Sciences Technical Sciences, 2015. **63**(3): p. 745-757.

[6] Fu, Y.L., et al. *Drop Dynamic Simulation for Landing Gear via SimMechanics*. in *Advanced Materials Research*. 2013. Trans Tech Publ.

[7] PTDI, *Design Document.* 2014.

[8] Giesbers, J., *Contact mechanics in MSC Adams-A technical evaluation of the contact models in multibody dynamics software MSC Adams.* 2012.

[9] Dunlop, *Tyre Application Chart*, 1989.

[10] MSC.Software, *Adams/Solver help.* 2010.

[11] Dassault.Systemes. *Motion Analysis*. 2017 [cited 2017 May 23]; Available from: solidworks.com/sw/products/simulation/motion-analysis.htm.

Buckling Analysis of Single and Multi Delamination In Composite Beam Using Finite Element Method

Hans Charles Simanjorang[1, a], Hendri Syamsudin[2, b], Muhammad Giri Suada[3, c]

[1, 2, 3] Lightweight Structure Research Group, Faculty of Mechanical and Aerospace Engineering

Institut Teknologi Bandung, Jl. Ganesha 10 Bandung 40132, Indonesia

[a]hans.simanjorang@students.itb.ac.id, [b]hendri.syamsudin@ae.itb.ac.id, [c]mgs@aero.pauir.itb.ac.id

Abstract. Delamination is one type of imperfection in structure which found usually in the composite structure. Delamination may exist due to some factors namely in-service condition where the foreign objects hit the composite structure and creates inner defect and poor manufacturing that causes the initial imperfections. Composite structure is susceptible to the compressive loading. Compressive loading leads the instability phenomenon in the composite structure called buckling. The existence of delamination inside of the structure will cause reduction in buckling strength. This paper will explain the effect of delamination location to the buckling strength. The analysis will use the one-dimensional modelling approach using two- dimensional finite element method.

Keywords: Delamination, buckling load, buckling mode, composite structure, finite element method

1. INTRODUCTION

The development of composite material has been successfully compete with another conventional materials such as metals in many applications. Composite material is material that has high specific strength, high specific stiffness, and low density [1]. Despite it has many advantages compared to metallic materials, composite materials also have weakness in its application. Composite materials especially composite laminates are prone to delamination. This will reduce the strength of composite material due to compressive loading. Sallam and Simitsess (1984) performed one dimensional buckling and post buckling analysis on cross-ply laminate that have delamination to calculate buckling loads. Parlapali, Shu, and Chai (2008) have developed the analytical model with Euler Bernoulli and Classical Lamination Theory basis to analyze buckling on composite beam with two delamination lower and upper bounds. Chen and Qiao (2011) have developed the analytical equation to analyze the buckling delamination on bi-layer column with the effect of transverse shear deformation and local tip deformation. This paper will explain the effect of delamination location, delamination length, and multi delamination to the buckling strength numerically using commercial finite element software.

2. LINEAR BUCKLING ANALYSIS

Numerical modelling using finite element is done with the linear buckling analysis – eigenvalue analysis. The purpose is to calculate the critical load and mode of buckling deformation. To calculate the critical load, it is required the equilibrium position of loads, external load, and internal load acting on the structure as in this following equation.

$$[K]\{d\} = ([K]_0 + \lambda[K]_\sigma)\{d\} = \{0\} \ldots\ldots\ldots\ldots (1)$$

Where [K] is total stiffness matrices, $[K]_0$ is the stiffness matrices calculated in the span of small displacement, $[K]_\sigma$ is the geometry stiffness matrices, λ is the eigenvalue and {d} is the displacement vector. The equation above is to solve the eigenvalue problem. If the displacement vector is not zero, then the solution of the equation can be obtained using the following characteristic equation.

$$det([K]_0 + \lambda[K]_\sigma) = 0 \ldots\ldots\ldots\ldots (2)$$

The equation can be solved for lambda. The smallest eigenvalue will take into consideration because this value will become the multiplier factor to calculate the critical load on the structure. Eigenvector also can be obtained and it shows the mode of buckling. Using this method, the structure can be assumed as linear until it reaches the instability point [2].

3. LAMINATE BEAM PLATE BUCKLING

In this paper, the thickness of the specimen is very small compared to its length and the force acting on the structure only in x-axis, and the lamina can be assumed as beam [3]. The following is linier differential equation for plate buckling.

$$\frac{\partial N_x^b}{\partial x} + \frac{\partial N_{xy}^b}{\partial y} = 0 \ldots\ldots\ldots\ldots (3)$$

$$\frac{\partial N_{xy}^b}{\partial x} + \frac{\partial N_y^b}{\partial y} = 0 \ldots\ldots\ldots\ldots (4)$$

$$\frac{\partial^2 M_x^b}{\partial x^2} + 2\frac{\partial^2 M_{xy}^b}{\partial x \partial y} + \frac{\partial^2 M_y^b}{\partial y^2} + N_x^i \frac{\partial^2 w}{\partial x^2} + 2N_x^i \frac{\partial^2 w}{\partial x \partial y} + N_y^i \frac{\partial^2 w}{\partial y^2} = 0 \ldots\ldots (5)$$

The superscript (b) shows there are moment and buckling forces and (i) shows the force shortly before buckling and pre-buckling. The complete details to obtain these equations can be seen at reference [4]. In this current paper, the force acts only in axial direction Nx and all differentiation in y is eliminated in the linear differential equation of plate buckling and it becomes

$$\frac{\partial N_x^b}{\partial x} = 0 \ldots\ldots\ldots\ldots (6)$$

$$\frac{\partial^2 M_x^b}{\partial x^2} + N_x^i \frac{\partial^2 w}{\partial x^2} = 0 \ldots\ldots\ldots\ldots (7)$$

For 1-D beam, $M_x^b = E_x I_y \frac{\partial^2 w}{\partial x^2}$ and $N_x^i = bN_x$, therefore the equation above becomes

$$\frac{\partial N_x^b}{\partial x} = 0 \ldots\ldots\ldots\ldots (8)$$

$$E_x I_y \frac{\partial^4 w}{\partial x^4} + N_x^i \frac{\partial^2 w}{\partial x^2} = 0 \quad (9)$$

$$\frac{\partial^4 w}{\partial x^4} + \frac{N_x^i}{E_x I_y} \frac{\partial^2 w}{\partial x^2} = 0 \quad (10)$$

Where $\lambda_b^2 = \frac{N_x^i}{E_x I_y}$

The general solution of the equation above is

$$w(x) = c_1 sin\lambda_b x + c_2 cos\lambda_b x + c_3 x + c_4 \quad (11)$$

C1, c2, c3, c4, obtained from the boundary conditions. To determine the critical buckling load, for example the length of the beam is L, thickness h, with the fixed supported in both ends, the equation become

$$w(0) = 0, \frac{\partial w}{\partial x}(0) = 0, w(L) = 0, \frac{\partial w}{\partial x}(L) = 0 \quad (12)$$

Therefore, the critical buckling load per unit thickness is

$$N_{cr} = \frac{\pi^2 E_x^b}{3}\left(\frac{h^3}{L^2}\right) \quad (13)$$

From the explanation above, it can be stated that the analysis of 1-D plate buckling is similar with beam/column buckling analysis.

4. MATERIAL PROPERTIES

Material that has been used in this research is fiberglass-epoxy that made from wet-lay-up method. The specimen contains 20 layers with cross-ply orientation. Material made following the standard ASTM 3039. The experiment to obtain the material properties is done by Mulia Minhat (2007) [5]. He conducted the tensile test on the specimens. Because of the limitation in the equipment tools, the property obtained from the experiment was only the young modulus. The poison ratio and shear modulus obtained from the literature.

Table 1. Material property of *Fiberglass-Epoxy Isotropic*

Fiberglass Property	
Longitudinal Modulus, E_1	9.25 GPa
Poisson Ratio, v_{12}	0.25
Shear Modulus, G_{12}	3.7 GPa

5. FINITE ELEMENT METHOD ANALYSIS

Due to the limitation of the experimental tools, the material will be assumed as isotropic in finite element model. Commercially finite element code MSC Patran Nastran 2011 will be used in the numerical analysis. MSC Patran will be used as pre-processing tool to model the geometry, load, boundary conditions, and input material properties and also to mesh the model. MSC Nastran will be used as processing tool to calculate and obtain eigenvalue and buckling mode. MSC Patran will be used again to show the calculated eigenvalue and mode of buckling. The composite beam will be modelled as 2-D shell type with CQUAD4 elements. The boundary conditions will be fixed supported at both ends with apply constant displacement in one side.

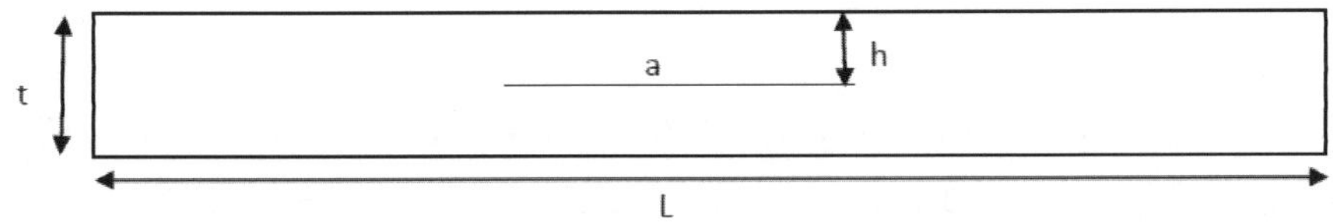

Figure 1. Single delamination model

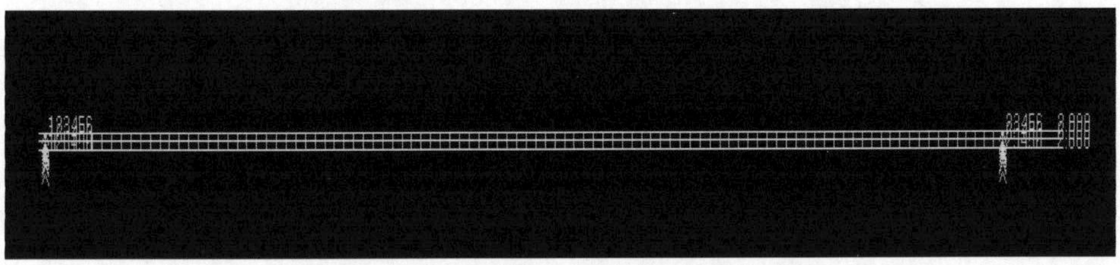

Figure 2. Single delamination model in FEM

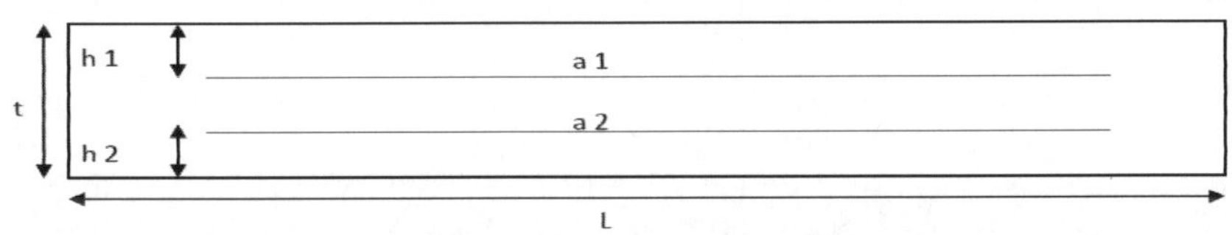

Figure 3. Multi delamination model

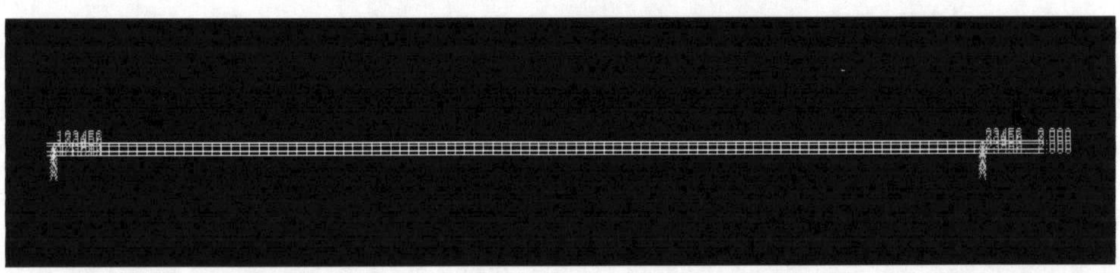

Figure 4. Multi delamination model in FEM

Table 2. Model Dimensions

Model dimensions	
Length, L	240 mm
Width, t	2.5 mm
Thickness, w	40 mm
Delamination length, a	Varying variable
Delamination location, h1	Varying variable
Delamination location, h2	Varying variable

IOP Conf. Series: Journal of Physics: Conf. Series **1005** (2018) 012008 doi:10.1088/1742-6596/1005/1/012008

Table 3. Variation of a/L to h/t in single delamination

Variation of a/L to h/t	
h/t	a/L
0.1	0.1 , 0.2 , 0.3 , 0.4 , 0.5 , 0.6 , 0.7 , 0.8 , 0.9
0.2	0.1 , 0.2 , 0.3 , 0.4 , 0.5 , 0.6 , 0.7 , 0.8 , 0.9
0.3	0.1 , 0.2 , 0.3 , 0.4 , 0.5 , 0.6 , 0.7 , 0.8 , 0.9
0.4	0.1 , 0.2 , 0.3 , 0.4 , 0.5 , 0.6 , 0.7 , 0.8 , 0.9
0.5	0.1 , 0.2 , 0.3 , 0.4 , 0.5 , 0.6 , 0.7 , 0.8 , 0.9

Table 4. Variation of a/L to h/t in multi delamination

Variation of a/L to h/t			
h1/t	h2/t	a1/L	a2/L
0.1	0.9	0.1 , 0.2 , 0.3 , 0.4 , 0.5 , 0.6 , 0.7 , 0.8 , 0.9	0.1 , 0.2 , 0.3 , 0.4 , 0.5 , 0.6 , 0.7 , 0.8 , 0.9
0.2	0.8	0.1 , 0.2 , 0.3 , 0.4 , 0.5 , 0.6 , 0.7 , 0.8 , 0.9	0.1 , 0.2 , 0.3 , 0.4 , 0.5 , 0.6 , 0.7 , 0.8 , 0.9
0.3	0.7	0.1 , 0.2 , 0.3 , 0.4 , 0.5 , 0.6 , 0.7 , 0.8 , 0.9	0.1 , 0.2 , 0.3 , 0.4 , 0.5 , 0.6 , 0.7 , 0.8 , 0.9
0.4	0.6	0.1 , 0.2 , 0.3 , 0.4 , 0.5 , 0.6 , 0.7 , 0.8 , 0.9	0.1 , 0.2 , 0.3 , 0.4 , 0.5 , 0.6 , 0.7 , 0.8 , 0.9
0.1	0.5	0.1 , 0.2 , 0.3 , 0.4 , 0.5 , 0.6 , 0.7 , 0.8 , 0.9	0.1 , 0.2 , 0.3 , 0.4 , 0.5 , 0.6 , 0.7 , 0.8 , 0.9

6. RESULTS AND ANALYSIS

Table 5. Critical load of a perfect structure

Equation	Critical Load (Newton)	% difference with theoretical value
Theoretical: column model	330.2	0.00 %
Theoretical: 1-D plate	330.2	0.00 %
Numerical FEM	330.3	0.03 %

It can be seen that the analysis using finite element model in perfect structure is valid with small difference with the theoretical values. This calculation is done to see the accuracy and also to know the critical load of a perfect structure that later on will be compared with the delaminated structure.

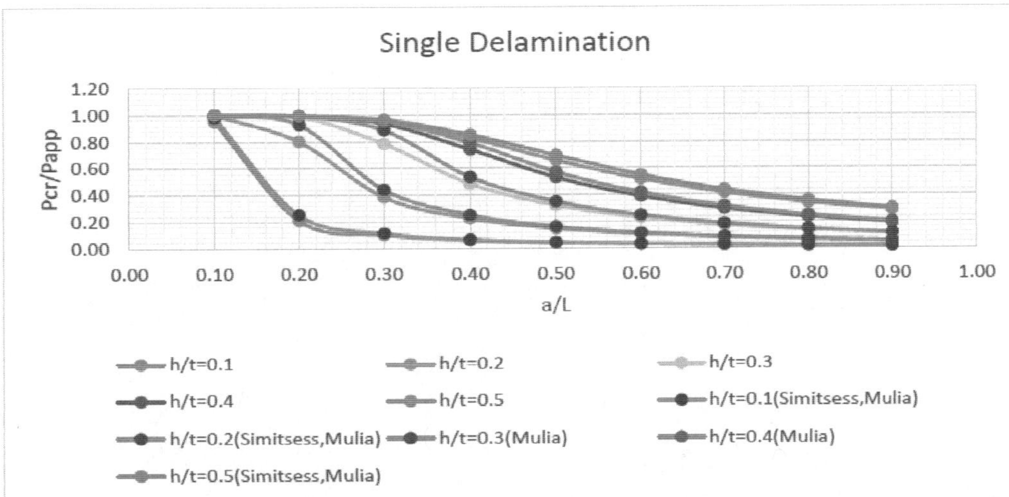

Figure 5. Pcr/P to the delamination length in single delamination

Table 6. Buckling mode in single delamination

a/L , h/t	0.1	0.2	0.3	0.4	0.5
0.1	Mixed	Global	Global	Global	Global
0.2	Local	Mixed	Mixed	Mixed	Global
0.3	Local	Mixed	Mixed	Mixed	Global
0.4	Local	Mixed	Mixed	Mixed	Global
0.5	Local	Mixed	Mixed	Mixed	Global
0.6	Local	Local	Mixed	Mixed	Global
0.7	Local	Local	Local	Mixed	Global
0.8	Local	Local	Local	Mixed	Global
0.9	Local	Local	Local	Local	Global

In the single delamination, it can be seen that the results are consistent with the results that obtained by previous researchers, Simitsess [6] and Mulia Minhat [5]. From this results, the longer the delamination, the buckling strength of the composite laminate structure will decrease quadratically. This results is match with the equation that derived by Euler that the longer the delamination, it will reduce the buckling strength quadratically.

Besides that, the delamination occur near to the upper or lower surface will reduce the buckling strength significantly and creates the local buckling phenomenon. This is caused by the small moment inertia in the delaminated area. It can be stated that for the delamination near to the structure surface, the structure will fail first by delamination and then by buckling when the compression force applied on it.

IOP Conf. Series: Journal of Physics: Conf. Series **1005** (2018) 012008 doi:10.1088/1742-6596/1005/1/012008

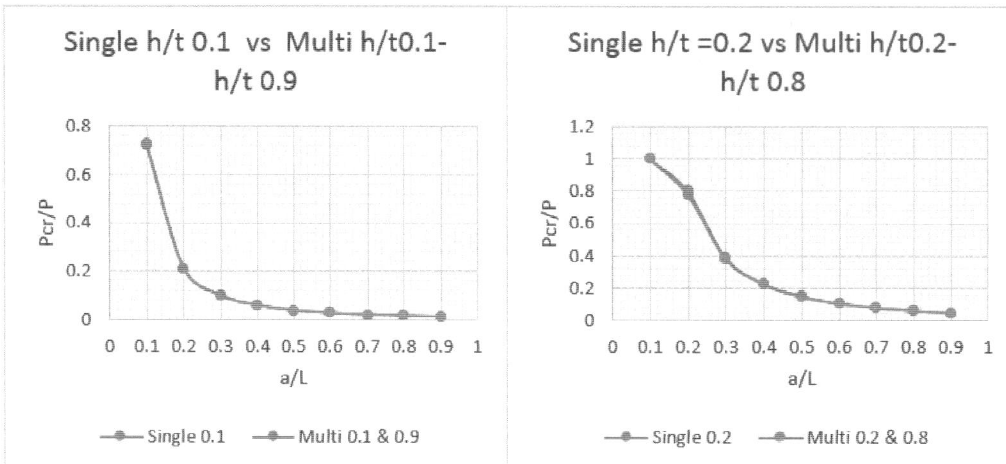

Figure 6 (left). Pcr/P to delamination length in single delamination h/t=0.1 and multi delamination h/t=0.1 and h/t=0.9

Figure 6 (right). Pcr/P to delamination length in single delamination h/t=0.2 and multi delamination h/t=0.2 and h/t=0.8

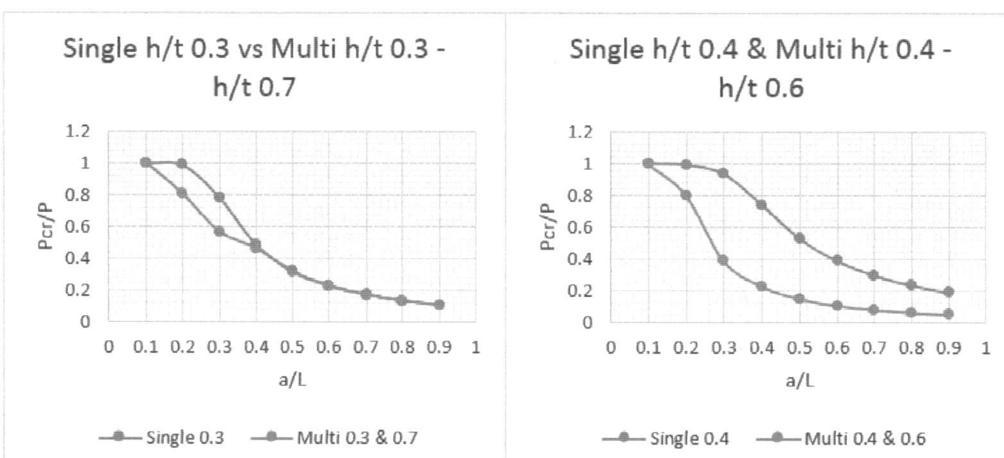

Figure 7 (left). Pcr/P to delamination length in single delamination h/t=0.3 and multi delamination h/t=0.3 and h/t=0.7

Figure 7 (right). Pcr/P to delamination length in single delamination h/t=0.4 and multi delamination h/t=0.4 and h/t=0.6

Table 7. Buckling mode in multi delamination

a/L , h/t	0.1 & 0.9	0.2 & 0.8	0.3 & 0.4	0.4 & 0.6	0.1 & 0.5
0.1	Mixed	Global	Global	Global	Mixed
0.2	Local	Mixed	S-shape	Mixed	Local
0.3	Local	Mixed	S-shape	Mixed	Local
0.4	Local	Mixed	Mixed	Mixed	Local
0.5	Local	Mixed	Mixed	Mixed	Local
0.6	Local	Local	Mixed	Mixed	Local
0.7	Local	Local	Mixed	Local	Local
0.8	Local	Local	Local	Local	Local
0.9	Local	Local	Local	Local	Local

From the results of multi delamination, it can be seen that the critical buckling strength will decrease with the presence of multi delamination. It can be seen at from figure 6 to 7. This parametric study shows if there are multi delamination in the composite laminate structure, it will make the buckling strength decreases more than with the single delamination.

Figure 6 shows that the results for single and multi-delamination with delamination near to the surface is close. This is because the similarity in the buckling mode which is local buckling that decreases the buckling strength drastically and also the location of delamination where near to the surface of the structure. It is different with figure 7, where the multi delamination decrease the structure buckling strength more than single delamination.

From this results we can see that the location of the delamination plays major role in reducing the buckling strength. If the delamination occur near to the surface whether it's single or multi delamination, serious concern should be taken on it. But if the delamination occur far from surface structure, it's also need to be checked whether the delamination is single or multi. If its multi delamination, serious concern should be taken because it will reduce the buckling strength more than single delamination.

The delamination near to the structure surface will make the buckling strength decreases significantly. This is happen because of the small moment inertia in the delaminated area. It can be seen that for the delamination near to the surface, the structure will fail first because of the delamination then by buckling.

Table 6 and 7 show the buckling mode of the structure in singe and multi delamination cases. There are three types of buckling mode namely local buckling, global buckling, and mixed mode buckling (local bucking and global buckling) [7].

Majority delamination near to the surface will lead to the local buckling, but when the delamination far from the surface the buckling mode will be in global or mixed buckling. From the results, we can conclude that the local buckling decreases the buckling strength of the structure drastically than global and mixed buckling mode.

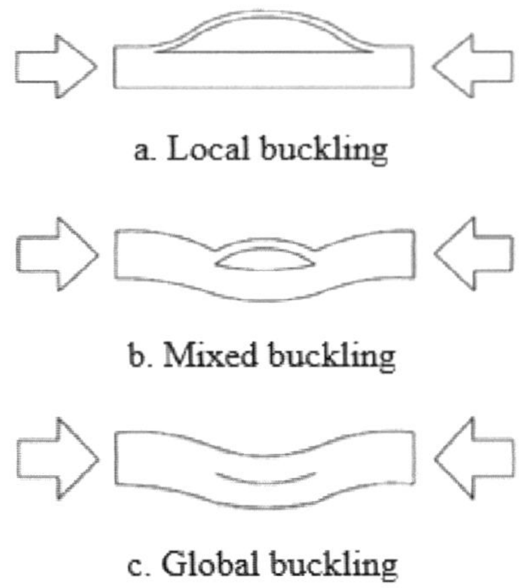

Figure 8. Modes of buckling [7]

7. CONCLUSIONS

The conclusions of this research are numerical modelling using finite element method with 2D Shell CQUAD4 element produce very good results that can show the decrease of buckling load phenomenon in the structure with delamination.

The longer the delamination in every location in the structure will lead to the decreasing of structure buckling load. The location of delamination will affect the buckling load of the structure, where the delamination near to the surface will make the buckling strength of structure decrease significantly.

The buckling strength of structure with multi delamination is lower than with single delamination. Local buckling makes the buckling load of the structure decreases significantly comparing to the other modes of buckling. The buckling phenomenon in the structure dominantly affected by geometrical factor, moment inertia and also the length of the structure.

There are three types of buckling namely local buckling, global buckling, and mixed mode buckling. In this research local buckling occur mainly when the delamination is located near to the surface. The global and mixed mode occur mainly when the delamination far from the surface of the structure. The local buckling mode play major role in decreasing the buckling strength significantly compared to the global and mixed buckling mode.

8. REFERENCES

[1] R.M. Jones, "Mechanics of Composite Materials", 2nd edition, Taylor & Francis, Inc, 1999.

[2] S. Sallam and G.J Simitsess, "Delamination Buckling and Growth of Flat, Cross-Ply Laminates", Composite Structures, Vol.4, pg 361-381, 1984.

[3] J.N. Reddy, "Mechanics of Laminated Composite Plates and Shells – Theory and Analysis", CRC Press, pg 165 – 182, 2004.

[4] M.J Sewell, "The Static Perturbation Techniques in Buckling Problems", Mech. Phys. Solids, 1965, Vol.13, pg. 247-265.

[5] Mulia Minhat, "Buckling Analysis of Delaminated Composite Beams Using Finite Element Method", Master Degree Thesis Insitut Teknologi Bandung, 2007.

[6] G.J Simitsess, S. Sallam, "Effect of Delamination of Axially Loaded Homogeneous Laminated Plates", AIAA Journal, pg 1437-1444, 1985.

[7] I. Sheinmann, M. Bass and O. Ishai, "Effect of Delamination on Stability of Laminated Composite Strip", Composite Structures, Vol. 11, pg 227-242, 1989.

IOP Conf. Series: Journal of Physics: Conf. Series **1005** (2018) 012009 doi:10.1088/1742-6596/1005/1/012009

Effect of mesh-peel ply variation on mechanical properties of E-glas composite by infusion vacuum method

K Abdurohman[1] and Mabe Siahaan[1]

1) Aeronautics Technology Center, National Institute of Aeronautics and Space (LAPAN), Indonesia
kosim.abdurohman@lapan.go.id, and mabesiahaan@yahoo.co.uk

Abstract. Composite materials made of glass fiber EW-135 with epoxy lycal resin with vacuum infusion method have been performed. The dried glass fiber is arranged in a mold then connected to a vacuum machine and a resin tube. Then, the vacuum machine is turned on and at the same time the resin is sucked and flowed into the mold. This paper reports on the effect of using mesh-peel ply singles on upper-side laminates called A and the effect of using double mesh-peel ply on upper and lower-side laminates call B with glass fiber arrangement is normal and ± 450 in vacuum infusion process. Followed by the manufacture of tensile test specimen and tested its tensile strength with universal test machine 100kN Tensilon RTF 2410, at room temperature with constant crosshead speed. From tensile test results using single and double layers showed that double mesh-peel ply can increase tensile strength 14% and Young modulus 17%.

1. Introduction

There is an increasing interest to use composite materials rather than conventional materials in structural applications for aircraft, aerospace, automotive, marine and other structures [1].
Due to the advantages of this composite material in its use has been widely used in the aerospace industry. The advantages of composites are among others have excellent mechanical properties and low structural weight. In fact, most of the costs associated with high performance composites are associated with manufacturing techniques. And along with the necessity of improving the nature, is the reason why some techniques are developed to automate the process. Much of the progress towards obtaining low-cost composite structures has occurred in the application of textile technology in its manufacturing methods. Traditional laminate structures generally exhibit poor interlaminar toughness, which makes them vulnerable to delamination while supporting interlaminar pressure. Where impact loading, manufacturing defects, geometric discontinuities such as free edges will result in delamination. The decrease in structural integrity is a result of the growth of delamination cracks under the influence of external loads resulting in catastrophic failure of the structure. [2].

During the past decade, the vacuum infusion or resin infusion processes have become popular for manufacturing structural polymer-based composites. Resin infusion processes have been identified as cost-effective alternatives to conventional autoclave manufacturing techniques. The resin infusion process is possible to produce or work on complex and thick parts with excellent mechanical properties and with less waste than traditional methods However, the process is somewhat difficult to control, firstly because the infusion-stage driving mechanisms are quite complex, and secondly, with existing industrial technology parameters such as thickness or resin in small dimensions inaccessible [3].The most popular terms to describe vacuum infusion processes are: Some types of infusion resin include Vacuum Assisted Resin Transfer Molding, Vacuum Assisted Resin Infusion Molding,

IOP Conf. Series: Journal of Physics: Conf. Series **1005** (2018) 012009 doi:10.1088/1742-6596/1005/1/012009

Composite Seemann Resin Infusion Process Molding, Vacuum Resin Transfer Molding Bag, Infusion Vacuum Assisted Resin Process, Liquid Resin Infusion [4]. All involve basically the same technology, and describe methods based on the impregnation of a dry reinforcement by liquid thermoset resin driven under vacuum.

From experimental studies of the tensile properties of hand lay up plain-weave woven glass fiber reinforced polymers (GFRP) panels [5] have resulted in characteristic tensile deformation of different lay-up arrangements and preservation pressures have shown relatively significant value.

Goren .A and Atas C. [1] has been studied manufacturing of polymer matrix composite using vacuum assisted resin infusion molding with double (upper and lower) peel ply but single mesh on the upper side

This paper describes about the effect of upper and lower side mesh-peel ply used in vacuum infusion or liquid resin infusion process to mechanical properties of the glass fiber reinforced polymer (GFRP). The objectives are to know the differences between single (upper) mesh-peel ply and double (upper and lower) mesh-peel ply used in vacuum infusion to tensile properties of the composite.

2. Experimental Method

2.1 Material

Woven glass fibers have been used in this research as reinforcing material, while epoxy resin (lycal type) has been utilized as the matrix material [6]. Fiberglass is a composite material consisting of glass fibers in a matrix of polymer [7]. It used e-glass cloth 135 (EW 135 cloth) from justus. The laminated consist of normal and $\pm 45^0$ cutted EW135 cloth fibers.

2.2 Vacuum Infusion Process

As has been explained above that to drive resin into laminate is by using vacuum pressure where this technique is called Vacuum Infusion Process. The material is placed dry into the mold and vacuum is applied before the resin is introduced. Once the complete vacuum is reached, the resin is literally sucked into the laminate through a carefully placed tube. This process is aided by a variety of supplies and materials [8].

Fig. 1 showed the normal procedure of composite manufacturing using vacuum infusion process. There are much runner in the upper side of normal procedure although the resin didn't fully wet through the laminates. To improve the resin absorption through fibers, it needs to add peel ply and flow mesh as resin distribution media in the lower side of fibers. The experiment manufactured composite with normal procedure vacuum infusion and by adding peel ply and flow mesh in the lower side to be compared then.

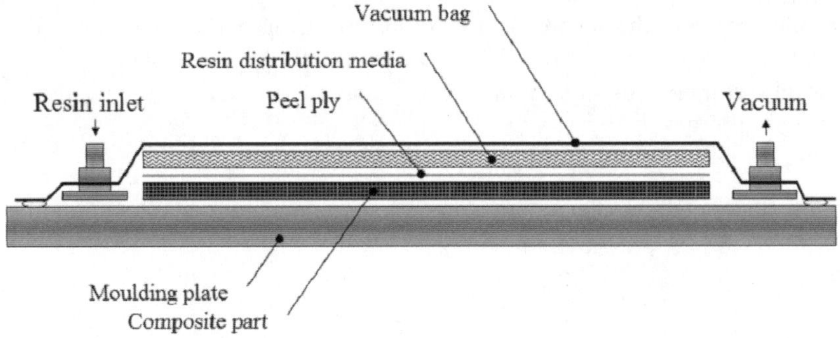

Figure 1. Normal procedure for vacuum infusion process [9]

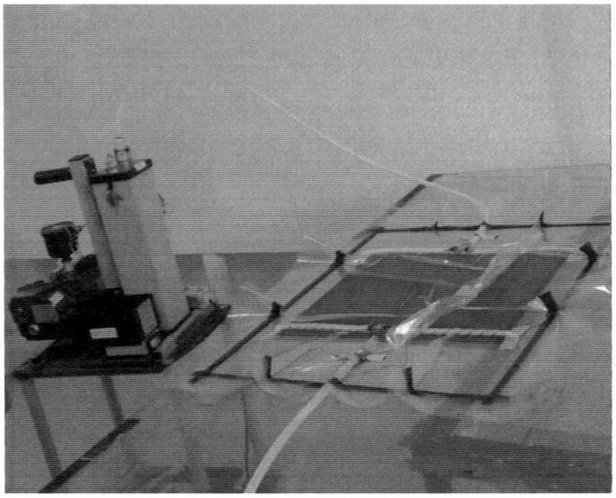

Figure 2. Composite manufactured by vacuum infusion process

3. Tensile Test

The woven glass fiber-reinforced lycal was used in this test; the total thickness of the laminate is ±2.5 mm. Tensile specimens were cut out of the laminates according to ASTM D3039. The number of tensile specimen is ten specimen manually cut out.

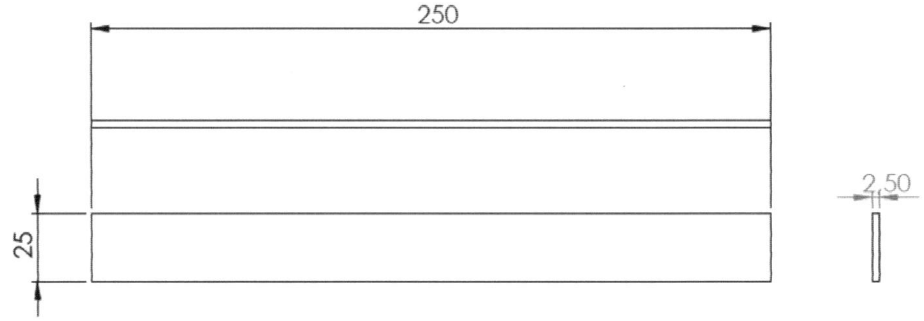

Figure 3. Geometry of the tensile test specimens

3.1 Testing Method

The tests was conducted from the beginning to the failure using a 100kN universal testing machine, Tensilon RTF 2410, at room temperature ±23°C with constant crosshead speed [10]. Load and strain measurement were automatically recorded during each test using a tensilon software. The displacement rate was 2 mm/min.

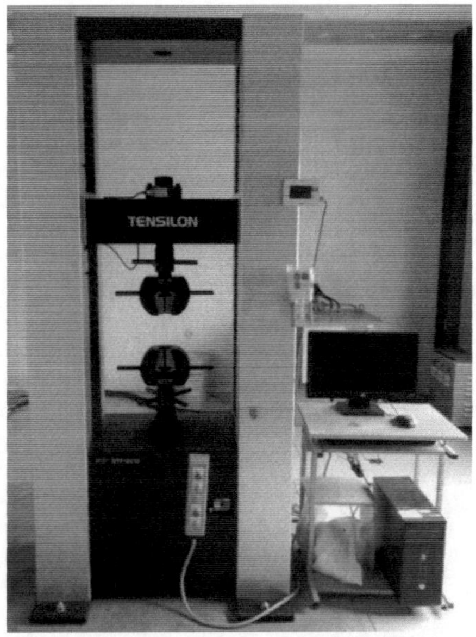

Figure 4. 100kN UTM Tensilon RTF 2410

4. Results and Discussions

The purpose of the static tensile tests was to determine tensile strength and Young modulus of wovwn glass laminates. Fig. 6 shows the typical stress-strain curve of woven glass laminate normal vacuum infusion (single mesh-peel ply) under tensile loads. Fig. 7 shows the typical stress-strain curve of woven glass laminate with adding mesh-peel ply in th lower side (double mesh-peel ply) under tensile loads.

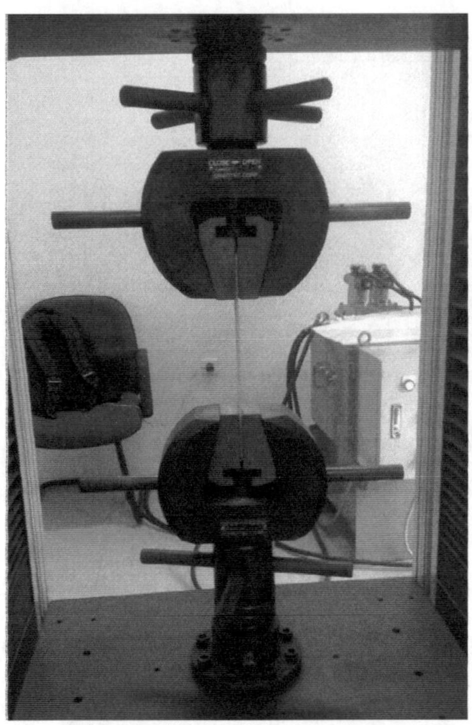

Figure 5. Tensile test set up

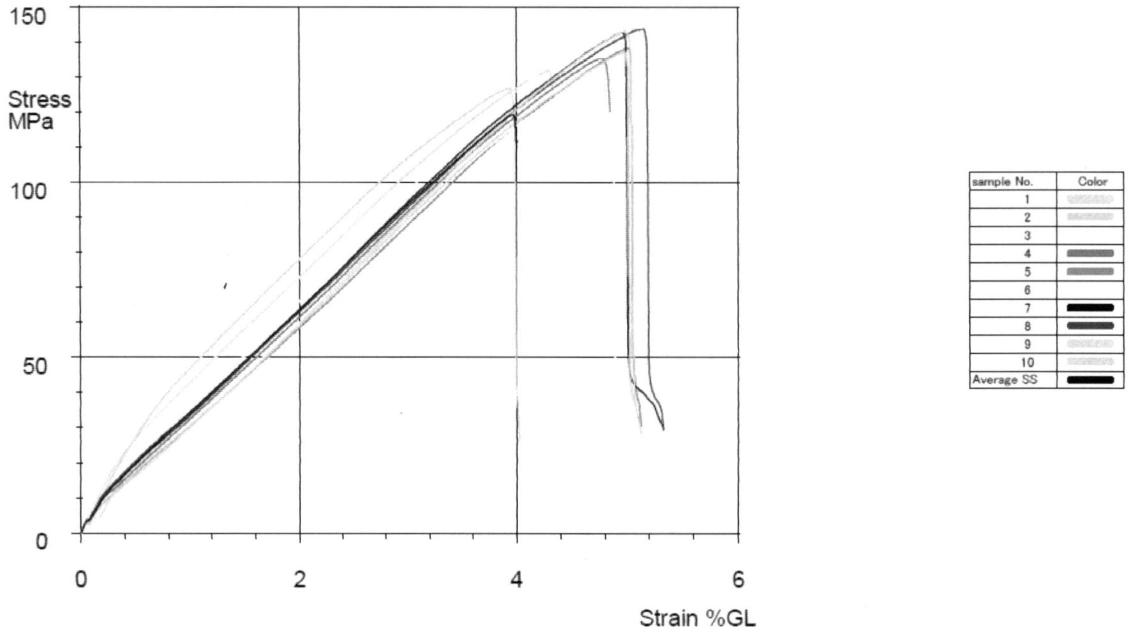

Figure 6. Tensile stress-strain curve of single mesh-peel ply composite

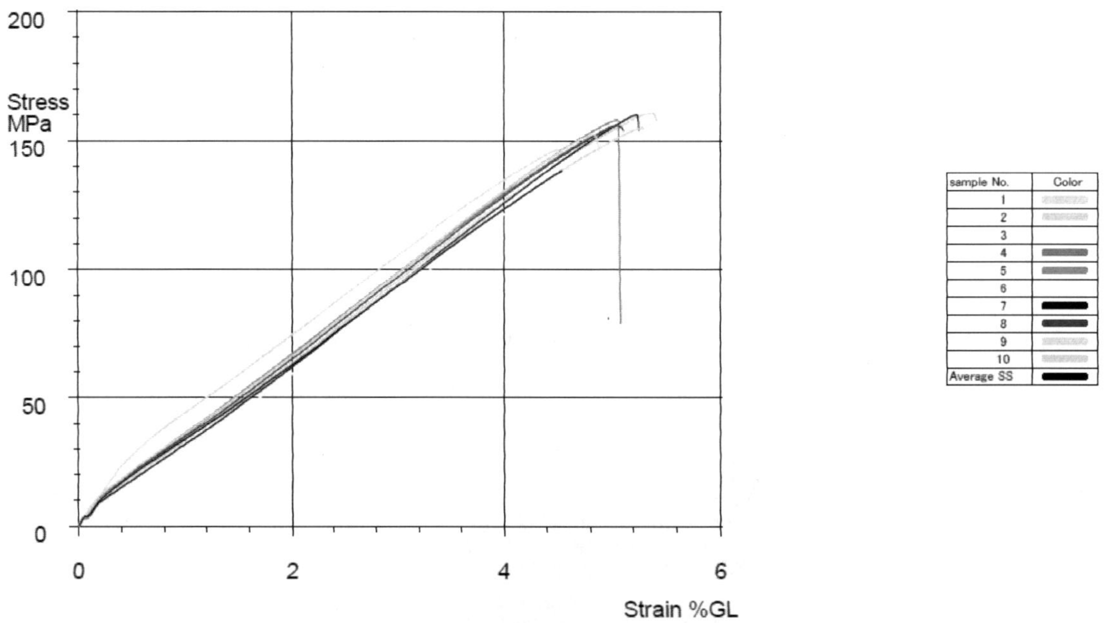

Figure 7. Tensile stress-strain curve of double mesh-peel ply composite

The tensile characteristic stress-strain results for single and double mesh-peel ply all have shown a similarity in trend growing from non-linearity and eventually develops into linear slops after about 0.3% amount strain. Mohd Aidy Faizal et.al. 2011 [5] said that the initial non-linearity is basically attributed to the deformation of the glass fibers. Thereafter, the characteristic linear slope predominantly reflects the deformation of the glass fibers. However, J. C. Chen et.al 2011 [11] said that for the certain curves displayed yielding and post yield strain hardening, indicating the reinforcing effect and structural homogenity for the composites. The slope between the yielding to failure point (called the post-yield modulus) represents the reinforcing efficiency of the composite. Therefore, the Young modulus value divided into two category; tensile modulus (before yield) and post-yield modulus (between yielding to failure point) [9]. This paper determined the Young modulus value in the 0.1% to 0.3% strain range according to ASTM D3039 for both single and double mesh-peel ply. In all tensile failure, the characteristics are very distinct and abrupt without plastic strain. The strain to failure is found to fall over a range 3.9% to 5% amount of strain for single mesh-peel ply and 4.5% to 5.4% amount strain for double mesh-peel ply, which can be explained as due to the resin absorption to the laminates effect. The stress to failure is also found to fall over a range 125 to 145 MPa for single and 135 to 160 MPa for double mesh-peel ply.

Figure 8. Specimens fracture for single mesh-peel ply

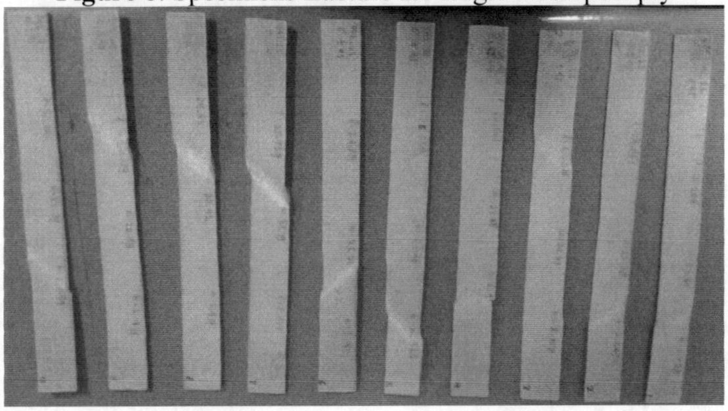

Figure 9. Specimens fracture for double mesh-peel ply

Fig. 8 and 9 showed the specimens after fracture. For both single and double mesh-peel ply had typically fracture region near the grip and in the gage length. Visual inspection showed that the fracture type is in the plus or minus 45^0 direction of fibers. It means the weakness of the specimens is in the $\pm45^0$ fiber direction. So the $\pm45^0$ fiber direction is not good for the longitudinal tensile loading. It could be reference for design a structue which subjected by tensile load.

The average value of tensile strength is 136.76 MPa for single and 155.78 for double mesh-peel ply. It means the average value of tensile strength increase 14% with double mesh peel ply. The average value of Young modulus is 5.06 GPa for single and 5.91 GPa for double mesh-peel ply. It means the double mesh-peel ply could improve the Young modulus for about 17%.

5. Conclusions

This paper showed the used of double mesh-peel ply in the vacuum infusion process for composite could improve the tensile properties of the composite. Tensile strength increased 14% and Young modulus increased 17%. However, both process had typically stress-strain curve and fracture type based on visual inspection.

Acknowledgement

This work is supported by Litbangyasa LSU-02 NG LD Program. The authors would like to thank the Director of Center for Aeronautical Technology LAPAN who has provided material and permission to use the laboratory facilities, so that this research can be done.

References

[1] Goren, A., & Atas, C. Manufacturing of Polymer Matrix Composite Using Vacuum Assisted Resin Infusion Molding. *Material Science and Engineering, Vol. 34, Issue 2, pp. 117-120, 2008.*

[2] Blas, F., & Fernandez I. Mechanical Characterisation of Carbon/Epoxy Composite Materials \ Manufactured by Resin Film Infusion Method with Stitching Reinforcement. *STO Meeting Proceedings, RTO-MP-069 (II), Vol. 5, No. 4, October 2001.*

[3] Wang, P., Drapier, S., Molimard, J., Vautrin, A., Minni, J. C. Numerical and Experimental Analyses of Resin Infusion Manufacturing Processes of Composite Materials. *Journal of Composite Materials, Vol. 46, No. 13, pp. 1617-1631, 2011.*

[4] Wang, P., Molimard, J., Drapier, S., Vautrin, A., Minni, J. C. Monitoring The Resin Infusion Manufacturing Process Under Industrial Environment Using Distributed Sensors. *Journal of Composite Materials, vol. 46, No. 6, pp. 691-706, August 2011.*

[5] Faizal, M. A., Beng, Y. K. and Dalimin, K. N. Tensile Property of Hand Lay-Up Plain-Weave Woven E-Glass/Polyester Composite: Curing Pressure and Ply Arrangement Effect. *Borneo Sience, The Journal of Science & Technology, 2011.*

[6] Torabizadeh, M. A. Tensile, Compressive and Shear Properties of Unidirectional Glass/Epoxy Composites Subjected to Mechanical Loading and Low Temperature Services. *Indian Journal of Engineering & Material Science, Vol. 20, pp. 299-309, 2013.*

[7] Bakir, B., & Hasim, H. Effect of Fiber Orientation for Fiber Glass Reinforced Composite Material on Mechanical Properties. *International Journal of Mining, Metallurgy & Mechanical Engineering (IJMMME), Vol. 1, Issue 5, 2013.*

[8] http://www.composites.ugent.be/home_made_composites/documentation/FibreGlast_Vacuum_infusion_process.pdf

[9] Perrin, H. F., D'acunto, A., Martin, P., Cauchois, J. P. Industrialization of Liquid Resin Infusion Processes (LRI) Considering to the Stress Transferred Influences. *The 10th International Conference on Flow Processes in Composite Materials (FPCM10), Monte Verita, Ascona, CH, July, 2010.*

[10] Abdurohman, K., & Marta, A. An Experimental Study of Polyester Composite Tensile Properties Reinforced Unidirectional Carbon Fiber Manufactured by Vacuum Infusion for LSU Materials. *Jurnal Teknologi Dirgantara, Vol. 14, No. 1, pp. 61-72, June 2016.*

[11] Chen, J. C., Wu, C. M., Pu, F. C., Chiu, C. H. Fabrication and Mechanical Properties of Self Reinforced Polyethylene Terepthalate Composite. *eXPRESS Polymer Letters, Vol. 5, No. 3, pp. 228-237, 2011.*

CFD Based Prediction of Discharge Coefficient of Sonic Nozzle with Surface Roughness

Agastya Bagaskara[*], **Mochammad Agoes Moelyadi**[**]

Faculty of Mechanical and Aerospace Engineering, Institut Teknologi Bandung, Indonesia
**agastya.bagaskara@students.itb.ac.id
**moelyadimochammad@gmail.com

Abstract. Due to its simplicity and accuracy, sonic nozzle is widely used in gas flow measurement, gas flow meter calibration standard, and flow control. The nozzle obtains mass flow rate by measuring temperature and pressure in the inlet during choked flow condition and calculate the flow rate using the one-dimensional isentropic flow equation multiplied by a discharge coefficient, which takes into account multiple non-isentropic effects, which causes the reduction in mass flow. Proper determination of discharge coefficient is crucial to ensure the accuracy of mass flow measurement by the nozzle. Available analytical solution for the prediction of discharge coefficient assumes that the nozzle wall is hydraulically smooth which causes disagreement with experimental results. In this paper, the discharge coefficient of sonic nozzle is determined using computational fluid dynamics method by taking into account the roughness of the wall. It is found that the result shows better agreement with the experiment data compared to the analytical result. |
Keywords: sonic nozzle, discharge coefficient, boundary layer, mass flow, roughness, Computational Fluid Dynamics

Nomeclature

A_t	Throat area, m^2
C^*	Critical flow function, dimensionless
C_d	Discharge coefficient, dimensionless
d	Throat diameter, m
m	Parameter defined by Geropp, dimensionless
$\dot{m}_{actual}$	Actual mass flow rate ($kg\ s^{-1}$)
$\dot{m}_{ideal}$	Isentropic one-dimensional mass flow rate ($kg\ s^{-1}$)
p_0	Inlet stagnation pressure, Pa
R	Specific gas constant (dry air, 287 $j\ kg^{-1}K^{-1}$)
Re	Throat Reynolds number, dimensionless
R_{wall}	Wall curvature radius, m
T_0	Inlet stagnation temperature, Kelvin
μ	Dynamic viscosity (dry air, 0.00001831 $Pa\ s$)
γ	Specific heat ratio, dimensionless (dry air, 1.4)
y^+	Dimensionless distance from wall

1. Introduction

Unlike other types of flow measurement devices such as turbines or variable area flow meters, sonic nozzle requires no moving part, thus extending its life and reliability. Since it operates at choked flow condition, any pressure changes downstream of the nozzle may not affect mass flow through the nozzle. Therefore, it is a suitable choice for accurate gas flow measurement and flow control.

The nozzle oprates at choked flow condition. Thus, when the inlet pressure and temperature is known (usually done by placing pressure and temperature sensor upstream of the nozzle), mass flow rate can then be calculated using isentropic one-dimensional flow equation multiplied by a discharge coeficient which serves as mass flow correction factor which takes into account non-isentropic and non-one dimensional effects. Accurate determination of the discharge coefficient is crucial in order to ensure accurate reading of mass flow.

Previous analytical discharge coefficient formulas derived by Hall [1] (which considers the effects of boundary layer in the nozzle wall, C_{d1}) and Geropp [2] (which considers multi-dimensional flow effects depending on geometry of nozzle, C_{d2}) assumed that the nozzle wall was hydraulically smooth [3], which can cause the formula to overestimate the experimental discharge coefficient.

In this study, Computational Fluid Dyanmics is used to determine the discharge coefficient of sonic nozzle with surface roughness. The inlet pressure boundary condition is varied in accordance with the experiment data in order to observe the effect of Reynolds number to the discharge coefficient, and the back pressure is lowered sufficiently to ensure critical flow at the throat section and to avoid shock formation in divergent section of the nozzle [4].

2. Theoretical Discharge Coefficient

The dimensions of the nozzle and the inlet pipe in milimetre is shown in figure 1.Discharge coefficient $C_d$is the ratio of actual critical mass flow rate $\dot{m}_{actual}$ to the ideal critical mass flow rate $\dot{m}_{ideal}$

$$C_d = \dot{m}_{actual}/\dot{m}_{ideal} \tag{1}$$

The critical mass flow rate $\dot{m}_{ideal}$ for one-dimensional isentropic flow of an ideal gas can be calculated by [5]

$$\dot{m}_{ideal} = \frac{A_t C^* p_0}{\sqrt{RT_0}} \tag{2}$$

Where C^* (critical flow function) is given by

$$C^* = \sqrt{\gamma\left(\frac{2}{\gamma+1}\right)^{\frac{\gamma+1}{\gamma-1}}} \tag{3}$$

The research of C_d is divided into three parts [3], namely:C_{d1} which considers mass flow reduction by viscous effects of boundary layers in the viscous region of the flow, C_{d2} which considers mass flow reduction in the inviscid region of the flow caused by the geometry of the nozzle, and C_{d3}(also called viral discharge coefficient) which considers the physical property of the gas, which is negligible when the gas is approximately ideal [4]. In this study, only C_{d1} and C_{d2} is considered. For Hall-Geropp nozzle, C_d can be expressed as

$$C_d = C_{d1}.C_{d2} \tag{4}$$

Formulas for determining C_{d1} for laminar and turbulent flows in smooth wall nozzle are as follows [2, 4, 6]

$$C_{d1} = \begin{cases} 1 - \frac{4}{\sqrt{Re.m}}\left(\frac{\gamma+1}{2}\right)^{\frac{1}{2(\gamma-1)}}\left(3\sqrt{2} - 2\sqrt{3} + \frac{\gamma-1}{\sqrt{3}}\right), \ laminar\ flow \\ 1 - 0.00525\left(\frac{d}{R_{wall}}\right)^{-0.4} Re^{-0.2}, \ turbulent\ Flow \end{cases} \tag{5}$$

Where

$$Re = \frac{4\dot{m}_{actual}}{\pi d\mu} \tag{6}$$

And

$$m = \sqrt{\frac{2d}{R_{wall}}\left(\frac{\gamma+1}{2}\right)^{\frac{3\gamma-1}{\gamma-1}}} \tag{7}$$

While the formula for determining C_{d2} is as follows [1]

$$C_{d2} = 1 - \frac{\gamma+1}{(2R_{wall}/d)^2}\left(\frac{1}{96} - \frac{8\gamma+21}{4608\left(\frac{2R_{wall}}{d}\right)} + \frac{754\gamma^2+1971\gamma+2007}{552960\left(\frac{2R_{wall}}{d}\right)^2}\right) \tag{8}$$

Therefore, since the R_{wall}/d of the nozzle is 2.2, $\gamma = 1.4$, and it is assumed (due the large Reynolds number regime in the experiment) that the boundary layer is turbulent, the theoretical discharge coefficient is given by

$$C_d = (1 - 0.0071966\ Re^{-0.2})\ .0.998978 \tag{9}$$

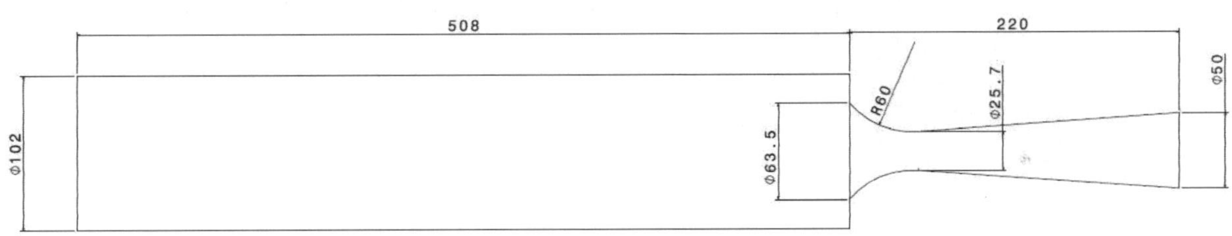

Figure 1 Inlet pipe and nozzle dimensions

3. CFD Simulation

3.1. Geometry and Meshing

3D model of the nozzle and inlet pipe is shown in figure 2. ICEM CFD software is used for meshing. Patch independent, quad dominant shell mesh is used as well as tetra/mixed volume meshing with octee mesh method. The cross section of the mesh is shown inFigure 3. The Prism mesh is used in the entire wall to ensure accurate simulation of boundary layer. The prism layers start in $y^+ = 5$, extending with a growth ratio of 1.2 for 17 layers as shown in figure 4.

Mesh quality values is used to assess the quality of the mesh and to ensure no negative volume exist in the mesh. Minimum element quality is 0.22 with maximum mesh quality of 1, averaging at 0.83 for a total of 9064479 elements. Therefore, the mesh is suitable for the simulation

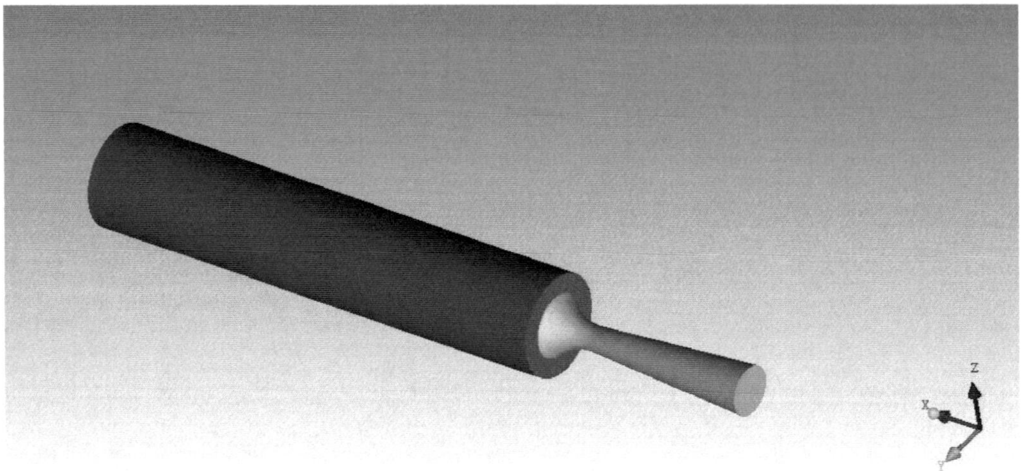

Figure 2 3D model

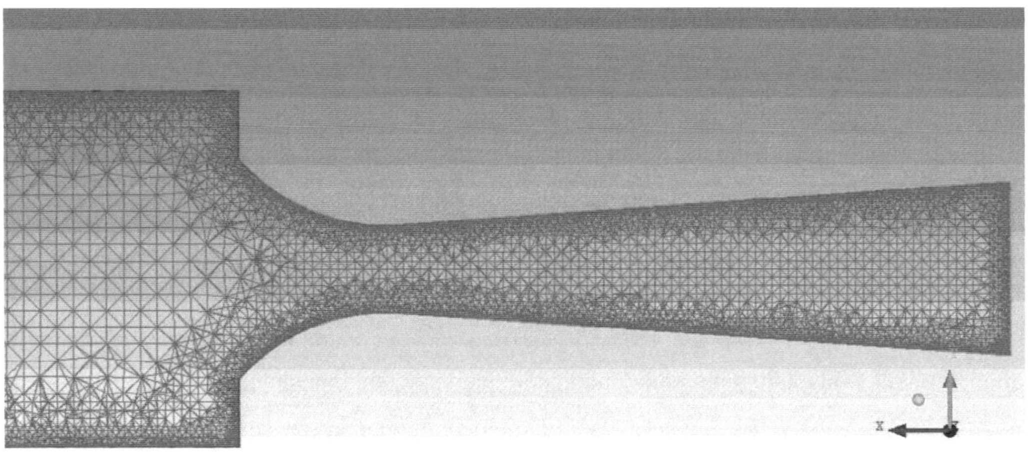

Figure 3 Cross section of the volume mesh

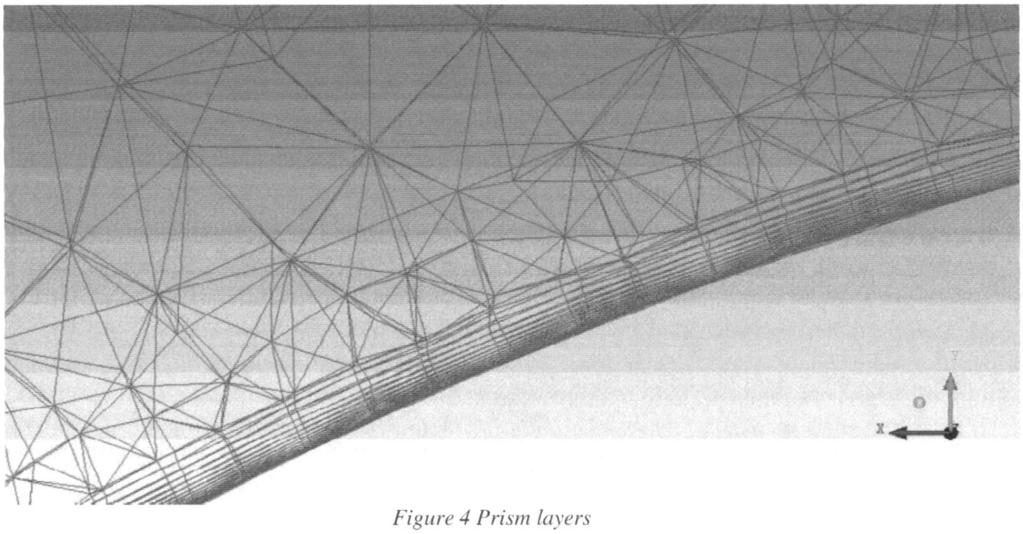

Figure 4 Prism layers

3.2. Solver Setting

CFX solver is used to solve Reynolds-Averaged steady state Navier-Stokes equation with shear stress transport turbulence model. It is assumed that the wall is adiabatic, and three different wall condition (smooth wall and rough wall with two different roughness) as well as inviscid flow condition is simulated. Wall roughness of both 1 micron and 2 micron is simulated to observe the effect of increased wall roghness to the discharge coefficient.

A total of five different inlet pressure is used in the simulation to observe the change of discharge coefficient with respect to change to Reynolds number. Aside from RMS residual target of 0.0001, conservation target of 0.001 is also used as convergence criterion to ensure no imbalances in the domain.

4. Results

All simulations satisfy the convergence criterion below the maximum number of iteration of 150. Figures 5, 6, 7 and 8 show the Mach number distribution for invsicid flow, viscous flow with smooth wall, viscous flow with wall roughness of 1 micron, and viscous flow with wall roughness of 2 micron, respectively. It can be seen that for viscous flows, both with smooth and rough walls, the viscosity effect cause reduction in Mach number near the walls of the nozzle, which forms the boundary layer.

Figure 9 shows the distribution of unit mass flow (product of flow density and velocity) at the wall of throat. As shown in the figure, there is a reduction of unit mass flow inside the boundary layer. This reduction in unit mass flow causes mass flow to be smaller compared to the ideal mass flow. As shown by the increasing wall distance in which the unit mass flow is equal to the freestream/core flow, the figure also shows that increasing wall roughness causes boundary layer thickness to increase as well, meaning that the boundary layer in the nozzle with rougher wall have larger displacement thickness, causing smaller effective throat radius compared to the smooth wall nozzle.

Figure 10 shows that this further reduction in effective throat diameter in nozzle with rough wall results in the decrease of discharge coefficient (thus in better agreement with the experiment result) compared to the nozzle with the smooth wall. For this simulation, a wall roughness of 2 micrometres causes the CFD result to better predict the experiment result by at least 43 percent compared to the smooth wall condition CFD.

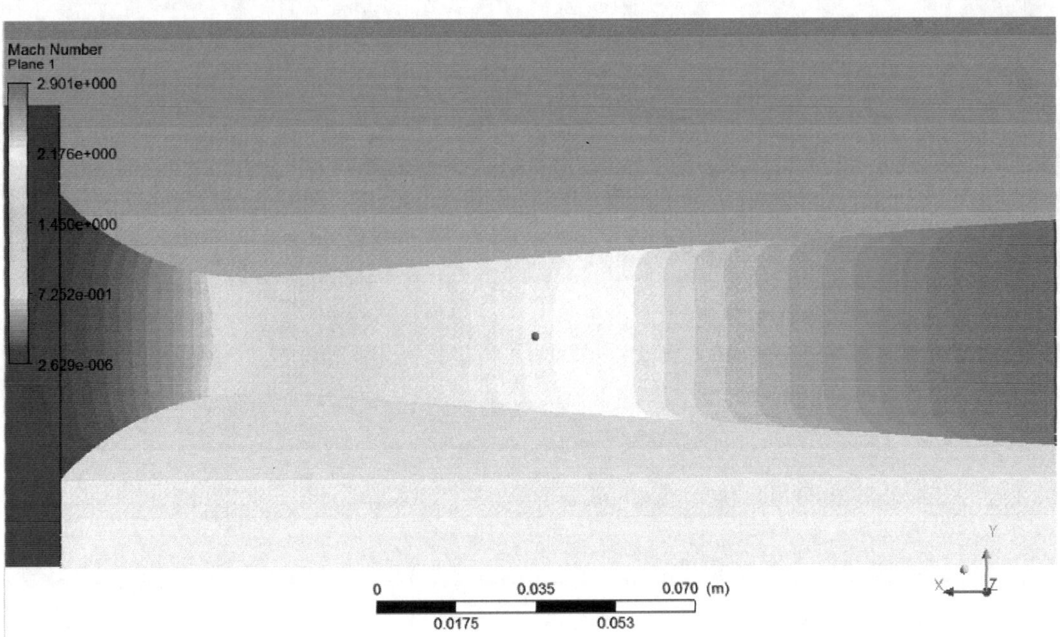

Figure 5 Inviscid flow Mach number distribution

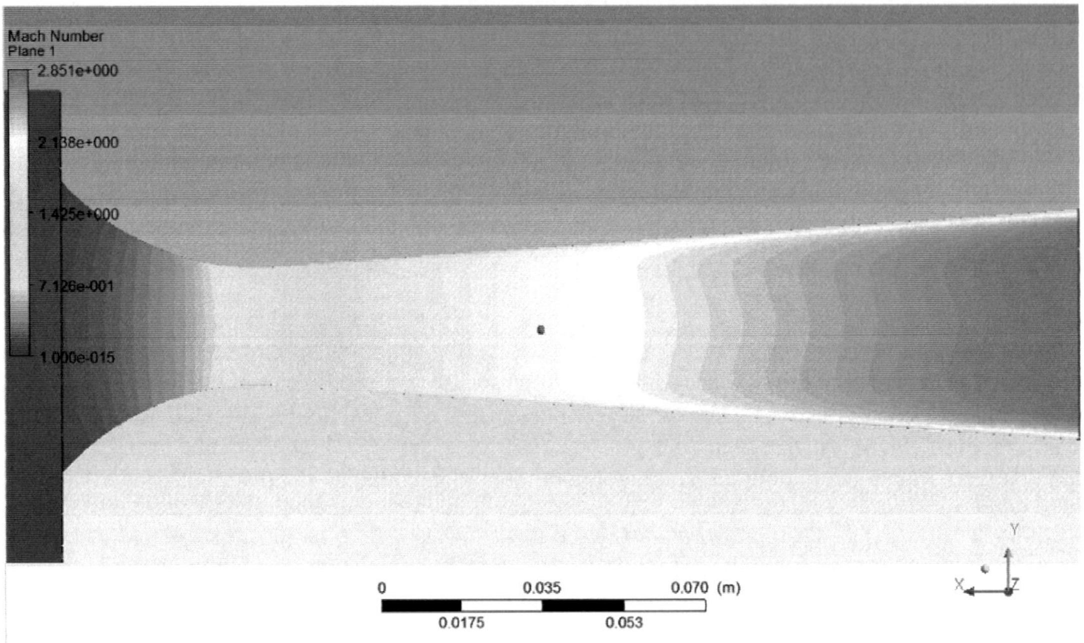

Figure 6 Viscous flow Mach number distribution with smooth wall

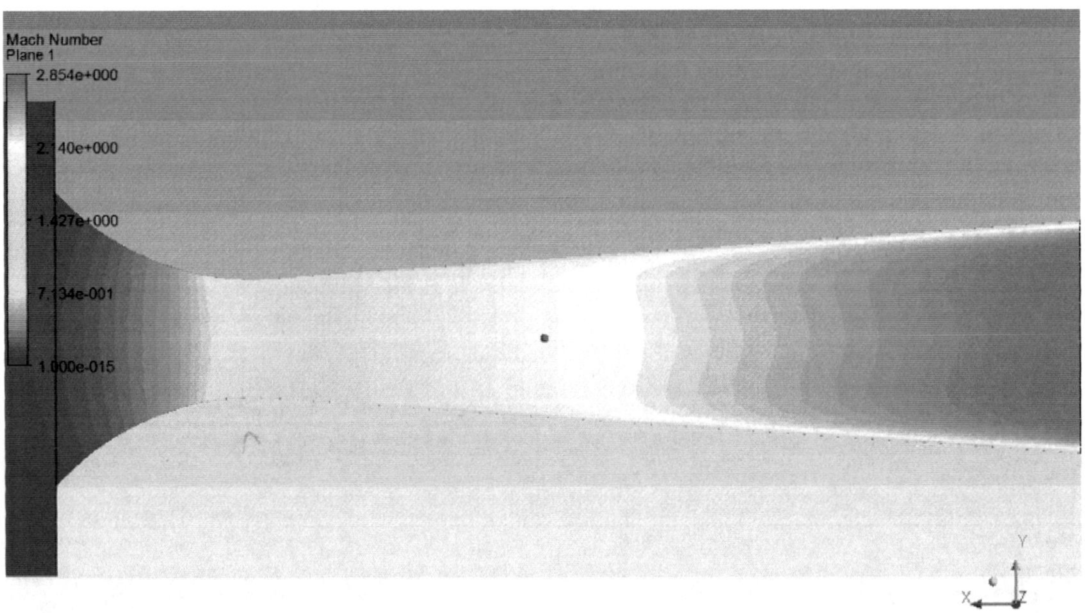

Figure 7 Viscous flow Mach number distribution with 1 micron roughness wall

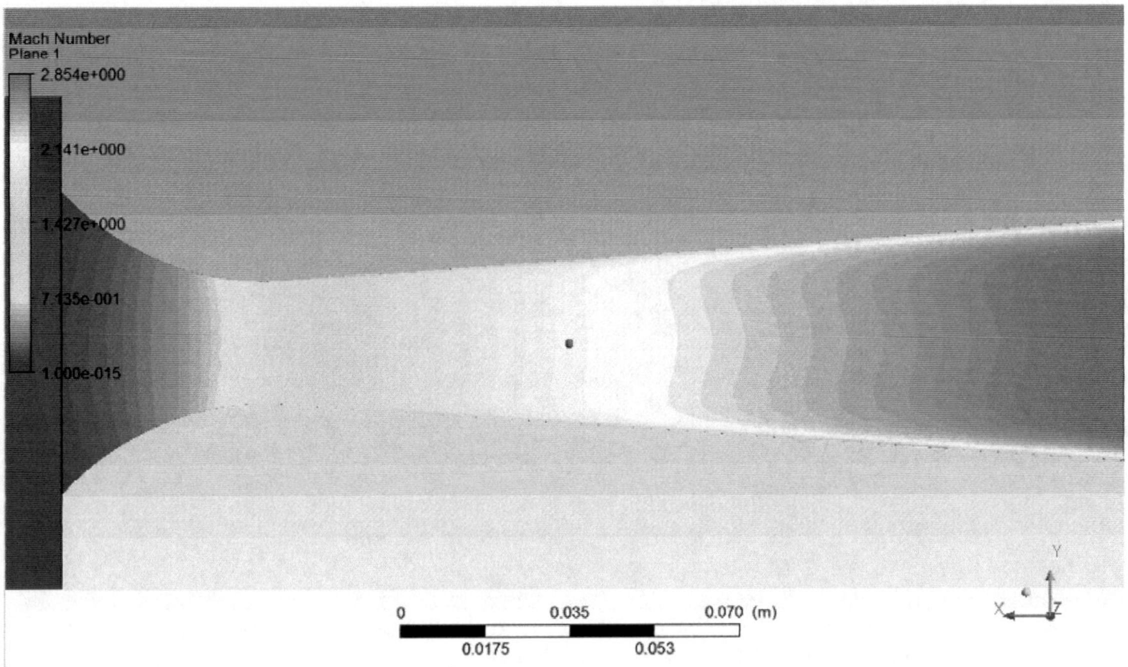

Figure 8 Viscous flow Mach number distribution with 2 micron roughness wall

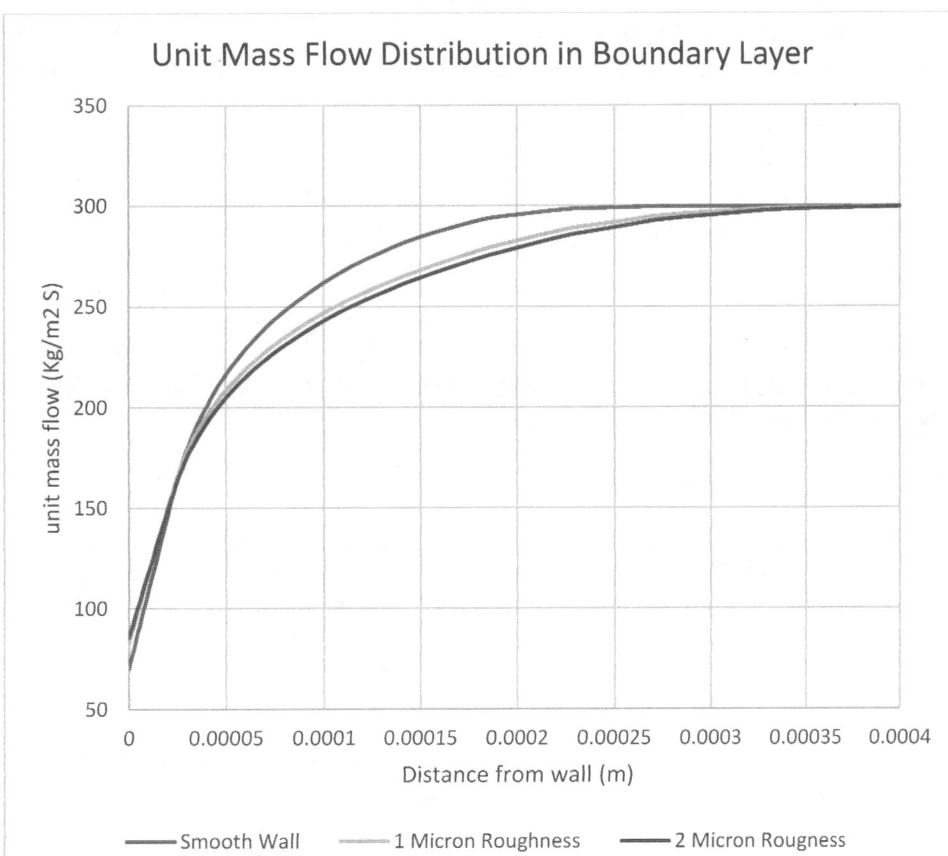

Figure 9 ***Unit mass flow*** *distribution inside boundary layer*

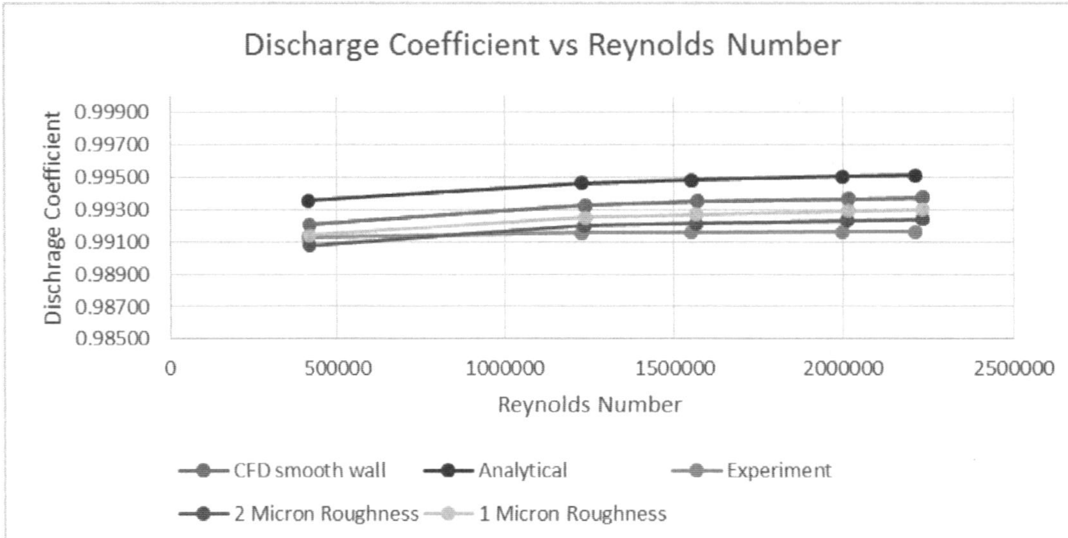

Figure 10 Dicharge coefficient vs Reynolds number

Conclusion

The effect of surface roughness on discharge coefficient of sonic nozzle is analyzed using CFD. Five different inlet pressure is simulated to show the change in discharge coefficient with respect to Reynolds number, as well as two wall conditions, smooth and rough. Two different roughness values of 1 and 2 microns are simulated to observe the effect of increased roughness. The outlet pressure is set to be sufficiently small to avoid formation of normal shocks in the diffuser section of the nozzle which can cause difficulty in achieving convergence criterion. Prism mesh is used in wall region to ensure accurate capture of boundary layer. The CFX Solver is used to solve Reynolds-Averaged Navier-Stokes equation with shear stress turbulence modelling. All simulation reaches both RMS and conservation target of 0.0001 and 0.001, respectively, within 150 iteration steps.

The results are compared with existing analytical formula which assumes that the surface of nozzle is hydraulically smooth and confirmed by comparing the results to experimental data. As shown in the CFD result, viscosity effect causes unit mass flow defect inside the boundary which in turn causes reduction in mass flow in the nozzle. The boundary layer in the nozzle with the rougher wall is thicker than the nozzle with smooth wall, causing it to have more unit mass flow defect. As a result, the nozzle with the rougher wall has lower discharge coefficient over the entire simulated region of Reynolds number, and the nozzle with 2 micron wall roughness is found to be in better agreement with the experiment result.

References

1. I.M. Hall, Transonic flow in two-dimensional and axially-symmetric nozzles, Q. J.Mech. Appl. Math. (Pt. 4) (1962) 487–508.
2. D. Geropp, Laminare Grenzschichten in ebenen und Rotation-Symmetrischen Lavaldüsen, Deutsche Luft-und Raumfahrt Forschungsbericht, German, 1971, pp.71–90.
3. H. Ding et al, Approximate solution for discharge coeffecient of the sonic nozzle with surface roughness, Flow Measurement and Instrumentation 52 (2016) 227–232
4. H. Ding et al, Influence of divergent sectionon flow fields and discharge coefficient of ISO 9300 toroidal-throat sonic nozzle, Flow Measurement and Instrumentation 40 (2014)19-27
5. ISO 9300. Measurement of gas flow by means of critical flow venturi nozzles; 2005.
6. B.S. Stratford, The calculation of the discharge coefficient of profiled choked nozzles and optimum profile for absolute air flow measurement, J. R. Aeronaut. Soc. 68 (1964) 237–245.

CFD Based Added Mass Prediction in Cruise Condition of Underwater Vehicle Dynamic

Mochammad Agoes Moelyadi, Bagus Bambang Riswandi
Department of Aeronautics and Astronautics,
Faculty of Mechanical and Aerospace Engineering, Bandung Institute of Technology, Indonesia

Email: moelyadi@ae.itb.ac.id, bagusbambang@students.itb.ac.id

Abstract. One of the unsteady flow behavior on the hydrodynamic characteristics of underwater vehicle is the presence of added mass. In cruising conditions, the underwater vehicle may require the addition of speed or experience the disturbance in the form of unsteady flow so that cause the hydrodynamic interaction between the surface of the vehicle with the surrounding fluid. This leads to the rise of local velocity of flow and the great changes of hydrodynamic forces which are very influential on the stability of the underwater vehicle. One of the result is an additional force called added mass. It is very useful parameter to control underwater vehicle dynamic.This paper reports the research on the added mass coefficient of underwater vehicles obtained through the Computational Fluid Dynmaic (CFD) simulation method using CFX software. Added mass coefficient is calculated by performing an unsteady simulation or known as transient simulation. Computational simulations are based on the Reynold Average Navier-Stokes (RANS) equation solution. The simulated vehicle moves forward and backward according to the sinus function, with a frequency of 0.25 Hz, a 2 m amplitude, a cruising depth of 10 m below sea level, and Vcruise 1.54 m / s (Re = 9.000.000). Simulation result data includes velocity contour, variation of force and acceleration to frequency, and added mass coefficient.

Keyword : Underwater vehicle; Unsteady hydrodynamic; Transient simulation; added mass coefficient.

1. Introduction

Added mass is a physical phenomenon that arises from the acceleration of objects in the fluid. Research on added mass is still being done because of its many benefits in underwater vehicle development. In fluid mechanics, added mass or virtual mass is the additional mass and inertia to the system due to acceleration or deceleration. This causes the fluid traversed by the object to be moved away because the object can not be occupied in the same time and space simultaneously. The movement of objects will accelerate the surrounding fluid with varying magnitude. The coefficient matrix added mass is a dimensionless value which can be expressed in the added mass coefficient divided by the mass of the displaced fluid. The added mass coefficient represents a force and moment component which is proportional to the acceleration. In general, this coefficient is in the form of a second order tensor where the value of each component depends on the vector of acceleration and the geometry of the object. The force vector added mass is always opposite the acceleration vector of the object.

The first added mass began to be investigated in 1776 by Dubua using a sperical pendulum experiment. Today, there has been a lot of research on this subject. Chen, *et al* [1] conducted an analytical and experimental study of added mass and damping on the vibrating rod. K. Vikestad *et al* [2] has calculated the value of added mass by experiment with a lightly damped elastically mounted rigid cylinder subjected to constant flow velocity. The analysis of added mass and damping on the circular cylinder, which oscillates in an air numerically has been performed by Uchiyama [3]. Ghassemi and Yari [4] have carried out the added mass computation of sphere, ellipsoid and marine propellers using boundary element method. Similar research about mass coefficient on underwater vehicles has also been done by Eng, et al [5]. They used experimental methods with a sinusoidal motion approach of scaled vehicle in water, known as Free Decay Test. The result through this experiment then compared well with the simulation results obtained from well-established computational fluid dynamics (WAMIT Software). Thus, the proposed approach can be use to find the added mass for other underwater vehicles.

This paper presents the added mass coefficient of the underwater vehicle maneuver in cruise condition. The simulation is performed at a frequency of 0.25 Hz and at a reynold number of 9,000,000. To model the motion, the axial position of the object changes with respect to time and the path of motion according to a sinus function that hase been set on CFX software. The flow model uses k-ε turbulence model. In addition, the computational domain around the underwater vehicle model has been made in multilayer domain to modeling simulated motion.

2. Computational Method

2.1 Geometry and Mesh Generation

Before perform the simulation, the definition of the computational domain, including the creation of the underwater vehicle geometry and mesh generation has to be carried out. The mesh generated in the computational domain has to accurately capture flow properties, such as velocity, pressure, and to properly model the underwater vehicle body. High quality (orthogonal) mesh is very important to prevent a negative volume during the simulation. For complex geometries such as the body,fin, and thruster of a vehicle, an unstructured mesh was created.

2.1.1 Geometry

Table 1. Underwater Vehicle Configuration

The length of the hull	4.3 m
Capacity (driver)	2 person
Maximum Payload	50 kg
Dry Empty Weight	1100 kg
Battery Capacity	9 kwh
Thrust Engine	182 kgf in bollard condition 109 kgf for cruising speed of 3.08 m/s
Diving Method	Automatic Bouyancy System
Cruise Depth	10 m
Maximum cruise depth	20 m
cruise speed	1.54 m/s
Maximum cruise speed	3.08 m/s

The geometry was created using CATIA software. The design adopts an existing mini submarine with little improvement on the hull and fin shape.

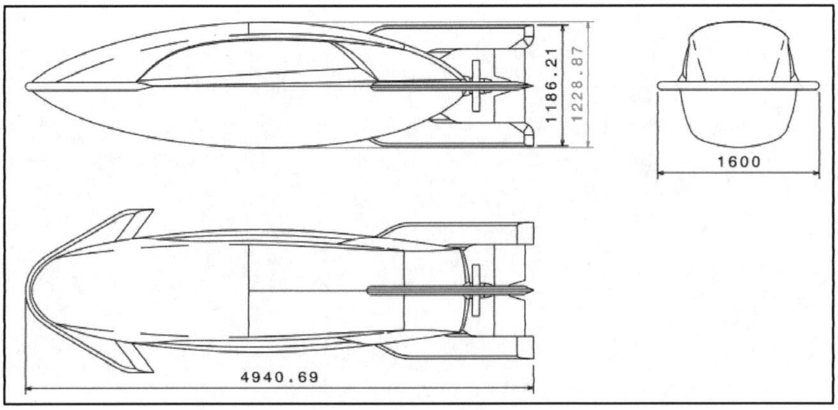

Figure 1. Geometry of Underwater Vehicle

2.1.2 Mesh Generation

Setting the size and quantity of elements and good mesh quality will improve the efficiency of the computing process. This process is performed using the Altair Hypermesh 13.0 software. Domain form that surrounds the object is cylinder-shaped with a diameter of 3.3 times the length of the object. The domain length in front of the nose is 2.1 times the length of the object, as well as the domain length behind the tail is 9 times the length of the object. In meshing process, the computational domain is divided into three parts, namely inner, outer and downstream domains. Inner domain is the domain that directly interact with the vehicle, consisting of 3 parts namely the boundary layer, layer 1, and layer 2. For the outter domain is divided into 2 parts namely layer 3 and layer 4. The front size for the inner domain or layer 2 is 0.5 times the length of the object, the measured rear distance from the tail is 3 times the length of the object, while for the diameter of layer 2 is 4 times the width of the object.

Inner domain is defined as a domain that is move simultaneously with the vehicle while the outter domain is defined as a fixed or immovable domain. The division of move and fixed domains is to avoid the occurrence of folded mesh when the displacement of the object is considerable. This will cause a negative volume on the calculation results. For this downstream domain measuring 9 time the object body length. This section serves to capture the properties of the eddy flow that occur due to the movement of object , so that the results of the calculation is closer to real conditions.

.

IOP Conf. Series: Journal of Physics: Conf. Series **1005** (2018) 012011 doi:10.1088/1742-6596/1005/1/012011

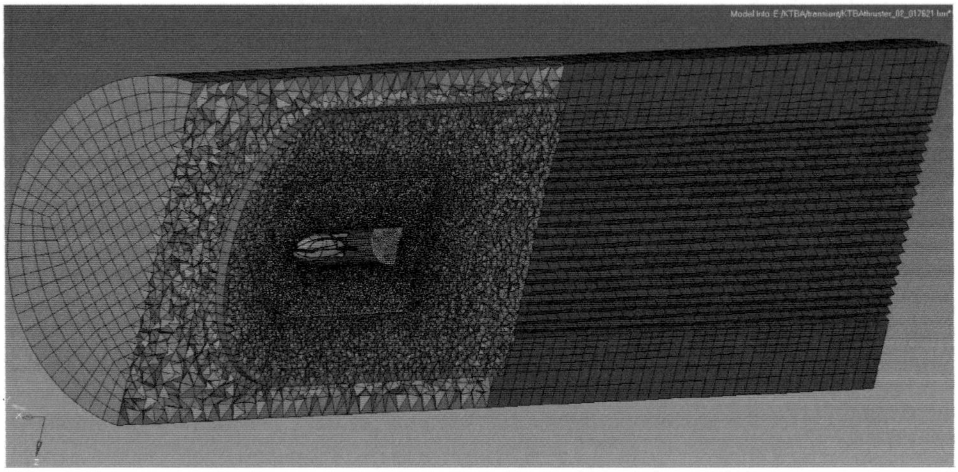

Figure 2. Mesh generation for Underwater Vehicle

Domain	Nodes	Elements
fluid 0	4171835	16650611

2.2 Motion Modeling

To obtain the added mass coefficient in cruise condition (axial direction), simulation is performed for one mode of motion. Therefore, simulation will be set so that the object moving back and forth or the position change on the positive and negative X-axis by time. The translational motion above is modeled by the following harmonic oscillation equation.

$$X(t) = X_{max}Sin\ (2\pi f t) \qquad (1)$$

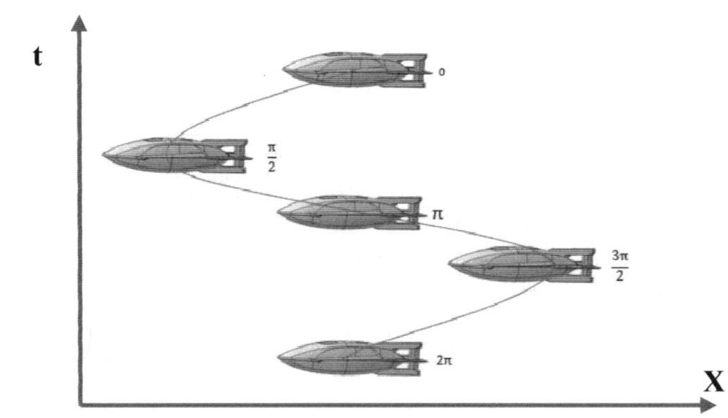

Figure 3. Motion illustration

Xmax [m]	ω [rad/s]	f [Hz]	Δt[s]	Total time Run
2	1.57	0.25	0.05	10 [s]

3. Results and Discussions

(a) Time Domain Data

The calculation result of force and moment for each timestep,Δt from the CFX is then plotted to see the resulting signal pattern. Below is presented plot of force and moment against time.

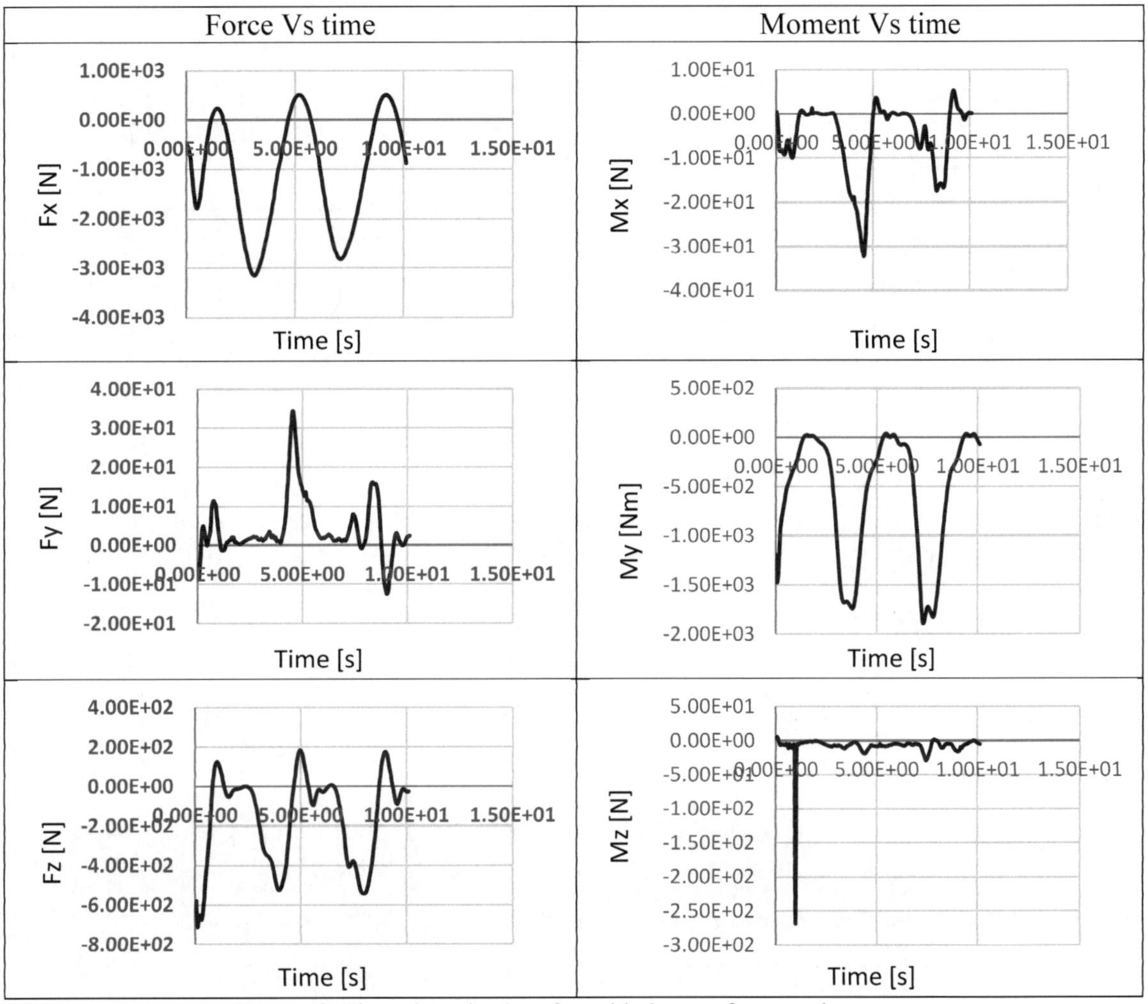

Figure 4. Time domain data for added mass force and moment

(b) Frequency Domain Data

The data set in point (a) is required as input data to be analyzed using fourier transforms. In its application, this fourier transform is already found in MATLAB software in the form of Fast Fourier Transform (FFT). Can be seen in **Figure 4**, the time-varying signal pattern of force and moment as a function of time is very complex, indicating that there are many combinations of harmonic wave with difference frequency and magnitude. **Figure 5** show the result of the fourier transformation of added mass force and moment using Fourier Transform Method.

IOP Conf. Series: Journal of Physics: Conf. Series **1005** (2018) 012011 doi:10.1088/1742-6596/1005/1/012011

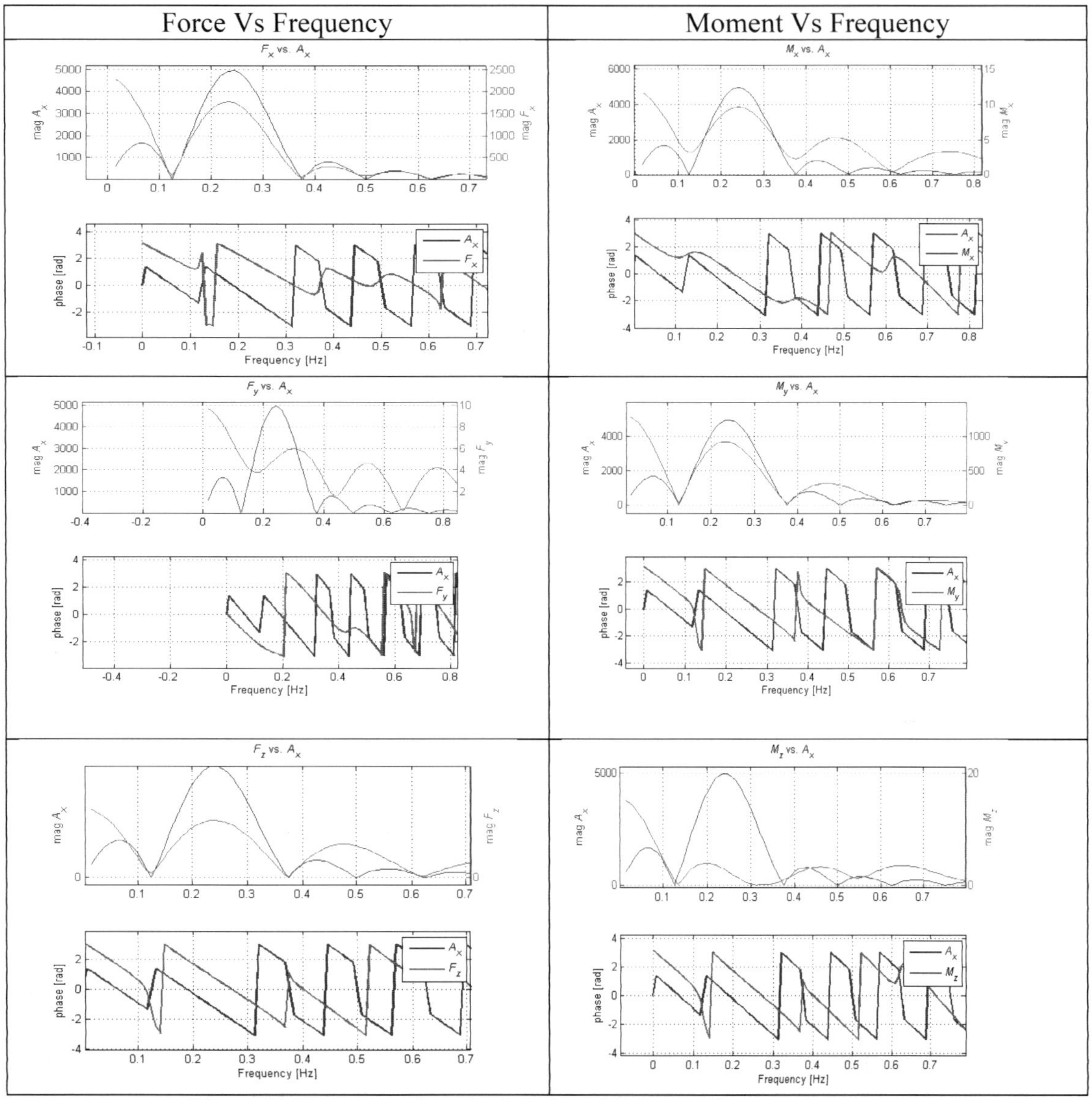

Figure 5. Frequency domain data for added mass force and moment

IOP Conf. Series: Journal of Physics: Conf. Series **1005** (2018) 012011 doi:10.1088/1742-6596/1005/1/012011

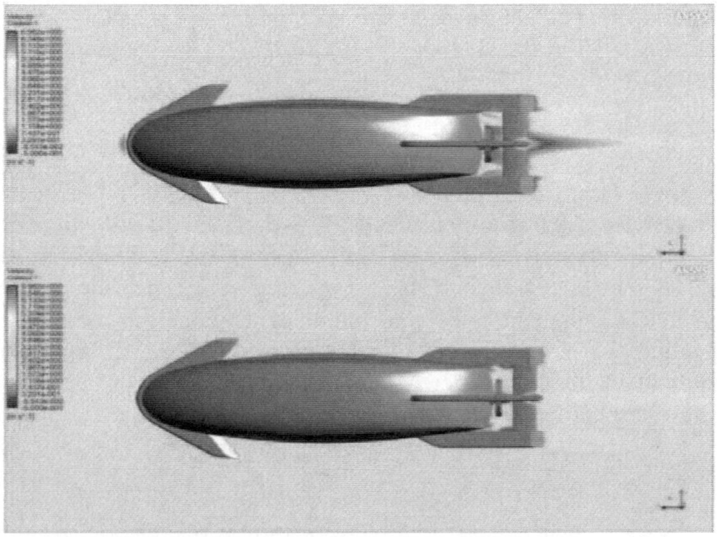

Figure 6. Velocity countur around the underwater vehicle
Top figure (x=0, t=4 s) dan bottom figure (upstroke, t=5 s)

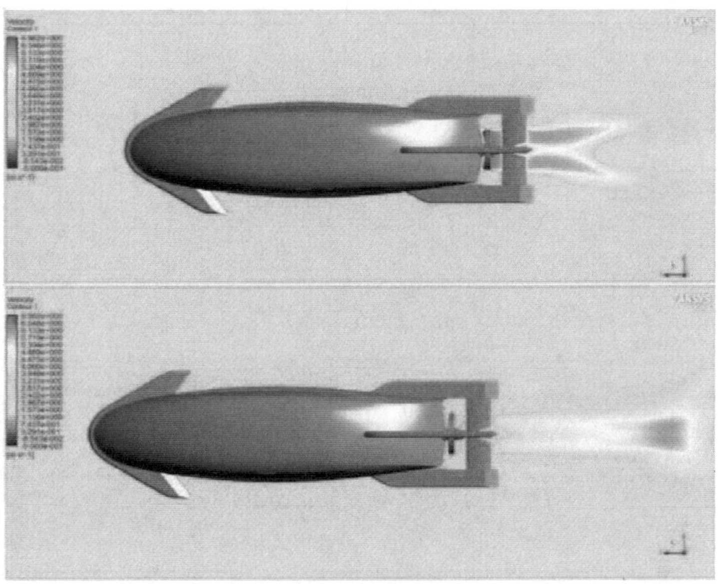

Figure 7. Velocity countur around the underwater vehicle
Top figure (t=6 s, x=0) dan bottom figure (t=7 s, downstroke)

(c) Added mass coefficient

Table 2 below shows the value of added mass force and moment component due to axial acceleration .

Table 2. Added mass Force and Moment Component

$X\dot{U}$ [N]	$Y\dot{U}$ [N]	$Z\dot{U}$ [N]	$K\dot{U}$ [Nm]	$M\dot{U}$[Nm]	$N\dot{U}$ [Nm]
-1714.70	-5.32	-252.12	-9.55	-897.03	-2.33

Based on reference [6], due to port and starboard symmetry of vehicle geometry, the added mass force in lateral direction (y-axis) can be assumed zero. Similarly, added mass moment in **x** and **z** axis for this vehicle can also be assumed zero. According to **Table 2**, there is still a value of force and moment in this axis but relatively small, so it can be ignored. This is due to the computational error on ANSYS CFX. Based on table 2, the significant added mass forces and moment are $X\dot{U}$, $Z\dot{U}$, dan $M\dot{U}$. The total added mass force and moment for vehicle with port and starboard symmetry expressed in **equation 2** [6].

$$
\begin{aligned}
X_{AM} &= -A_{11}\dot{u} - A_{13}\dot{w} - A_{15}\dot{q} - A_{31}U_0 q \\
Y_{AM} &= -A_{22}\dot{v} - A_{24}\dot{p} - A_{26}\dot{r} - A_{11}U_0 r + A_{31}U_0 p \\
Z_{AM} &= -A_{31}\dot{u} - A_{33}\dot{w} - A_{35}\dot{q} - A_{11}U_0 q \\
K_{AM} &= -A_{42}\dot{v} - A_{44}\dot{p} - A_{46}\dot{r} + A_{51}U_0 r - A_{31}U_0 v \\
M_{AM} &= -A_{51}\dot{u} - A_{53}\dot{w} - A_{55}\dot{q} + A_{31}\left(U_0^2 + 2U_0 u^*\right) + (A_{33} - A_{11})U_0 w + A_{35}U_0 q \\
N_{AM} &= -A_{62}\dot{v} - A_{64}\dot{p} - A_{66}\dot{r} - (A_{22} - A_{11})U_0 v - (A_{24} + A_{51})U_0 p - A_{26}U_0 r
\end{aligned}
\quad (2)
$$

The component of added mass and Moment due to only axial acceleration shows in **table 3**.

Table 3. Added Mass Component Due to Axial Acceleration ($\dot{U}$)

m_{11} [Kg]	m_{21} [Kg]	m_{31} [Kg]	m_{41} [Kgm]	m_{51} [Kgm]	m_{61} [Kgm]
349.65	1.08	51.41	1.95	182.92	0.48

$\dot{U}$ = 4.90 m/s2 (obtain from FFT)

Value of m_{21}, m_{41}, and m_{61} is relatively small, so it can be neglected and assumed to be zero magnitude. Added mass coefficient obtained from added mass component divided by mass displaced by the vehicle. The result shown in **table 4**.

Table 4. Added Mass Coefficient

m_{11}	m_{21}	m_{31}	m_{41}	m_{51}	m_{61}
0.095487	0.000296	0.01404	0.000887	0.083255	0.000216

$\rho_{sea\ water}$ = 1020 kg/m^3, Volume of the vehicle = 3.6 m^3

4. Conclusions

In this paper, a transient simulation of unsteady flow around the underwater vehicle was presented. The results includes time-dependent force and moment and unsteady flow behavior at upstroke and down-stroke of underwater vehicle motion. Based on the result on point 3(c), the significant added mass coefficient for cruise condition is A_{11}, A_{31}, and A_{51} with the largest added mass value is the one with the acceleration of the object that is A_{11}. This is appropriate with the existing reference below. The coefficient due to acceleration of direction x, $\dot{u}$ can be seen in point 3(c).

Acknowledgement
This research was supported by Program Magang Industri 2017 Fakultas Teknik Mesin dan Dirgantara, ITB. The authors would like to thank to director of Bhimasena Research and Development for providing research facilities and to supervisor of Aero-Naval division for sharing knowledge.

References

[1] Chen, S.S.,Wambsganss, M.W.,& Jendrzejczyk, J.A. (1976). *Added Mass and Damping of Vibrating Rod in Confined Viscous Fluid*, Journal of Applied Mechanic.
[2] Vikestad, K., Vandiver, J.K. & Larsen, C.M. (2000). *Added Mass and Oscillation Frequency for a Circular Cylinder Subjected to Vortex-Induced Vibrations and External Disturbance*, Journal of Fluid Structur.
[3] Uchiyama,Tomomi (2002). *Numerical prediction of added mass and damping for a cylinder oscillating in confined incompressible gas–liquid two-phase mixture*, Elsevier.
[4] Ghassemi, H., & Yari, E. (2011).*The Added Mass Coefficient computation of sphere, ellipsoid and marine propellers using Boundary Element Method*, Polish Maritime Research 1(68) 2011 Vol 18; pp. 17-26.
[5] Eng YH, Lau WS, Low E., Seet GGL and CS Chin (2008). Estimation of the Hydrodynamics Coefficients of ROV Using Free Decay Pendulum Motion. *Advance Online Publication*, 1-2.
[6] Lewandowski, Edward M : *The Dynamic of Marine Craft : Manuvering and Seakeeping.* Singapore, World Scientific Publishing, 2004

Effects of Turbulence Model and Numerical Time Steps on Von Karman Flow Behavior and Drag Accuracy of Circular Cylinder

E Amalia[1)] *, M A Moelyadi[1)], M Ihsan[1)]

1) Faculty of Mechanical and Aerospace Engineering, Institut Teknologi Bandung, Indonesia

*ema@ftmd.itb.ac.id

Abstract. The flow of air passing around a circular cylinder on the Reynolds number of 250,000 is to show Von Karman Vortex Street Phenomenon. This phenomenon was captured well by using a right turbulence model. In this study, some turbulence models available in software ANSYS Fluent 16.0 was tested to simulate Von Karman vortex street phenomenon, namely k-epsilon, SST k-omega and Reynolds Stress, Detached Eddy Simulation (DES), and Large Eddy Simulation (LES). In addition, it was examined the effect of time step size on the accuracy of CFD simulation. The simulations are carried out by using two-dimensional and three-dimensional models and then compared with experimental data. For two-dimensional model, Von Karman Vortex Street phenomenon was captured successfully by using the SST k-omega turbulence model. As for the three-dimensional model, Von Karman Vortex Street phenomenon was captured by using Reynolds Stress Turbulence Model. The time step size value affects the smoothness quality of curves of drag coefficient over time, as well as affecting the running time of the simulation. The smaller time step size, the better inherent drag coefficient curves produced. Smaller time step size also gives faster computation time.

1. Introduction

Flow around a circular cylinder is one of basic engineering problems and is extensively studied with the applications including heat exchanger, chimney, and offshore supporting feet [1]. From scientific perspective, flow around a circular cylinder is having some important phenomenon of fluid dynamics, such as transition to turbulent, separation, and vortex shedding. Vortex shedding gives important contributions to practical things such as lift and drag coefficients of the circular cylinder. Von Karman vortex street is one example of vortex shedding.

Moreover, the flow around a tube bundle is challenging from Computational Fluid Dynamics (CFD) point of view [2]. Its applications in nuclear power plant cooling system and heat exchanger makes study of flow around circular cylinder to be important to carry out. Both flows around a single singular cylinder and around a bundle of it are challenging to analyzed by CFD. These flows contain laminar and turbulent condition of flow. Analyzing such kind of flows by using CFD demands a right choice of technique and model of flow.

In this study, analyses of flow around a circular cylinder are carried out by using ANSYS Fluent 16.0, a software of CFD. The objectives of this study are: (1) to know which turbulence model that gives the best solution compared with experimental result; (2) to determine the time step size to provide efficient and accurate computational results. Time step size is also varied to know its effect to computational result. In addition, some conditions of flow are analyzed.

2. Method

Methodologies employed in this study are literature study and numerical study. The steps are as shown in figure 1. Step 1 is literature study of experimental data, and step 2 until 5 are numerical study.

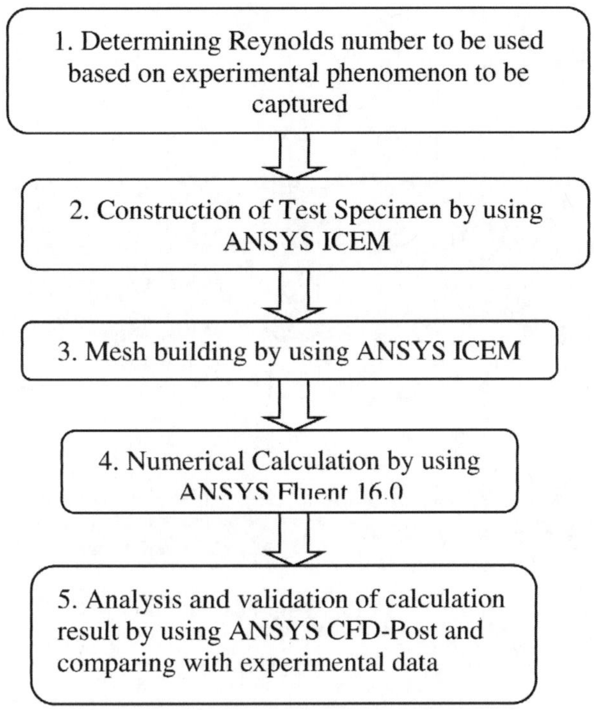

Figure 1. Methodology of this study

Purpose of Step 1 in figure 1 of determining Reynolds number is to assure validation of numerical calculation result. For flow around a circular cylinder, there are some ranges of Reynolds number with different experimental phenomenon as shown in figure 2. In this study, phenomenon chosen to be simulated is turbulent Von Karman Vortex Street, which happened in Range of Reynolds number between 300 and 300,000 or number 4 in Figure 2. Reynolds number chosen in this study is 250,000. Figure 3 shows turbulent Von Karman Vortex Street happened in cloud. This phenomenon is challenging in knowing which turbulence model in ANSYS Fluent 16.0 that best capturing the phenomenon of turbulent Von Karman Vortex Street. Also, it is interesting to know the effect of time step size variation to curve of lift coefficient of the circular cylinder in this phenomenon.

The step 2 (see figure 1) of this study is construction of test specimen of the circular cylinder by using software called ANSYS ICEM. There are two types of test specimen, which are two-dimensional and three-dimensional circular cylinder. The reason of why using two types of specimen is to know whether the same turbulence model that best in capturing the turbulent Von Karman Vortex Street in two-dimensional and three-dimensional condition of CFD model. Moreover, as shown in figure 4, one of turbulence model is LES (Large Eddy Simulation). This type of turbulence model is only usable in the condition of three-dimensional test specimen in ANSYS Fluent 16.0. Therefore, this study investigated both of two-dimensional and three-dimensional for the sake of complicity to check performance of some of turbulence model available in ANSYS Fluent 16.0 to case of flow around a circular cylinder. Diameter of cylinder is 72.6 mm for two-dimensional case and 100 mm for three-dimensional case. Different diameter is used because of different reference of determining suitable mesh for the two cases, which is Reference [3] for two-dimensional case, and reference [4] for three-dimensional case.

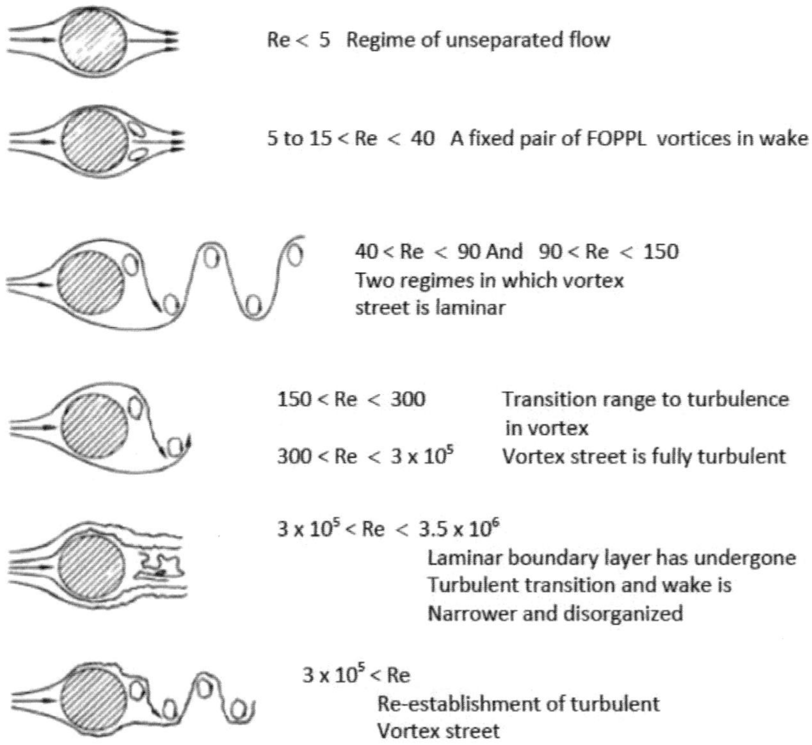

Figure 2. Regime of fluid flow across a smooth tube [5]

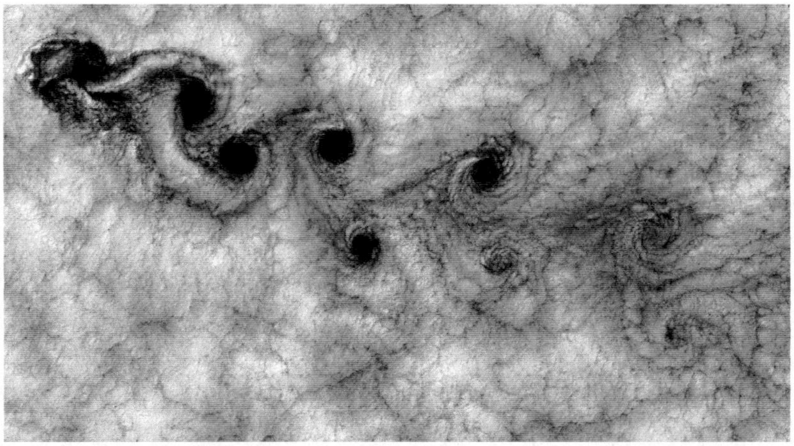

Figure 3. Turbulent Von Karman Vortex Happened in Cloud [6]

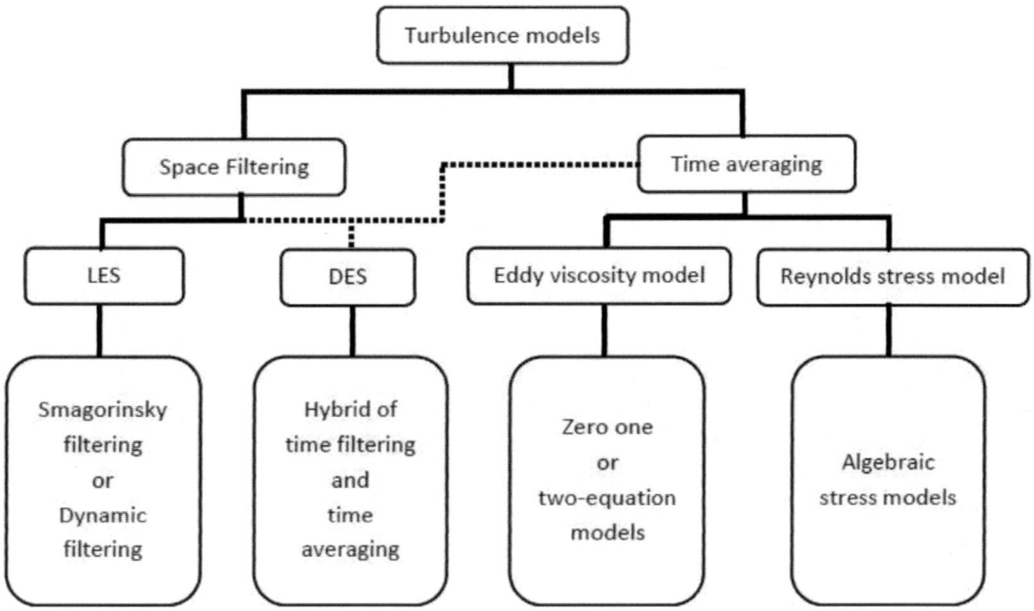

Figure 4. Classification of Turbulence Models [7]

Step 3 in figure 1 is carried out by considering the phenomenon that will be captured. To capture Turbulence Von Karman Vortex Street, fine grid near wall and downstream the circular cylinder is needed. To determine size of numerical calculation domain, we refer to Ref.[3] and Ref.[4]. Calculation domain for two-dimensional case is as illustrated in figure 5, while figure 6 shows calculation domain for three-dimensional case. Three-dimensional case is having 5 mm in thickness. To construct a good mesh for capturing turbulence Von Karman Vortex Street, distance of the first mesh from wall of circular cylinder should be determined first by using equation 1. In equation 1, ΔS is distance of first mesh from the wall, y+ is a constant value chosen based on turbulence model, ρ is density, and U is velocity of flow. By choosing y+ as 2, ρ as 1.225 kg/m^3, μ as 1.789e-5 kg/ms, and U as 36.51 m/s, we get ΔS for two-dimensional case is 0.0012 and for three-dimensional case is 0.017 mm. Mesh used in this study is structured grid type for the sake of accuracy. Mesh for two-dimensional case is as shown in figure 7, and for three-dimensional case is as shown in figure 8. More detailed of setting in mesh construction can be found in Ref. [8] chapter 3.

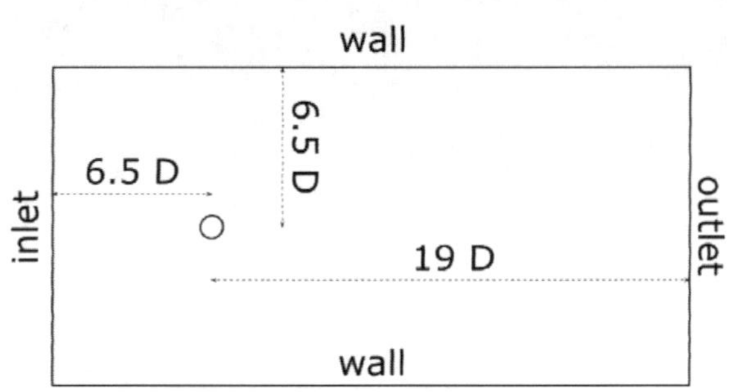

Figure 5. Calculation domain size for two-dimensional case

$$\Delta S = \frac{y^{+}\mu}{\rho U} \tag{1}$$

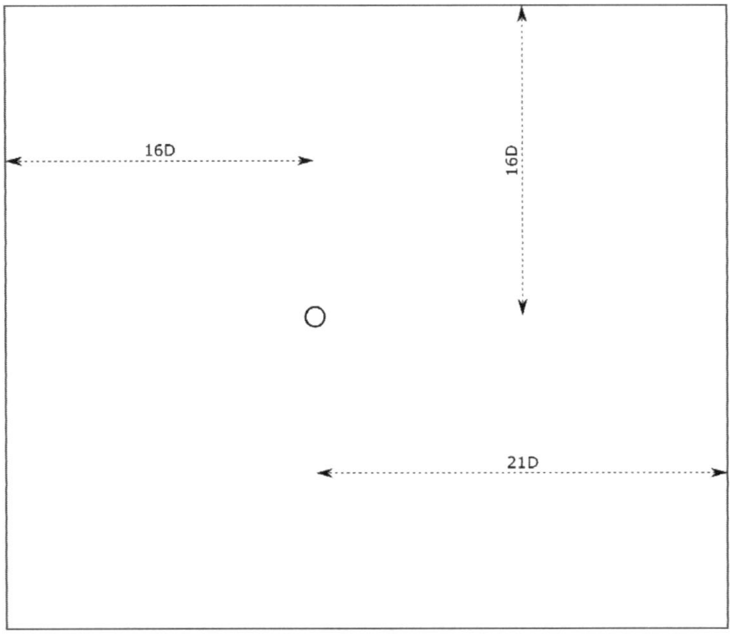

Figure 6. Calculation domain size for three-dimensional case

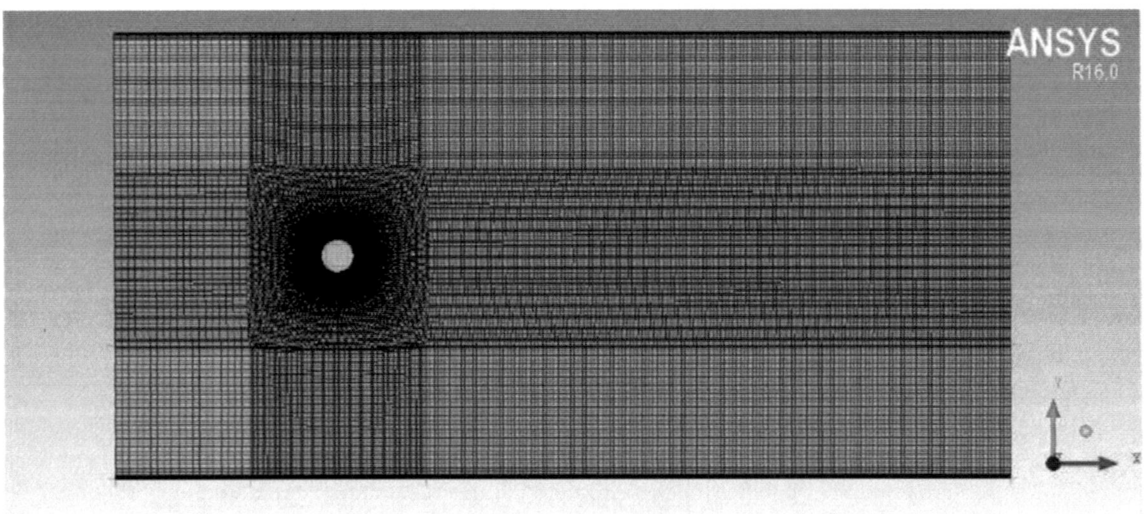

Figure 7. Mesh for two-dimensional case

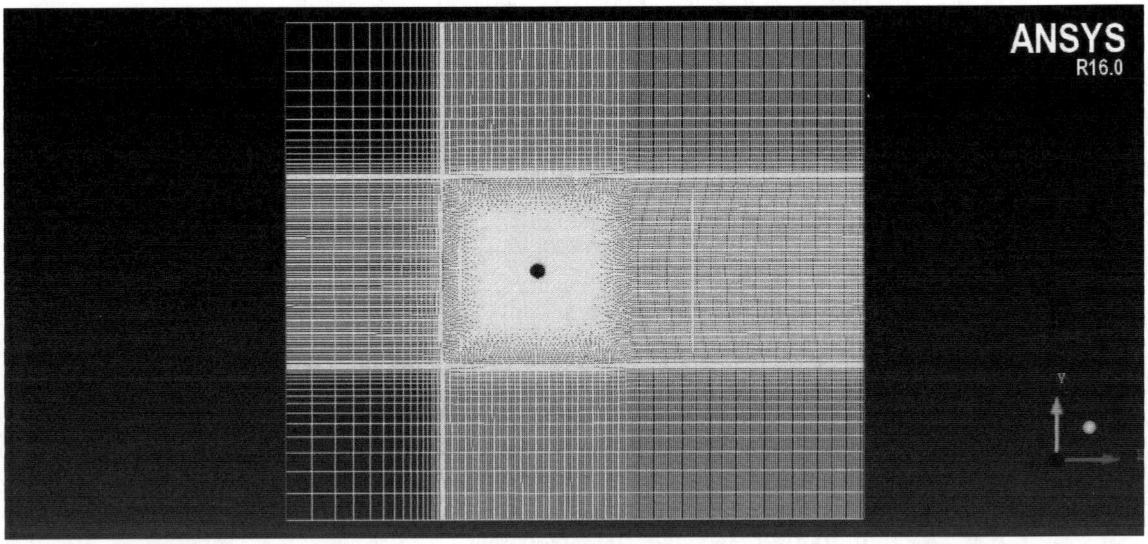

Figure 8. Mesh for three-dimensional case

Step 4 in figure 1 by using ANSYS Fluent 16.0 consists of exporting mesh file from ANSYS ICEM and doing many setting of solver as can be found in detailed in Ref. [8]. The main setting is to set reference values and turbulence models. Table 1 shows reference values for two dimensional and three dimensional cases. Table 2 shows turbulence models that can be utilized in ANSYS Fluent 16.0 for two-dimensional and three-dimensional cases. For two-dimensional case, turbulence models to be used are SST k-omega, k-epsilon, Reynolds Stress, and Detached Eddy Simulation (DES). As for three-dimensional case, turbulence model to be utilized are SST k-omega, k-epsilon, Reynolds Stress, Detached Eddy Simulation (DES), and Large Eddy Simulation (LES). Equations utilized in ANSYS Fluent 16.0 are explained in detailed in Ref. [8] chapter 2.

Results from step 4 are processed further by using ANSYS CFD-Post in Step 5 (see figure 1) to be easier to analyze and validate. In other word, step 5 is conducted so that the result of Step 4 will be more user-friendly and easy to compare with experimental data as figure 2 and 3.

Table 1. Reference values used in numerical calculation

Reference Values	Two-Dimensional Case	Three-Dimensional Case
Area (m^2)	0.004142563	0.00314
Density (kg/m^3)	1.225	1.225
Depth (m)	0.072644	-
Enthalpy (j/kg)	0	0
Length (mm)	72.644	100
Pressure (pascal)	0	101325
Temperature (k)	288.16	288.15
Velocity (m/s)	50.27	36.51
Viscosity (kg/m-s)	1.7894e-05	1.7894e-05
Ratio of Specific Heats	1.4	1.4

IOP Conf. Series: Journal of Physics: Conf. Series **1005** (2018) 012012 doi:10.1088/1742-6596/1005/1/012012

Table 2. Turbulence Models available in ANSYS Fluent 16.0

Turbulence Models	Two Dimensional Case	Three Dimensional Case
Inviscid	Available	Available
Laminar	Available	Available
Spalart-Allmaras (I eqn)	Available	Available
k-epsilon (2 eqn)	Available	Available
k-omega (2 eqn)	Available	Available
Transition k-kl-omega (3 eqn)	Available	Available
Transition SST (4 eqn)	Available	Available
Reynolds Stress (5 eqn)	Available	Not available
Reynolds Stress (7 eqn)	Not available	Available
Scale-Adaptive Simulation (SAS)	Available	Available
Detached Eddy Simulation (DES)	Available	Available
Large Eddy Simulation (LES)	Not available	Available

3. Results and Discussions

For case of two-dimensional circular cylinder, turbulence models that employed are SST k-omega, k-epsilon, Reynolds Stress, and Detached Eddy Simulation (DES). In this paper, only result of turbulence model of SST k-omega will be presented. For two-dimensional case, SST k-omega gives the best capturing of turbulent Von Karman Vortex Street (number 4 in figure 2 and figure 3). Figure 9 shows velocity contour of flow around two-dimensional circular cylinder with SST k-omega turbulence model, and figure 10 shows value of C_L of the two-dimensional cylinder with the same turbulence model. More completed result of other turbulence models could be seen in Ref. [8].

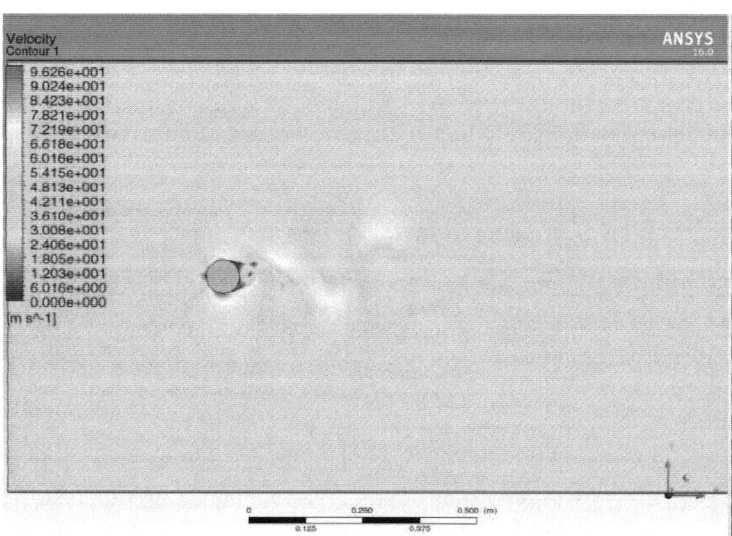

Figure 9. Velocity contour of flow around a two-dimensional circular cylinder for Reynolds number of 250,000 by using turbulence model of SST k-omega

IOP Conf. Series: Journal of Physics: Conf. Series **1005** (2018) 012012 doi:10.1088/1742-6596/1005/1/012012

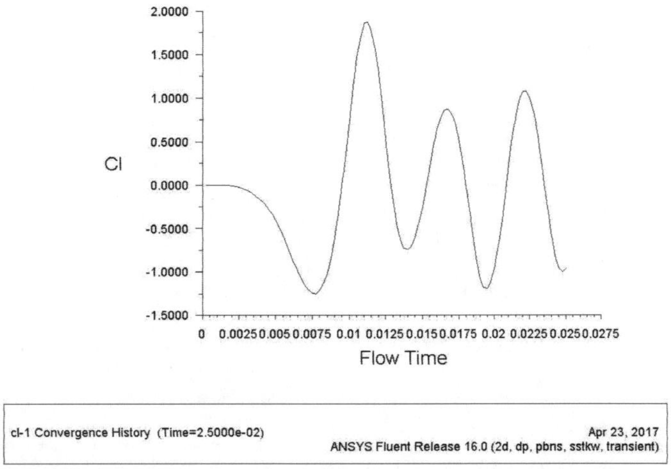

Figure 10. Lift Coefficient of the two-dimensional circular cylinder for Reynolds number of 250,000 by using turbulence model of SST k-omega

After getting result that turbulence model of SST k-omega gives the best result for two-dimensional case, investigation of effect of time step size are conducted only for this turbulence model. Figure 11 and 12 gives example of results for time step size of 0.25 ms and 4 ms. It could be seen that smaller time step size gives smoother result of drag coefficient curve over time.

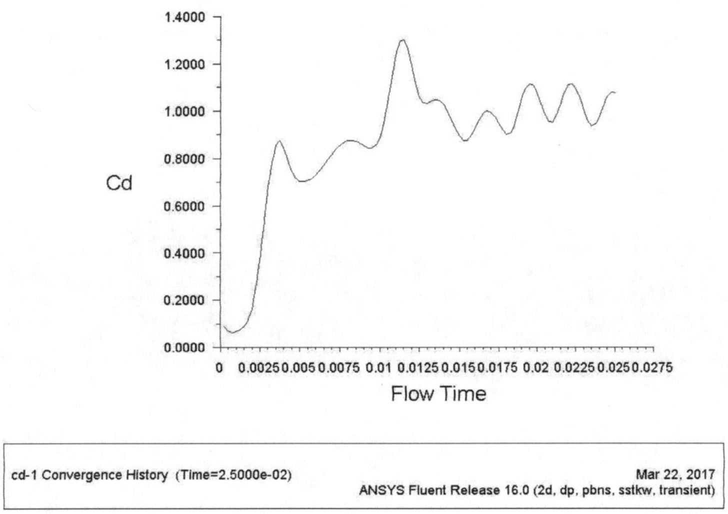

Figure 11. Drag coefficient value for two-dimensional circular cylinder with turbulence model of SST k-omega by using time step size of 0.25 ms

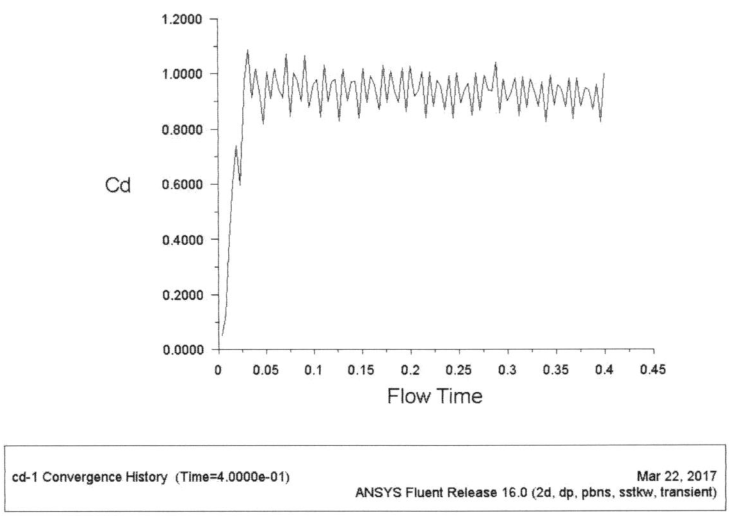

Figure 12. Drag coefficient value for two-dimensional circular cylinder with turbulence model of SST k-omega by using time step size of 4 ms

As for case of three-dimensional circular cylinder, turbulence models that employed are the same with them for two-dimensional case with addition of Large Eddy Simulation (LES). For three-dimensional case, turbulence model of Reynolds stress that gives the best result of capturing turbulent Von Karman Vortex Street. Figure 13 shows velocity contour result for three-dimensional circular cylinder by using turbulence model of Reynolds stress. It captured the same phenomenon as shown in figure 2 number 4 and figure 3. Figure 14 shows lift coefficient curve for the three-dimensional cylinder.

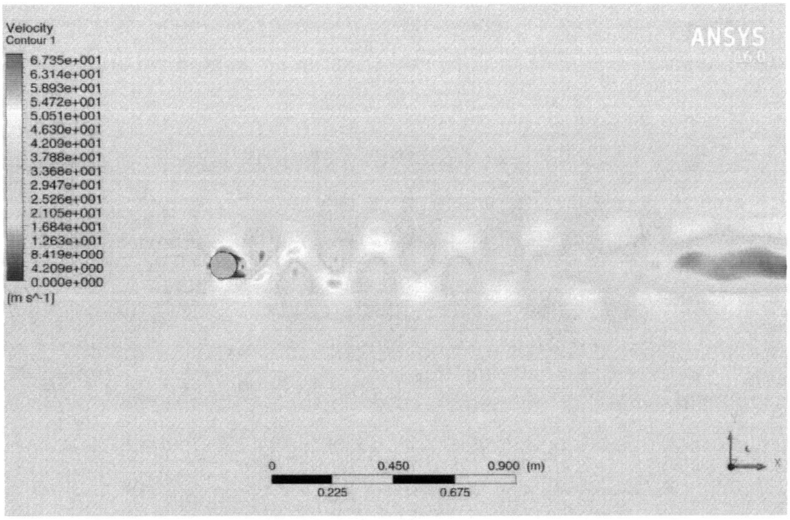

Figure 13. Velocity contour of three-dimensional circular cylinder for Reynolds number of 250,000 by using turbulence model of Reynolds Stress

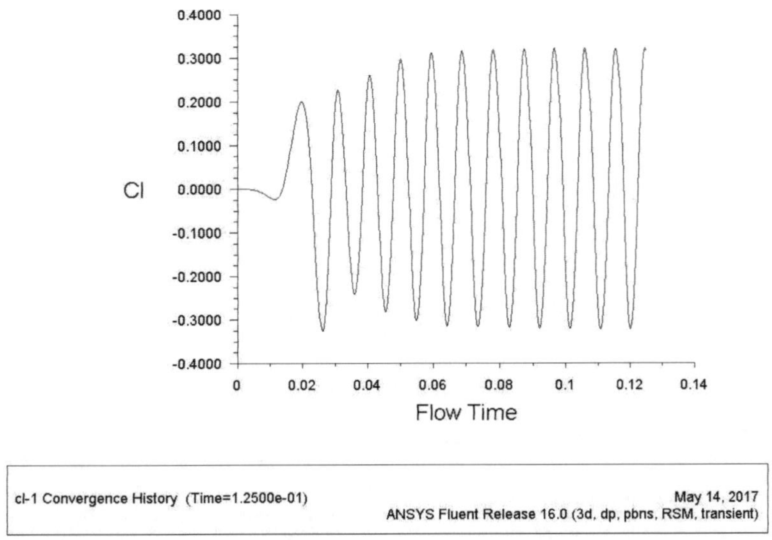

Figure 14. Lift Coefficient of three-dimensional circular cylinder for Reynolds number of 250,000 by using turbulence model of Reynolds Stress

After getting result that turbulence model of Reynolds stress gives the best result for three-dimensional case, investigation of effect of time step size are conducted only for this turbulence model. Figure 15 and 16 gives example of results for time step size of 0.25 ms and 0.5 ms. It could be seen that smaller time step size gives more correct result.

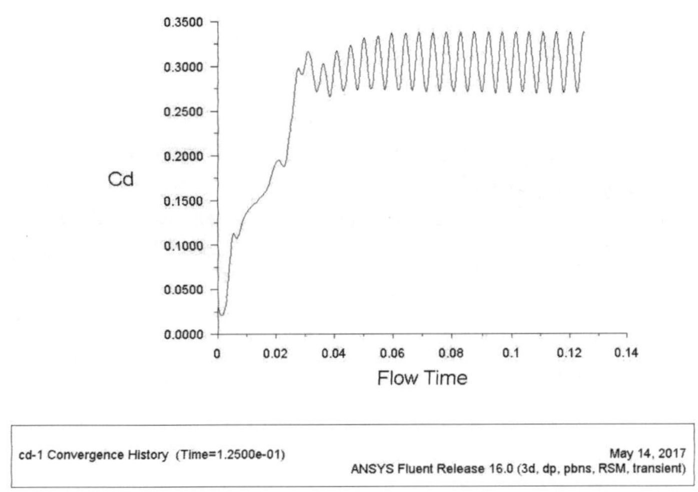

Figure 15. Drag coefficient value for three-dimensional circular cylinder with turbulence model of Reynolds stress by using time step size of 0.25 ms

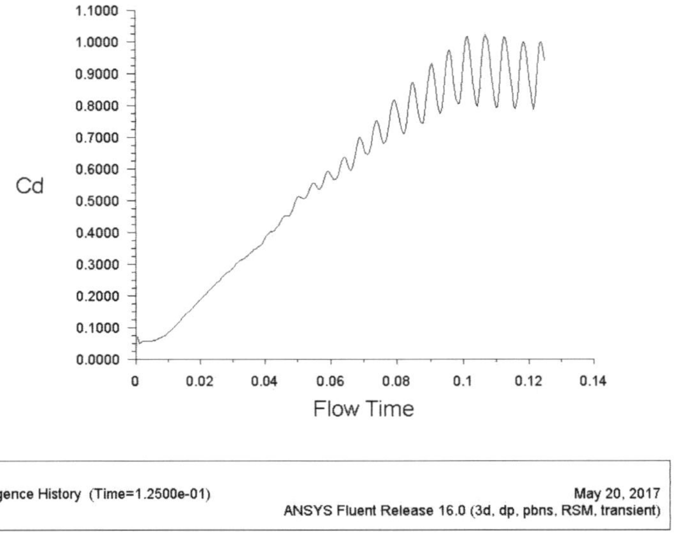

Figure 16. Drag coefficient value for three-dimensional circular cylinder with turbulence model of Reynolds stress by using time step size of 0.5 ms

4. Conclusions

From this study, it could be concluded that turbulence models used in a numerical calculation have crucial effect in getting a right result. For two-dimensional circular cylinder, turbulence models that gives results that the most agree with experiments is turbulence model of SST k-omega. As for the case of three-dimensional circular cylinder, turbulence models that gives the best result compared to experiments is Reynolds Stress. Moreover, time step size has very important effect to numerical calculation result. Smaller time step size gives faster and more accurate result although with more iteration number.

References

[1] Tutar, M., and HoldØ, A.,E., Computational Modelling of Flow around a Circular Cylinder in Sub-critical Flow Regime with Various Turbulence Models, *Journal of Numerical Methods in Fluids, Vol. 35, pp. 763-784, 2001.*

[2] Bouris, D., and Bergeles, G., Two Dimensional Time Dependent Simulation of The Subcritical Flow in a Staggered Tube Bundle Using a Subsgrid Scale Model, *Journal of Heat and Fluid Flow, Vol. 20, pp. 105-114.*

[3] Kensrud, J.R., Determining Aerodynamics Properties of Sport Ball in Situ, *Master Thesis, Washington State University, 2010.*

[4] Nidhul, K., Sunil, A.S., Benphil, C.P., Effect of Domain Size and Grid Spacing on Flow Past A Circular Cylinder at Low Reynolds Number, *International Journal of Engineering Research and Technology, 2014.*

[5] Blevins, R.D., Flow Induced Vibration, *2nd Edition, Van Nostrand Reinhold Co., 1990.*

[6] https://commons.m.wikimedia.org/wiki/File:Vortex-street-1-90deg.jpg

[7] Patel, Y., Numerical Investigation of Flow Past a Circular Cylinder and in a Staggered Tube Bundle Using Various Turbulence Models, *Master Thesis, Lappeenranta University of Technology, 2010.*

[8] Ihsan, M., Analysis of Turbulence Model and Effect of Time Step Size Value of Analysis of Flow around Circular Cylinder, *Bachelor Thesis, Institute of Technology Bandung, (to be published on 2017).*

IOP Conf. Series: Journal of Physics: Conf. Series **1005** (2018) 012013 doi:10.1088/1742-6596/1005/1/012013

Study of Swept Angle Effects on Grid Fins Aerodynamics Performance

G A Faza[1], H Fadillah[1], F Y Silitonga[1], Mochamad Agoes Moelyadi [2]

[1] Undergraduate Student, Department of Aerospace Engineering, Bandung Institute of Technology, Indonesia
[2] Head Lector, Department of Aerospace Engineering, Bandung Institute of Technology, Indonesia

Abstract. Grid fin is an aerodynamic control surface that usually used on missiles and rockets. In the recent several years many researches have conducted to develop a more efficient grid fins. There are many possibilities of geometric combination could be done to improve aerodynamics characteristic of a grid fin. This paper will only discuss about the aerodynamics characteristics of grid fins compared by another grid fins with different swept angle. The methodology that used to compare the aerodynamics is Computational Fluid Dynamics (CFD). The result of this paper might be used for future studies to answer our former question or as a reference for related studies.
Key Words: Grid Fin, Swept-Back, Lattice Fin, CFD Simulation.

1. Introduction

A Grid fin is an aerodynamic control surface that usually used on missiles and rockets (see Figure 1). Unlike the conventional planar fins, grid fins consist of an outer frame with internal lattice of intersecting thin walls with small chord instead of using single body supersonic airfoil (see Figure 2). The mounting of grid fins and conventional planar fins also different, unlike the conventional planar fins that aligned parallel to the direction of airflows the grid fins mounted perpendicular with the airflows.

Figure 1 Grid Fins mounted on SpaceX Falcon 9 Rocket, source: https://i.stack.imgur.com/rnHbB.jpg

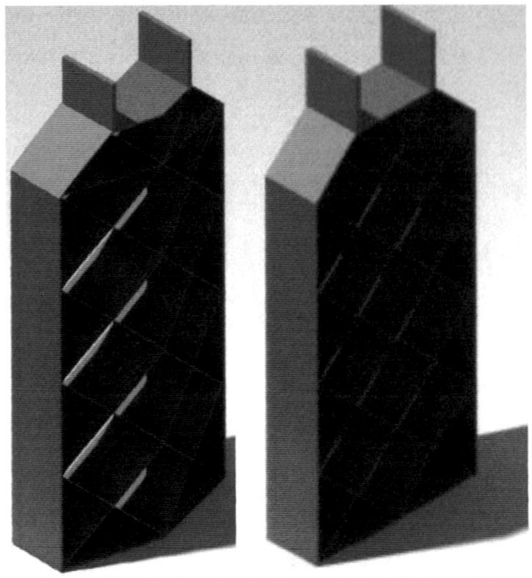

Figure 2 Grid Fin with Swept Back Angle (left) and Grid Fin without Swept Angle (right)

According to Marco Debiasi:

> The main advantage of grid fins is that they have significantly smaller chord than conventional planar fins. Thus, they generate smaller hinge moments which require smaller actuators to rotate them in a high-speed flow. Their small chord also makes them less likely to stall at high angles of attack which increases their control effectiveness compared to conventional planar fins. Another advantage of grid fins is that they can be folded easily against an aerodynamic body to make them more compact and convenient to store or transport. **[1]**

Zubairkhan Kittur and Amit Bahekar **[2]** wrote on their paper that grid fin is very efficient compared to planar fin in supersonic flow, but it's not efficient in transonic flow. When grid fin in transonic flow, it produces shockwaves at the front of the grid fin, this blocks the freestream air to going through the lattice and produce a high amount of drag. This statement also written by Salman Munawar **[3]** in his paper.

The other advantage of grid fin instead of planar fin is the stall angle of grid fin is very high. William D. Washington and Mark S. Miller **[4]** have conducted experiments of grid fin and the result was the grid fin did not stall at angle of attack of 30-degree.

There are many variations of grid fins design and one of them is the variations of swept angle. The use of swept angle on a grid fin either it's swept back or swept forward might improve the aerodynamics characteristics. Actually, there are many variations in grid fin geometry besides swept angle variations. But, this paper will only discuss about the effects between the aerodynamics characteristics of a grid fin and the swept angle on a grid fins. The aerodynamics characteristic that will be measured on this paper are drag coefficient, lift coefficient, and moment coefficient.

The method that will be conducted in this research is computational fluid dynamics (CFD). Computational fluid dynamics is able to simulate the fluid flow inside the lattice grid and also the fluid flow around the grid fin. DeSpirito and Sahu **[5]** have conducted research about viscous CFD calculations of grid fin using FLUENT. The result they obtained led to good agreement between the CFD data and aerodynamic coefficients measured by experiments. The geometries that will be evaluated to obtain aerodynamics coefficient are base configuration, swept back with 5°,15°, 20°, and 30° angles, and forward swept with 5°,15°, 20° angles.

To measure the aerodynamics performance criteria the magnitude of lift coefficient, drag coefficient, and pitching moment coefficient must be known. These coefficients can be calculated from force and moment obtained from CFD simulation. To determine which is the best configuration, the ratio between lift coefficient and drag coefficient (Cl/Cd) was used. The value of Cl/Cd indicates aerodynamic efficiency of an object. In other words, bigger Cl/Cd value is desired.

The main objective of this study is to compare the aerodynamics characteristics between grid fins with different configuration on its swept angle. It's expected that the variations in swept angle could improve the aerodynamic characteristics and the efficiency of the grid fins.

2. Geometry Configurations

The grid-fin geometry used in this study was based on the geometry that used by Zubairkhan Kittur [2], but with some modifications. All the dimensions that used in the geometry are relative to the spacing of the grid. Later we can define the spacing as "s". The width of the grid-fin is 2s, the length of the grid fin is 3s, the height of the grin fin is 0.625s, and the swept angle Λ. For this case, we have the value of s is 80 mm and Λ is 30 degree, therefore we have the dimension of the grid fin as stated in figure 3 below.

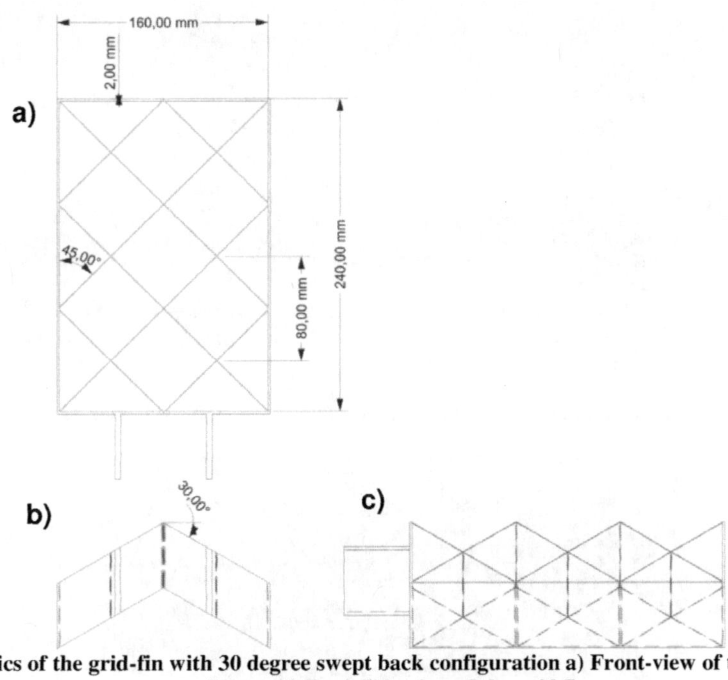

Figure 3 Schematics of the grid-fin with 30 degree swept back configuration a) Front-view of the grid fin b) Top view of the grid-fin c) Side view of the grid fin

3. Numerical Approach

In this study, the methodology that will be conducted is numerical simulation using computational fluid dynamics (CFD). Computational fluid dynamics is able to simulate the fluid flow inside the lattice grid and also the fluid flow around the grid fin, moreover this method provide the detailed flow fields including pressure contours and integrated aerodynamic coefficients **[6]**. ANSYS CFX will be used as the solver of CFD simulations. This solver software is based on Reynolds Averaged Navier-Stokes (RANS) method. This method was chosen because it delivers enough accuracy and faster processing rather than Large Eddy Simulation (LES) method that has higher accuracy but it takes a lot of memory and running time **[7]**.

RANS or Reynolds Averaged Navier-Stokes equations are the simplified form of the Navier-Stokes equations. These equations are simplified by separating the flow variables into the time-averaged components and the fluctuating components using Reynolds Decomposition. However, in order to achieve a good accuracy in turbulence flow modelling RANS equations need turbulence model equations. K-ε turbulence model were choosed to solve turbulence flow problems in this simulation. K-ε turbulence model is a two-equation model, that means this turbulence model adds two extra transport equations to represent the turbulence flow. This turbulence model is one of the most common turbulence model.

The CFD model have a quadrilateral prism domain with a perpendicular downstream boundary located 9,375s away from the main object, a perpendicular upstream boundary located 6,25s away from the main object, three slanted upstream boundaries, a symmetry boundary located on the grid fin's main arm root, and non-slip wall type for all solid surfaces. The boundaries were arranged to make analysis with multi angle of attack easier. The meshing type that will be used in this simulation are tetrahedrons and prisms. Tetrahedrons and prisms were choosed because grid fins geometry has so many acute angles that will be difficult to cover with hexahedrons meshing type.

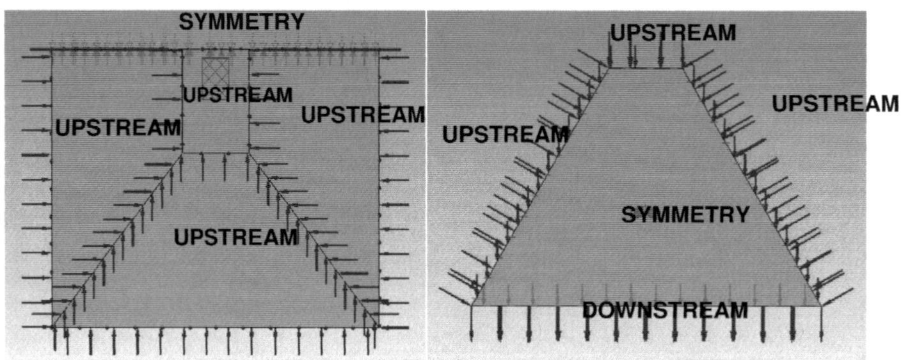

Figure 4 Boundary Conditions at the Domain: front view (left), side view (right)

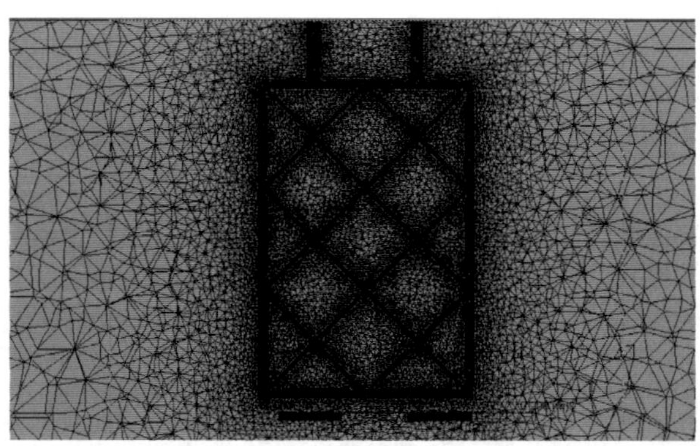

Figure 5 Mesh arounds the grid fin.

The number of mesh elements in the simulation was determined by using convergence test or grid sensitivity test. The test was done by adding the number of mesh elements until the drag coefficient in base configuration with alpha 0 reaches difference of 5 drag counts. Figure 6 below shows us that the number of mesh elements is about 3.5 million.

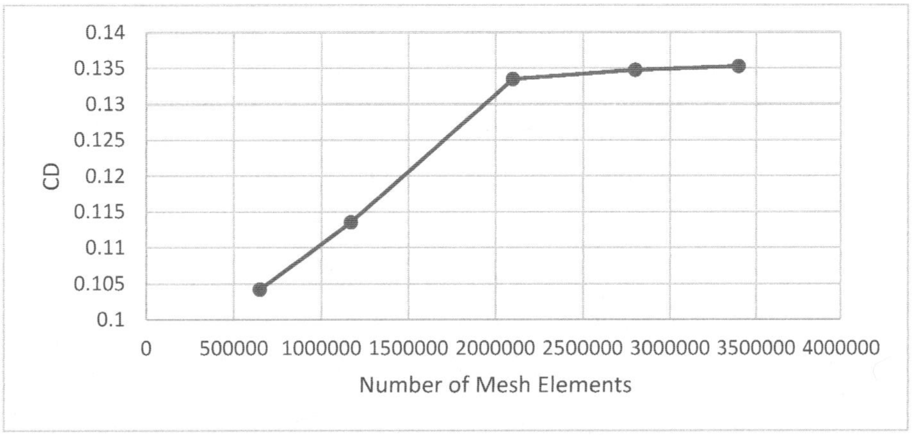

Figure 6 Grid Sensitivity Test.

In this simulation, the conditions were assumed that the grid fin is at the 10,000 meters ASL with ISA condition and the freestream speed is Mach 1.5. This conditions were chosen to represent the "actual" conditions of a missile with Mach 1.5 at the 10,000 meters above sea level. But, then again this "actual" conditions are just the writers' assumption.

After running the analysis. The lateral force, axial force, and the pitching moment can be measured using CFD-Post software. The lateral force represents the lift force and the axial force represent the drag force. In this case, lift force shows us the capability of a control surface to control the object's main body and the drag force shows us how much loss of momentum and energy while the grid fins control the object. On other hand, pitching moment shows us how much torque to be applied to the grid fin in order to control the main body.

4. Result

Coefficient of Drag v/s Angle of Attack:
According to figure 7 below, grid fin without any swept angle or the base model has the highest drag coefficient at any angle of attack among the other model. Meanwhile the 30-degree swept back model has the lowest drag coefficient at any angle of attack among the other model. The effect of swept back or swept forward also visible on the graph. Comparing swept back and swept forward with 5,15, and 20 degrees' angle, the swept forward configuration seems with higher drag coefficient than the swept back configuration with the same swept angle

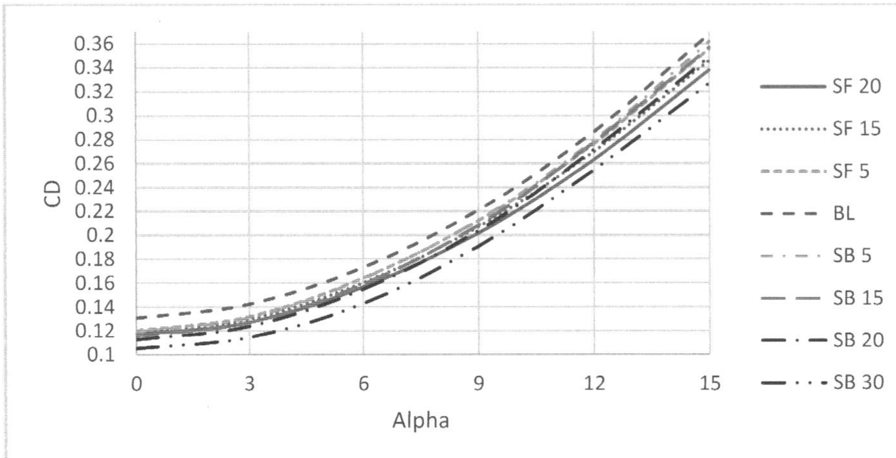

Figure 7 Graph of Drag Coefficient v/s AoA at Mach=1.5

Table 1 Drag Coefficient v/s AoA at Mach=1.5

Swept\alpha	0	3	6	9	12	15
-20	0.117270768	0.127146	0.156748	0.201526	0.26295	0.337844
-15	0.119459823	0.129895	0.160297	0.206798	0.269635	0.345825
-5	0.120397989	0.132013	0.164267	0.211957	0.276213	0.355692
0	0.13090545	0.142436	0.173039	0.220758	0.286261	0.367861
5	0.119459823	0.131118	0.163831	0.212386	0.278937	0.361602
15	0.11601988	0.127024	0.159157	0.208197	0.27756	0.356779
20	0.112767571	0.123833	0.154871	0.203581	0.27138	0.348636
30	0.105324786	0.114575	0.142918	0.190604	0.254462	0.326822

Coefficient of Lift v/s Angle of Attack:

Figure 8 shows us that the grid fin with 5 degree swept back configuration has the highest number of lift coefficient at any angle of attack between 0 to 15. Meanwhile the 30-degree swept back configuration and 20 degree swept forward configuration gives the two lowest lift coefficients among the other configuration.

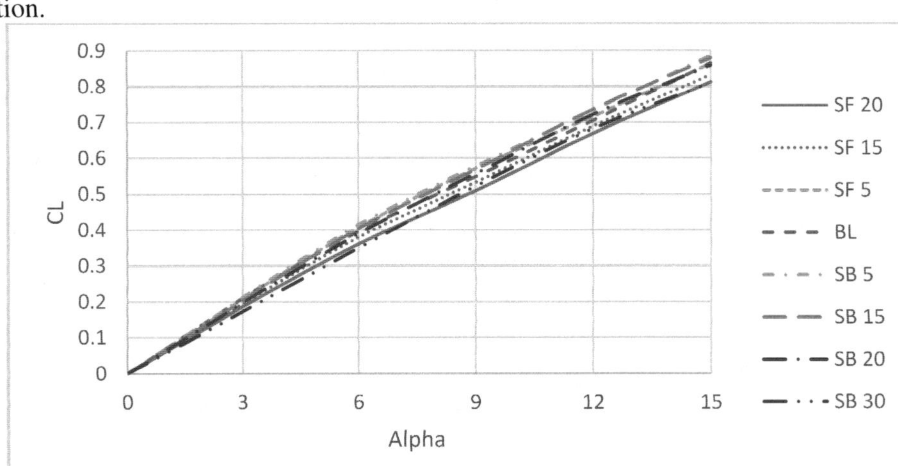

Figure 8 Graph of Lift Coefficient v/s AoA at Mach=1.5

Table 2 Lift Coefficient v/s AoA at Mach=1.5

Swept\alpha	0	3	6	9	12	15
-20	-2.68941E-05	0.186551	0.360481	0.508742	0.667928	0.8121
-15	0.000113831	0.194799	0.379289	0.532033	0.688247	0.831977
-5	3.70888E-06	0.207715	0.407487	0.570604	0.714984	0.863004
0	-0.000129467	0.210614	0.395811	0.547807	0.703896	0.864923
5	0.000235792	0.212146	0.414702	0.577312	0.732308	0.886025
15	-5.47264E-05	0.203342	0.403747	0.567653	0.735159	0.8789
20	-0.000150107	0.197434	0.389293	0.5561	0.721126	0.858419
30	0.000133845	0.172554	0.349168	0.521744	0.681242	0.809873

Coefficient of Pitching moment v/s Angle of Attack:

The pitching moment coefficient graph shows that at a particular angle of attack, grid fin with 30 degree swept back configuration has the most negative pitching moment. It indicates that this configuration need a motor with bigger torque to move it. Meanwhile the base model or 0 swept angle degree configuration has the most positive pitching moment among the other.

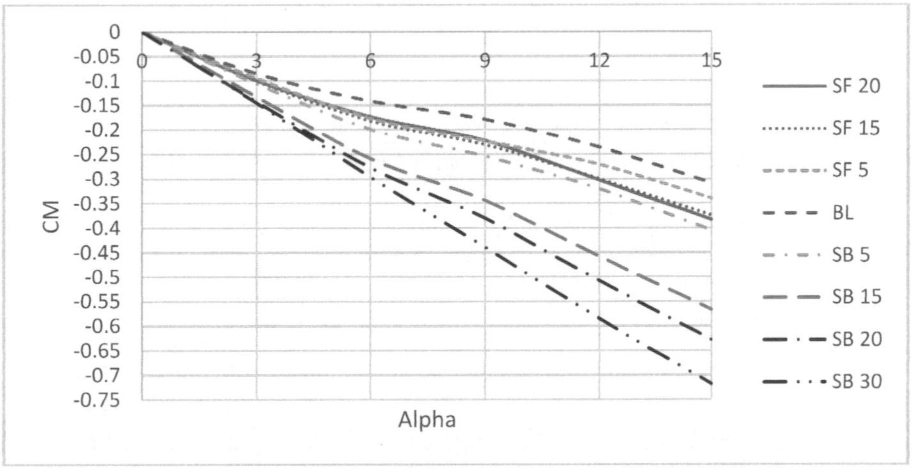

Figure 9 Graph of Pitching Moment Coefficient v/s AoA at Mach=1.5

Table 3 Pitching Coefficient v/s AoA at Mach=1.5

Swept\alpha	0	3	6	9	12	15
-20	0.00011258	-0.09943	-0.17325	-0.22178	-0.30284	-0.38352
-15	-4.44065E-05	-0.10073	-0.182	-0.23066	-0.30034	-0.37439
-5	-0.000132594	-0.09439	-0.17675	-0.22316	-0.27082	-0.34037
0	0.000187633	-0.08522	-0.14123	-0.17863	-0.23554	-0.30934
5	-0.000216529	-0.10603	-0.19902	-0.25406	-0.32048	-0.40566
15	0.000110078	-0.13209	-0.25893	-0.34387	-0.45808	-0.56715
20	0.000162615	-0.14235	-0.27757	-0.38002	-0.50786	-0.6287
30	-4.62829E-05	-0.14498	-0.29671	-0.43894	-0.58492	-0.71801

Aerodynamic Efficiency v/s Angle of Attack:
The aerodynamic efficiencies were determined by the ratio of lift coefficient over drag coefficient. From figure 4.4 below, the base model appears with the lowest aerodynamic efficiency at every angle of attack between 0 and 15 degrees. The highest number of aerodynamic efficiency at angle of attack up to 6 degree is reached by the 5-degree swept back configuration, meanwhile from angle of attack 9 degree up to 15 degree the 30-degree swept back configuration has the highest number of aerodynamic efficiency.

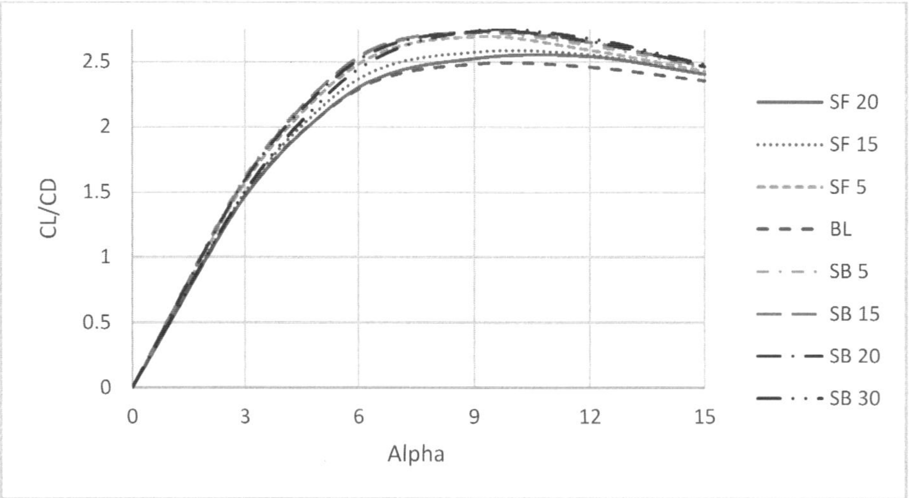

Figure 10 Graph of Aerodynamic Efficiency v/s AoA at Mach=1.5

Table 4 CL/CD v/s AoA at Mach=1.5

Swept\alpha	0	3	6	9	12	15
-20	0.000229	1.46722	2.299748	2.524452	2.540132	2.403775
-15	0.000953	1.499662	2.366164	2.572723	2.552515	2.405778
-5	3.08E-05	1.573449	2.480635	2.692075	2.588521	2.426266
0	0.000989	1.478655	2.287404	2.481483	2.45893	2.351219
5	0.001974	1.617985	2.531285	2.718218	2.625358	2.45028
15	0.000472	1.600821	2.53678	2.726524	2.648646	2.463432
20	0.001331	1.594358	2.513659	2.731595	2.657252	2.462221
30	0.001271	1.506035	2.44313	2.737324	2.67718	2.478025

Airflow Contour
The airflow contour of grid fin with swept back, base model, and swept forward configuration are shown in figure 11 and 12 below.

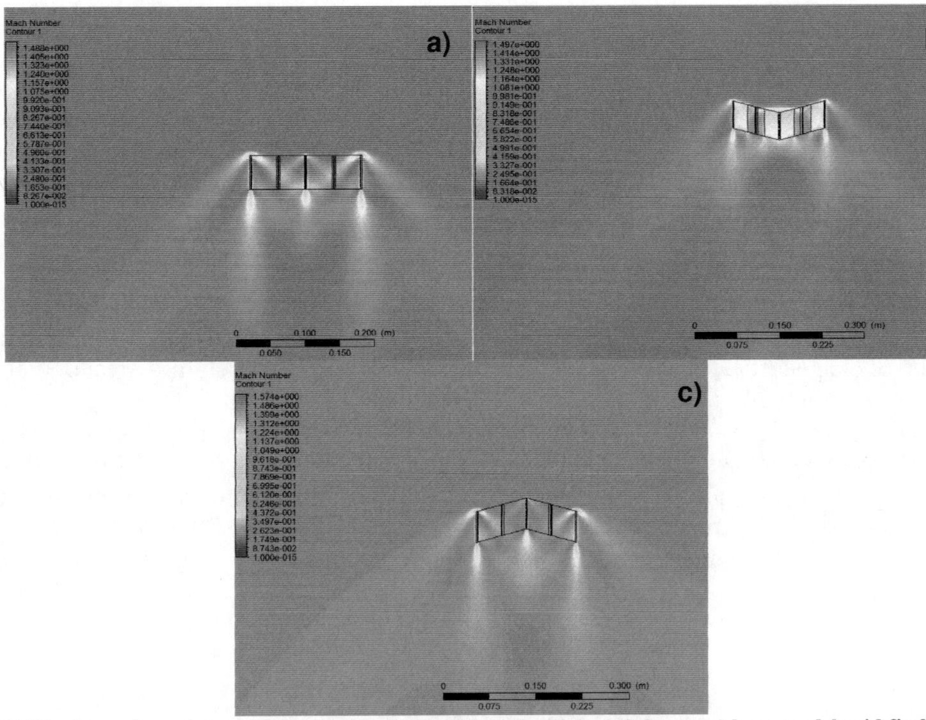

Figure 11 Mach number contour plot with freestream M= 1.5 and AoA 0 deg on a) base model grid fin b) 15 deg forward swept c) 15 deg swept back

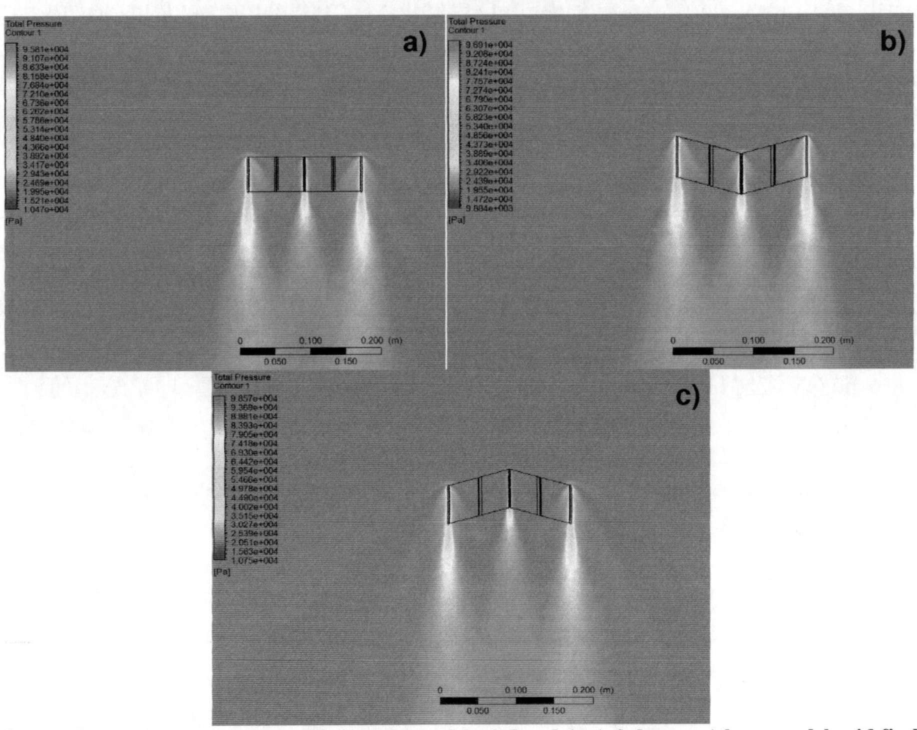

Figure 12 Total Pressure contour plot with freestream M= 1.5 and AoA 0 deg on a) base model grid fin b) 15 deg forward swept c) 15 deg swept back

5. Analysis and Discussion

According to the results section, the grid fin with 5 degree swept back angle has the highest aerodynamic efficiency at AoA from 0 to 6 degree, while the 30-degree swept back angle configuration has the highest aerodynamic efficiency at AoA from 9 to 15 degrees. The swept back configurations have the highest aerodynamic efficiency because of its shape that prevent energy loss caused by shockwave as much as the base model and the swept forward configuration model.

6. Conclusion

Using CFD simulation, calculation of a flow past a grid fin at supersonic speed can be obtained. The highest value of drag coefficient possessed by base model grid fin because it has greater total pressure loss than the other models. Meanwhile, at AoA from 0 to 15 grid fin with 5 degree swept back configuration has the highest lift coefficient among the other configuration. For pitching moment coefficient, the base model has the most positive pitching moment, contrary the swept back 30-degree model has the most negative pitching moment. At last, the grid fin with 5 degree swept back angle has the highest aerodynamic efficiency at AoA from 0 to 6 degree, while the 30-degree swept back angle configuration has the highest aerodynamic efficiency at AoA from 9 to 15 degrees. With these results, it can be concluded that generally, the swept back grid fin configuration is the most efficient configuration.

7. References

[1] M. Debiasi, "Development of New Grid-Fin Design for Aerodynamic Control," *AIAA paper 2012-2909,* 2012.

[2] Z. Kittur and A. Bahekar, "CFD Analysis of Grid FIn Application on Missile in Supersonic Flow Regime," Research Journal of Recent Sciences Vol 5(9), 2016.

[3] S. Munawar, "Analysis of Grid Fins as Efficient Control Surface in Comparison to Conventional Planar FIns," 27th International Congress of the Aeronautical Sciences, 2010.

[4] W. D. Washington and M. S. Miller, "Grid Fins A New Concept for Missile Stability and Control," AIAA Paper 93-0035, 31st AIAA Aerospace Sciences Meeting and Exhibit, Reno,NV, 1993.

[5] J. DeSpirito and J. Sahu, "Viscous CFD Calculations of Grid Fins Missile Aerodynamics in the Supersonic Flow Regime," AIAA Paper 2001-0257, 39th AIAA Aerospace Sciences Meeting and Exhibit., Reno,NV, 2001.

[6] Y.-q. ,. e. Deng, "Navier-Stokes Computation of Grid Fin Missile Using Hybrid Structured-Unstructured Grids," Chinese Journal of Aeronautics, 2006.

[7] B. Franzluebbers, "Drag Coefficient of Inclined Hollow Cylinders : RANS versus LES," Worcester Polytechnic Institute, 2013.

IOP Conf. Series: Journal of Physics: Conf. Series **1005** (2018) 012014 doi:10.1088/1742-6596/1005/1/012014

Application of CFD in Indonesian Research: A review

Ambarita H[1*], Siregar M R[2], Kishinami K[3], Daimaruya M[3], Kawai H[3]

[1]Sustainable Energy and Biomaterial Centre of Excellent, Faculty of Engineering, Universitas Sumatera Utara, Jl. Almamater Kampus USU Medan 20155, Indonesia
[2]Sarulla Operation Ltd., The Energy Building 51st floor, SCBD Lot 11A, Jl. Jend. Sudirman Jakarta 12190, Indonesia
[3]Mechanical Engineering Department, Muroran Institute of Technology, 27-1 Mizumoto-cho, Muroran, Hokkaido 050-8585, Japan

*Email: himsar@usu.ac.id

Abstract. Computational Fluid Dynamics (CFD) is a numerical method that solves fluid flow and related governing equations using a computational tool. The studies on CFD, its methodology and its application as a research tool, are increasing. In this study, application of CFD by Indonesian researcher is briefly reviewed. The main objective is to explore the characteristics of CFD applications in Indonesian researchers. Considering the size and reputation, this study uses Scopus publications indexed data base. All of the documents in Scopus related to CFD which is affiliated by at least one of Indonesian researcher are collected to be reviewed. Research topics, CFD method, and simulation results are reviewed in brief. The results show that there are 260 documents found in literature indexed by Scopus. These documents divided into research articles 125 titles, conference paper 135 titles, book 1 title and review 1 title. In the research articles, only limited researchers focused on the development of CFD methodology. Almost all of the articles focus on using CFD in a particular application, as a research tool, such as aircraft application, wind power and heat exchanger. The topics of the 125 research articles can be divided into 12 specific applications and 1 miscellaneous application. The most popular application is Heating Ventilating and Air Conditioning and followed by Reactor, Transportation and Heat Exchanger applications. The most popular commercial CFD code used is ANSYS Fluent and only several researchers use CFX.

1. Introduction

Computational Fluid Dynamics, hereafter named as CFD, is a numerical technique that solves a set of fluid flow and related governing equations to simulate the fluid flow and other parameters. Before the CFD era, the fluid flow problems were solved using the pure theoretical method and experimental method. The history of CFD started in the early 1970's and the beginning of CFD was triggered by the availability of increasing computational tools and CFD methodology its self. The first applications of the CFD methods include simulation of transonic flows based on solution non-linear potential equation. At the beginning of 1980's, the CFD has been used to solve two-dimensional and three-dimensional Euler equations. Forced by rapidly increasing speed of computers and development of numerical techniques, in the mid of 1980's, the solution of the viscous flows governed by Navier-Stokes equations become feasible. In parallel, a variety of turbulence models developed with different degree of numerical complexity and accuracy [1].

CFD has now matured and has been widely used not only in aerodynamics application but also in many multidisciplinary subjects. In the present time, CFD has become an integral part of the engineering design and analysis environment of many industrials. This is because of its ability to predict the performance of a new designs or process before they are ever manufactured or implemented. It is now easily found in the literature that CFD plays an important role as one of the key tools in many applications such as Hydrodynamics, Powerplant, Turbomachinery, electronic cooling device, chemical process, Heating Ventilating & Air Conditioning (HVAC), Biomedical Engineering, etc. There is no strict rule in dividing all of the CFD applications in literature. However, several researchers have been reported application of CFD in a particular area. Xia and Sun [2] published a review article on applications of CFD in the food industry. In the food industry, CFD is applied in several main processes such as drying, sterilization, refrigeration and mixing. It was shown that CFD has already proven to be a valuable tool to the food processing industry. However, there are still some problems that block the widespread use of CFD. Some are the materials in food processing differ in many ways from those is conventionally applied (air, water, and any fluids) and the specific knowledge in the food processing is extremely needed to be coupled with the CFD expert. Norton et al. [3] reviewed the applications of CFD in the modeling and design of ventilation systems in the agricultural industry. The results showed that advances in CFD technology have meant that field solutions can include the biological response of housed entities, and thereby enhance the realism of simulations. Bhutta et al. [4] reported their review of the CFD applications in various heat exchanger design. The results showed that CFD has been replacing the conventional methods used for the design, analysis, and optimization of heat exchangers. Easily accessible general purpose CFD commercial software such as FLUENT and CFX can fulfill the requirements of CFD analysis of a various type of heat exchangers. One of the most discussed issues in the heat exchanger analysis is the modeling of turbulent flow. In this issue, the turbulence model has been most widely used for heat exchanger design and optimization. The simulation results generally show a good agreement with experimental with discrepancy varies from 2% to 10%. Al-abidi et al. [5] reviewed the applications of CFD in the latent heat thermal energy storage. The results show that numerical solution in the analysis of Phase Change Material (PCM) is more accurate than the analytical solution and CFD software is used successively to simulate the application PCM as the latent heat thermal energy storage. Various CFD commercial software is used in PCM engineering includes ANSYS Fluent, Comsol Multiphysic, and Star-CCM. Among these the most widely used is ANSYS Fluent.

The above literature shows that the CFD which was originally developed for the aerodynamic purpose, it is now in present spread out to multidisciplinary applications. Today, CFD plays an important role in the design, analysis, and optimization of all related fluid engineering problems. There is significant growth in the CFD application in research and industry in the world includes in Indonesia. The objective of the present work is to review the applications of CFD published by Indonesian researchers that are reported in the literature. The results are expected to supply the necessary information in the development of research in Indonesia that related to CFD application.

2. Methods

The present work is divided into two sections. In the first section, the principle of computational fluid dynamics will be briefly presented including the governing equations and methodology of the solutions. In the second section, the growth of research paper related to CFD in the world and selected countries include Indonesia will be examined. In addition, the research articles, where at least one of Indonesian researcher as the author, are further reviewed. It is realized that there are many published research papers are found in the literature. In order to make clear and more focus discussion, only the research articles that are indexed in Scopus [6] will be used. The data were drawn from Scopus database that accessed on September 1, 2017.

3. Results and Discussions

3.1. Principles of CFD

The definition of CFD is the methodology that converts the governing equations into a set of linear equations and solves them iteratively. The fluid flow can be formulated using partial differential equations and named as governing equations. There are several different types of governing equations depend on the assumptions that have been made. The hierarchy of the governing equations is shown in

Figure 1. The real flow typically found in the real problem is turbulent flow regime. Here the energy is transferred from large scale motions to very smaller eddies until the scale become so small that the motion is dissipated by viscosity. In order to solve such problem numerically, in three-dimensional case, the computational domain needs to be divided into grids with a number with the order of Re9/4. Where Re is a Reynold number of the flow. For the case with big dimension, for instance, an aircraft with typical Re number is 30 million, that grid number is beyond the range of any current of the foreseeable computer. Thus, several assumptions have been made to simplify the real governing equations. The simplest to the most complex flow equations are shown in the figure. In the lowest hierarchy, the simplest flow is inviscid flow equation. In the inviscid calculations combined with boundary layer equation can result in accurate estimation of lift and drag coefficient. In the top of the hierarchy is Reynolds averaged Navier-Stokes (RANS) equations. This is a time-averaging of the Navier-Stokes equation and the continuity equation for incompressible flow.

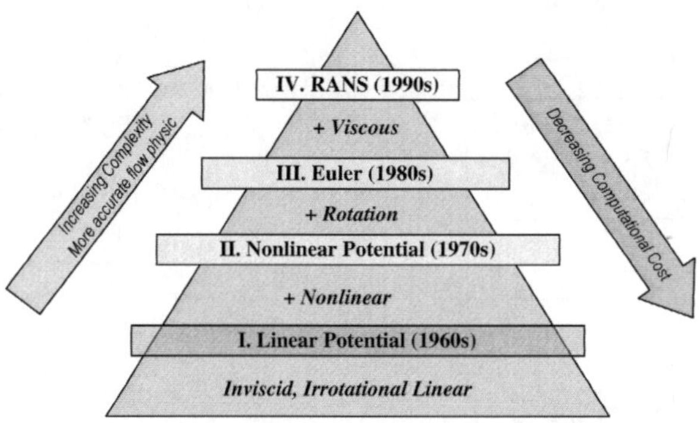

Figure 1. The hierarchy of governing equation [7]

Introducing the time averaging mean value of velocity, as formulated in equation (1)

$$\overline{v}_i = \lim_{T\to\infty}\frac{1}{T}\int_t^{t+T} v_i dt \tag{1}$$

where $\overline{v}_i$ is the mean value of velocity, the RANS governing equations of continuity and momentum equations can be written as follows.

$$\frac{\partial\overline{v}_i}{\partial x_i}=0 \tag{2}$$

$$\rho\frac{\partial\overline{v}_i}{\partial t}+\rho\overline{v}_j\frac{\partial\overline{v}_i}{\partial x_j}=-\frac{\partial\overline{p}}{\partial x_i}+\frac{\partial}{\partial x_j}\left(\overline{\tau}_{ij}-\rho\overline{v_i'v_j'}\right) \tag{3}$$

In this equation, the additional term of $\rho\overline{v_i'v_j'}$ is known as Reynolds-stress tensor represents the transfer of momentum due to turbulent fluctuations.

$$-\rho\overline{v_i'v_j'}=-\rho\left(\overline{v_i'v_j'}-\overline{v}_i\overline{v}_j\right) \tag{4}$$

In the equation (3), the laminar viscous stresses $\overline{\tau}_{ij}$ can be calculated using the following equation:

$$\overline{\tau}_{ij}=2\mu\overline{S}_{ij}=\mu\left(\frac{\partial\overline{v}_i}{\partial x_j}+\frac{\partial\overline{v}_j}{\partial x_i}\right) \tag{5}$$

In order to close the above equations, there are several turbulence models proposed by researchers.

The governing equations presented above are solved numerically by converting those equations into linear equation system in every grid. In general, the steps of CFD method are depicted in Figure 2. The fluid engineering problem needs to be modeled includes the dimensions, boundary, and other related parameters. After decided, the selected problem must be transferred into computational domain. Here, any available software can be used such as any CAD software, GAMBIT, ICEM or ANSYS Design

Modeler. By using these software, the computational domain will be divided into grids and meshes. The mesh will be exported to the solver. There are several commercially CFD solvers are available in the market. They are ANSYS Fluent, CFX, PHOENICS, and CFD2000. The self-developed code also can be used to solve the governing equations in every grid.

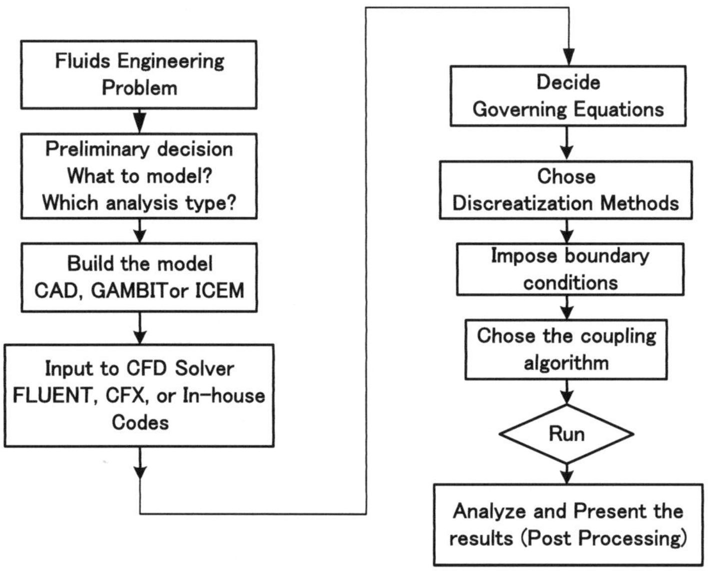

Figure 2 Typical Algorithm of CFD

The method of converting the governing equations into a system of linear equations is known as spatial discretization method. There are several spatial discretization methods typically available in the commercial software. They are First and Second Order Upwind, Power Law, QUICK and Third-Order MUSCL. These discretization methods are ordered from the cheapest to the most expensive computational cost. The highest cost related to high accuracy. As a note, the governing equations consists of several fields such as pressure and velocity fields also temperature field, turbulent kinetic energy field depends on the governing equations. These fields are coupled iteratively. There are several coupling schemes are typically available in the commercial software such as SIMPLE scheme, SIMPLEC scheme, and PISO scheme. The most commonly used is SIMPLE (Semi-Implicit Method for Pressure-Linked Equations) proposed by Patankar and Spalding [8].

Even though at present the commercial CFD code has been matured and has been widely used in many industrial applications, however, there are several challenges in the applications CFD. In the present worldwide commercial and government codes are based on the algorithms developed in the 80s and 90s. These codes can handle complex geometry but are generally limited to the second order of accuracy. The turbulences are handled using modeling such as first and second order RANS and for the present highest performances computers, the turbulence can be solved using LES and DNS. The unsteady simulations are still very expensive and several questions on accuracy still remain. These are several challenges in the future development of CFD [7].

3.2. Growth of Research Paper on CFD

The database of Scopus has been used to explore research paper related on the CFD. The growth of research paper indexed by Scopus in the last 22 years, during 1996 to 2017 is shown in Figure 3. There are two research documents shown in the figure. The first is the document related to all CFD topics (shown by the red line and its number on the right). The second is the document related to CFD in a particular area of aerodynamics (shown by the blue line and its number on the left). As a note, the number of document in the year 2017 is lower than in the year 2016. This is because the year 2017 is not yet finished. It will higher. The figure shows that the research paper increases with increasing time. The research paper related to CFD increases 10.2% in every year. This fact reveals that application of CFD in all research areas is increasing. In the specific application the aerodynamics, as the original application of CFD, the application of CFD is increasing with the averaged growth of 10.5%. The share of the research paper in the aerodynamic application is around 9.6%. This fact suggests that 9.6 of research papers out of 100 research paper related to CFD are related to aerodynamic applications.

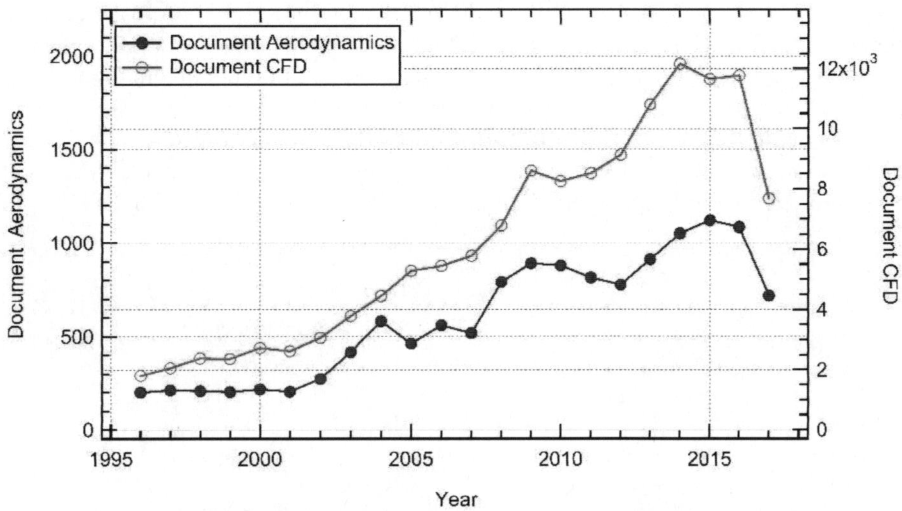

Figure 3 Worldwide document related to CFD indexed by Scopus

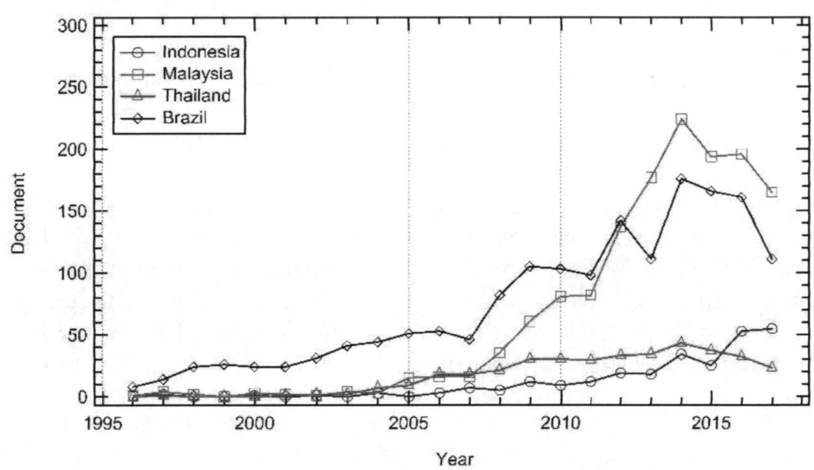

Figure 4 Document related to CFD indexed in Scopus

Several countries also show the same trend with the worldwide trend. Figure 4 shows the research document related to CFD published by researchers with affiliation from Malaysia, Brazil, Thailand, and Indonesia. It can be seen the similar trend by these countries. In cumulative, the total document published by Malaysia, Brazil, Thailand and Indonesia researchers are 1419, 1641, 370, and 260 documents, respectively. Brazil is the most productive among these countries. During the last ten years (2006 to 2016), Brazilian researchers published 110 documents per year and followed by Malaysia, Thailand, and Indonesia with the averaged productivity of 110.8 documents, 29.5 documents, and 17.9

documents, respectively. It reveals that Indonesia shows the lowest productivity among these countries. However, there is a significant growth from Indonesian researchers. In the last ten years, Indonesia shows the average growth of document; it is 48.06% per year. The second is Malaysia with a growth of 33.75% and followed by Brazil and Thailand with yearly growth are 15.9% and 7.2%, respectively. This fact reveals that there is an increasing growth of application CFD in Indonesia in the last ten years.

3.3. Indonesia Research Articles related to CFD

For Indonesia case, the first document related on CFD published by Indonesian researcher was in June 1996 by Kindangen et al. [9]. Here, the CFD was used to investigate natural ventilation, and the results were compared with wind tunnel result. After this article, there are 260 documents indexed by Scopus produced by Indonesian researchers during period 1996 to 2017. The type of those documents consists of conference paper with 135 documents and followed by 125 research articles, one book, and one review. The distribution of those documents is presented in Figure 5(a). In this work, the 125 of research article are reviewed to explore the characteristics of CFD application published by Indonesian researcher.

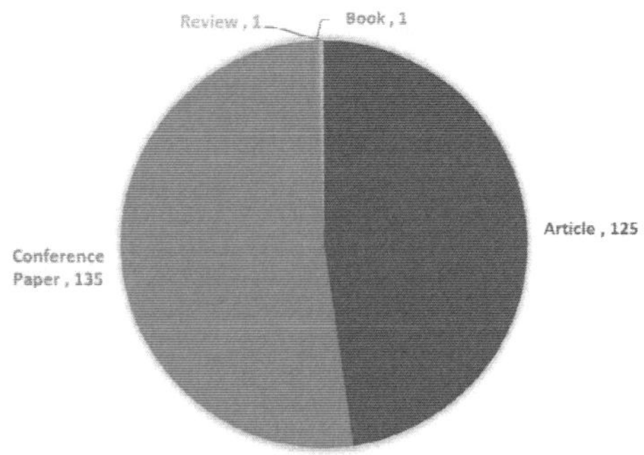

(a) Number of documents

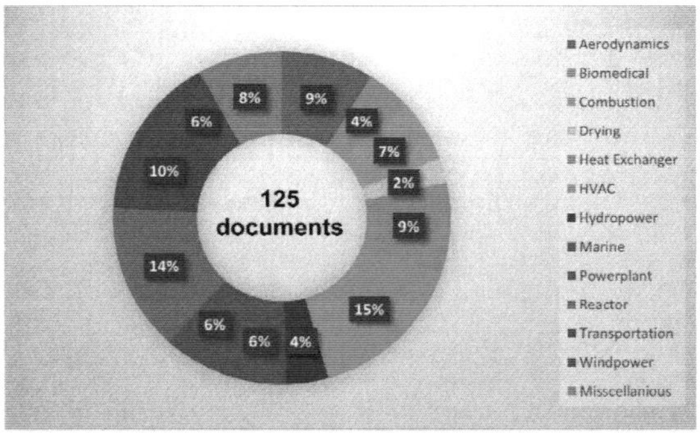

(b) Article distributions

Figure 5 Number of document and article distribution

The review shows that CFD research papers can be divided into two modes. The first mode is the research related to the CFD method and the second mode is the application of CFD to solve the engineering problem. In the first mode, typically a new method (algorithm, scheme, discretization technique, etc.) is proposed and compared with the previous one. There are 8 of research articles deal with the first mode. Febrianto and Zuhal [10] proposed vortex in cell [VIC] method to predict flutter phenomenon of 2D bridge deck model. The performance of the proposed method to simulate moving body is tested by simulating fluid-structure interaction and flutter phenomena of long-span bridge. The result of the proposed mode showed good agreement with another numerical method and experimental results. Tsai et al. [11] presented a coupled wave-vegetation simulation for the moving object of the coastal vegetation on tsunami wave height damping. Solitary wave propagation on a group of emergent cylinders was employed to idealize the problem. To model the turbulence, general RANS equation combined with renormalization group turbulent closure. The general moving object (GMO) model is employed to simulate the coupled motion of vegetation with wave dynamically. Djojodihardjo and Safari [12] reported a development the foundation for the computational scheme for the calculation of the influence of the acoustic disturbance to the aeroelastic stability of the structure. Wahyudi et al. [13] proposed a new method to study the gas-solids flow and heat transfer in fluidized beds with an immersed tube. The new method is a fully three-dimensional model combined CFD and discrete element method. At first, a critical bed thickness is determined at which the bed can be regarded as fully 3D. The result shows that the model can successfully reproduce the typical relationship between pressure drop and gas velocity, and flow and heat transfer characteristics such as the four distinct stages of bubble transit through the tube and peak of heat transfer coefficient between the tube and the bed for certain gas velocity. This phenomenon was well predicted by the proposed method which is not well-predicted by previous 2D CFD-DEM and 2D CFD-3D DEM models. van Groesen and Andonowati [14] proposed a new approach to solve fully nonlinear treatment of the dynamic wave-ship interaction for potential flows. Hamilton description of the inviscid free surface wave was extended in a straightforward consistent way to include ship-wave interaction. By describing the fluid flow motion in horizontal variables only, computational complexity was reduced. The internal fluid motion is modeled in a consistent approximative way. The Lagrangian variational principle is used to develop the equations for the wave-ship interaction. The results show how to obtain consistently the extension of known Hamiltonian formulation for free surface waves to a Hamiltonian formulation of the fully nonlinear wave-ship interaction while retaining the dimension reduction of dealing only with the fluid potential at the free surface and wetted ship hull. The main observation of their work is that the Hamiltonian formulation avoids the calculation of the dynamics part of the pressure which becomes part of a canonical momentum variable that is updated by the hydrostatic forces.

Gun Gun et al. [15] investigated performance of several turbulence models by comparing the numerical results and experimental one. Three models were tested, they are k-epsilon model, renormalization group (RNG) k-epsilon and Reynolds stress model (RSM). The results revealed that those three turbulence models provide good results to predict the distribution of speed and pressure of the fluid flow in the wind tunnel. In the case of predicting the turbulent kinetic energy, turbulent dissipation rate and turbulent effective viscosity, the k-epsilon model showed a good agreement in comparison with RSM model. Nugroho et al [16] proposed an analytical solution to three-dimensional incompressible Navier-Stokes equation with the continuity equation. For simplicity, the spatial and temporal coordinates are transformed into a single coordinate ξ. The form of solution $V = \nabla\Phi + \nabla\times\Phi$ was proposed. Here Φ is a potential function that is defined as $\Phi = P(x,\xi)R(\xi)$. As a result, more general solutions are also obtained based on the particular solution of P and the explicit analytical solutions are found to be mathematically similar for the cases of zero and constant pressure gradient. The conclusion was the selection of variables for the potential function can be interchanged from the beginning, resulting in similar explicit solutions.

In the second mode of the research, the CFD code, typically the commercial one, is used as a tool to provide analysis, design, and optimization of an engineering problem or process. As expected, this mode is the most used mode by Indonesian researchers. Figure 5(b) shows the application distribution of research articles related to CFD. It consists of Aerodynamics (9%), Biomedical (4%), Combustion (7%), Drying (2%), Heat Exchanger (9%), Heating Ventilating and Air Conditioning, hereafter as HVAC (15%), Hydropower (4%), Marine (6%), Powerplant (6%), Reactor and Separation Process (14%), Transportation (10%), Wind Power (6%), and Miscellaneous (8%). This distribution shows that for Indonesian researchers the CFD most used in HVAC application with 19 research articles and followed by Reactor and Transportation applications with 18 and 13 research articles, respectively. The typical research article for each application is reviewed below.

3.3.1 Aerodynamics applications

In the aerodynamics application, there are 11 research articles resulted by Indonesian researcher indexed in Scopus. Bangga and Sasongko [17] provided CFD predictions of a pitching NACA 0012 air foil at a reduced frequency of 0.1 and at small Reynolds number vane of 135,000. The turbulence flow was modeled using adjusted the k-epsilon URNAS model. Here the dumping factor was introduced as a function of wall distance in the buffer zone region. Parametric studies on the involving variables were conducted. Their study success to demonstrate the fluid flow analysis using the CFD ANSYS Fluent. Effendy et al. [18] investigated the cooling performance of blade trailing edge cutback of the gas turbine that typically used in the aircraft. Effects of the lip thickness to slot height (t/H) ratio was explored. Their study was aimed to enrich the information on the dynamic flow interaction process between the internal coolant fluid and external warm mainstream fluid in comprehensive manner. Figure 6 shows the isometric view of computational domain (a), schematic view of trailing edge cutback, and (c) turbulent flow structure at several t/H ratios. The CFD with detached-eddy simulation (DES) combined with SST k-omega turbulence model can be used to simulate the dynamic flow interaction process of trailing edge cutback of a gas turbine. The observed vortex shedding and its characteristics in the near wake region (as shown in Figure 6c) are found to play an important role in determining the dynamics process of the cod and warm airflow mixing, which in turn have significant influences on the prediction accuracy of the near-wall heat transfer performance. Those papers are only two out of 11 articles Scopus indexed published Indonesian researchers.

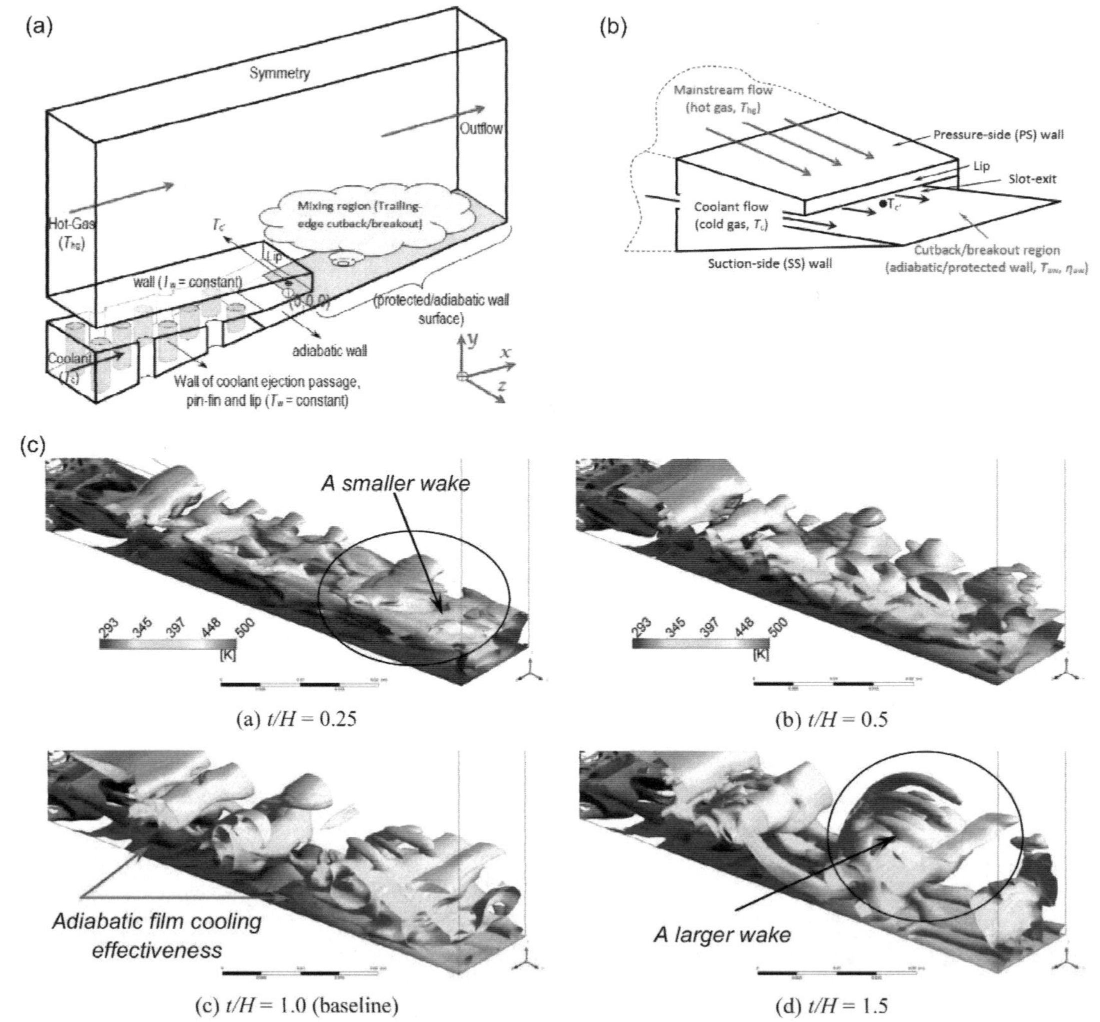

Figure 6 the isometric view of computational domain (a), schematic view of trailing edge cutback, and (c) turbulent flow structure at several *t/H* ratios [18]

3.3.2 Biomedical engineering

Biomedical engineering also uses CFD as a tool to provide analyses. There are five research articles have been found in literature indexed in Scopus. Three of those works are briefly presented here. Tenekecioglu et al. [19] employed CFD ANSYS Fluent to simulate blood flow in a porcine coronary artery model. The objective was to evaluate the impact of the scaffold design on the local hemodynamic microenvironment. Figure 7 shows the CFD results of the analysis. The conclusion of the study is that the CFD assessment can be used to guide improvements in the scaffold design for a more "hemo-compatible" geometry. Sanka et al. [20] employed CFD to simulate the water flow over the surface of the crab carapace and the interactions of water with the micro-protuberant setae of the carapace surface. The objective of the modeling was to determine how the micro-protuberant setae of the crab carapace may instigate micro-organism nucleation by altering fluid motion. In their work Fiji-Image and Rhino 3D (CAD modeling software) were used to develop the computational domain and Comsol Multi-physics was used to execute the numerical analyses. Figure 8 shows the velocity field and streamline over Tiarinia cornigera carapace. In total, there are five research articles related to the application of CFD in biomedical engineering. In this work only these two papers are briefly reviewed.

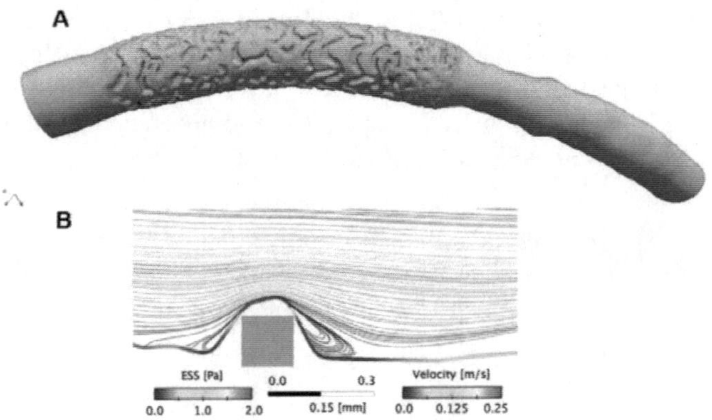

Figure 7 CFD analysis of blood flow in a porcine coronary artery model [19]

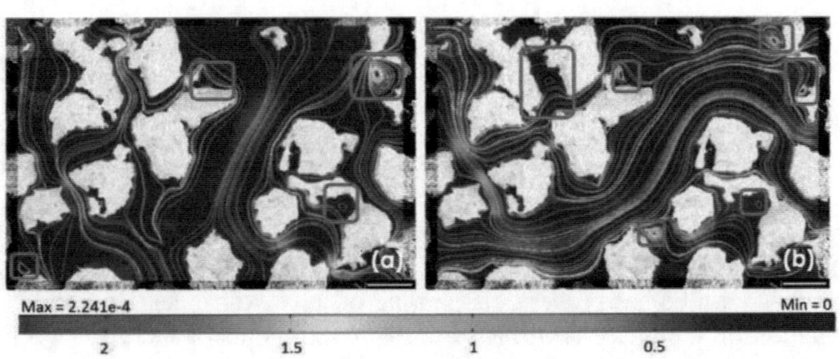

Figure 8 CFD analysis of velocity field and streamline over *Tiarinia cornigera* carapace [20]

3.3.3 *Combustion*

The combustion process is also a field in engineering that typically employs CFD to simulate the fluid flow during the combustion process. Even though combustion process is a perfect application of CFD, however only 7% of the research articles are related to this application published by Indonesian researcher. This might be because the CFD application in combustion needs more numerical effort. Also, the combustion process is integrated with many familiar CFD code, such as ANSYS Fluent, only in a few years later. One of these paper is the study by Ambarita et al. [21] which employed ANSYS Fluent commercial code to simulate combustion process in a small compression ignition engine run on duel-fuel (diesel-fuel) mode. In the case of thermal efficiency, the CFD and experimental results showed a good agreement.

3.3.4 Drying

In the drying process, CFD is typically used to simulate simultaneous heat and mass transfer in the fluid flow of drying medium. For instance, Simanjuntak et al. [22] explored the drying process of coal in swirl fluidized bed coal drying. Here, CFD commercial code was employed to investigate the effect of different angles of guide van in a transient three-dimensional computational domain. The results showed that CFD commercial code could be used to simulate the drying process.

3.3.5 Heat Exchanger

Recently, heat exchanger industry and researchers employ CFD commercial code to provide analysis, design, and optimization to the heat exchanger. The literature review shows that 9% of the research papers deal with heat exchangers. Afrianto et al. [23] used CFD to investigate LNG flow and heat transfer characteristic in a heat exchanger. In their work, the commercial code Fluent was used to simulate the LNG flow and heat transfer characteristics in a shell and tube heat exchanger. The shear stress transport (SST), the k-omega turbulence model, was adopted. The CFD code successfully depicts the flow characteristics inside the heat exchanger as shown in Figure 9.

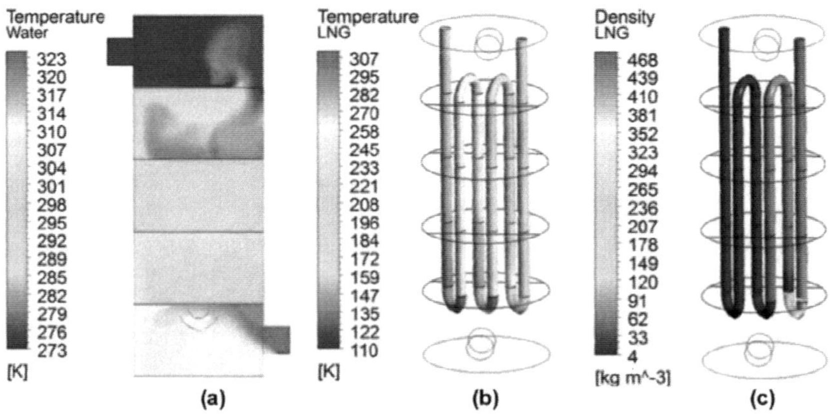

Figure 9 Contour of fluid distributions: (a) shell side wall temperature, (b) tube side LNG temperature, and (c) tube side density distribution [23]

Santosa et al. [24] employed CFD code Fluent to investigate local refrigerant and air heat transfer coefficient in plain fin-and-tube gas cooler coils, a type of heat exchanger. The objective is to fill the absence of investigation on the analysis of heat transfer coefficients of CO2 gas coolers in literature. The CFD result is shown in Figure 10. The k-epsilon model was used to model the turbulence model. The results reveal that CFD modeling was found to be able to adequately represent the heat transfer characteristics of the gas cooler and be an effective simulation tool for the determination of local air and refrigerant-side heat transfer coefficient in the coil.

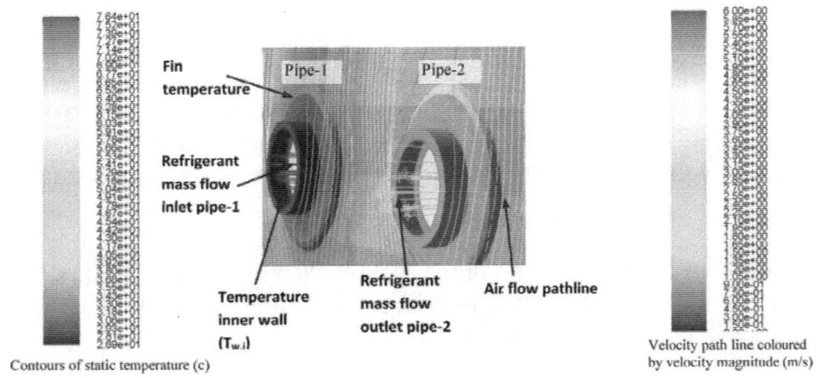

Figure 10 Contour of temperature and velocity path at a segment of finned-tube heat exchanger [24]

3.3.6 Heating Ventilating and Air Conditioning

This application is the most popular for CFD applications by Indonesia researchers. About 15% of the research articles dealt with HVAC. In this application, CFD is employed to simulate temperature and velocity distributions. Using these fields, the ergonomic parameters can be estimated. The applications are not aimed only for the human, but also for livestock such a study reported by Yani et al. [25]. In the paper, CFD is used to design the housing of the broiler. Solid Work Flow simulation as a CFD code was used.

3.3.7 Hydropower

There are five research articles related to the application of CFD to hydropower found in literature indexed in Scopus affiliated by Indonesian researcher. Budiman et al. [26] employed CFD ANSYS Fluent to calculate the load for various opening angles of the guide vane. To model the turbulence mode, the k-epsilon standard was used. The CFD results of the study are shown in Figure 11. Arini et al. [27] employed CFD to simulate fluid flow around vertical axis tidal turbine to explore the operating performance such as torsion and output power. The profile of the turbine is NACA0015, and the turbulence flow was modeled using RANS.

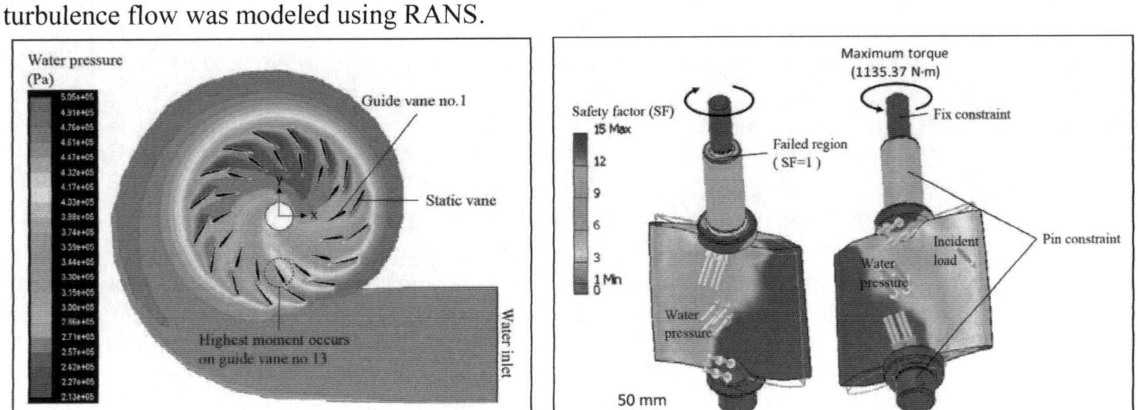

Figure 11 (a) Pressure contour and (b) over load case on guide vane [26]

3.3.8 Marine

Marine also employs CFD to simulate fluid flow. There are eight research papers have been found in the literature. Tsai et al. [28] used open-source CFD IHFOAM to investigate wave-damping performance that considering vegetation motion coupled with tsunami wave. The fractional area/volumes obstacle representation technique is used to simulate rigid body motion dynamically coupled with fluid flow. The free water surface tracking in the model is accomplished by using the volume of fluid (VOF) method. And the turbulence flow is modeled using k-epsilon model. The simulated results show that the damping of wave height and turbulent kinetic energy by the moving cylinders. Also, the wave decay by the coastal vegetation may be overestimated if the vegetation was represented as a stationary state. Wulp et al. [29] reported a study on numerical simulations of river discharges, nutrient flux and nutrient dispersal in Jakarta Bay, Indonesia. The Delf3D flow model was used to simulate flow characteristics by solving the shallow water equations for flow. The method also incorporates nutrient loads such as total nitrogen and total phosphorus. The nutrition distributions during flood flow and during ebb flow are shown in Figure 12.

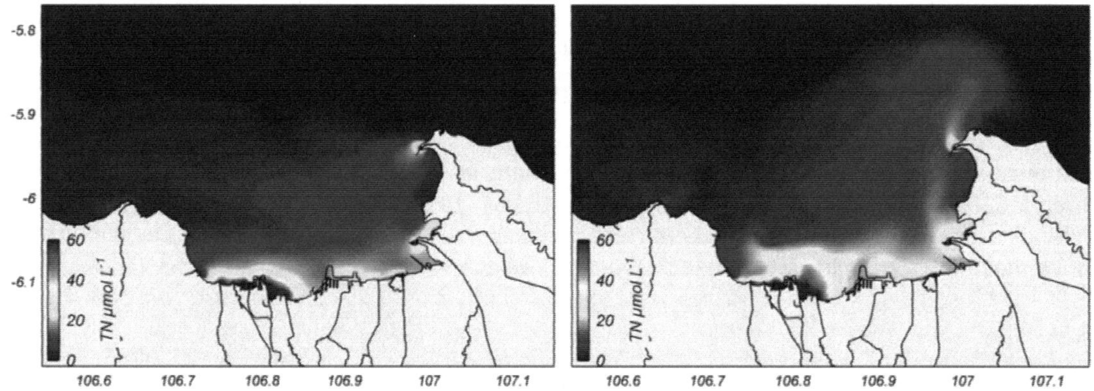

Figure 12 Nutrition distribution, total nitrogen gradient during (a) flood flow and (b) ebb flow

3.3.9 Powerplant

Several research articles related to application of CFD in powerplant have been found in Scopus indexed literature. Typically, the CFD codes are used to explore the characteristic of fluid flow during energy conversion in the powerplant. Daryus et al [30] reported a study on a micro gas turbine bioenergy proto X-3 which is known as a prototype turbine designed to be able to operate on any kind of fuel, include renewable and it is aimed as a powerplant for green building application. In order to explore the flow phenomena in the combustor of the turbine. In this study two type of turbulence models were tested, STD k-epsilon and RNG k-epsilon modes.

3.3.10 Reactor

In this In this application, the reactor includes the apparatus that maintain and control a nuclear reaction and also the container to hold chemicals undergoing a chemical reaction. There are 18 research articles found in the literature. Wahyudi et al. [31] used CFD to analyze fluid flow and heat transfer mechanism in low temperature packed and fluidized beds. In this case, there are two flows will be analyzed, the fluid flow and the solid flow. In order to combine these two flows, the CFD-DEM model is employed. The typical results of flow fields and solid flow are presented in Figure 13. It was demonstrated that both the gas and solids flow are 3D in nature. The unique feature of 3D orientations of a gas velocity vector field around the bubble plays an important role in generating in a consistency of the size and upward movement of the bubble. Muharam et al. [32] used CFD to model hydrotreating reactor to produce renewable diesel from non-edible vegetable oils. The reactor is packaged with spherical catalyst particles. The model was two-dimensional axisymmetric. Simanungkalit and Rinaldi [33] reported a study on CFD analysis of the fast pyrolysis for empty fruit bunches in Fluidized-Bed Reactor. In order to model the two-phase flow, the Eulerian approach was used. The approach considers that discrete secondary phase (solid) and continuum primary phase (fluid) are similar and illustrated for the primary phase. ANSYS Fluent was used to perform the analysis. The contours of solid phase distribution, reactor temperature, and bio-oil vapor concentration were presented. Ramadhan et al. [34] employed CFD analysis to perform a preliminary study for design core of nuclear research reactor of Triga Bandung using fuel element plate MTR. Here porous and non-porous models of the fuel element plate were considered. The model was created using CAD software based on a three-dimensional picture of the fuel plates. However, the governing equations and type of CFD code are not declared in the report. The CFD results of the velocity field and temperature distributions were presented. The above works are only a few samples of the research papers affiliated by Indonesian researchers.

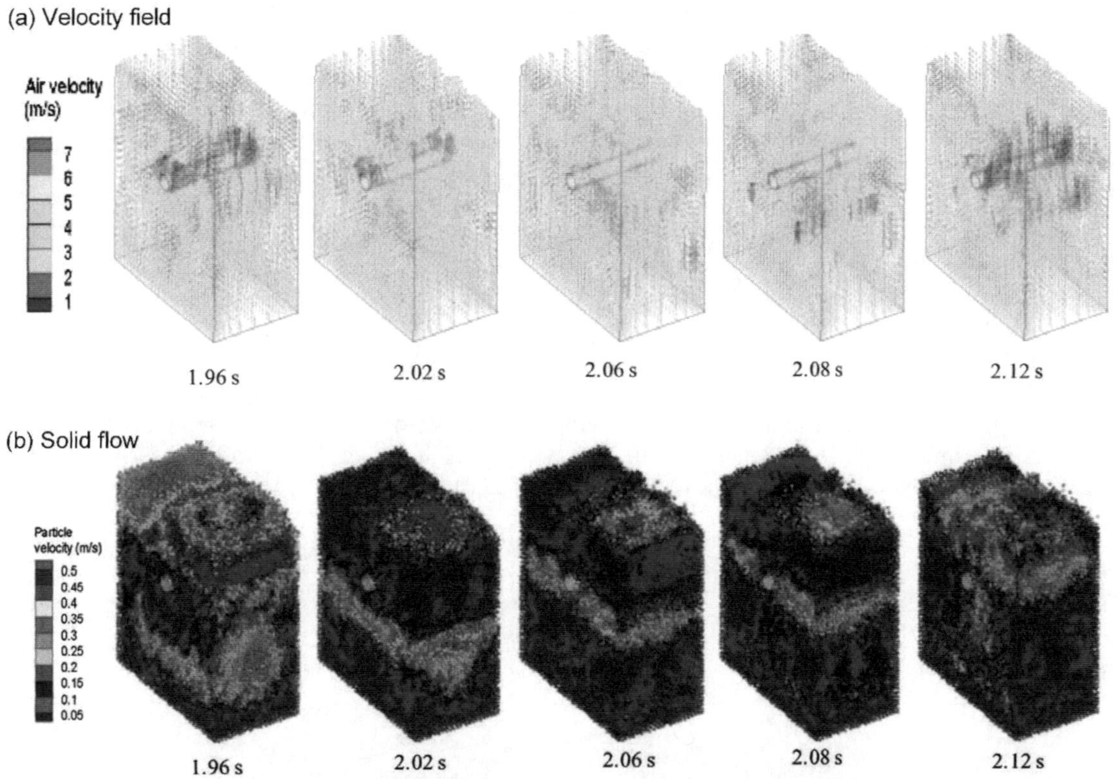

Figure 13 (a) Fluid flow distribution and solid flow distribution in fluidized bed [31]

3.3.11 Transportation

Transportation industry and research is a traditional costumer of CFD commercial code. This trend is also shown in the research articles affiliated by Indonesian researchers. There are 13 research articles found in literature, some of those reviewed here. Even though road transportation is the most used transformation, however, in Indonesia case, water transportation has the biggest share in comparison with road transportation. Taha et al. [35] used CFD code ANSYS Fluent to simulate fluid flow analysis to investigate aerodynamics characteristics of "Merdeka 2" a prototype of a solar vehicle that participated in the world solar challenge. The design concept of the car was based on box fish which was also claimed the concept design of Mercedes Benz minivan. Reynold-averaged Navier-Stokes with the k-epsilon model was used to model the turbulent flow. The CFD results and experimental data do not agree well. Syamsuar et al. [36] reported a study on using CFD to investigate outer wing of flying boat remote control model. The objective was to explore the aerodynamics and hydrodynamics characteristics of the model during hydro planning. Samuel et al. [37] used CFD to investigate the resistance components of converting a traditional monohull fishing vessel into Cataraman form. In the analysis, the real Cataramans size was modeled into 1:10 model.

3.3.12 Wind power

In this section, CFD is typically used to explore the fluid flow characteristics and comparison with the experimental results. About 6% of the research articles deal with wind power. Sanusi et al. [38] employed CFD code ANSYS Fluent to simulate the fluid flow of three blades models of Savonius Wind Turbines. The objective was to improve the performance of the wind turbine by combining a conventional blade with an elliptical blade into a combined blade rotor. The type of governing equations was not explored. The velocity and pressure contours of conventional and proposed design resulted by CFD code were plotted, and examinations were made based on these CFD results. Bangga et al. [39] employed CFD code to investigate a single bladed vertical axis wind turbine under a dynamic stall. The blade was developed based on NACA 0012 profile and was operated under stalled conditions at tip speed ratio of 2. The analysis was performed using ANSYS Fluent, and the turbulence was modeled using SST turbulence model. Soebiyan et al. [40] explored the potency of wind energy in high-rise buildings in humid tropical climate using CFD as a tool. The wind speed data were used to perform the CFD analysis. It was concluded that in the case of Bina Nusantara University wind energy could replace 3.27% of main building energy consumption and this is comparable to 127,650 kgCO2 emission reduction.

3.3.13 Miscellaneous

Several research articles related to miscellaneous such as sport, gun, bearing, etc. are also found in the literature. One of the specific studies was published by Mubin et al. [41]. In the study, CFD code is used to explore the drag and lift coefficient of sepak takraw ball at different face orientations. The sepak takraw ball was claimed to be unique since it is not enclosed like other balls such as golf, tennis and soccer balls. In the study, CFD code of CFX was employed to perform the analysis. The turbulent flow was modelled using RANS turbulence model. The typical results of the CFD software are shown in Figure 14. It was shown that CFD is a powerful tool for determining the drag and lift of a sepak takraw ball.

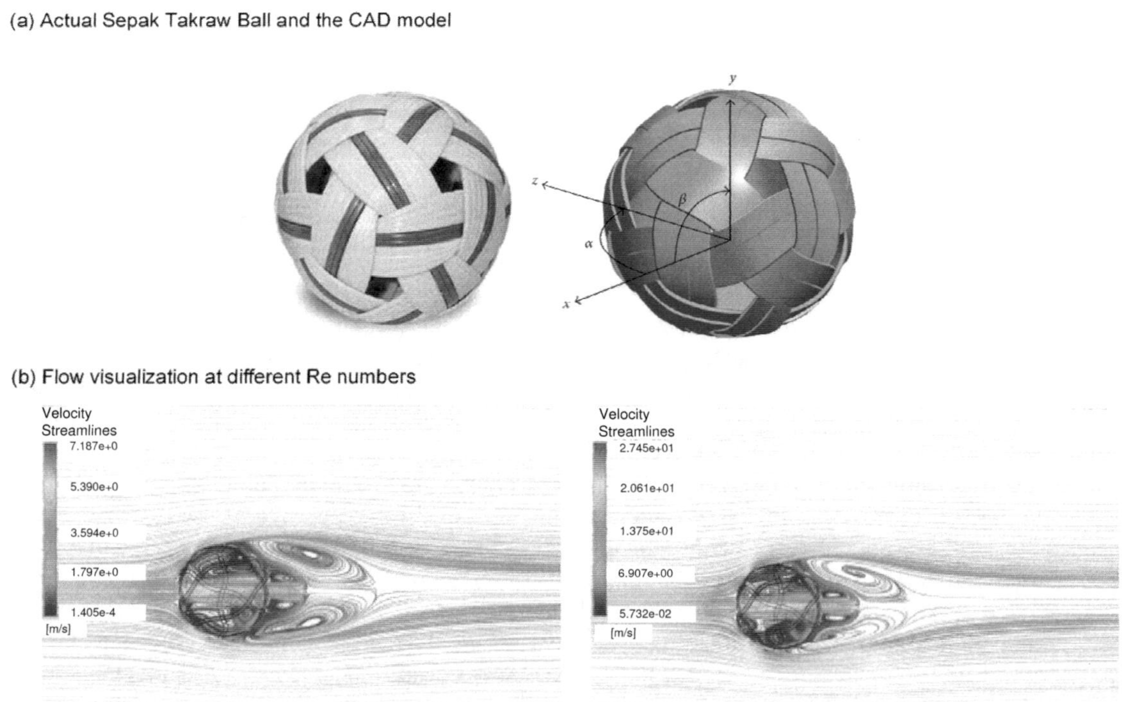

Figure 14 The actual and model of the ball (a) and flow visualization (b)

4. Conclusions

In this work, the document related to CFD indexed in Scopus affiliated by Indonesian researcher has been briefly reviewed. The following conclusions are drawn in this study.

- To the present, there are 260 documents related to CFD affiliated by Indonesian researchers found in literature indexed by Scopus. These documents divided into research article 125 titles, conference paper 135 titles, book one title and review one title.
- In comparison to neighbor countries such as Malaysia and Thailand also with Brazil, a country with a population closer to Indonesia, the document productivity is relatively lower. Indonesia is comparable with Thailand. However, Indonesia shows a significant growth, especially in the last five years.
- In the research articles, only limited researchers focused on the method development. The research articles, almost all, focus on using the specific application or using CFD as a research tool in the specific application such as aircraft and heat exchanger.
- The topics of the 125 research articles can be divided into 12 specific applications and one miscellaneous application. The most popular applications are HVAC followed by Reactor, Transportation and Heat Exchanger applications.
- The most popular commercial CFD code used by Indonesia researcher is ANSYS Fluent, and only several researchers use CFX.

References

[1] Blazek J 2001 *Computational Fluid Dynamics: Principles and Applications* (Amsterdam: Elsevier)
[2] Xia B and Sun DW 2002 *Computers and Electronic in Agriculture* **34**, 5-24
[3] Norton T, Sun DW, Grant J, Fallon R and Dodd V 2007 *Bioresource Technology* **98**, 2386-2414
[4] Bhutta M M A, Hayat N, Bashir M H, Khan A R, Ahmad K N and Sarfaraz K 2012 *Applied Thermal Engineering* **32**, 1-12
[5] Al-abidi A A, Mat S B, Sopian K, Sulaiman M Y and Mohammed A Th 2013 *Renewable and Sustainable Energy Reviews* **20**, 353-363
[6] www.scopus.com [Accessed September 1, 2017]
[7] https://pdfs.semanticscholar.org/de25/08faa97d1144ee1185d9a322c821803e4e96.pdf
[8] Patankar S V and Spalding D V 1972 *International Journal of Heat and Mass Transfer* **15**, 1787-1806
[9] Kindangen J, Krauss G and Kindangen J 1996 *Architectural Science Review* **39**, 113-120
[10] Febrianto E V and Zuhal L R 2017 *ARPN Jurnal of Engineering and Applied Science* **12(10)**, 3040-3045
[11] Tsai C P, Chen Y -C, Sihombing T O and Lin C 2017 *Natural Hazards and Earth System Sciences* **17(5)**, 693-702
[12] Djojodihardjo H and Safari I 2013 *CMES-Computer Modeling in Engineering and Sciences* **91(3)**, 205-234
[13] Wahyudi H, Chu K, and Yu A 2016 *International Journal of Heat Transfer* **97**, 521-537
[14] van Groesen E and Andonowati 2017 *Applied Mathematical Modeling* **42**, 133-144
[15] Gun Gun R G, Siswantara A I, Budiarso, Daryus A, and Pujowidodo H 2016 *International Journal of Technology* **7(8)**, 1362-1371
[16] Nugroho G, Ali A M S, and Abdul Karim Z A 2009 *Applied Mathematics Letters* **22(11)** 1639-1644
[17] Bangga G and Sasongko H 2017 *Journal of Applied Fluid Mechanics* **10(1)**, 1-10
[18] Effendy M, Yao Y F, Yao J and Marchant D R 2016 *Applied Thermal Engineering* **99**, 434-445

[19] Tenekecioglu E, Torii R, Bourantas C, Abdelghani M, Cavalcante R, Sotomi Y, Crake T, Su S, Santoso T, Onuma Y and Serruys P W 2017 *International Journal of Cardiology* **227**, 467-473

[20] Sanka I, Suyono E A, Rivero-Muller A and Alam P 2016 *Acta Biomaterialia* **41**, 52-59

[21] Ambarita H, Widodo T I and Nasution D M 2017 *Journal of Physic: Conference Series* **801(1)**, 012095

[22] Simanjuntak M E, Prabowo, Ichsan D and Widodo W A 2016 *JP Journal of Heat and Mass Transfer* **13(4)**, 497-51

[23] Afrianto H, Tanshen Md. R, Munkhbayar B, Suryo U T, Chung H and Jeong H 2014 *International Journal of Heat and Mass Transfer* **68**, 110-118

[24] Santosa I M C, Gowreesunker B L, Tassou S A, Tsamos K M and Ge Y 2017 *International Journal of Heat and Mass Transfer* **107**, 168-180

[25] Yani A, Suhardiyanto H, Erizal and Purwanto B P 2014 *Media Peternakan* **37(1)**, 17-23

[26] Budiman B A, Suharto D, Djodikusumo I, Aziz M and Juangsa F B 2016 *Advances in Mechanical Engineering* **8(7)**, 1-8

[27] Arini N R, Arywibowo T H and Nugroho S 2014 *International Journal of Applied Engineering Research* **9(22)**, 17793-17800

[28] Tsai C P, Chen Y C, Sihombing T O and Lin C 2017 *Natural Hazards and Earth System Sciences* **17**, 693-702

[29] van der Wulp, Damar A, Ladwig N and Hesse K J 2016 *Marine Pollution Bulletin* **110**, 675-685

[30] Daryus A, Siswantara A I, Darmawan S, Gunadi G G R, Camalia R 2016 *International Journal of Technology* **2**, 204-211

[31] Wahyudi H, Chu K and Yu A 2016 *International Journal of Heat and Mass Transfer* **97**, 521-537

[32] Muharam Y, Nugraha O A, Leonardi D 2017 *Chemical Engineering Transactions* **56**, 1561- 1566

[33] Simanungkalit S P and Rinaldi N 2013 *Journal of Engineering and Applied Scince* **8 (9-12)**, 290-294

[34] Ramadhan A I, Suwonon A, Umar E and Tandian N P 2017 *Engineering Journal* **21(3)**, 173-181

[35] Taha Z, Passarella, Sugiyono, Abd Rahim N, Md Sah J and Ahmad-Yazid N 2011 *Advanced Science Letter* **4**, 2807-2811

[36] Syamsuar S, Djatmiko E B, Mujahid A S, Erwandi and Subchan 2015 *Jurnal Teknologi (Science & Engineering)* **76(1)**, 221-228

[37] Samuel, Iqbal M and Utama I K A P 2015 *International Journal of Technology* **3**, 432-441

[38] Sanusi A, Soeparman S, Wahyudi S and Yulianti L 2017 International Journal of Fluid Machinery and Systems 10(1), 54-62

[39] Bangga G, Hutomo G, Wiranegara R and Sasongko H 2017 *Journal of Mechanical Science and Technology* **31(1)**, 261-267

[40] Soebiyan V, Saragih J F B and Tedja M 2017 *Chemical Engineering Transaction* **56**, 241-246

Aerodynamic Characteristics of Tube-Launched Tandem Wing Unmanned Aerial Vehicle

Nurhayyan H. Rosid[1)*], Irsyad Lukman E.[1)], and Ahmad Fadlillah M.[1)] , M. Agoes Moelyadi[1)]
[1)] Faculty of Mechanical and Aerospace Engineering, Institut Teknologi Bandung, Indonesia

*nurhayyan.halim.rosid@s.itb.ac.id

Abstract. Tube Launched UAV with expandable tandem-wing configuration becomes one of the most interesting topic to be investigated. Folding wing mechanism is used due to the requirements that the UAV should be folded into tubular launcher. This paper focuses on investigating the aerodynamics characteristics because of the effects of folding wing mechanism, tandem wing configuration, and rapid deploying process from tube launcher. The aerodynamic characteristics investigation is conducted using computational fluid dynamics (CFD) at low Reynolds numbers (Re < 200000). The results of the simulation are used for the development of ITB Tube-launched UAV prototype and for future studies.

Keywords : tandem wing, folding wing, expandable wing, tube-launched UAV, aerodynamic, CFD

1. Introduction

Unmanned Aerial Vehicles (UAVs) have been widely used both in military and civil all across the world in recent year. Many types of UAV technlogies have been developed to answer the global needs, such as for aerial mapping, aerial surveilllence, atmospheric sensing, payload delivering, and for war missions. One of the latest technologies in aerial vehicle is morphing shape which can manipulate it's external shape to achieve better flight performance and to gain other good advantages. Among the various kinds of morphing vehicles, there are certain types of UAV with folded wing that could launch from tubular launcher. The folding wing technology can reduce space usage so the carriage of the UAV can be easier and the packaging can be simplified. On the other hand, the tubular launcher has higer security and faster take off speed than other conventional launching method[1].These adventages are needed in military application helping the troops to carry and to deploy UAV easier. During launching process, the rapid tranformation from folded wing to expanded wing can have significant influence on aerodynamic characteristics,stability and maneuverability. The numerical simulation will be conducted to investigate the aerodynamic characteristic both in transition process and steady level flight.

The small allowable space of tube launcher requires the folding wing mechanism, and the available volume must be utilized to achieve the wing area as large as possible, so the tandem wing configuration is the best choice. Rival D[2] has discovered that tandem-wing system has better better fuel efficiency. The interaction between forward and rear wing generates suction bubble creating thrust addition. Another research done by Rhodes shows that there are some benefits using dual-wing

IOP Conf. Series: Journal of Physics: Conf. Series **1005** (2018) 012015 doi:10.1088/1742-6596/1005/1/012015

configuration compared to conventional aircraft design. Dual-wing configuration has better cruise performance and lower fuel consumption than single-wing configuration[3].

Aircraft Configuration

The deploying process of tandem-wing UAV with folding mechanism is shown in figure 1. This configuration contains canard (front lifting surface), wing (rear lifting surface), dual vertical tail, and pusher electric propulsion. This configuration is designed to achieve desired lift from 2 lifting surfaces, canard and wing. The wing span is 190 mm longer than canard span. This difference is designed to make aircraft neutral point (NP) moves backward. In order to achieve good longitudinal stability, the center of gravity must be located in front of the neutral point. The center of gravity of this configuration is about 20 mm in front of NP. Aircaft wing is located below the fuselage to anticipate the collision between vertical stabilizer and wing when in rapid deploying proccss. Downwash effect of the canard will be investigated later.

There are some parameters that affect the performance of tandem wing UAV such as the vertical gap between canard and wing denoted as y_L and y_R, distance between canard-wing leading edge denoted as d, small vertical gap between left and right lifting surfaces shown in figure 2, and the size of vertical stabilizer that significantly affects the lateral-directional stability of aircraft.

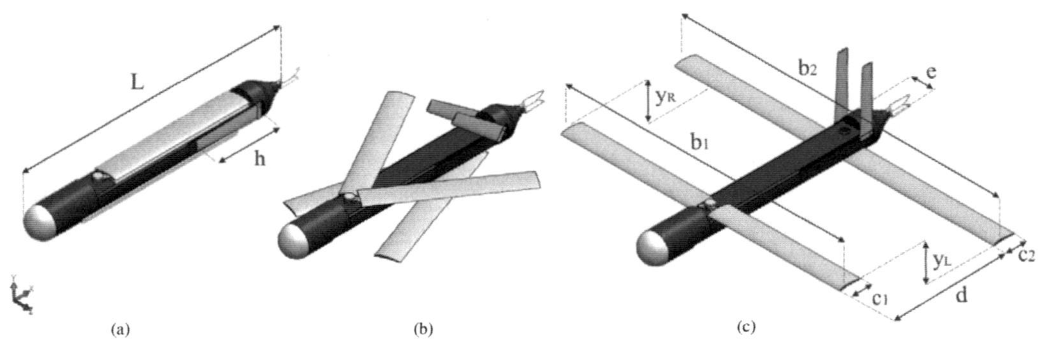

Figure 1. Deploying process : (a) Folded Condition (b) Transition Condition (c) Expanded Condition

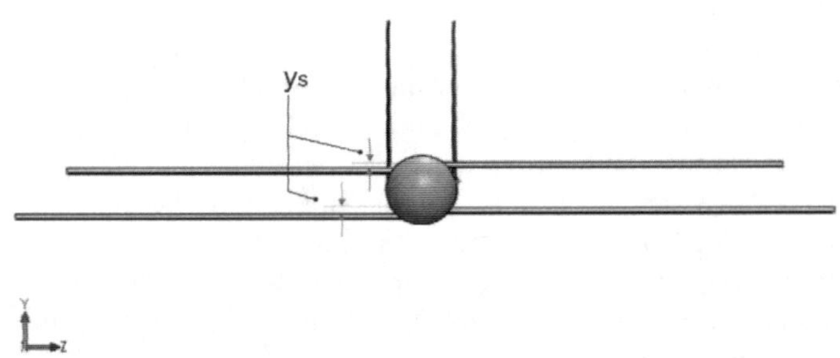

Figure 2. Front view : small vertical gap between left and right lifting surfaces

Table 1. UAV Specification

Definition	Value
Length ***L***	1124 mm
Chord Length ***c1***	100 mm
Chord Length ***c2***	100 mm
Canard Span ***b1***	1318 mm
Wing Span ***b2***	1508 mm
Vertical Stabilizer ***h***	300 mm
Canard-Wing LE distance ***d***	635 mm
Canard-Wing Gap (Left) y_L	75 mm
Canard-Wing Gap (Right) y_R	75 mm
Left-Right Lifting Surfaces Vertical Gap y_S	11 mm
Distance among Vertical stabilizers	118 mm
Canard Airfoil	NACA 6408
Wing Airfoil	NACA 6408
Vertical Stabilizer Airfoil	NACA 0010
MTOW	3 kg
Cruise Speed	25 m/s
Altitude	100 m
S ref	2.69 m^2

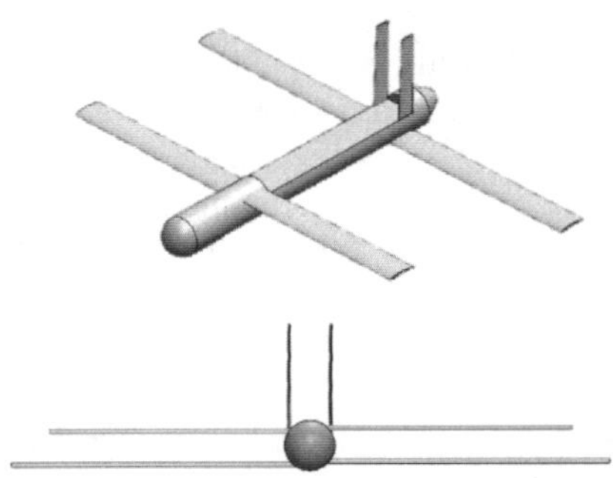
(a)

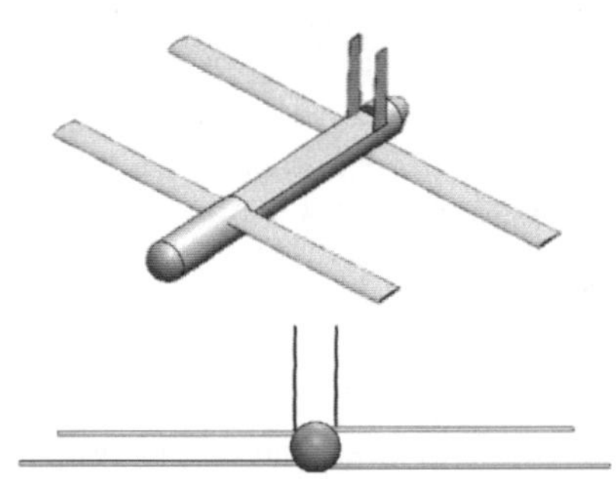
(b)

Figure 3. Two types UAV configuration: (a) Model 1 (b) Model 2.

Because of folding wing mechanism, the wing and canard must be separated at symmetric plane of UAV. The left and right wings are joined through wing spar and folding mechanism. Fig. 3 shows two types of configuration in placing lifting surfaces. Model 1 places left canard higher than right canard and also left wing higher than right wing. Meanwhile, model 2 places left wing lower than right wing with the same canrd position as model 1. In this study, the effect of these two different configurations are investigated whether there is difference on lift distribution between left and right section.

IOP Conf. Series: Journal of Physics: Conf. Series **1005** (2018) 012015 doi:10.1088/1742-6596/1005/1/012015

2. Numerical Simulation Method

The numerical simulation is conducted using Ansys Software with CFX solver and based on finite volume method. General equation commonly used to represent compressible flow behavior is Navier-Stokes equation which is written as

$$\rho\left[\frac{\partial\mu}{\partial t}+(u\nabla)u\right]=-\nabla p+\mu\nabla^2 u+\frac{1}{3}\mu\nabla(\nabla u)+\rho F \quad (1)$$

Where ρ is the fluid density, μ is the dynamic viscosity, t is the time , u is the fluid velocity vector, and F is the body force vector acting on unit mass of fluid[1]. Because of the estimated cruise speed is about 25 m/s with operating altitude 100 m, the value of Mach number is about 0.07. This small Mach number drives to incompressible flow assumption. For incompressible flow the density is constant

$$\frac{\partial\rho}{\partial t}=0 \quad (2)$$

The continuity equation of fluid can be expressed as

$$\nabla(\rho u)+\frac{\partial\rho}{\partial t}=0 \quad (3)$$

From equation (1),(2), and (3) , the general form of Navier-Stokes equation for incompressible flow can be written as

$$\rho\left[\frac{\partial\mu}{\partial t}+(u\nabla)u\right]=-\nabla p+\mu\nabla^2 u+\rho F \quad (4)$$

In order to achieve good accuracy of drag due to turbulent viscosity , two transport equation are solved for turbulence kinetic energy, k, and turbulence dissipation rate,ε[4]. k- ε turbulence model is the most common model used in CFD to simulate mean flow characteristics for turbulent flow condition.

Rectangular bounding box is chosen for computational domain. The size of the domain is 4.2 m x 3 m x 6 m and the maximum box length is approximately 60 times of the chord. In order to choose appropriate node number, the grid test is performed for several nodes number. The node numbers are from 6.5 x 10^5, 8.2 x 10^5, 1.12 x 10^6,1.13 x 10^6, 1.133 x 10^6, 1.19 x 10^6. Figure 5 shows the effect of node number on lift to drag ratio. Greater node number results in converging value and more accurate.

The computation domain and mesh are shown in fig. 4. In order to capture boundary layer flow, the inflation method is used. Further, in this simulation, flow material is defined as ideal gas. The boundary condition (BC) setup contains inlet BC with 25 m./s of flight velocity at zero angle of attack, boundary condition at ooutlet is defined using static pressure 99700 Pa defined from UAV operating altitude, boundary condition at wall is defined as no slipping wall, and far field regions are defined as opening.

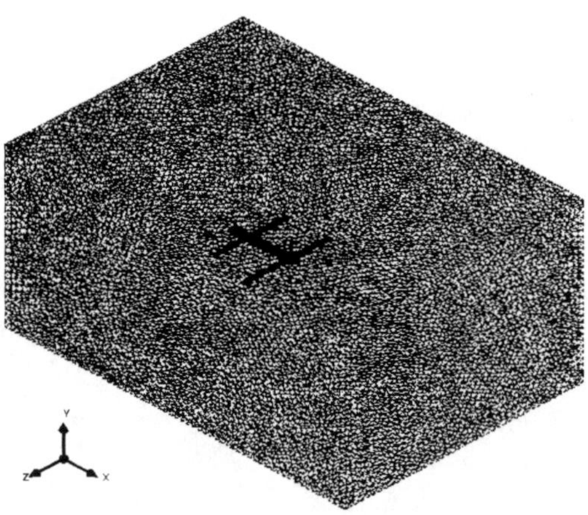

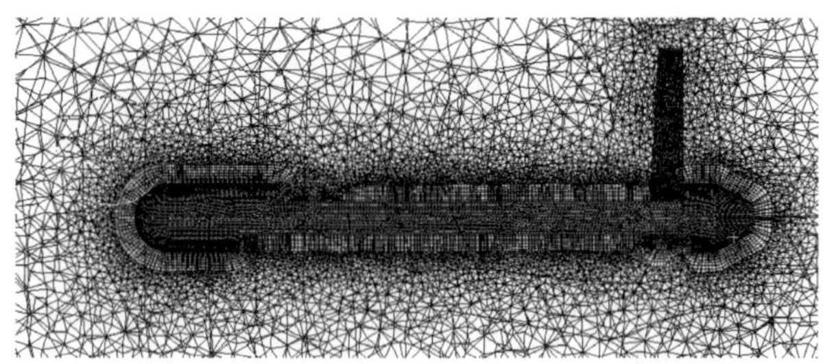

Figure 4. Computation domain and mesh

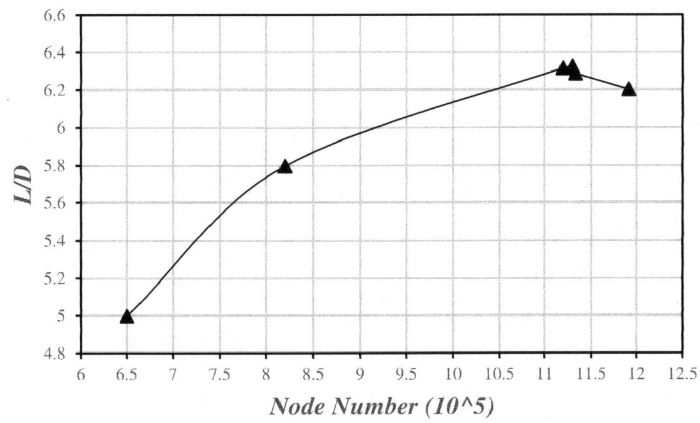

Figure 5. The effect of node number on Lift to drag ratio

3. Results and Discussions

The numerical simulation on aerodynamic characteristics of two different models with varying angle of attack and with the same boundary condition are presented in figure 6 (a) and (b).The lift and drag coefficient both for two models are almost overlapping. The small difference of lifting surface gap has negligible effect on the lift and drag characteristic of the aircraft. In contrast, there is difference value of roll moment between two models as shown in figure 6 (c). Roll moment of model 1 tends to have greater value than model 2. Both two models have positif roll moment. Roll moment of model 1 keeps increasing until it reaches 12 degree with maximum value 0.36 N.m. Meanwhile, the roll moment of model 2 reaches the maximum value at 6 degree and the magnitude is about 0.25 N.m. As a whole view, model 2 has lower roll moment. At this perspective, lifting surfaces arrangement of model 2 is better than model 1 because it is closer to zero roll moment. Huge roll moment can cause lateral-directional instability of an aircraft.

At small angle of attack, both configurations do not visualize any downwash effect generated by canard flow as shown in figure 7. Figure 8 shows streamline velocity for high AoA (9 deg). It is clearly seen that the flow from canard does not give any wash flow effect to the wing. From this result, both configuration is good enough to minimize the canard-wing flow interaction due to downwash effect.

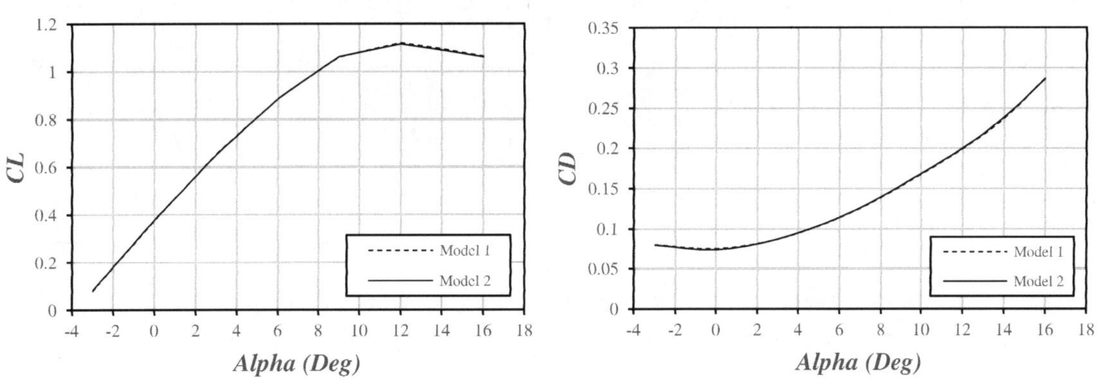

(a) (b)

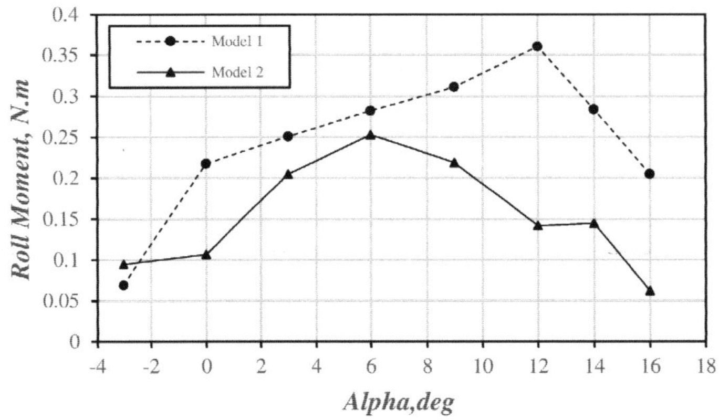

(c)

Figure 6. Comparison of the Aerodynamic characteristic for both model 1 and model 2 : (a) Lift Coefficient (b) Drag coefficient (c) Roll moment

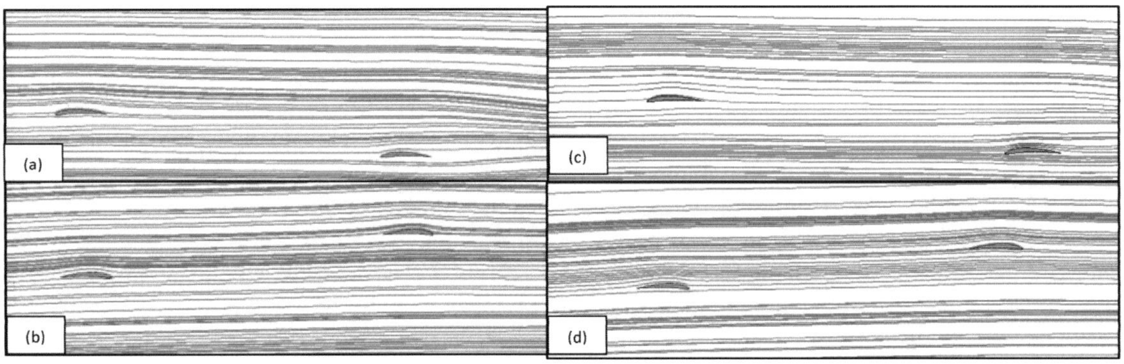

Figure 7. Comparison of the Velocity streamline for both model 1 and model 2 at zero AoA: (a) Model 1 left view (b) Model 1 right view (c) Model 2 left view (d) Model 2 right view

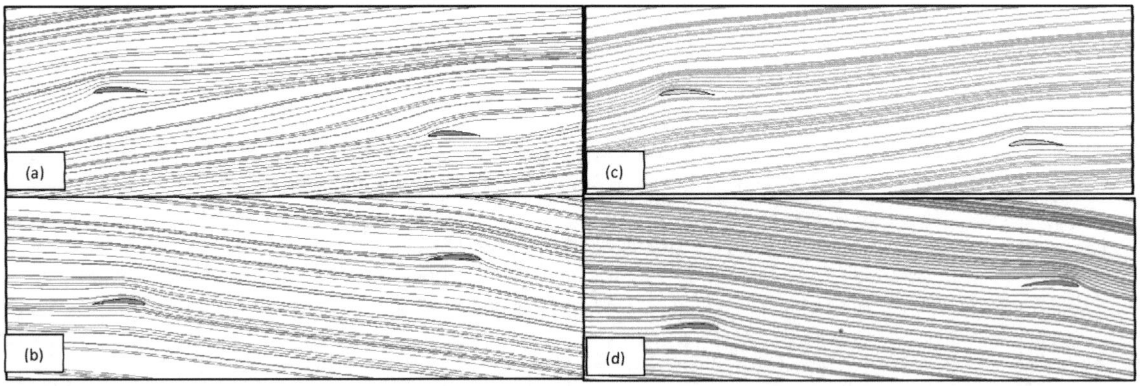

Figure 8. Comparison of the Velocity streamline for both model 1 and model 2 at 9 degree AoA: (a) Model 1 left view (b) Model 1 right view (c) Model 2 left view (d) Model 2 right view

4. Conclusions or Concluding Remarks

For small gap (11 mm) between left and right lifting surfaces, there is no significant differences in lift and drag coefficient for various angle of attack. Model 2 is better than model 1 because it generates less roll moment.

Acknowledgement

The authors woud like to thank The Indonesian Ministry of Education Directorate General of Higher Education (DIKTI) for financial support and Aksantara UAV Research & Development ITB for providing research laboratory.

References

[1] Liang Gao, Changle Li, Hongzhe Jin, Yanhe Zhu, Jie Zhao and Hegao Cai. Aerodynamic Characteristics of a Novel Catapult Launched Morphing Tandem-Wing Unmanned Aerial Vehicle. *Advances in Mechanical Engineering* 2017, vol. 9.

[2] Y. Zhou, A.M. Anvar. Design and Simulation of New Versions of Tube Launched UAV. *21st International Congress on Modelling and Simulation,* Australia, 2015.

[3] Justin Ott, Daniel Biezad. Design of a Tube-Launched UAV. *AIAA 3rd "Unmanned Unlimited" Technical Conference*,Chicago,2004.

[4] Tony S. Tao. *Design and Development of a High-Altitude, In-Flight-Deployable Micro-UAV.* Cambridge : MIT Lincoln Laboratory, 2010.

[5] Rival D, Hass G and Tropea C. Recovery of energy from leading-and trailing-edge vortices in tandem-airfoil configurations. J Aircr 2011; 48: 203–211.

[6] Rhodes MD and Selberg BP. Benefits of dual wings over single wings for high-performance business airplanes. J Aircr 1984; 21: 116–127.

IOP Conf. Series: Journal of Physics: Conf. Series **1005** (2018) 012016 doi:10.1088/1742-6596/1005/1/012016

Aerodynamic Efficiency Analysis on Modified Drag Generator of Tanker-Ship Using Symmetrical Airfoil

Starida Moranova[1) *], Rahmat Hadiyatul A., S.T.[1)], Indra Permana S., S.T.[2)]

[1)] Faculty of Mechanical and Aerospace Engineering, Institut Teknologi Bandung, Indonesia
2) PT Regio Aviasi Industri, Indonesia

*staridamaronova@gmail.com

Abstract. Time reduction of tanker ship spent in the sea should be applied for solving problems occured in oil and gas distribution, such as the unpunctuality of the distribution and oil spilling. The aerodynamic design for some parts that considered as drag generators is presumed to be one of the solution, utilizing our demand of the increasing speed. This paper suggests two examples of the more-aerodynamic design of a part in the tanker that is considered a drag generator, and reports the value of drag generated from the basic and the suggested aerodynamic designs. The new designs are made by adding the NACA airfoil to the cross section of the drag generator. The scenario is assumed with a 39 km/hour speed of tanker, neglecting the hydrodynamic effects occured in the tanker by cutting it at the waterline which separated the drag between air and water. The results of produced drag in each design are calculated by Computational Fluid Dynamic method.

1. Introduction

A lot of problems happenned when distributing offshore and oil by a tanker ship from one land to another, these are the off-shedule distribution and the large amount of oil spill during shipping. If the time spent in the water could be reduced, hopefully there is such an efficient way to minimalized these problem. One solution to decrease the wasted time of tanker ship in the water is by reducing the drag that could be occured around the body of the ship.

Angle-edged body like prism have a 1.14 coefficient of drag, larger compared to streamlined body which is 0.45 [11]. A part, located in the deck of the basic tanker design in figure 1, considered as drag generator regarding to its sharp edges of shape. This motivates the making of new design, smooth-edged shape, in order to reduce drag coefficient of the tanker.

Figure 1 Deck of tanker ship basic design

The objective of this paper is to present the effect of aerodynamic shapes in the deck of a tanker ship to the total drag produced in the entire body of tanker. The drag generator will be replaced by a streamline-shaped airfoil to reach the need of aerodynamic efficiency. The drag generated on each design will be analysed using ANSYS software. Tanker ship model used is a basic oil tanker, created by Harris Nubly (downloaded from grabcad.com) [12], length 500 m.

2. Design and Simulation

Assumptions

Assumptions used in the design process and simulation are:

1. The basic model used is a 500 meter length tanker, scaled 1:500. The result calculated is not the real value, but this paper will only focus on the comparison of drag generated between each design.
2. Simulation will only analyse the part of tanker body which in contact with the air as fluid. The tanker is cut at the imaginary line which separate between surface contacted with air and water (next will be called *waterline*) [6].
3. The analysis uses a condition when the tanker ship floats on height of water 27.5 m above the lowest surface of tanker body. In this scaled 1:500 analysis, the waterline is 55 mm height.

Airfoil Selection

Thinner camber and closer camber position towards the leading edge of the airfoil will result to lower drag coefficient obtained [1]. These will be put in consideration of deciding airfoil parameter. We choose the lowest number maximum camber and maximum camber position as possible, which is zero. The thickness is then adjusted to the deck of the tanker. Finally, the selected airfoil is NACA 0024, which parameters are shown in table 1.

Table 1. Airfoil Parameter

Max Camber	0%
Max camber position	0%
Max thickness	24%

IOP Conf. Series: Journal of Physics: Conf. Series **1005** (2018) 012016 doi:10.1088/1742-6596/1005/1/012016

Design Modelling

The design concept is to insert the geometry of airfoil and align the shape of the drag generator constraining the shape of NACA 0024 using CATIA V5. The three-view drawing of basic tanker ship design is shown in figure 2.

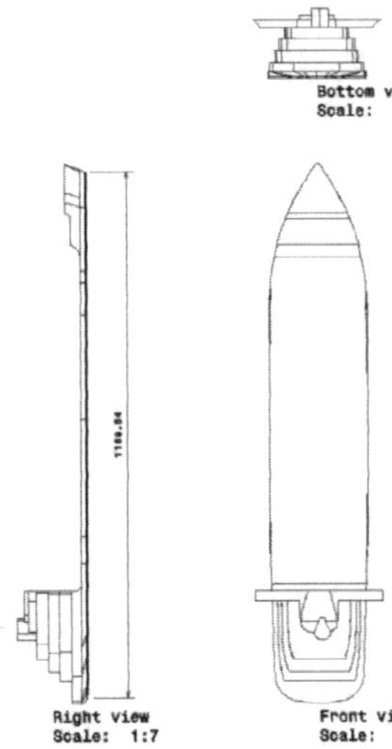

Figure 2. Three view drawing of basic tanker design.

There are two suggested designs. The first design uses airfoil aligning the whole drag generator surface (next will be mentioned as Design 1). Three view drawing of Design 1 is shown in figure 3 below.

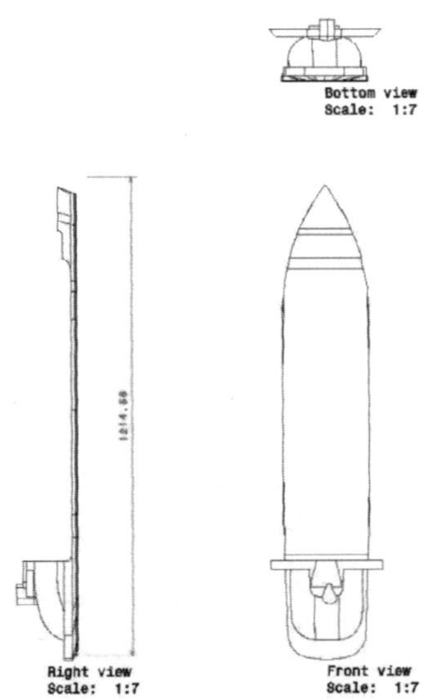

Figure 3. Three view drawing of Design 1

The second design uses airfoil aligning each angle-edged shape, like three stairs on the deck (next will be mentioned as Design 2). Three view drawing of Design 2 is shown in figure 4 below.

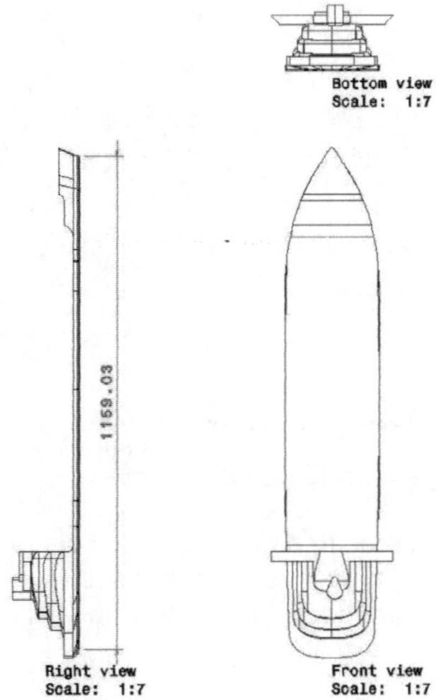

Figure 4. Three view drawing of design 2.

Simulation

The simulation process begins by importing the CAD model as the geometry from CATIA V5 into ANSYS Workbench. The geometry will be given an enclosure as a boundary layer to define the airflow around the tanker. In analyzing the force acting on the vehicle in an unbound fluid domain, ideally an infinite large fluid domain is needed, which is not practicable in CFD and even in experiment. According to reference [2], [3], the assumption used for an unbound fluid domain is 20 times the length, width, and height of the object. Thus, the dimensions of enclosure surrounding the tanker are shown in table 2.

Table 2. Dimensions of enclosure (fluid domain)

Length, y axis (m)	20
Width, x axis (m)	4
Height, z axis (m)	2.5

The negative of z axis is not included, since the hydrodynamic effect is neglected in this analysis. The enclosure is then coincide the symmetry plane x-y, as shown in figure 3.

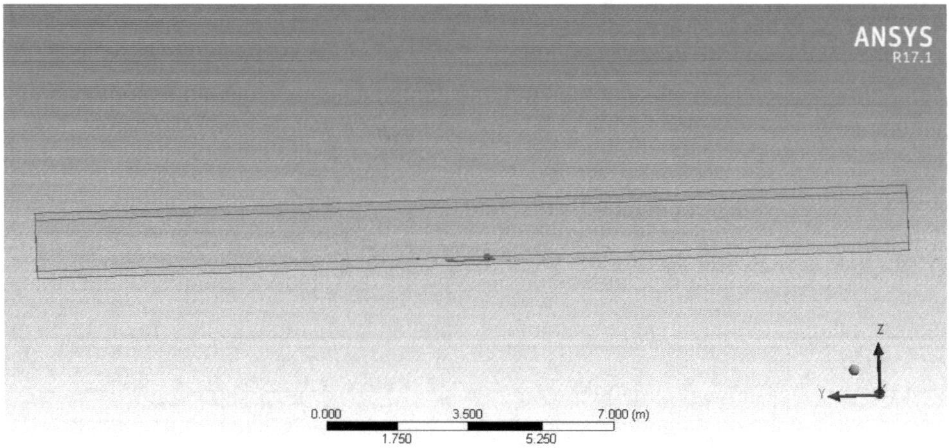

Figure 5 Boundary layer

To start the simulation, the model needs to be meshed finely to get the most accurate result. The mesh is a collection of elements which serve to break-up the CAD model into smaller structural pieces - each piece (or element) can be thought of as a little spring with a three-dimensional spring equation, and the collection of elements represents a structure.

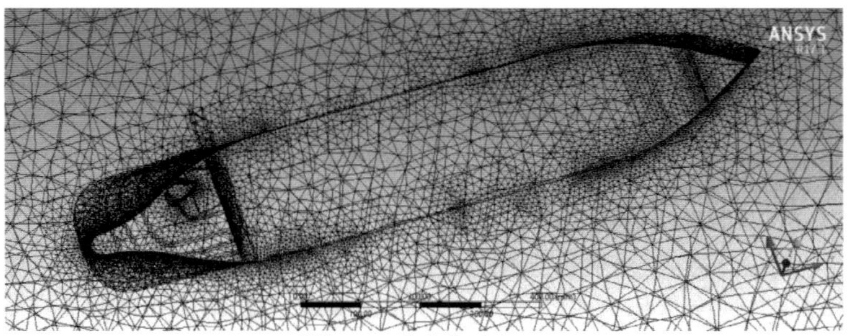

Figure 6 Meshing

If all the above processes are done, setup the simulation with the configurations that needed to be changed, listed in Table 3.

Table 3. Setup configuration

Fluid Material	Air Ideal Gas
Reference Pressure	0 atm
Normal Speed	10.833 ms^{-1}
Turbulence	Medium (Intensity = 5%)
Opening Relative Pressure	1 atm
Relative Average Static Pressure	1 atm

When the setup is finished, the model is ready for simulation using CFX Solver Manager. There are three types of models that are analyzed and each model is set equitably according to all the processes described above.

3. Results and Discussions

The final results of drag generated on each design are shown below:

- Basic tanker model : 12.767 Newton
- Design 1 : 3.491 Newton
- Design 2 : 2.021 Newton

Or the ratio of 63:17:10.

Validation

As shown in this paper, we use CFX from ANSYS 17.0 to solve and analyze the flow of the fluid surrounding the tanker ship that we modify. ANSYS 17.0 is a Computer Aided Engineering software that is used for analyzing variety of disciplines such as finite element analysis, structural analysis, computational fluid dynamics, and many others. In our study, we are using this sofware because it is commonly used to define the characteristics and properties of the air flow surrounding the model in a virtual environment which we want to analyze.

There are two evidences that can validate the method of utilizing ANSYS CFX that we use in this analysis. First, there are references with similar method that already proven and used worldwide with many varieties in terms of models [5], [6], [7]. Secondly we can see from the result of the CFX whether it is convergent or divergent.

From the result of simulations, the drag plots during iteration are shown in figure 7, 8, and 9.

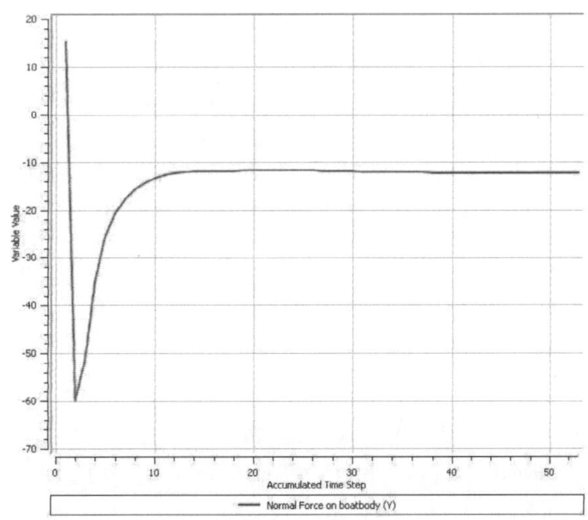

Figure 7. Normal force (drag) generated on tanker body for basic design

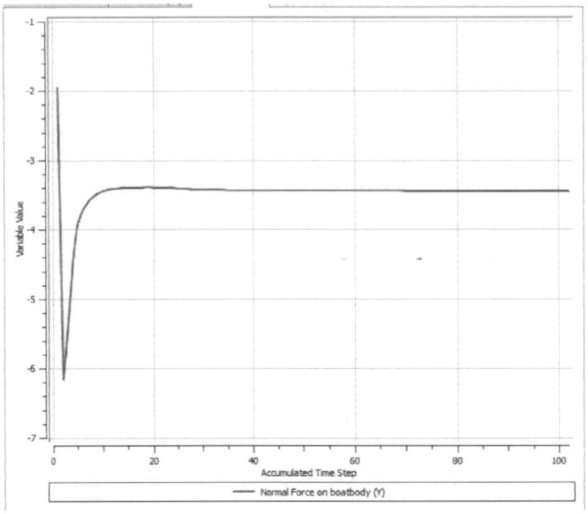

Figure 8 Normal force (drag) generated on tanker body for Design 1

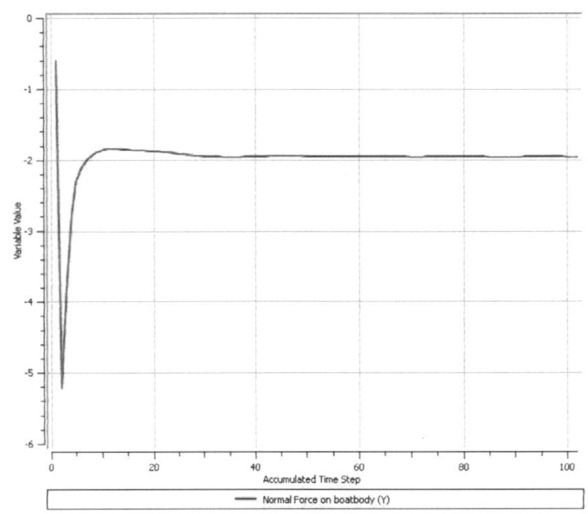

Figure 9 Normal force (drag) around tanker body for Design 2

4. Conclusions or Concluding Remarks

Conclusion

Results of calculation describes that the ratio of drag generated in Basic Design, Design 1 and Design 2 is 63:17:10. In other words, with the modified drag generator using symmetrical airfoil, the tanker ship with Design 1 configuration manages to gain 72,65% reduction from Basic Design (or 9,3 Newton), and with Design 2 configuration reaches 84,17% reduction from Basic Design (or 10,746 Newton). This percentage shows that the modified tanker ship can get less burden from the airflow which will lead to a more profitable cruise in a particular depth of water.

Acknowledgements

This work would have not been finished without the support of everyone that helped the authors along the way. We would like to thank Mr. Ony Arifianto, Ph.D. as our mentor. We would also like to thank FTMD ITB who allowed us to use the computational facility of the department and all of our friends from Aeronotika & Astronotika ITB such as Inggi, Rio, Fadli, Njet, Darda,, Tasya, Horas, Bembi, and Agas for providing a full support in our work.

References

[1] Abbott, Ira H. 1959. *Theory of Wing Sections Including Summary of Airfoil Data.* New York: Dover Publications, Inc.

[2] Chin, Cheng Siong. 2013. *Computer-Aided Control Systems Designs: Practical Applications Using MATLAB and Simulink.* Boca Raton: Taylor & Francis Group.

[3] Patel, Rocky and Ramani, Satyen. Determination of Optimum Domain Size for 3D Numerical Simulation in ANSYS CFX. *International Journal of Innovative Research in Science, Engineering and Technology, Vol. 4, Issue 6, June 2015.*

[4] Priyadarshini, J.Sita, Abhinav, A.V.S., Chandra, B. Sharath, and Swathi, R.S. Use of Aerodynamic Lift in Increasing the Fuel Efficiency of Heavy Vehicle. *IOSR Journal of Mechanical and Civil Engineering (IOSR-JMCE), Vol. 12, Issue 3 Ver. I, July-August 2015.*

[5] Latorre, Dr. Robert. Ship Hull Drag Reduction Using Bottom Air Injection. *Ocean Engineering, Vol. 24, No. 2, November 1995.*

[6] Choi, Jin-Keun, and Chahine, Georges L. Numerical Study on the Behavior of Air Layers Used for Drag Reduction. *28th Symposium on Naval Hydrodynamics, September 2010.*

[7] Padagannavar, Praveen and Bheemanna, Manohara. Automotive Computational Fluid Dynamics Simulation of a Car Using ANSYS. *International Journal of Mechanical Engineering and Technology (IJMET) Vol. 7, Issue 2, March-April 2016.*

[8] Tejaas, Vishan Ravi. Low Reynolds Number Flow Over Thick Symmetrical Airfoil with Dimple (NACA 0018). *International Journal of Scientific and Research Publications, Vol. 6, Issue 9, September 2016.*

[9] Mosaad, M. A., Gaafary, M.M., El-Kilani, H., and Amin, I.A.. Effect of Airfoil Camber on WIG Aerodynamic Efficiecy. *Port Said Engineering Research Journal.*

[10] http://airfoiltools.com/airfoil/naca4digit

[11] https://www.grc.nasa.gov/www/k-12/airplane/shaped.html

[12] https://grabcad.com/library/simple-oil-tanker-vessel-2

IOP Conf. Series: Journal of Physics: Conf. Series **1005** (2018) 012017 doi:10.1088/1742-6596/1005/1/012017

Implementation of Dryden Continuous Turbulence Model into Simulink for LSA-02 Flight Test Simulation

Teuku Mohd Ichwanul Hakim[1], Ony Arifianto[2]

Abstract. Turbulence is a movement of air on small scale in the atmosphere that caused by instabilities of pressure and temperature distribution. Turbulence model is integrated into flight mechanical model as an atmospheric disturbance. Common turbulence model used in flight mechanical model are Dryden and Von Karman model. In this minor research, only Dryden continuous turbulence model were made.

Dryden continuous turbulence model has been implemented, it refers to the military specification MIL-HDBK-1797. The model was implemented into Matlab Simulink. The model will be integrated with flight mechanical model to observe response of the aircraft when it is flight through turbulence field. The turbulence model is characterized by multiplying the filter which are generated from power spectral density with band-limited Gaussian white noise input.

In order to ensure that the model provide a good result, model verification has been done by comparing the implemented model with the similar model that is provided in aerospace blockset. The result shows that there are some difference for 2 linear velocities (v_g and w_g), and 3 angular rate (p_g, q_g and r_g). The difference is instantly caused by different determination of turbulence scale length which is used in aerospace blockset. With the adjustment of turbulence length in the implemented model, both model result the similar output.

Introduction

Along with national needs in the area of airborne remote sensing and surveillance [1], Aeronautics Technology Center of LAPAN in cooperation with TU Berlin, Germany, develop a project to investigate the implementation of an Electronic Flight Control System (EFCS) into a light utility aircraft. The project itself started from 2014 and planned to be finish in 2019 [1]. Goal of the EFCS development in the future is to provide the control system that can help and even replace the pilot in missions that are extremely difficult, long or dangerous. In order to fulfill that goal, the EFCS have to have full authority and highly reliable. Furthermore, the EFCS shall be able to fly the aircraft to follow predefined trajectories with high precision, to stabilize the aircraft and measurement system during flight in calm and turbulence atmosphere. In the project a demonstrator aircraft which called as an LSA-02 is developed in order to investigate the flight control function and flight control laws [2]. The LSA-02 will

[1] LAPAN, Bandung Institute of Technology, TU Berlin
[2] Bandung Institute of Technology

use to perform research in the area of UAV technology, especially for new flight control functions. During the testing, safety pilot will be on board for safety reason and conduct the experiment.

In the aircraft, the EFCS is installed to supplement the basic mechanical flight control system. When engaged, the EFCS uses the mechanical linkage of the mechanical FCS to command control surfaces via electro-mechanical actuators. The commands are computed by Flight Control Laws (FCL). Basic FCL are implemented in the flight control computer (FCC) of the Basic Electronic Flight Control System (BEFCS). The BEFCS provides all standard autopilot modes [3].

Flight control laws (FCL) testing is an important part in the development of Electronic Flight Control System (EFCS) of an LSA-02. The testing is used to verify and validate the developed FCL against the requirements in order to ensure the functionality and safety. Verification and validation of the FCL can be performed through several type of test e.g. software in the loop simulation (SILS), hardware in the loop simulation (HILS) and flight testing (FT).

On ground test facility such as SILS and HILS, the dynamic behavior of the aircraft is simulated in flight mechanical model. The environmental disturbances e.g. wind, gust and turbulence are also simulated to provide real condition that may be faced by the aircraft during flight testing. It is not sufficient to simulate and to evaluate the aircraft only in calm condition.

Objective of the research is to provide atmospheric turbulence model in the form of Simulink model which has been verified to support the simulation. Later, the model will be integrated to the flight mechanical model to observe response of the aircraft when it is flight through turbulence field and how the FCL response to stabilize the aircraft. Figure 1 shows environment for FCL test, where the atmospheric disturbance integrated to flight mechanical model block.

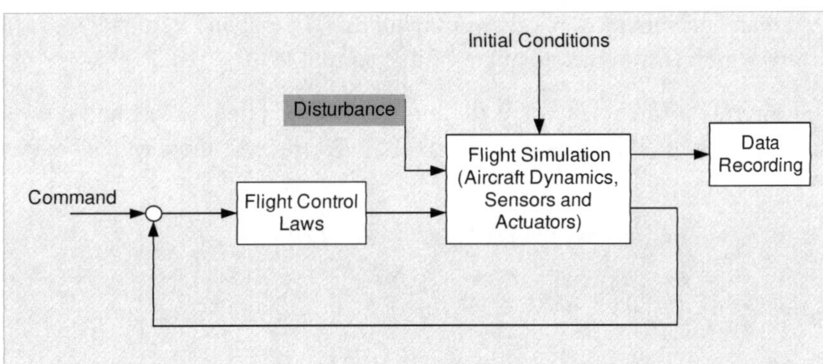

Figure 1 Environment for FCL test [4]

By definition, turbulence is a movement of air on small scale in the atmosphere that caused by instabilities of pressure and temperature distribution in clouds, near ground and in the jet-stream region. Based on [5] and [6], turbulence is defined as a stochastic process which are determined by velocity spectra. The turbulence field is assumed to be visualized as frozen in time and space

There are 2 common continuous turbulence models that are usually used in flight mechanical model, those are:

- Dryden Continuous Turbulence Model; and
- Von Karman Continuous Turbulence Model

IOP Conf. Series: Journal of Physics: Conf. Series **1005** (2018) 012017 doi:10.1088/1742-6596/1005/1/012017

In practical, Dryden turbulence model is used more frequently, it is caused by the Dryden model has simpler form and the ease to produce [7]. The Dryden wind turbulence model, or it is also called as the Dryden gusts, is a mathematical model of continuous gusts which is accepted by the United States Department of Defense to be used in aircraft design and simulation applications.

Atmospheric Turbulence Model Theory

Dryden and Von Karman model are the common turbulence model that are used in flight mechanics. In this case, Dryden model will be used because it is easier to implement in simulation even though both of them tends to similar result. According to [8] there are four assumptions are taken in turbulence modeling for statistical properties of the turbulence

- stationary (independent of time)
- homogeneous (independent of location in space)
- isotropic (independent of direction)
- larger than the airplane (all parts of the airplane are equally affected)

In Dryden turbulence model, the linear and angular velocity components of continuous gusts treated as spatially varying stochastic processes and specifies each component's power spectral density. It is applying shaped noise with known spectral properties as velocity and angle rate perturbations to the body axes of the vehicle, the effect of turbulence is captured during discrete time simulations [9]. The noise spectrum for each of the perturbations is described by a turbulence scale length (L), the airspeed (V), and the turbulence intensity (σ). The turbulence scale length L is the length of turbulence field which expressed in longitudinal, lateral and vertical axis (L_u, L_v and L_w). Turbulence intensity σ is the magnitude of the turbulence which expressed in longitudinal, lateral and vertical (σ_u, σ_v and σ_w). It has dimension ft/s or m/s which is root-mean-square of its magnitude.

Figure 3 shows the aircraft with velocity of V fly into a turbulence field with sine-type vertical velocity distribution with the length of L in longitudinal axis (L_u). To traverse the turbulence, the aircraft takes time

$$T = \frac{L}{V} \tag{1}$$

The aircraft experiences the turbulence wave with the temporal frequency of

$$\omega = \frac{2\pi}{T} = \frac{2\pi}{(L/V)} \tag{2}$$

The turbulence wave not just consist of one frequency, whole spectrum are exist from low to high frequency. Turbulence spectral density that are modeled in Dryden function expressed in non-dimensional frequency form as:

$$\omega T = \omega \frac{L}{V} \tag{3}$$

IOP Conf. Series: Journal of Physics: Conf. Series **1005** (2018) 012017 doi:10.1088/1742-6596/1005/1/012017

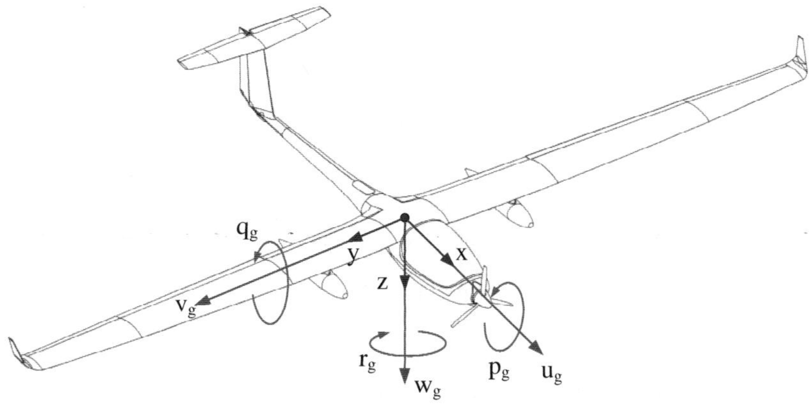

Figure 2 Axes definition of turbulence linear velocity and angular rate follows body axis of the aircraft

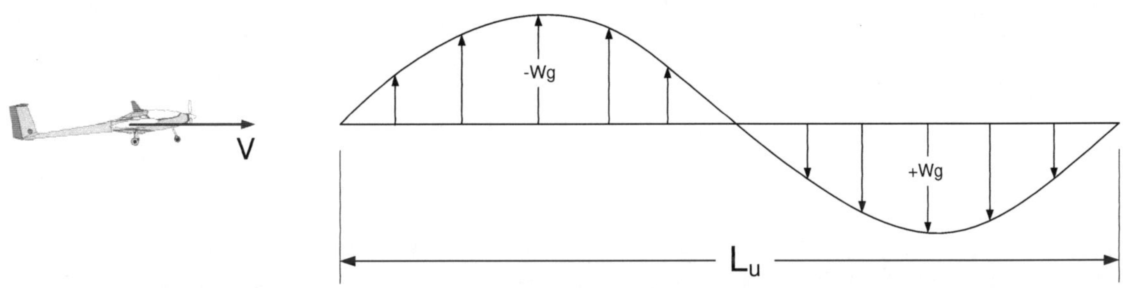

Figure 3 Illustration of the aircraft flies into sine-type turbulence field

Power Spectral Densities

Dryden turbulence model is characterized by following power spectral densities (Φ) as the function of temporal frequency. Power spectral densities for three linear velocity component (u_g, v_g, w_g) are:

$$\Phi_u(\omega) = \frac{2\sigma_u^2 L_u}{\pi V} \cdot \frac{1}{1+\left(L_u \frac{\omega}{V}\right)^2} \tag{4}$$

$$\Phi_v(\omega) = \frac{2\sigma_v^2 L_v}{\pi V} \cdot \frac{1+12\left(L_v \frac{\omega}{V}\right)^2}{\left[1+4\left(L_v \frac{\omega}{V}\right)^2\right]^2} \tag{5}$$

$$\Phi_w(\omega) = \frac{2\sigma_w^2 L_w}{\pi V} \cdot \frac{1+12\left(L_w \frac{\omega}{V}\right)^2}{\left[1+4\left(L_w \frac{\omega}{V}\right)^2\right]^2} \tag{6}$$

Power spectral densities for three angular rate components (p_g, q_g, r_g) are:

$$\Phi_p(\omega) = \frac{\sigma_w^2}{VL_w} \cdot \frac{0.8\left(\frac{2\pi L_w}{4b}\right)^{1/3}}{1+\left(\frac{4b\omega}{\pi V}\right)^2} \tag{7}$$

$$\Phi_q(\omega) = \frac{\pm\left(\frac{\omega}{V}\right)^2}{1+\left(\frac{4b\omega}{\pi V}\right)^2} \cdot \Phi_w(\omega) \tag{8}$$

$$\Phi_q(\omega) = \frac{\pm\left(\frac{\omega}{V}\right)^2}{1+\left(\frac{3b\omega}{\pi V}\right)^2} \cdot \Phi_v(\omega) \tag{9}$$

where:

- $\sigma_u, \sigma_v, \sigma_w$ indicate turbulence intensities
- L_u, L_v, L_w indicate scale length
- b indicates wing span

Continuous Dryden Filter

Continuous Dryden filters are derived from the square roots of the Dryden power spectrum equations as it is expressed in equation 4 - 9. In order to generate turbulence signal that has similar frequency spectrum as Dryden power spectral density in the simulation, the continuous filter is used. To excite waves with full spectrum, white noise (band-limited Gaussian white noise) signal is used. It is passed through the appropriate filters. White noise is a random signal that having equal intensity at different frequencies, giving it a constant power spectral density.

Based on [10] Dryden continuous filter for velocity spectra are expressed as following transfer function:

$$H_u(s) = \sigma_u\sqrt{\frac{2L_u}{\pi V}}\frac{1}{1+\frac{L_u}{V}s} \tag{10}$$

$$H_v(s) = \sigma_v\sqrt{\frac{2L_v}{\pi V}}\frac{1+\frac{2\sqrt{3}L_v}{V}s}{\left(1+\frac{2L_v}{V}s\right)^2} \tag{11}$$

$$H_w(s) = \sigma_w\sqrt{\frac{2L_v}{\pi V}}\frac{1+\frac{2\sqrt{3}L_w}{V}s}{\left(1+\frac{2L_w}{V}s\right)^2} \tag{12}$$

Dryden continuous filter [10] for angular rate spectra expressed as following transfer function:

$$H_p(s) = \sigma_w\sqrt{\frac{0.8}{V}}\frac{\left(\frac{\pi}{4b}\right)^{1/6}}{(2L_w)^{1/3}\left(1+\frac{4b}{\pi V}s\right)} \tag{13}$$

$$H_r(s) = \frac{\mp\frac{s}{V}}{\left(1+\frac{3b}{\pi V}s\right)} \cdot H_v(s) \tag{14}$$

$$H_q(s) = \frac{\mp\frac{s}{V}}{\left(1+\frac{4b}{\pi V}s\right)} \cdot H_w(s) \tag{15}$$

Continuous Dryden filter is a kind of low-pass filter, where the frequency that higher than certain cut-off frequency is eliminated. Cut-off frequency of the filter is determined by the ratio of turbulence scale length to the airspeed.

Application of turbulence model is clearly defined in [5] [6]. In the references there are distinction of the altitude, which are low-altitude and medium/high altitude.

Low-altitude region is defined as the altitude equal to and less than 1000 feet. In this region, following values are defined:

- Turbulence scale length is the function of altitude. According to [6], turbulence scale length for low altitude is defined as follows:

$$2L_w = h \tag{16}$$

$$L_u = 2L_v = \frac{h}{(0.177+0.000823h)^{1.2}} \tag{17}$$

- Turbulence intensities

$$\sigma_w = 0.1W_{20} \tag{18}$$

$$\sigma_u = \sigma_v = \sigma_w \frac{1}{(0.177+0.000823h)^{0.4}} \tag{19}$$

W_{20} is typical values for wind speed at height of 20 feet, it is defined in [5] and [6]. Typical values for W_{20} defined in the Table 1.

Table 1 Typical value for W_{20}

Turbulence Level	**W_{20}**
Light	15 knots
Moderate	30 knots
Severe	45 knots

Medium/high altitude is defined for altitude higher than or equal to 2000 feet. In this region turbulence is assumed isotropic, therefore following values are defined

- Turbulence scale length in medium/high altitude is assumed to be constant. Relation for turbulence scale length in each axes is defined as follow:

$$L_u = 2L_v = 2L_w = 1750\,ft \tag{20}$$

- Turbulence intensities in [5] and [6] are defined in a diagram that contains root-mean-square (RMS) turbulence intensities as the function of altitude and probability of exceedance as it is shown in Figure 4.
  Relation for turbulence intensities for each axes is defined as follow:

$$\sigma_u = \sigma_v = \sigma_w \tag{21}$$

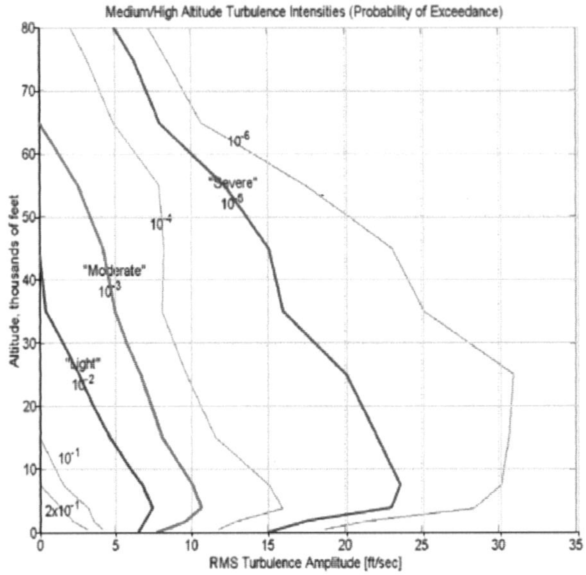

Figure 4 Medium/high altitude turbulence intensities plot compared with [10]

Output of the filter are 3 linear velocity component (u_g, v_g, w_g) and 3 angular rate component (p_g, q_g, r_g). Those output will goes to the flight mechanical model as a disturbance signal.

Model Development

Model for Dryden continuous turbulence was made in Simulink. It consist of input subsystem, scale length subsystem, turbulence intensity subsystem, velocity filter subsystem and angular rate subsystem. Input values and plotting commands are defined in Matlab script. High level structure of the developed turbulence model is shown in Figure 5. Detail information of the developed model can be found in [4].

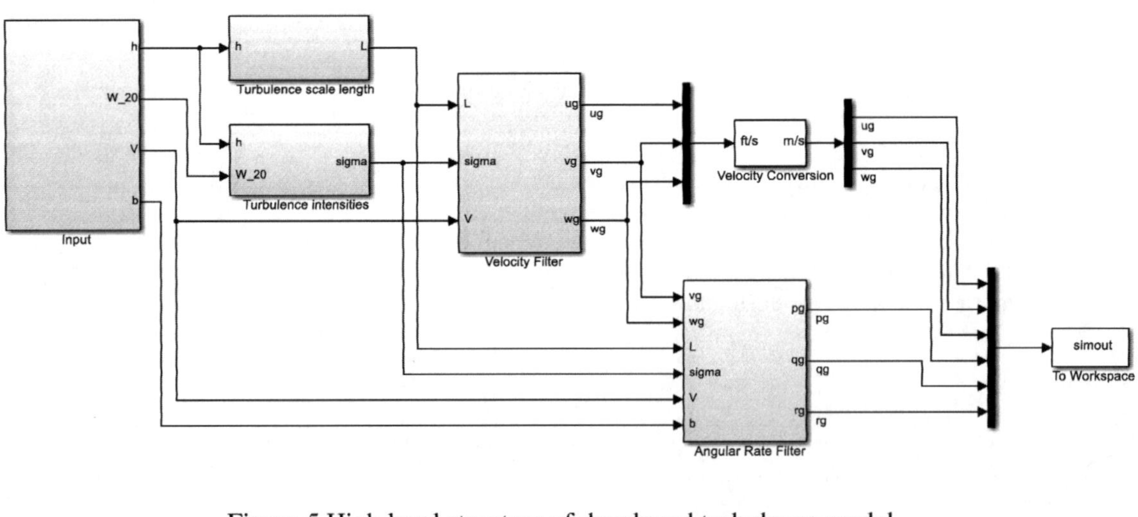

Figure 5 High level structure of developed turbulence model

IOP Conf. Series: Journal of Physics: Conf. Series **1005** (2018) 012017 doi:10.1088/1742-6596/1005/1/012017

In order to ensure that the Simulink models provide a good results, it is need to verify the model with other models that has been proven. Therefore the developed model is compared with Dryden continuous turbulence model which are available in Matlab Aerospace Blockset. Block structure comparison between developed models with Dryden model can be seen in Figure 6. Both model use the same input in order to provide comparable output. Those model are calculated using [6].

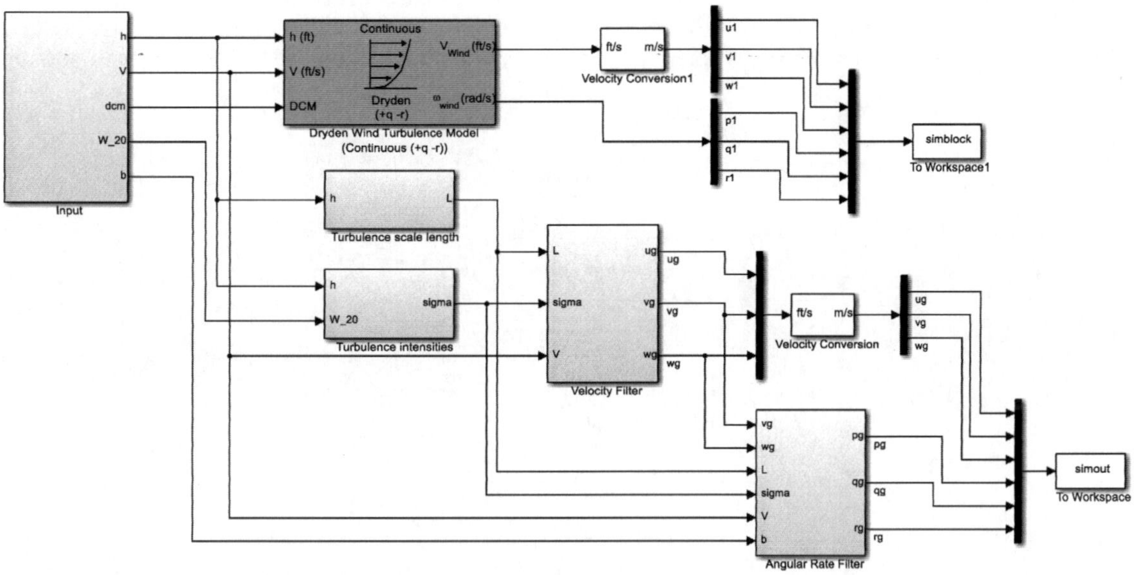

Figure 6 High level structure of comparison model

Result and Discussion

Turbulence model is simulated with the time of 100 seconds. Result of the simulation is described as follow.

Output from both model stored in the workspace and then compared in the same plot to show how close the developed model to the reference. Figure 7 shows the comparison of linear velocity (u_g, v_g and w_g) between developed model and aerospace blockset model. The result shows that output of linear velocity u_g is fit to the output that results from aerospace blockset. But it is not the same for output of v_g and w_g, the magnitude are did not fit to the output of the reference. The same condition also occur for angular rate p_g, q_g and r_g as it is shown in Figure 8. Magnitude of three angular rate of developed model are different from magnitude of reference.

With the same input values given, both model should result the same output. From further analysis finally it is found that the difference between the models are the value of turbulence scale length (Lv and Lw) that are used in calculation. Even though both models are use the same standard [6], in fact [10] use turbulence scale length as it is defined in [5]. Therefore only u_g that fit together, because [5] and [6] has the same definition for Lu.

Reference [10] state that the use of different Lu as it is defined in standard [5] and [6] is acceptable, despite it would result the difference turbulence characteristic. Turbulence length scale has direct impact to the power spectral density and magnitude of the turbulence.

IOP Conf. Series: Journal of Physics: Conf. Series **1005** (2018) 012017 doi:10.1088/1742-6596/1005/1/012017

Furthermore, to ensure that the developed model result the same output as aerospace blockset, then the turbulence scale length in developed model were adjusted. After adjustment in turbulence scale length, the result are shown in Figure 9 and Figure 10. In those figure all output are fit together. As well as for the power spectral density (PSD) for linear velocity and angular rate are fit together. Figure 11 shows the sample for single sided PSD with FFT for u_g and q_g.

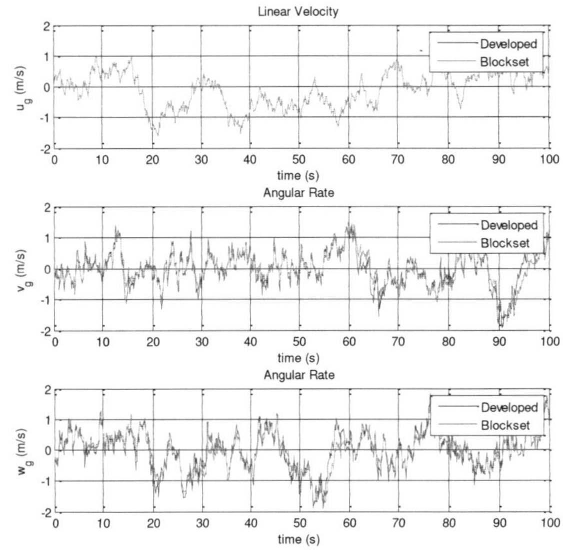

Figure 7 Comparison results for longitudinal linear velocity u_g, v_g and w_g

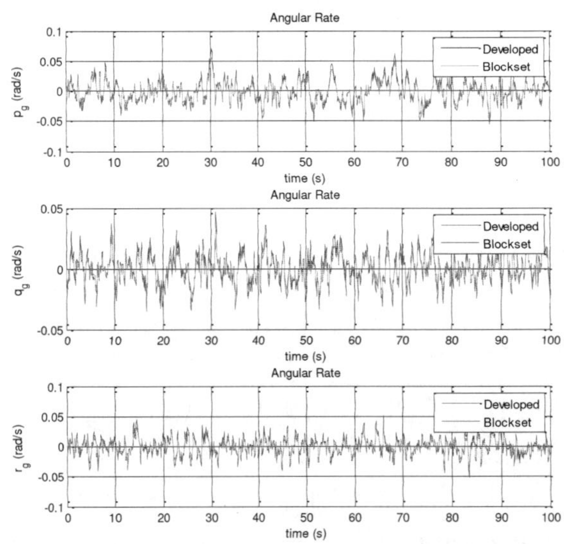

Figure 8 Comparison results for longitudinal angular rate p_g, q_g and r_g

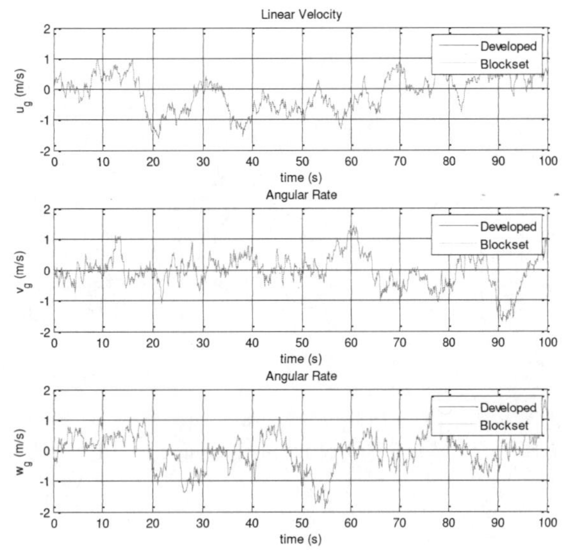

Figure 9 Comparison of linear velocity for adjusted Lv and Lw

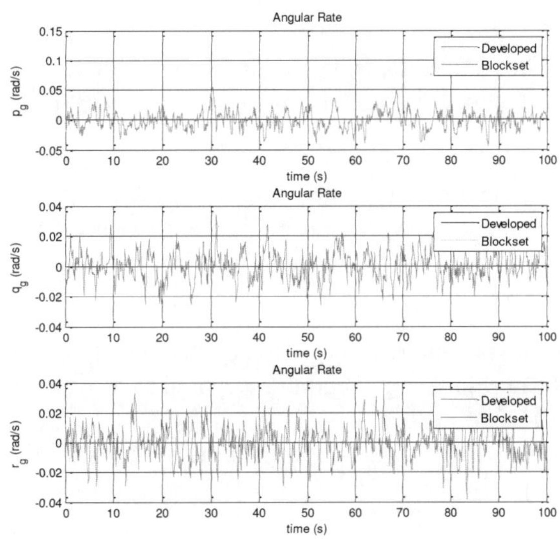

Figure 10 Comparison of angular rate for adjusted Lv and Lw

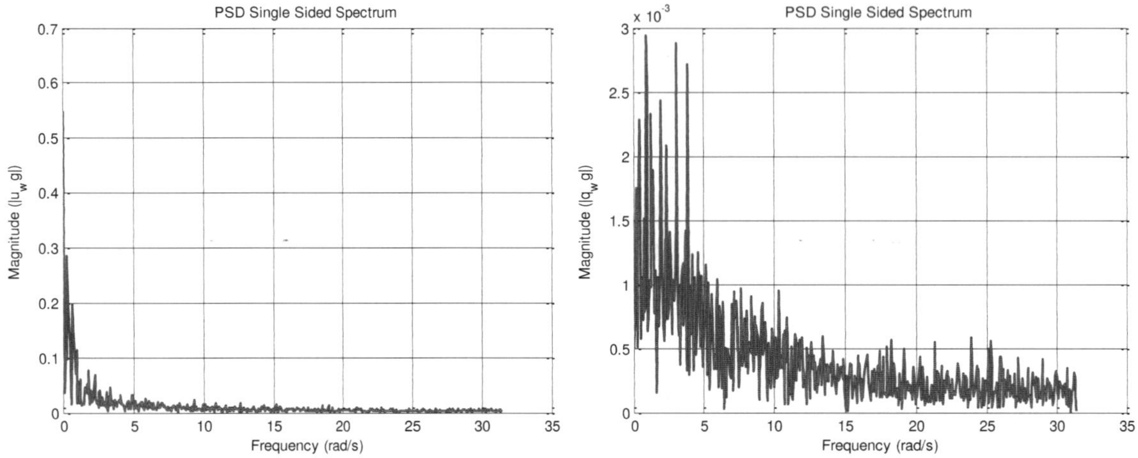

Figure 11 Single sided PSD plot for u_g and q_g using FFT method

Conclusion

Dryden continuous turbulence model based on [6] has been successfully developed for low altitude and medium/high altitude.

Linear interpolation has been introduced to calculate turbulence scale length and turbulence intensities for altitude between 1,000 ft and 2,000 ft to provide value of L and σ for turbulence calculation in those altitude range.

Verification of the Simulink model result the same output in u_g as the model from aerospace blockset. Other 2 linear velocity and 3 angular rate results in different magnitude. The differences caused by different turbulence scale length definition for Lv and Lw as it is determined in [6]. The [10] choose Lv and Lw as it is determined in [5].

The developed turbulence model is quite reasonable to be integrated to the flight mechanical model for FCL testing, because it has been result a good output compared to the reference.

References

[1] Pustekbang LAPAN, "Program Pengembangan LAPAN Light Surveillance Aircraft (LSA)," Pustekbang, 2017. [Online]. Available: http://pustekbang.lapan.go.id/index.php/subblog/pages/2014/13/Light-Surveillance-Aircraft-LSA. [Accessed 27 June 2017].

[2] S. Bahri, "TR LSA 02: Specification and Concept of LSA-02 (Technology Demonstrator)," LAPAN, Berlin, 29.06.2016.

[3] T. M. I. Hakim, "TR LSA 02: BEFCS Definition, Requirements and Concept," LAPAN, Berlin, 2016.

[4] T. M. I. Hakim, "Minor Research Report: Dryden Continuous Turbulence Model," Bandung Institute of Technology, Bandung, 2017.

[5] MIL-F-8785C, " US Military Specification: Flying Qualities of Piloted Airplanes," 5 November 1980.

[6] MIL-HDBK-1797, "US Military Handbook: Flying Qualities of Piloted Aircraft," 19 December 1997.

[7] T. R. Beal, "Digital Simulation of Atmospheric Turbulence for Dryden and Von Karman Models," *Journal of Guidance, Control, and Dynamics,* Vols. 16, No. 1, January - February 1993.

[8] P. H. Zipfel, Modeling and Simulation of Aerospace Vehicle Dynamics, Blacksburg, Virginia : American Institute of Aeronautics and Astronautics, Inc. , 2007.

[9] E. T. King, "Distributed Coordination and Control Experiments on a Multi-UAV Testbed," Massachusetts Institute of Technology, August 20, 2004.

[10] S. Gage, "Creating a Unified Graphical Wind Turbulence Model from Multiple Specifications," in *AIAA Modeling and Simulation Technologies Conference and Exhibit*, Austin, Texas, 11-14 August 2003.

Design of Flight Control Panel Layout using Graphical User Interface in MATLAB

A Wirawan[1, 3] *, T Indriyanto[2]

[1)] Aeronautic Technology Centre, National Institute of Aeronautics and Space (LAPAN), Indonesia
2) Faculty of Mechanical and Aerospace Engineering, Institut Teknologi Bandung, Indonesia
3) Institut für Luft- und Raumfahrt, Technische Universität Berlin, Germany

*adi.wirawan@lapan.go.id

Abstract. This paper introduces the design of Flight Control Panel (FCP) Layout using Graphical User Interface in MATLAB. The FCP is the interface to give the command to the simulation and to monitor model variables while the simulation is running. The command accommodates by the FCP are altitude command, the angle of sideslip command, heading command, and setting command for turbulence model. The FCP was also designed to monitor the flight parameter while the simulation is running.

1. Introduction

In general, an aircraft is controlled by the following principal modes: mechanical control mode and electronic control mode. The electronic control mode can support two types of control, which are automatic control and manual control. In automatic control mode, pilot sends input commands to Electronic Flight Control System (EFCS) using Flight Control Panel (FCP). The FCP also display flight plan sequentially as a function of aircraft position.

The Flight Control Panel (FCP) also has functioned as Human Machine Interface (HMI). In this case, the FCP is placed in the avionic panel in front of the pilot. The FCP has three main functions:

1. To load and transmit initialization data to Flight Control Computer (FCC) to be read by Flight Control Law (FCL). The loaded data are aircraft and weight data, waypoints or flight path pattern and mission data, as well as runway data.
2. To allow the pilot to input data or to manually give commands to EFCS. Commands given by pilot to EFCS are engaged or disengage EFCS in normal condition, various switching command modes, and flight profile commands (airspeed, height, and heading).
3. To display information that the pilot has to observe during flight such as flight condition, EFCS status and visual warnings [9].

The FCP in these research is required to sends input commands to the flight control model and turbulence model and also to display aircraft data model.

2. Study on FCP Layout

LAPAZ Project

LAPAZ Project (abbreviation for ***L**uft-**A**rbeits-**P**lattform für die **A**llegemeine **Z**ivilluftfahrt in German*, translated as an aerial work platform for general civil aviation). The project is intended to develop and demonstrate a reliable, high-precision automatic flight control system for the S15 aircraft to support in flight geo-exploration and surveillance tasks. The partners are German aircraft manufacturer STEMME, the Universität Stuttgart and Technische Universität Berlin (TUB).

Automatic Flight Control Panel (AFCP) of LAPAZ project is an example of human machine interface in aircraft. LAPAZ AFCP allows the pilot to engage and disengage AFCS, to set target values for various flight states (airspeed, altitude, course, etc.), to switch between flight control modes as well as to load predefined waypoints and runway data. [3]

The AFCP consists of a central processor unit, a display, push buttons and rotary encoders as shown in **Figure 1**.

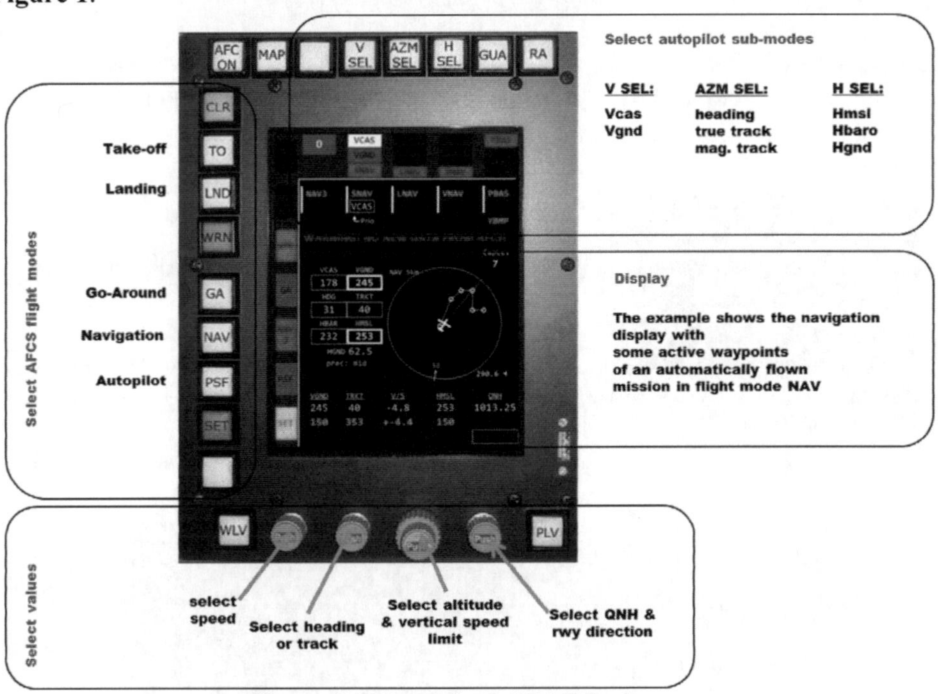

Figure 1. FCP of the LAPAZ Project

TEMO HMI

Another example of dedicated HMI platform is TEMO HMI for cockpit touch screen display as shown in **Figure 2**. The TEMO HMI consists of a Primary Flight Display (PFD), Navigation Display (ND), Vertical Situation Display (VSD) and a Time & Energy Display (TED) [8].

Generally, two types of cockpit applications distinguished by symbology applications and interactive applications. Symbology applications present one-way information to pilot. Typically, symbology displays are data-driven applications. Interactive applications enable the pilot to react on the information shown by means of, for example, a toggle button, push button or combo box. This type of information exchange has an event-driven character [12].

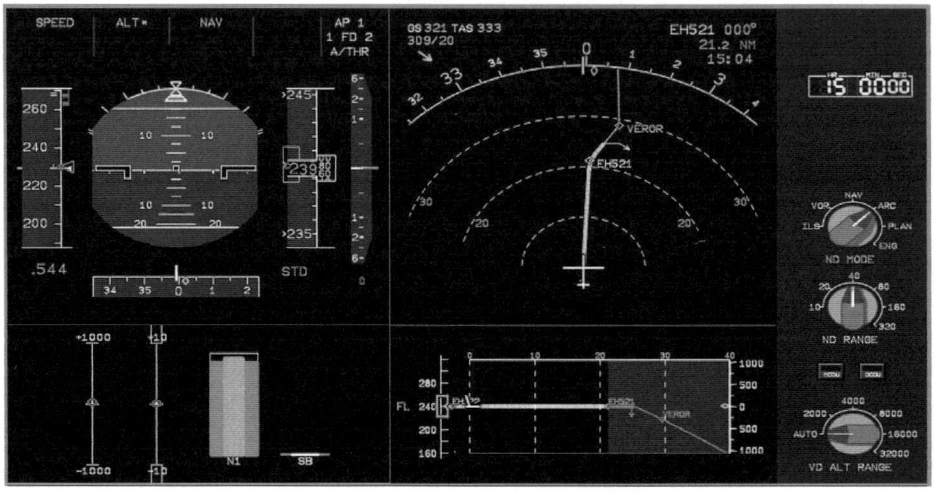

Figure 2. *TEMO cockpit HMI [8]*

HMI for A380 flight test

Another example of HMI (**Figure 3)** is a unique HMI used in flight test and On-board computing for Airbus A380 that has been designed to integrate all test monitoring applications [2].

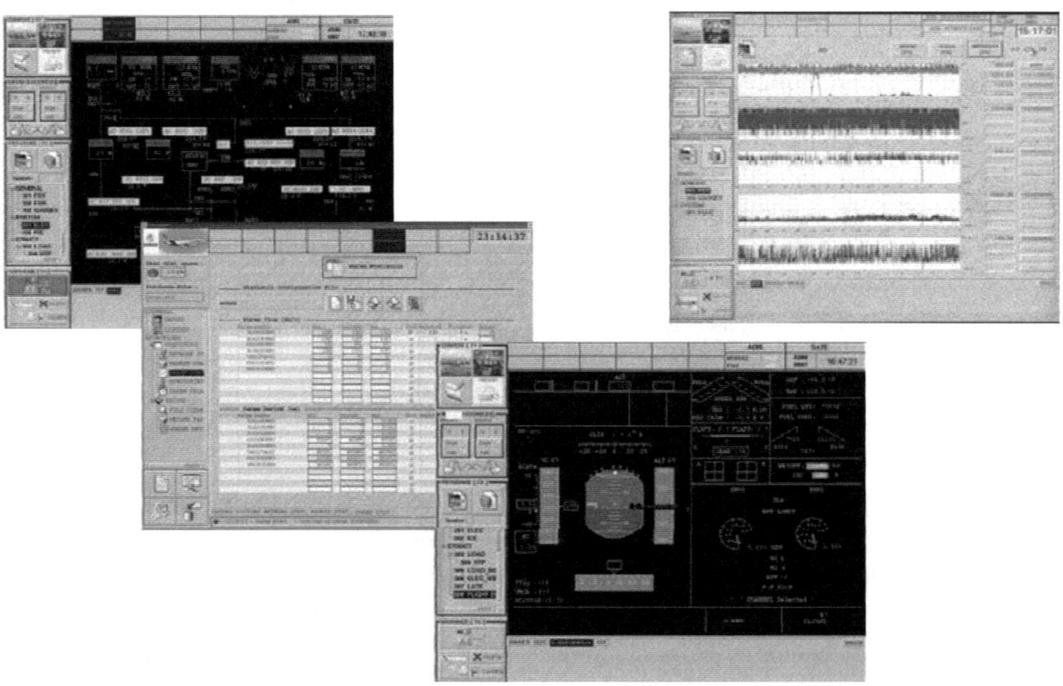

Figure 3. HMI for Flight Test of Airbus A380 [2]

3. Requirements and Design Process

The method used within the research are: define the key function of FCP, design the layout, then implement the layout by using GUIDE Matlab, iterate the design to fulfil all requirements, and finally test the communication within the FCP program and with simulation program in X-Plane and Simulink.

Key maps for the initial design of FCP in this research consist of three major functions as required in FCP requirements, i.e. input command, display flight parameter and load and transmit initialization. The FCP key maps function is shown in Table 1.

Table 1. FCP key maps function

Input Command	Initialization	Display
Trimming command Velocity command Heading command Beta command	Set initial condition Set loop mode Set turbulence condition	Velocity Altitude Heading Beta Visualization of aircraft heading

Layout

Because of the similarity of the functions, the layout of FCP in this research is designed similar to LAPAZ AFCP. For current task, the design will follow Matlab GUIDE in [6].

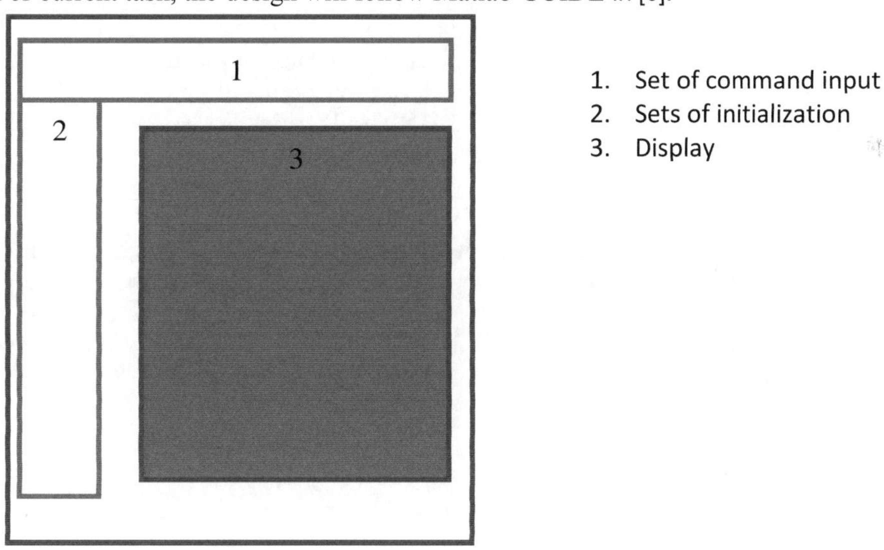

Figure 4. Initial FCP layout design

Initial FCP layout design as shown in **Figure 4** consists of three main block as defined in FCP key maps function in Table 2. The layout consists of command input block, initialization block, and display block. After several evaluations and iterations, the final layout is as shown in **Figure 5**.

IOP Conf. Series: Journal of Physics: Conf. Series **1005** (2018) 012018 doi:10.1088/1742-6596/1005/1/012018

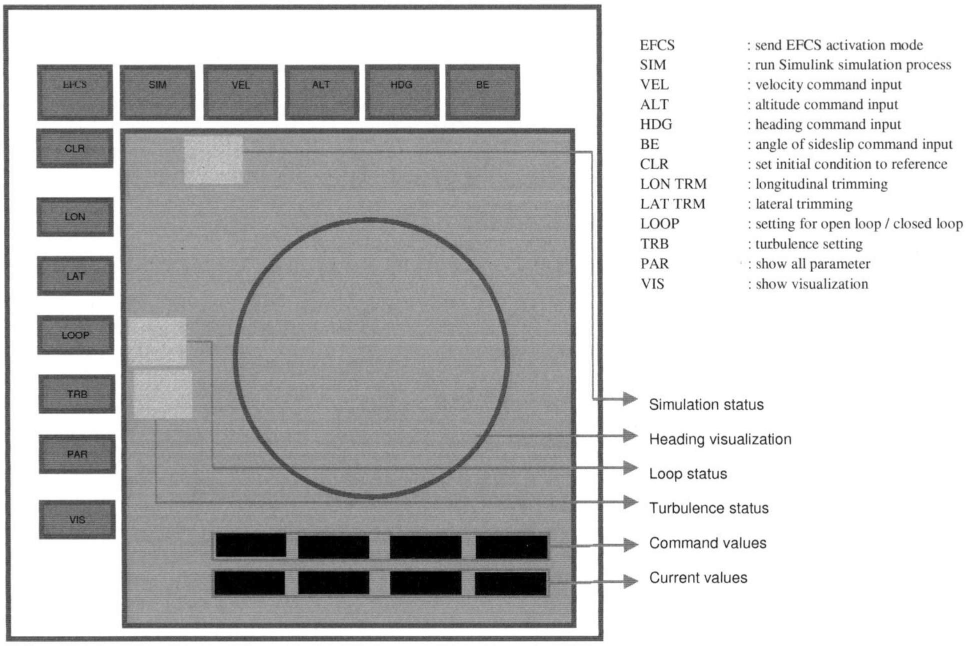

Figure 5. Final FCP Layout design

MATLAB GUIDE

MATLAB has a built-in Graphical User Interface Development Environment (GUIDE), which can be used to lay out GUI graphically and generate MATLAB code automatically [6]. The GUI is a modern tool found in MATLAB versions 7.0 or later, and not automatically compatible with version 6.5 and earlier [11]. GUIDE automatically generates two kinds of Matlab files: one file for Matlab interface figures and another M-file used to store command function of Matlab program [5]. In previous Matlab version, users had to write Matlab algorithm in an M-file and if there were any changes, they had to go back to M-file to make the changes.

General procedure to design GUI in MATLAB are shown in **Figure 6**. The design starts with user interface design where programmer put button, text, editor or other GUI objects and then continue with programming callback function for each GUI objects to define the behavior and responses of the objects. After programming callback functions, users then continue to build M-file as editor for user interface design. In the build M-file process if the process success then continues to the final process that is the creation of user interface for program application, but if still, any function needs to add then the process back to the callback functions.

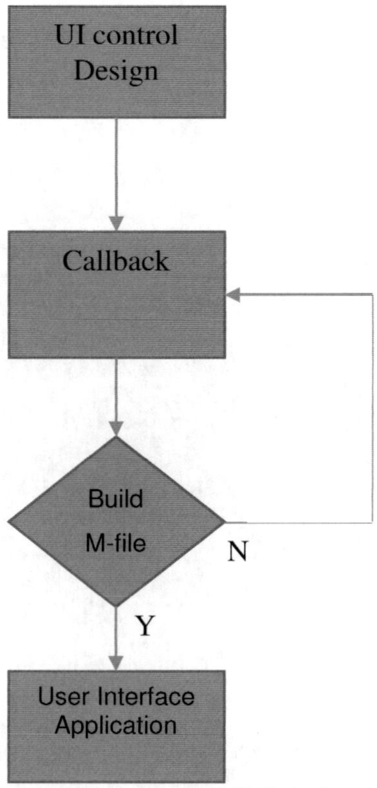

Figure 6. General flowchart of GUI design with Matlab

Figure 7 shows an example of Quick Start window to create a new GUI in MATLAB.

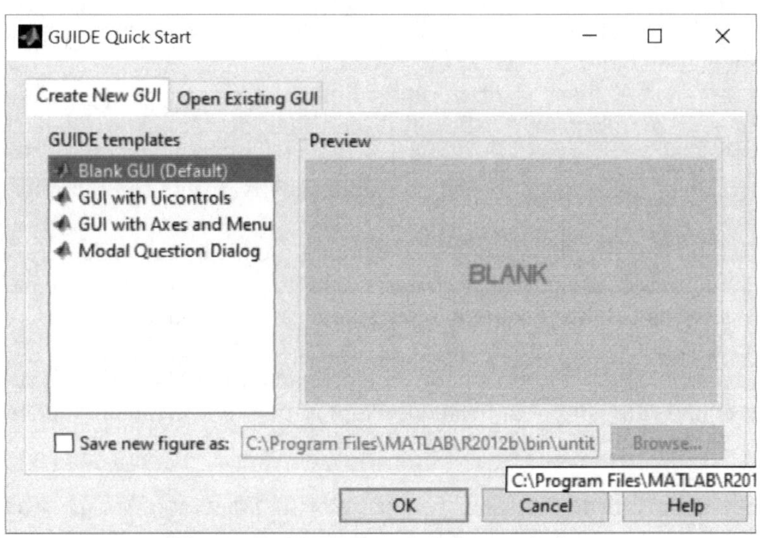

Figure 7. GUIDE Quick Start window

By using the GUI Layout Editor, the user can populate a GUI by clicking and dragging GUI components — such as axes, panels, buttons, text fields, sliders, into the layout area. From the Layout Editor, the size of GUI interface can be changed. GUIDE automatically generates an M-file that controls how the operation of the GUI. This M-file provides code to initialize the GUI and contains a framework for the GUI click-backs, i.e. the routines that execute when a user interacts with a GUI component. Using the M-file editor, the user can add code to the click-backs to perform required functions [9]. The GUI for the development process of FCP design is shown in **Figure 8**.

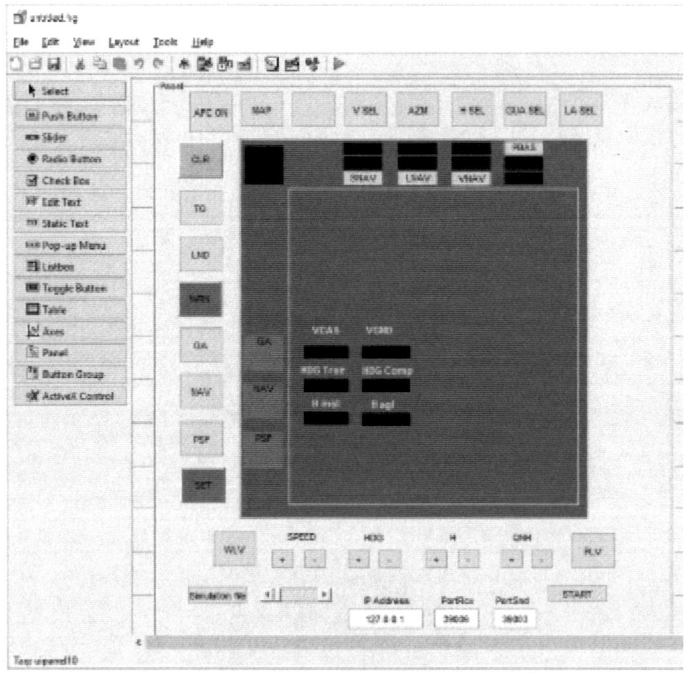

Figure 8. GUI development process for FCP

Dynamic Data Input

To get dynamic data as the input for GUI FCP, data from the flight simulation program were taken. X-plane is the flight simulation program used in this research.

The principle of operation of X-Plane is based on reading the geometric shape of an aircraft to predict its flight dynamics and performance. X-Plane is certified by the U.S. FAA (Federal Aviation Administration) to train pilots because its method ensures a reliable system since it is more detailed, flexible and advanced than flight model based on stability derivatives used by other flight simulators [1].

X-plane standard method for communicating with external processes and machines is User Datagram Protocol (UDP). UDP provides a minimal overhead communication method for passing data between nodes. Unlike Transmission Control Protocol (TCP), UDP is a non-guaranteed protocol and gives no assurances that data packets will arrive in order or at all. UDP is designed to minimize bandwidth usage but presents a possible problem resulting from corrupted data. Although this problem does exist, degradation of control and simulation has been unseen in X-plane experimentations [4].

X-Plane UDP Data

On this research, the communication established between X-Plane and Matlab was performed on a single computer using the IP address "127.0.0.1", which is the network card's internal address. In this type of configuration, it is necessary to use four access points, two for Simulink, two for X-Plane and a single IP address as shown in **Figure 9**.

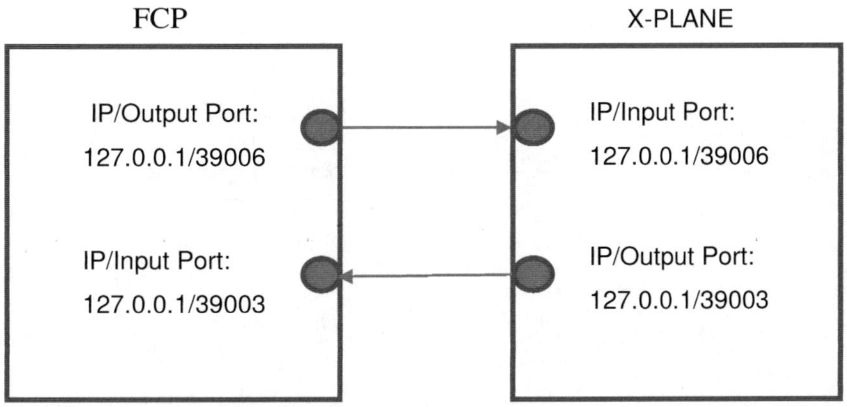

Figure 9. UDP ports connection between FCP and X-Plane

To receive data from UDP user should define IP Port and local port first. Because the program is run on the same computer, remote IP address is set to 127.0.0.1 and local IP Port to 39006. M-file code for receiving UDP data test is shown in **Figure 10**.

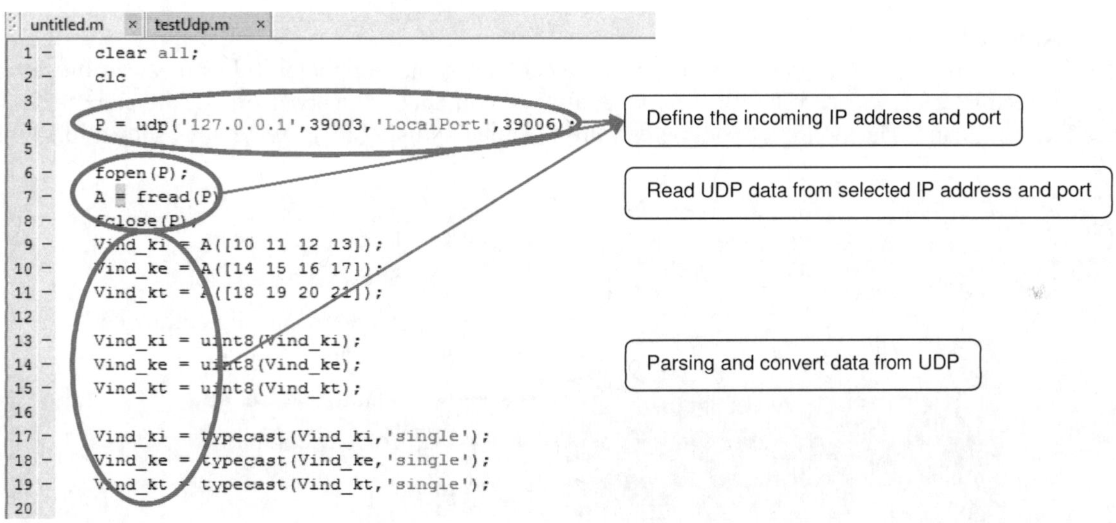

Figure 10. M-file for UDP Receive

To send data from UDP user should also define IP Port and local port. Because the program is run in the same computer remote IP address is set to 127.0.0.1 and local IP Port to 39003 as was set in X-plane. M-file code for sending UDP data test is shown in **Figure 11**.

IOP Conf. Series: Journal of Physics: Conf. Series **1005** (2018) 012018 doi:10.1088/1742-6596/1005/1/012018

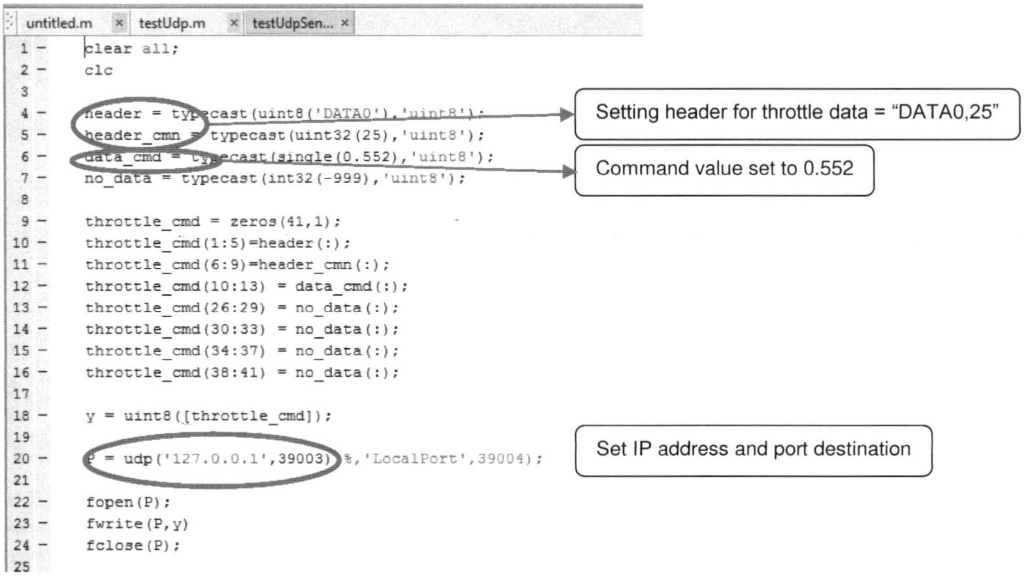

Figure 11. M-file for UDP Send

4. Results and Discussion

To demonstrate the integration according to the scenario, some parameter values need to be set up. Those parameter values are velocity command, altitude command, the angle of sideslip command and heading command. The setting of parameter values is done by using FCP function as shown in **Figure 12**.

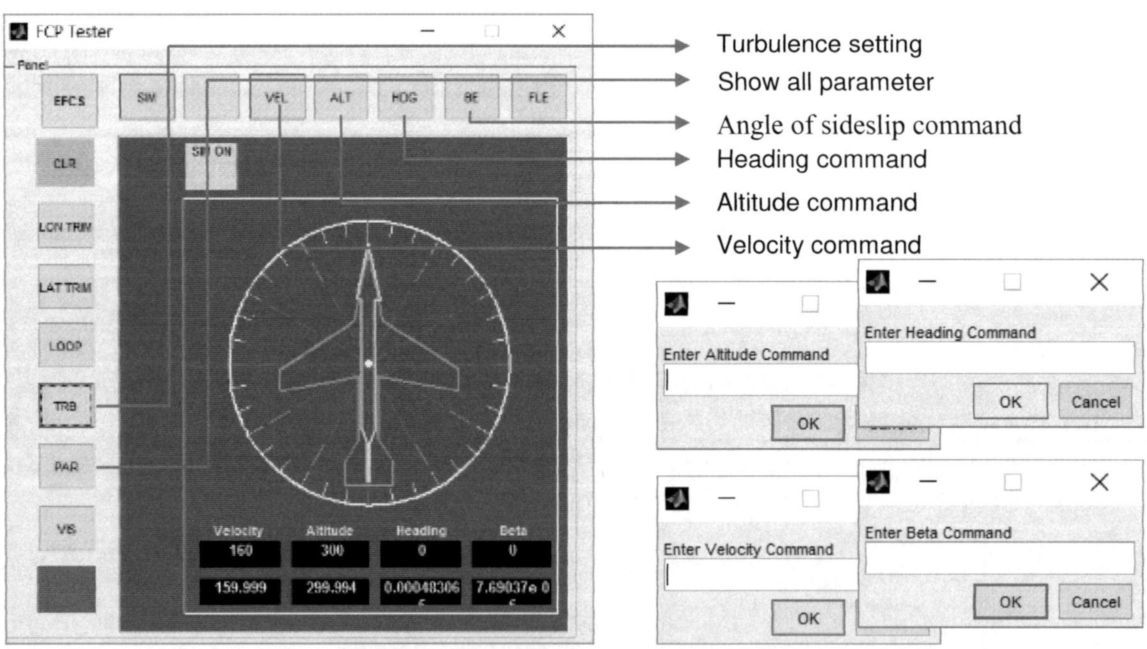

Figure 12. Parameter values setting in FCP

IOP Conf. Series: Journal of Physics: Conf. Series **1005** (2018) 012018 doi:10.1088/1742-6596/1005/1/012018

Turbulence setting consists of turbulence intensity selection, current turbulence intensity mode run on Simulink models as well as a button to activated and deactivated turbulence models. The turbulence setting on FCP is shown in **Figure 13**.

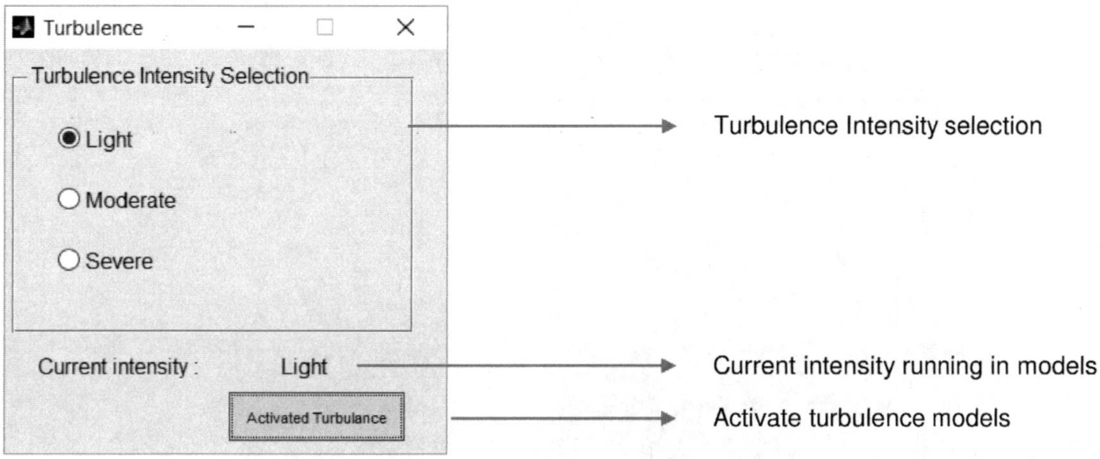

Figure 13. Turbulence models setting in FCP

While the simulation in Simulink is run, flight parameters observe by using PAR function on the FCP as shown in **Figure 14**.

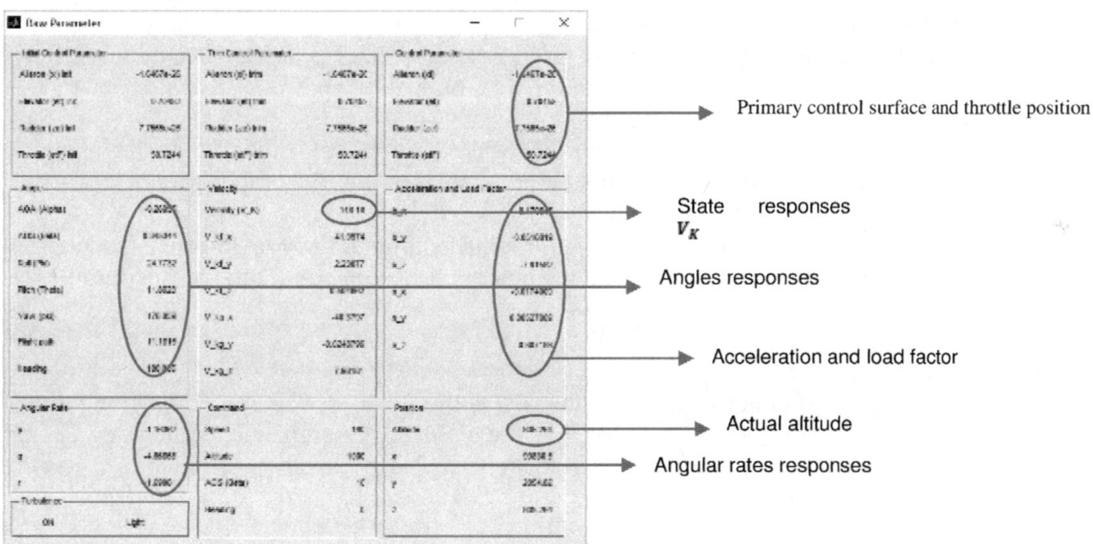

Figure 14. Current Parameters values

Final layout of FCP is shown in **Figure 15** below with description and function of each button and display.

IOP Conf. Series: Journal of Physics: Conf. Series **1005** (2018) 012018 doi:10.1088/1742-6596/1005/1/012018

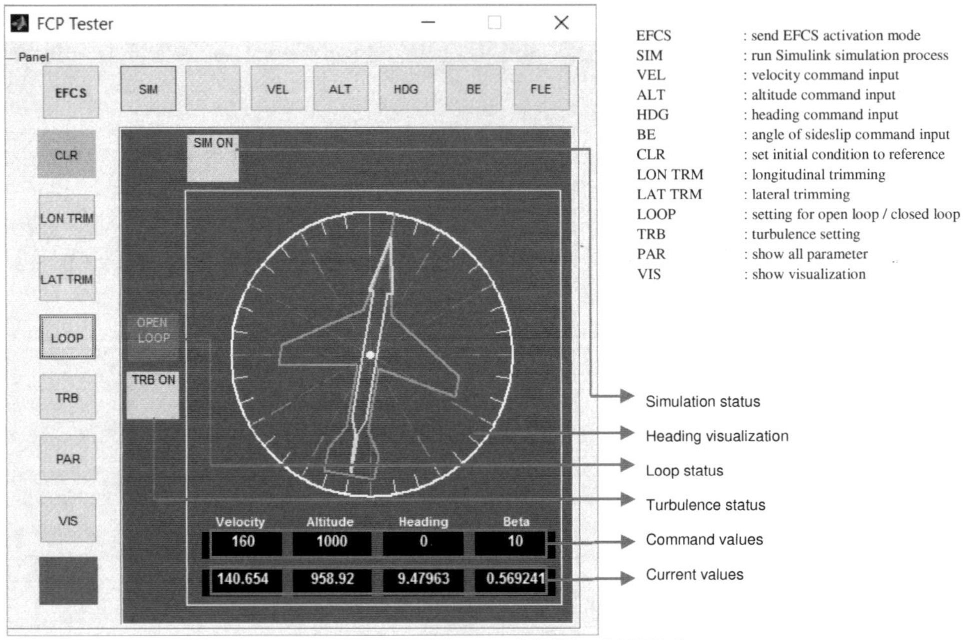

Figure 15. Final FCP layout

5. Conclusions and Future Works

Flight Control Panel Layout using Graphical User Interface MATLAB has been successfully designed. The FCP is used as interface and monitor for command input and aircraft data model including an input for turbulence model setting.

The setting of command can also be done using FCP function with velocity, altitude, the angle of sideslip and heading command values. While simulation in Simulink is run, flight parameter can be observed by using PAR function on the FCP.

Later on, the FCP will be combined with the flight mechanical model and disturbance model. The FCP will be the is used as an interface to command the input to the flight control model and turbulence model and to be an interface for aircraft data model.

Acknowledgements

This work is the final report of minor research of the first author, held in ITB Bandung. The authors would like to thank Drs. Gunawan Setyo Prabowo, MT the director of Aeronautic Technology Center (Pustekbang) LAPAN for providing administrative support, and also Dr. Ony Arifianto, Dr. Rianto Adhi Sasangko and Dr. Yazdi Ibrahim Jenie for supporting this research.

References

[1] A. Bittar, H. Figuereido, P. Guimaraes and A. Mendes, "Guidance Software-in-The-Loop Simulation Using X-Plane and Simulink for UAVs," *International Conference on Unmanned Aircraft Systems (ICUAS),* pp. 993-1002, 2014.

[2] D. Lafourcade, "Conduct of Flight Tests and On-Board Computing for the A380," in *Meeting Proceedings RTO-MP-SCI-162*, France, 2005.

[3] M. Lamp, "Konzept und hardware fur das Automatic Flight Control Panel," FMRA, TU Berlin, Berlin, 2008.

[4] R. Garcia and L. Barnes, "Multi-UAV Simulator Utilizing X-Plane," *Intelligent and Robot system,* pp. 393-406, 2010.

[5] R. C.Gonzalez, R. E. Woods and S. L. Eddins, Digital Image Processing Using Matlab, Gatesmark Publishing, 2009

[6] H. Silver, "Creating Graphical User Interfaces with MATLAB," in *Conference of the American Society of Engineering Education*, 2013.

[7] T. MathWorks, "Creating Graphical User Interfaces," 2014.

[8] R. Verhoeven, F. Bussink, P. X. and D. R., "FLIGHT TESTING TIME AND ENERGY MANAGED OPERATIONS (TEMO)," in *Society of Flight Test Engineers - European Chapter Symposium*, 2016.

[9] T. Hakim, "BEFCS Definition, Requirements and Concept," Berlin, 2016.

[10] B. Jähne, Digital Image Processing, Springer, 2002

[11] I. Pitas, Digital Image Processing Algorithms and Applications, Willey, 2000

[12] R.P.M. Verhoeven and A.J.C. de Reus, Prototyping interactive cockpit applications, NLR, 2005

IOP Conf. Series: Journal of Physics: Conf. Series **1005** (2018) 012019 doi:10.1088/1742-6596/1005/1/012019

Prediction of Flutter Boundary Using Flutter Margin for The Discrete-Time System

Angga Dwi Saputra[1)] *, R. Wibawa Purabaya[2)]

[1, 2)] National Laboratory for Aerodynamics, Aeroelastics, and Aeroacoustics Technology (BBTA3), Indonesia

*angga.dwi@bppt.go.id

Abstract. Flutter testing in a wind tunnel is generally conducted at subcritical speeds to avoid damages. Hence, The flutter speed has to be predicted from the behavior some of its stability criteria estimated against the dynamic pressure or flight speed. Therefore, it is quite important for a reliable flutter prediction method to estimates flutter boundary. This paper summarizes the flutter testing of a wing cantilever model in a wind tunnel. The model has two degree of freedom; they are bending and torsion modes. The flutter test was conducted in a subsonic wind tunnel. The dynamic data responses was measured by two accelerometers that were mounted on leading edge and center of wing tip. The measurement was repeated while the wind speed increased. The dynamic responses were used to determine the parameter flutter margin for the discrete-time system. The flutter boundary of the model was estimated using extrapolation of the parameter flutter margin against the dynamic pressure. The parameter flutter margin for the discrete-time system has a better performance for flutter prediction than the modal parameters. A model with two degree freedom and experiencing classical flutter, the parameter flutter margin for the discrete-time system gives a satisfying result in prediction of flutter boundary on subsonic wind tunnel test.

1. Introduction

The most important aeroelastic phenomenon is flutter. Flutter is self-excited oscillation of the elastic structure under the action of the aerodynamic loads. Flutter instabilities often exhibit an explosive behavior that cause a sudden change in stability despite a small change in flight condition. Further, the aeroelastic vibrations that occur at all flight regimes have a strong impact on the fatigue life of the structure . Since flutter is the self excited vibration which causes a fatal damage to a wing, engineers have to pay attention to the occurrence of flutter in the airplane design and development. At the final stage of development, therefore, they have to conduct the flutter tests to check that flutter does not occur in the flight envelope, and to evaluate the flutter boundary. Since the flutter tests are carried out only in the safety range far below the critical point, the flutter boundary is predicted from the behavior of some stability measure plotted against the flight speed or the dynamic pressure [1].

Flutter boundary prediction during wind tunnel and flight tests is one of the most important, but very difficult tasks imposed in the process of development of a new airplane [2]. One difficult aspect of flutter testing is that a flutter mode can suddenly become unstable with a small increase in dynamic pressure. Furthermore, there may be little or no indication of the approach of the instability in a plot of modal damping against dynamic pressure. Although such a sudden onset of instability is not always the case, it occurs in practice frequently enough that flutter prediction based on the projection of modal damping may not be adequate [3]. As an alternative approach, several researchers have attempted to propose more suitable parameters for the flutter prediction than the modal damping. Among them, the flutter margin introduced by Zimmermann and Weissenburger [4] is the most famous and successful one.

The flutter margin decreases almost monotonically toward zero as the dynamic pressure increases, which is a very suitable property for the flutter prediction, though it is applicable only to binary-flutter [5]. Currently, measurement and analysis of test data are generally performed on a digital computer, where sampled data are processed to define the characteristics of the systems. System identification gives powerful and useful procedures to construct a linear discrete-time model, known as an Auto Regressive Moving Average (ARMA) model. The ARMA model consists of autoregressive (AR) terms of measurement, that is linear combination of present and the finite number of past data, and the moving average (MA) of white noise. The AR part correspond to the characteristic polynomial of the system and gives information on stability [6]. In the discrete-time domain, Matsuzaki and Ando have proposed to use Jury's stability parameter as the indicator of stability margin, which is the stability criterion for the discrete-time system and are calculated from the Auto Regressive Moving Average (ARMA) model identified. Then Jury's parameters and introduced the new parameter called the Flutter Margin for the Discrete-time System (FMDS), which is approximately equivalent to the flutter margin introduced. Though the FMDS has superior properties as the flutter prediction parameter, it is applicable to the data which include only two coupling modes [1]. They have shown that the stability parameter is a more effective index for the flutter prediction than modal damping through numerical and experimental studies [7].

Generally, flutter testing on a wind tunnel carried out at subcritical speeds to avoid facilities damage. Hence, the flutter speed has to be predicted from the behavior some of its stability criterions estimated against the dynamic pressure or flight speed. Therefore, it is quite important for a reliable flutter prediction method to estimates flutter boundary speed. A flutter prediction method has to estimates the flutter boundary speed as accurate as possible. This paper summarize flutter testing of a wing cantilever model in a wind tunnel. The model has two degree of freedom, are bending and torsion modes. The flutter test was conducted in the subsonic wind tunnel. The measurement was repeated by increasing the wind speed stepwise. The dynamic responses was measured by two accelerometers that were mounted on leading edge and center of wing tip. Dynamic responses were used to determine parameter stability FMDS. Flutter boundary of the model were estimated using extrapolation of FMDS against dynamic pressure. The predicted flutter speed using FMDS were compared with actual flutter speed to evaluate the accurary of FMDS method. The goal of this study is to determine alternative method that is accurate and effective for predicting flutter boundary in a wind tunnel.

2. Method

2.1. Flutter Margin for Discrete-Time System (FMDS)

FMDS defined based on the time-series model, hence that it is suitable for the digital data processing, and it doesn't need the modal damping and frequency. After preprocessed through the band-pass filter so as to include only the coupling mode, sampled data is identified by the ARMA model

$$y_1 + \alpha_1 y_{t-1} + \alpha_2 y_{t-2} + \alpha_3 y_{t-3} + \alpha_4 y_{t-4} = e_t + \gamma_1 e_{t-1} + \gamma_2 e_{t-2} + \gamma_{t-m} e_{t-m} \quad (1)$$

where y_t is the data observed at time t, and et a white noise. The order, m, in the right hand side of Eq.(1) is determined by the Akaike Information Criteria (AIC). The left hand side corresponds to the characteristic equation of the system:

$$G(z) = z^4 + \alpha_1 z^3 + \alpha_2 z^2 + \alpha_3 z + \alpha_4$$

$$= \prod_{i=1}^{2} (z - z_i)(z - z_i^*)$$

Where z_i* is complex charateristic roots. It is obvious that *i* represents the number of the modes consisting of the aeroelastic system and *T* is sampling rate of measurement. The modal frequencies, ω_i, and the damping ratio, ξ_i, of the i-th mode are respectively given as [8]:

$$\xi_i = -(\ln|z_i| / |\ln z_i|) \tag{2}$$

$$\omega_i = \frac{1}{T} |\ln z_i| \tag{3}$$

The stability margin of the system is measured by [1]

$$F_z = \frac{\det(X_3 - Y_3)}{(1 - \alpha_4)^2} \tag{4}$$

Where

$$X_3 = \begin{bmatrix} 1 & \alpha_1 & \alpha_2 \\ 0 & 1 & \alpha_1 \\ 0 & 0 & 1 \end{bmatrix}, \; Y_3 = \begin{bmatrix} \alpha_2 & \alpha_3 & \alpha_4 \\ \alpha_3 & \alpha_4 & 0 \\ \alpha_4 & 0 & 0 \end{bmatrix}$$

The value of Fz decreases monotonically and almost linearly toward zero in the subcritical range with the increase of the dynamic pressure [1].

2.2. *Wind-Tunnel Flutter Testing*

The Flutter test was conducted at Indonesia Low Speed Wind Tunnel (ILST). ILST is a close circuit wind tunnel. This tunnel has a close rectangular test section with size 4 x 3 m. Installation the model in the wind tunnel is depicted in Figure 1. The wind speeds around model was measured by a pitot tube that was installed close to the model.

Figure 1. The Flutter Model in wind tunnel

Flutter model is depicted in Figure 2. The model is cantilever wing with two degree of freedom, are bending and torsion mode. Data responses were measured by two accelerometers that placed on leading edge and center of wing tip. Both are inside the wing. So that they don't disturb the flow over model

IOP Conf. Series: Journal of Physics: Conf. Series **1005** (2018) 012019 doi:10.1088/1742-6596/1005/1/012019

surface. During each measurement, the wind speed was kept constant and the wing was excited by impulsive disturbance at the same time.

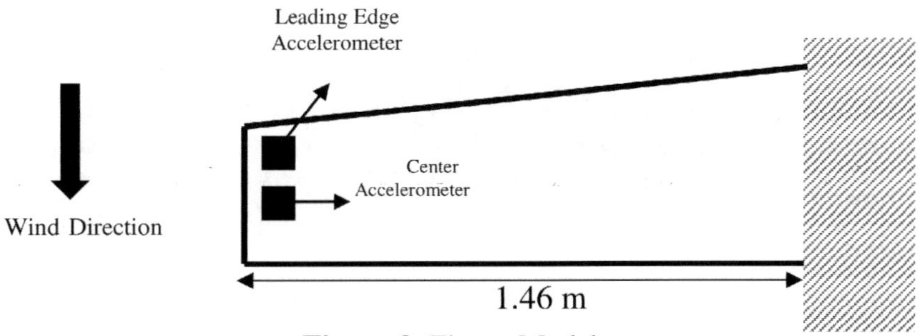

Figure 2. Flutter Model

Figure 3 describes the flow chart of flutter test. The flutter test conducted in the subsonic wind tunnel. The measurement was repeated by increasing the wind speed. The model excited by an impulsive force. The purpose of excitation are to get sufficient response to distinguish modes from noise. The data response was measured by accelerometers and then was digitized in a data acquisition. The digital data are then processed to eliminate low-frequency noise and higher modes using digital filter. Hence, the filtering data that consists only the lowest two modes. Then, filtering data identified with ARMA model, to estimate ARMA coefficients. Flutter margin parameter Fz were determined using ARMA coefficients, and then the flutter boundary speed was predicted by linear fitting of flutter margin Fz [8].

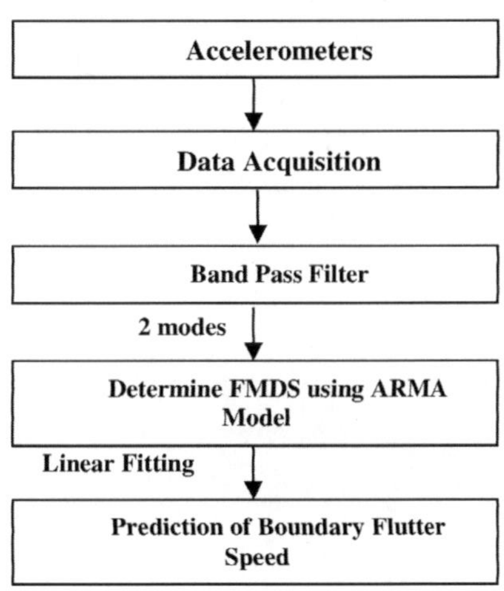

Figure 3. Flow Chart of Flutter Test

3. Results and Discussions

3.1. Modal Testing

The Modal testing was conducted to determine natural frequency of the flutter modes. Bending and torsion modes are the flutter modes. Attention must be paid to those frequencies in the wind tunnel test. The Modal testing was carried out using single input multiple outputs (SIMO) method. Excitation or inputs was provided by an impact hammer that is equipped by a force sensor on the tip. The dynamic response were measured by eight accelerometers. Those accelerometers were installed on the model surface. Installation of sensor is depicted in Figure 4.

IOP Conf. Series: Journal of Physics: Conf. Series **1005** (2018) 012019 doi:10.1088/1742-6596/1005/1/012019

Figure 4. Modal Testing of Model

The result of modal testing is presented in Table 1. Each vibrational mode shape corresponds to a single frequency. The first vertical bending and torsion mode are the important modes contributed to the occurrence of flutter.

Table 1. Natural Frequencies

No.	**Mode shape**	Frequency (Hz)
1	***First Vertical Bending***	3.2
2	***First Inplane Bending***	5.03
3	***First Torsion***	8.6

3.2. Wind Tunnel Test

In the wind tunnel test, impulsive disturbance was applied to the model. The model was impacted by a hammer to create impulsive force and model oscillated freely toward a stable condition afterward. Transient responses of the model is depicted in Figure 5, the disturbance effect to the model was damped out and it returned to the original stable condition.

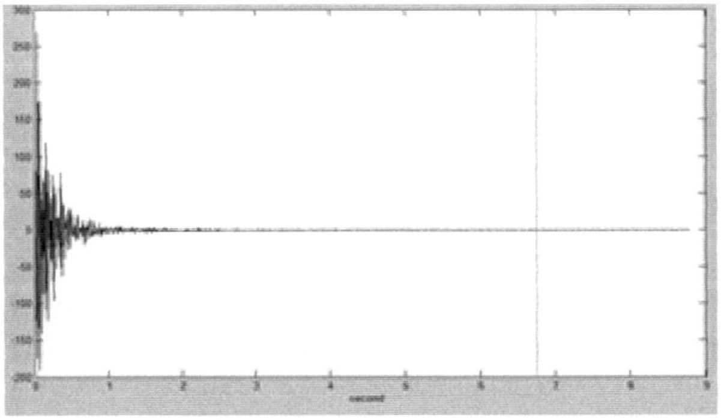

Figure 5. Transient Response

Modal parameters were extracted at each wind speed. The modal dampings that were extracted at each test point are given in Figure 6(a). The damping data indicates that that one mode is becoming less stable, whereas the other mode is more stable. The critical speed for flutter is happened if the total damping of the model is going to zero [9]. As shown in Figure 6(a), the modal damping is not possible to predict flutter boundary speed. The modal frequencies that were extracted at each test point are given in Figure 6(b). These data shows the indication that the model is experiencing a classical bending-torsion flutter. The natural frequencies appear to be coalescing, as expected for a classical flutter until a possible coalescence at the airspeed very close to the onset of flutter [10].

•

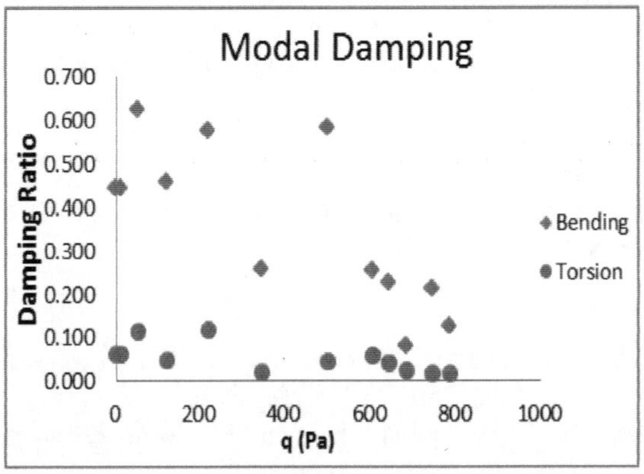

Figure 6(a). Measured Modal Damping

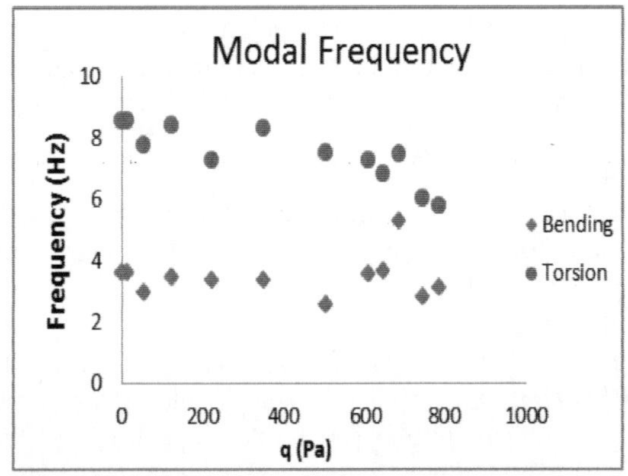

Figure 6(b). Measured Modal Frequency

The flutter boundary is predicted by parameter flutter margin discrete systems (*Fz*). Parameters *Fz* was determined using Eq.(4). As shown in Figure 7, the curve *Fz* against dynamic pressure (q) is almost linear. Therefore, a linear fit is convenient to predict flutter boundary for this case. The parameters *Fz* decrease monotonically toward zero with increasing dynamic pressure.

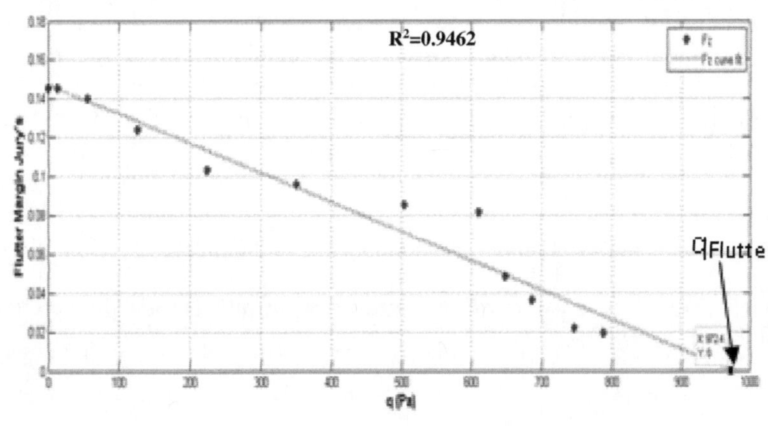

Figure 7. The paramaters *Fz* against Dynamic Pressures (q)

The parameter *Fz* has significant linear correlation against dynamic pressure (q) that is shown by value correlation coefficient (R^2) close to one. The critical speed or dynamic pressure occurs when the value of *Fz* is zero or the point of fitted line crossing the horizontal coordinate. The Applying extrapolation the fitted line gives the critical pressure dynamic of q_f= 972.5 Pa or the wind speed of v_f = 41.7 m/s. Based on the experiment , the actual flutter boundary speed is v_{fact}= *40.5m/s*, hence the error predicted flutter is *error*=3%.

4. Conclusions

The flutter test of a half wing model at Indonesia Low Speed Wind Tunnel (ILST) had been carried out. This experiment has proved that the modal damping and frequency cannot predict flutter boundary accurately. The model was experiencing a classical bending-torsion flutter, It seems that the natural frequencies appear to be coalescing, as expected for a classical flutter until a possible coalescence at the airspeed very close to the onset of flutter. The parameter flutter margin for discrete-time systems (FMDS) has a significant correlation against dynamic pressure. Therefore, the parameter FMDS has the *better performance* for flutter prediction than the modal parameters. The application to a model with two degree freedom that experiencing a classical flutter, the FMDS gives a satisfying result in prediction of flutter boundary on subsonic wind tunnel test.

Acknowledgment
The authors would like to thank The National Laboratory for Aerodynamics, Aeroelastics, and Aeroacoustics Technology (BBTA3) for providing research support facilities.

References

[1] Torii, H. The Extension of Discrete-Time Flutter Margin . *Proceedings of 28th Congress of The Intertional Council of The Aeronautical Sciences, ISBN 9780956533319, 2012*
[2] Bisplinghoff, R.L., Ashley, Holt, and Halfman, L, R. Aeroelasticity. Second Edition, *Addison-Wesley Publishing Company, Massachusetts, 1955.*
[3] Bennett, R.,M. Aplication of Zimmerman Flutter Margin Criterion to a Wind Tunnel Model. *NASA Technical Memorandum 84545, Virginia, 1982*
[4] Zimmermann, N., and Weissenburger, J. Prediction of Flutter Onset Speed Based on Flight Testing at Subcritical Speeds. *Journal of Aircraft, Vol. 1, No. 4, pp 190-202, 1964*
[5] Price, S.J.,and Lee. B.H.K. Evaluation and Extension Flutter Margin Method for Flight Flutter Test. *Journal of Aircraft, Vol.30, No.3,1993, pp.395-402*
[6] Torii, H., dan Matsuzaki,Y. Flutter Margin Evaluation for Discrete-Time Systems. *Journal of Aircraft, Vol.38, No.1, January-February 2001.*
[7] Matsuzaki,Y., and Ando,Y. Responce Characteristics of a Two-Dimensional Wing Subject to Turbulence Near the Flutter Boundary. *Journal of Sound and Vibration, Vol.136, No.2, 1990, pp.187-199*
[8] Matsuzaki,Y. An Overview of Flutter Prediction In Test Based on Stability Criteria in Discrete- Time Domain. *International Journal of Aeronautical and Space Sciences, Vol. 12, No.4, 2011, pp.305-317*
[9] Fariduzzaman and Leonardo Gunawan. Flutter Margin Determination of A Long span Bridge. *4th Pacific International Conference on Aerospace Science and Technology Kaohsiung, Taiwan, 21-23 May 2001*
[10] Lind, Rick. Flight Test Evaluation of Flutter Prediction Methods. *Journal Of Aircaft Vol. 40, No. 5, September–October 2003.*

Hinge Moment Coefficient Prediction Tool and Control Force Analysis of Extra-300 Aerobatic Aircraft

Chandra Nurohman[1)*], **Ony Arifianto**[1)], **Agra Barecasco**[2)]
[1)]Faculty of Mechanical and Aerospace Engineering, Bandung Institute of Technology, Indonesia
[2)]PT T&E Simulation, Bandung, Indonesia

*chandranrhmn@gmail.com

Abstract. This paper presents the development of tool that is applicable to predict hinge moment coefficients of subsonic aircraft based on Roskam's method, including the validation and its application to predict hinge moment coefficient of an Extra-300. The hinge moment coefficients are used to predict the stick forces of the aircraft during several aerobatic maneuver i.e. inside loop, half cuban 8, split-s, and aileron roll. The maximum longitudinal stick force is 566.97 N occurs in inside loop while the maximum lateral stick force is 340.82 N occurs in aileron roll. Furthermore, validation hinge moment prediction method is performed using Cessna 172 data.

Keywords: hinge moment coefficient, prediction tool, control force, Extra-300, Microsoft Excel

1. Introduction

When an aircraft flies, the air passes through its control surfaces will change the load on them and generate force and moment around the hinge line. This moment has to be countered by pilot through cockpit control devices to deflect the corresponding control surfaces **[1]**. There are two kinds of aircraft control concept, reversible and irreversible **[2]**.

In reversible system, the pilot controls are directly connected using combinations of pushrods, cables, and pulleys to the control surface. As the consequence, if pilot controls are deflected, the corresponding control surface is deflected and at the same time if the control surface is deflected, then the pilot controls will be deflected. In irreversible system, there are additional power augmentation system that support pilot to counter the hinge moment. The augmentation system consists usually of hydraulic system or pneumatic (electro-mechanical actuators).

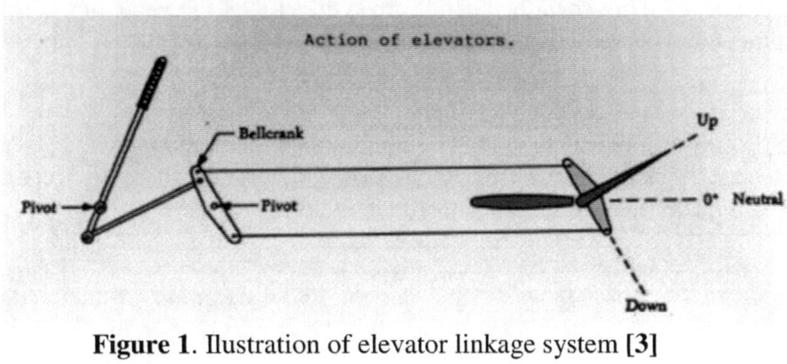

Figure 1. Ilustration of elevator linkage system **[3]**

Hinge moment usually expressed as coefficient of hinge moment (C_h). Generally, specific aircraft has unique hinge moment coefficient as its aerodynamic characteristic. Hinge moment analysis has three main following purposes:

1. To compute stick, wheel, and pedal cockpit control forces so they can be checked against airworthiness requirements.
2. To compute actuator force level so that hydraulic or electro-mechanical actuators can be properly sized **[4]**.
3. To design proper artificial feel or control loading system.

Roskam's method of hinge moment derivative is a method that is applicable to predict hinge moment coefficient of subsonic aircraft. Hinge moment coefficient is function of control surface geometry, aircraft specification, angle of attack, and surface deflection.

A hinge moment prediction tool is developed based on the Roskam's method to calculate hinge moment coefficient. This tool is designed to calculate hinge moment coefficient in ailerons and elevators and these values will be validated using reference data of Cessna 172S. The advantages of this tool is the convenience to inspect the calculation steps and to debug if any problem occurs so it is suitable for both educational, design, or other analysis.

The tool is applied to Extra-300 aerobatic aircraft which operates in high load factor region during its operation. There is various level of aerobatic maneuvers known i.e. basic maneuver, basic combination maneuver, and basic sequence maneuver **[5].** In this work, inside loop, half cuban 8, split-s, and aileron roll maneuver is selected. Longitudinal and lateral control forces are analyzed during those maneuvers. Control force is one kind of cueing aspect in aircraft. In complex, high consequence environments such as aviation, the capacity to acquire, integrate, and respond to task-related cues is critical **[6]**. Pull back stick that generate forward control force is considered as positive control force **[7]**.

To conclude the introduction statements, the purpose of this work is to explain the development of a hinge moment prediction tool based on Roskam's method. The tool is applied to analyze hinge moment coefficient of Extra-300. Furthermore, flight simulation is conducted and the control forces are analyzed during some aerobatic maneuvers.

2. Hinge Moment Prediction

Roskam's method is used as basic theory for the developed tool. The following is the general equation of hinge moment coefficient (C_h)

$$C_h = C_{ho} + C_{h\alpha}.\alpha + C_{h\delta}.\delta + C_{h\delta t}.\delta t \quad (1)$$

Hinge moment coefficient is separated into three derivatives i.e. angle of attack, control surface deflection, and trim tab deflection. Besides, it's also contribution of C_{ho} which is function of airfoil feature. For Extra-300 in which the airfoils are symmetrical, the value of C_{ho} is 0.

Hinge moment derivative due to angle of attack can be determined using this equation

$$C_{h\alpha} = \frac{A\cos\Lambda_{\frac{c}{4}}}{A+2.\cos\Lambda_{\frac{c}{4}}}.C_{h\alpha M} + \Delta C_{h\alpha} \quad (2)$$

The term $C_{h\alpha M}$ is the contribution of 2D hinge moment derivative which is function of Mach number and balance ratio factor as shown in the equation below. Balance ratio represents the ratio between forward and backward part of control surface in respect to the hinge line.

$$C_{h\alpha M} = \frac{(C_{h\alpha})_{balance}}{(1-M^2)^{\frac{1}{2}}} \quad (3)$$

$C_{h\alpha M}$ can be obtained through some steps and the first is examining trailing edge (TE) angles by checking whether following equation (4) is satisfied. TE angle notation explained in the Figure 4

$$\tan(\frac{\Phi'_{te}}{2}) = \tan(\frac{\Phi''_{te}}{2}) = \tan(\frac{\Phi_{te}}{2}) = \frac{t}{c} \quad (4)$$

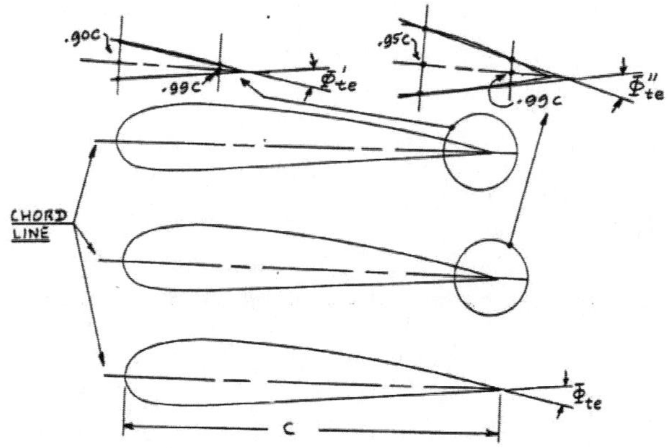

Figure 2. trailing edge definition

The second step is calculating $C'_{h\alpha}$ using equation (5) and the value of $\frac{C'_{h\alpha}}{(C_{h\alpha})_{theory}}$ and $(C_{h\alpha})_{theory}$ can be obtained from Figure 2.

$$C'_{h\alpha} = \frac{C'_{h\alpha}}{(C_{h\alpha})_{theory}} \cdot (C_{h\alpha})_{theory} \tag{5}$$

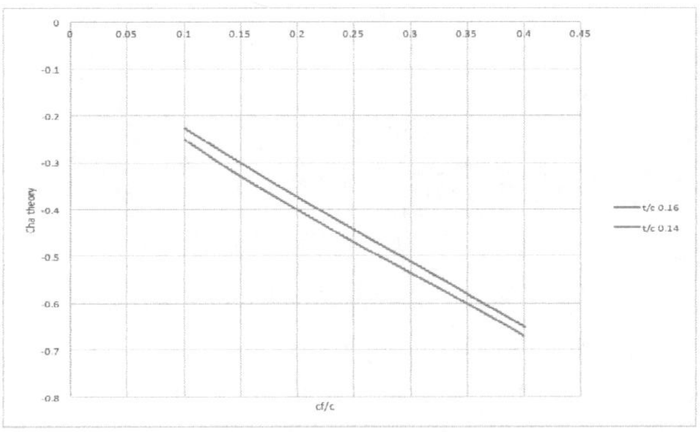

(a)

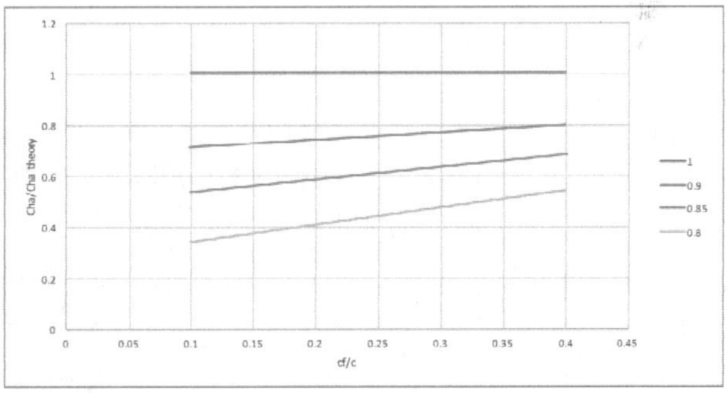

(b)

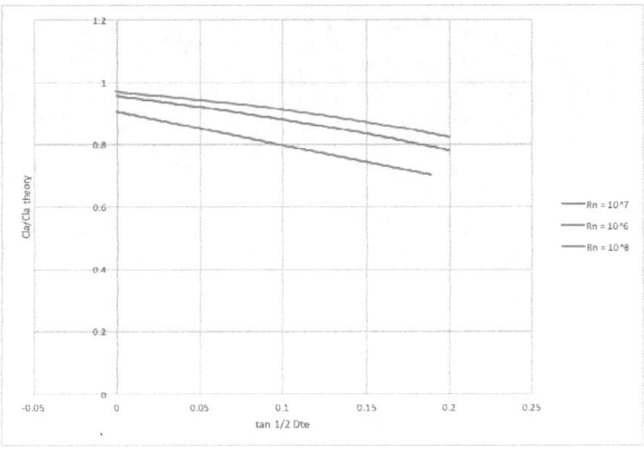

(c)

Figure 3. (a) Graph $(C_{h\alpha})_{theory}$ vs cf/c, (b) Graph $\frac{C'_{h\alpha}}{(C_{h\alpha})_{theory}}$ vs cf/c, (c) $\frac{C_{l\alpha}}{(C_{l\alpha})_{theory}}$ vs $\tan(\frac{\Phi_{te}}{2})$

If equation (4) is satisfied, then $C''_{h\alpha} = C'_{h\alpha}$, but if it is not , $C''_{h\alpha}$ can be calculated using equation (6)

$$C''_{h\alpha} = C'_{h\alpha} + 2.(C_{l\alpha})_{theory}.[1 - \frac{C_{l\alpha}}{(C_{l\alpha})_{theory}}].[\tan\frac{\Phi_{te}}{2} - \frac{t}{c}] \quad (6)$$

Then the next step is calculating $(C_{h\alpha})_{balance}$ using following equation

$$(C_{h\alpha})_{balance} = C''_{h\alpha}.\frac{(C_{h\alpha})_{balance}}{C''_{h\alpha}} \quad (7)$$

As the feature of aircraft's control surface, balance ratio equation is stated below

$$BR = \sqrt{(\frac{c_b}{c_f})^2 - (\frac{t}{2c_f})^2} \quad (8)$$

Then, balance ratio information is applied to Figure 3(a) to obtain the value of $\frac{(C_{h\alpha})_{balance}}{C''_{h\alpha}}$

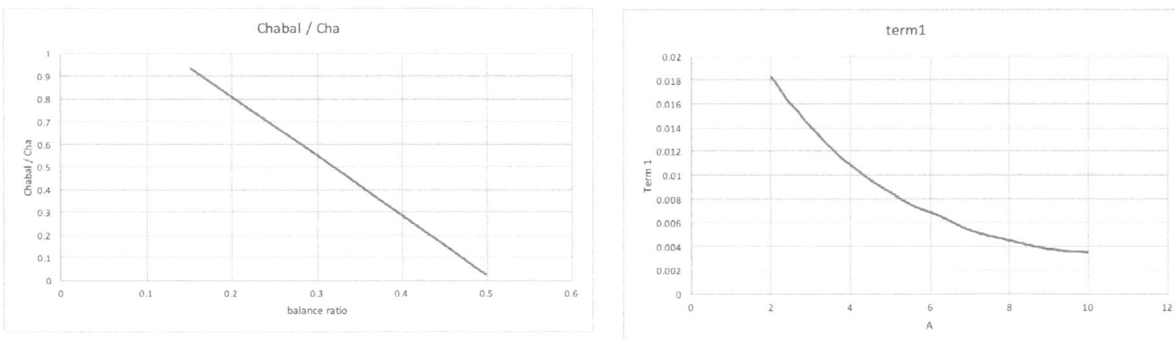

(a) (b)

Figure 4. (a) Balance ratio graph, (b) $\frac{\Delta C_{h\alpha}}{C_{l\alpha}.B2.K_a.\cos\Lambda_{\frac{c}{4}}}$ graph

After completing previous steps, the value of C_{haM} can be calculated from equation (2). The next step is examining $\Delta C_{h\alpha}$. The equation is stated below

$$\Delta C_{h\alpha} = [\frac{\Delta C_{h\alpha}}{C_{l\alpha}.B2.K_a.\cos\Lambda_{\frac{c}{4}}}].C_{l\alpha}.B2.K_a.\cos\Lambda_{\frac{c}{4}} \quad (9)$$

The term $\frac{\Delta C_{h\alpha}}{C_{l\alpha}.B2.K_a.\cos\Lambda_{\frac{c}{4}}}$ can be evaluated from figure 3(b) while the values of $C_{l\alpha}$ and $\cos\Lambda_{\frac{c}{4}}$ are the property of airfoil and wing or HTP plane respectively. K_a can be calculated using the equation and graph 4(a) below:

$$K_a = \frac{K_{ai}.(1-\eta_i) - K_{ao}.(1-\eta_o)}{\eta_o - \eta_i} \quad (10)$$

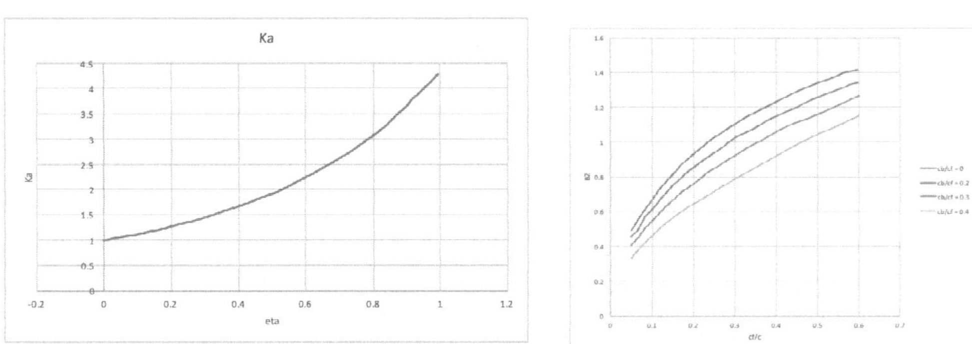

(a) (b)

Figure 5. (a) Graph Ka vs η, (b) Graph B2 vs cf/c

The variable η_i and η_o refer to inboard and outboard span of control surface as fraction of the main surface semi span respectively. The value of B2 can be obtained from the graph 4(b). Return to equation (1), all variables have been known hence value of $C_{h\alpha}$ can be calculated by this equation

$$C_{h\alpha} = \frac{A\cos\Lambda_{\frac{c}{4}}}{A+2.\cos\Lambda_{\frac{c}{4}}} . C_{h\alpha M} + \Delta C_{h\alpha} \tag{11}$$

This method is applicable for both ailerons in wings and elevator in horizontal tail. The value of angle of attack in wing is not equal to in tail. There is a correction due to the presence of downwash is computed using the equation below

$$\frac{d\alpha_H}{d\alpha} = \frac{[A^2.(1-M^2)+4]^{0.5}-2}{[A^2.(1-M^2)-4]^{0.5}+2} \tag{12}$$

The next step is calculating hinge moment derivative due to control surface deflection. The general formula of hinge moment derivative due to control surface deflection is stated below

$$C_{h\delta} = \cos\Lambda_{\frac{c}{4}} . \cos\Lambda_{hl} . [C_{h\delta M} + \alpha_\delta . C_{h\alpha M} . \frac{2\cos\Lambda_{\frac{c}{4}}}{A+2\cos\Lambda_{\frac{c}{4}}} + \Delta C_{h\delta}] \tag{13}$$

From the previous steps, the value of $C_{h\alpha M}$ is already known. The value of α_δ can be obtained from the following graph

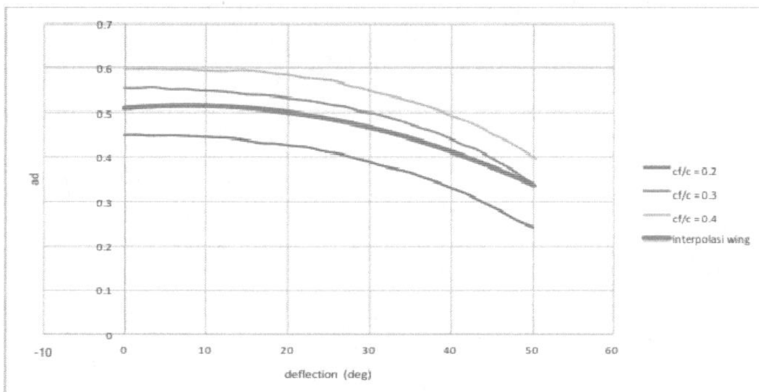

Figure 6. Graph α_δ vs surface deflection

Equivalent to the hinge moment derivative due to angle of attack, the deflection factor is also started from 2D hinge moment derivative formulated as

$$C_{h\delta M} = \frac{(C_{h\delta})_{balance}}{(1-M^2)^{\frac{1}{2}}} \tag{14}$$

After examining the trailing edge angles which has been done before in angle of attack discussion, the second step is calculating $C'_{h\delta}$ using this equation :

$$C'_{h\delta} = \frac{C'_{h\delta}}{(C_{h\delta})_{theory}} . (C_{h\delta})_{theory} \tag{15}$$

IOP Conf. Series: Journal of Physics: Conf. Series **1005** (2018) 012020 doi:10.1088/1742-6596/1005/1/012020

The value of $\frac{C'_{h\delta}}{(C_{h\delta})_{delta}}$ and $(C_{h\delta})_{theory}$ can be obtained from Figure 6.

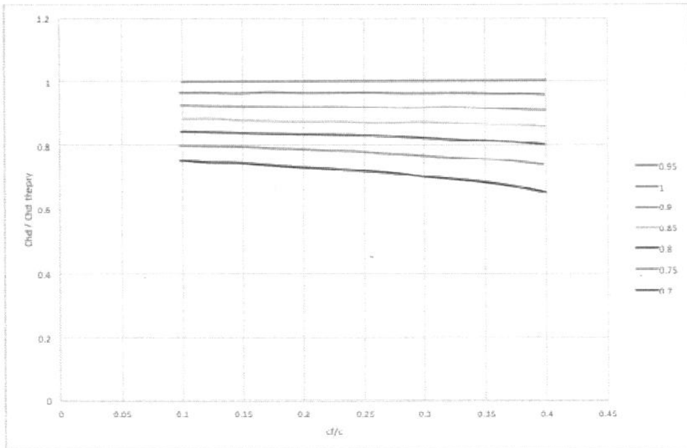

(a)

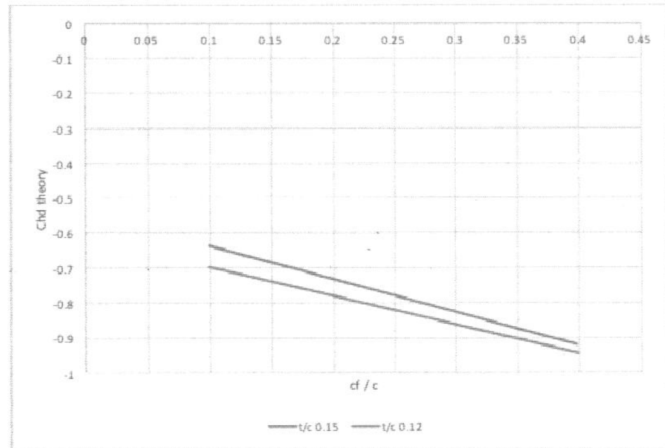

(b)

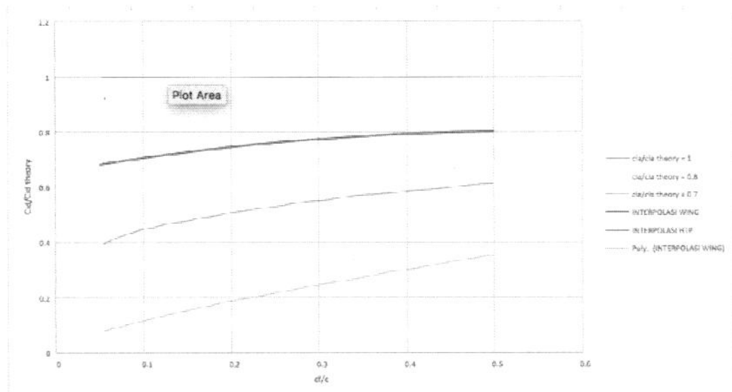

(c)

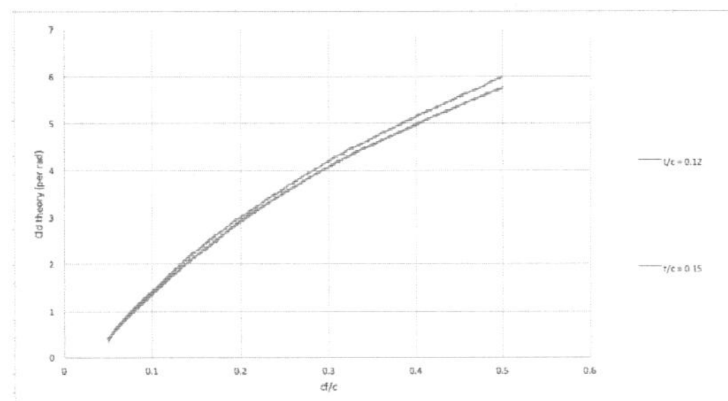

(d)

Figure 7. (a) Graph $\frac{C'_{h\delta}}{(C_{h\delta})_{delta}}$vs cf/c, (b) Graph $(C_{h\delta})_{theory}$vs cf/c, (c) Graph $\frac{C_{l\delta}}{(C_{l\delta})_{theory}}$ vs $\tan(\frac{\Phi_{te}}{2})$, (d) Graph $(C_{l\delta})_{theory}$ vs cf/c

If equation (4) is satisfied then $C''_{h\delta} = C'_{h\delta}$, but if not $C''_{h\delta}$ can be calculated by using equation (16)

$$C''_{h\delta} = C'_{h\delta} + 2.(C_{l\delta})_{theory}.[1 - \frac{C_{l\delta}}{(C_{l\delta})_{theory}}].[\tan\frac{\Phi_{te}}{2} - \frac{t}{c}] \quad (16)$$

$(C_{l\delta})_{theory}$ can be obtained from graph 6(d). The next step is calculating $(C_{h\delta})_{balance}$ using following equation

$$(C_{h\delta})_{balance} = C''_{h\delta}.\frac{(C_{h\delta})_{balance}}{C''_{h\delta}} \quad (17)$$

Next, balance ratio equation is stated below

$$BR = \sqrt{(\frac{c_b}{c_f})^2 - (\frac{t}{2c_f})^2} \quad (18)$$

IOP Conf. Series: Journal of Physics: Conf. Series **1005** (2018) 012020 doi:10.1088/1742-6596/1005/1/012020

Then, balance ratio information is applied to this graph to obtain the value of $\frac{(C_{h\delta})''_{balance}}{C''_{h\delta}}$

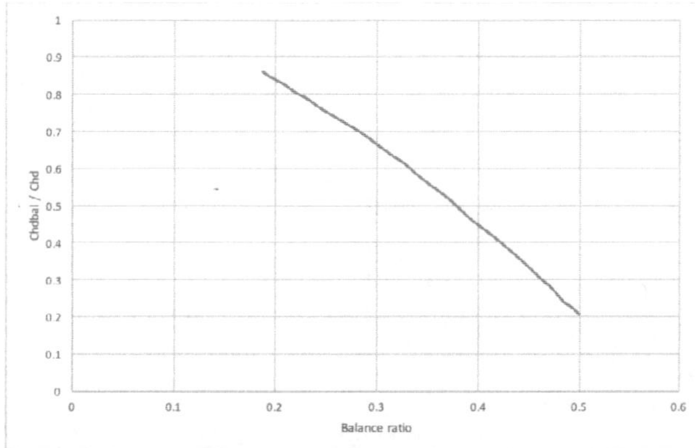

Figure 8. Balance ratio graph

The next step is examining the value of $\Delta C_{h\delta}$ in which the steps are similar to the calculation of $\Delta C_{h\alpha}$. Return to the general equation:

$$C_{h\delta} = \cos\Lambda_{\frac{c}{4}}.\cos\Lambda_{hl}.\left[C_{h\delta M} + a_{\delta}.C_{h\alpha M}.\frac{2\cos\Lambda_{\frac{c}{4}}}{A+2\cos\Lambda_{\frac{c}{4}}} + \Delta C_{h\delta}\right] \quad (20)$$

All the variables can be completely obtained from the steps explained before. Therefore, the value of hinge moment coefficient derivative as the effect of control surface deflection can be calculated. The method of calculation the effect of tab is similar to the surface deflection calculation.

At this point, can be concluded that an aircraft has its unique hinge moment coefficient. The coefficient depends on aircraft geometry, angle of attack, and control surface deflection and proportional to angle of attack and surface deflection. The more advance step is calculating control force of an aircraft during its operation or maneuvers. Control force is function of gearing ratio (*G*), hinge moment coefficient (C_{he}), HTP efficiency (η_{htp}) which is the dynamic pressure ratio between HTP and wing, air density (ρ), airspeed (v), surface area (S), and surface mean chord (c) as shown in equation (21)

$$F_e = \frac{1}{2}.G.C_{he}.\eta_{htp}.\rho.v^2.S_e.c_e \quad (21)$$

Those steps above have been implemented in the hinge moment prediction tool based on MS. Excel. User needs to input aircraft specification including general data, wing and aileron data, and HTP and elevator data. The output of the tool is hinge moment derivative due to angle of attack ($C_{h\alpha}$) and control surface deflection ($C_{h\delta}$).

The hinge moment calculation tool has some features as shown in the Figure 8, 9, and 10.

1. Main sheet as the input and output center of the tool.
2. Airfoil analyzer to visualize the airfoil and calculate its angle especially for trailing edge.
3. Graph sheets, contains graph data and interpolation feature.

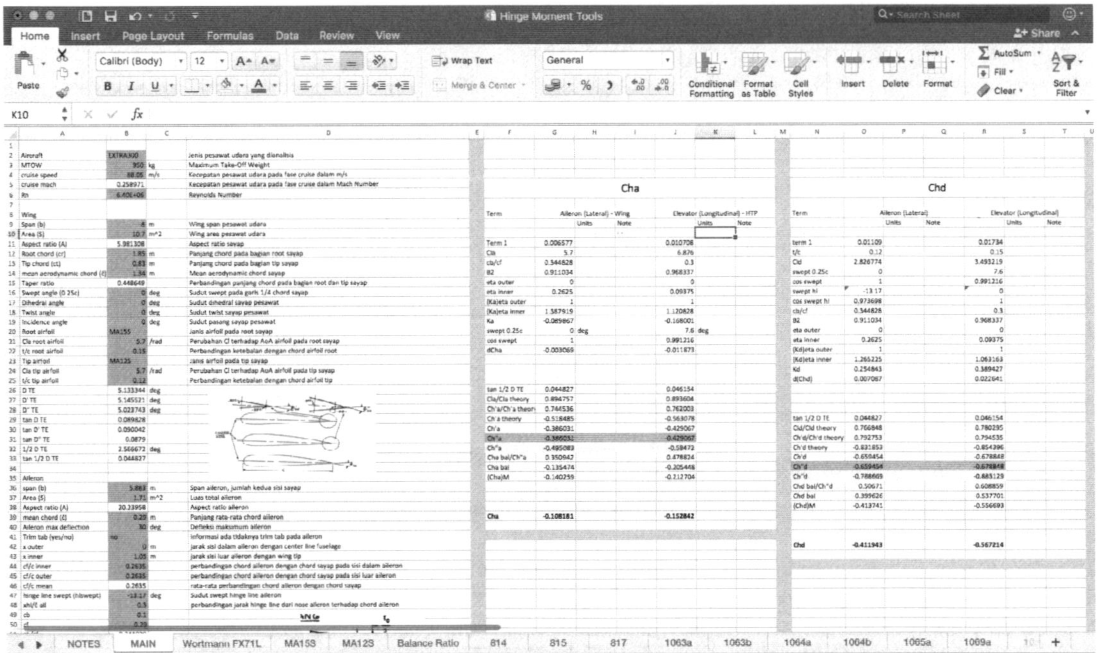

Figure 9. Main sheet of the prediction tool

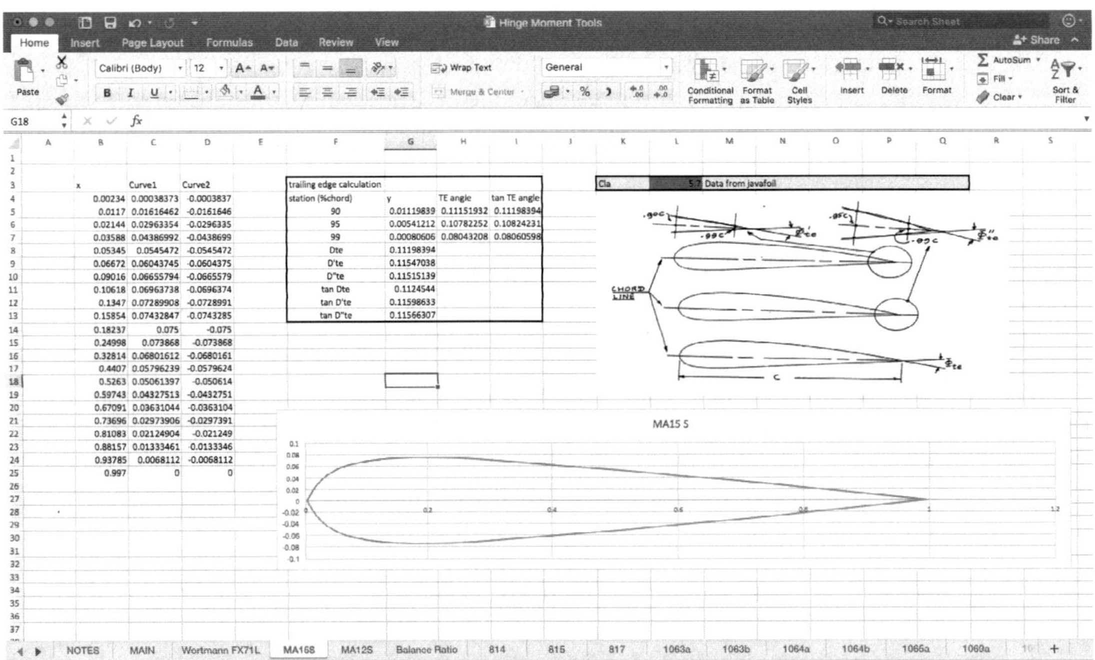

Figure 10. Airfoil sheets as the trailing edge analyzer

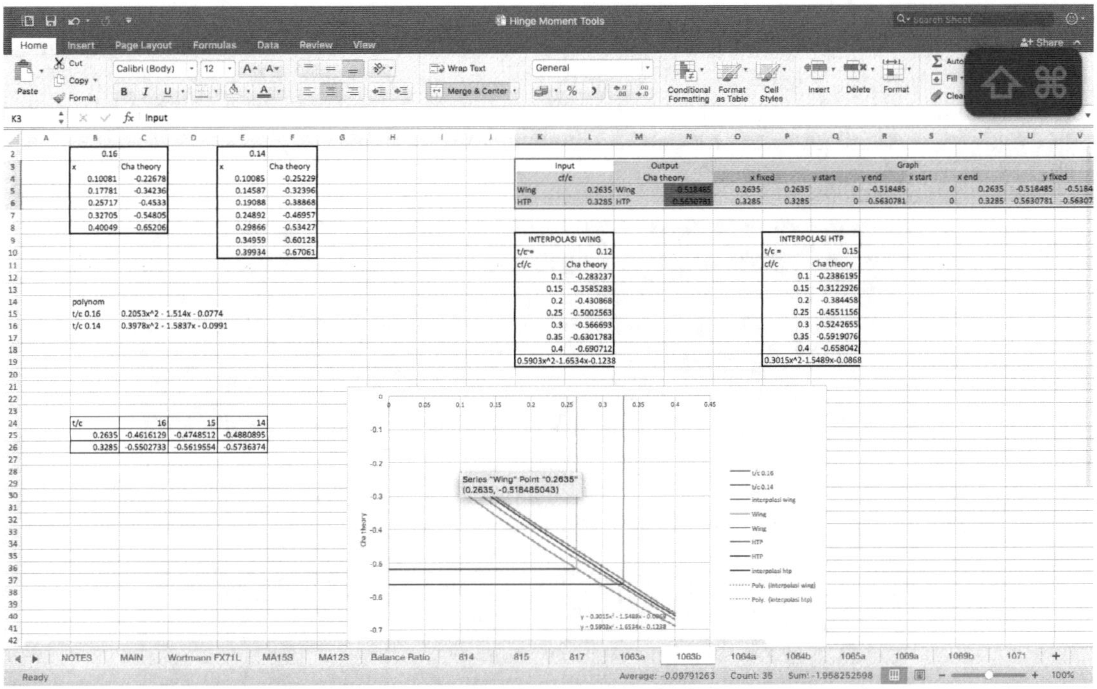

Figure 11. Graph sheet to calculate variable values

3. Result and Discussion

The following table shows the calculation result and the reference data for Cessna 172S

Cessna 172		reference	calculation	error
aileron	Cha	-0.05	-0.078	55.38
	Chd	-0.1	-0.115	14.88
elevator	Cha	-0.295	-0.283	4.04
	Chd	-0.59	-0.566	3.99

Table 1. Comparison between calculation result and reference for Cessna 172S

There are slight differences between reference data and calculation result of elevator for both angle of attack (4.04%%) and control surface deflection (3.99%) derivatives. However, the errors are quite bigger in aileron with 55.38% for angle of attack and 14.88% for surface deflection. But these conditions happen because the coefficients based on reference data are relatively small. Furthermore, there are no contribution of angle of attack in aileron control force because the effect is canceling each other between right and left aileron.

Extra-300 aerobatic aircraft is the object of this research. Table below shows hinge moment coefficient of Extra-300.

Aileron	Cha	-0.10818
	Chd	-0.41194
Elevator	Cha	-0.15284
	Chd	-0.56721

Table 2. Hinge moment coefficient of Extra-300

IOP Conf. Series: Journal of Physics: Conf. Series **1005** (2018) 012020 doi:10.1088/1742-6596/1005/1/012020

Aircraft model of Extra-300 has been simulated using XPlane-10 doing normal flight and maneuvers. Each maneuvering flight consist of several maneuver executions. In this part, the most critical control force condition for each maneuver are presented. The maximum stick force in inside loop is 566.97 N while the most critical forces for half cuban 8 and split-s are 561 N and 480.05 N respectively as shown in the following graphs

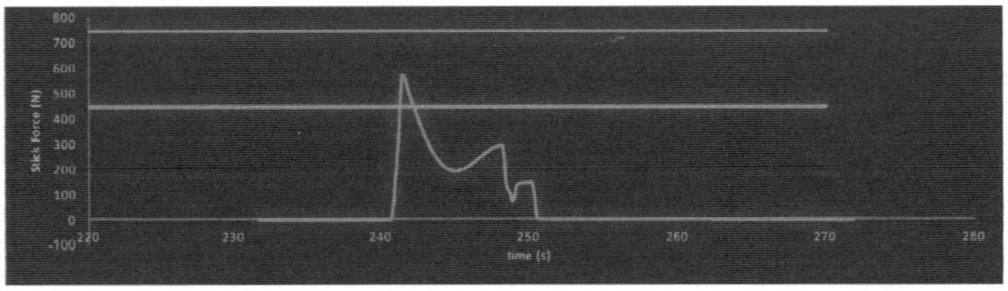

Figure 12. Stick force in inside loop maneuver

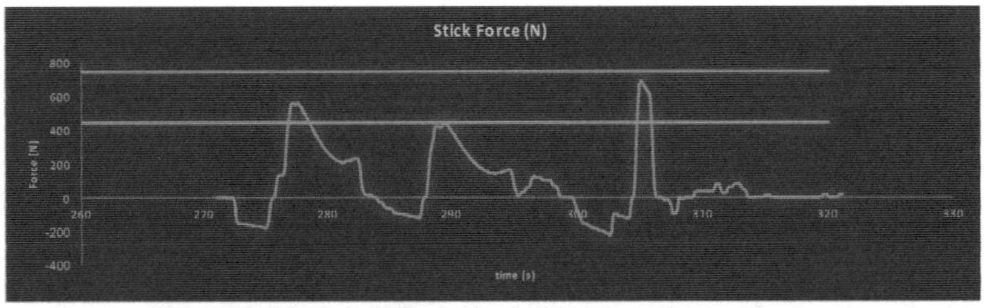

Figure 13. Stick force in half cuban 8 maneuver

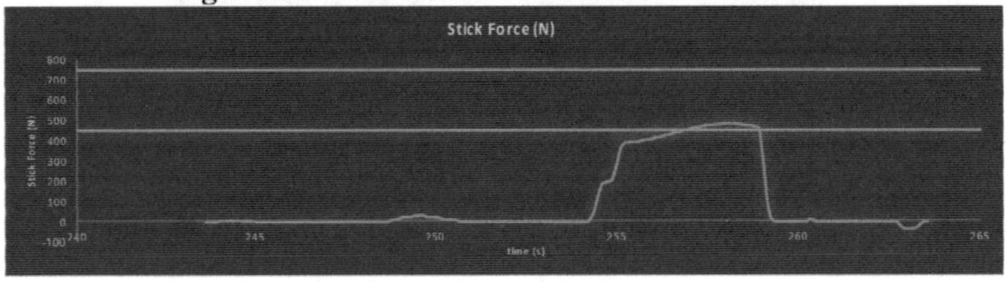

Figure 14. Stick force in split-s maneuver

Furthermore, since aileron roll involves aileron only, the critical force happen in aileron or called wheel force. The maximum wheel force is 340.82 N as shown in the figure below

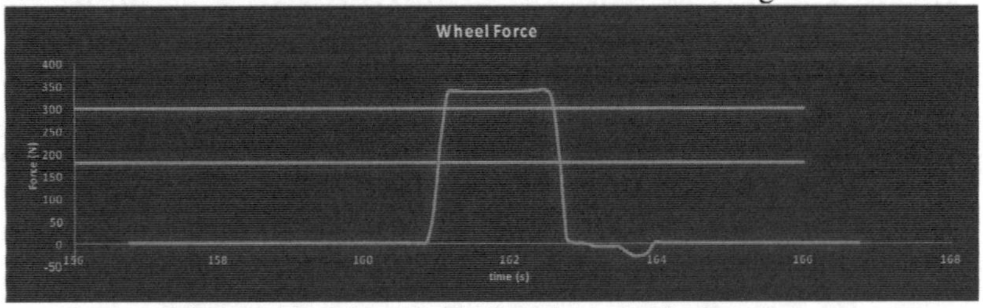

Figure 15. Wheel force in aileron roll maneuver

Control force is a function several parameters including airspeed and control surface deflection. Force is proportional to airspeed hence higher airspeed results in higher force. Besides, surface deflection is proportional to hinge moment coefficient which is also proportional to force. The following split-s graph shows the relation between airspeed and force.

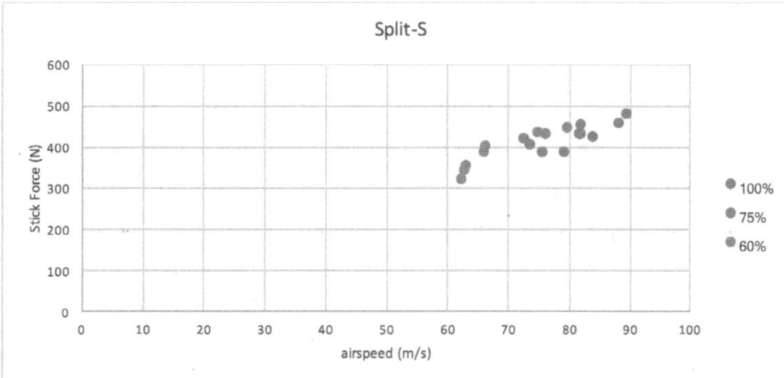

Figure 16. Relation between stick force and airspeed

The following inside loop maneuver presents the relation between surface deflection and force.

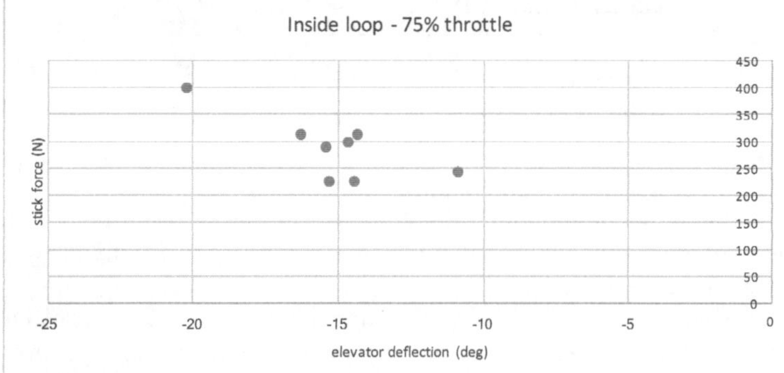

Figure 17. Relation between stick force and elevator deflection

Besides, there is correlation between airspeed or deflection to maneuver shape i.e. loop diameter. Elevator deflection and entry airspeed are inversely proportional to loop diameter. Bigger deflection will generate more extreme inside loop with smaller diameter and greater load factor. Extra-300 can afford until 10 gravities **[8]**.

For further analysis, the maneuvering flight data are recorded and tabulated as follow:

throttle	100%					
no	entry airspeed	entry altitude	max altitude	loop diameter	max elevator deflection	max stick force
	m/s	m	m	m	deg	N
1	87.5	565.2	884.1	318.9	-11.28	316.49
2	90.71	663.06	1074.51	411.45	-8.43	297.72
3	87.4	853.3	1197.63	344.33	-10.63	301.6
4	89.43	894.17	1152.84	258.67	-14.91	332.64
5	93.57	799.75	973.04	173.29	-23.72	566.97
6	86.62	796.31	1097.06	300.75	-11.16	315.73
7	90.04	875.39	1026.88	151.49	-20.46	504.07

Table 3. Flight data during inside loop maneuver

throttle	100%									
no	entry airspeed	entry altitude	exit altitude	min altitude	max altitude	max altitude change	max elevator deflection	max aileron deflection	max stick force	max wheel force
	m/s	m	m	m	m	m	deg	deg	N	N
1	92.43	702.57	1103.09	670.29	1117.6	447.31	-8.67	-16.24	195.28	-89.52
2	81.68	602.74	756.96	572.44	775.62	203.18	-13.26	-30	309.09	-234.8
3	89.77	910	1109.52	790.7	1147.02	356.32	-9.22	-17.4	302.8	-146.7
4	89.21	732	737.83	564.95	822.43	257.48	-25	-18.47	484.32	-153.64
5	94.38	574.51	472.42	368.29	592.8	224.51	-23.64	-10.93	561	-146.52

Table 4. Flight data during half cuban 8 maneuver

throttle	100%							
no	entry airspeed	entry altitude	min altitude	curve diameter	max elevator deflection	max aileron deflection	max stick force	max wheel force
	m/s	m	m	m	deg	deg	N	N
1	88.21	570.39	243.16	327.23	-11.78	-5.3	456.89	-194.95
2	79.23	590.18	179.38	410.8	-8.6	-5.58	387.14	-154.37
3	81.71	555.31	194.44	360.87	-10.85	6.93	430.31	190.82
4	89.42	488.66	177.19	311.47	-13.1	-6.49	480.05	-235.73
5	83.87	402.99	45.91	357.08	-9.93	12.28	423.13	292.88
6	81.99	397.71	137.52	260.19	-14.48	9.32	431.71	238

Table 5. Flight data during split-s maneuver

throttle	75%			
no	entry airspeed	entry altitude	max aileron deflection	max wheel force
	m/s	m	deg	N
1	78.13	461.9	-5.9	-201.72
2	78.4	450.89	6.39	196.68
3	76.07	488.06	-6.93	-202.97
4	79.29	465.67	8.13	243.37
5	80.1	433.69	-10.62	-318.32
6	79.27	418.22	14.36	340.82
7	80.99	407.92	-7.76	-251.9
8	79.65	448.52	8.76	258.04
9	83.45	409.93	-7.09	-250.78

Table 6. Flight data during aileron roll maneuver

4. Conclusion

This research includes development, validation, and application of hinge moment prediction tool. The tool can be used to calculate coefficient of hinge moment of a subsonic aircraft. It has been validated using comparison data of Cessna-172.

Application on Extra-300 aerobatic aircraft generates hinge moment coefficient derivatives due to angle of attack of -0.108 for aileron and -0.152 for elevator. Besides hinge moment coefficient derivatives due to control surface deflection are -0.411 for aileron and -0.567 for elevator. The maximum stick force occurs in inside loop maneuver up to 566.97 N, while the maximum wheel force is 340.82 N in aileron roll. Control force is proportional to airspeed and control surface deflection.

5. Acknowledgement

The work is supported by Faculty of Mechanical and Aerospace Engineering and PT T&E Simulation. The authors would like to thank President Director of PT T&E Simulation for the support and Department of Aeronautics and Astronautics FTMD ITB for the facility.

Reference

[1] Nelson, Robert C.1998.*Flight Stability and Automatic Control*.Singapore: McGraw-Hill Book Co

[2] http://www.dept.aoe.vt.edu/~lutze/AOE3134/Stickfreecharacteristics.pdf

[3] https://www.slideshare.net/nyinyikyaw/basic-aircraft-control-system-7376945

[4] Roskam, Jan. 1985. *Airplane Design*. Kansas: Roskam Aviation and Engineering Corporation

[5] Williams, Neil. 1975. *Aerobatics*. Ramsbury: Airlife Publishing

[6] Wiggins, Mark W et al. 2014. *Cue-utilization typologies and pilots' pre-flight and in-flight weather decision-making*. www.elsevier.com

[7] Corda, Stephen. 2017. Introduction to Aerospace Engineering with a Flight Test Perspective. West Sussex: John Wiley & Sons Ltd.

[8] Extra-300 Information Manual.

Flight Dynamic Simulation of Fighter In the Asymmetric External Store Release Process

Imam Safi'i *****, **Ony Arifianto, Chandra Nurohman**

Faculty of Mechanical and Aerospace Engineering, Institut Teknologi Bandung, Indonesia

*asy.syafii16@gmail.com

Abstract. In the fighter design, it is important to evaluate and analyze the flight dynamic of the aircraft earlier in the development process. One of the case is the dynamics of external store release process. A simulation tool can be used to analyze the fighter/external store system's dynamics in the preliminary design stage. This paper reports the flight dynamics of Jet Fighter Experiment (JF-1 E) in asymmetric Advance Medium Range Air to Air Missile (AMRAAM) release process through simulations. The JF-1 E and AIM 120 AMRAAAM models are built by using Advanced Aircraft Analysis (AAA) and Missile Datcom software. By using these softwares, the aerodynamic stability and control derivatives can be obtained and used to model the dynamic characteristic of the fighter and the external store. The dynamic system is modeled by using MATLAB/Simulink software. By using this software, both the fighter/external store integration and the external store release process is simulated, and the dynamic of the system can be analyzed.

Keywords : fighter, external store, flight dynamic simulation

1. Introduction

Generally, a fighter is designed to paralyze the enemies. For this purpose, a fighter usually equipped with weapons or external stores such as missiles, bombs, rockets, etc. In case of external store release to attack enemies, a fighter undergoes a huge disturbance due to the loss of components with a large mass and a considerable aerodynamics contribution. If left unchecked, this process could interfere the fighter's combat operation. Therefore, the dynamics of a fighter in the external store release process is a case that needs to be analyzed and evaluated in the beginning of the development process of the fighter.

One of the method that can be used to perform the analysis is simulation. Simulation has several advantages compared with flight test method. By using this method, it spent only 40% of the fund compared with the flight test method. It is safer to use simulation due to zero percent accident risk. It also prevent losses due to aircraft damage. Simulation is also a flexibel method as it can be done under various conditions [1].

One of the tool that can be used in simulation is Computational Fluid Dynamics (CFD). CFD can be used to do a lot of simulation and analysis. By using CFD, the static stability derivatives can be computed and then the stability of the aircraft can be modelled and predicted [2]. It can be used to simulate and analyze the separation characteristics of a missile, either using Cartesian grids [3] or using Grid-Free Euler Solver [4].

IOP Conf. Series: Journal of Physics: Conf. Series **1005** (2018) 012021 doi:10.1088/1742-6596/1005/1/012021

Another tool that can be used in the dynamics simulation and analysis is MATLAB/Simulink software. Compared to CFD, the numerical calculation of MATLAB/Simulink is lighter on computer and takes a relatively short time.

Institut Teknologi Bandung (ITB), that has been involved in the development process of IF-X, a fighter that will be produced by Indonesian government in cooperation with South Korea Government, is developing a simulation model and facility for a fighter. The fighter model being developed is named Jet-Fighter Experiment (JF-1 E) [1][5]. In the previous works [1][5], besides modeling of JF-1E, air combat maneuvers simulation has also been done. In the next development, it is important to analyze the dynamic response of the fighter for various external store release scenarios. Therefore, a dynamic system that consist of fighter and external store model is needed to be developed.

Figure 1. The illustration of JF-1 E model [1]

The present paper will specifically discuss the simulation of JF-1 E's dynamics in the asymmetric external storage release process using MATLAB/Simulink software.

2. Method

The modeling of JF-1 E has been done by using Advanced Aircraft Analysis (AAA) software, so the stability and control derivatives of its aircraft has been obtained, both in longitudinal [1] and lateral-directional [5] dimension. By using Missile Datcom software, the stability derivatives of AIM 120 AMRAAM has been obtained, too [6]. Both AAA and Missile Datcom are suitable softwares for calculating aerodynamic parameters of an aircraft and a missile at the preliminary design stage. Table 1 shows the stability and control derivatives of JF-1 E as the result of the previous works [1][5] :

Table 1. JF-1 E stability derivatives

Parameters		**h = 3000 m**		
		V = 240 m/s	**V = 300 m/s**	**V = 400 m/s**
C_X	C_{X_0}	-0,0149	-0,0149	-0,0149
	C_{X_α} (/rad)	0,1658	0,123	0,1454
	$C_{X_{\alpha^2}}$ (/rad)	0,0055	0,0055	0,0055
	C_{X_M}	0,0027	0,003	0,0035
	$C_{X_{\delta_F}}$ (/rad)	0	0	0
	$C_{X_{iH}}$ (/rad)	0,0033	-0,0038	0
C_Y	C_{Y_β} (/rad)	-0,6685	-0,6685	-0,6685
	C_{Y_0}	0	0	0

	C_{Y_r} (/rad)	0,5952	0,5894	0,5786
	$C_{Y\delta_r}$ (/rad)	0,1281	0,0829	0,0049
	$C_{Y\delta_a}$ (/rad)	0	0	0
C_Z	C_{Z_α} (/rad)	3,0401	3,5298	3,7369
	C_{Z_0}	0,1078	0,1059	-0,0548
	C_{Z_q} (/rad)	5,7818	5,5973	5,8977
	$C_{Z_{\alpha^2}}$ (/rad)	-0,0027	-0,0027	-0,0027
	$C_{Z\delta_f}$ (/rad)	0	0	0
	$C_{Z_{iH}}$ (/rad)	-0,2506	-0,1128	-0,0619
C_l	C_{l_0}	0	0	0
	C_{l_β} (/rad)	-0,0891	-0,0894	-0,0895
	C_{l_p} (/rad)	-0,2588	-0,222	-0,2231
	$C_{l_{\delta_r}}$ (/rad)	0,015	0,0097	0,0006
	$C_{l_{\delta_a}}$ (/rad)	0,1112	0,1158	0,422
C_m	C_{m_0}	-0,049	-0,0497	-0,0619
	C_{m_α} (/rad)	-0,9263	-1,1154	-1,144
	C_{m_T}	0,0039	0,0044	0,0037
	C_{m_q} (/rad)	-4,6024	-4,0754	-4,3082
	$C_{m_{\delta_f}}$ (/rad)	0	0	0
	$C_{m_{iH}}$ (/rad)	0,3347	0,1501	0,0086
C_n	C_{n_0}	0	0	0
	C_{n_β} (/rad)	0,2371	0,2329	0,2269
	C_{n_r} (/rad)	-0,3722	-0,3592	-0,3471
	$C_{n_{\delta_r}}$ (/rad)	-0,0843	-0,0546	-0,0032
	$C_{n_{\delta_a}}$ (/rad)	-0,0032	-0,003	0,0006

One of the weapon or external store of JF-1 E is a missile named AIM 120 AMRAAAM. It is a flexible missile that can be used in either air-to-air or ground-to-air scenario [6]. Figure 2 shows the stability derivatives of AIM 120 AMRAAM as the result of the previous works [6] :

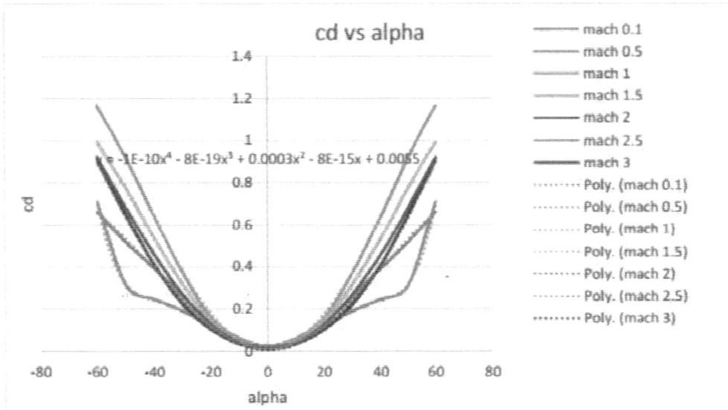

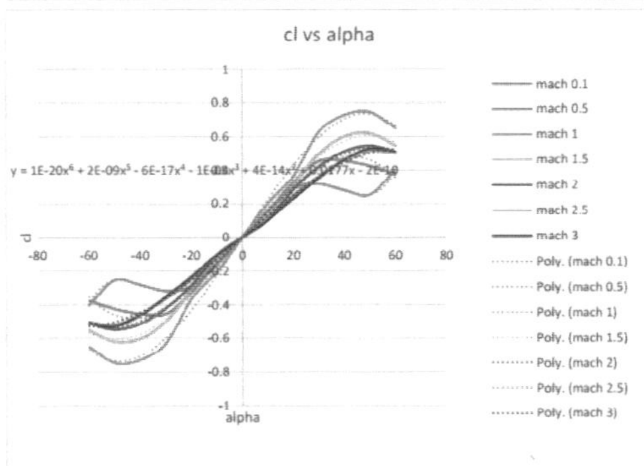

(a) (b)

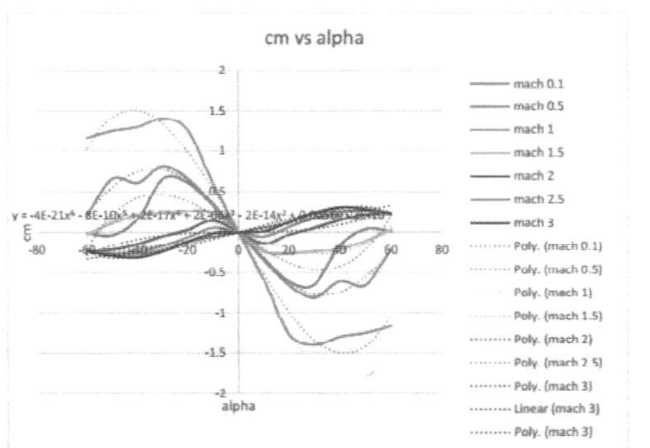

(c)

Figure 2. (a) C_d vs alpha graph, (c) C_l vs alpha graph, C_m vs alpha graph, of AIM 120 AMRAAM model

The works to build the dynamic system can be classified into several parts :

a. modeling JF-1 E's airframe and control coefficients ;
b. modeling AIM 120 AMRAAM coefficients ;
c. modeling aerodynamic forces and moments of JF-1 E and AMRAAM ;
d. modeling the transients equations of external store release process ;
e. modeling the equations of motion.

The general schematic diagram of the flight dynamic simulation of JF-1 E in the asymmetric AIM 120 AMRAAM external store release process is represented in Figure 3.

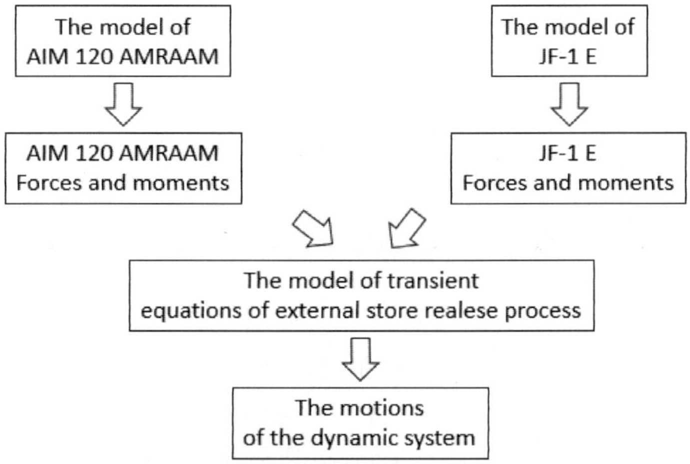

Figure 3. Schematic diagram of the simulation

The model of JF-1 E is built to obtain its airframe and control coefficients. By using all of the stability derivatives that has been obtained from AAA, the fighter's airframe and control calculation blocks was made in Simulink. The total aerodynamic coefficients of the fighter is the sum of the airframe and the control coefficients.

The model of AIM 120 AMRAAM is built to obtain only its airframe coefficients. The airframe coefficients calculation block in Simulink can be made by using stability derivatives that has been obtained by using Missile Datcom software.

Both JF-1 E and AIM 120 AMRAAM coefficients are used to obtain their aerodynamic forces and moments. Since the stability derivatives of JF-1 E and AIM 120 AMRAAM have different references, it will be easier to combine their forces and moments.

To model the fighter/external stores integration and the external store release process, some equations are needed to be developed, called the transients equation of external store release process. The equations must satisfy all of the changed properties of the fighter/external store system caused by the attached or released external stores. The attached or released external stores can change the total mass, total inertia, center of gravity (CG) location, total forces, and total moments of the system. But the effect of the changes of those properties to the equations of motion of the aircraft can be ignored, since the equation of motion are derived relative to the CG.

Generally, the total magnitude of the fighter/external store system's properties is the superposition of the fighter's and the external store's properties. For example, the total mass of the system is the sum of the fighter's mass and the attached external stores's mass. To represent either the integration or the release process, it is needed to add an additional constant to the equation. This constant's magnitude will change over time depending on each external store condition. If the external store is attached to the fighter's body or wing, this constant's magnitude is one. But, if the external store is released, this constant's magnitude will be zero. This additional constant is similar with the binary digit (bit) in the computer. Bit only has two possible values, it is always one or zero.

The transients equation below is used to model the external store release process. For these equations, it is assumed that four AIM 120 AMRAAMs are attached on the wing of JF-1 E. It is also assumed that the aerodynamic forces acting to JF-1 E's airframe is always distributed in a balance condition between the left and the right side, so the total aerodynamic force always located in the middle. Here it is the detailed equations

$$m_{Total} = m_{Aircraft} + m_{Missile}\,(C_1 + C_2 + C_3 + C_4) \quad (1)$$

$$X_{CG_{Total}} = \frac{X_{CG_{Aircraft}} m_{Aircraft} + m_{Missile}\,(X_1C_1 + X_2C_2 + X_3C_3 + X_4C_4)}{m_{Aircraft} + m_{Missile}(C_1 + C_2 + C_3 + C_4)} \tag{2}$$

$$Y_{CG_{Total}} = \frac{Y_{CG_{Aircraft}} m_{Aircraft} + m_{Missile}\,(Y_1C_1 + Y_2C_2 + Y_3C_3 + Y_4C_4)}{m_{Aircraft} + m_{Missile}\,(C_1 + C_2 + C_3 + C_4)} \tag{3}$$

$$Z_{CG_{Total}} = \frac{Z_{CG_{Aircraft}} m_{Aircraft} + m_{Missile}(Z_1C_1 + Z_2C_2 + Z_3C_3 + Z_4C_4)}{m_{Aircraft} + m_{Missile}(C_1 + C_2 + C_3 + C_4)} \tag{4}$$

$$\begin{aligned} I_{xx_{Total}} = I_{xx_{Aircraft}} &+ \left(\left(Y_{CG_{Total}} - Y_{CG_0}\right)^2 + \left(Z_{CG_{Total}} - Z_{CG_0}\right)^2\right) m_{Aircraft} + C_1(I_{xx_{Missile}} \\ &+ \left(\left(Y_1 - Y_{CG_{Total}}\right)^2 + \left(Z_1 - Z_{CG_{Total}}\right)^2\right) m_{Missile}) + C_2(I_{xx_{Missile}} \\ &+ \left(\left(Y_2 - Y_{CG_{Total}}\right)^2 + \left(Z_2 - Z_{CG_{Total}}\right)^2\right) m_{Missile}) + C_3(I_{xx_{Missile}} \\ &+ \left(\left(Y_3 - Y_{CG_{Total}}\right)^2 + \left(Z_3 - Z_{CG_{Total}}\right)^2\right) m_{Missile}) + C_4(I_{xx_{Missile}} \\ &+ \left(\left(Y_4 - Y_{CG_{Total}}\right)^2 + \left(Z_4 - Z_{CG_{Total}}\right)^2\right) m_{Missile}) \end{aligned} \tag{5}$$

$$\begin{aligned} I_{yy_{Total}} = I_{Aircraft} &+ \left(\left(X_{CG_{Total}} - X_{CG_0}\right)^2 + \left(Z_{CG_{Total}} - Z_{CG_0}\right)^2\right) m_{Aircraft} + C_1(I_{yy_{Missile}} \\ &+ \left(\left(X_1 - X_{CG_{Total}}\right)^2 + \left(Z_1 - Z_{CG_{Total}}\right)^2\right) m_{Missile}) + C_2(I_{yy_{Missile}} \\ &+ \left(\left(X_2 - X_{CG_{Total}}\right)^2 + \left(Z_2 - Z_{CG_{Total}}\right)^2\right) m_{Missile}) + C_3(I_{yy_{Missile}} \\ &+ \left(\left(X_3 - X_{CG_{Total}}\right)^2 + \left(Z_3 - Z_{CG_{Total}}\right)^2\right) m_{Missile}) + C_4(I_{yy_{Missile}} \\ &+ \left(\left(X_4 - X_{CG_{Total}}\right)^2 + \left(Z_4 - Z_{CG_{Total}}\right)^2\right) m_{Missile}) \end{aligned} \tag{6}$$

$$\begin{aligned} I_{zz_{Total}} = I_{zz_{Aircraft}} &+ \left(\left(X_{CG_{Total}} - X_{CG_0}\right)^2 + \left(Y_{CG_{Total}} - Y_{CG_0}\right)^2\right) m_{Aircraft} + C_1(I_{zz_{Missile}} \\ &+ \left(\left(Y_1 - Y_{CG_{Total}}\right)^2 + \left(X_1 - X_{CG_{Total}}\right)^2\right) m_{Missile}) + C_2(I_{zz_{Missile}} \\ &+ \left(\left(Y_2 - Y_{CG_{Total}}\right)^2 + \left(X_2 - X_{CG_{Total}}\right)^2\right) m_{Missile}) + C_3(I_{zz_{Missile}} \\ &+ \left(\left(Y_3 - Y_{CG_{Total}}\right)^2 + \left(X_3 - X_{CG_{Total}}\right)^2\right) m_{Missile}) + C_4(I_{zz_{Missile}} \\ &+ \left(\left(Y_4 - Y_{CG_{Total}}\right)^2 + \left(X_4 - X_{CG_{Total}}\right)^2\right) m_{Missile}) \end{aligned} \tag{7}$$

$$\begin{aligned} J_{xy_{Total}} = J_{xy_{Aircraft}} &+ \left(X_{CG_{Total}} - X_{CG_0}\right)\left(Y_{CG_{Total}} - Y_{CG_0}\right) m_{Aircraft} \\ &+ C_1\left(J_{xy_{Missile}} + \left(X_{CG_{Total}} - X_1\right)\left(Y_1 - Y_{CG_{Total}}\right) m_{Missile}\right) \\ &+ C_2\left(J_{xy_{Missile}} + \left(X_{CG_{Total}} - X_2\right)\left(Y_2 - Y_{CG_{Total}}\right) m_{Missile}\right) \\ &+ C_3\left(J_{xy_{Missile}} + \left(X_{CG_{Total}} - X_3\right)\left(Y_3 - Y_{CG_{Total}}\right) m_{Missile}\right) \\ &+ C_4\left(J_{xy_{Missile}} + \left(X_{CG_{Total}} - X_4\right)\left(Y_4 - Y_{CG_{Total}}\right) m_{Missile}\right) \end{aligned} \tag{8}$$

$$\begin{aligned} J_{xz_{Total}} = J_{xz_{Aircraft}} &+ \left(X_{CG_{Total}} - X_{CG_0}\right)\left(Z_{CG_{Total}} - Z_{CG_0}\right) m_{Aircraft} \\ &+ C_1\left(J_{xz_{Missile}} + \left(Y_1 - Y_{CG_{Total}}\right)\left(Z_1 - Z_{CG_{Total}}\right) m_{Missile}\right) \\ &+ C_2\left(J_{xz_{Missile}} + \left(Y_2 - Y_{CG_{Total}}\right)\left(Z_2 - Z_{CG_{Total}}\right) m_{Missile}\right) \\ &+ C_3\left(J_{xz_{Missile}} + \left(Y_3 - Y_{CG_{Total}}\right)\left(Z_3 - Z_{CG_{Total}}\right) m_{Missile}\right) \\ &+ C_4\left(J_{xz_{Missile}} + \left(Y_4 - Y_{CG_{Total}}\right)\left(Z_4 - Z_{CG_{Total}}\right) m_{Missile}\right) \end{aligned} \tag{9}$$

$$\begin{aligned} J_{yz_{Total}} = J_{yz_{Aircraft}} &+ \left(Y_{CG_{Total}} - Y_{CG_0}\right)\left(Z_{CG_{Total}} - Z_{CG_0}\right) m_{Aircraft} \\ &+ C_1\left(J_{yz_{Missile}} + \left(X_{CG_{Total}} - X_1\right)\left(Z_1 - Z_{CG_{Total}}\right) m_{Missile}\right) \\ &+ C_2\left(J_{xz_{Missile}} + \left(X_{CG_{Total}} - X_2\right)\left(Z_2 - Z_{CG_{Total}}\right) m_{Missile}\right) \\ &+ C_3\left(J_{xz_{Missile}} + \left(X_{CG_{Total}} - X_2\right)\left(Z_3 - Z_{CG_{Total}}\right) m_{Missile}\right) \\ &+ C_4\left(J_{xz_{Missile}} + \left(X_{CG_{Total}} - X_4\right)\left(Z_4 - Z_{CG_{Total}}\right) m_{Missile}\right) \end{aligned} \tag{10}$$

$$F_{x_{Total}} = F_{x_{Aircraft}} + F_{x_{Missile}}\left(C_1 + C_2 + C_3 + C_4\right) \tag{11}$$

$$F_{y_{Total}} = F_{y_{Aircraft}} + F_{y_{Missile}}\left(C_1 + C_2 + C_3 + C_4\right) \tag{12}$$

$$F_{z_{Total}} = F_{z_{Aircraft}} + F_{z_{Missile}}\left(C_1 + C_2 + C_3 + C_4\right) \tag{13}$$

$$\begin{aligned} L_{Total} = L_{Aircraft} &+ L_{Missile}C_1 + L_{Missile}C_2 + L_{Missile}C_3 + L_{Missile}C_4 \\ &+ F_{Z_{Missile}}\left(Y_1 - Y_{CG_{Total}}\right)C_1 + F_{Z_{Missile}}\left(Y_2 - Y_{CG_{Total}}\right)C_2 \\ &+ F_{Z_{Missile}}\left(Y_3 - Y_{CG_{Total}}\right)C_3 + F_{Z_{Missile}}\left(Y_4 - Y_{CG_{Total}}\right)C_4 \\ &- F_{Y_{Missile}}\left(Z_1 - Z_{CG_{Total}}\right)C_1 - F_{Y_{Missile}}\left(Z_2 - Z_{CG_{Total}}\right)C_2 \\ &- F_{Y_{Missile}}\left(Z_3 - Z_{CG_{Total}}\right)C_3 - F_{Y_{Missile}}\left(Z_4 - Z_{CG_{Total}}\right)C_4 \\ &- \left(Y_{CG_{Total}} - Y_{CG_0}\right)F_{Z_{Aircraft}} + \left(Z_{CG_{Total}} - Z_{CG_0}\right)F_{Y_{Aircraft}} \end{aligned} \tag{14}$$

$$\begin{aligned} M_{Total} = M_{Aircraft} &+ M_{Missile}C_1 + M_{Missile}C_2 + M_{Missile}C_3 + M_{Missile}C_4 \\ &+ F_{Z_{Missile}}\left(X_1 - X_{CG_{Total}}\right)C_1 + F_{Z_{Missile}}\left(X_2 - X_{CG_{Total}}\right)C_2 \\ &+ F_{Z_{Missile}}\left(X_3 - X_{CG_{Total}}\right)C_3 + F_{Z_{Missile}}\left(X_4 - X_{CG_{Total}}\right)C_4 \\ &+ F_{X_{Missile}}\left(Z_1 - Z_{CG_{Total}}\right)C_1 + F_{X_{Missile}}\left(Z_2 - Z_{CG_{Total}}\right)C_2 \\ &+ F_{X_{Missile}}\left(Z_3 - Z_{CG_{Total}}\right)C_3 + F_{X_{Missile}}\left(Z_4 - Z_{CG_{Total}}\right)C_4 \\ &+ \left(X_{CG_{Total}} - X_{CG_0}\right)F_{Z_{Aircraft}} - \left(Z_{CG_{Total}} - Z_{CG_0}\right)F_{X_{Aircraft}} \\ &- \left(Z_{CG_{Total}} - Z_{CG_0}\right)F_{Propulsion} \end{aligned} \tag{15}$$

$$\begin{aligned} N_{Total} = N_{Aircraft} &+ N_{Missile}C_1 + N_{Missile}C_2 + N_{Missile}C_3 + N_{Missile}C_4 \\ &- F_{X_{Missile}}\left(Y_1 - Y_{CG_{Total}}\right)C_1 - F_{X_{Missile}}\left(Y_2 - Y_{CG_{Total}}\right)C_2 \\ &- F_{X_{Missile}}\left(Y_3 - Y_{CG_{Total}}\right)C_3 - F_{X_{Missile}}\left(Y_4 - Y_{CG_{Total}}\right)C_4 \\ &- F_{Y_{Missile}}\left(X_1 - X_{CG_{Total}}\right)C_1 - F_{Y_{Missile}}\left(X_2 - X_{CG_{Total}}\right)C_2 \\ &- F_{Y_{Missile}}\left(X_3 - X_{CG_{Total}}\right)C_3 - F_{Y_{Missile}}\left(X_4 - X_{CG_{Total}}\right)C_4 \\ &+ \left(Y_{CG_{Total}} - Y_{CG_0}\right)F_{X_{Aircraft}} + \left(Y_{CG_{Total}} - Y_{CG_0}\right)F_{Propulsion} \\ &+ \left(X_{CG_{Total}} - X_{CG_0}\right)F_{Y_{Aircraft}} \end{aligned} \tag{16}$$

where m is mass, X_{CG}, Y_{CG}, and Z_{CG} are CG location, X, Y, and Z are the location of each external store, I_{xx}, I_{yy}, and I_{zz} are moment of inertia, J_{xy}, J_{xz}, and J_{yz} are product of inertia, F_x, F_y, and F_z are forces, L, M, and N are moments, and C is the additional coefficient of each external store.

The dynamic system's motions is modeled using six-degree-of-freedom (6DOF) equation of motion, so both longitudinal and lateral-directional motions of the system can be simulated and analyzed simultaneously.

3. Results and Discussions

Some scenarios are simulated to analyze the dynamic of the fighter in the external store release process. One of the release process of AIM 120 AMRAAM from JF-1 E's wing is simulated in the following conditions :

- four external stores are attached to the fighter, the one placed in the most left side is released ;

- the system is set in the steady symmetric level flight condition at the speed of 200 m/s and the altitude of 3.000 m;
- the time for simulation is 300 seconds, the external store is released in the 25th second;
- the mass of the fighter is assumed to be constant at 24.000 kg during the simulation ;
- the results of the simulation will be shown from 15th second.

It will be easier to analyze the system if it is in a trim condition before the external store is released. Table 3 shows the magnitude of the inputs given to the fighter to make the system in the steady symmetric level flight condition :

Table 2. Input to the dynamic system

Setting angle of horizontal tail (°)	-22,2151
Flaps deflection (°)	0
Aileron deflection (°)	0
Rudder deflection (°)	0
Thrust (N)	51779,2981

Before the model is simulated, the poles of the linearized system can be analyzed to see the characteristic of the system. Table 3 shows the poles of the dynamic system :

Table 3. The dynamic system's poles

Dimension	Poles
Longitudinal	-0.0102 + 0.0666i
	-0.0102 - 0.0666i
	-0.9170 + 4.6902i
	-0.9170 - 4.6902i
Lateral-Directional	-0.2747 + 3.3275i
	-0.2747 - 3.3275i
	-0.0193
	-1.0486

From table 3 above, the poles show that the system is stable, both in longitudinal and lateral-directional dimension. In the longitudinal dimension, the poles show that the dynamic system has short period mode with a high enough natural frequency. Otherwise, the natural frequency of phugoid mode is quite low. In the lateral-directional dimension, the poles show that the dutch roll mode has a high enough natural frequency. The spiral and roll damping mode show a stable behaviour.

Some of the obtained results of the simulation are as follow :

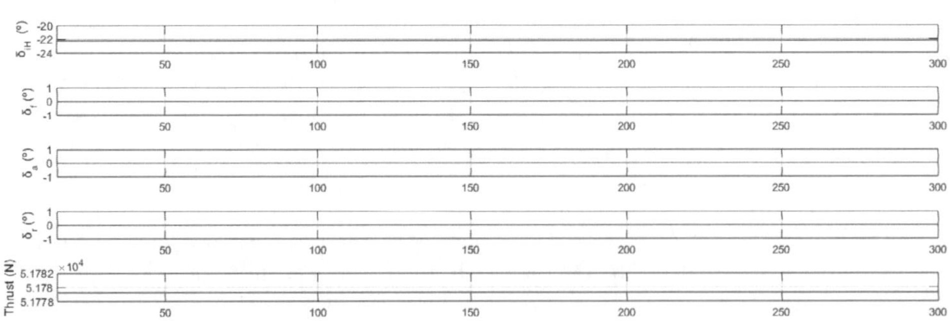

Figure 4. Inputs given to the fighter during the simulation

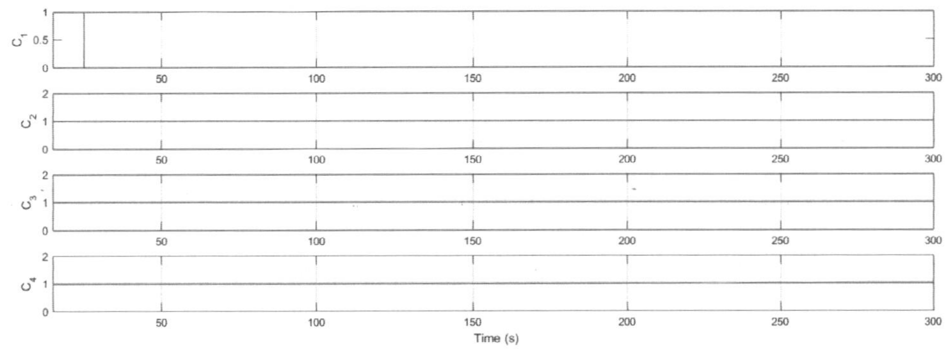

Figure 5. Inputs to the external store release constants

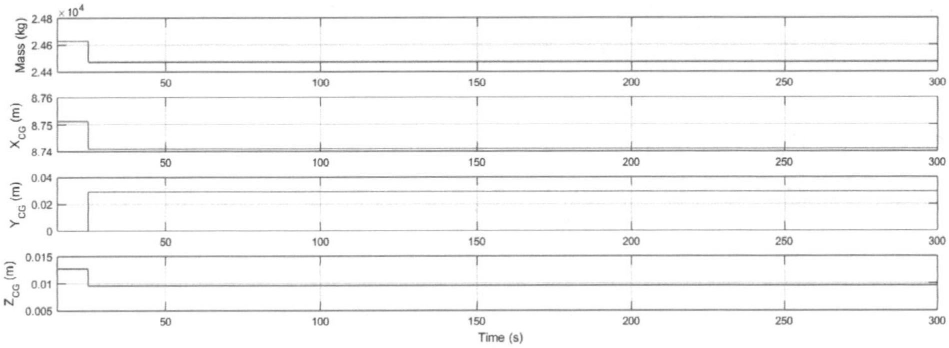

Figure 6. Mass and CG location

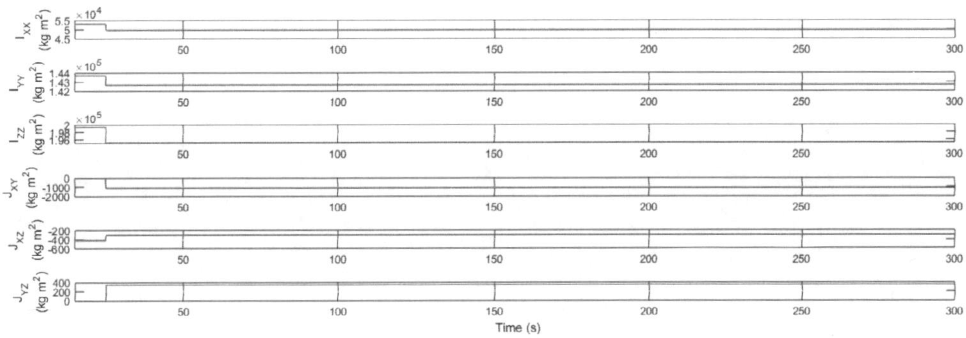

Figure 7. Inertia

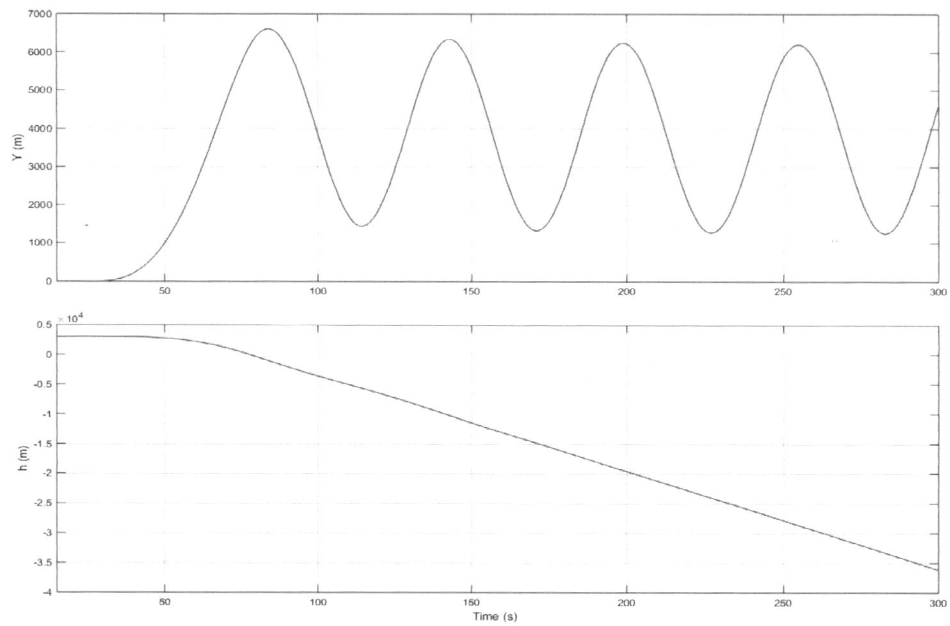

Figure 8. Positions

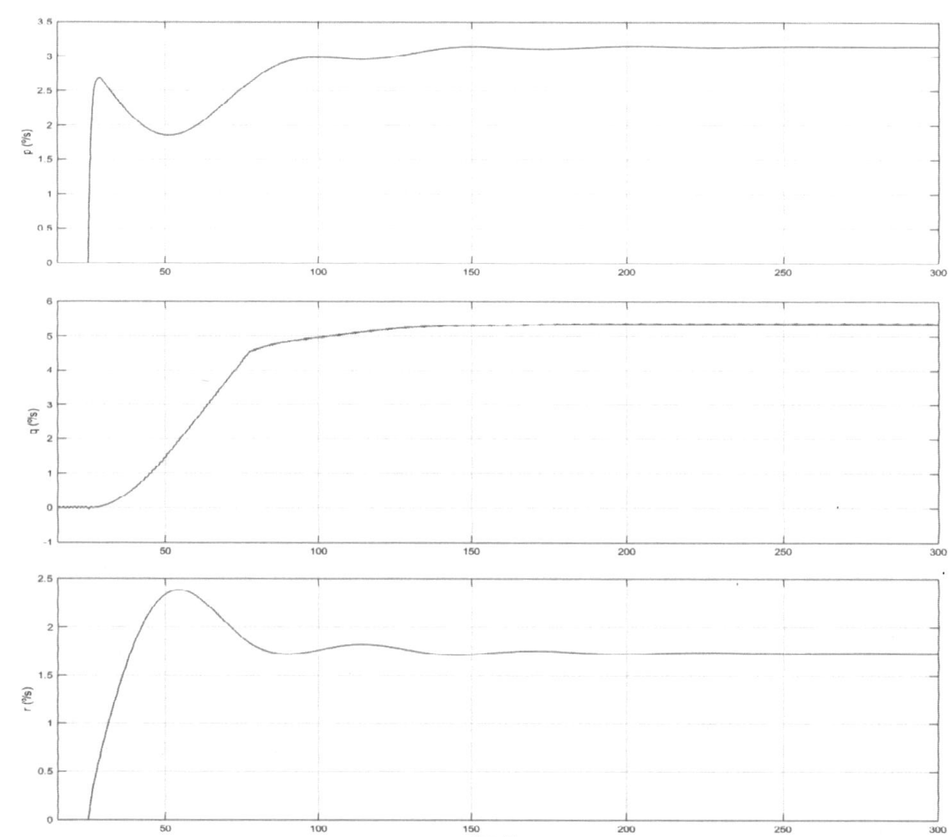

Figure 9. Angular velocity

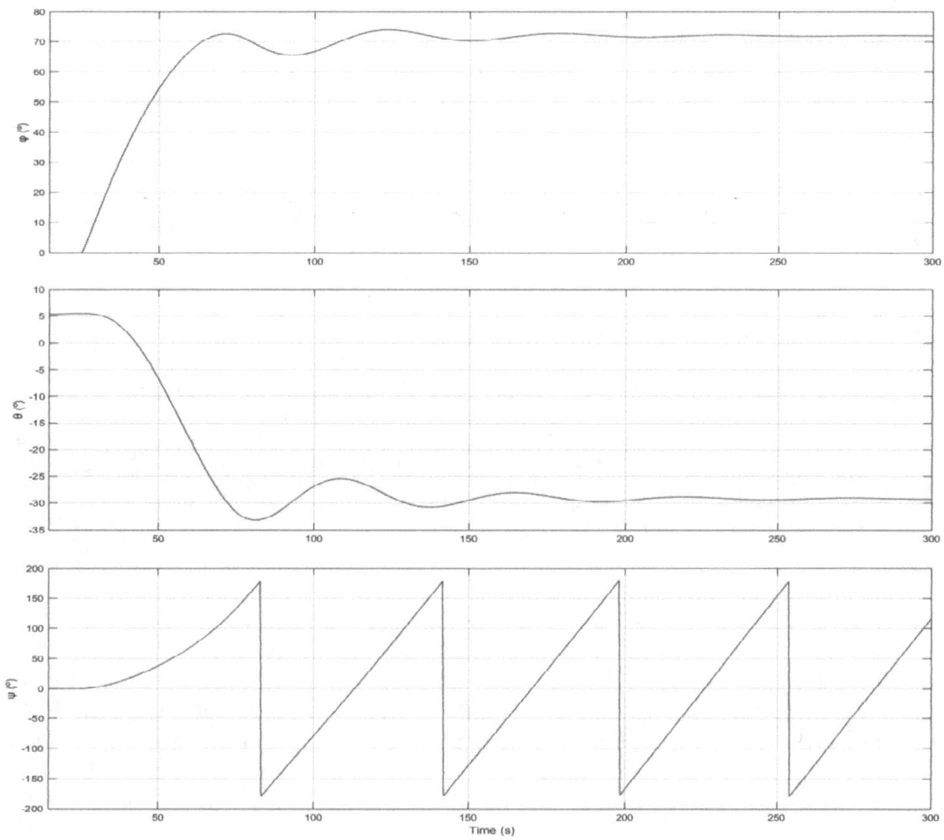

Figure 10. Attitude Angles

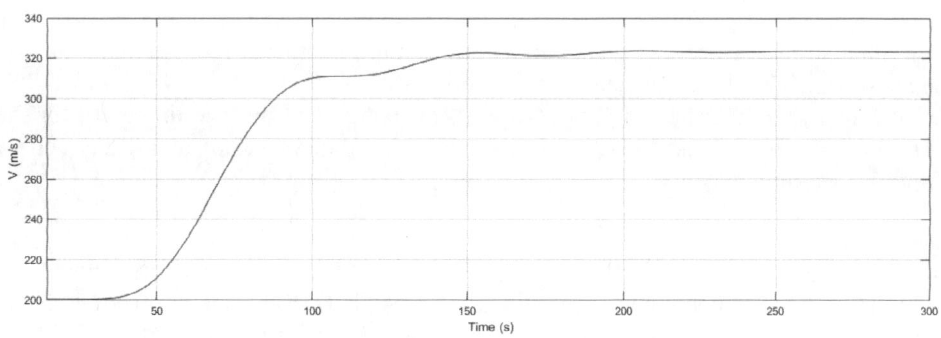

Figure 11. Speed of the dynamic system

IOP Conf. Series: Journal of Physics: Conf. Series **1005** (2018) 012021 doi:10.1088/1742-6596/1005/1/012021

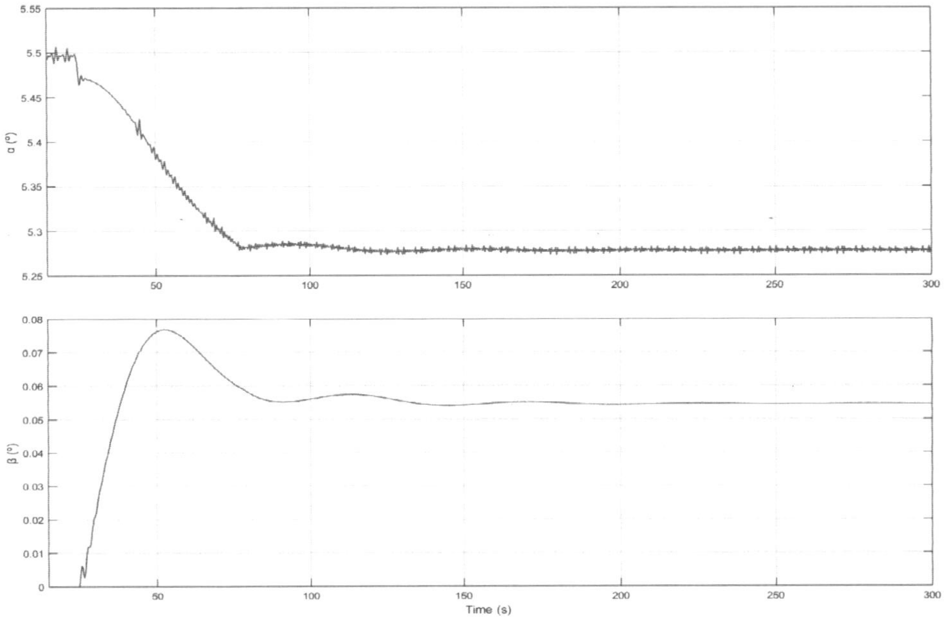

Figure 12. Aerodynamic Angle

According to the simulation results above, it can be seen that the external store release process gives a large enough effect to the dynamics of the fighter. As the external store released, there are some changes in the dynamic system's properties, such as mass, inertia, and center of gravity (CG) location. The total mass of the system is decreased and cause the CG location of the system shift to the right. The mass changes also lead to changes in inertia. All of the changes on these properties will affect the magnitude of the forces and moments of the dynamic system.

The results of the simulation show that the spiral mode of the system is more dominant than the other modes. The system move in a circle to form a spiral while continuing to lose its altitude. The magnitude of the yaw angle show that the system undergoes a turnaround, since it moves from -180° to 180° and back again. The Y-position of the system show a sinusoidal pattern, it indicates that the system move in a circular path with a certain diameter. The dynamic system's speed increases due to the dive. A small side slip angle also arises as the result of spiral motion.

Since the spiral mode has a long enough time constant, it presents no difficulty for the pilot to control the fighter even if it is unstable [7]. To reduce the spiral mode effect, a yaw damper system can be installed to the fighter.

4. Conclusion

The dynamic simulation of JF-1 E in the AIM 120 AMRAAM release process has been done in this paper. The dynamic system model has been built using MATLAB/Simulink and tested. The results of the simulation show that external store release process can affect the dynamic of the fighter. It shows that the spiral mode is more dominant than the other natural responses. Actually, the pilot will find no difficulty to control the fighter since the spiral mode take a long time. To reduce the spiral mode effect, a yaw damper system can be installed to the fighter.

Acknowledgement
This work is supported by Faculty of Mechanical and Aerospace Engineering, Institut Teknologi Bandung.

IOP Conf. Series: Journal of Physics: Conf. Series **1005** (2018) 012021 doi:10.1088/1742-6596/1005/1/012021

References

[1] Megantara, K. 2016. "Simulasi Manuver Tempur Pesawat "JF-1 E' Pada Matra Lateral-Direksional Menggunakan MATLAB/Simuink". *Essay*. Faculty of Mechanical and Aerospace Engineering, Institut Teknologi Bandung.

[2] Dean, J. P., Morton, S. A., McDaniel, D. R. and Gortz, S., "Efficient High Resolution Modelling of Fighter Aircraft with Stores for Stability and Control Clearance," *AAIA Paper 2007-1652*, 2007.

[3] Murman, S. M., Aftosmis, M. J. and Berger, M. J., "Simulations of Store Separation from an F/A-18 with a Cartesian Method," *Journal of Aircraft, Vol. 41, No. 4,* 2004.

[4] Anandhanarayanan, K., Arora, K., Shah, V., Khrishnamurty, R. and Chakraborty, D., "Separation Dynamic of Air-to-Air Missile Using a Grid-Free Euler Solver," *Journal of Aircraft, Vol. 50, No. 3,* 2013.

[5] Prasetyo, K. Simulasi Manuver Tempur Pesawat "JF-1 E' Pada Matra Longitudinal Menggunakan MATLAB/Simuink. *Essay*. Faculty of Mechanical and Aerospace Engineering, Institut Teknologi Bandung.

[6] Laksmana, Y. D. Perancangan Dan Simulasi Sistem Panduan, Sistem Kendali Serta Dinamika Terbang AIM 120 AMRAAM. *Essay*. Faculty of Mechanical and Aerospace Engineering, Institut Teknologi Bandung.

[7] Caughey, D. A. 2011. *Introduction to Aircraft Stability and Control Course Notes for M&AE 5070*. New York: Cornell University.

IOP Conf. Series: Journal of Physics: Conf. Series **1005** (2018) 012022 doi:10.1088/1742-6596/1005/1/012022

Development of Nonlinear Flight Mechanical Model of High Aspect Ratio Light Utility Aircraft

S Bahri[1),2),3)] * and R A Sasongko[2)]

1) Aeronautic Technology Center, National Institute of Aeronautics and Space (LAPAN), Indonesia
2) Faculty of Mechanical and Aerospace Engineering, Bandung Institute of Technology (ITB), Indonesia
3) Institute of Aeronautics and Astronautics, Berlin Institute of Technology (TU Berlin), Germany
*sayr.bahri@lapan.go.id

Abstract. The implementation of Flight Control Law (FCL) for Aircraft Electronic Flight Control System (EFCS) aims to reduce pilot workload, while can also enhance the control performance during missions that require long endurance flight and high accuracy maneuver. In the development of FCL, a quantitative representation of the aircraft dynamics is needed for describing the aircraft dynamics characteristic and for becoming the basis of the FCL design. Hence, a 6 Degree of Freedom nonlinear model of a light utility aircraft dynamics, also called the nonlinear Flight Mechanical Model (FMM), is constructed. This paper shows the construction of FMM from mathematical formulation, the architecture design of FMM, the trimming process and simulations. The verification of FMM is done by analysis of aircraft behaviour in selected trimmed conditions.

1. Introduction

The development of Electronic Flight Control System (EFCS) is conducted inside the cooperation between LAPAN and TU Berlin whose purpose is to build a demonstrator aircraft named LAPAN Surveillance Aircraft (LSA-02). The aircraft selected is a light utility single engine aircraft with payload capability up to 350 kg (see Figure 1). The original mechanical flight control system is modified with EFCS to provide full authority to reduce pilot work load and enhance control performance especially during extreme tasks such long endurance flight and high accuracy maneuver. The development of Flight Control Law (FCL) for the Basic EFCS (BEFCS) is conducted to realize the attitude control and flight path control for the LSA-02[8][9]. Other development that is navigation control shall be assigned as the experimental FCL. The development of basic FCL employs a strict development process namely V model development process [1].

The development of basic FCL can be only realized when an accurate model of the aircraft dynamics is available. Assumed as a rigid body, the real aircraft has six degrees of freedom and possess dynamic nonlinearity. Hence, to mimic the real aircraft response and characteristics, the flight mechanical model (FMM) needs to be supported by accurate data such as mechanical properties data (mass, inertias, CG. position, etc.), propulsion system data, and also aerodynamic and stability derivatives coefficients. Since at initial stage some data may be confidential or proprietary which are difficult to obtain, then prediction techniques and assumptions will be used for reconstructing the required data.

The development of FMM considers numerous aspects but it usually starts with the definition of the equations of motion. In addition, the accurate representation of the atmosphere variables should also be considered to provide realistic atmospheric effects to the aircraft response during flight. Furthermore, the contribution of disturbance may be added to the aircraft motion. The formulation of forces and moments coefficients are essential and unique for each aircraft especially for aerodynamic forces and moments coefficients. These may be the most valuable data for an aircraft and often they are confidential. However, some prediction tools may be used to estimate the aerodynamic coefficients including the stability derivatives, which to some degree may help at the early stage of the FMM development. Later in the development process, the data from wind tunnel tests or ultimately the flight tests shall be used for validating the FMM. The propulsion parameters may be easier to acquire from the engine data. However, some parameters can only obtained by means ground tests or flight tests. Other variables like acceleration and load factor shall be included as well in the FMM to provide valuable information especially if the acceleration or load factor exceed the threshold of airframe structural strength, which may be a threat for the safety of human on board.

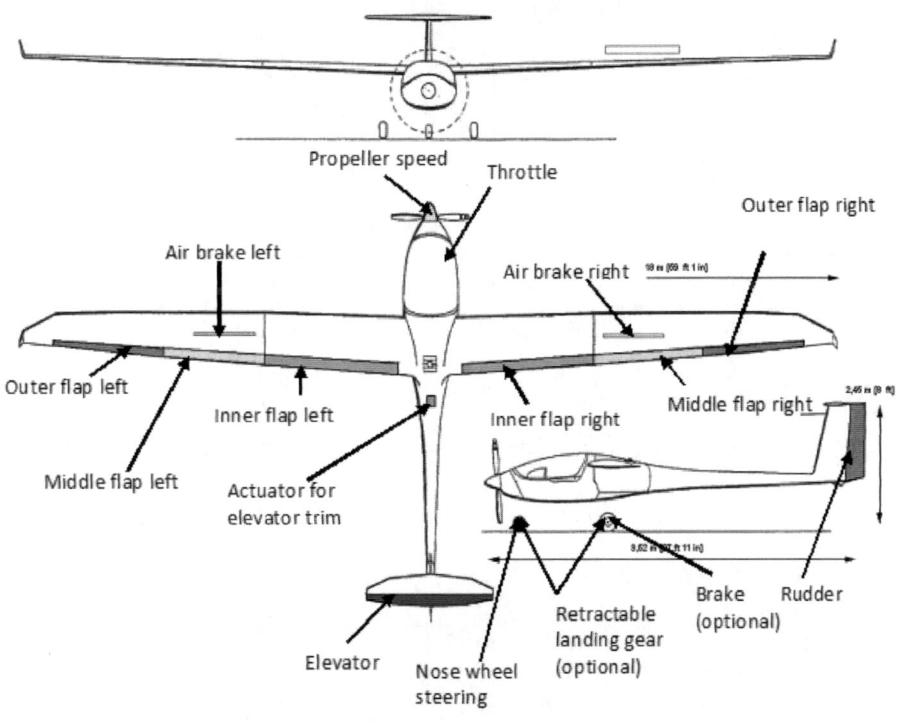

Figure 1. Aircraft STEMME ES15 which is modified with EFCS to become LSA-02 [1].

The main structure of a FMM may be generic and can be adjusted for some different types of aircraft. . However, as the aircraft possesses huge number of parameters which are related to its specific characteristics and the flight regimes it may cover, , the development of FMM shall be carried out independently and exclusively for a specific aircraft.

The research objectives of the work presented in this paper are to develop a nonlinear flight mechanical model of LSA-02 aircraft and to verify the flight dynamic characteristics of the aircraft, especially the natural Eigen mode responses from the flight simulation. This paper shows the construction of FMM, including the definition its dynamic variables, the mathematical formulation, and the architecture design of FMM numerical representation using SIMULINK. In addition, a trimming method for longitudinal and lateral motions and the dynamic response of the FMM are also simulated and presented.

2. Formulation of FMM

It is important to firstly define clear terminology used in FMM development [2]. The formulation of FMM involves four main coordinate systems, i.e. geodetic, aerodynamic, kinetic and body-fixed coordinate systems (see Figure 2).

The equations of motion mostly is described in body-fixed coordinate system. The forces equation is the summation of the forces come from the weight of the aircraft, the thrust of the engine and the aerodynamic forces. The thrust is assumed to work coincidently in body-fixed axis x_f which generates no moments. Hence, the moments equation only comes from aerodynamic moments. The airspeed is obtained from the flight path velocity subtracted with the wind velocity. The Euler angles or the attitude angles are calculated from angular rates. The geodetic position is obtained from calculated ground speed.

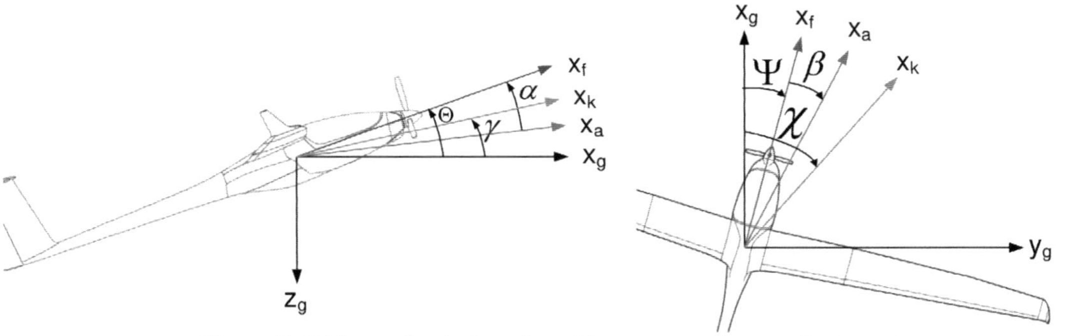

Figure 2. Flight mechanical angles with respect to the coordinate systems.

There are six aerodynamic forces and moments coefficients which are firstly calculated before computing the aerodynamic forces and moments. These coefficients derivatives are explained in [3] and can be written to be

$$C_W = C_{W0} + \frac{C_A^2}{\pi \Lambda e} \tag{1}$$

$$C_Q = C_{Q\beta}\beta + C_{Qp}\left(\frac{p_f s}{V}\right) + C_{Qr}\left(\frac{r_f s}{V}\right) + C_{Q\xi}\xi + C_{Q\zeta}\zeta \tag{2}$$

$$C_A = C_{A0} + C_{A\alpha}\alpha + C_{A\alpha_K}\eta_K \tag{3}$$

$$C_l = C_{l\beta}\beta + C_{lp}\left(\frac{p_f s}{V}\right) + C_{lr}\left(\frac{r_f s}{V}\right) + C_{l\xi}\xi + C_{l\zeta}\zeta \tag{4}$$

$$C_m = C_{m0} + C_{m\alpha}\alpha + C_{m\eta}\eta \tag{5}$$

$$C_n = C_{n\beta}\beta + C_{np}\left(\frac{p_f s}{V}\right) + C_{nr}\left(\frac{r_f s}{V}\right) + C_{n\xi}\xi + C_{n\zeta}\zeta \tag{6}$$

Here, the coefficient index W is drag, Q is side, A is lift. The control surfaces deflection of aileron, elevator, rudder and flaps are denoted respectively with ξ, η, ζ and η_K. Other coefficients e.g. $C_{W0}, \Lambda, e, C_{Q\beta}, C_{Qp}$, etc. on Equation (1)-(6) are defined following the description in [6].

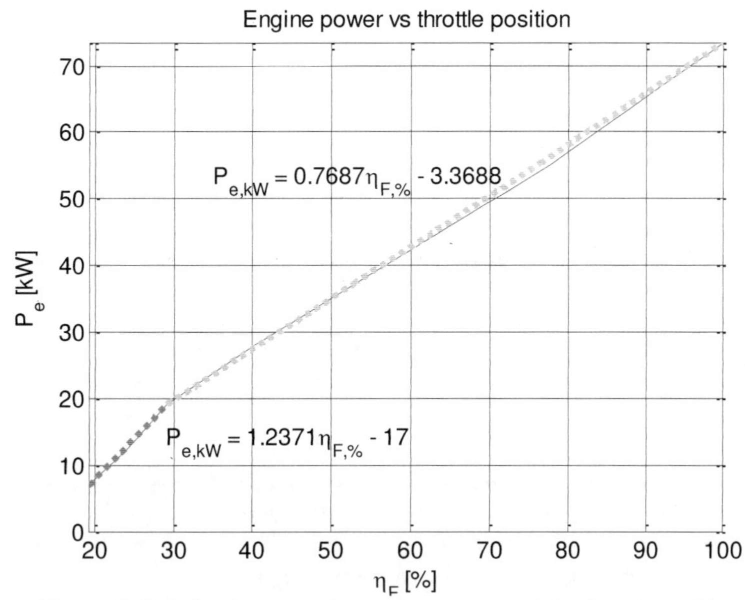

Figure 3. Relation between the engine power and the throttle position.

The thrust force equation is derived from the engine data [4]. Here, it is assumed that the throttle position is proportional to the fuel consumption. Mapping the engine rotation with the throttle position will result the engine power. The thrust force is calculated by equation (**7**).

$$F = \frac{\bar{\eta}_p P_e}{V} \tag{7}$$

where $\bar{\eta}_p$ is propeller efficiency, P_e is engine power and V is airspeed. The propeller efficiency is assumed to be constant and is given to be 0.85. The construction of relation chart between throttle position and the engine power is displayed in **Error! Reference source not found.**.

The atmospheric variables are assumed to follow the international standard atmosphere (ISA). The variables required are geopotential height, static pressure, temperature and air density. The geometric height or the altitude is obtained from geodetic position which may be computed by solving the navigational parts of the equations of motion.

3. Architecture of FMM

The FMM consists of several subsystems which will be continuously developed. The existing subsystems are namely the command subsystem, trim control subsystem, aircraft subsystem and output subsystem. However, the FMM emphasizes more to the aircraft subsystem or the aircraft model. The original aircraft model receives command signals from the input model and produces responses to the output model.

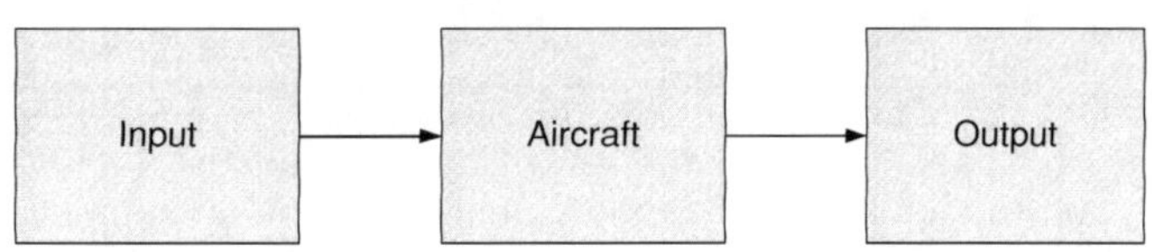

Figure 4. The aircraft model with input and output models.

The environment model is included inside the aircraft model or the aircraft subsystem while the disturbance model is included in the input subsystem. The FMM is realized by means of Simulink / MATLAB applications. The block diagram of the aircraft model and its architecture are depicted in Figure 5.

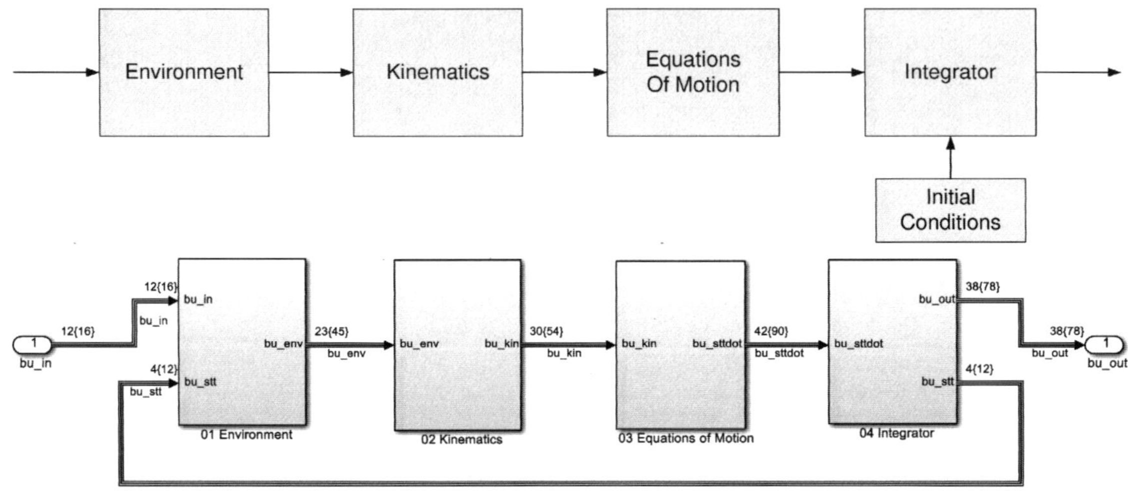

Figure 5. Block diagram of the aircraft model (top) and the Simulink model of the aircraft (bottom).

All sub-subsystems of the aircraft model are derived based on the formulation of FMM which has been previously developed. Environment sub-subsystem comprises wind speed and ISA atmosphere data computation. Kinematics block calculates airspeed and ground speed. Equations of motion sub-subsystem block consists of forces, moments and states computation. The forces and moments block include the weight, propulsion, aerodynamics and inertial data and computation. The integrator sub-subsystem integrates the state derivatives resulted from the equations of motion. There are four derivatives obtained from the integrator sub-subsystem, i.e. flight path speed, angular rate, attitude and position derivatives.

4. Trimming the Aircraft Model

The next phase after building the aircraft model is to compute the trimmed condition of the aircraft. Trimmed condition is when the resultant of forces and moments are zeros. This means the aircraft is on steady flight condition either in longitudinal or lateral motion. There are many techniques to trim the aircraft, even Simulink MATLAB offers the automatic trim application directly from the developed model. Here, the trimming method is proposed by means of FCL implementation. For this purpose, the FCL for trimming the aircraft model is called by the trim control subsystem and form a closed system. Therefore, the FMM is expanded by introducing the command and trim control subsystems. The previous input subsystem is included inside the trim control subsystem.

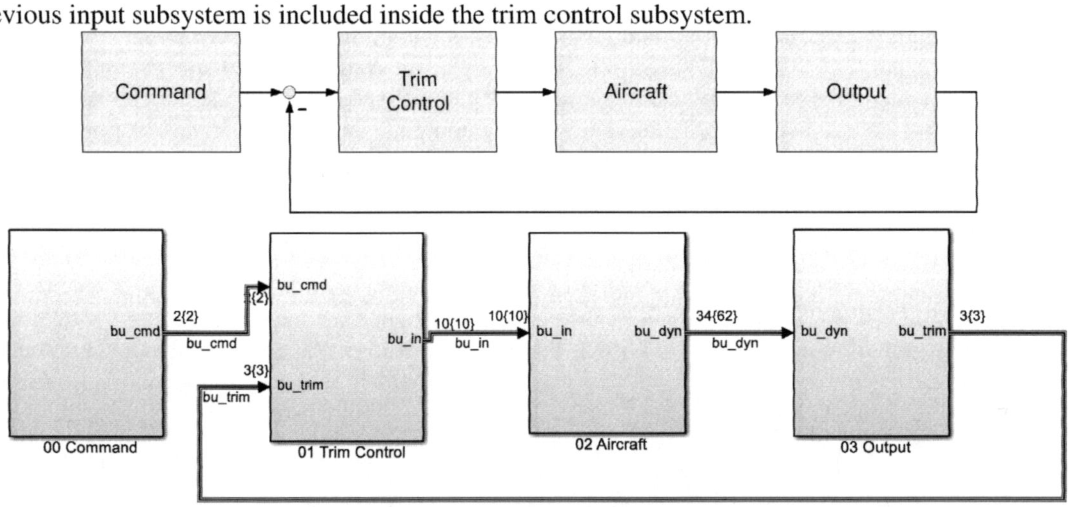

Figure 6. The block diagram of expanded FMM with introduction of trim control subsystem (top) and the Simulink model (bottom).

The trim control subsystem has purpose of computing the required commands for the control surfaces to trim the aircraft in longitudinal and lateral motion. The trim control in this report is essentially the cascade of the flight-path control and attitude control, which are employed as a means to trim the aircraft, hence the required control surfaces deflection are obtained. Therefore, their implementations only concern the steady state responses of the aircraft. The trim control, based on its function, is divided into two controls, i.e. longitudinal motion control and lateral motion control. The longitudinal control consists of the flight-path velocity control and the altitude control, while the lateral control comprises the angle of sideslip control and heading control. Here, the proportional and integral gains are used as controller. The block diagram of trim control is displayed from Figure 7 to Figure 10.

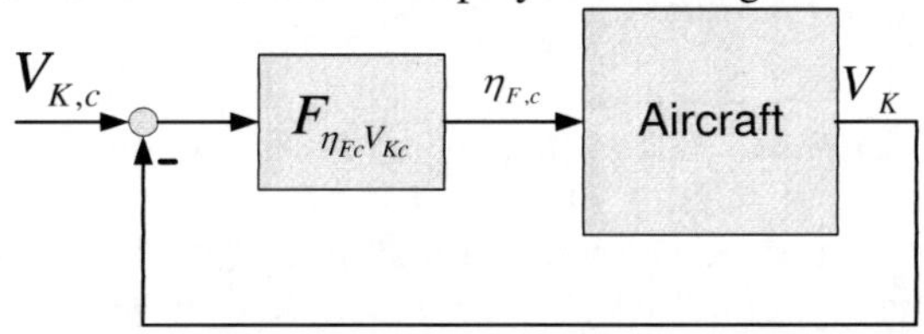

Figure 7. The block diagram of flight path velocity control.

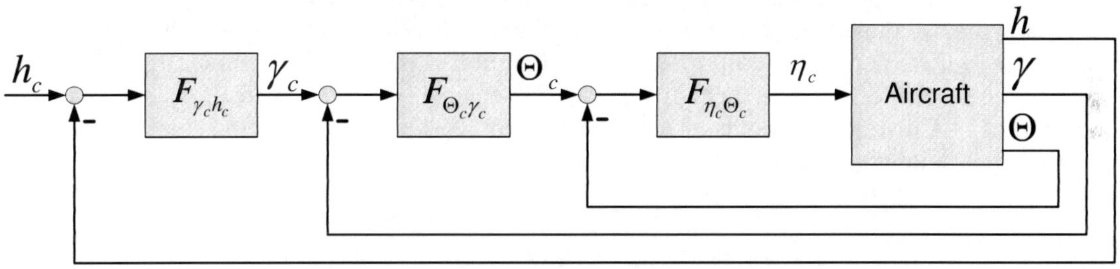

Figure 8. The block diagram of altitude control.

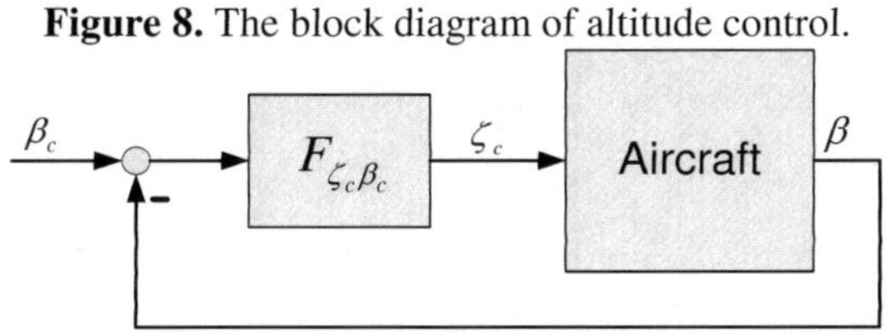

Figure 9. The block diagram of angle of sideslip control.

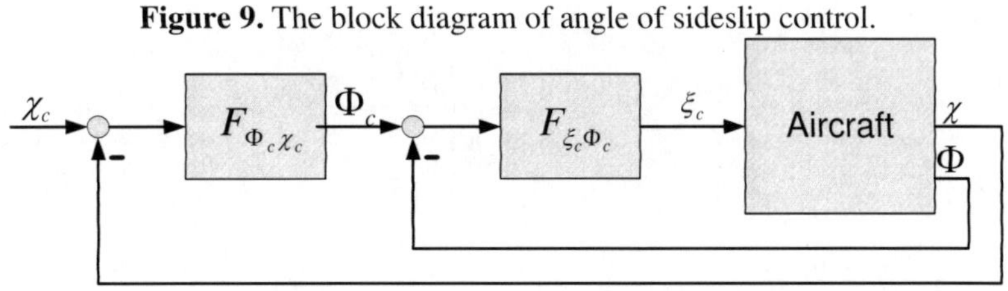

Figure 10. The block diagram of heading control.

The integration between the flight-path velocity control, altitude control, angle of sideslip control, and azimuth control is implemented in trim control subsystem and is displayed in Figure 11.

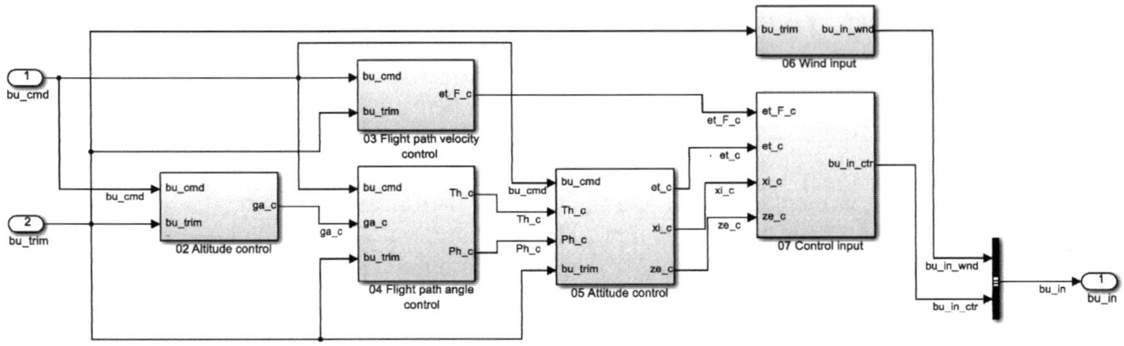

Figure 11. The Simulink model of the trim control.

Initially, the trimming of the aircraft subsystem is carried out for the longitudinal motion, and then is continued for the lateral motion. The longitudinal motion is trimmed by maintaining flight path velocity and altitude. The flight path velocity is realized by the throttle position while the altitude is realized by the elevator. The lateral motion is trimmed by maintaining the angle of sideslip. The angle of sideslip is realized by the rudder while the zero heading angle is realized by the aileron.

In this work, for the longitudinal trimming, the flight path velocity chosen are 130, 160, and 190 km/h while the altitude is fixed at 1000 m. The lateral motion is trimmed by maintaining the angle of side slip and the heading angle. The angle of side slip chosen are -10, -5, 0, 5, and 10 degrees, while the heading is fixed at 0 degree. The results of longitudinal and lateral trimming are the steady state values and required control inputs (control surface deflection).

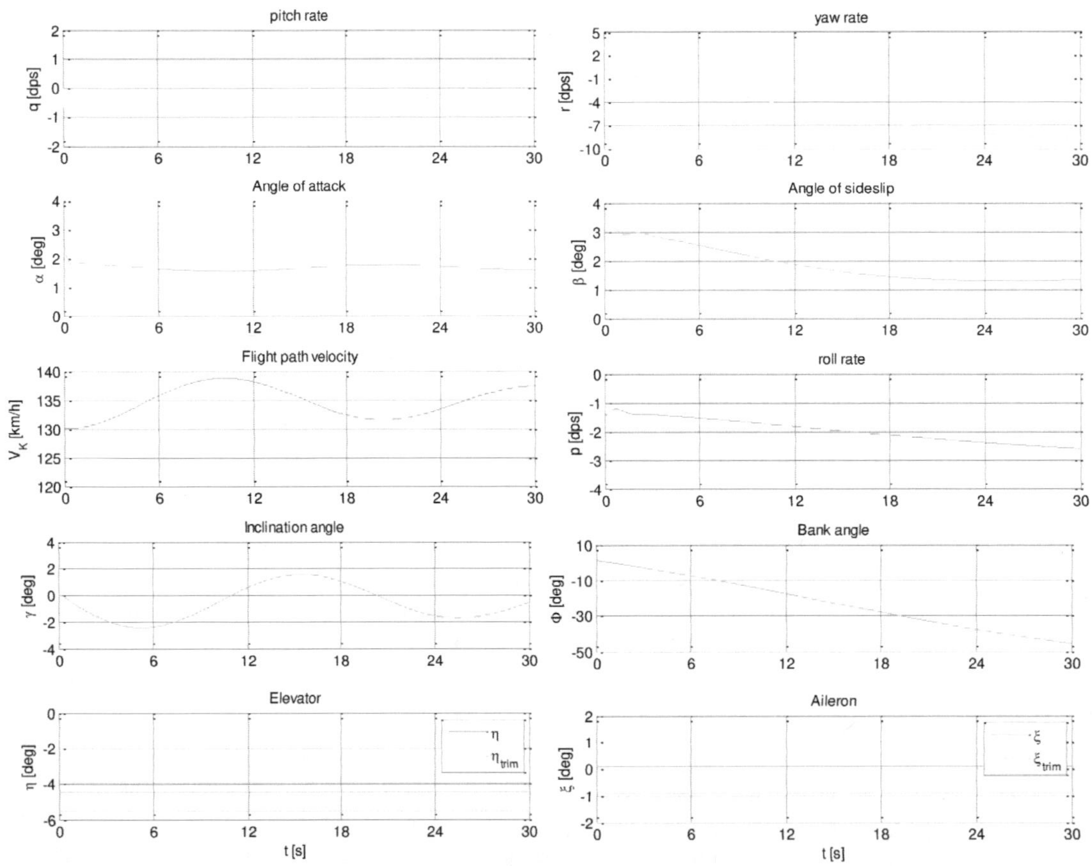

Figure 12. Example of flight simulation. Open loop responses of longitudinal motion (left) and lateral motion (right).

After that, the open loop response is examined to identify the natural behavior of the aircraft model. From trimming values, the disturbance is excited so that the response of the aircraft can be observed. The simulation result for disturbed trimming condition can be seen from Figure 12.

From Figure 12 on the left column, the step input of 1 degree is given to the trimmed elevator deflection. The longitudinal motion consists of two Eigen modes, i.e. short period mode and phugoid mode. The short period can be clearly observed from the pitch rate response around the first two seconds after elevator excitation. The angle of attack although slower response than the pitch rate, but the short period can be distinguished as well within the same time. After two seconds the phugoid appears in all states. From this simulation we can qualitatively estimate that the short period has frequency of 0.5 Hz or around 3 rad/s (because the period is around 2 seconds) while the phugoid has frequency of 0.05 Hz or around 0.3 rad/s (because the period is around 20 seconds). For further examination, the Eigenvalues and Eigenvectors of the linearized system can be computed and compared to the nonlinear simulation results

Figure 12 on right column shows the state variables response from the step input of 1 degree from the aileron. The aileron excites the rolling motion in the aircraft which is represented by the roll rate and the bank angle responses. The nature of the aircraft is unstable typically because of the spiral mode. The spiral mode is mainly reflected by, from the simulation, the bank angle response. This explains that the bank angle keeps diverging away from zero. The roll mode is depicted from the roll rate but surprisingly it shows slight oscillation which probably is caused by low roll damper C_{lp} value. . The Dutch-roll can be clearly seen from the response of the yaw rate around the first two seconds. The weaker Dutch-roll can be observed as well from the angle of sideslip response.

After investigating the open loop responses, the next step is to investigate the trimming curve. The longitudinal and lateral trimming results are observed. The relationships selected for the longitudinal motion are elevator versus flight-path velocity and angle of attack versus flight-path velocity. On the other hand, the relationships selected for the lateral motion are aileron versus angle of sideslip and rudder versus angle of sideslip. The result of trimming curves are given in Figure 13.

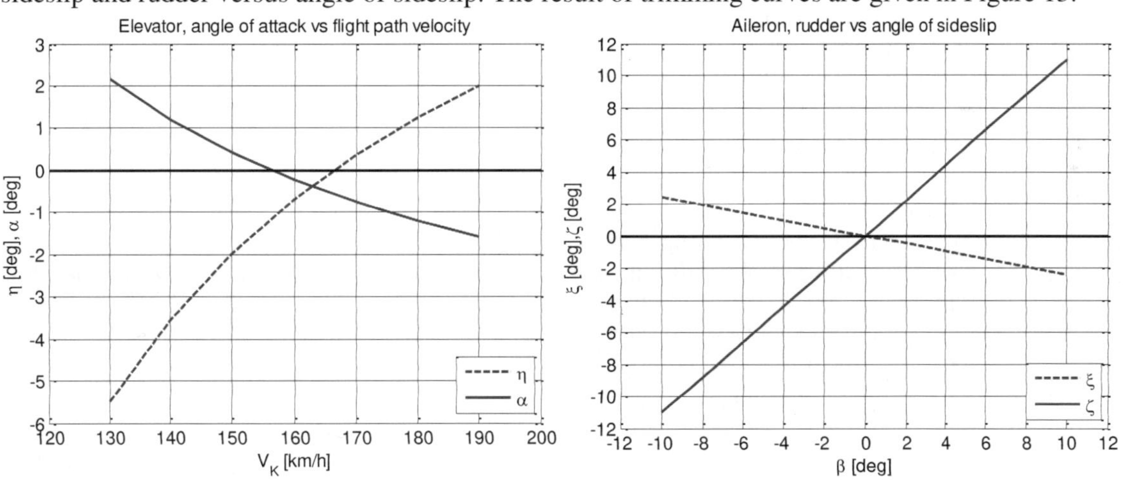

Figure 13. Trimming curve of the elevator, angle of attack with flight path velocity (left) and Trimming curve of the aileron, rudder with the angle of sideslip (right).

From Figure 13 on the left, the relation for elevator and angle of attack with the flight path velocity in longitudinal motion is non-linear. The angle of attack is positive when the flight path velocity is low which explains that the aircraft needs more lift force to fly in low speed. The angle of attack gradually decrease to reduce the lift contribution when the flight path velocity is increased. The elevator has purpose to hold the pitch angle so that the flight path inclination stays at zero. When the flight path velocity is low, the elevator is negative to produce positive pitching moment until the altitude is maintained. If elevator is kept fixed while the flight path velocity is increased, the aircraft will climb

IOP Conf. Series: Journal of Physics: Conf. Series **1005** (2018) 012022 doi:10.1088/1742-6596/1005/1/012022

and the altitude command will fail. Therefore, to maintain the altitude the elevator will increase (from negative to positive) as the flight path velocity is increased.

From Figure 13 on the right, the relation for aileron and rudder with the angle of sideslip in lateral motion seems to be linear. The aileron has purpose to compensate the yawing moment caused rudder deflection. The rudder deflection also causes the rolling moment which is needed to be anticipated by aileron. The aileron has smaller deflection because of long lever arm that is wing. On the other hand, to achieve angle of sideslip command, the rudder is seen to have relatively proportional deflection. In addition, the flight path velocity in longitudinal motion does not affect the relationship in lateral motion. This is because the aircraft is laterally commanded when the longitudinal motion is already trimmed.

5. Conclusions

Some conclusions drawn from this paper are

The development of the nonlinear flight mechanical model of high aspect ratio light utility aircraft has been done. This is shown by the formulation and architecture design of the nonlinear flight mechanical model

Trimming points aircraft model have been selected and the flight simulation verifies the natural Eigen mode responses of the real aircraft. The real aircraft has all stable Eigen modes except for spiral mode which is slightly unstable.

For further investigation, some work can be suggested, i.e. utilization of more comprehensive and accurate stability and control derivatives, computation of trimmed aircraft on more variation of flight condition.

References

[1] Bahri, S. Specification and Concept of LSA-02 (Technology Demonstrator). *Technical Report TR_LSA02_LSA_1100_01, 2016.*
[2] Deutsches Insitut für Normung (DIN). Begriffe, Größen und Formelzeichen der Flugmechanik. *DIN 9300, October 1990.*
[3] Luckner, R. Flugmechanik 1 – Grundlagen und stationäre Flugzustände. *Flight Mechanics lecture script, 2007.*
[4] Rotax. Operators Manual For Rotax Engine Type 914 Series, *2014.*
[5] Stemme AG. Maintenance Manual Stemme ASP Model S15-1. *2016.*
[6] Bahri, S. Development of Nonlinear Flight Mechanical Model of LSA-02 Technology Demonstrator. *Minor Research Report AE 6003 FTMD ITB, 2017.*
[7] Mathworks. MATLAB and Simulink, *R2013b, 2013.*
[8] Hakim, T.M.I. BEFCS Definition, Requirement and Concept. *Technical Report TR_LSA2_LSA_1500_01, 2016*
[9] Bahri, S. Requirement, Definition, and Concept of FCL for BEFCS. *Technical Report TR_LSA02_LSA_3500_01, 2014.*
[10] Allerton, D. Principles of Flight Simulation. *John Wiley & Sons, 2009.*

Study of Required Thrust Profile Determination of a Three Stages Small Launch Vehicle

A Fariz[1)], R A Sasongko[1)*] R E Poetro[1)]

[1)] Faculty of Mechanical and Aerospace Engineering, Institut Teknologi Bandung, Indonesia

*r.a.sasongko@gmail.com

Abstract. The effect of solid rocket motor specifications, i.e. specific impulse and mass flow rate, and coast time on the thrust profile of three stages small launch vehicle is studied. Solid rocket motor specifications are collected from various small launch vehicle that had ever been in operation phase, and also from previous study. Comparison of orbital parameters shows that the radius of apocenter targeted can be approached using one combination of solid rocket motor specifications and appropriate coast time. However, the launch vehicle designed is failed to achieve the targeted orbit nor injecting the satellite to any orbit.
Keywords: thrust profile, small launch vehicle, specific impulse, mass flow rate, coast time

1. Introduction

One of the main factors in space exploration is the capability to transport objects from earth surface to specific positions in space or orbit. The development of launch vehicle plays key roles in providing this capability. Hence the technology to design, develop, and operate launch vehicle is obligatory.

Indonesia, as one the countries that is developing its capability for space exploration, has also an objective to develop launch vehicle for delivering payloads to orbits, in order to support the implementation of some maritime surveillance application in Indonesia. In order to achieve the objective, it is planned to develop a three stages solid small launch vehicle that capable of carrying 50 kg payload to low Earth orbit in 2019 [1]. Some studies related to the planned program has been carried out, one of them is the works by Utama [2] that explores on the design of multi-stage launch rocket and its trajectory control system. However, due to the limitations of the study, iteration of rocket motor specifications has not done yet.

The objective of this paper is to study the possibility to approximate reference trajectory, which is shaped by reference thrust profile. Approximation is done by varying the specifications of solid rocket motor, i.e. specific impulse and mass flow rate, and the duration of coast time. Various solid rocket motor specifications data are collected from various operated small launch vehicle and also from previous research by Utama [2,3]. Two-point boundary value problem developed by Tewari [4] is used to generate optimal trajectory of desired initial and final flight condition. The trajectory generated is used as reference trajectory. The deviation of trajectory to the reference leads a study of how coast time duration affect the flight, as Pontani and Teoffilato [5] use this method to evaluate launch vehicle flight performance. To determine the orbit achieved by the launch vehicle, the orbit parameters, i.e. semi-major axis and eccentricity, are calculated from terminal condition of position, velocity, and flight path angle.

2. Launch Vehicle Configurations and Modelling

2.1. Multistage Rocket

In this paper we distinguish all nomenclature system, except the one used by Cornelisse [6]. The term sub-rocket define a complete rocket vehicle consisting of one or more stages together with the payload and guidance control system. Besides, the term stage defines a complete propulsion unit together with control equipment, which is discarded completely when staging process has occurred. Here is the example of three stages rocket. The total mass of rocket m_0 is divided into payload mass m_u, useful propellant mass m_p, structural mass m_c. In case of three-stages rocket with, the formula of mass should be as following.

$$m_{01} = m_{c1} + m_{p1} + m_{02}$$

$$m_{02} = m_{c2} + m_{p2} + m_{03}$$

$$m_{03} = m_{c3} + m_{p3} + m_u \quad (1)$$

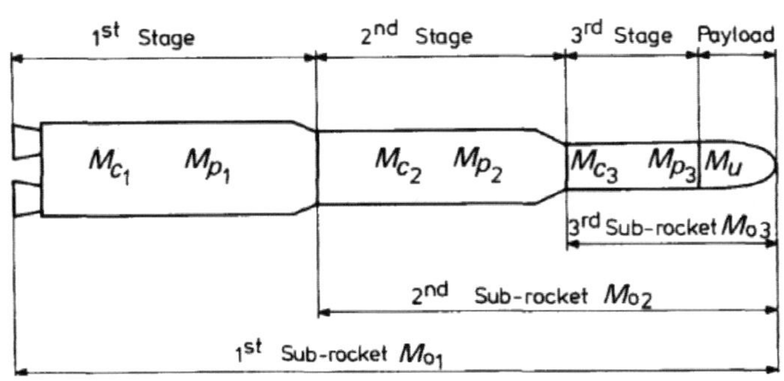

Figure 1. Nomenclature for a three-stages rocket [6]

2.2. Mass Parameters of Rocket

Given mass parameters of rocket [6] which has range, for single stage rocket in general, presented in Table 1. Here are ratio of masses defined, i.e. Mass Ratio *MR*, Payload Ratio *PR*, Structural Efficiency *SE*, and Propellant Ratio *PrR*. The payload mass, structural mass, and propellant mass of rocket should fullfil the range.

Table 1. Mass Parameters of Single Stage Rocket [6]

Parameters	*PR*	*PrR*	*SE*	*MR*
Formula	$PR = m_u/m_0$	$PrR = m_p/m_0$	$SE = m_c/(m_c + m_p)$	$MR = m_0/m_e$
Range	$0{,}01 < PR < 0{,}2$	$0{,}5 < PrR < 0{,}9$	$0{,}2 < SE < 0{,}5$	$2 < MR < 10$

2.3. Solid Rocket Motor

The thrust profile of launch vehicle depends on the rocket motor specifications. A rocket motor, which generates thrust during burning period, delivers a total impulse I. The term specific impulse I_{sp} is used to define impulse per unit weight of propellant. Mass flow rate $\dot{m}$ is the rate of propellant mass flowing through the nozzle. Often, steady-state mass flow assumption is used in convergent-divergent nozzle analysis to deal with ideal motor case [6]. In this case the density, flow speed, and cross-sectional area of exit nozzle remains constant.

$$I_{sp} = \int_0^{t_b} F(t)dt \,/(m_p \, g_0) \tag{2}$$

$$\dot{m} = \rho \, V \, A = constant \tag{3}$$

Consider one solid rocket motor with its propellant has one value of specific impulse I_{sp} and constant mass-flow rate, the thrust should be continuous and can not be throttled over burning period. Thrust T can be calculated in terms of specific impulse, mass-flow rate, and sea-level gravity acceleration g_0 as :

$$T = \dot{m} \, g_0 \, I_{sp} \tag{4}$$

Various solid rocket motor specifications data, referred to previous research and operated small launch vehicle [2,3], are compiled in Table 2. In this study, we consider only the specific impulse and mass-flow rate.

Table 2. Various Solid Rocket Motor Specifications per Stage [2, 3]

Stage	SRM	I_{sp} (s)	$\dot{m}$ (kg/s)
1	1A	192	79.719
	1B	240	128.46
	1C	253	202.71
	1D	263	380.228
	1E	286	573.213
2	2A	193	41.2
	2B	203	625.961
	2C	250	35.184
	2D	250	85.888
	2E	255	48.052
	2F	267	102.041
	2G	280	178.571
	2H	285	173.613
3	3A	211	14.218
	3B	258	3.080
	3C	260	120
	3D	277	33.412
	3E	278	10.021
	3F	280	89.286
	3G	292	41.306
	3H	301	99.668

2.4. Coast Time

Coast time t_c defined as time between burnout of nth stage to ignition of $(n+1)^{th}$ stage. In coast time period, there is no thrust generated because the burning does not occured. Thus, the gravity turn will change the flight path direction, and thus convert the potential energy to kinetic energy [5].

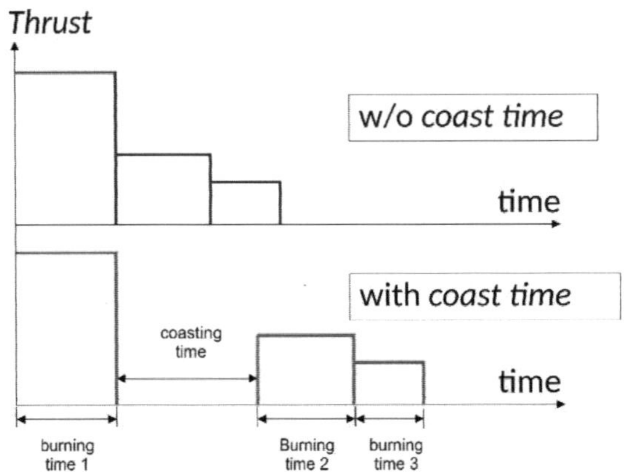

Figure 2. Thrust generated (a) w/o coast time, (b) with coast time

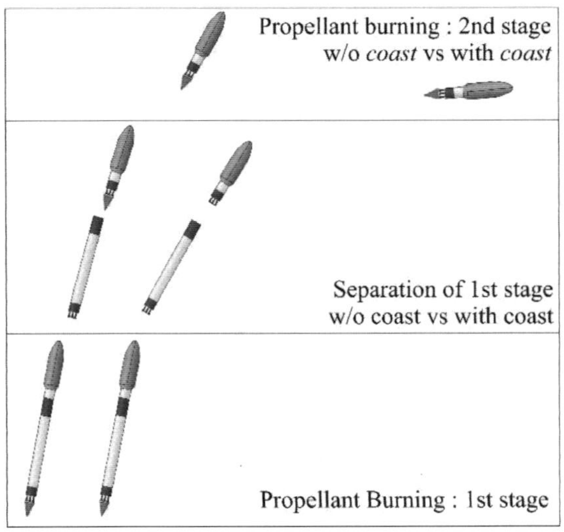

Figure 3. Illustration of flight (a) w/o coasting time, (b) with coasting time

2.5. Equation of Motion

The ascent of launch vehicle from planetary surface through the atmosphere can be modelled as flight in vertical plane. It is usually follows a gravity-turn trajectory, where the angle of attack α is maintained close to zero, leading to negligible lift force. The duration of launch vehicle flight is generally small enough to neglect the effects of planetary rotation. In this study the plane of Earth is assumed to be flat. As a point mass, the equation of motion are written as follows :

$$\dot{x} = V \cos \gamma \tag{5}$$

$$\dot{h} = V \sin \gamma \tag{6}$$

$$\dot{V} = u_1 - g \sin \gamma \tag{7}$$

$$\dot{\gamma} = (u_2 + (V^2/r - g) \cos \gamma)/V \tag{8}$$

IOP Conf. Series: Journal of Physics: Conf. Series **1005** (2018) 012023 doi:10.1088/1742-6596/1005/1/012023

$$u_1 = (T - D)/m \tag{9}$$

$$u_2 = T_c/m \tag{10}$$

$$\dot{m} = -\dot{m}_p \tag{11}$$

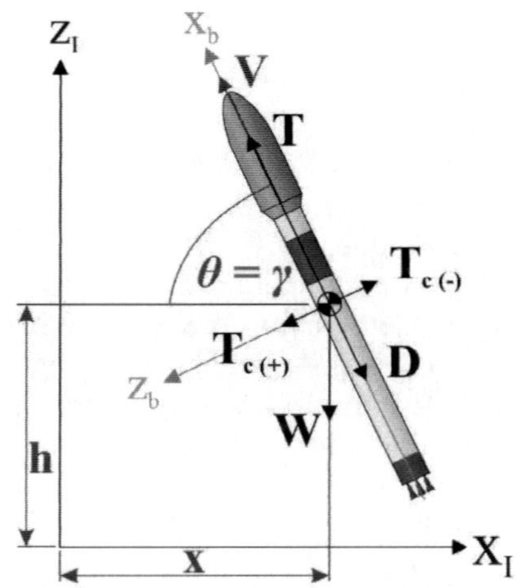

Figure 4. Flight in Vertical Plane

with state variables flight path angle γ, velocity V, and radius r, input variables thrust T and input control thrust T_c, gravity acceleration g, and drag D. In this study, the drag coefficient is assumed to be constant due to short duration of flight in atmosphere. If the shape of nose cone is a half-round, the C_D should be around 0.5 [3]. The drag is formulated as :

$$D = 0.5\, D = 0{,}5\, \rho\, V^2\, S\, C_D\, V^2\, S\, C_D \tag{12}$$

2.6. Orbital Mechanics

In gravitational condition, the two-body problem assumption is used [7]. Gravity force exerted on the body is the function of radius from the inertial frame. On the other hand, Velocity of body on orbit is obtained using Vis-Viva Integral of energy. The formula is as follows :

$$g = -\mu/r^2 \tag{13}$$

$$V = \sqrt{\mu(2/r - 1/a)} \tag{14}$$

To determine the orbit achieved by the launch vehicle, parameters of orbit, i.e. semi major axis, angular momentum, and eccentricity are calculated in terms of final altitude, velocity, and flight path angle. The formulas are as follows [8].

IOP Conf. Series: Journal of Physics: Conf. Series **1005** (2018) 012023 doi:10.1088/1742-6596/1005/1/012023

$$a = \left(\frac{2}{r} - \frac{V^2}{\mu}\right)^{-1} \tag{15}$$

$$|\boldsymbol{h}| = |\boldsymbol{r}\ x\ \boldsymbol{V}| = |\boldsymbol{r}||\boldsymbol{V}|\sin\left(\frac{\pi}{2} - \gamma\right) \tag{16}$$

$$1 - e^2 = \frac{|\boldsymbol{h}|^2}{\mu}\left(\frac{2}{r} - \frac{V^2}{\mu}\right) \tag{17}$$

2.7. Payload and Mission Specifications

In this study, the three-stages solid launch vehicle is adopted from previous research by Utama [2], where the launch vehicle has total mass 12,800 kg. The allocation of total mass from 1st stage to 3rd stage is 91%, 7.5%, and 1.5%, which are determined to fullfil the mass parameters of rocket [6] and has structural efficiency assumed to be 0.2. The payload mass is 50 kg. The launch vehicle is targeted to inject the payload to low earth orbit with semi major axis a = 6528.14 km. The flight begin from the ground, with initial velocity at liftoff is 0.001 km/s and elevation angle 70^0 [2]. The satellite is planned to be injected from apocenter r_a = 6578.14 km. In order to get into the orbit, the velocity of satellite injected should be 7.75439 km/s and the final flight path angle should be 0^0.

Table 3. Mission Specifications

Parameters	r (km)	V (km/s)	γ (^{0})	$\|\boldsymbol{h}\|$ (km^2/s)	a (km)	e	r_p (km)
Initial	6378.14	0.001	70	-	-	-	-
Target/final	6578.14	7.75439	0	51009.48	6528.14	0.0077	6478.14

3. Thrust Profile Shaping

3.1. Method Description

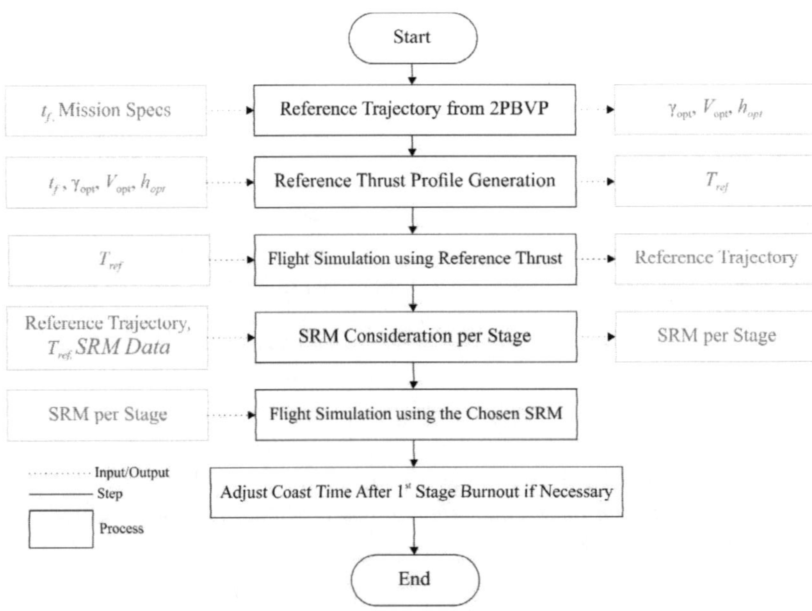

Figure 5. Flow Chart of Method

The method of thrust profile shaping is illustrated in Figure 5. The reference trajectory is generated from flight simulation of launch vehicle using optimal longitudinal acceleration from 2PBVP solution. Approximation to reference trajectory, and thus the reference thrust profile, is done by considering the solid rocket motor and the duration of coast time. To determine the orbit achieved by the launch vehicle, the orbit parameters, i.e. angular momentum, semi-major axis, and eccentricity, are calculated from terminal conditions. The terminal conditions cosidered are position, velocity, and flight path angle.

3.2. Reference Trajectory

The reference trajectory generated in this paper is obtained from two-point boundary value problem (2PBVP) solution developed by Tewari [4]. It is done by running bvp4c function in MATLAB [9]. The longitudinal acceleration u_1 is the optimized input. Optimization is based on quadratic cost $L = u^2$ and without terminal penalty, which yields the Hamiltonian equation H. The necessary condition of the derivation of Hamiltonian equation by input variables yields the extremal longitudinal acceleration input. Below presented the state and co-state equation, $\boldsymbol{f}$, and the optimal longitudinal acceleration input, u_1^*.

$$H = L + \lambda \boldsymbol{f} \tag{18}$$

$$dH/du = dL/du + \lambda df/du = 0 \tag{19}$$

$$\boldsymbol{f} = \begin{Bmatrix} \dot{r} \\ \dot{v} \\ \dot{\gamma} \\ \dot{\lambda}_2 \\ \dot{\lambda}_3 \end{Bmatrix} = \begin{Bmatrix} v \sin\gamma \\ u_1 - (\mu/r^2)\sin\gamma \\ (v/r - (\mu/(r^2 v)))\cos\gamma \\ -(2\mu \sin\gamma)\,\lambda_3/r^3 \\ -(\sin\gamma)\,\lambda_2 \end{Bmatrix} \tag{20}$$

$$u_1^* = -\lambda_3/(2\,R_1) \tag{21}$$

The function bvp4c required time of flight t_f as the time interval $0 \le t \le t_f$ and boundary condition from table 2. The time of flight $t_f = 555$ s is chosen as it generates optimal trajectory with initial value of flight path angle $70{,}01^0$, which is considered as intial condition of the flight in this study. The optimal trajectory defined by optimal flight state variables and input is depicted in Figure 6. Note that the longitudinal acceleration input will determine the thrust optimal needed as the launch vehicle mass is known during the simulation.

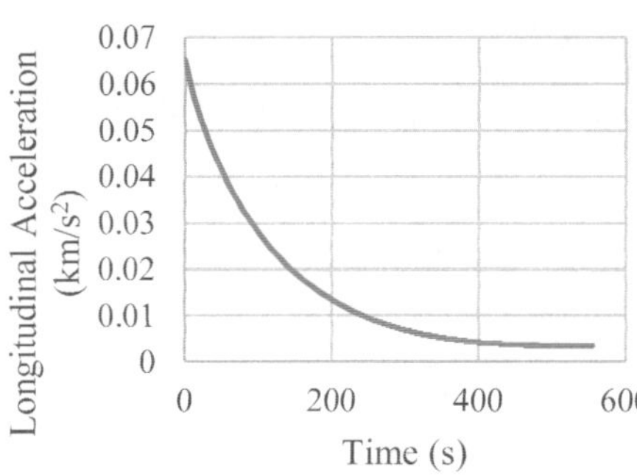

(a)

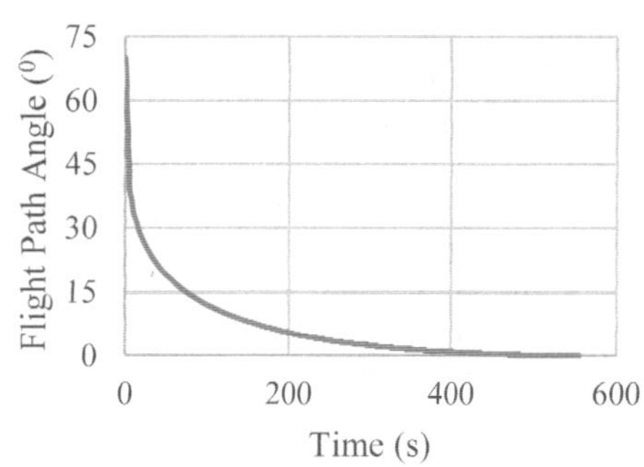

(b)

IOP Conf. Series: Journal of Physics: Conf. Series **1005** (2018) 012023 doi:10.1088/1742-6596/1005/1/012023

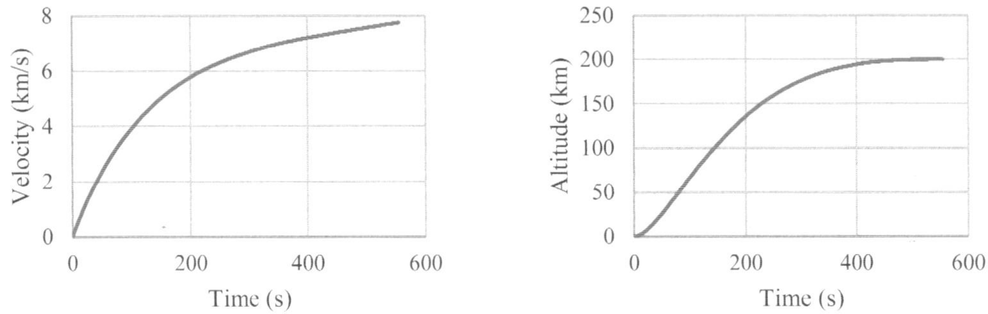

(c) (d)

Figure 6. (a) Optimal Longitudinal Acceleration Input , (b) Optimal Flight Path Angle, (c) Optimal Velocity, (d) Optimal Height solutions by two-point boundary value problem

4. Flight Simulation and Analysis

4.1. Simulation Platform

To perform flight simulations, a flight simulation platform is build in Simulink environment. The Simulink model consist of equations of motion depicted in Figure 7.

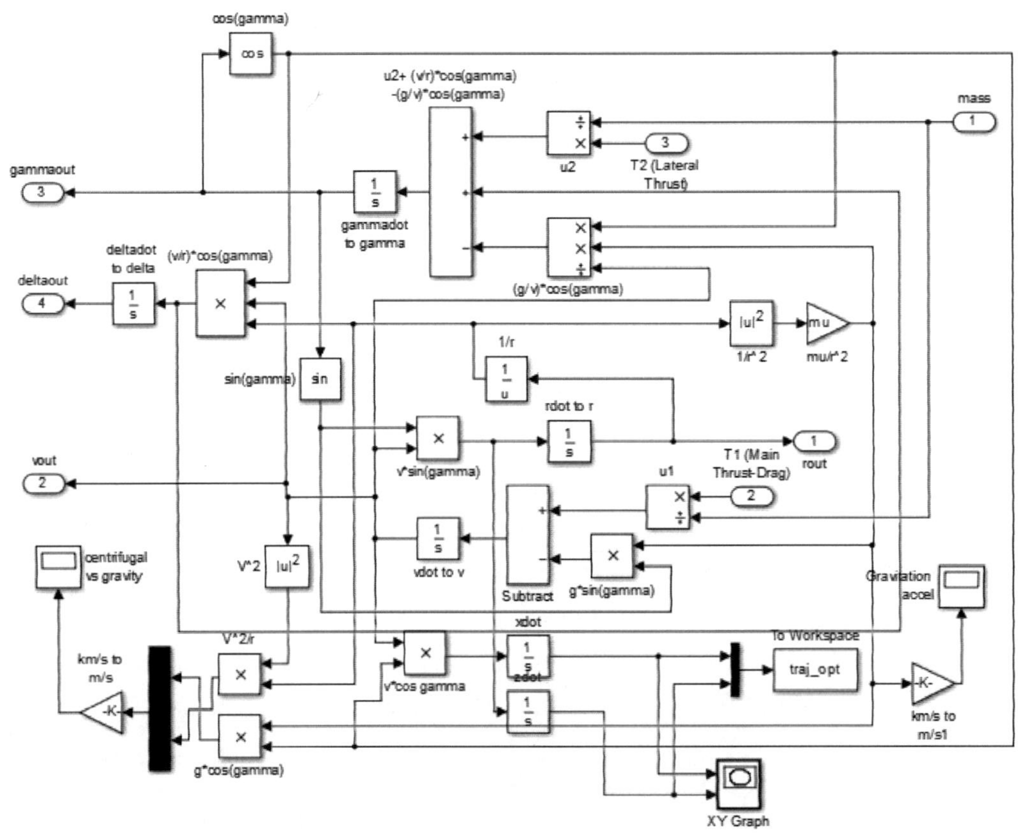

Figure 7. Equation of Motion in Simulink

IOP Conf. Series: Journal of Physics: Conf. Series **1005** (2018) 012023 doi:10.1088/1742-6596/1005/1/012023

4.2. Simulating Reference Trajectory

By using optimal longitudinal acceleration input from 2PBVP and initial guess of specific impulse and mass flow rate, the reference trajectory is obtained from flight simulation. The reference thrust profile is also generated. The orbital parameters achived are presented in Table 3.

Table 4. Orbital Parameters of Launch Vehicle

Parameters	r (km)	V (km/s)	γ (^{0})	$\|h\|$ (km^2/s)	a (km)	e	r_p (km)
Target	6578.14	7.75439	0	51009.48	6528.14	0.0077	6478.14
Reference	6580.793	7.75276	0.21	51018.96	6530.656	0.0085	6575.208

Reference trajectory could be shaped only if the thrust profile is similar to the reference. The reference thrust generated could be achieved if the launch vehicle equipped with liquid rocket motor. Thus, in this section considerations of solid rocket motor and coast time are done to approximate the the reference trajectory, and thus the reference thrust profile.

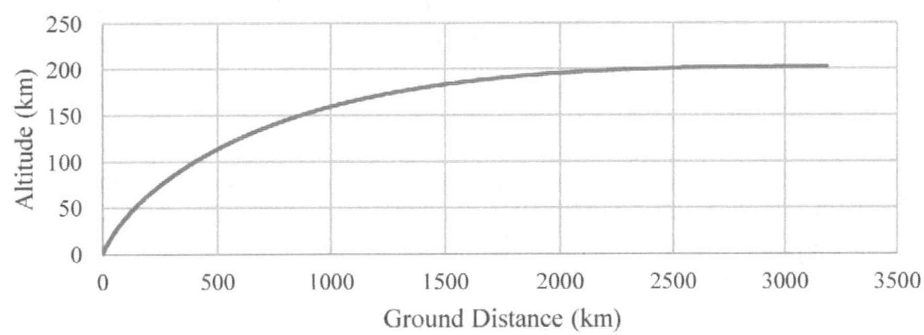

Figure 8. Reference Trajectory

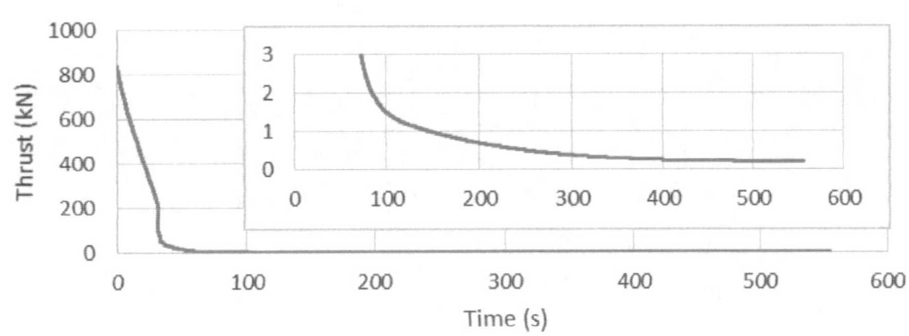

Figure 9. Reference Thrust Profile

4.3. Solid Rocket Motor Specifications Consideration

The steps begin with choosing the specifications of 1st stage rocket motor, followed by the 2nd stage and the 3rd stage. The consideration is based on the trajectories formed by three-stages launch vehicle with

IOP Conf. Series: Journal of Physics: Conf. Series **1005** (2018) 012023 doi:10.1088/1742-6596/1005/1/012023

various specifications of solid rocket motors from Table 2. The trajectories and thrust plotted in Figure 11 until Figure 16 are analysed in order to get the nearest one to the reference.

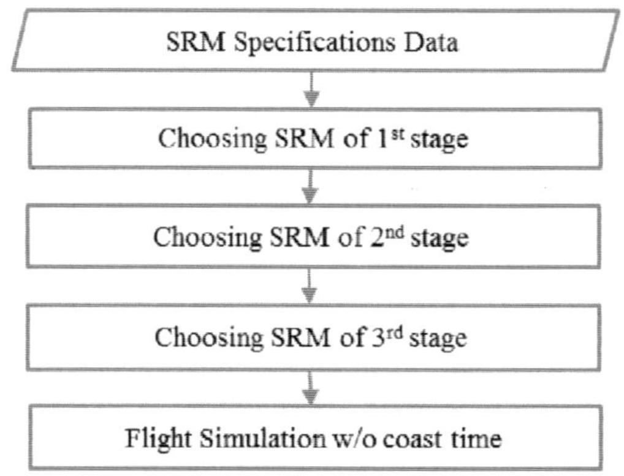

Figure 10. Flow Chart of Solid Rocket Motor Specifications consideration

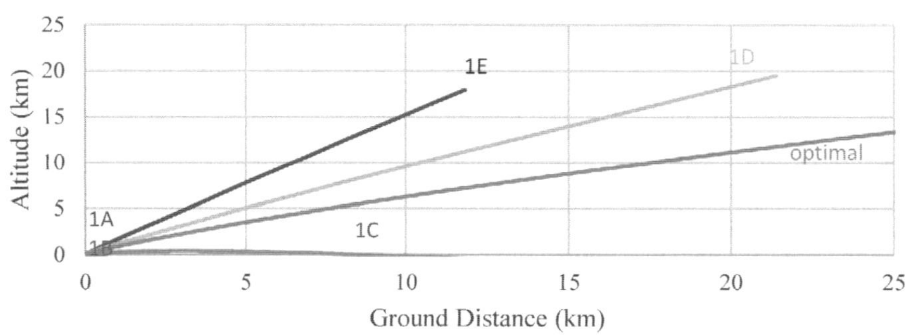

Figure 11. Trajectories After 1st Stage Burnout

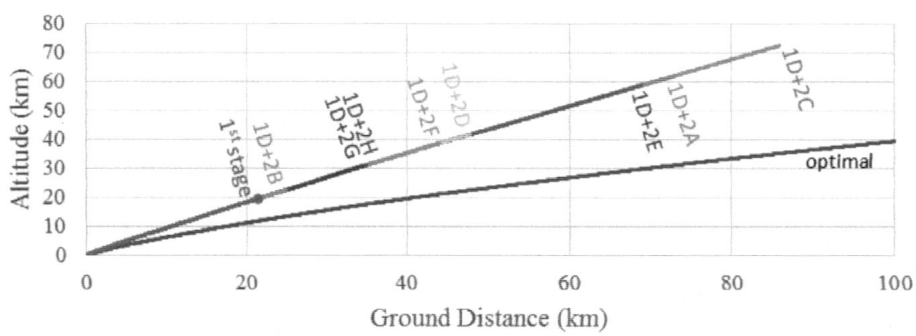

Figure 12. Trajectories After 2nd Stage Burnout

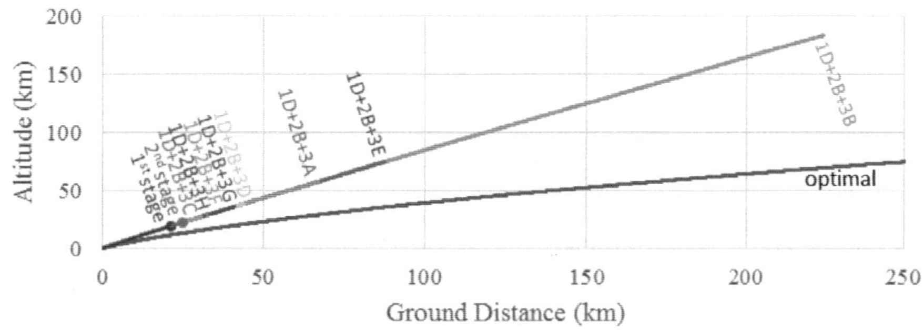

Figure 13. Trajectories After 3rd Stage Burnout

For the 1st stage, solid rocket motor with code 1A, 1B, and 1C could not propel the launch vehicle during the burning time. On the other hand, code 1E propels the launch vehicle to the higher altitude than the reference and code 1D trajectory at the same ground distance. Thus, code 1D is chosen as solid rocket motor for 1st stage. For the 2nd stage and 3rd stage, code 2B and 3C are chosen as it increases the altitude of launch vehicle less than the other solid rocket motor.

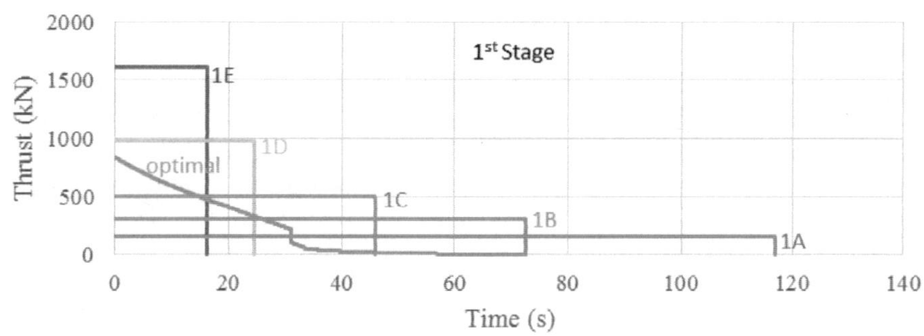

Figure 14. Thrust Profile Variation of 1st Stage

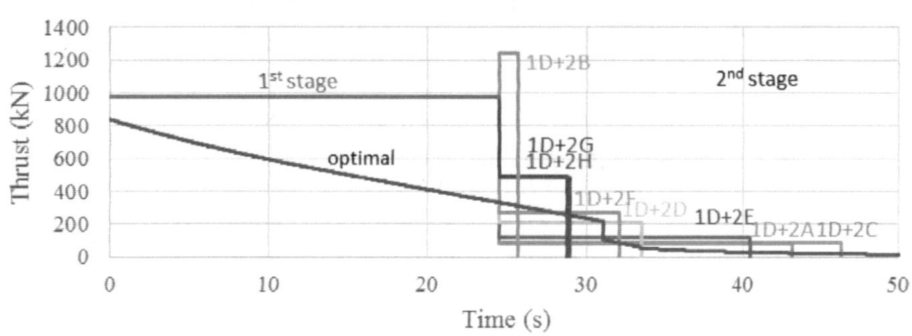

Figure 15. Thrust Profile Variation of 2nd Stage

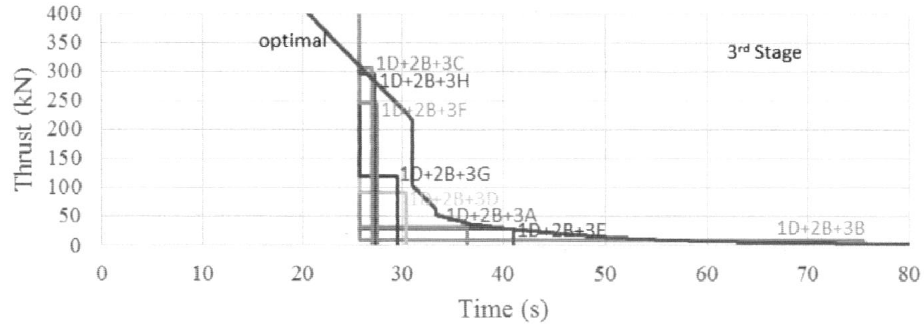

Figure 16. Thrust Profile Variation of 3rd Stage

Thrust of each stage plotted in Figure 14 until Figure 16 shows how the thrust affecting the trajectory shaped. 1st stage with specifications code 1A, 1B, and 1C could not propel the launch vehicle along burning time because the thrusts provided are lower than the reference thrust, and thus the impulses generated could not fullfil the minimum to propel the launch vehicle. Thrust generated by code 1D is actually higher than the reference, but it is lower than the code 1E. Thrust generated by code 2B and 3C are the highest among each stage, but the duration is the shortest. This condition leads to the increment of altitude less than the others. Thus, the chosen combination of solid rocket motor specifications is 1D+2B+3C.

4.4. Flight simulation

After choosing the combination of solid rocket motor, flight simulation is conducted in 555 s as the reference trajectory takes 555 s to achieve the orbit. The trajectory and thrust profile of launch vehicle is plotted along the reference. The final conditions and the orbital parameters are presented in Table 4.

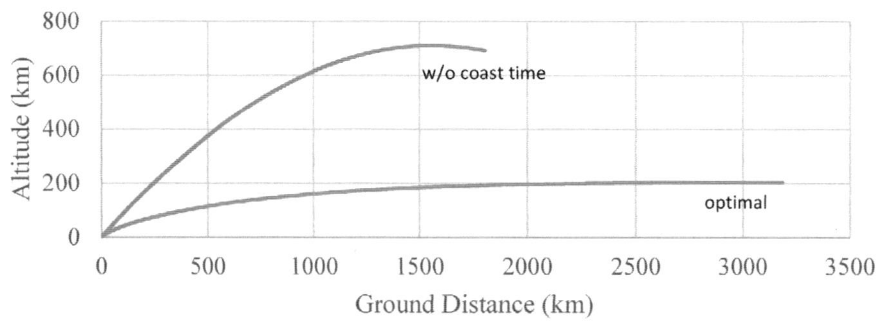

Figure 17. Trajectory of Launch Vehicle

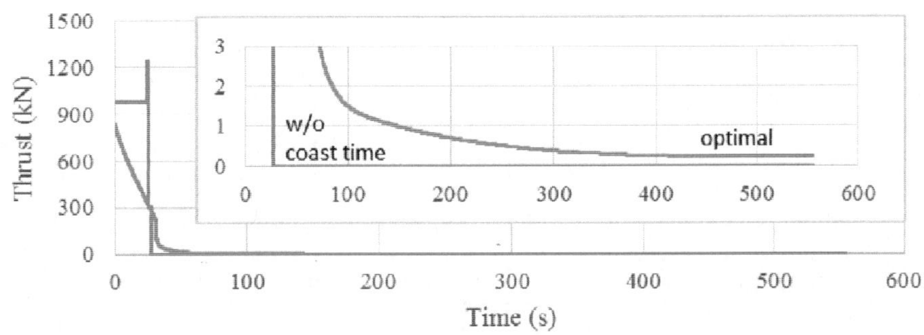

Figure 18. Thrust Profile of Launch Vehicle

Table 5. Orbital Parameters achieved by Launch Vehicle

Parameters	r (km)	V (km/s)	γ (^{0})	$\|\boldsymbol{h}\|$ (km^2/s)	a (km)	e	r_p (km)	r_a (km)
Target	6578.14	7.75439	0	51009.48	6528.14	0.0077	6478.14	6578.14
Reference	6580.79	7.75276	0.21	51018.96	6530.66	0.0085	6575.21	6586.10
Coast 0 s	7068.65	3.27162	-8.7	22860.59	3904.92	0.8150	722.37	7087.48

From the flight simulation of launch vehicle without coast time, the launch vehicle fly through a trajectory higher than the reference. From the data presented in Table 4, the radius of pericenter achieved is 722.367 km, which is lower than radius of Earth (6378.14 km). In this case the launch vehicle will fly into the atmosphere and crashing the ground. On the other hand, the radius in final condition is 7068.646 km, higher than the orbit target or reference. Also, the final velocity is slower than the velocity targeted. This shows that it is important to change the direction of flight to convert potential energy to kinetic energy, which can be done by gravity turn. Thus, for the following flight simulation, the duration of coast time after burnout of 1st stage is adjusted.

4.5. Coast Time Consideration

To get the optimal coasting duration after burnout of 1st stage, various coast time are considered. Then the radius of pericenter and radius of apocenter are compared, and the nearest one to the referenece is chosen as the appropriate coast time. The orbital parameters of launch vehicle with various coast time are presented in Table 5. The radius of pericenter and apocenter of various coast time plotted in Figure 16.

Table 6. Orbital Parameters with Various Coast Time

Parameters	r (km)	V (km/s)	γ (^{0})	$\|\boldsymbol{h}\|$ (km^2/s)	a (km)	e	r_p (km)	r_a (km)
Target	6578.14	7.75439	0	51009.48	6528.14	0.0077	6478.14	6578.14
Reference	6580.79	7.75276	0.21	51018.96	6530.66	0.0085	6575.21	6586.10
Coast 0 s	7068.65	3.27162	-8.7	22860.59	3904.92	0.8150	722.37	7087.48
Coast 10 s	8116.16	4.74713	26.9	34345.3	5266.34	0,6619	1780,746	8751,93
Coast 30 s	8054.67	5.03499	26.1	36425.83	5414.11	0,6206	2053,99	8774,22
Coast 50 s	7885.11	5.15608	22.9	37459.34	5349.13	0,5847	2221,43	8476,83
Coast 80 s	7623.65	5.37042	17.8	38976.27	5263.58	0,5253	2498,68	8028,48
Coast 110 s	7355.62	5.62449	12.5	40386.7	5193.84	0,4606	2801,64	7586,04
coast 150 s	6995.39	6.01810	5.1	41930.21	5127.14	0,3738	3210,68	7043,59
Coast 200 s	6566.79	6.56428	-4.4	42978.21	5090.1	0,2993	3566,48	6613,72

Coast 205 s	6526.94	6.61972	-5.4	43016.41	5089.43	0,2964	3580,86	6598
Coast 210 s	6487.86	6.67496	-6.3	43041.72	5089.34	0,2946	3590,18	6588,49
Coast 211 s	6480.14	6.68598	-6.5	43045.22	5089.39	0,2943	3591,43	6587,35
Coast 212 s	6472.45	6.69698	-6.7	43048.22	5089.46	0,2942	3592,46	6586,46
Coast 213 s	6464.80	6.70797	-6.9	43050.69	5089.56	0,2940	3593,29	6585,83
Coast 214 s	6457.17	6.71858	-7.1	43050.3	5089.36	0,2940	3593,32	6585,40
Coast 215 s	6449.59	6.71828	-7.3	42979.59	5079.68	0,2961	3575,61	6583,75
Coast 220 s	6414.74	4.67776	-8.7	29658.44	3892.77	0,6581	1330,90	6454,64
Coast 230 s	6393.66	0.24777	-54.5	919.60	3198.40	0,9997	1,06	6395,74
Coast 250 s	6382.42	0.1040	-89.8	2.22	3191.49	1	6,18E-06	6382,98

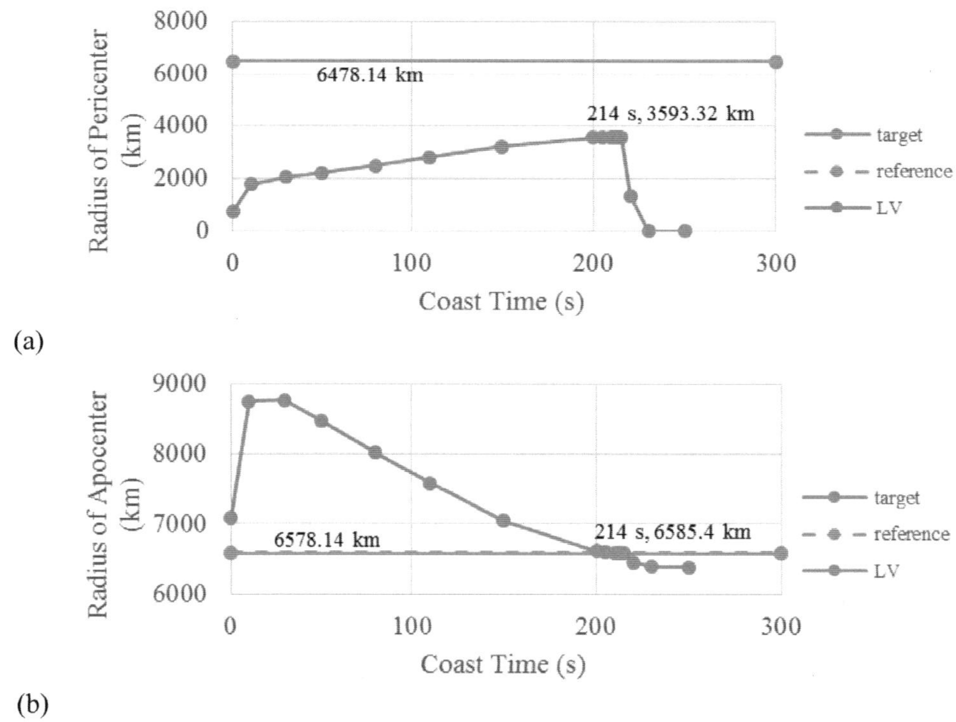

Figure 19. Radius of (a) Pericenter and (b) Apocenter of Launch Vehicle with Various Coast Time

From Figure 16, the highest radius of pericenter occured when the duration of coasting is 214 s. The radius of pericenter is 3593.32 km, thus can be concluded as lower than the target and reference. This also shows that the launch vehicle will fail to inject the satellite into the orbit target nor bring the satellite orbiting the Earth. The radius of apocenter of reference and target can be approximated, which is 6585.4 km.

4.6. Analysis

Section 4.3, i.e. choosing the solid rocket motor specifications, shows that the altitude achieved is based on the specifications chosen. Higher mass flow rate will lead to shorter duration of burning, thus the altitude will be lower at burnout. Compared to the reference trajectory, the altitude of launch vehicle rises faster with 1st stage rocket motor specifications chosen (code 1D). Thus, it is reasonable to choose the 2nd and 3rd stage rocket motor specifications (code 2B and 3C) which increases the altitude less than by the other codes.

By simulating the launch vehicle without coast time, it can be seen from Figure 14 that the launch vehicle will achieve altitude higher that the apocenter of the target (shown by blue marker). This happens because the thrust generated is higher than the reference thrust. It means the impulse is higher than the referenece and propel the launch vehicle to a higher trajectory. It is also shown from Table 3 that the radius of pericenter is 722.37 km, which is lower that Earth's radius. It means that the launch vehicle and the satellite will crash the ground and counted as a failed mission. It is necessary to change the direction of flight, which can be done naturally by gravity turn.

Figure 17 shows that there is change of direction of flight by using coast time. This means that gravity turn effect will be dominant in a situation where there is no thrust generated, and thus convert the potential energy to kinetic energy. From Figure 18, the thrust profile of launch vehicle is adjusted by the coast time, thus in the coast phase the trajectory following gravity turn trajectory. Unfortunately, in this case the highest radius of pericenter is 3593.32 km. It shows that the launch vehicle with the chosen specifications of rocket motor and adjusted coast time could not inject the satellite to any orbit.

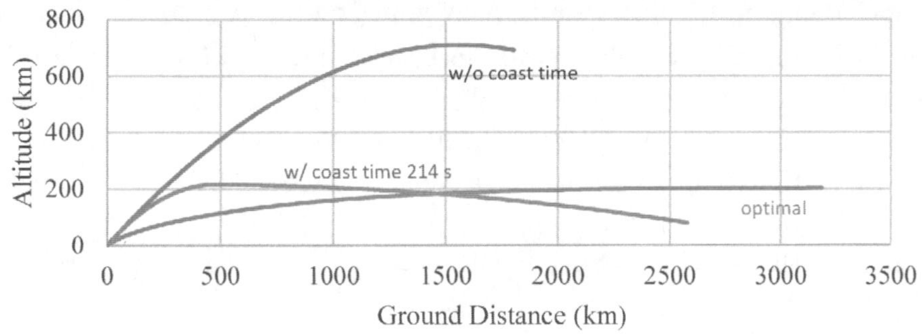

Figure 20. Trajectory of Launch Vehicle with Coast Time

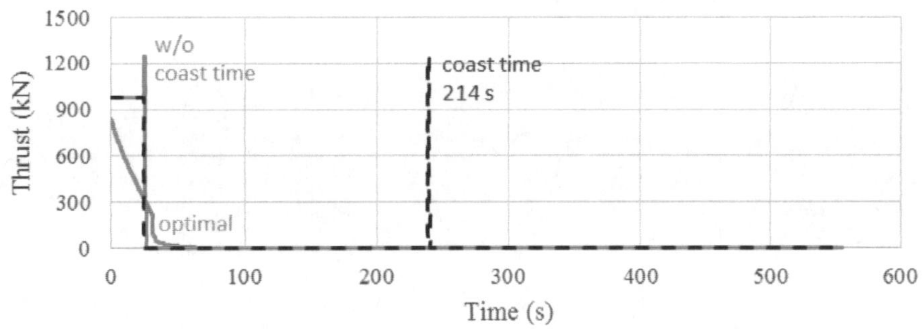

Figure 21. Thrust Profile of Launch Vehicle

5. Conclusions and Future Work

In this paper, a study of thrust profile shaping of small launch vehicle based on solid motor specifications and coast time cosideration was carried out. Varying the solid rocket motor configuration of each stage yields the variety of altitude achieved. The consideration of choosing solid rocket motor code 1D for the 1st stage is done based on the altitude achieved is slightly higher than the reference trajectrory, but lower that the code 1E. Solid rocket motor code 2B and 3C are chosen for 2nd and 3rd stage to overcome the

IOP Conf. Series: Journal of Physics: Conf. Series **1005** (2018) 012023 doi:10.1088/1742-6596/1005/1/012023

higher altitude achieved. Applying the appropriate coast time yields the higher radius of pericenter, which means that the direction of flight is changing as effect of gravity turn become dominant. This lead to the lower altitude achieved.

For the future work, it is necessary to do some steps. First, the design optimization is using existing multidisiplinary optimization method to solve the design problems. The methods chosen must be applicable to launch vehicle design problems. Second, the propellant grain should be designed in order to generate mass flow rate in which the thrust generated is approximating the reference thrust. Third, configuration of mass of the launch vehicle should be iterated to find the optimal one.

References

[1] LAPAN, “Fokus Pengembangan Roadmap Roket Peluncur Satelit,” [Online]. Available: http://pustekroket.lapan.go.id/index.php/subblog/pages/2014/17/Fokus-Pengembangan-Roadmap-Roket-Peluncur-Satelit. [Diakses 10 Juni 2017].

[2] S. Utama, “Perancangan Konfigurasi dan Sistem Kendali Wahana Pengorbit Satelit,” ITB, Bandung, 2015.

[3] “Wikipedia,” [Online]. Available: https://en.wikipedia.org/.

[4] A. Tewari, Advanced Control of Aircraft, Spacecraft, and Rockets, Wiley, 2011.

[5] M. Pontani, P. Teofilatto, “Simple method for performance evaluation of multistage rockets,” *Acta Astronautica,* vol. 94, no. 1, pp. 434-445, 2014.

[6] J.W. Cornelisse, H.F.R. Schoyyer. K.F. Wakker, Rocket Propulsion and Spaceflight Dynamics, Pitman, 1979.

[7] R.G. Bate, D.D. Mueller, J.E. White, Fundamentals of Astrodynamics, New York: Dover Publications, 1971.

[8] R.E. Poetro, "Astrodinamika," ITB, Bandung.

[9] Mathworks, *MATLAB SIMULINK Documentation.*

Development of Ground Test System For RKX-200EB

Herma Yudhi Irwanto

Rocket Technology Center, Indonesian National Institute of Aeronautics and Space (LAPAN)
herma.yudhi@lapan.go.id

Abstract. After being postponed for seven years, the development of RKX-200EB now restarts by initiating a ground test, preceding the real flight test. The series of the development starts from simulation test using the real vehicle and its components, focusing on a flight sequence test using hardware in the loop simulation. The result of the simulation shows that the autonomous control system in development is able to control the X tail fin vehicle, since take off using booster, separating booster-sustainer, making flight maneuver using sustainer with average cruise speed of 1000 km/h, and doing bank to maneuver up to ±40 deg heading to the target. The simulation result also shows that the presence of sustainer in vehicle control can expand the distance range by 162% (12.6 km) from its ballistic range using only a booster.

1. Introduction

Referring to 2010 Program Manual Control System related to the development of RKX-200EB (Figure 1), a two-stage guided rocket, the development in 2016 aims at mastering the technology of control system with medium sub sonic vehicle using a solid rocket booster-sustainer, which is applied in the ground test included in several tests in laboratory using the hardware in the loop simulation method [1][2] and booster-sustainer static test to ensure the characteristic of rocket motor and to measure delay time when booster and sustainer are separated.

Figure 1. RKX-200EB in 2010

Not only focusing on the system development of RKX-200EB ground test, the test also directly involves the autonomous control system [3] to make it simpler and enable it to implement in the hardware control system of RKX-200EB.

In brief, the development is expected to revive RKX-200EB program, which has been stalled since 2012. The target of the year is making an optimum simulation and a ground test, as a preparation of the real flight test. The simulation is done using flight sequence step by step, including necessary maneuvers.

Similar TNI rockets, such as the C-802 and C-705 that imported from Chine are also rocket 2 stages, rely on solid rocket booster and turbo jet engines sustainer to fly and maneuver toward the target [4][5]. This RKX-200EB was developed in an effort to reduce the dependence of rocket technology from abroad.

2. Target of RKX-200EB Development

There were at least three big agenda related to the development of RKX-200EB, ensuring that the entire system in the test works well before the expected flight test.

- *Flight Sequence Test and Hardware in the Loop Simulation System (HILS)*
  The flight simulation test of RKX-200EB used HILS method. In the simulation, the Micro Controller that would interact in the real flight test later was directly used.
- *Roll Stability Test using Mini Wind Tunnel System*
  A mini wind tunnel was used to test RKX-200EB concerning its roll stability in the real time. The target smart roll became the purpose of the system development using control techniques to stop roll and later to maneuver to the determined target. It was forced to connect with HILS system point 1 if possible.
- *Characteristic and Real Delay Measurement on Static Test System*
  It was the static test of booster-sustainer motor rocket of RKX-200EB that included characteristic test and measurement of delay time when the booster and the sustainer were separated.

3. Methodology

As there were three big targets to achieve in the system development of RKX-200EB ground test, each target had particular methodology in formulating the system to develop. Exclusively for the first target, namely Flight Sequence Test, it was completed using methods :

- Hardware in the Loop Simulation (HILS) [6] test in a computer to make sure that RKX-200EB could be controlled since ignition of booster, ignition of sustainer, and maneuver control to the target.
- Removing the controller entirely by real controller (NI MyRIO), replacing the programming in the computer. This used Integrated Simulation System (ISS) [7] that had been developed earlier.
- Combining ISS, making the separation test using the real RKX-200EB.

4. Component Test of RKX-200EB

The paper solely explains the first point of the three targets mentioned above, namely Flight Sequence Test and Hardware in the Loop Simulation System (HILS). In the meantime, the other two points would be explored in a separated paper. The test involved almost all components that would be attached in the real flight test. Figure 2 shows RKX-200EB that has been integrated and ready in a laboratory test.

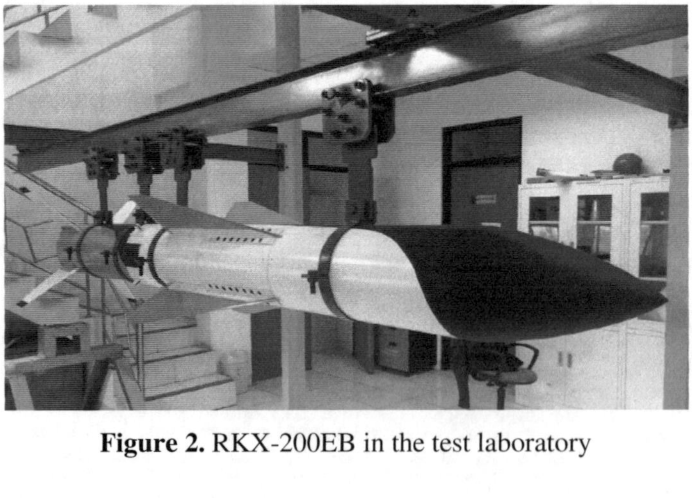

Figure 2. RKX-200EB in the test laboratory

The activities in the simulation system development mainly involved modules and tools, so that the result of the simulation might be as close as possible to the reality. Block diagram of the development is seen in the following Figure 3.

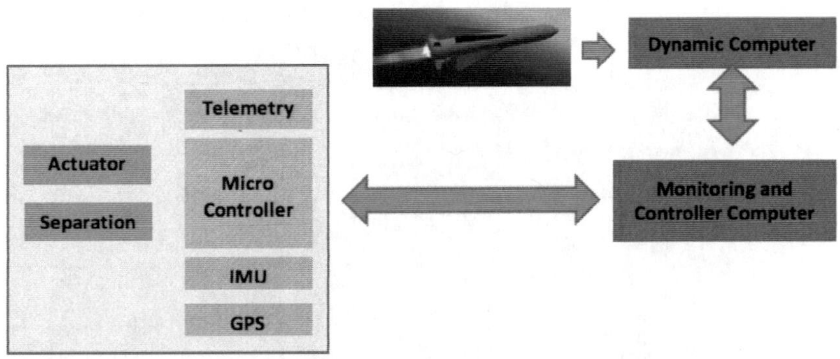

Figure 3. Block diagram of the development system

4.1. Real Payload System

Payload RKX-200EB in the simulation was designed and produced in its real form. It means that all structures and components, such as micro controller, inertial measurement unit (IMU) and navigation sensor (GPS) were real components, which would be used in the real flight test.

Meanwhile, the technical specifications of micro controller and peripherals used in the payload were:

- Single Board NI MyRIO (Xilinx FPGA and Dual-core ARM® Cortex™-A9 processor), Figure 4 left.
- IMU Micro Strain 3DM-GX3-25 (is a high-performance, miniature Attitude Heading Reference System (AHRS), utilizing MEMS sensor technology), Figure 4 right.

NI MyRIO Microstrain 3DM-GX3-25

Figure 4. Onboard controller components

- Peripherals needed in the onboard controller were:
  - 1 chn TTL serial for IMU
  - 1 chn serial TTL for telemetry
  - 1 chn serial for GPS, including converter RS-232 to TTL
  - 1 chn for detecting separation
  - 1 chn DIO for igniting rocket sustainer
  - 4 chn output PWM for servo motor actuator
  - 4 chn DIO (digital input output) for other needs
  - 4 chn AI (analog input) for reading encoder actuator
  - 2 chn AI (analog input) for other needs

Figure 5. Assembled onboard controller

4.2. Power Module

RKX-200EB needed two units of power modules. The first one was LiPo 3S 11.1V 5000mAH, which was used for micro controller, IMU, GPS and other modules located outside the payload block. The second one was LiPo 2S 7.4V 5000mAH (Figure 6) for actuator and encoder.

Exclusively for the laboratory test, external power modules with capacity 12V and 5V were also prepared to replace the earlier modules, as seen in Figure 7. There was a switch to enable choosing either internal or external power.

Figure 6. LiPo as the internal power module

Figure 7. AC-DC converter as the external power module

4.3. Real Telemetry System

The telemetry radio in use in the laboratory test of RKX-200EB was identical with the real module telemetry commonly used in flight tests. It was MaxStream xTend 900 MHz, hopping in frequency 902 – 928 MHz.

Telemetry radio and GPS structure block was located at the nose cone, the front part of the rocket. All communication data was sent wirelessly in real time from the rocket to the ground station, as in the real flight test.

IOP Conf. Series: Journal of Physics: Conf. Series **1005** (2018) 012024 doi:10.1088/1742-6596/1005/1/012024

4.4. Real Actuator System

RKX-200EB had four units of actuators that were controlled independently. They applied the control system X-Fin Tail Control, as seen in Figure 8. Each actuator was composed of two units of servo motors, as the torque enhancer, and one unit of encoder, as the sensor for the actuator's fin position. High voltage torque, steel gear servo motor product of Hitec, HS-7954SH had the maximum torque capacity 29 kg.cm and speed 0.12 sec / 60 deg. It was completed with analog encoder product from US Digital.

Figure 8. Actuator RKX-200EB

4.5. Real Body RKX-200EB

The system development of RKX-200EB ground test intentionally used a metal body frame, as commonly used in the real flight test. This aimed at making accurate placement of all components and modules attached in RKX-200EB, as in the real flight. In the development of RKX-200EB before flying, several things could become reference and warning points, including wiring complexity in the payload module, actuator placement in a narrow room in the nozzle area, types and uses of cables and connectors for external connection and internal connection between the two stages, and many other things (Figure 9).

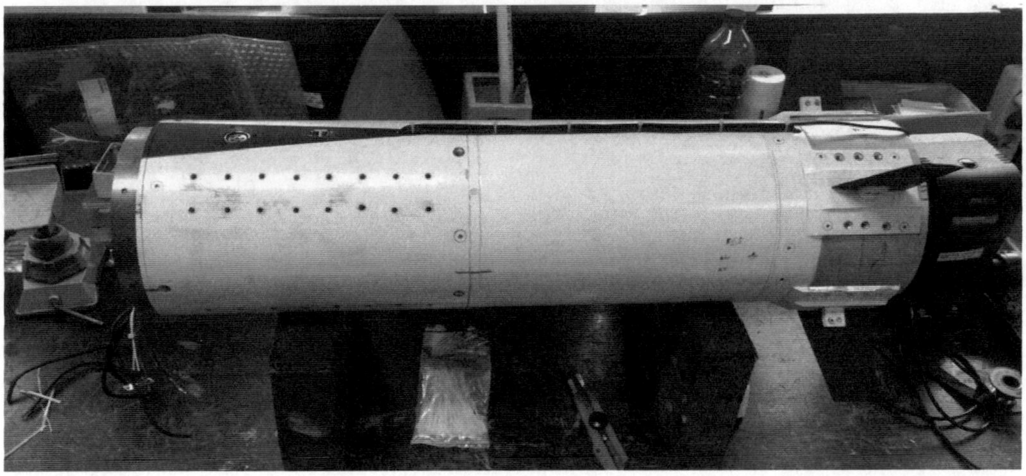

Figure 9. Structure of RKX-200EB

IOP Conf. Series: Journal of Physics: Conf. Series **1005** (2018) 012024 doi:10.1088/1742-6596/1005/1/012024

4.6. Flight Simulator

In conducting the simulation, the system was assisted by XPlane flight simulator [8][9]. Vehicle model of RKX-200EB developed in XPlane (Figure 10) was considered realistic to be the model of this simulation.

Figure 10. RKX-200EB in simulation

Connection between XPlane and Controller Computer was done using UDP protocol. In this case, XPlane would produce data on aerodynamic, attitude and vehicle position in flight. On the other hand, XPlane received data of motion field (moving X fin) to control the movement of the vehicle. The close loop process could be used by PID control or other control strategies to simulate the vehicle flight using the method of hardware in the loop simulation (HILS).

XPlane software was completed with XPlane Maker, for designing the vehicle and its equipment, including engine rocket, wing, fin, booster (using JATO / Jet Assisted Take Off facility) [10], C_G setting and others so that RKX-200EB vehicle (Figure 11) could fly in the simulation.

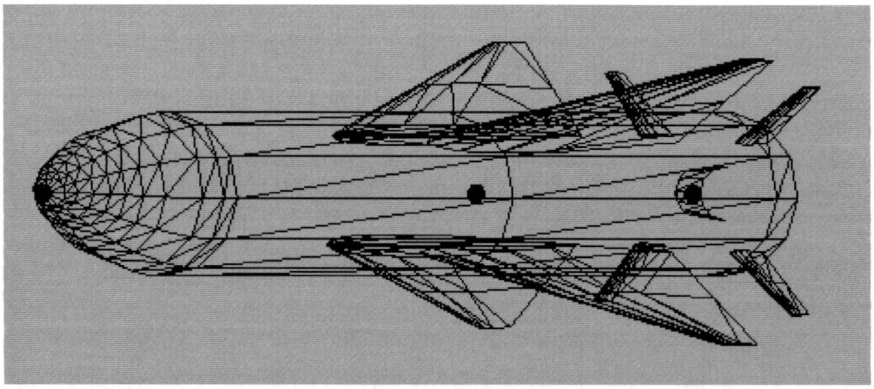

Figure 11. Model design using XPlane Maker

5. Simulation Test

The simulation involved tools having been built earlier, namely Integrated Simulation System (ISS), in order to simplify and to shorten the test in laboratory. In the test, all components mentioned above were involved entirely, along with XPlane as the flight simulator. Several tests done in the simulation were:

5.1. Measuring thrust in simulation for each stage (booster and sustainer)

This was necessary to ensure that RKX-200EB model had a thrust corresponding to the real design. Booster ignition was done using JATO facility, which was followed by sustainer that might later become a very significant feedback in setting the ignition time delay between both of them.

In RKX-200EB design, the booster had thrust of 500 kgf, and burning time 3.8 sec while the sustainer had 80 kgf, and burning time 24 sec. Based on flight test in 2010, there had been separation time delay around 7 sec. Compared with its design, the flight simulation of RKX-200EB was expected to have not only the same frame dimension but also the same thrust. The following figure shows the result of thrust simulation and speed compared to time.

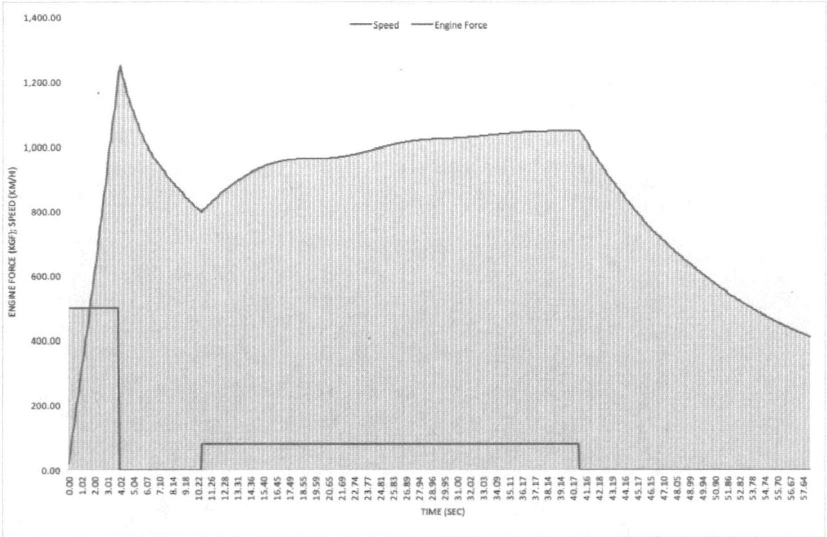

Figure 12. Simulation result of thrust booster, delay time separation and thrust sustainer

Based on the chart seen in Figure 12 above, the flight simulation of RKX-200EB was done with thrust booster force 499 kgf within 3.9 sec, followed by delay in sustainer ignition (which was assumed as delay of booster-sustainer separation) for 6.5 sec, and continued by sustainer ignition with constant thrust of 79.45 kgf for 30 sec. The combination of the three variables resulted in maximum speed of 1,247 km/h, equivalent to 1.16 Mach. Sustainer ignition was able to boost the vehicle up to 1,047.6 km/h, until it finally went gliding to the target.

5.2. Vehicle control

The development of autonomous control system for RKX-200EB was significantly different from that of RKX-200TJ in terms of programming, although both used X tail fin control strategy. Control program for RKX-200TJ could not be applied directly to RKX-200EB, because of big difference in speed. RKX-200EB had cruise speed 800 - 1000 km/h, when sustainer was turned on, and its control fin functioned well within the speed. However, when the sustainer was off, the control fin barely functioned except only for stability and was unable to maneuver though the speed was still above 400 km/h.

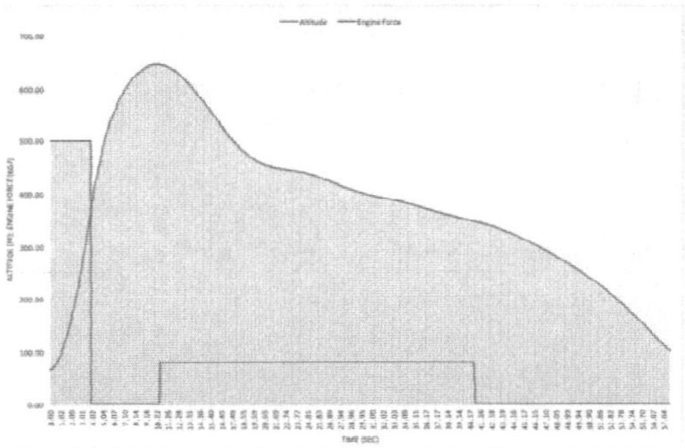

Figure13. Simulation of thrust and altitude

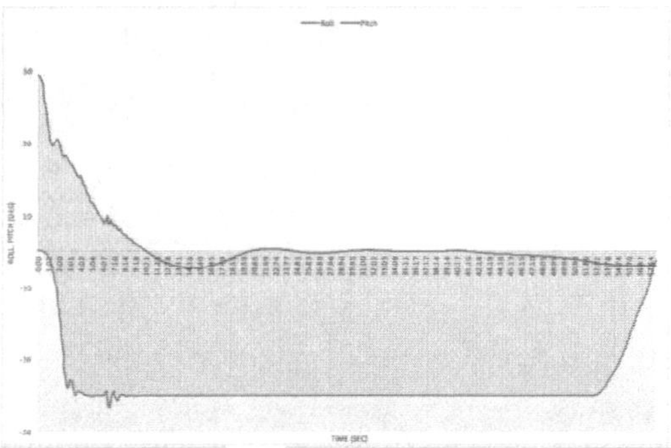

Figure14. Simulation of roll and pitch

IOP Conf. Series: Journal of Physics: Conf. Series **1005** (2018) 012024 doi:10.1088/1742-6596/1005/1/012024

As seen in Figure 13, the simulation result showed that when sustainer was on and making maneuver to the target, the fall of RKX-200EB was successfully delayed although its altitude had declined. As a comparison, if it relied only on booster, the vehicle would reach not more than 4.8 km distance with launching angle of 50 deg. However, with sustainer and its control system, the distance range of the vehicle could expand to 12.6 km, with initial heading of 154 deg and final heading of 67 deg. The condition of X tail fin control with sustainer, as seen in Figure 14, showed that the vehicle's pitch in the program was maintained at 0 deg, and at the same time, the vehicle made maximum roll of -40 deg (according to the program), to maneuver to the left, heading to the target. The complete simulation result can be seen in Figure 15, in the format of KML file in Google Earth software.

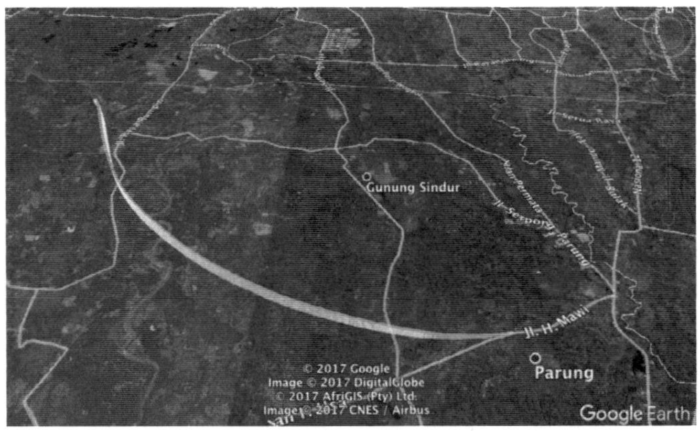

Figure 15. Simulation of RKX-200EB maneuver

5.3. Separation system test

The current separation system was the first design since its debut flight in 2010, and there had been no change. This research aimed at testing the real design in the real component series that showed separation between booster and sustainer. The simulation of the separation was done by adding a piston in the sustainer part, which forced the booster backward (Figure 16). Moreover, the separation time was a moment before the sustainer ignition.

Figure 16. Separation test between booster and sustainer

6. Conclusion and Recomendation

The first point of targets in development of RKX-200EB ground test was related to laboratory test of sequence test program using HILS facility. It led to several conclusions that the autonomous control system was able to make series of programs since booster ignition, separation between booster and sustainer, ignition of sustainer and maneuvering to the target based on the planned program. RKX-200EB was able to make banking maneuver up to ±40 deg to the target while the sustainer was still on. The control system worked well using thrust sustainer, followed by X tail fin directing the vehicle to the target as far as 12.6 km with change in heading maneuver up to 87 deg. The success in the simulation system would encourage further tests in the sequence of RKX-200EB development, as a long term development of RPS.

Acknowledgement

I would like to show my gratitude to Deputy of Aerospace Technology and Head of Rocket Technology Center for the facilities and supports in completing this research activity. I would also like to thank my colleagues in Control System Program for all help provided both directly and indirectly.

References

[1] Kowalchuk, S. (2012). MODEL-BASED DESIGN STRATEGIES FOR REAL-TIME HARDWARE-IN-THE-LOOP ROCKET SYSTEM SIMULATIONS. *AIAA Modeling and Simulation Technologies Conference*, (August), 1–9. http://doi.org/doi:10.2514/6.2012-4627

[2] Cho, J. M., Hwang, D. H., Lee, K. C., Jeon, J. W., Park, D. Y., Kim, Y. J., & Joh, J. S. (2001). Design and implementation of HILS system for ABS ECU of commercial vehicles. *ISIE 2001. 2001 IEEE International Symposium on Industrial Electronics Proceedings (Cat. No.01TH8570)*, *2*, 1272–1277. http://doi.org/10.1109/ISIE.2001.931663

[3] Hawley, P. A., & Blauwkamp, R. A. (2010). Six-degree-of-freedom digital simulations for missile guidance, navigation, and control. *Johns Hopkins APL Technical Digest (Applied Physics Laboratory)*, *29*(1), 71–84. http://doi.org/10.1155/2015/829473

[4] Siouris, G. M. (2004). Missile Guidance and Control Systems. *Applied Mechanics Reviews*, *57*(6), B32. http://doi.org/10.1115/1.1849174

[5] Platzer, M. F. (2010). Missile Aerodynamics. In *Encyclopedia of Aerospace Engineering*. http://doi.org/10.1002/9780470686652.eae017

[6] Sufendi, Trilaksono, B. R., Nasution, S. H., & Purwanto, E. B. (2013). *Design and implementation of hardware-in-the-loop-simulation for uav using pid control method.* In Proc. of 2013 3rd Int. Conf. on Instrumentation, Communications, Information Technol., and Biomedical Engineering: Science and Technol. for Improvement of Health, Safety, and Environ., ICICI-BME 2013 (pp. 124–130).

[7] Herma Yudhi Irwanto. *Integrated Simulation System for High Speed UAV.* In Bunga Rampai Hasil Litbangyasa: Teknologi pada Pesawat Terbang, Roket dan Satelit, LAPAN 2015.

[8] Laminar Research. (2016). Laminar Research | Creators of the X-Plane Flight Simulator. Retrieved from http://www.laminarresearch.com/

[9] Bittar, A., De Oliveira, N. M. F., & De Figueiredo, H. V. (2014). Hardware-in-the-loop simulation with X-plane of attitude control of a SUAV exploring atmospheric conditions. *Journal of Intelligent and Robotic Systems: Theory and Applications*, *73*(1–4), 271–287. http://doi.org/10.1007/s10846-013-9905-8

[10] Meyer, A. (2011). X-Plane Operation Manual. *X-Plane User Manual*, 1–224. Retrieved from papers2://publication/uuid/4E5FDDE4-13E1-443C-A0F6-9D79D0F8A2DF

Design of High Altitude Long Endurance UAV: Structural Analysis of Composite Wing using Finite Element Method

Khodijah Kholish Rumayshah [*], Aditya Prayoga, and Dr. Ing. Mochammad Agoes Moelyadi

Faculty of Mechanical and Aerospace Engineering, Bandung Institute of Technology, Indonesia
*kkrumayshah@gmail.com

Abstract. Research on a High Altitude Long Endurance (HALE) Unmanned Aerial Vehicle (UAV) is currently being conducted at Bandung Institute of Technology (ITB). Previously, the 1st generation of HALE UAV ITB used balsa wood for most of its structure. Flight test gave the result of broken wings due to extreme side-wind that causes large bending to its high aspect ratio wing. This paper conducted a study on designing the 2nd generation of HALE UAV ITB which used composite materials in order to substitute balsa wood at some critical parts of the wing's structure. Finite element software ABAQUS/CAE is used to predict the stress and deformation that occurred. Tsai-Wu and Von-Mises failure criteria were applied to check whether the structure failed or not. The initial configuration gave the results that the structure experienced material failure. A second iteration was done by proposing a new configuration and it was proven safe against the load given.

1. Introduction

In recent decades, the use of Unmanned Aerial Vehicles (UAV) in both military and civilian applications have been increased. The potential to substitute satellite to High Altitude Long Endurance (HALE) as a relay for communications is one such application. Indeed, to achieve the same capability as an orbital satellite which can cover a large area, the aircraft must be able to stay aloft for a long time in a very high altitude.

Bandung Institute of Technology (ITB) is now doing research on HALE UAV. In order to achieve low drag and long flight endurance, an aircraft must be designed having high aspect ratio wing. The 1st generation of HALE UAV ITB has been done developed with 12 m wing span. Most of its structure used balsa wood and polyurethane foam [1]. Hasan [2] had performed numerical analysis on its structure and gave many alternative designs on it. However, even though finite element analysis and static test had proven that the structure could withstand the designed load, flight test showed otherwise due to unforeseen strong side wind and wrong launch technique.

Castellani et al [3] stated that high aspect ratio wing is susceptible to many structural drawbacks. Its large span results in heavy structural weight. To overcome this problem, a very lightweight structure is needed, which in turn leads to a very flexible structure so that a large displacement cannot be neglected anymore. Romeo [4] found that a good structural wing solution to enhance bending stiffness could be represented by an opportunely reinforced high modulus carbon fiber leading edge, besides, the main spar could be subdivided into some parts which could be connected to each other. The structural configuration of the 2nd generation of HALE UAV ITB is expected to solve the problems that occurred in the previous generation. It is expected that by replacing some parts originally made from balsa wood to composite may strengthen the wing structure but by considering the target mass not to be too heavy. Many airframe structures currently using composites due to its ideal strength to weight ratio [5].

For the 2nd generation, HALE UAV ITB is designed having 16 m wing span with 0.4 m chord length. The wing is built modular with the length of each segment is 2 m. Every wing segment will be joined using a glass-fiber joiner at the front spar. The structural configuration is shown in **Figure 3**. It has a rectangular front spar which is located at 36% chord and made from a composite carbon fiber. The rear spar is an aluminum hollow tube located at 70% chord. There are 42 ribs which are made from carbon fiber along one-half span wing. Hasan [2] suggestion was to use a non-structural skin for a wing with large deformation in order to avoid buckle at the upper skin. So the 2nd generation of HALE UAV ITB is designed having skin made from plastic film, except skin at leading edge region which is strengthened using carbon fiber and a rectangular beam made from balsa.

Figure 1 First generation of HALE UAV ITB. (Source: Personal documentation)

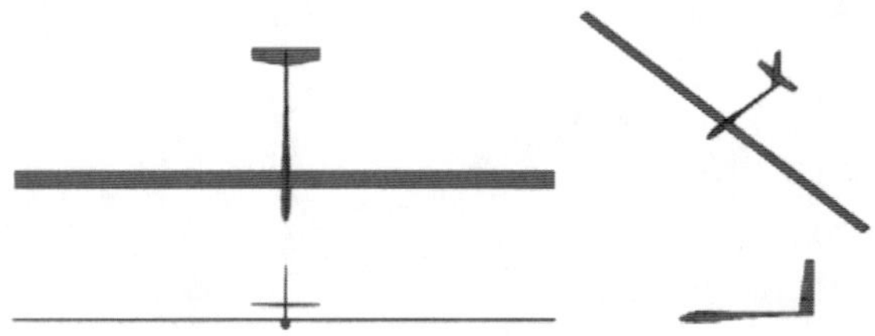

Figure 2 Design drawing of the 2nd generation of HALE UAV ITB.

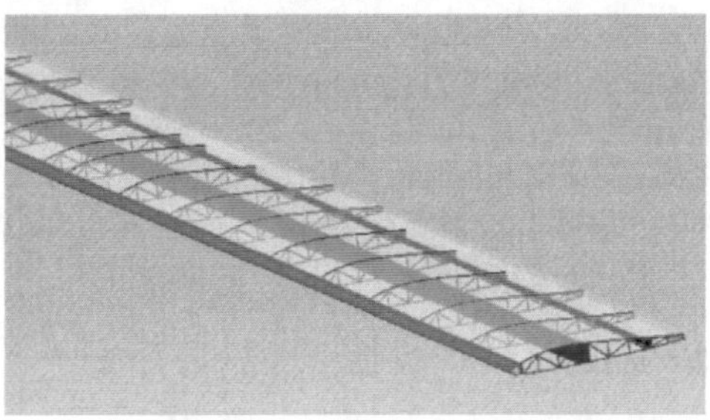

Figure 3 Wing structural configuration of the 2nd generation of HALE UAV ITB.

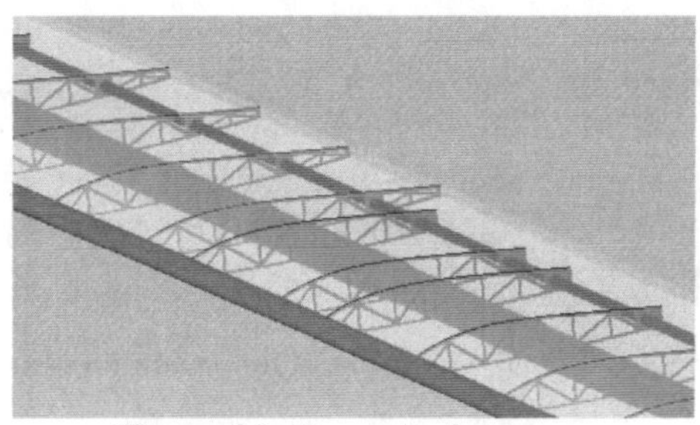

Figure 4 Joiner at the front spar.

Numerical analysis will be conducted for predicting the response of the structure to the loads given, i.e. the lift force which is distributed along wing span following non-linear Prandtl lifting line theory, the structural weight which is uniformly distributed, and the weight of the motor propeller which is located at a distance of 2 m from the airplane's axis of symmetry. Since its symmetric geometry, the numerical analysis will be conducted only for one-half wing span in order to reduce the computing time [5]. The desired output is ensuring whether the whole structure can withstand the given loads. There will be improvements in some critical parts so as to obtain a more optimal performance.

In the next development plan, HALE UAV ITB is going to be featured with a solar panel system as its additional energy source. By innovating this hybrid fuel, the aircraft is expected can stay aloft for a longer time (up to months) since it will be automatically charged its battery by the solar energy at day and it will use its reserve energy stored in its battery at night.

2. Finite Element Method and Simulation Setup

The finite element method (FEM) is a tool for analyzing certain structural response which comes to overly complex analytical hand-calculation. FEM divides the structures that being analyzed into finite numbers of elements each associated with some nodal points. The compatibility equations of each

IOP Conf. Series: Journal of Physics: Conf. Series **1005** (2018) 012025 doi:10.1088/1742-6596/1005/1/012025

element are then combined so that being appropriated for each node [6]. The static structural analysis in this simulation is performed using finite element software ABAQUS/CAE. ABAQUS is chosen since it has pre-processing features that make composite material property definition become easier.

2.1. Geometry Model

The CAD model was first generated using CATIA V5 in solid 3D. The geometry is translated into ABAQUS/CAE per part and simplified into shell 2D using mid surface technique. Broekaart [7] and Wysocki et al [8] explained why creating a mid-surface shell model is still favorable compared to analyze the whole model in solid 3D.

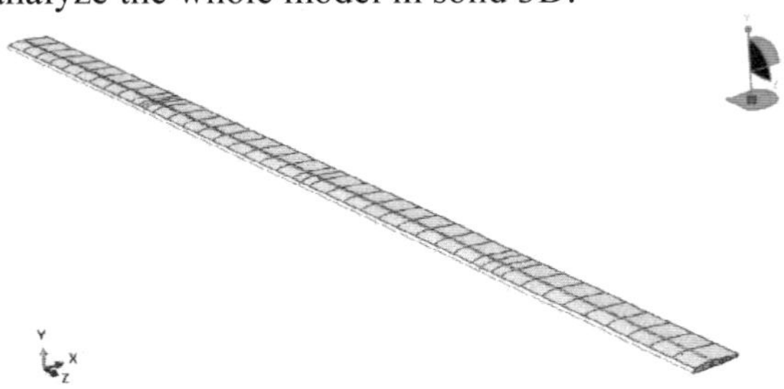

Figure 5 Simplified geometry model using shell element (full model).

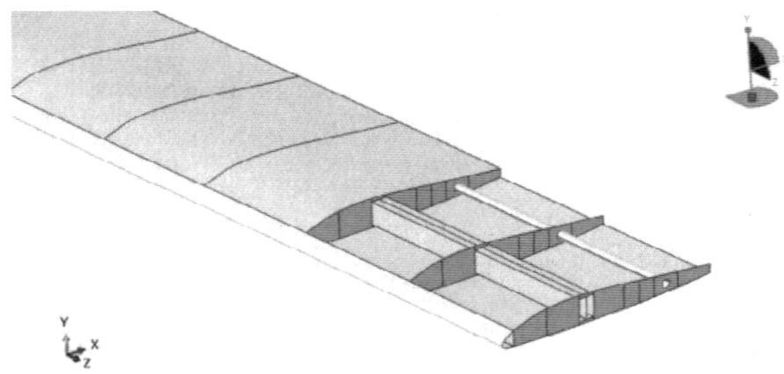

Figure 6 Simplified geometry model using shell element (zoom-in).

2.2. Material Properties

The wing structure of HALE UAV ITB consists of various materials on each part. For composite materials, the ply stacking sequence is defined by firstly predicting the major load direction acting at a certain region. Then the fiber orientation can be adjusted to its direction so that the load transfer can take place optimally. Rajadurai [9] states that [0/90/45/-45/90/0] is the best ply arrangement for aircraft composite wing.

Table 1 Material, ply orientation (for composite materials), and thickness of each part in the model.

Part	Material	Ply Orientation	Thickness (mm)
Upper Skin	Plastic Film	-	1
Lower Skin	Plastic Film	-	1
Front Spar	Woven Carbon Epoxy	$[90_2/0_2/45_2/-45_2]_s$	0.2 @ply
Joiner	Woven S-Glass Epoxy	$[90_2/0_2/45_2/-45_2]_s$	0.2 @ply
Rear Spar	Aluminum 6061 T6	-	2
LE (curved)	Woven Carbon Epoxy	$[90_2/0_2/45_2/-45_2]_s$	0.2 @ply
LE (flat)	Balsa	-	5
Ribs	Woven Carbon Epoxy	$[90/0/45/-45]_s$	0.2 @ply

The mechanical properties of each material are obtained from various references. The properties of woven carbon epoxy and woven s-glass epoxy are obtained from Performance Composites Ltd. website [10]. The aluminum tube is manufactured using drawing technique with the mechanical properties obtained from Military Handbook [11]. The mechanical properties of balsa wood are obtained from Wood Handbook [12]. WS Hampshire Inc. [13] provides data for plastic film.

IOP Conf. Series: Journal of Physics: Conf. Series **1005** (2018) 012025 doi:10.1088/1742-6596/1005/1/012025

Table 2 Mechanical properties of balsa wood.

Young's Modulus	**E**	3	GPa
Poisson Ratio	**ν**	0.38	-
Density	**ρ**	0.13	gr/cm^3

Table 3 Mechanical properties of aluminum 6061 T6.

Young's Modulus	**E**	9.90E+03	ksi
Poisson Ratio	**ν**	0.33	-
Density	**ρ**	0.098	lb/in^3

Table 4 Mechanical properties of woven carbon epoxy.

Young's Modulus	**E_{11}**	70	GPa
	E22	70	GPa
Shear Modulus	**G_{12}**	5	GPa
Poisson Ratio	**ν**	0.1	-
Density	**ρ**	1.6	gr/cm^3
Tensile Strength	**σ_T**	600	MPa
Compressive Strength	**σ_C**	-570	MPa
Shear Strength	**τ_{12}**	90	MPa

Table 5 Mechanical properties of woven s-glass epoxy.

Young's Modulus	**E_{11}**	25	GPa
	E22	25	GPa
Shear Modulus	**G_{12}**	4	GPa
Poisson Ratio	**ν**	0.2	-
Density	**ρ**	1.9	gr/cm^3
Tensile Strength	**σ_T**	440	MPa
Compressive Strength	**σ_C**	-425	MPa
Shear Strength	**τ_{12}**	40	MPa

Table 6 Mechanical properties of plastic film.

Young's Modulus	**E**	125000	psi
Poisson Ratio	**ν**	0.3	-
Density	**ρ**	0.93	gr/cm^3

2.3. Loads and Boundary Conditions

Silitonga et al [14] had done a comparative study of wing lift distribution for HALE UAV ITB. It came up to the conclusion that the use of Non-linear Prandtl Lifting Line Theory (NLLT) is recommended for preliminary analysis of HALE UAV ITB. **Figure 7** shows the spanwise lift distribution along one-half wing span by following NLLT. Besides lift force, there are also gravity force and any other concentrated loads working on the wing structure. The weight of the structure is automatically calculated by considering the material density input into ABAQUS pre-processing. While the concentrated force comes from 2 kg motor propeller which is located at a distance of 2 m from the aircraft's symmetry axis. ENCASTRE (fixed support) boundary condition is applied to the ribs at the wing root which will prevent translation and rotation in all direction. While the wing tip is free from constraint in all degree of freedoms. Overall loads and boundary conditions acted on the model is shown in **Figure 8**.

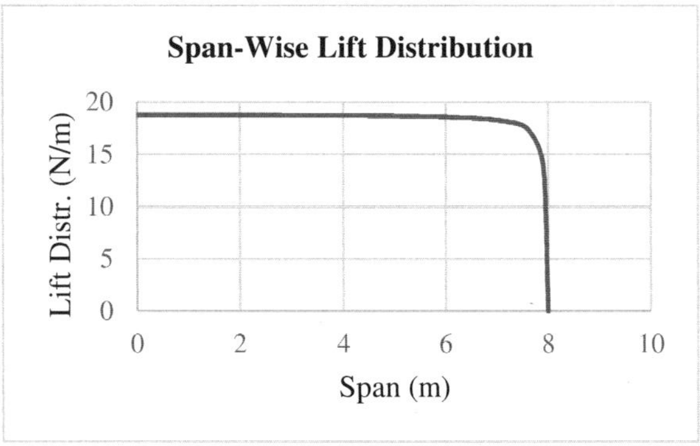

Figure 7 Span-wise lift distribution using lifting line method.

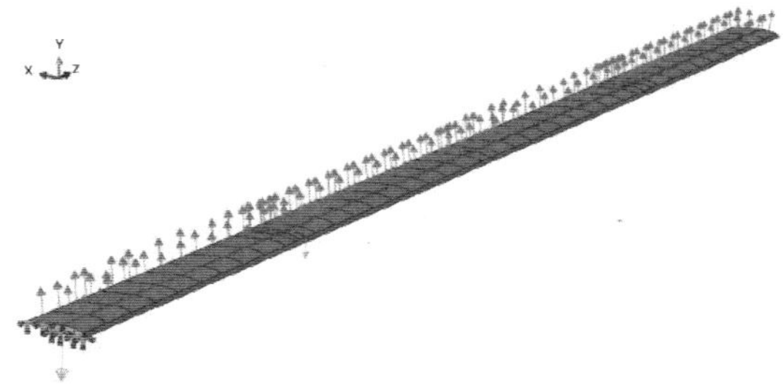

Figure 8 Loads and boundary conditions.

2.4. Meshing

The global mesh seed is generated for the whole model with the size of approximately 5 mm. *Mesh control* is activated for regulating the shape of the element and the meshing technique which is being used. For all of the parts except ribs, *quad-dominated* shape element and *structured* meshing technique is used. While for ribs, since the airfoil shape gives many curved edges, the meshing technique is *free*. The mesh result is shown by **Figure 9** below.

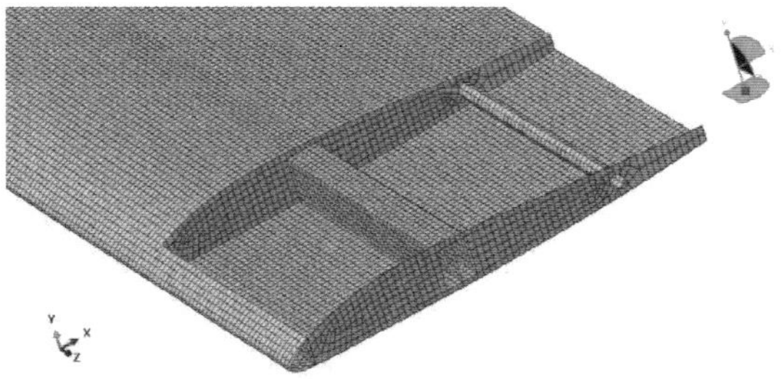

Figure 9 Mesh result.

3. Failure Criteria

Failure criteria are used to ensure the ability of a structure standing to the loads given, whether it fails or not. There are many different kinds of failure criteria based on the material characteristic. In this paper, the authors use Tsai-Wu failure criterion for parts made of composite materials and use Von-Mises failure criterion for isotropic or homogeneous materials.

Numerous failure criteria have been proposed to examine a composite material. Rajanish [15] conducted a study on it and found that Tsai-Hill, Tsai-Wu, and Hashin failure criterion give the best fit with the experimental result compared to the other failure criteria widely used. Further, Kaw [16] stated that Tsai-Wu failure criterion is more general than the Tsai-hill failure criterion since it distinguishes between compressive and tensile strength of a lamina. Tsai-Wu failure criterion is developed based on Beltrami failure theory concerning strain energy. Lamina (composite layer) in a structure is called fail if it does not satisfy **Equation 1** as follows [16].

$$H_1\sigma_1 + H_2\sigma_2 + H_6\tau_{12} + H_{11}{\sigma_1}^2 + H_{22}{\sigma_2}^2 + H_{66}{\tau_{12}}^2 + 2H_{12}\sigma_1\sigma_2 < 1 \qquad \textbf{Equation 1}$$

Which σ_1 is stress in fiber direction, σ_2 is stress in transverse direction, and τ_{12} is shear stress. While the value of H_1, H_{11}, H_2, H_{22}, H_6, and H_{66} can be obtained using **Equation 2** up to **Equation 7** as follows.

$$H_1 = \frac{1}{({\sigma_1}^T)_{ult}} - \frac{1}{({\sigma_1}^C)_{ult}} \qquad \textbf{Equation 2}$$

$$H_2 = \frac{1}{(\sigma_2{}^T)_{ult}} - \frac{1}{(\sigma_2{}^C)_{ult}}$$ **Equation 3**

$$H_{11} = \frac{1}{(\sigma_1{}^T)_{ult}(\sigma_1{}^C)_{ult}}$$ **Equation 4**

$$H_{22} = \frac{1}{(\sigma_2{}^T)_{ult}(\sigma_2{}^C)_{ult}}$$ **Equation 5**

$$H_6 = 0$$ **Equation 6**

$$H_{66} = \frac{1}{(\tau_{12})^2{}_{ult}}$$ **Equation 7**

Which $(\sigma_1{}^T)_{ult}$ is tensile strength in fiber direction, $(\sigma_1{}^C)_{ult}$ is compressive strength in fiber direction, $(\sigma_2{}^T)_{ult}$ is tensile strength in transverse direction, $(\sigma_2{}^C)_{ult}$ is compressive strength in transverse direction, and $(\tau_{12})_{ult}$ is shear strength. While the value of H_{12} can be obtained from Mises-Hencky criterion as follows.

$$H_{12} = -\frac{1}{2}\sqrt{\frac{1}{(\sigma_1{}^T)_{ult}(\sigma_1{}^C)_{ult}(\sigma_2{}^T)_{ult}(\sigma_2{}^C)_{ult}}}$$ **Equation 8**

Whereas to examine homogeneous material parts, Von-Mises failure criterion is used. The magnitude of Von-Mises stress can be obtained from general Von-Mises equation as follows [17].

$$\sigma_{VM} = \frac{1}{\sqrt{2}}[(\sigma_1 - \sigma_2)^2 + (\sigma_2 - \sigma_3)^2 + (\sigma_3 - \sigma_1)^2]^{\frac{1}{2}}$$ **Equation 9**

Which σ_1, σ_2, and σ_3 respectively are normal stresses that occur in the x-axis, y-axis, and z-axis. Plane stress assumption is used in the analysis so that σ_3 can be neglected and the equation comes into

$$\sigma_{VM} = \sqrt{\sigma_1{}^2 - \sigma_1\sigma_2 + \sigma_2{}^2}$$ **Equation 10**

The comparison between Von-Mises stress (σ_{VM}) and allowable stress of the material (yield stress/σ_Y) is defined as safety factor (SF).

$$SF = \frac{\sigma_Y}{\sigma_{VM}}$$ **Equation 11**

The structure is safe if the safety factor is greater than 1. Otherwise, safety factor less than 1 means the structure has performed a plastic deformation which is unpreferable [18].

4. Results and Discussion

Figure 10 to **Figure 15** below provides visualization of the numerical simulation results.

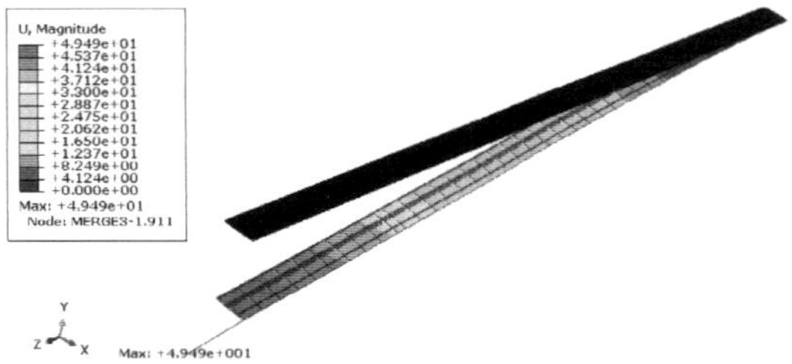

Figure 10 Deformation (U) plot.

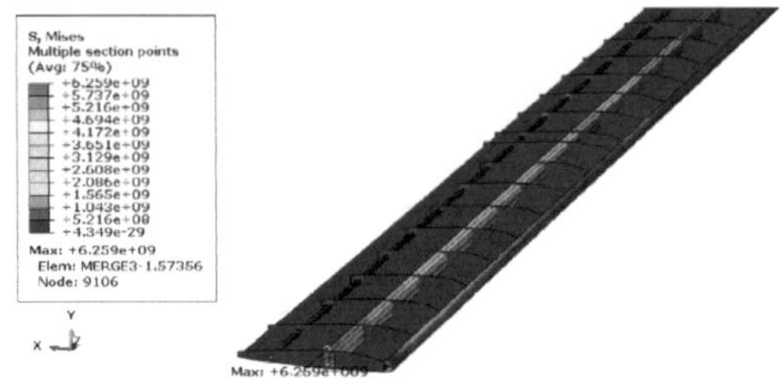

Figure 11 Von-Mises stress plot.

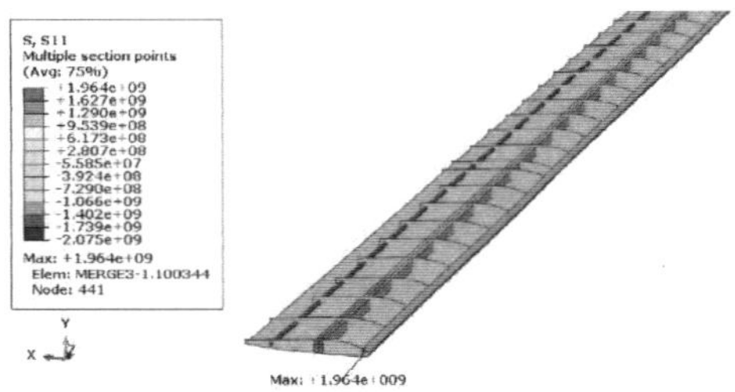

Figure 12 Stress plot in fiber direction (S11).

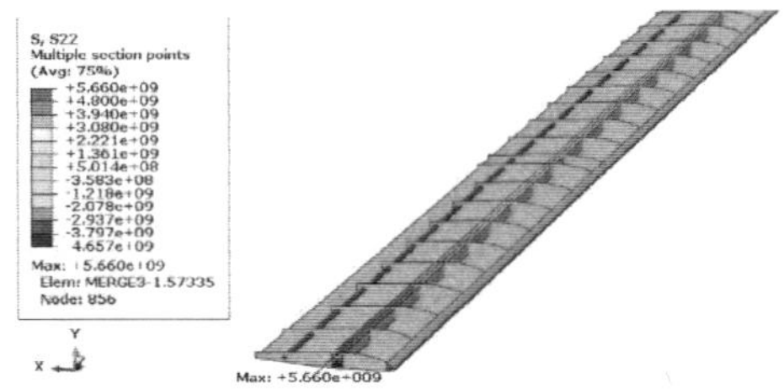

Figure 13 Stress plot in transverse direction (S22).

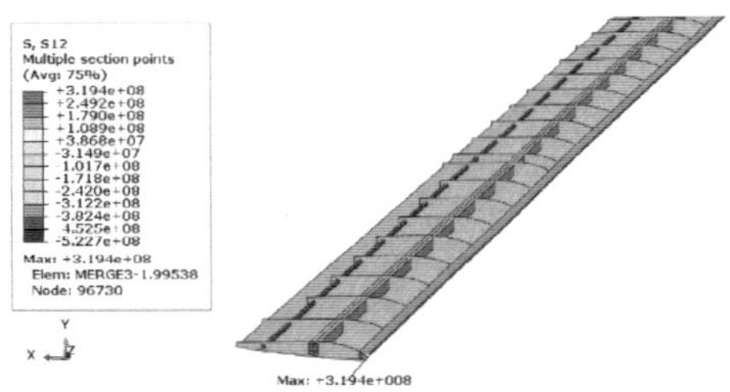

Figure 14 Shear stress (S12) plot.

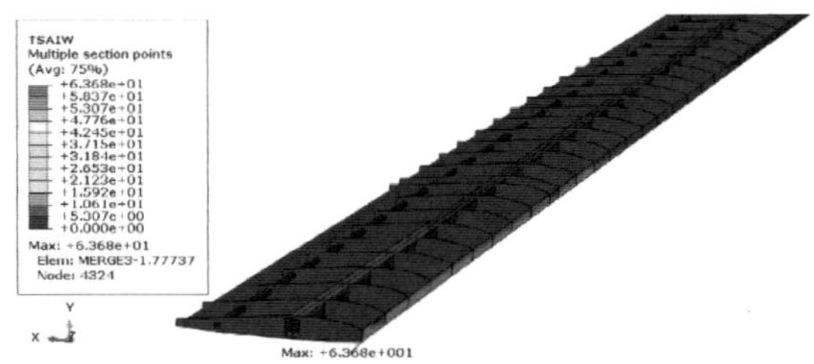

Figure 15 Tsai-Wu failure criterion (TSAIW) plot.

The numerical calculation of Tsai-Wu failure criterion gives the result as shown in **Table 7** below. The structure is failed since the calculation of Tsai-Wu equation gives the number of 63.68 so that **Equation 1** is not fulfilled. The most critical part is leading edge since it gives the highest number of Tsai-Wu failure criterion.

Table 7 Tsai-Wu failure criterion calculation for parts using composite material.

	Tsai-Wu Criteria	**Conclusion**
Front Spar	9.656	Fail
Joiner	4.117	Fail
LE (curved)	8.075	Fail
LE (flat)	63.68	Fail
Ribs	5.632	Fail

The safety factors of homogeneous parts are calculated in **Table 8** below. Von-Mises failure criterion gives the result that the structure is failed since the value of safety factor is less than 1 in the entire model which means the stress that occurs in the structure has caused the material to be yielding. The most critical part is rear spar since it has the lowest value of safety factor.

Table 8 Von-Mises failure criterion for parts using homogenous material.

	Von-Mises Stress (Pa)	Material	Yield stress (Pa)	Safety Factor	Conclusion
Upper Skin	1.06E+08	Plastic Film	2.14E+07	0.2016	Fail
Lower Skin	1.46E+08	Plastic Film	2.14E+07	0.1468	Fail
Rear Spar	2.13E+09	Aluminum 6061 T6	3.E+08	0.1297	Fail

The second iteration is then conducted in order to achieve a better structural configuration. The principle is to strengthen parts that receive high load either by means of material replacement or part thickening. Changing the ply orientation also quite helpful in doing this optimization. Outward wing (4th segment from wing root) only accepts small portions of load, so that parts of this segment can be reduced in strength in order to reduce weight until a satisfactory result is reached. **Table 9** and **Table 10** below show the new configuration offered.

Table 9 A new configuration in the use of material for 1st to 3rd wing segment.

Part		Material	Ply Orientation	Thickness (mm)
Upper Skin	Skin	Woven Carbon Epoxy	$[90_2/0_2/45_2/-45_2]_s$	0.2 @ply
	Core	Balsa	-	4
Lower Skin	Skin	Woven Carbon Epoxy	$[90_2/0_2/45_2/-45_2]_s$	0.2 @ply
	Core	Balsa	-	4
Front Spar	Horizontal Surface	Woven Carbon Epoxy	$[90_3/0_3/45_2/-45_2]$s	0.2 @ply
	Middle Splitter	Woven Carbon Epoxy	$[90_2/0_2/45_2/-45_2]_s$	0.2 @ply
	Vertical Surface	Woven Carbon Epoxy	$[90_2/0_2/45_3/-45_3]_s$	0.2 @ply
Joiner	Hor&Ver Surface	Woven Carbon Epoxy	$[90_2/0_2/45_2/-45_2]_s$	0.2 @ply
Rear Spar		Woven Carbon Epoxy	$[90/0/45/-45]_s$	0.2 @ply
LE (curved)		Woven Carbon Epoxy	$[90_3/0_3/45_2/-45_2]$s	0.2 @ply
LE (flat)		Woven Carbon Epoxy	$[90/0/45/-45]_s$	0.2 @ply
Ribs		Woven Carbon Epoxy	$[90/0/45/-45]_s$	0.2 @ply

Table 10 Material changing for 4th wing segment.

		Material	Ply Orientation	Thickness (mm)
Front Spar	Hor, Mid, & Ver Surface	Woven Carbon Epoxy	[-45/45]	0.2 @ply
Joiner	Hor&Ver Surface	Woven Carbon Epoxy	[-45/45]	0.2 @ply
Rear Spar		Woven Carbon Epoxy	[-45/45]	0.2 @ply
LE (curved)		Woven Carbon Epoxy	[-45/45]	0.2 @ply
LE (flat)		Woven Carbon Epoxy	[-45/45]	0.2 @ply
Ribs		Woven Carbon Epoxy	[90/0/45/-45]	0.2 @ply

The numerical analysis results of the new configuration offered are shown by **Figure 16** up to **Figure 20**. It is proven that the structure succeeds as it does not experience failure against the given load. The Tsai-Wu failure criterion for the overall model has the value of 0.865. In the new configuration, there is no part using homogeneous material anymore so as Von-Mises failure criterion is no longer used.

The maximum displacement is 3.265 m occurs at the wing tip. It can be reduced by locating the concentrated load to a point further away from wing root. The 2 kg motor propeller is previously located at a distance of 2 m from wing root. Simulation has been done to put the motor propeller at a distance of 6 m and give a result of 3.217 m maximum deflection at wing tip. This changing in motor propeller location needs further consideration including the flight control aspect.

However, the use of woven carbon epoxy in almost all of the parts makes this configuration heavier than the initial configuration. In the actual model, ribs have holes which are not modeled in this simulation. The existence of holes in ribs can reduce the mass of ribs by 53% in comparison with ribs without holes. It is an important consideration for an overweight problem that arises. The further suggestion offered is to make the 4th wing segment become tapered.

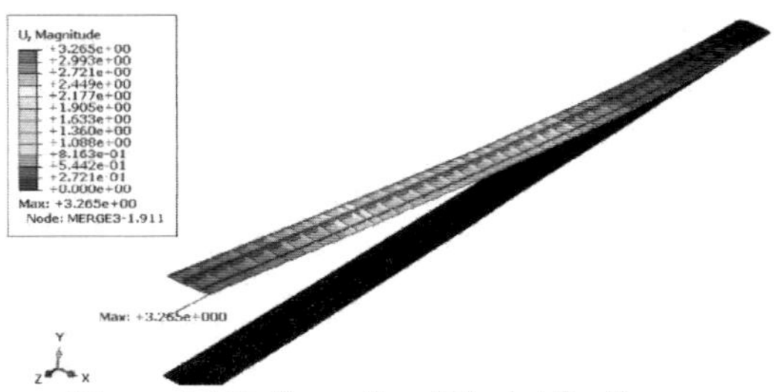

Figure 16 Deformation (U) plot for the new configuration with motor propeller located at 2 m from root.

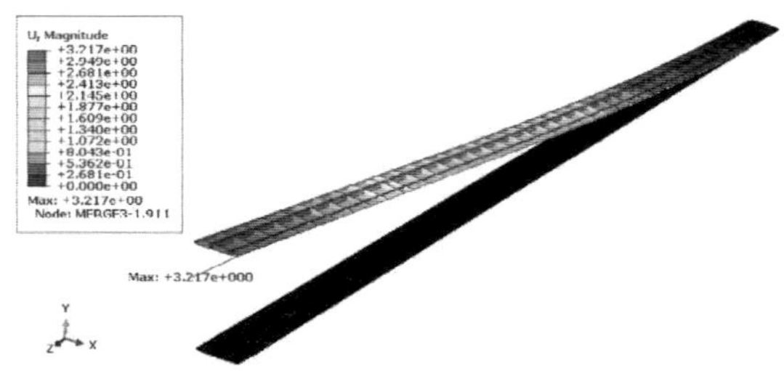

Figure 17 Deformation (U) plot for the new configuration with motor propeller located at 6 m from root.

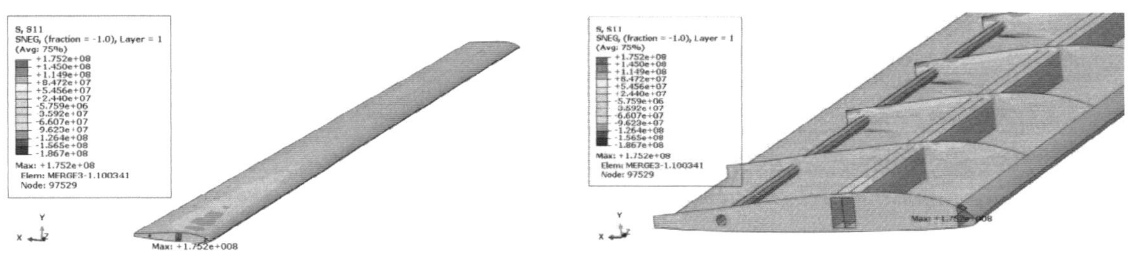

Figure 18 Stress plot in fiber direction (S11) for the new configuration.

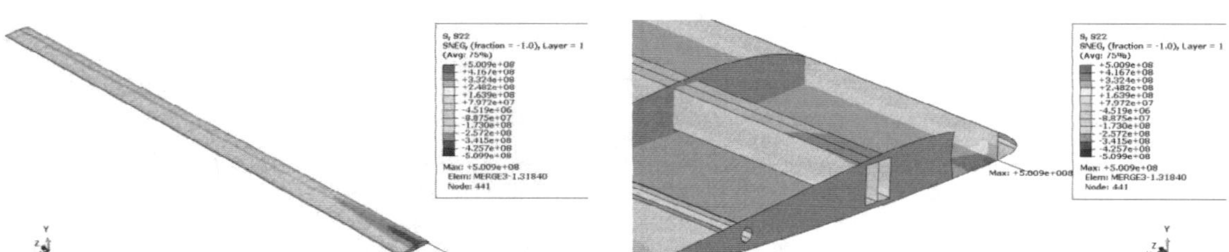

Figure 19 Stress plot in transverse direction (S22) for the new configuration.

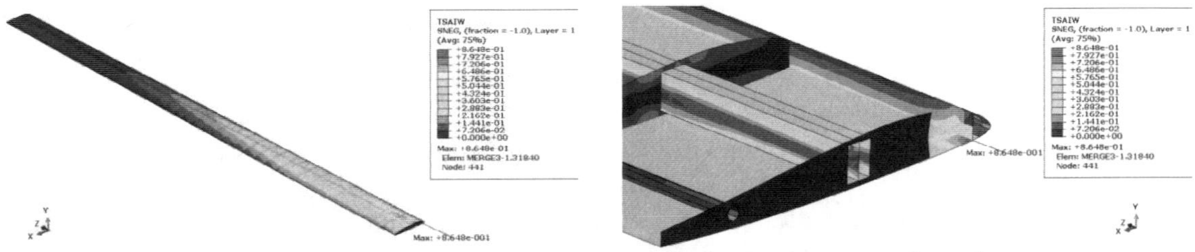

Figure 20 Tsai-Wu failure criterion (TSAIW) plot for the new configuration.

5. Conclusions

Structural analysis on the 2nd generation of HALE UAV ITB is done using finite element software ABAQUS/CAE. The results prove that the initial configuration fails as evidenced by Tsai-Wu failure criterion for composite parts is 63.68 and Von-Mises failure criterion for homogeneous parts is less than 1 for the entire model. The second iteration is made by proposing a new configuration which done some changes of the material, ply direction, and thickening in some parts. This new structural configuration is proved succeeds against the given loads with Tsai-Wu failure criterion is at the value of 0.865.

Acknowledgement

This work is a collaboration between Center for Unmanned System Studies (CentrUMS)-ITB and industry partner engaged in UAV fields. It is funded by the Ministry of Research, Technology and Higher Education of the Republic of Indonesia (*Kemenristekdikti*) through its research program *Riset Andalan Perguruan Tinggi dan Industri (RAPID).*

References

[1] M. A. Moelyadi, R. Zain, D. Widagdo and R. A. Sasongko, "Laporan Akhir Tahun RAPID Pengembangan Wahana Udara Tanpa Awak HALE Guna Keperluan Aerial Surveillance Sumber Daya Alam dan Batas Wilayah NKRI," ITB, Bandung, 2016.

[2] A. Hasan, "Analisis Desain Sayap Komposit HALE UAV dengan Metode Elemen Hingga," ITB, Bandung, 2014.

[3] M. Castellani, J. E. Cooper and Y. Lemmens, "Flight Loads Prediction of High Aspect Ratio Wing Aircraft Using Multibody Dynamics," *International Journal of Aerospace Engineering,* 2016.

[4] G. Romeo, G. Frulla and E. Cestino, "Design of a High-Altitude Long Endurance Solar-Powered Unmanned Air Vehicle for Multi Payload and Operations," *Proceedings of the Instituion of Mechanical Engineer, Part G: Journal of Aerospace Engineering,* vol. 221, pp. 199-216, 2007.

[5] M. R. A. Hutagalung, A. A. Latif and H. Israr, "Structural Design of UAV Semi-Monoque Composite Wing," *Journal of Transport System Engineering,* vol. 3, no. 1, pp. 26-34, 2016.

[6] D. A. P. Ketti, "Kaji Desain Struktur Komposit Filament Winding Tabung Motor Roket RHAN-450," ITB, Bandung, 2017.

[7] D. Broekaart, "5 Reasons Why You Should Use a Mid-Surface Shell Mesh for Thin-Walled Parts," Simuleon, 6 September 2016. [Online]. Available: http://info.simuleon.com/blog/5-reasons-why-your-fea-simulations-should-be-setup-with-a-mid-surface-shell-mesh-for-thin-walled-parts. [Accessed 5 August 2017].

[8] A. Wysocki and D. Soteropoulos, "Modeling a Composite Rib-Section by Mid-Surfacing using Conventional Shell Elements in Abaqus/CAE," University of Massachusetts, Lowell.

[9] M. Rajadurai, P. Vinayagam, G. M. Priya and K. Balakrishnan, "Optimization of Ply Orientation of Different Composite Materials for Aircraft Wing," *International Journal of Advanced*

IOP Conf. Series: Journal of Physics: Conf. Series **1005** (2018) 012025 doi:10.1088/1742-6596/1005/1/012025

Engineering Research and Science (IJAERS), vol. 4, no. 6, 2017.

[10] "Performance Composites Ltd.," [Online]. Available: http://www.performance-composites.com/carbonfibre/mechanicalproperties_2.asp. [Accessed 7 July 2017].

[11] Military Handbook, Metallic Materials and Elements for Aerospace Vehicle Structures, Department of Defense United States of America, 1998.

[12] Wood Handbook: Wood as an Engineering Material, Forest Product Laboratory, USDA Forest Service, 1999.

[13] WS Hampshire Inc., "Ultra High Molecular Weight Polyethylene (UHMW PE)," [Online].

[14] F. Y. Silitonga and M. A. Moelyadi, "Comparative Study of Wing Lift Distribution Analysis for High Altitude Long Endurance (HALE) Unmanned Aerial Vehicle," (in submission), Bandung, 2017.

[15] Rajanish, Nanjundaradhya, R. S. Sharma and B. pal, "A Review of Failure of Composite Materials," *International Journal of Engineering Research and Applications (IJERA),* vol. 3, no. 2, pp. 122-124, March-April 2013.

[16] A. K. Kaw, Mechanics of Composite Materials, Florida: Taylor & Francis Group, 2006.

[17] A. Andriyana, "Failure Criteria for Yielding," Centre de Mise en Forme des Materiaux, CEMEF UMR CNRS 7635 Ecole des Mines de Paris, 06904 Sophia Antipolis, France, 2008.

[18] FAA (Federal Aviation Administration), FAR (Federal Aviation Regulations) Part 23 Airworthiness Standards: Normal, Utility, Acrobatic and Commuter Category Airplanes, vol. Amandement 55.

IOP Conf. Series: Journal of Physics: Conf. Series **1005** (2018) 012026 doi:10.1088/1742-6596/1005/1/012026

Design of Single Stage Axial Turbine with Constant Nozzle Angle Blading for Small Turbojet

Putra Adnan F[1)*], Firman Hartono[2)]

Aeronautics and Astronautics Study Program
Faculty of Mechanical and Aerospace Engineering
Institut Teknologi Bandung

*putraadnanf@gmail.com

Abstract. In this paper, an aerodynamic design of a single stage gas generator axial turbine for small turbojet engine is explained. As per design requirement, the turbine should be able to deliver power output of 155 kW at 0.8139 kg/s gas mass flow, inlet total temperature of 1200 K and inlet total pressure of 335330 Pa. The design phase consist of several steps, i.e.: determination of velocity triangles in 2D plane, 2D blading design and 3D flow analysis at design point using Computational Fluid Dynamics method. In the determination of velocity triangles, two conditions are applied: zero inlet swirl (i.e. the gas flow enter the turbine at axial direction) and constant nozzle angle design (i.e. the inlet and outlet angle of the nozzle blade are constant from root to tip). The 2D approach in cascade plane is used to specify airfoil type at root, mean and tip of the blade based on inlet and outlet flow conditions. The 3D approach is done by simulating the turbine in full configuration to evaluate the overall performance of the turbine. The observed parameters including axial gap, stagger angle, and tip clearance affect its output power. Based on analysis results, axial gap and stagger angle are positively correlated with output power up to a certain point at which the power decreases. Tip clearance, however, gives inversely correlation with output power.

1. Introduction

Small turbojet engine used for Unmanned Aerial Vehicle (UAV) has being developed in our laboratory. This turbojet engine consist of three main components, namely compressor, combustor and turbine. In the design of turbojet engine, a turbine is considered as a component which can stand high temperature loads of heated air from combustion chamber. The kinetic energy of the turbine in rotation is used to drive the compressor so that the engine attains idle condition and to produce thrust output. For small turbojet engine, the use of a single axial turbine is to give simplicity and to provide lower cost in development and manufacture. In addition, using the simple turbine, the engine is easy in maintenance. In this study of design axial turbine, there are several previous research that become a basis and reference, for example D.G Ainley and G.C.R. Matheison mentions in Ref. [1] that the information obtained from the cascade method (which gives two-dimensional of blade section) is not enough because the three- dimensional effect are so powerful. Donald E. Holeski and Samuel M. Futral predict turbine efficiency decreased linearly with increasing rotor tip clearance [2]. James Allen, Ref [3], predict that axial clearance (axial gap) can affect turbine efficiency and torque produced by the turbine. In this work,

IOP Conf. Series: Journal of Physics: Conf. Series **1005** (2018) 012026 doi:10.1088/1742-6596/1005/1/012026

the design of the axial turbine of small turbojet is carried out by considering constant nozzle. Two methods namely parametric cycle analysis (analytical) and computation fluid dynamics (numerical) approaches are used to obtain the axial turbine design which fulfills the turbine design requirement and objectives. With this study of design a simple axial turbine, be expected can contribute to turbomachinery field in develop a high performance turbine with low cost in manufacture and maintenance.

2. Design Point and Method

The designed turbine must fulfills its design points. The design points is summarized in Table 1. The steps that used to obtain design point (that have been appointed before), firstly use analytical method and then result of the analytical method is used as input value of numerical method to obtain the form of the turbine. The designed turbine and then simulated with CFD to assess its performance. For the analytical method in this paper use parametric cycle analysis, parametric cycle analysis is one of method to approximate thermodynamic properties through gas turbine engine primary components such as compressor, combustion chamber, and turbine, based on Brayton cycle. The result from analytical calculation with parametric cycle and then used to be input value to calculate the velocity triangle with numerical calculation. After obtaining the velocity triangle data, and then use the data to generate the airfoil profile and furthermore build the 3D model of the designed turbine. As previously discuss, that the last step is simulate the designed turbine with CFD to assess the designed turbine performance.

Table 1. Design Requirement and Objective of Turbine

RPM turbine	80000 rev/min
Mass Flow Inlet Turbine	0.8139 Kg/s
Temperature Inlet Turbine	1200 deg K
Total Pressure Inlet Turbine	335.33 KPa
Total Pressure Outlet Compressor	372.589 KPa
Coefficient Pressure Ratio	3.677
Compressor Efficiency	67.1 %
Turbine Efficiency	70 %
Mechanical Efficiency	99 %
Fuel Flow	0.01723 Kg/s

The following are the brief explanation of the steps from the method that used (analytical method and analytical method)

2.1 Analytical Method

2.1.1 Parametric Cycle analysis

The Figure 1 is summarized of the parametric cycle analysis, as previously explained that parametric cycle analysis is analytical method used to obtain the performance of the gas turbine (start from compressor, combustion chamber, and then turbine).

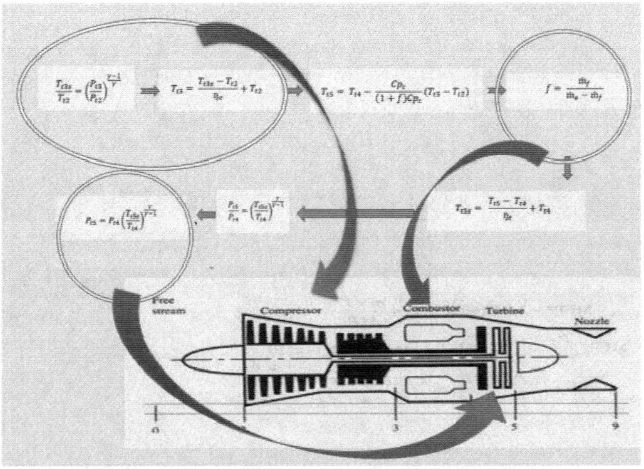

Figure 1. Brief review of parametric cycle equations

2.1.2 Choice of Blading Design

There are some kinds of blading design that commonly applied such as free vortex, constant nozzle angle, and constant mass flow [4]. Free vortex blading commonly have stage characteristic such as the stagnation enthalpy h_0 is constant over the annulus, the axial velocity is constant over the annulus, and the whirl velocity is inversely proportional to the radius. Constant nozzle angle blading is one of blading type with specific form of stator (NGV) blade is not twisted. Blading design that used in this study is constant nozzle angle type, this type is selected due to the simple geometry in stator (NGV) contour and with simple geometry contour will be affected in manufacturer cost.

2.1.3 Velocity Triangle

The determination of angles of the turbine blades (stator blade and rotor blade) uses velocity triangle method as shown in Figure 2 (left). The angles such as alpha-2, beta-2 and beta-3 are used to generate the profile of airfoil at each section along span. As a result the full configuration turbine blade are created. The typical *zero inter stage swirl* velocity triangle used in the design process of turbine blade is shown in Figure 2 (right).

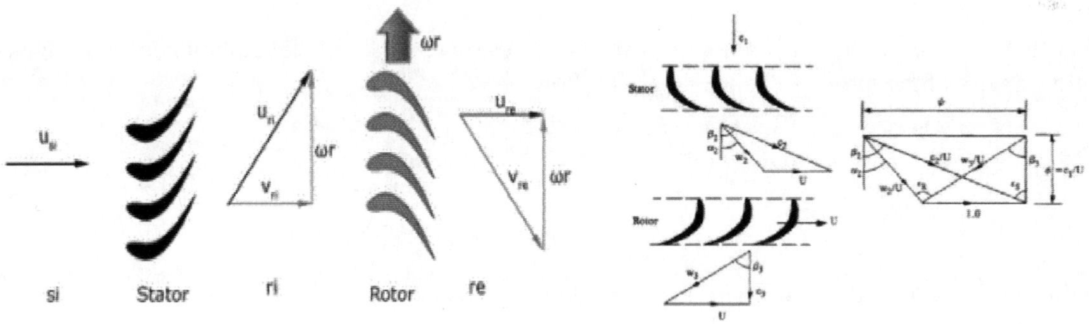

Figure 2. Characteristics of typical velocity triangle through turbine blade (left) and Zero Inter Stage Swirl Velocity Triangle (right) [4]

As shown in the Figure 3, the velocity triangle of each section is changed from tip section to the hub section of the turbine blade. Changes on the flow angle is significantly can be seen at the tip section to the mean section of the blade (the flow angle is reversed). The flow angle for mean section to the hub section is increased from 22.4 degree to 41 degree. The following are the illustration of velocity triangle of the turbine at the tip, middle (mean), and hub section.

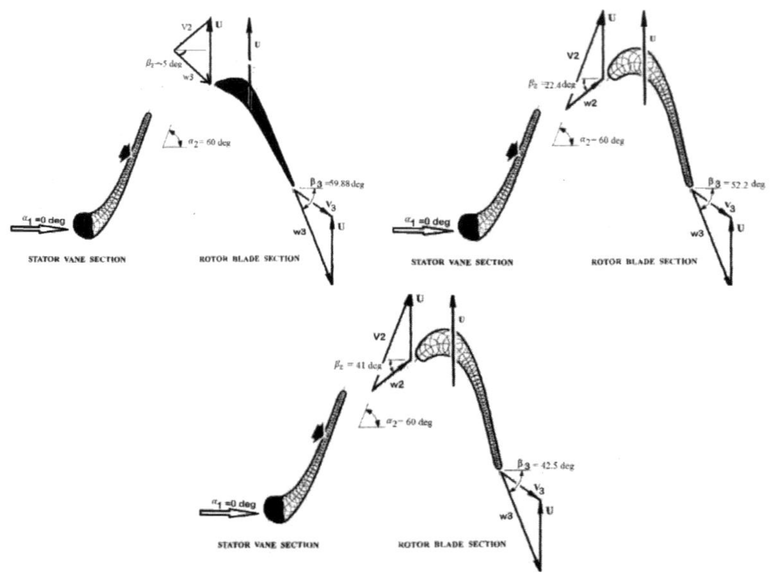

Figure 3. Illustration of velocity triangle of each section of designed turbine blade (tip (left), mean (middle), and hub (right) [4]

The following are the detail calculation to derive the specific flow angle at each blade section (tip, mean and hub section). From the velocity triangle, it can be derived the equation to determine flow angle (such as alpha-2, beta-2 and beta-3. The derivation of flow angle equations can be started by defining other parameters, namely flow coefficient, ϕ and work coefficient, ψ as follows:

$$\phi = \frac{C_x}{U} \quad (1)$$

$$\psi = \frac{\Delta h_0}{U^2} \quad (2)$$

From equations 1 and 2, by performing some mathematical manipulation, the relation between velocity and flow angles of the blade can be obtained as follows:

$$\frac{U}{C_a} = tan\alpha_2 - tan\beta_2 \quad (3)$$

$$\frac{U}{C_a} = tan\beta_3 - tan\alpha_3 \quad (4)$$

And then the equations of the flow angles $\alpha 2$, $\beta 2$ and $\beta 3$ can be obtained as follows:

$$\alpha_2 = \varepsilon_s = arc\tan(\psi/\phi) \quad (5)$$

$$\beta_2 = arc\tan\left(\frac{\psi-1}{\phi}\right) \quad (6)$$

$$\beta_3 = arc\ tan\left({}^{1}/_{\phi}\right) \quad (7)$$

The axial velocity variable, Ca in Equations 3 and 4 can be derived by taken constant nozzle angle shown in the following equations:

IOP Conf. Series: Journal of Physics: Conf. Series **1005** (2018) 012026 doi:10.1088/1742-6596/1005/1/012026

$$C_a \frac{dC_a}{dr} + C_W \frac{dC_W}{dr} + \frac{C_W^2}{r} = \frac{dh_0}{dr} \quad (8)$$

$$\frac{C_{a2}}{C_{w2}} = cot\alpha_2 = constant \quad (9)$$

$$C_{w2}cot^2\alpha_2 \frac{dC_{w2}}{dr} + C_{w2}\frac{dC_{w2}}{dr} + \frac{C_{w2}^2}{r}$$

$$\frac{dC_{a2}}{dr} = \frac{dC_{w2}}{dr}cot\alpha_2 \quad (10)$$

$$(1 + cot^2\alpha_2)\frac{dC_{w2}}{dr} + \frac{C_{w2}}{r} = 0 \quad (11)$$

$$\frac{dC_{w2}}{C_{w2}} = -sin^2\alpha_2 \frac{dr}{r} \quad (12)$$

$$C_{w2}r^{sin^2\alpha_2} = constant \quad (13)$$

$$C_{a2}r^{sin^2\alpha_2} = constant \quad (14)$$

$$Ca_2r = Ca_2m\left(\frac{r_m}{r}\right)^{sin^2\alpha_2}$$

$$Ca_2m = Ca_3m = Cw_2m\, cot\, \alpha_2 \quad (15)$$

$$Cw_2m = \frac{C_p\Delta T}{U_m} \quad (16)$$

2.1.4 Turbine Blade Cross Section

Based on the analytic calculation discussed at the previous section, the profile of the airfoil is obtained by using some numerical methods. The calculated angle from analytical method is imported as an input value and target value to obtain the desired flow angle whether flow inlet angle or flow outlet angle from each blade sections. Also, the variable such as airfoil maximum camber location, camber angle theta, stagger angle, pitch chord ratio, and airfoil profile thickness is arranged in such a way, that the input value (flow inlet angle) can provide desired output angle (correspond with analytical calculation). Furthermore it also can provide the proper shock free beta angle (wherever possible the beta angle and shock free beta angle gives the same value). The detailed airfoil profile geometry parameter of each section (NGV, hub rotor, mean rotor, and tip rotor) is summarized in Table 2.

Table 2. Specification of NGV and rotor blade turbine design

NGV Airfoil Profile	
Airfoil Type	C4 base profile
Profile Thickness Scale	1
Camber line type	Parabolic
x/l of maximum camber	0.2
Camber angle theta	-90 degree
Stagger angle lamda	39.6 degree
Pitch/chord ratio t/l	1
Fluid inlet angle beta 1	0 degree
Rotor Hub Airfoil Profile	
Airfoil Type	C4 base profile
Profile Thickness Scale	1.2
Camber line type	Parabolic
x/l of maximum camber	0.3
Camber angle theta	107.7 degree
Stagger angle lamda	-8 degree
Pitch/chord ratio t/l	1
Fluid inlet angle beta 1	41 degree

Table 2. Specification of NGV and rotor blade turbine design (cont.)

Rotor Mean Airfoil Profile	
Airfoil Type	C4 base profile
Profile Thickness Scale	1.2
Camber line type	Parabolic
x/l of maximum camber	0.3
Camber angle theta	100 degree
Stagger angle lamda	-22 degree
Pitch/chord ratio t/l	1
Fluid inlet angle beta 1	22.4 degree
Rotor Tip Airfoil Profile	
Airfoil Type	C4 base profile
Profile Thickness Scale	1.2
Camber line type	Parabolic
x/l of maximum camber	0.3
Camber angle theta	82 degree
Stagger angle lamda	-38 degree
Pitch/chord ratio t/l	1
Fluid inlet angle beta 1	-5 degree

Illustration of the stator (NGV) – rotor configuration both of with airfoil profiled NGV and flat plate profiled NGV is depicted in Figure 4. In the flat plate profiled NGV configuration, the flat plate is generated by following the main camber line of the airfoil that previously used, the flat plate thickness that used in the configuration is 2 mm (will be discussed at the following chapter).

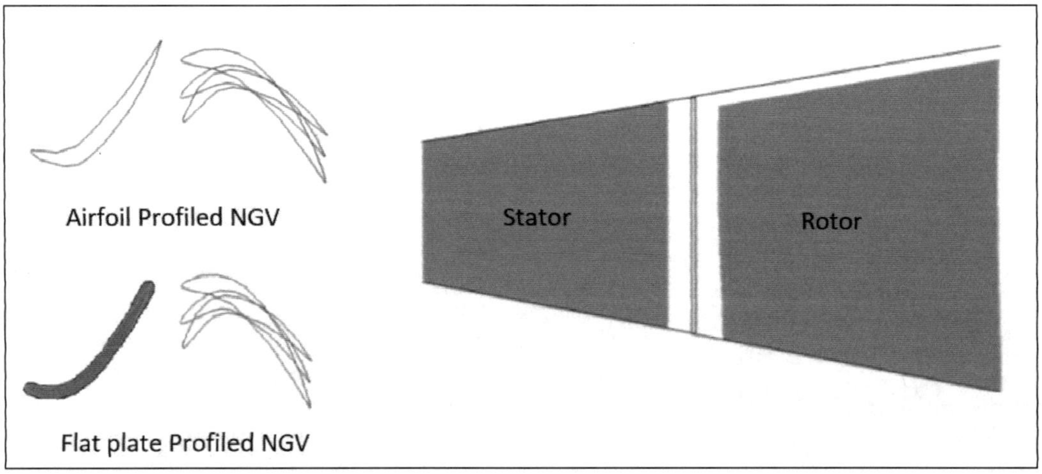

Figure 4. Turbine airfoil arrangement (stator (left) and rotor (right))

2.2 CFD Simulation

2.2.1 3D Model of Designed Turbine

The generation of three dimensional blade model as full configuration of the turbine is used for numerical simulation and turbine performance computation. The single blade model and full NGV – rotor model are shown in Figure 5.

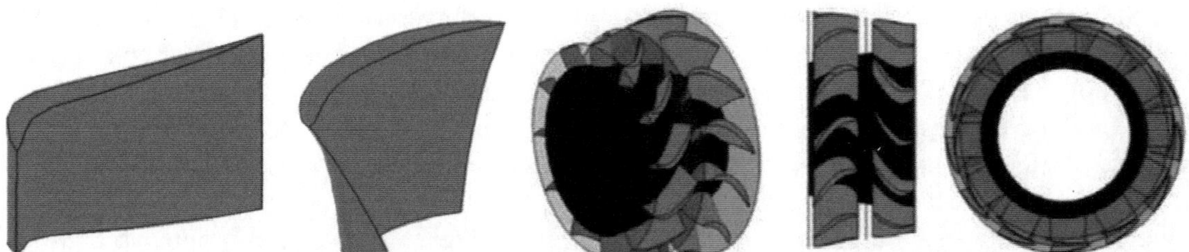

Figure 5. Three-dimensional blade model (left) and full three dimensional designed turbine (right)

2.2.2 Numerical Model

The equation that used in the simulation is Reynolds Averaged Navier-Stokes with K-epsilon turbulence model. The number of mesh that applied on the computational domain geometry is around 5 million to 6 million mesh. The type of the simulation is quasi-steady simulation. This simulation type uses relative move of the flow in the domain to simulate the rotational movement of the rotor blade. The rotor is remain at rest but the flow in the rotor domain move with certain rpm that inputted. This conditions is known by applying frozen rotor interface.

2.2.3 Boundary Condition

The computational simulation starts with the generation of computational domain in which the thermodynamic properties such as velocity, pressure and temperature are computed. The domain is defined with some boundaries such as inlet, shroud, hub, and interfaces as shown in Figure 6. The boundary conditions that applied in the pre-setting of the simulation steps are inlet (in this boundary conditions use mass flow to be input value and total temperature), wall (use default setting of the wall boundary conditions), and outlet (in this boundary use static pressure as input value). The place of the boundary conditions that applied at the computational domain is depicted in Figure 6. In the simulation, two type of the convergence criteria i.e., error convergence criteria and residual result convergence criteria, were used to check the convergence. The final step of the simulation is running numerical process to get convergence solution

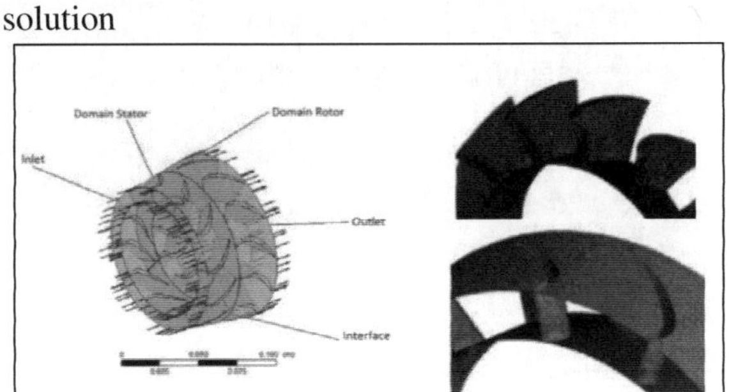

Figure 6. Computational domain (left) and meshing (right) that used in this design study

3 Convergence of the Simulation

The criteria of convergence used in the simulation is residual growth of mass and momentum as well as the curves of normal force of blade as shown in Figure 7. The residual is one of the most fundamental measures of an iterative solutions convergence, as it directly quantifies the error in the solution of the system equation. In a CFD analysis, the residual measures the local imbalance of a conserved variable in each control volume. In an iterative numerical solution, the residual will never be exactly zero. For CFD, RMS residual levels of 1E-4 are considered to be loosely converged, level 1E-5 are considered to be well converged, and the last level of 1E-6 are considered to be tightly converged. For complicated problems, however, it is not always possible to achieve residual levels as low as 1E-6 or even 1E-05. As shown in the Figure 7 (residual error curve) that the curve is already achieve convergence level of 1E-4.In a steady state analysis, the solution field should not change iteration to iteration for an analysis to be deemed converged. Monitoring integrated quantities such as force, drag, or average temperature can help judge if the simulation have been converged or not. In the Figure 7 (quantities of interest curve) is shown that the curve is already converged since iteration 40.

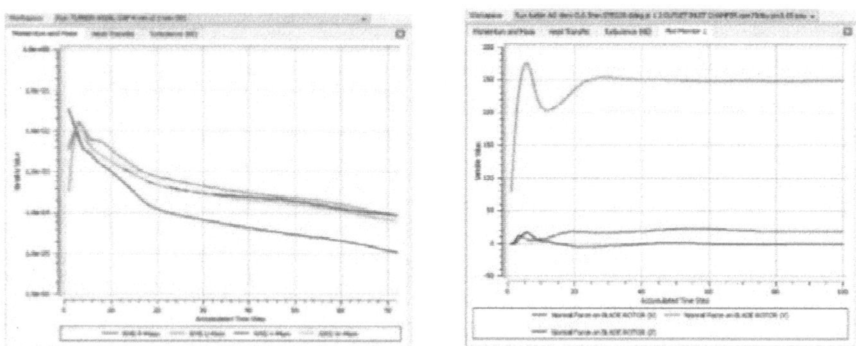

Figure 7. Error curve (left) and quantities of interest curve (right) from running process

4 Result (On Design Performance)

4.1 Case 1: Turbine with Airfoil-Profiled NGV

Figure 8 below shows the detailed dimension of the turbine for simulation. Figure 9 shows the effect of axial gap and clearance on turbine output power resulted from simulations. From Figure 9, it can be seen that the maximum power for various clearance is resulted for the axial gap of 20 % of stator (NGV) chord. The turbine power increases about 20% by changing the tip clearance from 10% to 2.5% of the blade span. The configuration of the axial turbine designed with methodologies as explained before, specifically with 4 mm axial gap (20 % stator chord) and airfoil-profiled stator blade is presented in Figure 8 as follows. The detailed sizes of the designed axial turbine is also presented in Figure 8.

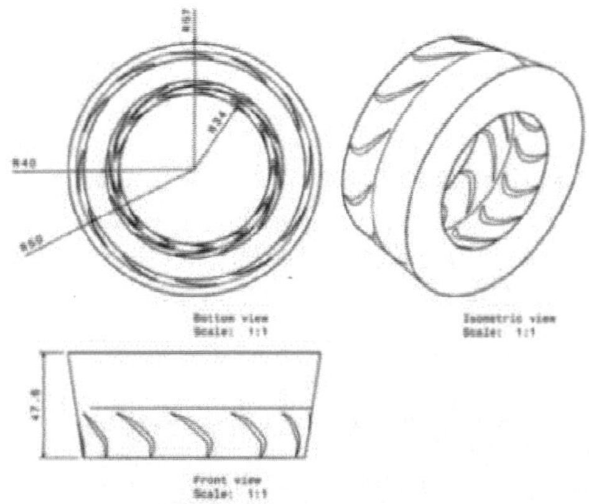

Inlet outer radius	50 mm
Inlet inner radius	40 mm
Interface outer radius	53 mm
Interface inner radius	37 mm
Outlet outer radius	57 mm
Outlet inner radius	34 mm
Axial gap	4 mm
Clearance	0.5 mm or 1 mm

Figure 8. Designed Turbine (left) detailed of turbine size specification (right)

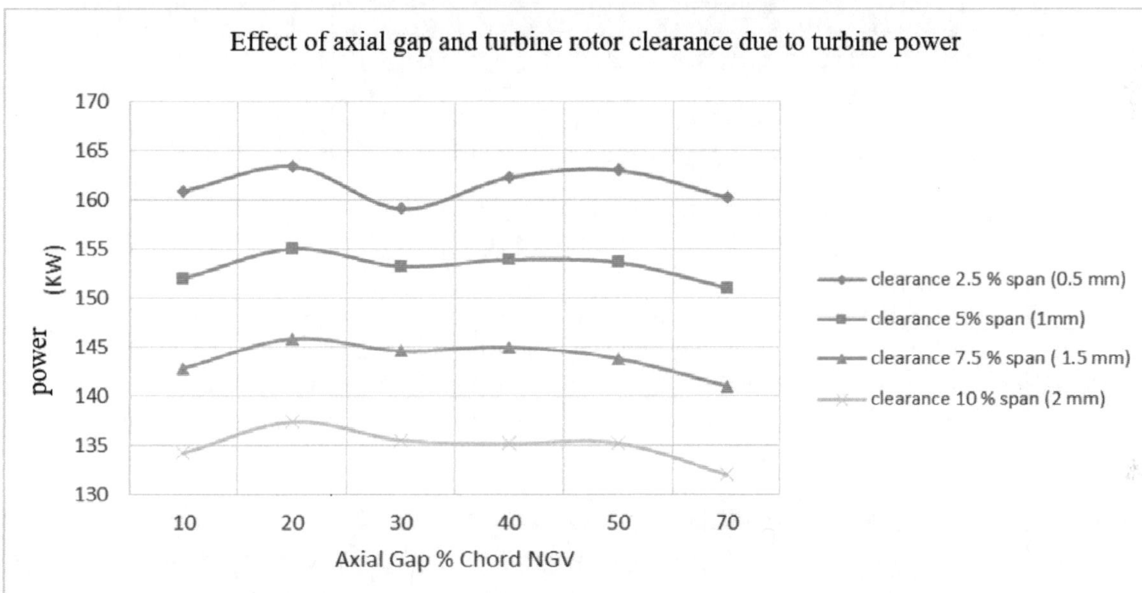

Figure 9. Effect of axial gap and turbine rotor clearance due to turbine power

The effect of tip clearance is examined further for 4 mm axial gap turbine case. Table 3 shows the performance of the axial turbine with the axial gap of 4 mm for the tip clearances of 0.5 mm (left) and 1.0 mm (right), respectively. As can be seen from Table 3, the output power increases by reducing clearance.

Table 3. Performance of two type designed turbine, turbine with 4 mm axial gap-0.5 mm rotor clearance (left) and turbine with 4 mm axial gap-1 mm rotor clearance (right)

0.5 mm rotor clearance		1 mm rotor clearance	
Total Pressure Inlet	373.2 KPa	Total Pressure Inlet	356.2 KPa
Total Pressure Outlet	156.1 Kpa	Total Pressure Outlet	158.4 KPa
Total Temperature Inlet	1200 deg K	Total Temperature Inlet	1200 deg K

Total Temperature Outlet	1022 deg K	Total Temperature Outlet	1023 deg K
Mass Flow	0.8139 Kg/s	Mass Flow	0.8139 Kg/s
Power	160196 watt	Power	155121 watt

4.2 Case 2: Turbine with Flat Plate Profile that applied at the Stator (NGV) Section

Considering manufacture cost and simplicity, the airfoil-profiled stator blade might possibly be substituted by a curved flat plate profile which follow the camber line of the airfoil as shown in Figure 10. The curved plate axial turbine is then simulated with the axial gap of 4 mm for two different clearances, 0.5 mm and 1 mm. The thickness of the plate is set to 2 mm.

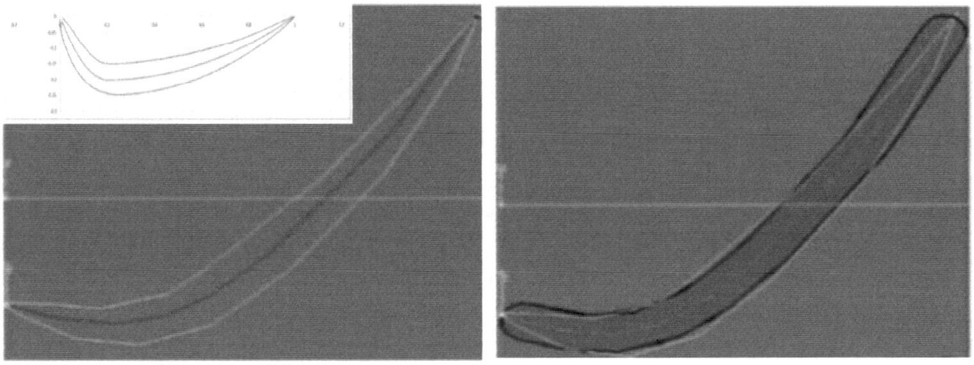

Figure 10. Stator profile using airfoil profile (left) and using flat plate (right)

The performances of the designed turbine with curved plate stator blade are presented in Table 4. Based on the simulation result between two kind of turbine models (turbine with NGV blade is airfoil profiled and turbine model with flat plat profiled), the output power of turbine model with flat plate profiled NGV is decreasing around 3 % from the previous NGV model. Although the flat plate NGV model give lower output power but the reduction is not significant, with simple NGV geometry turbine with flat plate NGV is still worth to be developed.

Table 4. Performance of turbine with flat plate stator-rotor clearance 0.5 mm (left) and rotor clearance 1 mm (right)

Rotor clearance 0.5 mm		Rotor clearance 1 mm	
Total Pressure Inlet	3.812e+05 Pa	Total Pressure Inlet	3.817e+05 Pa
Total Pressure Outlet	1.586e+05 Pa	Total Pressure Outlet	1.588e+05 Pa
Total Temperature Inlet	1.200e+03 K	Total Temperature Inlet	1.200e+03 K
Total Temperature Outlet	1.028e+03 K	Total Temperature Outlet	1.031e+03 K
Mass Flow	0.8319 kg/s	Mass Flow	0.8139 kg/s
Power	156831 watt	Power	152082 watt

4.3 The Effect of Stagger Angle and Flat Plate Thickness to Turbine's Output Power

4.3.1 The Effect of Turbine Stagger Angle to Turbine Output Power

The influence of stagger angle on turbine output power is depicted in Figure 11. The simulation results show the turbine output power are lower than 155 KW for the turbine models with clearance 1.5% and 2%. The turbine power output can be increased by setting up stagger angle of the blades to the right position. As shown from the simulation, the higher angle of stagger gives more power to the turbine.

IOP Conf. Series: Journal of Physics: Conf. Series **1005** (2018) 012026 doi:10.1088/1742-6596/1005/1/012026

However, the increase in stagger angle that produces higher power output required higher inlet pressure and decreases turbine efficiency also as shown in Figure 11.

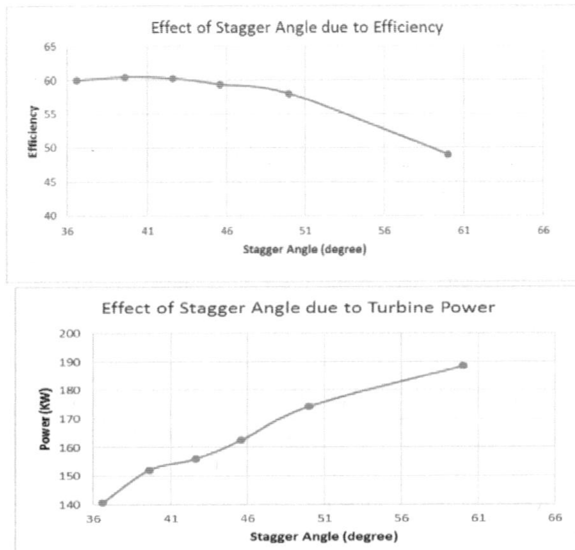

Figure 11. Effect of Stagger angle due to turbine efficiency (left) and turbine power (right)

4.3.2 Flat Plate Thickness Effect

The change of turbine power output due to the thickness of the flat plate is shown in Figure 12. In this study, the flat plate thickness is varied as 1 mm, 2 mm and 3 mm. From simulation results, the maximum turbine power output is obtained for the flat plate thickness of 2 mm. The layout of the turbine power output for various thickness is also depicted in Figure 12.

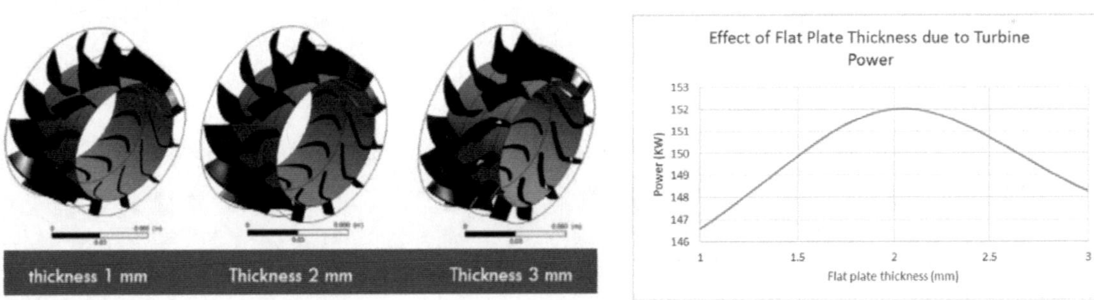

Figure 12. Three models of designed turbine (left) and effect of flat plate due to turbine power (right)

5. Conclusion

In the design of axial turbine there are several things that affect output power of the turbine such as stagger angle, axial gap, clearance, and the profile of the airfoil (both the stator and rotor profile). Based on the simulation results there is a certain distance of axial gap that provide optimal power, and the output power of the designed turbine is tend to decrease with increasing clearance (distance between rotor blade and shroud wall). The design model that closest with design point (the design point are : turbine that can produce 155 KW in power wiht 0.8139 kg/s mass flow at 80000 RPM) is the turbine model with 20% axial gap, 1 mm clearance between tip rotor blade and shroud wall and still use airfoil profile at the NGV section. The flat plate NGV model also can achieve the design point with change the clearance from 1 mm to 0.5 mm or may be can also change the stagger angle of the NGV's blade.

References

[1] D.G. Ainley and G.C.R. Mathieson, A Method of Performance Estimation for AxialFlow Turbines, Aeronautical Research Council Reports and Memoranda, 1952
[2] Donald E. Holeski and Samuel M. Futral, Effect of Rotor Tip Clearance on the Performance of A 5-Inch Single Stage Axial Flow Turbine, National Aeronautics and Space Administration, 1969
[3] James Allen Messegee, Influence of Axial and Radial Clearance on the Performance of a Turbine Stage with Blunt Edge Non-Twisted Blades,Dudley Knox Library, 1967,
[4] Lewis, Turbomachinery Performance Analysis, Elsevier Science & Technology, 1996
[5] H. Cohen, Gas Turbine Theory, Longman Group Limited, 1996
[6] Jack D. Mattingly, Elements of Propulsion: Gas Turbines and Rockets, AIAA Education Series, 1989
[7] Anderson Jr., John D., Fundamental of Aerodynamics, McGraw-Hill Series
[8] Hartono, Firman, Aero Propulsion Presentation Chapter Axial Turbine
[9] Hartono, Firman, Aircraft Propulsion Chapter Parametric Cycle Analysis
[10] Hartono, Firman, Aircraft Propulsion Chapter Turbomachinery

Review of Design Aspects and Challenges of Efficient and Quiet Amphibious Aircraft

Rhea P. Liem, Ph.D

Department of Mechanical and Aerospace Engineering
Hong Kong University of Science and Technology (HKUST), Hong Kong

rpliem@ust.hk

Abstract. Apart from the commercial and military aviation sectors, the general aviation (GA) sector is expected to experience a rapid growth, especially in Asia. The increasing economic activities in the region would demand for more efficient and convenient transportation, which would open door to more GA services. This development would require sufficient infrastructure supports, including airports. However, insufficient land area has often imposed limitations in airport development. As such, some areas (e.g., remote islands) are not easily accessible by air. One implication is that travels can only be done via land or water, which might prolong the travel time. This applies to business travels, with the significant increase in business and economic activities, which in turns demands for more efficient and faster mobility. In other cases, this involves some rural areas where the infrastructures are not very well-developed, and where the geographical terrains are too challenging to build a pad for vertical takeoff and landing (VTOL) air vehicles. Under such circumstances, it would be imperative to enable air travels to carry critical logistics such as medical supplies, food, and even sick patients. In this regard, we propose to develop a low-payload, low-altitude amphibious aircraft, which can takeoff and land on both water and land. Aircraft design process is a complex procedure and multidisciplinary in nature, and for amphibious aircraft design we need to consider the two takeoff and landing modes, which imposes further challenges to the design. In this paper we present two preliminary design projects, for two-seater and ten-seater aircraft. To design an efficient and quiet amphibious aircraft, we conduct some experiments on noise shielding mechanisms to reduce the propeller noise. The challenges and resulting designs are briefly discussed in this paper. Amphibious aircraft development will be very relevant to Indonesia, which is the world's largest archipelago with thousands of islands. More efficient inter-island transportation and mobility would be crucial in the overall economic development in the country.

1. Introduction

The air transportation has experienced a rapid growth, both in military and civil aviation sectors. The civil aviation sector can generally be categorized into two, namely the commercial air transport and general aviation [1]. The commercial air transport involves all operations that transport passengers, cargo, or mail for remuneration or hire. Scheduled services (revenue) are flights scheduled and performed for remuneration according to a published timetable, or recognizably systematic. A charter flight is a non-scheduled operation using a chartered aircraft. Air taxi revenue flights include on-demand, non-scheduled flights on short notice for the carriage by air of passengers, freight or mail, or any combination thereof for remuneration. General aviation, on the other hand, can be classified into instructional, business, and pleasure flying, as well as aerial work. Its activities include air ambulance, crop dusting, traffic monitoring, aerial photography, law enforcement, forest-fire fighting, gliding, etc.

Commercial airlines have been the fastest growing mode of passenger transport. The revenue passenger kilometer (RPK) of international air travel has grown by around 9% annually since 1960 [2, 3, 4, 5]. The passenger traffic in the United States alone has more than tripled since 1970. However, the energy used in air transportation has only increased by 43% [6]. The increase in aircraft efficiency came from several factors, including the introduction to jet engines [7], as well as improvements in avionics, wing design, and materials [8]. The doubling in aircraft efficiency has led to a very rapid decline in air transport price, especially between 1959 and 1972, when jet engines were introduced [8]. This in turn further promotes commercial aviation, and its demand is expected to continue increasing at an average annual rate of 4.8% through 2036 [9].

The general aviation (GA) sector is expected to experience a substantial growth. According to the General Aviation Manufacturers Association (GAMA) [10], the total number of GA aircraft flying worldwide today has reached 362,000 in 2016, ranging from two-seater training aircraft and utility helicopters to intercontinental business jets. Of all these aircraft, 204,000 are based in the United States and 110,000 are based in Europe. In the United States, the GA sector supports $2,219 billion in total economic output and 1.1 million total jobs. The number of GA airports in the United States and Europe has exceeded 5,000, and GA fleet can access over 4,200 airports. In terms of the total number of GA aircraft, Asia-Pacific is still lagging behind, with only 25,000 GA aircraft in operation.

One important aspect of the GA sector is the amphibious aircraft, i.e., air vehicles that can takeoff and land from both ground and water. It is different from seaplane, which can only takeoff and land on water. Amphibious planes have been around since the early age of aviation history. Numerous models of amphibious aircraft or seaplanes were developed for general aviation, military and rescue. Amphibious aircraft have often been used for exploration purposes, due to the versatility in their landing mechanism. In particular, they are favorable in exploring rural areas with rough terrain or with limited areas for proper runways. At present, amphibious aircraft found their applications in leisure, military, and rescue/emergency purposes. With the continued growth of the amphibious aircraft market, we foresee a demand for commercial amphibious aircraft in service for passenger transports.

Needless to say, the dual landing mechanisms and versatility of amphibious aircraft could offer some competitive advantages that conventional, ground-based aircraft could not. These unique characteristics can provide more accessibility. Some of the potential applications for the amphibious aircraft include: (1) short-haul business aviation development to meet the demand for customized high-efficiency travel services, (2) high-end seasonal tourism activities and transport to remote areas, (3) public services, such as search and rescue, fire-fighting, disaster relief, and medical aid, (4) industrial, agricultural, and forestry production activities, (5) sport activities (e.g., diving), and (6) leisure and point-to-point travel between residential properties, resorts, hotels, etc.

The unique characteristics and competitive advantages of amphibious aircraft, however, come with more design challenges than those of conventional aircraft. These challenges will be further discussed in Section 4. This paper presents some preliminary work on designing efficient and quiet amphibious aircraft, as part of students' final year design project (FYDP). At this preliminary stage, we select three primary design challenges to focus on, namely the noise impact, hydrofoil design, and landing gear mechanism.

The remaining of this paper is organized as follows. Section 2 will summarize the market research for amphibious aircraft, focusing on China and Asian region. A brief overview of amphibious aircraft design and current technology, including some popular examples, is then given in Section 3. We then discuss about some challenges on the amphibious aircraft development, including the design challenges (Section 4) and regulations (Section 5). The resulting preliminary designs of two aircraft (two-seater and ten-seater) are then presented in Section 6. This paper is then consolidated with conclusion and some future works in Section 7.

2. Emerging market for amphibious aircraft in Asia

This section summarizes the market research and analysis for our preliminary design projects. Focus is given to the Pearl River Delta region, which covers Hong Kong and some regions in China. The results from this market research are used to determine the mission requirements for the amphibious aircraft considered in this project. The design process and entailing discussion, however, will also be applicable to other uses of amphibious aircraft. Some other potential markets for similar amphibious aircraft designs, e.g., in the Indonesian archipelago, will be discussed briefly at the end of this section.

In the recent years, Asian countries have seen a healthy economic growth. While the US still tops the list in terms of the gross domestic product (GDP), China has been catching up to be the second place, according to the World Bank Data Catalog [11]. At present, the GDP of China is roughly 60% of that of the US. In terms of the GA sector, however, we still observe a stark difference between the two countries, despite the rapid growth observed in the GA market in China in the past decade. This difference can be clearly seen in the summary presented in Table 1.

Category	US	China
Number of aircraft	216,000	2,250
Number of airport	13,513	approx. 300
Number of flown hours	23 millions	0.7 millions

Table 1. General aviation sector comparison between the US and China.

Despite the recent development in military and commercial passenger aircraft in China, the GA production in the country has not been very active [12]. In 2006, the market share of foreign aircraft in GA industry domestically was 56%. However, the China's domestic GA manufacturing quantity is currently growing [13]. By the end of 2014, there were only 2,533 licensed pilots to fly GA aircraft [14]. In 2013, the US operated more than 300,000 GA aircraft, with at least 24,000 landing sites. In China, on the other hand, only 1 654 GA aircraft and 399 landing sites were available, according to industry statistics [14]. Moreover, there is still a lack of GA pilots in the region. However, China is expected to see an increase in the number of GA's pilots in the near future. The Civil Aviation Administration has relaxed standards in theoretical exams, flight tests, and physical conditions set for flight students [14]. With all this development in sight, we can optimistically expect to see further development in GA aircraft, including amphibious aircraft, in the Asian region.

Geographically, Hong Kong is one of the most economically dynamic cities in China. It is located within the Pearl River Delta zone, which allows for a close connection to cities such as Macau, Zhuhai, Guangzhou, and Shenzhen, that are located 65 km, 60 km, 132 km, and 30 km from Hong Kong, respectively. China has a coastline of 18,888 km in length, a total of 1,500 rivers with catchment area of over 1,000 km^2. There are over 24,800 lakes, where 2,800 of them have water surface areas larger than 1 km^2. There is a high concentration of middle-class and wealthy populations in the coastal regions such as Beijing-Tianjin-Hebei region, the Yangtze River Delta, and the Pearl River Delta. In these regions, land is expensive and increasingly sparse. On the other hand, there are a large number of lakes, waterways, rivers, coastal lines, and islands. Moreover, the waterfronts are likely to be adjacent to high-end housing estates. Within this distance range, a light GA aircraft is an ideal means for inter-city transportation. An amphibious configuration that can takeoff and land from water would offer a greater advantage on focusing on the aforementioned potential high-end markets. Amphibious aircraft also offer an advantage superior to other conventional air transportation vehicles in terms of docking space. In general, the water landing mechanism leads to amphibious aircraft requiring a minimum land area compared to that of airport and railway station. Being an island in nature and a harbor in function, Hong Kong could take an advantage of the surrounding sea which constitutes a great source for the development of GA industry. Figure 1 illustrates the areas covered within an 800 km and 1,000 km radius from Hong Kong, which could be the potential coverage of the GA flights from Hong Kong.

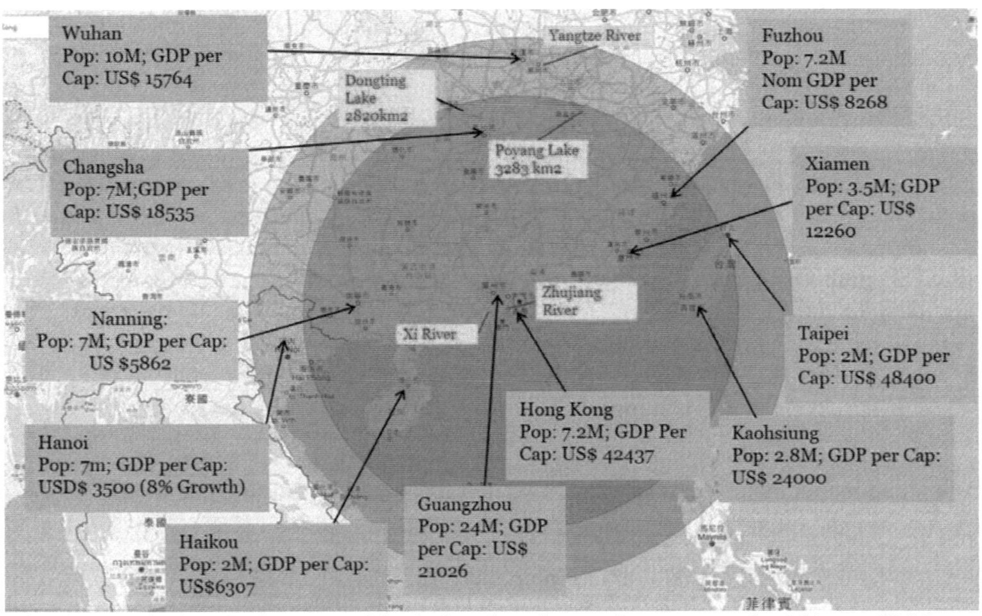

Figure 1. The Pearl River Delta area with circles showing regions covered within an 800 km and 1,000 km radius from Hong Kong, respectively.

Though the above discussion is mostly focused in the China region, the amphibious aircraft development will be crucial and beneficial in other countries and regions. Indonesia, in particular, would benefit significantly with it being the world's largest archipelago. The reported total number of islands in Indonesia varies with different sources, ranging from 13,000 to 19,000, due to tidal activities, remoteness, small island sizes, etc. A survey conducted by Indonesian's National Institute of Aeronautics and Space (LAPAN) in 2002 stated that there are 18,307 islands in Indonesia. Among all these islands, 922 of them are permanently inhabited [15]. However, not all of them are already well-developed and easily accessible. Moreover, they might not have the infrastructure required to build proper airports. Most of those islands would rely on transportation by sea, which could be significantly more time consuming than by air. Needless to say, amphibious aircraft would provide more transportation means between islands, which can further boost economic development, support logistics to more rural and remote areas, improve medical facilities and mobility.

In addition to the amphibious aircraft's role as the main means of transportation and logistics, it can also further boost the tourism and sport industries. Diving tourism, for instance, often includes several diving sites that are spread around in the region. Indonesia boasts some of the world's best diving sites, which are spread around in different islands. Boats or yachts are currently the most common means to transport divers from one site to another, which might be a bit slow and limits the number of sites that can be visited in one day. An amphibious aircraft will offer an excellent solution in this situation.

3. Amphibious Aircraft Overview and Current Technology

An amphibious aircraft is an aircraft that can takeoff and land on water and on land. It typically features a hull-type fuselage that can assist in the water operation and retractable landing gear to enable landing on land [16].

Some of the earlier generations of amphibious aircraft from the 1950's and 1960's include de Havilland Canada DHC-3 Otter, Viking Air DHC-6 Series 400, and Viking Air DHC-6 Series 400S Seaplane. The Dornier Seastar was introduced in 1984, which is a turboprop-powered amphibious aircraft. Composite materials form the bulk of its structures. There have been some recently developed amphibious aircraft designs in the market, such as the ICON A5, MVP Model 3, and LISA Akoya. All three of them are designed to carry two passengers. These three promising designs are shown in Figure 2. The design for each aircraft is described briefly below.

Figure 2. Some existing amphibious aircraft in the market.

ICON A5 is a high-wing monoplane with its wing and fuselage made up of carbon-fibre material. It is powered by a single 100hp Rotax 912iS engine driving three-bladed push propeller. Its Dornier-style sponsons provide hydrodynamic stability and acts as a step for the crew (for access and egress). One main advantage of ICON A5 is that its wings can be folded for ground transport and storage purposes. MVP Model 3 looks very similar to ICON A5 but it is designed to operate on snow and ice, in addition to water. It has a high-wing structure with foldable wings, made up of carbon fibre and is powered by 115hp Rotax 914 engine driving three-bladed push propeller. One unique feature of the MVP Model 3 is that its floor panels can be rearranged to accommodate fishing or camping. LISA Akoya looks different from the previous two designs. Its unique features include: the wing that can be folded almost 90° , the single engine (100hp Rotax 912 ULS) mounted high on the tail, and its carbon-fibre-reinforced polymer composites body and wing structures. It has a very high aspect ratio of 18, which is uncommon for a powered aircraft, and uses unique trailing edge extensions rather than the conventional hinged flaps.

4. Design and technical challenges for amphibious aircraft

Since an amphibious aircraft need to perform both on water and in the air, there are some operational issues that need to be carefully considered in the design process. It often requires some tradeoffs between the aerodynamics and hydrodynamics performances. For instance, the fuselage now needs to be designed as a hull, to be able to operate on water. This shape, however, is not aerodynamically efficient [16]. The higher drag in turns reduces the rate-of-climb and cruise speed of amphibious aircraft, when using the same power as a conventional aircraft. The relatively low operational altitude of amphibious aircraft makes noise impact a more important consideration in the design compared to that of

IOP Conf. Series: Journal of Physics: Conf. Series **1005** (2018) 012027 doi:10.1088/1742-6596/1005/1/012027

conventional aircraft. Therefore, any noise shielding or mitigation procedures need to be considered in the design.

The harsher operating environment also imposes some challenges to amphibious aircraft. For instance, its constant contact with seawater requires a more rigorous corrosion protection and maintenance, including the special selection of the materials and coatings used for the aircraft components. The hull is also more susceptible to damage as it slams into waves or floating debris, or when sliding up on a beach. This requires a strong and sturdy hull and as a consequence, an amphibious aircraft is typically heavier than a conventional aircraft for the same mission requirements. As an alternative to the hull design, some amphibious aircraft use hydrofoils, e.g., LISA Akoya. The hydrofoils can help reduce drag, and thus allows for higher rate-of-climb and cruise speed. Hydrofoils are also adopted in one of our designs, which will be further discussed Section 6.2.

In addition, there are some basic requirements of amphibious aircraft that do not exist in conventional aircraft. The aircraft needs to be watertight, especially for doors, windows, and any panels that give access to dry components. The lower aerodynamic efficiency might call for additional features to control the aircraft when there are not enough aerodynamic forces. The aircraft stability needs to be ensured both during flight and when it is on water, and the aircraft must be controllable in water at all speeds. Hydrostatic stability refers to the tendency of the aircraft to return to its at-rest position, upon application of any external forces that tilt it to one side. For this consideration, we need to look into the buoyancy, which is not required when designing a conventional aircraft. Some additional features, that are not typically used in a conventional aircraft, become necessary when designing an amphibious aircraft. For instance, a keel can help guide the aircraft to move in a straight line, a chine can help direct the water spray away from the hull in addition to increasing the hydrodynamic lift, and spray rails can help reduce the water spray on the propulsive system, since it could be destructive to the propeller.

Pilots need special training before they can operate and fly amphibious aircraft. First, they need to be familiar with some nautical terms used that are not common in any conventional aircraft. For instance, port and starboard are used instead of left and right, windward and leeward to refer to the upwind and downwind, bow and stern to refer to the nose and the bottom of the empennage, respectively. Landing on different water conditions (e.g., a normal rippled water, glassy or mirror-like water, and rough water) also need different techniques and procedures that the pilots need to be familiar with. Moreover, the cockpit design should provide a good visibility for pilots, especially to judge the altitude above water.

5. Challenges in Amphibious Aircraft Development

In addition to the technical challenges, there are three main challenges on the development of the amphibious aircraft industry. They include the regulatory restrictions, restrictive airspace, and weak support/insufficient supplies of equipment. Focus will be given to the China region, but these impeding factors are common for GA sectors in many countries. Each of them will be described briefly below:

1. **Regulatory restriction**

   China's shortcoming of coordinated regulations and policies across all shareholders made regulators overly cautious in the GA industry. Moreover, China lacks an efficient regulatory framework that encourages the GA development and ensures necessary safety and security [12]. The approval process of GA activities is lengthy, time consuming, and costly. Things have slowly improved since GA is included as one of the key areas of development in the five-year plan of China [17]. The country has attempted to improve the regulatory framework, as well as organize conferences and exhibitions specialized in the development of GA that involve foreign experts and manufacturers [18].

2. Restricted airspace

The airspace in China is divided into classes A, B, C, and D which are tightly controlled and thus restrict the GA aircraft operations. This limited access to airspace and insufficient air traffic management have curtailed GA growth [12]. However, China is gradually moving towards the liberalization of airspace. In November 2010, the State Council and Central Military Commission of China issued "Considerations on the deepening of reform in control mode of airspace at low altitudes" [19], which presented a plan to open the airspace below 4,000 m for amphibious aircraft. Since 2011, China reduced the time needed to obtain flight permissions and in 2014, China took a significant step towards liberalizing airspace below 1,000 m (3,280 ft above mean sea level) upon a joint approval from the State Council, the Air Traffic Control Commission and Central Military Commission [20]. This regulation provides classifications such as (1) those under control and requiring prior permission, (2) areas that do not require prior permission but remain under surveillance, and (3) areas where aircraft can fly freely after filing flight plans [20].

3. Weak support and insufficient supplies of equipment

China still lacks the domestic after-sale support in the GA industry. As such, aircraft owners need to rely heavily on foreign overhaul and supports. To address this problem, China has been trying to establish joint ventures with a majority of the world's leading manufacturers of GA equipment in the past ten years. In 2007, China's first GA aircraft shop and service center "4S" opened in Hangzhou [12]. Clearly, more companies will be required to support the expected growth of the GA industry development. Moreover, in order to create more favorable conditions for the operators and private owners of GA aircraft, the Civil Aviation Adminstration of China (CAAC) needs to continue to help reduce operating costs. MacCorkle and Wong stated that CAAC planned to reduce the import duties and value added tax (VAT) for GA aircraft and parts [13]. In fact, the import duties on GA aircraft were reduced from 23% to 5% in 2010, but the VAT still remains at a high 17% [12].

6. Preliminary Designs for Amphibious Aircraft

We have started the preliminary design projects for two amphibious aircraft at the Hong Kong University of Science and Technology (HKUST). The two designs have different basic payload requirement: one is for a two-seater aircraft and the other for a ten-seater aircraft. These design projects are designed to suit Final Year Design Project (FDYP) scopes and requirements for undergraduate students. Each design will be described briefly below. Due to the different payload requirement, the resulting designs are vastly different from each other, as can be observed in the following. Considering the limited time (less than a year) and that the projects were done mainly by undergraduate students, the designs are still at a very preliminary stage and there are still a lot to be done for completed designs.

6.1 Two-seater amphibious aircraft design

The proposed design for the two-seater amphibious aircraft is shown in Figure 3. This configuration features sponsons and an H-tail with two vertical stabilizers. The propeller is located at the back of the tail. This placement would shift the aircraft center of gravity (CG) rearward, which will affect the stability of the aircraft where the aircraft might tip over backwards in water. To counter this rearward CG shifting, we need to adjust the size and positions of the floats. The vertical alignment of the main wings, horizontal stabilizers, and the propeller would affect the aerodynamic performance, due to the interaction of wake produced by the wing during flight. The wing wake is a narrow band of reduced dynamic pressure resulting from the interaction between the upper and lower boundary layers. If the wing wake impinges the stabilizer or the propeller, the overall aerodynamic efficiency will be reduced, and the flight stability will be adversely affected as well. This issue is resolved by shifting the propeller upward. This design change comes with a tradeoff between the aerodynamic efficiency and total gross weight, by means of the additional structure to attach the propeller to the empennage. Table 2 summarizes the key design parameters and performance of the proposed two-seater amphibious aircraft. The Rotax 912ULS engine is used, which has previously been implemented in LISA Akoya aircraft.

Figure 3. Proposed design for the two-seater amphibious aircraft.

One of the key design elements in the proposed two-seater amphibious aircraft is the noise shielding, to reduce the noise impact to people on the ground. Aircraft noise (acoustics) is an inherent problem in any aircraft operations, with airframe and aircraft engines being the two major noise sources. This problem is even more prominent during takeoff and landing, when the aircraft is still at a relatively low altitude.

In this project, we conduct experiments to test different noise shielding configurations, including the vertical shield and circular shield. The experiments also include the unshielded propeller configuration for comparison purposes. Several sizes for the noise shields (both circular and vertical) are tested, as shown in Table 3. This table summarizes the detailed experimental setup and apparatus. The configurations for the vertical and circular shields are illustrated in Figure 5. For the experimental setup, we use scaled models of propellers and shields. Acrylic is used as the noise shielding materials. The experimental setup is shown in Figure 4. Four microphones are placed at various locations, all with the same height. Microphone 1 is located directly intersecting the propeller axis of rotation, microphone 4 is located within the plane of propeller rotation, whereas microphones 2 and 3 are located at equal angular distances between microphones 1 and 4. To measure the far field noise level of the propeller, a distance between the microphone and propeller is set to be 1.5 m. The motor and microphones are connected to a data acquisition system using the LabVIEW System Design Software.

The experiment results suggest that the proposed method of reducing propeller noise by using the vertical stabilizer as noise shields is ineffective, and could even potentially increase the noise level. On the other hand, the circular cover shield result in reduced noise levels by as much as 9 dB. To further improve the effectiveness of the noise shielding, we can further explore the modifications of the propeller geometry and material selection. Selecting a suitable material for the shield would be quite challenging. The selected material needs to be water-resistant as the amphibious aircraft would consistently be exposed to water. The material also needs to be smooth so as not to increase the aerodynamic drag, which could in turn affect the aerodynamic efficiency of the aircraft. Another critical material property would be the strength of the material, as the shield would be experiencing constant force from the aerodynamic drag. Identifying and integrating an appropriate noise absorbing material would greatly enhance the noise reducing property of the shield.

Table 2. Summary of the design parameters and performance for the two-seater amphibious aircraft.

Design Factors		Design Values
Certification	Certification type	FAA Light-Sport Aircraft Airworthiness Certification
Propulsion system	Engine	Rotax 912ULS
	Propeller	Constant speed 6-blade propeller, D = 1.7m
	Fuel tank capacity	30gal MOGAS (80.63kg)
Wing	Area	11.7 m^2
	Span	10.515 m
	Aspect ratio	9.45
	Sweep	No sweep
	Taper ratio	0.5 – 1.0
	Airfoil shape	NASA LANGLEY LS(1)-0417 (GA(W)-1)
	Angle of incidence	2.34°
Tail	Horizontal tail area	2.99 m^2
	Horizontal tail span	3.458 m
	Horizontal tail mean chord	0.865 m
	Vertical tail area	0.78 m^2
	Vertical tail span	1.249 m
	Vertical tail mean chord	0.625 m
	Airfoil shape	NASA LANGLEY LS(1)-0417 (GA(W)-1)
	Horizontal tail angle of incidence	-5°
	Horizontal tail aspect ratio	4
	Vertical tail aspect ratio	2
	Taper ratio	0.5 – 1.0
	Sweep	0 - 20°
Material selection	Fuselage	Carbon-fibre composite
	Wing and tail surface	Alclad alloys
	Wing and tail internal structures	A1 7475-T7361
Aircraft performance	Payload weight	180 kg
	Gross weight	643 kg
	Empty weight	397 kg
	Range	1,015.6 km
	Engine endurance	7.6 hours
	Takeoff distance (on land)	315.53 m
	Takeoff distance (on water)	410.19 m
	Cruise stall speed	22.37 m/s

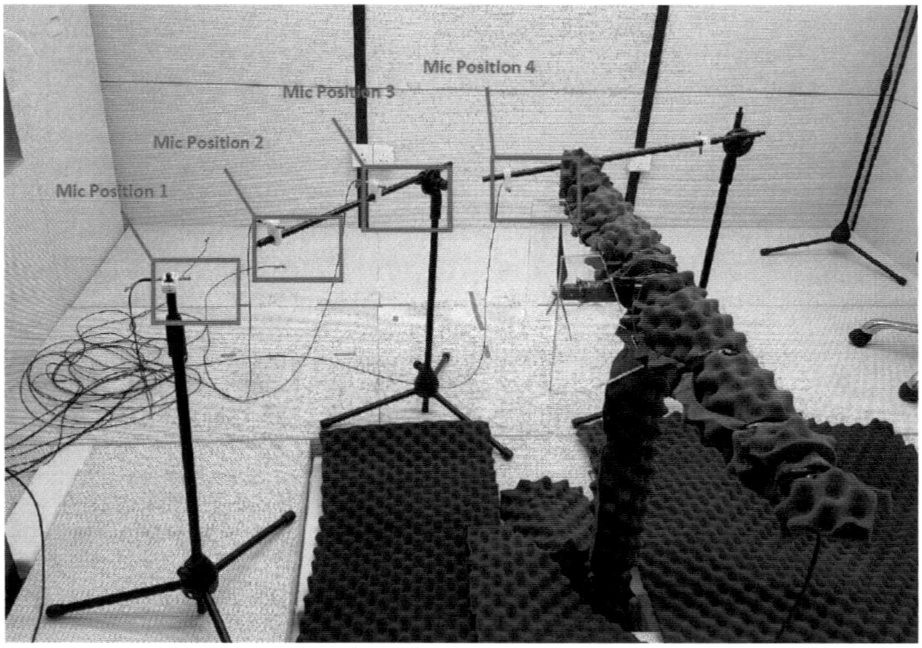

Figure 4. Noise shielding experimental setup.

Table 3. Apparatus setup for the noise shielding experiments.

Apparatus	Specification/model	Amount
Motor	E305 2312E (960 kV, CCW)	1
2-blade propeller	Diameter: 250 mm	1
3-blade propeller	Diameter: 250 mm	1
Microphones	-	4 sets
DC power supply	GPC3030DN	1
Anechoic chamber	AVIC Advanced Noise Technology Center	-
Data acquisition system	LabView System Design Software	-
Circular shield	Diameter: 260 mm, 270 mm, 280 mm	1 each
Vertical stabilizer shield	Height: 250 mm, 270 mm, 290 mm, 310 mm	1 set each

(a) Vertical stabilizer shield (b) Circular cover shield

Figure 5. Noise shielding configurations.

6.2 Ten-seater amphibious aircraft design

The second student project focuses more on designing a ten-seater amphibious aircraft, with eight passengers and two crews. The resulting design is notably different from the two-seater one, which shows that an aircraft design is significantly driven by the mission requirements. Table 4 summarizes the mission requirements for this ten-seater amphibious aircraft. These requirements were decided after performing a market research, to suit the regional needs where we focused on the Pearl River Delta's region.

Table 4. Mission requirements for the ten-seater amphibious aircraft.

Mission requirement	Value
Maximum range	1,000 km
Number of passengers	8
Number of crews	2
Baggage allowance	20 kg baggage per passenger 10 kg baggage per crew
Total payload	950 kg
Cruising speed	170 knots (approx. 314 km/h)
Maximum cruising altitude	3,048 m (10,000 ft)

Figure 6. Proposed design for the ten-seater amphibious aircraft.

Figure 6 shows the final proposed design for the ten-seater amphibious aircraft, and the selected design parameters are tabulated in Table 5. This configuration is a high-wing, T-tail, twin-engine aircraft with hydrofoil mounted at the bottom of the fuselage. The landing gear mechanism is also specially designed, though not shown in the figure, but will be briefly described shortly. This design is decided upon performing some empirical and numerical analyses on aerodynamic, structures, and aircraft stability. We mainly use OpenVSP [21] as the main aircraft conceptual design software tool. OpenVSP is an open-source software developed by a group of NASA engineers led by J. R. Gloudemans. Using this software, users can construct preliminary geometry models of aircraft, and perform aerodynamic analysis on the model using the vortex lattice method or panel method with an embedded aerodynamic solver, VSPAero.

Table 5. Design parameters for the ten-seater amphibious aircraft.

Design parameter	Value
Total length	16.5 m
Span	20.8 m
Height	7.07 m
Maximum takeoff weight	5,740 m
Powerplant	2 Trace OE600 radial engine, 600 hp each
Aspect ratio	9.08
Taper ratio	0.4
Wing skin material	Carbon-fibre reinforced epoxy
Wing internal structure	A1 6061

The aircraft structure, including the internal structures, are modeled in Solidworks which enables the estimation of the aircraft mass, center of gravity, and center of buoyancy (for the hydrostatic stability analysis). The landing gear mechanism design and the aircraft structural integrity are analyzed via the finite element method (FEM) performed in ANSYS. The inverted-T variable sweep hydrofoil is designed using Xfoil and Solidworks. The aircraft stability and control were analyzed in MATLAB. To analyze the aircraft dynamic stability, we look into five modes of dynamic responses. They include the longitudinal stability (short and phugoid modes), and lateral stability (roll, Dutch roll, and spiral modes).

As mentioned, we need to design our own landing gear. This is due to the hydrofoil placement under the fuselage, which requires more ground clearance than other standard-sized landing gears. In designing the landing gear mechanism, it is important to make sure that it would not impair the streamline geometry, so as not to sacrifice the aerodynamic performance. We decide on the common tricyle landing gear design with oleo shock absorbers for good ground stability, relatively light in weight, and its efficiency in energy absorption. Figures 7 and 8 show the main and nose landing gear mechanisms, respectively. The landing gear positioning, which is illustrated in Figure 9, is determined upon analyzing the ground stability (tip-back angle) and the aircraft operation and movement during takeoff and landing, to ensure enough clearance and safety. Load analysis is performed to size the wheels and shock-absorber, to ensure that they can withstand the aircraft weight and loads.

An amphibious aircraft needs to be able to takeoff and land from water. Hydrofoil, which is similar to airfoil but works underwater, can provide extra lift and reduce drag during the water takeoff. A hydrofoil is a lift-generating device mounted at the bottom of the fuselage. The hydrofoil can get the hull out of the water, allowing for a faster takeoff with a shorter distance. By lifting the hull out of the water, the aircraft needs to only overcome the drag on the foils instead of all the drag on the hull. There are two major considerations for designing hydrofoil of an amphibious aircraft. The first one is to prevent cavitation, which occurs when the water pressure drops to the point where the water starts to boil. When cavitation occurs, the foil no longer generates enough lift, which might risk the aircraft crashing down onto the water. Just like airfoil, when a hydrofoil encounters a free-stream velocity, the pressure on the upper surface drops. With the acceleration of the aircraft, this pressure might drop below the vapor pressure of the working liquid such that cavities form. As pressure of downstream of upper surface increases, the bubbles collapse when pressure rises above the vapor pressure. After the bubbles collapse, high density of energy (the latent heat of condensation) is released so that the local pressure and temperature also increase. As cavities form and implode in a very short duration continuously, the phenomenon negatively affects the structural integrity of airfoil and increases the hydrodynamic drag. Therefore, preventing cavitation is the primary consideration in designing the hydrofoil system.

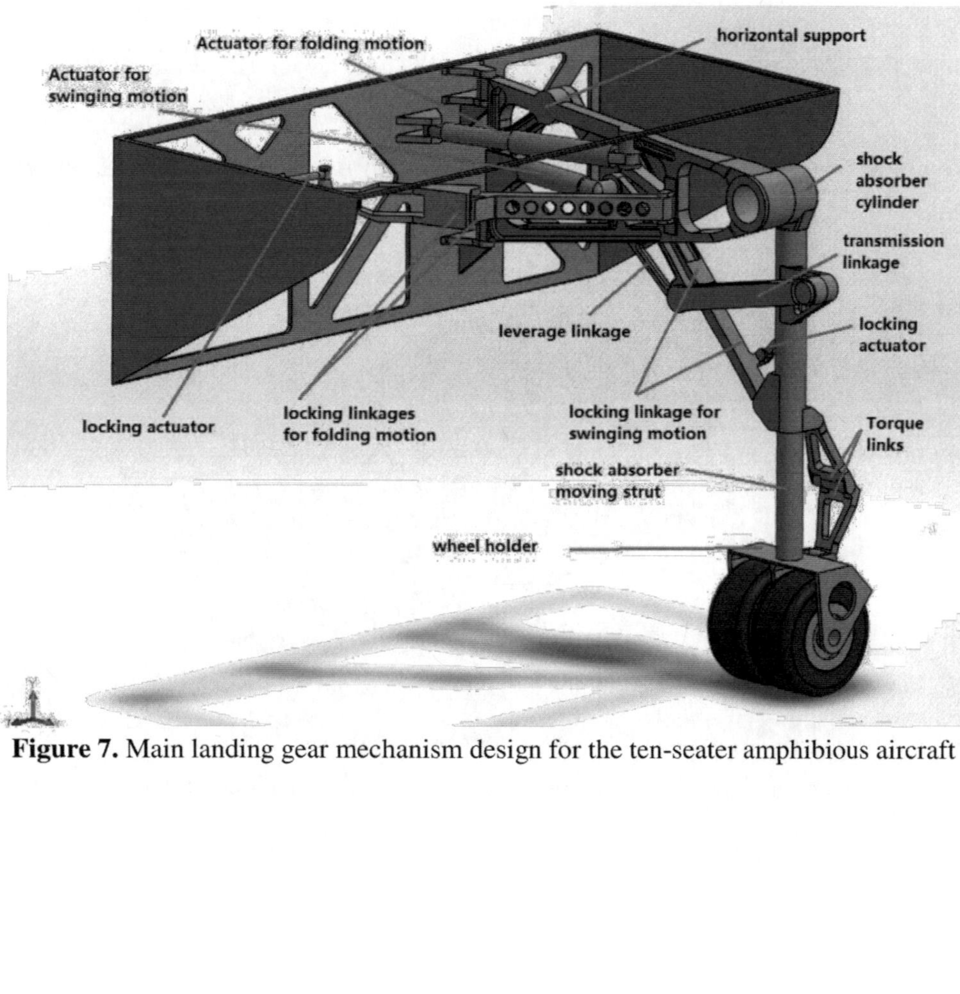

Figure 7. Main landing gear mechanism design for the ten-seater amphibious aircraft

IOP Conf. Series: Journal of Physics: Conf. Series **1005** (2018) 012027 doi:10.1088/1742-6596/1005/1/012027

Figure 8. Nose landing gear mechanism design for the ten-seater amphibious aircraft.

Figure 9. Main landing gear mechanism design for the ten-seater amphibious aircraft.

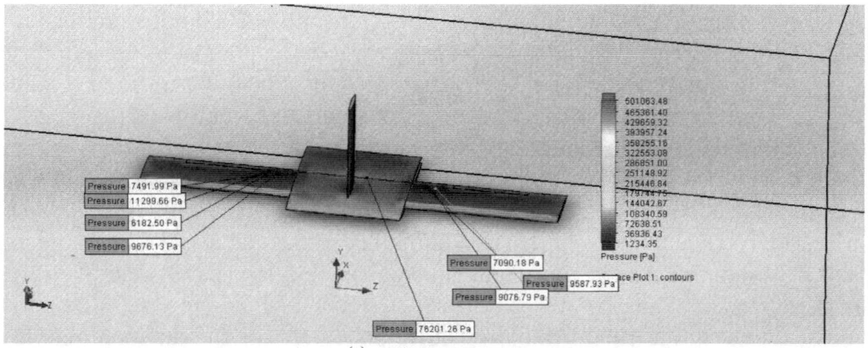

(a) Sweep angle: 0°

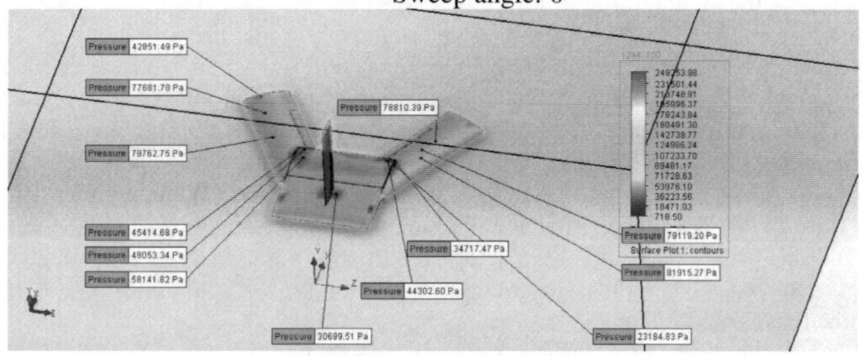

(b) Sweep angle: 60°

Figure 10. The pressure differences with variable sweep hydrofoils, measured at a water flow rate of 25 m/s.

The second one is the lift transfer from hydrofoil to airfoil. Before takeoff, hydrofoil and airfoil both provide lift to the aircraft. However, upon reaching the takeoff speed, the airfoil needs to provide all the required lift as the hydrofoil leaves the water and therefore does not provide lift anymore. This phenomenon would lead to the porpoising of the vehicle before it actually takes off. Therefore, a gradual transfer of lift from the hydrofoil to the airfoil becomes another important design requirement for the hydrofoil system.

To accommodate these two design considerations, an innovative variable sweep hydrofoil design is proposed. Figure 10 shows that changing the sweep angle of the hydrofoils can help increase the pressure acting on them, thus preventing cavitation. This figure shows that when the hydrofoil is unswept, its upper surface is covered by low-pressure region, which corresponds to cavitation. However, when the hydrofoil is swept, the surface pressure dramatically increases and thus prevents cavitation. At the same time, lift is transferred to airfoil.

7. Conclusion

In this paper we have discussed the prospect of developing amphibious aircraft, focusing more on the Asian region. The initial market research has revealed its vast potentials, spanning from business, logistics, safe and rescue, and tourism industry. The unique takeoff and landing mechanisms of amphibious aircraft would solve the problem of limited airport infrastructure in some areas, since they can takeoff and land from both water and land. Moreover, amphibious aircraft fly at a lower altitude than the typical commercial passenger and cargo aircraft, and thus their introduction would not further add to the air traffic congestion.

Aircraft design process is a complex procedure, since we need to take into account the different disciplines and their interrelations (i.e., how changing one parameter in one discipline would affect other disciplines). Some disciplines include aerodynamics, structures, stability and control, energy and propulsive system, and aeroacoustics, to name a few. Each discipline is complex on its own, and taking into account the interdisciplinary relations between them could make the aircraft design problem intractable. An aircraft design process is primarily driven by the specified mission requirements, e.g., payload (passenger, cargo, or both), mission range, cruise speed and altitude, etc.

We discuss some of the design challenges, both the technical and non-technical (e.g., regulatory and certification), and present two preliminary design projects. Most of the mission requirements for these two designs are similar, which were derived based on the market research, with the mission ranges aimed to cater for the Pearl River Delta region. The two amphibious aircraft, however, are designed to cater for different payload requirements. In particular, two-seater and ten-seater aircraft are designed and analyzed. From these design examples we can clearly see how one mission requirement can result in very different designs, and not just in term of size difference. Despite the different requirements, both design projects follow the similar aircraft design procedure.

For the two-seater amphibious aircraft, we look more deeply into the aeroacoustics aspect, i.e., to reduce aircraft noise. While there are multiple sources of aircraft noise (e.g., from airframe and engine), in this work we only focus on reducing the noise impact from the engine propeller. A series of experiments are setup and performed, from which we observe that the circular noise shielding has the potential for the propeller noise reduction. For the ten-seater aircraft, we focus more on the hydrofoil design, landing gear mechanism, and the stability and control analyses. We come up with the variable sweep hydrofoil mechanism to ensure that the aircraft has enough lift upon leaving the water during takeoff. The hydrofoil placement under the fuselage requires a special landing gear mechanism design. The stability and control analyses look into five dynamic response modes, including those in lateral and longitudinal directions.

While still at a preliminary stage, the work and discussion presented in this paper provide reviews on design aspects and challenges of quiet and efficient amphibious aircraft. Due to the complexity of an aircraft system, there are still a lot to be done to produce the final, integrated designs. To design a truly efficient amphibious aircraft, we need to solve other inherent problems of amphibious aircraft, namely higher drag, heavier aircraft, and to obtain the right balance between aerodynamic and hydrodynamic performance. These have been some of the bottleneck problems in designing an amphibious aircraft. In order to revolutionize amphibious aircraft in general, more research efforts need to be invested to overcome the aforementioned challenges.

As mentioned before, the amphibious aircraft development would not only be beneficial to the China region (where most of the background study for this project is based on), but to other countries and regions in Asia as well. Indonesia, in particular, would benefit a lot with the thousands of islands scattered around in the vast archipelago. We strongly believe that stronger collaboration among different institutions in Asia can help propel this project into a greater success.

Acknowledgement

I would like to acknowledge the work of several people who contribute significantly to this project, in particular Prof. Xin Zhang from HKUST who co-supervise this project. The students who work on the two-seater aircraft include Won-Kyu Lee, Jeng-hun Lee, Chung Hee Kim, and Tak Shing Yiu; and for the ten-seater aircraft Wai Kong Liu, Ming Chak Chan, Cheuk Hei Chue, and Chung Hang Sze. I would also acknowledge the involvement of some undergraduate research opportunity program (UROP) students: Yu Him Lai, Kiko Muramatsu, Zhenyang Dong, Cheuk Hang Wong, Rui Hong Chen, Hsun-yun Tang, Zhiyuan Fan, and Anish Deva. Lastly, I would like to express my gratitude to my student Jefry Yanto, who has helped convert this document from LaTeX typesetting format to Microsoft Words.

References

[1] ICAO, "Review of the Classification and Definitions Used for Civil Aviation Activities," Working Paper STA/10-WP/7, International Civil Aviation Organization, 2009.

[2] Becken, S., "Analyzing International Tourist Flows to Estimate Energy Use Associated with Air Travel," Journal of Sustainable Tourism, Vol. 10, No. 2, 2002, pp. 114–131. doi:10.1080/09669580208667157.

[3] Button, K., Michalski, W., and Weiss, P., The Future of International Air Transport Policy: Responding to Global Change, Organization for Economic Cooperation & Devel (OECD), Paris, France, January 1997.

[4] Oppermann, M. and Cooper, M., "Outbound Travel and Quality of Life: The Effect of Airline Price Wars," Journal of Business Research, Vol. 44, No. 3, March 1999, pp. 179–188. doi:10.1016/S0148-2963(97)00199-9.

[5] Schafer, A. and Victor, D. G., "Global passenger travel: implications for carbon dioxide emissions," Energy, Vol. 24, No. 8, August 1999, pp. 657–679. doi:10.1016/S0360-5442(99)00019-5.

[6] Greene, D. L., "Energy-efficiency improvement potential of commercial aircraft," Annual Review of Energy and the Environment, Vol. 17, 1992, pp. 537–574.

[7] Smith, J. B., "Trends in Energy Use and Fuel Efficiency in the U.S. Commercial Airline Industry," Transportation Research Record 870, Transportation Research Board, Washington, DC, USA, 1981.

[8] Hummels, D., "Transportation Costs and International Trade in the Second Era of Globalization," Journal of Economic Perspectives, Vol. 21, No. 3, 2007, pp. 131–154.

[9] ICAO, "Aviation and Climate Change," International Civil Aviation Organization (ICAO) Environmental Report, 2010.

[10] General Aviation Manufacturers Association (GAMA), "2015 General Aviation Statistical Databook & 2016 Industry Outlook," 2016.

[11] World Bank Data Catalog, GDP Ranking Table. http://data.worldbank.org/data-catalog/GDP-ranking-table.

[12] Araslanov, S. A., "GA Market in China," Tech. rep., General Aviation Scientific & Technical Center Ltd., 2011.

[13] MacCorkle, J. and Wong, T., "General Aviation in China: Seizing Growth opportunities," Tech. rep., Booz & Company Inc., 2009.

[14] Zhao, L., "Number of pilots rises, but more are needed," http://www.chinadaily.com.cn/china/2015-03/03/content19699370.htm, March 2015, (last retrieved:-June 27, 2017).

[15] "Seminar Nasional Penetapan Nama Pulau-pulau Kecil Dalam Perspektif Sejarah" or "National Seminar on Small Island Identification from the Historical Perspective", 16 to 18 July 2008 in Palembang, South Sumatra, Indonesia.

[16] Gudmundsson, S., General Aviation Aircraft Design: Applied Methods and Procedures, Appendix C3: Design of Seaplanes, Butterworth-Heinemann, 1st ed., 2013.

[17]
National Business Aviation Association (2015). China's Draft Five-Year Plan Envisions GA Infrastructure Boost. Retrieved from: https://www.nbaa.org/ops/intl/mid/20150806-chinas-draft-five-year-plan-envisions-general-aviation-infrastructure-boost. php (last retrieved: June 28, 2017).

[18] Shanghai International Business Aviation Show (SIBAS). Retrieved from: https://www.acnnewswire.com/press-release/english/6168/it's-shanghai-international-business-aviation-show's-time (last retrieved: June 28, 2017).

[19] National Business Aviation Association (2011). China Opens More Airspace for General Aviation. Retrieved from: https://www.nbaa.org/ops/intl/mid/20110815-china-opens-more-airspace-for-general-aviation.php (last retrieved: June 28, 2017).

[20] Lynch, K., "China to Continue Airspace Relaxation," Retrieved from: http://www.ainonline.com/aviation-news/business-aviation/2015-04-13/china-continue-airspace-relaxation (last retrieved June 28, 2017), 2015.

[21] OpenVSP Aircraft Conceptual Design Software. http://www.openvsp.org/ (last retrieved: June 30, 2017)

IOP Conf. Series: Journal of Physics: Conf. Series **1005** (2018) 012028 doi:10.1088/1742-6596/1005/1/012028

The Comparison Of In-Flight Pitot Static Calibration Method By Using Radio Altimeter As Reference with GPS and Tower Fly By Methods On CN235-100 MPA

Derajat[1)]* and Hindawan Hariowibowo[2)]

[1)] Flight Test Data Analyst, Flight Test Engineering Operation, Flight Test Center Division, Indonesian Aerospace (IAe), Indonesia
[2)] Flight Test Engineer and Head of Flight Test Center Division, Indonesian Aerospace (IAe), Indonesia

*derajat@indonesian-aerspace.com

Abstract. The new proposed In-Flight Pitot Static Calibration Method has been carried out during Development and Qualification of CN235-100 MPA (Military Patrol Aircraft). This method is expected to reduce flight hours, less human resources required, no additional special equipment, simple analysis calculation and finally by using this method it is expected to automatically minimized operational cost. At The Indonesian Aerospace (IAe) Flight Test Center Division, the development and updating of new flight test technique and data analysis method as specially for flight physics test subject are still continued to be developed as long as it safety for flight and give additional value for the industrial side. More than 30 years, Flight Test Data Engineers at The Flight Test center Division work together with the Air Crew (Test Pilots, Co-Pilots, and Flight Test Engineers) to execute the flight test activity with standard procedure for both the existance or development test techniques and test data analysis. In this paper the approximation of mathematical model, data reduction and flight test technique of The In-Flight Pitot Static Calibration by using Radio Altimeter as reference will be described and the test results had been compared with another methods ie. By using Global Position System (GPS) and the traditional method (Tower Fly By Method) which were used previously during this Flight Test Program (Ref. [10]). The flight test data case are using CN235-100 MPA flight test data during development and Qualification Flight Test Program at Cazaux Airport, France, in June-November 2009 (Ref. [2]).

1. Introduction

This paper is proposed to introduce a new in-flight pitot static calibration method by using Radio Altimeter as a reference. Many pitot static in-flight calibration methods are available to determine position error correction, such as GPS (Global Position System) Method, Tower Fly By Method, Pacer Aircraft Method, Trailing Cone Method, Trailing Boom Method etc (Ref. [5]). This new method is still under development and in case it is applicable, it is going to be applied in Indonesian Aerospace for the

IOP Conf. Series: Journal of Physics: Conf. Series **1005** (2018) 012028 doi:10.1088/1742-6596/1005/1/012028

development and certification flight of the new N219 aircraft, the CN235-220 Modified Version aircraft, the NC212i aircraft etc.

When we mesure some physical quantity with an instrument and obtain a numerical value, usually we are concerned with how close this value may be to the "true" value (Ref. [1]). Therefore precise flight test data in altitude and airspeed position error calibration is needed in determining aircraft flight performance and flight dynamics especially in type certification process. The altitude and airspeed indicator are actuated by pressure and sensed by pitot static systems (Ref. [4]). The position of installed pitot static tube to free airstream resulted an error that effects the value indicated by altitude and airspeed instrument. These error are called position error. Therefore beside corrected from instrument error, the indicated altitude and airspeed should be corrected from position error (Ref. [3]). Therefore by using this new method, it is expected to get the position error accurately.

The In-flight Pitot Static Calibration Using Radio Altimeter as Reference test technique is performed by flying the aircraft at height of out of ground effect (recommended at about 100 - 150 feet above ground level for CN235 airplane) with the acceleration speed at specified test configuration and condition. Then repeat for other flaps and landing gear configurations setting. By recording height of Radio Altimeter during acceleration flight together with indicated altitude, indicated airspeed and outside air temperature via OBDAS (On Bord Data Aquition System), the position error correction of altitude and airspeed can be calculated.

As previously mentioned, this method is still under development, the test data are using CN235-100 MPA flight test data during Tower Fly By Method with several constant speeds for each flaps configuration.

2. Aircraft Description

Figure 1. CN235-100 MPA Military Version

The test aircraft was using CN-235-100M MPA version as depicted in Figure1. The aircraft external shape has been changed due to installation of several additional equipment, which did not exist on the original CN-235-100M, in order to meet a certain mission requirements.
Modifications of CN-235-100M MPA from CN-235-100M (Ref. [2]) are:

- OM Radar Radome (belly),
- FLIR/TV (belly),
- Bubble Windows (rear fuselage),
- Different Nav/Com Antennas,
- ADFU Antennas (front upper fuselage and rear fuselage),
- UV Sensors (front upper and lower fuselage, landing gear fairing and tail cone),
- MAD Boom (tail cone).
- Chaff & Flare dispensers (front and rear fuselage)

The pitot static system location are shown in Figure 2 and Figure 3 for Pilot and Co-Pilot side systems respectively. Figure 4 shows the installation location of radio altimeter antenna on the test aircraft CN235-100 MPA

Figure 2. Pilot Pitot Static Probe location (Ref. [2])

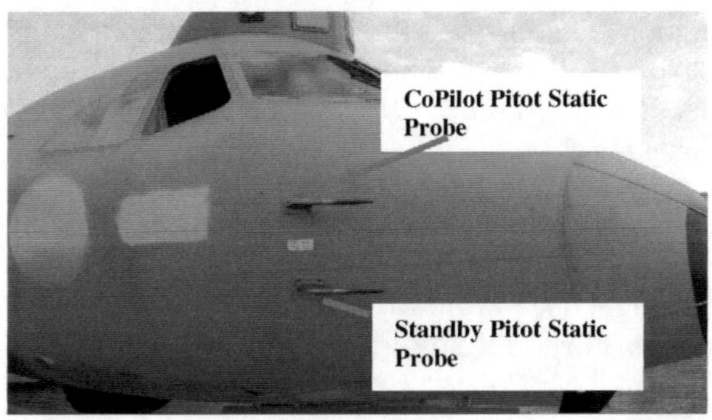

Figure 3. Co-Pilot and Standby Pitot Static Probe location (Ref. [2])

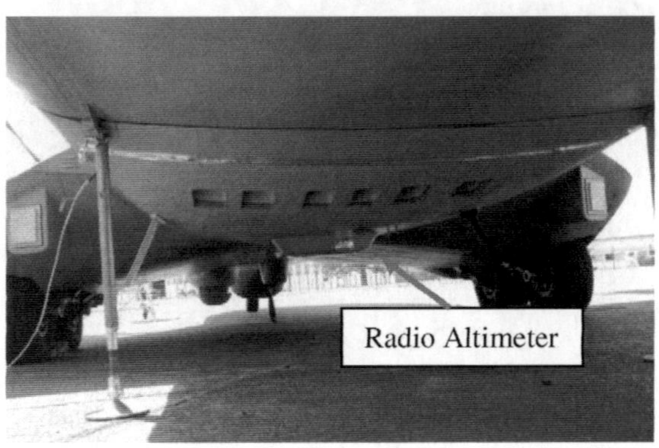

Figure 4. Radio Altimeter location (Ref. [11])

IOP Conf. Series: Journal of Physics: Conf. Series **1005** (2018) 012028 doi:10.1088/1742-6596/1005/1/012028

3. Mathematical Model Approximation

3.1 General Calibration Equation

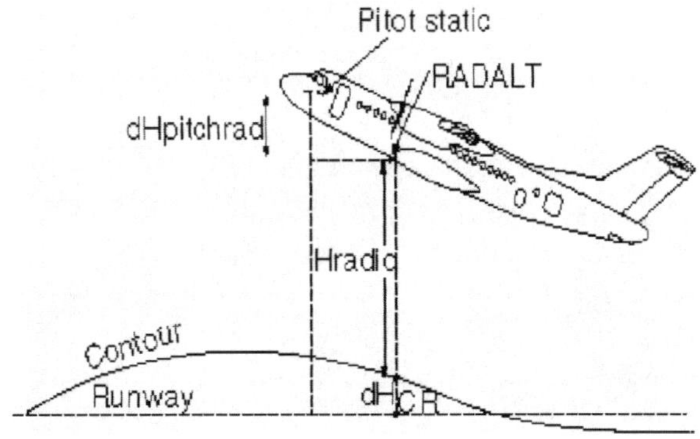

Figure 5. Aircraft illustration flies passed on the runway

By using radio altimeter calibration method, when the aircraft is passing the ruway as shown in figure 5, it can be shown that:

$$Href = Hradio + dHpichrad + dH_{CR} + Hairport \quad (1)$$

Where *Href* : Reference Height
Hradio : Radio Altimeter height
dHpichrad : The different height of pitot static to the Radio Altimater corrected by pitch and bank angle
dH_{CR} : Vertical height due to Runway countur
Hairport : Elevation of Airport Runway

The correction are then diveded into :
- Pilot system position error correction and
- Co-Pilot system position error correction

These correction are described below on the next paragraph

4. Pilot System Position Error Correction

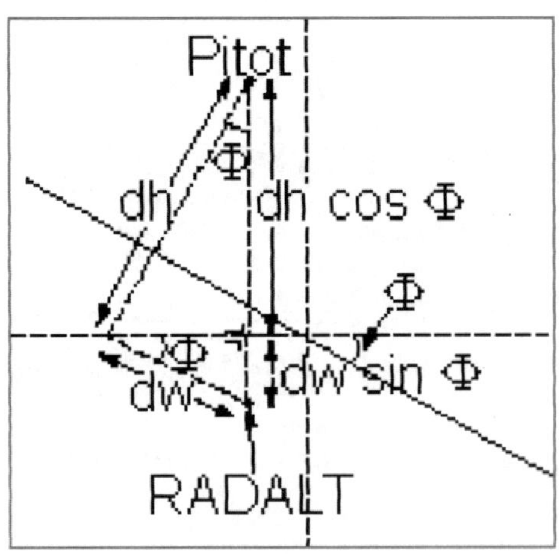

Figure 6. Pilot Pitot System Position Error Correction due to bank angle correction

$$H\Phi L = dh\ cos\Phi + dw\ sin\Phi \tag{2}$$

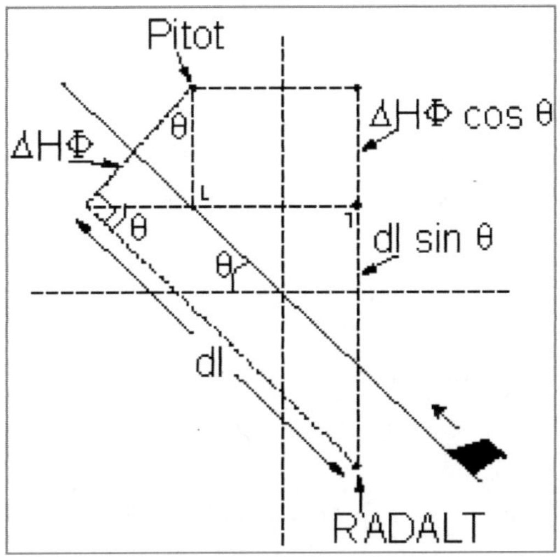

Figure 7. Pilot System Position Error Correction due to bank and pitch angle correction

From the above figure 7 it can be seen that

$\Delta H\theta\Phi L = \Delta H\Phi L\ cos\theta + dl\ sin\theta$, and by using equation *(2)*

$$= (dh\ cos\Phi + dw\ sin\Phi)cos\ \theta + dl\ sin\theta$$

$$= dh\ cos\Phi cos\theta + dw\ sin\Phi cos\theta + dlsin\theta \tag{3}$$

5. Co-Pilot System Position Error Correction

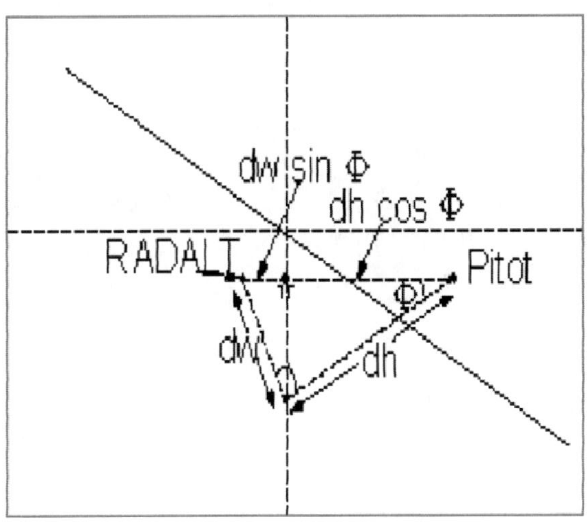

Figure 8. Co Pilot System Position Error Correction due to bank angle correction

From the above figure 8 it can be seen that

$$\Delta H\Phi R = dh\ cos\Phi - dw\ sin\Phi \tag{4}$$

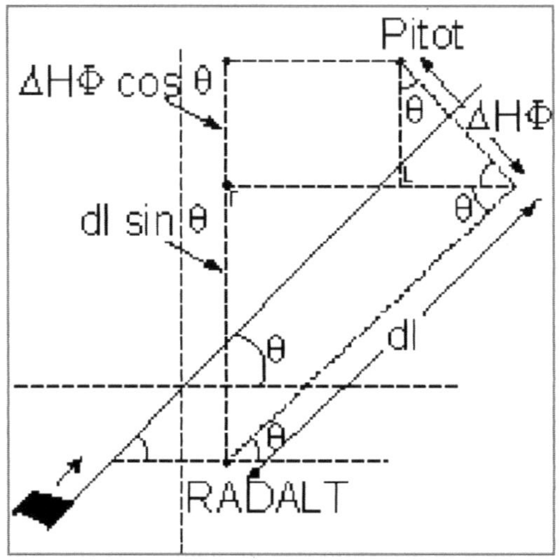

Figure 9. Co Pilot System Position Error Correction due to bank and pitch angle correction

From the above figure 9 it can be seen that

$$\Delta H\theta\Phi R = \Delta H\Phi R\ cos\theta + dl\ sin\theta$$

$$= (dh\ cos\Phi - dw\ sin\Phi)cos\theta + dl\ sin\theta, \text{ and by using equation } (4)$$

$$= dh\ cos\Phi\ cos\theta - dw\ sin\Phi\ cos\theta + dl\ sin\theta \quad (5)$$

Finally for pilot and co-pilot system it can be determined that

$$Hpref = Hradio + dHpitchrad + dH_{CR} + Hairport$$

$$= Hradio + dh\ cos\Phi cos\theta + dw\ sin\Phi cos\theta + dlsin\theta + dH_{CR} + Hairport \text{ (pilot system)} \quad (6)$$

And

$$Hpref = Hradio + dHpitchrad + dH_{CR} + Hairport$$

$$= Hradio + dh\ cos\Phi\ cos\theta - dw\ sin\Phi\ cos\theta + dl\ sin\theta + dH_{CR} + Hairport \text{ (co-pilot system)} \quad (7)$$

6. Calculation of dHpc and dVpc

Altitude position error correction is now :

$$dHpc = Hpref - Hic_{A/C} \quad (8)$$

The pressure change surrounding basic pitot static probe is :

$$dP = -\rho_A * g * dHpc \quad (9)$$

Where ρ_A-density at test altitude
g-constant of gravity = 9.80665 m/s^2

Calibrated airspeed now can be calculated by :

$$Vc = \sqrt{\frac{-2dP}{\rho_0} + V^2_{icA/C}} \tag{10}$$

With ρ_0-standard sea level density = 1.225 kg/m3

Position error correction for airspeed is :

$$dVpc = Vc - V_{icA/C} \tag{11}$$

For standardization and correction, which is dedicated for supporting Aircraft Flight Manual (AFM) chart, the formulation is start by calculation lift coefficient at test actual weight. Then plotting pressure coefficient is function of lift coefficient by using a polynomial regression of second order. By using this equation, the altitude and airspeed position error correction can be determined for the other weight and altitude (Ref. [5]).

7. Method Verification

CN235-100 MPA had finished the qualification test program (Ref. [2]). The method used during in-flight pitot static calibration test were Tower Fly By Method and GPS Method (Ref.[8], [9]). These flight test data from Tower Fly By Method are used to verify this method.
In the following table there are flight test data which is used for verification this method.

Table 1. Flight Test Data F00, LGUP, 100%RPM, PLF (Ref. [2])

Hi [feet]	Vi [knots]	Hrad [feet]	OAT [deg.C]	THETA [deg.]	PHI [deg.]
279.59	197.41	148.25	22.16	1.53	-0.5
255.15	171.94	137.1	21.64	2.99	6.17
246.99	148.51	115.03	20.27	4.53	-0.9
263.29	128.32	143.7	19.75	-6.06	1.75
255.14	118.15	139.15	19.49	-6.95	2.04

Table 2. Flight Test Data F10, LGUP, 100%RPM, PLF (Ref. [2])

Hi [feet]	Vi [knots]	Hrad [feet]	OAT [deg.C]	THETA [deg.]	PHI [deg.]
255.14	160	148.03	20.44	0.3	-1.36
255.14	131.21	137.56	19.49	3.65	-0.63
246.99	120.18	144.6	19.84	3.39	1.63
246.99	103.64	132.78	19.32	8.78	0.23

Table 3. Flight Test Data F15, LGUP,100%RPM, PLF (Ref. [2])

Hi [feet]	Vi [knots]	Hrad [feet]	OAT [deg.C]	THETA [deg.]	PHI [deg.]
444.31	158.51	340.54	21.13	-3.49	4.32
246.99	141.43	138.01	20.27	0.74	0.92
255.14	121.06	147.34	19.75	2.53	0.74
246.99	106.77	141.2	19.67	4.84	5.26

Table 4. Flight Test Data F23, LGUP, 100%RPM, PLF (Ref.[2])

Hi [feet]	Vi [knots]	Hrad [feet]	OAT [deg.C]	THETA [deg.]	PHI [deg.]
322.53	153.81	229.04	20.61	-3.7	6.43
287.74	132.7	191.72	20.18	-1.79	-0.59
238.84	117.86	131.19	19.41	-1.15	1.15
230.69	98.07	123.45	19.24	4.65	2.41

By using all equations in mathematical model approximation, and with *dl* (distance between Pitot Static and Radio Altimeter, longitudinal direction) 6.693 feet, *dh* (distance between Pitot Static and Radio Altimeter, vertical direction) 3.642 feet, and *dw* (distance between Pitot Static and Radio Altimeter, lateral direction) 3.297 feet, the altitude and airspeed position error correction for flaps 0, 10, 15, and 23 can be determined then it can be shown as the following figures respectivelly.

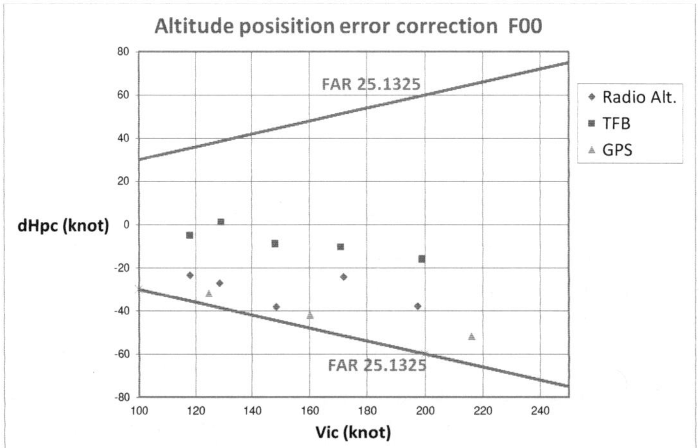

Figure10. Altitude Position Error Correction F00,LGUP, 100%RPM, PLF

IOP Conf. Series: Journal of Physics: Conf. Series **1005** (2018) 012028 doi:10.1088/1742-6596/1005/1/012028

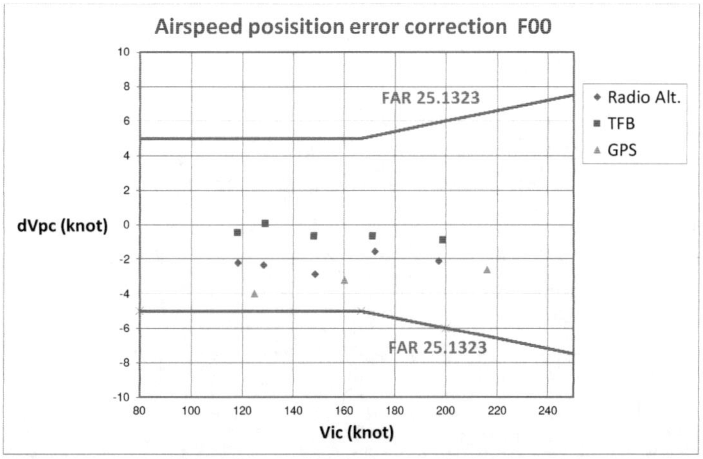

Figure 11. Airspeed Position Error Correction F00, LGUP, 100%RPM, PLF

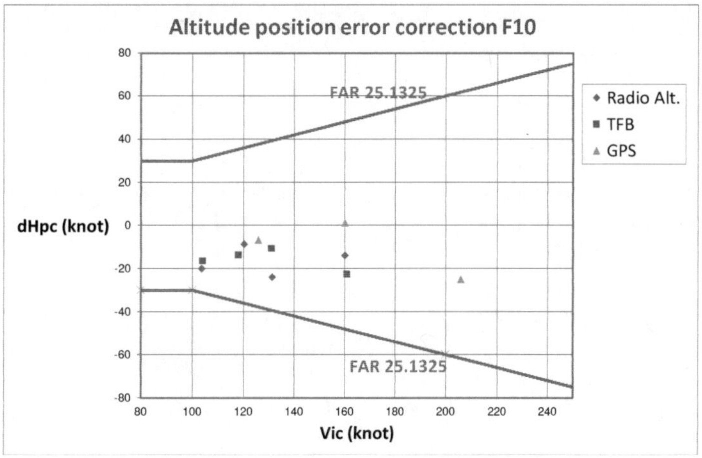

Figure 12. Altitude Position Error Correction F10, LGUP, 100%RPM, PLF

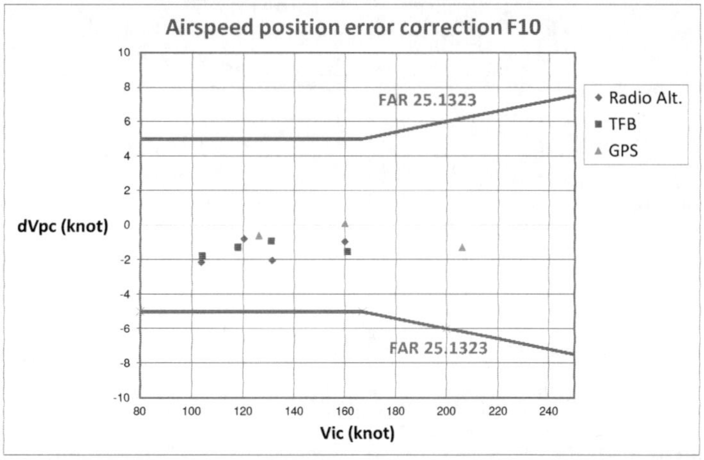

Figure 13. Airspeed Position Error Correction F10, LGUP, 100%RPM, PLF

IOP Conf. Series: Journal of Physics: Conf. Series **1005** (2018) 012028 doi:10.1088/1742-6596/1005/1/012028

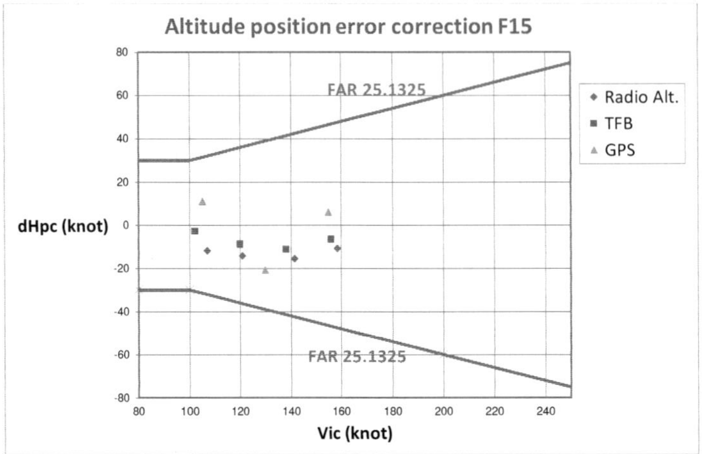

Figure 14. Altitude Position Error Correction F15, LGUP, 100%RPM, PLF

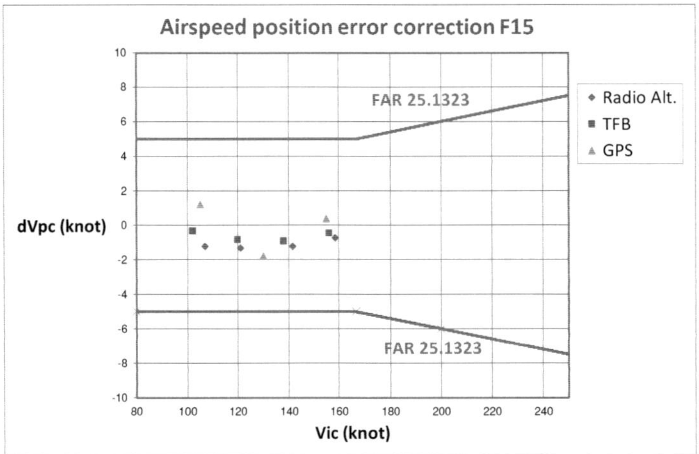

Figure 15. Airspeed Position Error Correction F15, LGUP, 100%RPM, PLF

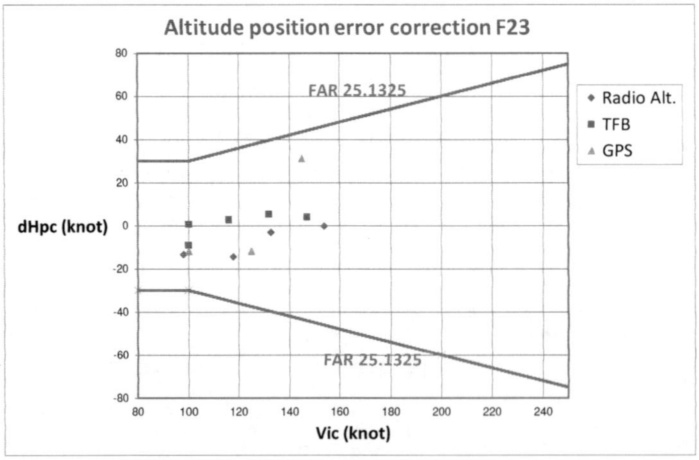

Figure 16. Altitude Position Error Correction F23, LGDW, 100%RPM, PLF

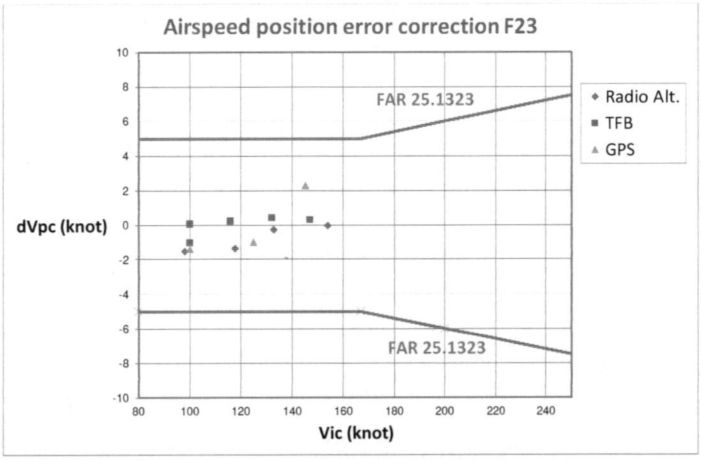

Figure 17. Airspeed Position Error Correction F23, LGDW, 100%RPM, PLF

From all that figures above, it can be seen that all altitude and airspeed position error correction by using Radio Altimeter for flaps 0, 10, 15 and 23 are well within the boundaries as stated by FAR 25.1325 and FAR 25.1323 (Ref. [6] and Ref. [7]). Comparing to Tower Fly By Method and GPS Method the test results are closed.

For the future, by using accereleration flight during the test for each flap configuration, it is expected to reduce flight hours due to all speed range can be taken in the same time. This method is also less human resources required compare to the Tower Fly By Method and take more efficient time comparing to GPS method.

8. Conclusions

From the description above it can be concluded that the altitude and airspeed position error correction by using Radio altimeter as reference are well within the the boundaries of FAR 25.1325 and FAR 25.1323 regulation. By using acceleration airspeed flight test technique during the test, this Radio Altimeter Method is more efficient in term of reducting flight hours, less human resources required, and not require any additional special equipment compare to the Tower Fly By and GPS Method.

Acknowledgement

This development is supported by Directorate of Technology and Development, Indonesian Aerospace (IAe). The authors would like to thank Director of Technology and Development for providing flight test data and research opportunity, Turkish Navy for test aircraft, and Thales Airborne System (TAS) for supporting test facilities at Cazaux.

References

[1] "Measurement System Application & Design", Ernest O. Doeblin 3rd, Mc Grow Hill Book Company, 1983.

[2] "CN235-100M MPA1 Position Error Calibration (PEC) Flight Test Results", Qualification Document, Indonesian-Aerospace (Iae), July 21th 2009.

[3] "Light Aircraft Performance For Test Pilots And Flight Test Engineers", Sean C. Roberts, Flight Research. INC., Mojave, California 93501.

IOP Conf. Series: Journal of Physics: Conf. Series **1005** (2018) 012028 doi:10.1088/1742-6596/1005/1/012028

[4] "Measurement of Aircraft Speed and Altitude", William Gracey, Aeronautical Research Scientist, NACA/NASA (Ret.)

[5] "Flight Test Analysis Description Manual", Flight Test Engineering Analysis Department, Flight Test Center, Indonesian-Aerospace, Doc. No. D000NA12113, June 1996.

[6] "Federal Aviation Regulation (FAR), Part 25".

[7] "Advisory Circular 25-7A, Flight Test Guide for Certification of Transport Category Airplane".

[8] "Flight Test Log, Airspeed Calibration Using GPS", Doc. No. TASFR00636990-F011.

[9] "Flight Test Log, Airspeed Calibration Using TFB", Doc. No. TASFR00636990-F011.

[10] "CN235-100 MPAP Development Flight Test Program – Flight Physic With Clean Wing", Flight Test Plan for first phase of CN235-100 MPA Program, Doc. No. 3M0006021, January 2009.

[11] "CN235-100M MPA1 Drag Polar Flight Test Results", Qualification Document, Indonesian-Aerospace (IAe), July 23th 2009.

IOP Conf. Series: Journal of Physics: Conf. Series **1005** (2018) 012029 doi:10.1088/1742-6596/1005/1/012029

Research Amplitudo Vibration On Holder Due To The Process Of Lathe Nozzle Rocket Rx 450

A.B.Djatmiko[1], Ediwan[1], E.D.Arisandi[1], H Purnomo[1]
[1] Rocket Technology Center, National Institute of Aeronautics and Space (LAPAN), Indonesia **Email : agusbudi60@gmail.com**

Abstract. The main function of the rocket nozzle is to convert the enthalpy efficiency from combustion gas to kinetic energy and also to make high velocity out of the gas. The rocket nozzle usually consists of a converging and diverging part. With a smaller area on the neck and enlarged at the exit area. The velocity flow through the nozzle enlarges into the speed of sound through the neck and then becomes super sonic in the divergent part.
Nozzle making or machining using conventional lathes, first performed is drilling on a massive metal that is bonded to the veneer, then after a sufficient gap is done deep-boring. At the time of the process of lathe in the nozzle RX 450 there is an obstacle that is vibrating tool holder chisel or holder so it is worried about not precision of the process of lathe. This should not happen because it can cause failure in the latter for it needs to be studied and studied further so that the lathe process goes accordingly.
The holder material of ST 60 with a modulus of elasticity 200 GPa and a nozzle material of AISI 4340 alloy steel with σ_{yield} = 470 MPa, Shear Modulus G = 80 GPa.
The purpose of this research is to observe the amplitude of vibration on the holder due to RX-450 nozzle lathe processing for the purpose of amplitude that occurs in accordance with the desired so that the nozzle structure is no damage process.
The result of the research was obtained holder with length (L) 80cm, profile width (B) 5 cm, height of profile (H) 10 cm, turning machine ω = 8.98 rad / sec and natural holder frequency ω_n = 89.8 rad / second, Amplitude of vibration of δ = 1.21 mm, while the amplitude of the design X = 1.22 mm
From the results of this study it can be said that the holder of a chisel or holder can be used as a tool at the time of RX nozzle retrieval process and is quite safe because it works under the condition ω/ω_n <0.3 (Plans and Specifications of A Suborbital Rocket Payload "AKPV Engineering University of Wyoming 2009)

1. Introduction

The main function of the rocket nozzle is to convert the enthalpy efficiency of the combustion gas to kinetic energy and also to make high exit speeds of the gas. The nozzle is a very efficient tool for gas acceleration to supersonic speed. The rocket nozzle usually consists of a converging and diverging part. With a smaller area on the neck and enlarged in the exit area. The velocity flow through the nozzle enlarges into the speed of sound through the neck and then becomes supersonic on the divergent part. [3]

The pressure occurring in a rocket motor is channeled through the nozzle, giving rise to a thrust force coming out of the rocket nozzle. For that required precision in making the nozzle so that the maximum results obtained from the rocket.

Making or machining nozzles using conventional lathe. The first performed is a drilling on a massive round metal that is bonded to the veneer, then after getting a sufficient gap to be done deep-boring. In the deep-lathe process to create a nozzle embossed force that causes vibration in the holder. Vibration on the chisel can cause damage to the inner surface of the nozzle. For that need to be studied further so that the holder of a chisel or holder does not vibrate during the nozzle turner.[1,2,6]

The machining process of the lathe is the occurrence of relative motion between the chisel and the workpiece will result in the variation of the chip resulting in a change in force, so that the vibration amplitude continues to expand rapidly. Enlarged amplitude will create a high-pitched sound coming from the chisel that cuts off the workpiece. Increased amplitude is also caused because the energy used cutting will be absorbed by the system. The energy that excites the vibration comes from the cutting process itself is called self-excitation vibration or better known as the chatter. Chatter is the vibration that occurs when the chisel moves toward the workpiece on the cutting process. Chatter should be avoided as it can degrade surface quality even resulting in geometry errors. The causes of chatter vary, one of which is the cutting parameter. [9,12]

The purpose of this study is to observe the amplitude of vibration in the holder due to the RX-450 nozzle retention process which is aimed to amplitude that occurs in accordance with the desired so that the process of making nozzle structure no damage.

2. THEORETICAL BASIS

2.1. Force of lathe

To find out the phenomenon that occurs at the time of the nozzle, it is necessary to find the size of the lathing force that is:

Material nozzle structure made of stainless steel AISI 4340 with the following specifications: σ_y = 470 MPa, Shear Modulus G = 80 GPa,

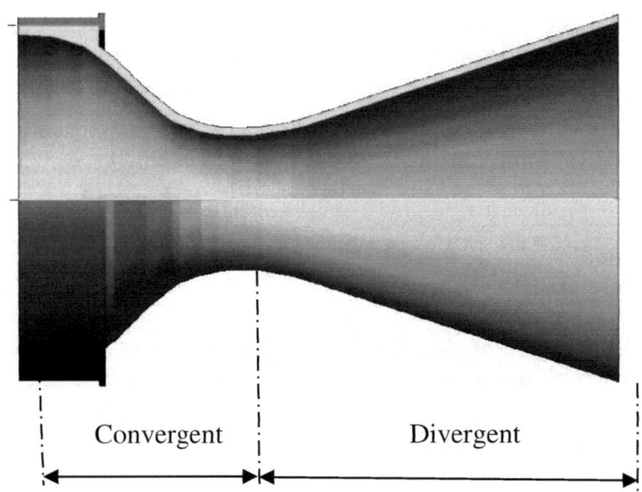

Figure 1. Rocket Nozzle

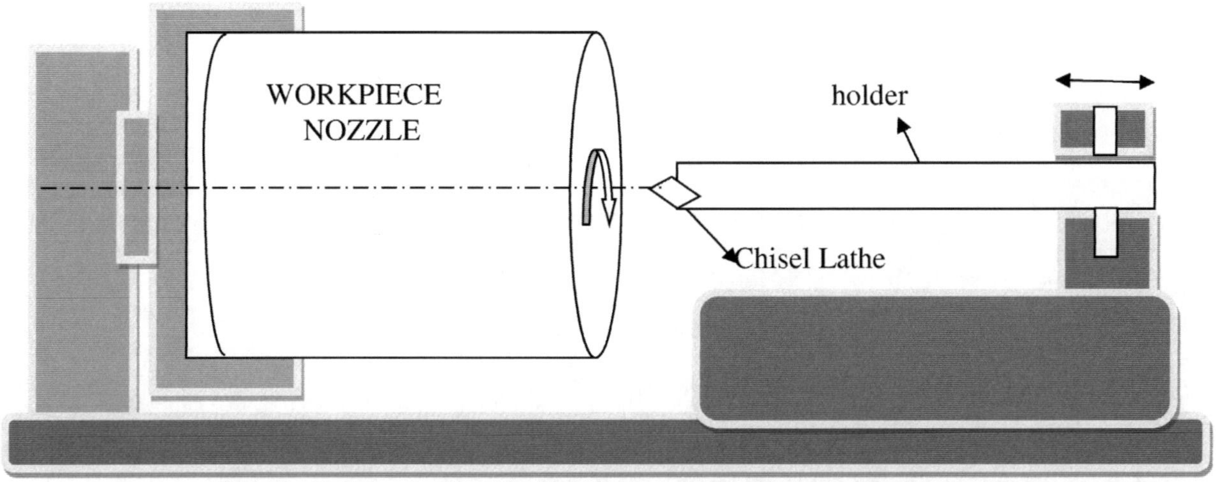

Figure 2. Schematic of lathes for forming rocket nozzle form

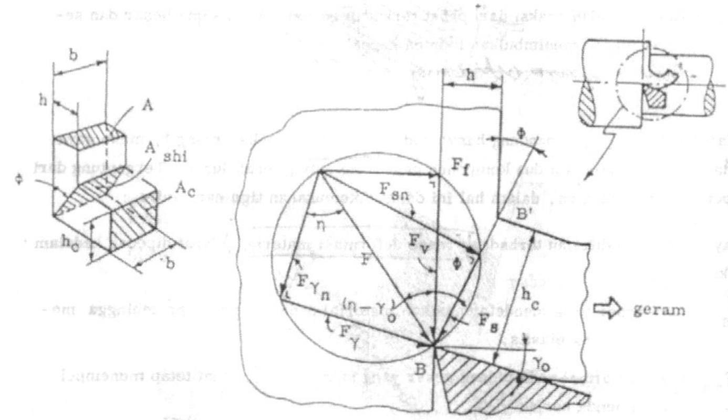

Figure 3. Total cutting force is resolved into two component, horizontal component vertical component

The relationship between furious with cutting force can be seen in the following formula [7] : $F = \tau_{shi}.b.h\dfrac{\cos(\eta - \gamma_o)}{\sin\phi\cos(\phi + \eta - \gamma_o)}$ [7]..(1)

Information :

F = Cutting force (N), τ_{shi} = Shear stress in shear plane (N / mm2), b = cutting width (mm), h = Thickness before cutting (mm), η = Friction angle, γo = Angle growled, ϕ = Shear angle From the above formula can be seen that the resulting fury is directly proportional to the cutting force imposed on the workpiece.

The force generated from the cutting process by the chisel causes the chisel to vibrate. Vibration on the chisel will increase as the resulting cut style increases.[8]

IOP Conf. Series: Journal of Physics: Conf. Series **1005** (2018) 012029 doi:10.1088/1742-6596/1005/1/012029

2.2. Rigidity of the Cross Direction Column

Considered a holder structure with the tip of the tip and L length, given a force load of F at one end as shown above.

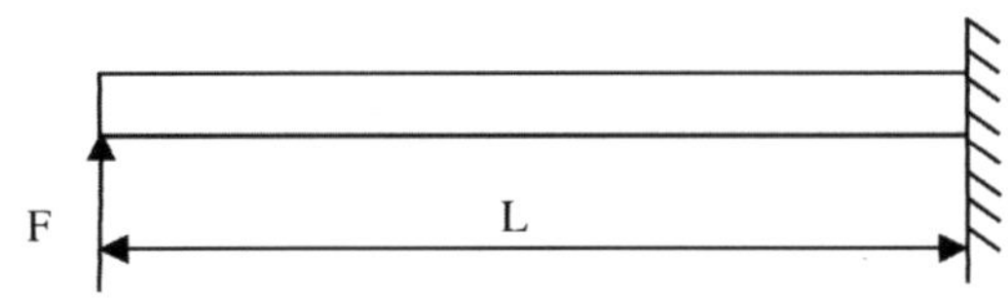

Figure 4. Modeling the holder

F = the lathing force, L = the length of the holder
The magnitude of the column stiffness of the holder structure is

$\delta = \dfrac{FL^3}{3EI}$, and the rigidity of the column holder is $k = \dfrac{F}{\delta}$, then

$$k = \frac{3EI}{L^3} \quad [2,5] \ldots\ldots (2)$$

Where: $I = (b.h^3/12)$ = moment of profile inertia, b = column width, h = column height E = Modulud elasticity of holder material [6]

2.3. Natural Frequency

To obtain the natural frequency of the holder, the equation of motion of the holder system (see figure 5) is $m\ddot{x} + c\dot{x} + kx = 0$ where the dampening system is ignored, the natural frequency is [1]:

$$\omega_n = \sqrt{k/m} \quad [4] \ldots\ldots (3)$$

Where: k = the column stiffness of the holder (N / m) and m = analogous as the lathe force (kg)

2.4. Vibration amplitude

To obtain the amplitude that occurs in the holder structure in the nozzle-bolting process, the following equations are used:

$$X = \frac{F/k}{\sqrt{\left[1 - (\omega/\omega_n)^2\right]^2 + (2\xi\,\omega/\omega_n)^2}} \quad [4] \ldots\ldots (4)$$

And the relationship between the amplitude ratio with the ratio of frequency to damping factor can be explained in figure 5 [4,5,11], :

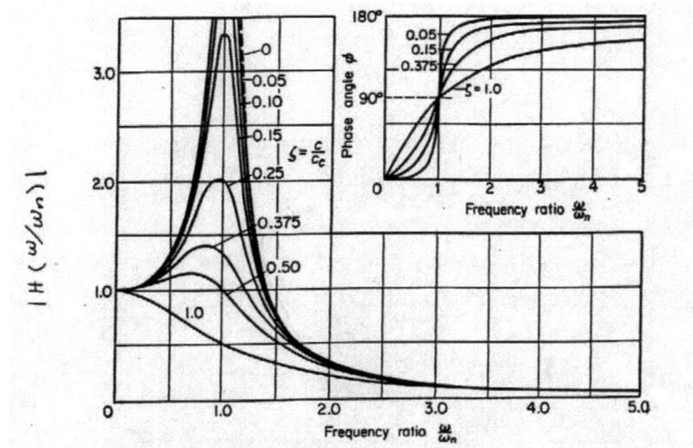

Figure 5. frequency ratio

For structural strength as well as holder, the designed frequency ratio is ω / ω_n <0.3, in order that the structure is not damaged due to large amplitude or resonance state. [11]

3. METHODOLOGY

The methodology used in designing the holder on RX 450 rocket nozzle was firstly searched for holder and nozzle data, large cut style, holder column elasticity modulus (ST 60), σ_{yield} from nozzle material, and long holder (1) , And then calculate the stiffness of the column holder (k) to find the equation of motion, after which the natural frequencies of the holder structure and vibration amplitude are calculated.

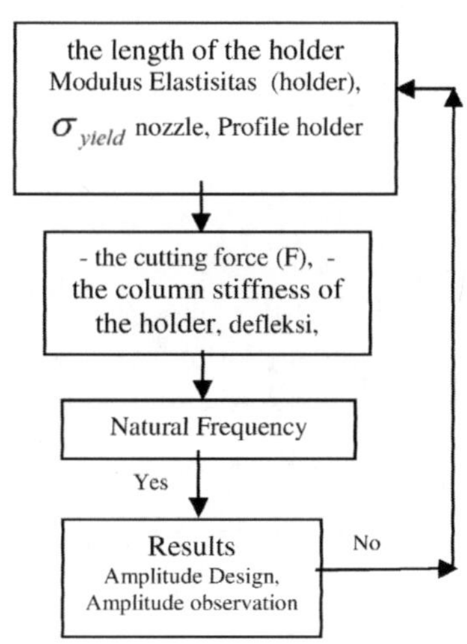

Figure 6. Flow Chart Designing Holder Structure

4. DATA PROCESSING

4.1. Calculating cutting force on lathe

To calculate the magnitude of the cutting force in the lathe process, equations (1)
Taken by cutting width (b) = 3 mm and the thickness of the flutter before cutting (h) = 2 mm 0,5 x 470

IOP Conf. Series: Journal of Physics: Conf. Series **1005** (2018) 012029 doi:10.1088/1742-6596/1005/1/012029

MPa = 235 MPa = 2,35 e2 N / mm^2, $\eta - \gamma_o = 35^0$, $\phi = 20^0$ then the cutting force (F) Is,

$$F = 2{,}35e2x3x2\frac{\cos 35}{\sin 20\cos 55} = 5887.68 \text{ N}$$

With the help of a simple microsoft excel program, there is a large number of cutting force (F) with some cutting width (b) and thickness before cutting (h), can be seen in table 1 below.

Where : $\eta - \gamma_o = 35^0$, $\phi = 20^0$ dan $\phi + \eta - \gamma_o$

Table 1. Result of Calculating cutting force on lathe

τ_{shi} (N/mm^2)	b (mm)	h (mm)	$\cos(\eta - \gamma_o)$	$\cos(\phi + \eta - \gamma_o)$	$\sin\phi$	F (N)
2.35E+02	3	2	7.66E-01	*0.50045969*	0.341	5887.68
2.35E+02	4	2	7.66E-01	*0.50045969*	0.341	7850.24
2.35E+02	5	2	7.66E-01	*0.50045969*	0.341	9812.80
2.35E+02	6	2	7.66E-01	*0.50045969*	0.341	11775.36

4.2. Calculates column rigidity

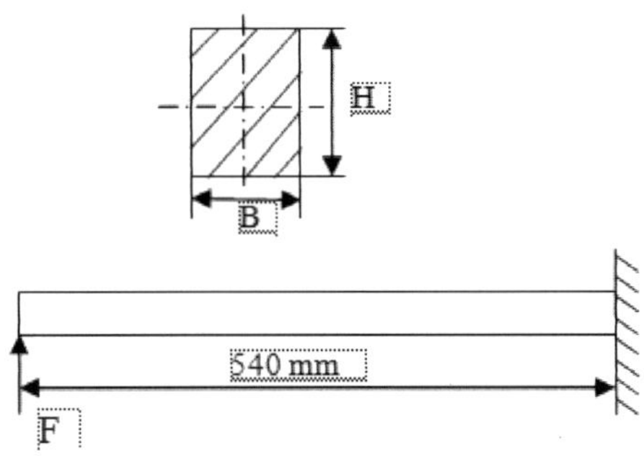

Figure 7. Profile holder form

I = moment of profile inertia $I = (B.H^3/12)$ [10], where:

B = width of holder profile, H = height of holder profile E = Modulud elasticity holder (st 60 steel = 200 GPa), then the stiffness of the base cross direction column k1 with L = 80 cm is:

$$k = \frac{3EI}{L^3}, \text{ Then } k = \frac{3x200e9x9e-7}{0.8^3} = 10.e6 \text{ N / m}$$

From the calculation of the cutting force of F = 5887.68 N, anf F = 5935 N approach is taken as the calculation data for completion in the design of the holder.

4.3. Deflection of observation results and Calculate the vibration amplitude in the holder

Figure 8. Observing the vibration amplitude of the holder in the LAPAN structure laboratory

The result of observation at the time of RX 450 nozzle lathe was obtained for observation with condition B = 0,06 m H = 0,12 m L = 0,8 m and k = 10 e6 N / m2
Obtained deflection on measuring instrument δ = 1,21 –e3 m and deflection by the calculation method obtained:

In this plan the holder structure is planned to work in the area of frequence ratio (see figure 5) and the attenuation factor is ignored or, then the above 3 equation becomes:

$$X = \frac{F_o / k}{\sqrt{\left[1-(0{,}1)^2\right]^2}} \text{ , Then } X = \frac{5995./10e6}{\sqrt{\left[1-(0{,}1)^2\right]^2}} = 0.00122 \text{ m}$$

he design result obtained by the value of vibration amplitude (X), not too far with deflection () result of observation that happened, with difference of error value equal to

$$\Delta = \frac{0{,}00059 - 0{,}00058}{0{,}00059} x100\% = 1\%.$$

And the above calculation and observation results with several holder profile sizes, length holder L = 0.8 m, and Modulus of elasticity E = 200 GPa, can be seen in table 2 below

Table 2. Result of holder design

F (N)	B (m)	H (m)	I (m^4)	K (N/m)	ω rad/det	δ (m)	n (rpm)	X (m)
5935	*0.03*	*0.06*	*5E-07*	632812.	32.341	9.38E-03	31,6	0.00947
5935	*0.04*	*0.08*	*2E-06*	2000000	5,75	2.97E-03	56,18	0.00299
5935	*0.05*	*0.1*	*4E-06*	4882813	8.98	1.21E-03	87,77	0.00122
5935	0.06	0.12	9E-06	10e6	12,93	5.86E-04	126,38	0.00059

δ = Deflection of observations
X = amplitude calculated

5. RESULTS AND DISCUSSION

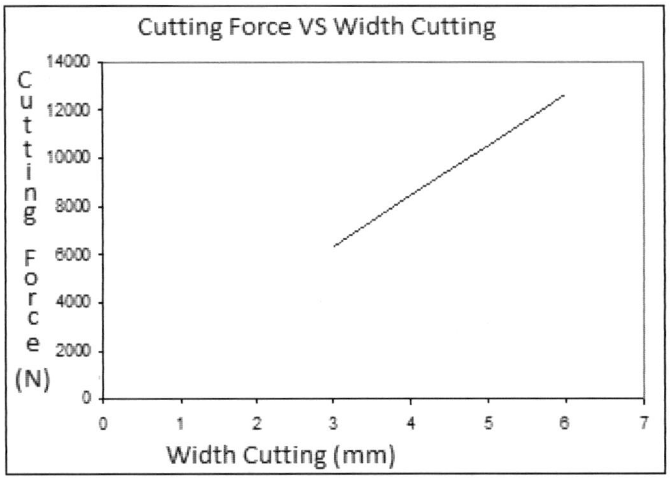

Figure 9. The relationship between the cutting force against cutting width

Figure 9: Describes the relationship between the cutting width (b) to the cutting force F (N) on the RX 450 nozzle making. The larger cutting width is also the required cutting force. In this design nozzle material of AISI 4340 allloy steel with σ_{yield} = 470 MPa, Slide Modulus G = 80 GPa, cutting width b = 3 s / d 6 mm while 2 mm thick bolt, where τ_{shi} = Shear stress in shear plane (N / Mm2) $\tau_{shi} = \frac{1}{2}\sigma_{yield}$ = 2.35E + 02 N / mm2, friction angle $\eta = 50^0$, shear angle $\phi = 20^0$, and angle of angle $\gamma o = 15^0$, the design result by taking cutting width b = 3 mm obtained by cutting force F = 5887, 68 N.
In the design of RX 450 nozzle lathe, 3 mm width cut data and 2 mm thickness of bolster are required because the smallest cutting force is required to maintain the stability of the lathe during the cutting process. (See the design result in table 1) In the design of RX 450 nozzle lathe, 3 mm width cut data and 2 mm thickness of bolster are required because the smallest cutting force is required to maintain the stability of the lathe during the cutting process. (See the design result in table 1) [1,2,8,12]

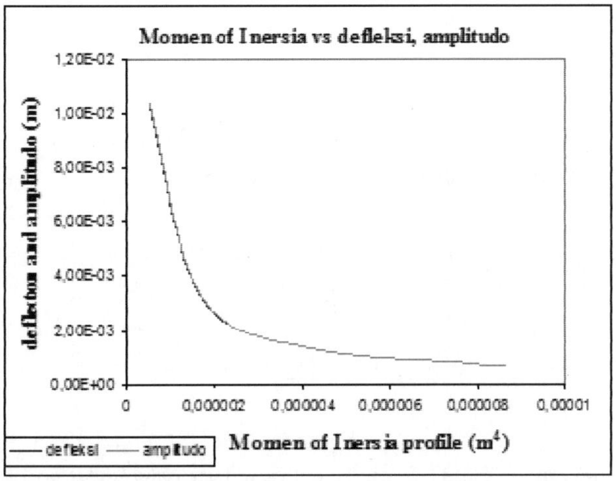

Figure 10. The relationship between the moment of profile inertia against deflection and amplitude

Figure 10: Describes the deflection and amplitude occurring in the holder of the chisel lathe, it is seen that the greater the moment of profile inertia of the holder the smaller the deflection and amplitude occur. The deflection and amplitude are almost the same because at the time of designing the vibration amplitude is taken condition $\omega/\omega_n = 0{,}1$.

Results of holder sculpture holder design in RX 450 rocket nozzle processing with holder profile size as follows, width (B) = 0.06 m, height (H) = 0.12 m and length L = 0.8 m, with steel material ST 60 which has E = 200 GPa and cutting force of F = 5587 N, obtained a deflection on the holder of X = 0.00122 m and the result of observation δ = 1.21E-03 m..

In the planning of vibration amplitude, the holder structure is planned to work in the area of the frequency ratio and the damping factor is negligible or, with K = 10.E6 N / m, the vibration amplitude X = 0.00122 m is obtained.

The design result obtained by the value of vibration amplitude (X), not too far with deflection (δ) result of research or observation with difference of value equal to Δ = 1%.

From the moment graph of the profile inertia moment of the holder to the deflection and deflection (graph 2) seen the two graphs of deflection and amplitude coincide for all values of moment of profile inertia, this shows that the holder structure design is very safe because the frequency ratio is ω/ω_n <0.3 (see figure 5) [1,2,6,8,11]

Deflection = 1,21 mm and amplitude X = 1,22 mm, with the size of the holder specified above can be used for holder design for the manufacture of RX 450 LAPAN nozzle.

6. CONCLUSION

From the results of research and design of holder holder chisel on the process of making nozzle RX 450 LAPAN can be concluded as follows:
Because the nozzle making by using conventional lathe requires a high degree of stability of the lathe tool, hence based on the result of the study selected holder with ST 60 material where the profile width (B) = 5 cm, height profile (H) = 10 cm and length L = 80 Cm with a frequency ratio $\omega/\omega_n = 0{,}1$ with a rotational speed of 87,77 rpm, resulting in a vibration amplitude of 0.58 mm, quite safe because in addition to small amplitudes, the cutting process works on a fairly safe area that is in the frequency ratio $\omega/\omega_n < 0{,}3$ (recommendation from AKPV Engineering University of Wyoming).
The design result of RX 450 nozzle making process or with AISI 4340 alloy steel with σ_{yield} = 470 MPa required cutting force of F = 5935 N.
The observed amplitude on the holder due to the cutting force of δ = 1,21 mm, the value is almost the same as the amplitude of design is X = 1,22 mm with deviation of Δ = 1%. [8,9,11,12].
From the results of research can be said that the design of the holder in accordance with the planned and holder can be used as a tool holders chisel on the process of making the RX 450 LAPAN nozzle and can reduce vibration due to amplitude and deflection that occur in safe area for the strength of the holder structure.
The holder has been used on the lathe for the process of making the RX 450 LAPAN nozzle with satisfactory results

References :

[1] NBV Lakshmi Kumari*1, S. Irfan Sadaq2, G. Prasana Kumar3 " **Analysis of Single Point Cutting Tool of a Lathe Machine Using FEA** " Mechanical Engg. Dept., Muffakham Jah College of Engg. & Tech., Hyderabad, Telangana, India ISSN: 2231-5381 http://www.ijettjournal.org Page 214 ***International Journal of Engineering Trends and Technology (IJETT) – Volume 20 Number 5 – Feb 2015***

[2] K. Khalili, M. Danesh " **Investigation of overhang effect on cutting tool vibration for tool condition monitoring** "The University of Birjand, Birjand, Iran 2, Corresponding author **E-mail:** *1kkhalili@birjand.ac.ir, 2mdanesh_ir@yahoo.com (Received 20 September 2013; accepted 26 September 2013)* VIRBROENGINEERING. ISSN PRINT 2335-2124, ISSN ONLINE 2424-4635, KAUNAS, LITHUANIA

[3] J.W. Cornelisse, H.F.R. Schoyer, K.F. Wakker " Rocket Propulsion and Spaceflight Dynamics " Pitman Publishing Limited London 1979.

[4] William T. Thomson " **Theory Of Vibration With Aplications** " 2nd Edition Prentice-Hall Inc. California 1981

[5] Singiresu S. Rao " **Mechanical Vibrations** " Addison-Wesley Publishing Company, Third Edition California 1995

[6] Bala Raju J, Leela Krishna J, Tejomurthy P " **EFFECT AND OPTIMIZATION OF MACHINING PARAMETERS ON CUTTING FORCE AND SURFACE FINISH IN TURNING OF MILD STEEL AND ALUMINUM** " *, Dept. of Mechanical Engineering, Gudlavalleru Engineering College, Vijayawada, A.P, India.* ***jakkulabalraj@gmail.com, jleelakrishna5@gmail.com, tejomurthy@gmail.com*** **International Journal of Research in Engineering and Technology eISSN: 2319-1163 | pISSN: 2321-7308 Volume: 02 Issue: 11 | Nov-2013, Available @ http://www.ijret.org**

[7] Rochim, Taufiq. " Teori dan Teknologi Proses Pemesinan ". Jakarta : Higher Education Development Support Project, 1993

[8] Kassab S. Y., Khoshnaw Y. K. " **The effect of cutting tool vibration on surface roughness of workpiece in dry turning operation** " Eng. Tech., Vol. 25, Issue 7, 2007, p. 879-889. lubrication

in high-speed milling of hardened steel, International Journal of Machine Tools & Manufacture 47, 2007. Pp. 1660–1666

[9] S.B.SALVI , R.R.DESHMUKH , DR.S.D.DESHMUKH, " **ANALYSIS OF SURFACE ROUGHNESS IN HARD TURNING BY USING TAGUCHI METHOD** " Department of Mechanical Engineering, MGM's Jawaharlal Nehru Engineering College, Aurangabad. salvisunilkumar@gmail.com International Journal of Engineering Science and Technology (IJEST) ISSN : 0975-5462 Vol. 5 No.02 February 2013

[10] R.S Khurmy , J.K. Gupta " Machine Design " Eurasia Publishing House, New Delhi second Edition 1980

[11] Luke Voss, Tony Allais, Sean King, Jeff Parkins " Plans and Specifications of A Suborbital Rocket Payload "AKPV Engineering University of Wyoming 2009

[12] M. Siddhpura and R. Paurobally " **Experimental Investigation of Chatter Vibrations in Facing and Turning Processes** " World Academy of Science, Engineering and Technology International Journal of Mechanical, Aerospace, Industrial, Mechatronic and Manufacturing Engineering Vol:7, No:6, 2013.

PHOTO TESTING

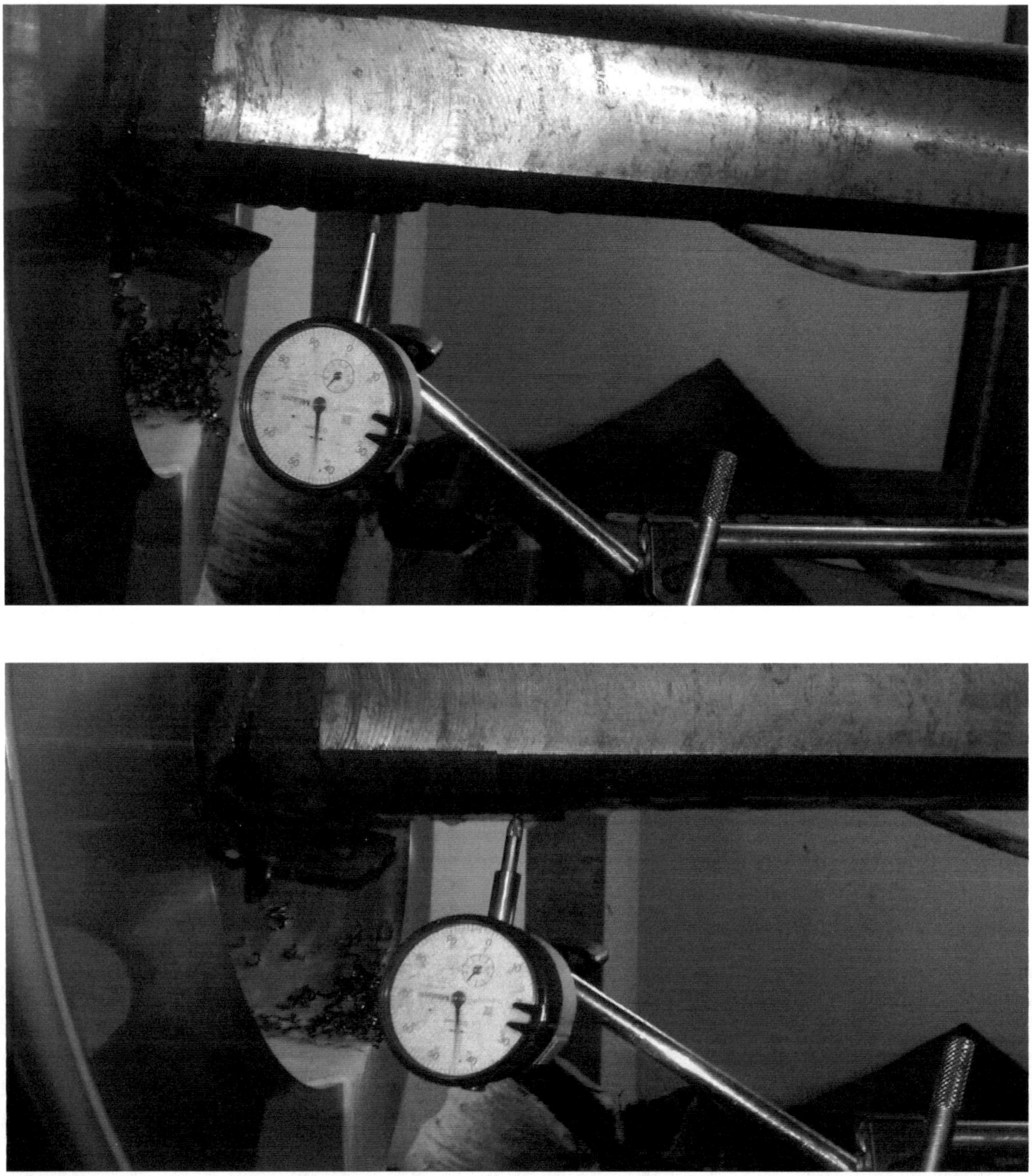

Figure 11 : observing the vibration amplitude of the holder in the LAPAN structure laboratory

The Design of the Longitudinal Autopilot for the LSU-05 Unmanned Aerial Surveillance Vehicle

Muhammad Fajar[1,2] **and Ony Arifianto**[1]
[1]Department of Aeronautics and Astronautics, Faculty of Mechanical and Aerospace Engineering, ITB, Jl. Ganesha No. 10, Bandung, Indonesia
[2]Aeronautics Technology Center, LAPAN, Jl. Raya Lapan, Rumpin, Bogor, Indonesia

Abstract. Longitudinal autopilot design for the LAPAN Surveillance Vehicle LSU-05 will be described in this paper. The LSU-05 is the most recent Unmanned Aerial Vehicle (UAV) project of the Aeronautics Technology Center (Pusat Teknologi Penerbangan – Pustekbang), LAPAN. This UAV is expected to be able to carry 30 kg of payload, four surveillance purposes. The longitudinal autopilot described in this paper consists of four modes, those are Pitch damper, Pitch Attitude Hold, Altitude Hold, and Speed Hold. The Autopilot of the UAV will be designed at four operating speeds, namely 15 m/s, 20 m/s, 25 m/s, and 30 m/. The Athena Vortex Lattice software is used to generate the aerodynamic model of the LSU-05. Non-linear longitudinal aircraft dynamics model is then developed in MATLAB/SIMULINK environment. Linearization of the non-linear model is performed using the linearization tool of SIMULINK. The controller is designed, based on the linear model of the aircraft in the state space form. A Proportional-Integral-Derivative (PID) controller structure is chosen, using root locus method to determine mainly the proportional (P) gain. Integral (I) and derivative (D) gain will only be used if the proportional gain can not achieve the desired target or if an overshoot / undershoot reduction is required. The overshoot/undershoot should not exceed 5% and settling time is less than 20 seconds. The controller designed is simulated using MATLAB and SIMULINK. Preliminary analysis of the controller performance shows that the controller can be used to stabilize the aircraft and to automatize the speed and altitude control throughout the considered speed range.
Keywords: longitudinal, autopilot, PID, UAV

1. Introduction

LSU-05 is an Unmanned Aerial Vehicle (UAV) designed by Pustekbang, LAPAN [1]. The UAV is designed as a multipurpose aircraft. The main mission of the UAV is surveillance. However, it can also be used for aerial photography, disaster evacuation, and as a research platform. The UAV is mainly controlled remotelly, with radio signals from the ground station. It can be controlled manually by a human pilot on the ground, or autonomously using an autopilot system which will be discussed in this paper.

Generic longitudinal autopilot system for this type of UAV consists of altitude-hold/select, speed-hold/select, and a glide-slope control for auto take-off and landing (ATOL) system [2, 3]. However, for semi-automatic control mode, it is necessary to be able to control pitch attitude using the control stick. This pitch attitude control can also serves as the inner loop of either the speed-hold or the altitude-hold control system. The Pitch Attitude Hold is a control system to keep the pitch angle of the aircraft at the commanded value. The Speed Hold is a control system to keep the aircraft at a certain speed, and the Speed Select is a controller to drive the aircraft to go to the commanded airspeed. The Altitude Hold is a system to keep the aircraft at a certain altitude, and the altitude select is a controller that drives the aircraft to the required altitude. Essentially, the "hold" and the "select" autopilots are the same controller, however, the way of operation is different.

Control systems used on many autopilot systems are PID controller [4, 5, 6]. The PID control system is explained in detail in reference [7]. Several methods are described in sufficient detail such as the root locus method and Frequency-Response method (Nyquist, Bode, Nichols, etc.). In addition to the PID control system, which has been developed in the world is a hybrid control system that is a combination of several control systems such as neural network - PID, neural network - fuzzy logic, fuzzy logic - PID, etc. [8, 9, 10]. Several papers compare the methods [8, 9], the results can be complementary to one another.

In this paper, the controller used was a PID controller with root locus method to get proportional gain. Integral and derivative gain will only used if the proportional gain did not achieve desired target or to reduce overshoot/undershoot. The overshoot/undershoot should not exceed 5% and settling time is less than 20 second. The model of the UAV was linearized model of longitudinal equations motion. It is represented by a state-space model for each operating trim speed. The controller designed will be simulated using MATLAB and Simulink.

2. Modeling of UAV

LSU-05 is a fixed-wing aircraft with tail boom connected to its tail. The UAV is powered with piston engine and 28 inch propeller. As many conventional aircraft, the control surfaces used in this UAV are elevator, rudder, and aileron. For longitudinal motion, the elevator had much effect to this motion. Three view drawing of the UAV is presented on Figure 1.

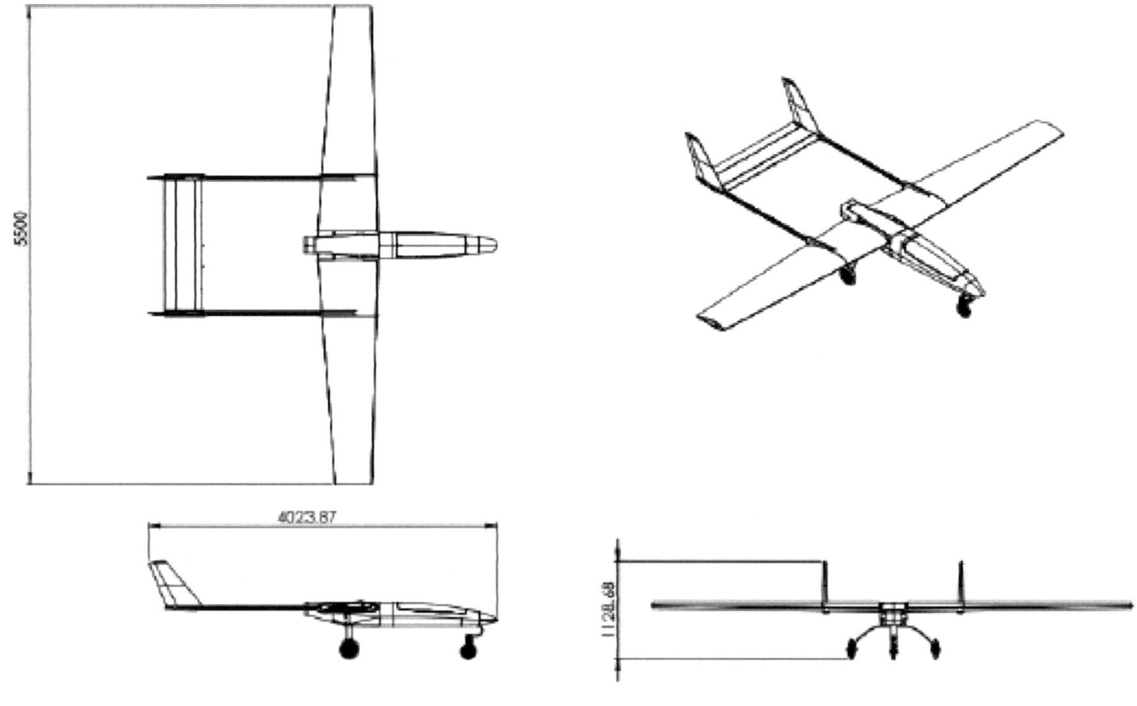

Figure 1 Three-view drawing of LSU-05

The UAV was designed to carry up to 30 kg payload with maximum take-off weight 75 kg. It has 5.5 m wing span and 3.49 m^2 wing area. The UAV configuration is presented in Table 1.

IOP Conf. Series: Journal of Physics: Conf. Series **1005** (2018) 012030 doi:10.1088/1742-6596/1005/1/012030

Table 1 UAV configuration

Parameter	Value	Unit
MTOW	75	kg
S	3.4925	m^2
b	5.5	m
c	0.6	m
AR	8.6614	

The aerodynamics characteristics of the aircraft obtained by vortex lattice method. Athena Vortex Lattice (AVL) [11] was used to get aerodynamic coefficients and the derivative of stability and control of the aircraft. Longitudinal aerodynamics characteristics and stability derivatives of the UAV is presented in Table 2.

Table 2 Longitudinal aerodynamics characteristics and stability derivatives of LSU-05

Parameter	Value	Unit
$C_{L\alpha}$	5.568	[per rad]
$C_{m\alpha}$	-1.295	[per rad]
C_{Lq}	9.876	[per rad/s]
C_{mq}	-13.815	[per rad/s]
$C_{L\delta e}$	0.0081	[per deg]
$C_{D\delta e}$	0.0013	[per deg]
$C_{m\delta e}$	-1.36	[per deg]

Aircraft dynamics modeling is described in many references to both longitudinal dynamics and lateral-directional dynamics [12, 13, 14, 15]. The references described how to derive aircraft equations of motion, triming, static and dynamic stability, derivative of stability and control, stability augmentation system [12], and automatic control [14, 15]. To facilitate the design of automatic control systems, the motion equations are linearized by small disturbance theory. This makes the equations can be decoupled into two modes of motion, longitudinal and lateral-directional.
Trim point observed at four operating speed in cruise phase, which is steady straight flight (γ=0). The velocity observed was 15 m/s, 20 m/s, 25 m/s, and 30 m/s. The altitude of operation is 1000 m. Trim condition at each operating trim point is presented in Table 3.

Table 3 Trim condition for steady straight flight (γ=0)

Velocity (m/s)	α (deg)	Thrust (N)	δe (deg)	L/D
30	1.3052	55.84	-4.44	13.16
25	3.3602	43.75	-6.39	16.79
20	7.1262	38.82	-9.96	18.96
15	15.1137	47.5	-17.55	15.78

From Table 3, the optimum operating speed is 20 m/s for maximum range, it has highest lift to drag ratio. The critical speed is 15 m/s, it has high angle of attack (α) which is close to stall condition. From these trim point, the state space for each trim point can be generated from equation (1) and equation (2). The derivations of equations is presented in reference [14]. Matrix A is a state matrix and matrix B is a control matrix. Matrix A consist of velocity component in X-axes (u), velocity component in Z-axes (w), pithc rate (q), and pitch angle (θ). Matrix B consist of elevator deflection (δe) and delta thrust (δT). In many references, delta thrust is modeled in delta throtle.

$$\bar{\dot{x}} = \boldsymbol{A}\bar{x} + \boldsymbol{B}\bar{u} \quad (1)$$

$$\begin{bmatrix}\dot{u}\\ \dot{w}\\ \dot{q}\\ \dot{\theta}\end{bmatrix}=\begin{bmatrix}X_u & X_w & 0 & -g\\ Z_u & Z_w & u_0 & 0\\ M_u+M_{\dot{w}}Z_u & M_w+M_{\dot{w}}Z_w & M_q+M_{\dot{w}}u_0 & 0\\ 0 & 0 & 1 & 0\end{bmatrix}\begin{bmatrix}u\\ w\\ q\\ \theta\end{bmatrix}+\begin{bmatrix}X_{\delta e} & X_{\delta T}\\ Z_{\delta e} & Z_{\delta T}\\ M_{\delta e}+M_{\dot{w}}Z_{\delta e} & M_{\delta T}+M_{\dot{w}}Z_{\delta T}\\ 0 & 0\end{bmatrix}\begin{bmatrix}\delta_e\\ \delta_T\end{bmatrix} \quad (2)$$

The stability of the UAV can be analyzed from the eigenvalues of matrix A. If all eigenvalues are negative, then the UAV is stable. If there is a positive eigenvalue then the UAV is unstable. The eigenvalues of each trim speed is presented in Table 4 and Table 5. For longitudinal motion, the modes of motion can be analyzed in two modes, short period mode and phugoid mode.

Table 4 Eigenvalue, damping, and frequency of matrix A for each trim speed for short period mode

Velocity (m/s)	Eigenvalue	Damping	Frequency (rad/s)
30	-2.52 ± 1.44i	0.869	2.91
25	-2.14 ± 1.25i	0.863	2.48
20	-1.77 ± 1.10i	0.848	2.08
15	-1.43 ± 1.03i	0.811	1.76

It is presented in Table 4, for short period mode, the eigenvalues are negative. The damping for each trim point is positive means the short period mode are damped.

Table 5 Eigenvalue, damping, and frequency of matrix A for each trim speed for phugoid mode

Velocity (m/s)	Eigenvalue	Damping	Frequency (rad/s)
30	0.0152 ± 0.383i	-0.0398	0.383
25	0.0409 ± 0.447i	-0.091	0.449
20	0.0801 ± 0.527i	-0.15	0.533
15	0.1290 ± 0.615i	-0.205	0.628

To make the UAV stable, the phugoid mode must have negative eigenvalue. From Table 5, the eigenvalue of each operating point is positive. The phugoid mode of the UAVs is unstable. It will makes the UAV also unstable, although the short period mode is stable.

3. Autopilot Design and Simulation

This section described the design of basic longitudinal autopilot consist of Pitch Attitude Hold, Altitude Hold, and Speed Hold. The sensors and actuators of the UAV was not modeled in designing of the autopilot.

3.1. Pitch Attitude Hold

Pitch attitude hold is an autopilot to hold aircraft attitude at desired pitch angle (θ). Pitch attitude hold feedback pitch angle (θ) and compare to reference pitch angle. It used pitch damper as inner loop. Pitch damper is a system to control pitch rate (q). As described in previous sections, the UAV is unstable. With pitch damper, the UAV could be stable. The design of pitch damper is presented on Figure 2.

IOP Conf. Series: Journal of Physics: Conf. Series **1005** (2018) 012030 doi:10.1088/1742-6596/1005/1/012030

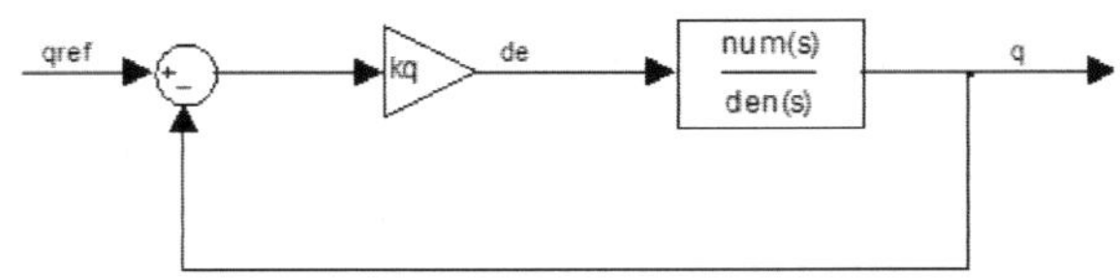

Figure 2 Design of pitch damper with feedback gain kq.

Using root locus method, feedback gain for each trim point is presented in Table 6. The root locus used is negative root locus. The gain is chosen based on location of the pole which has well damping and and miminum overshoot.

Table 6 Feedback gain and damping chosen for pitch damper

Velocity (m/s)	Gain	Damping
30	-0.242	0.775
25	-0.292	0.801
20	-0.277	0.85
15	-0.214	0.999

Pitch damper is used as inner loop for pitch attitude hold. The pitch angle is integration of pitch rate. From the pitch damper designed above, the output pitch rate is integrated with integrator addition to get pitch angle. The pitch angle is fedback to make the closed-loop system. The design of pitch attitude hold is presented on Figure 3.

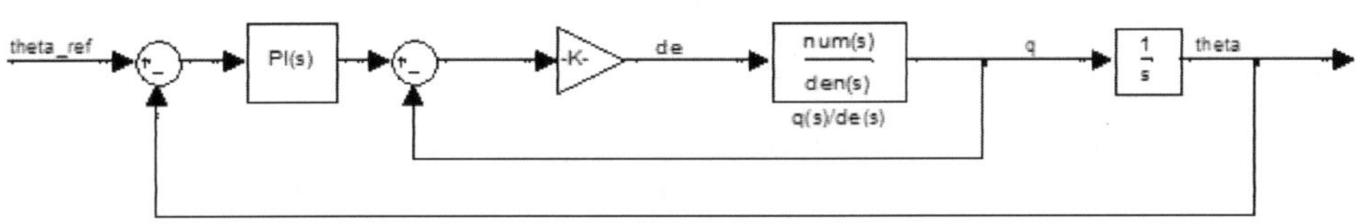

Figure 3 Design of pitch attitude hold with pitch damper as inner loop and PI controller used to control pitch angle.

Root locus of pitch attitude hold for each operating trim point is presented on Figure 4.

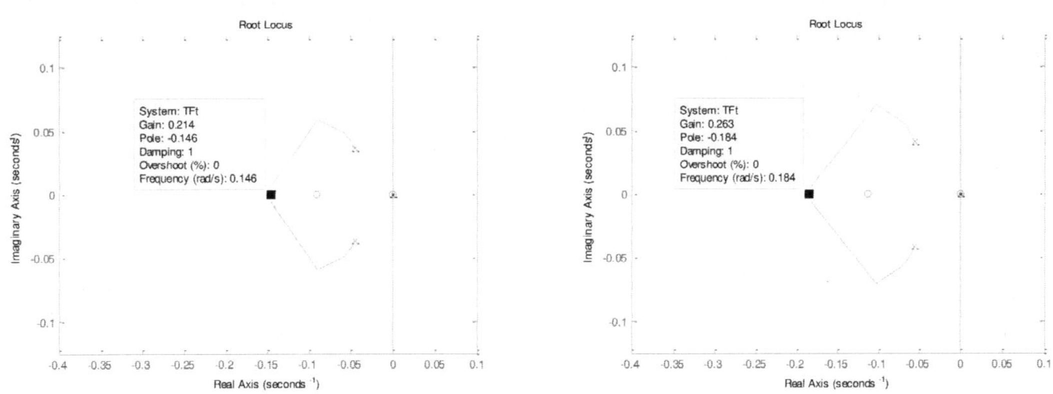

(a) (b)

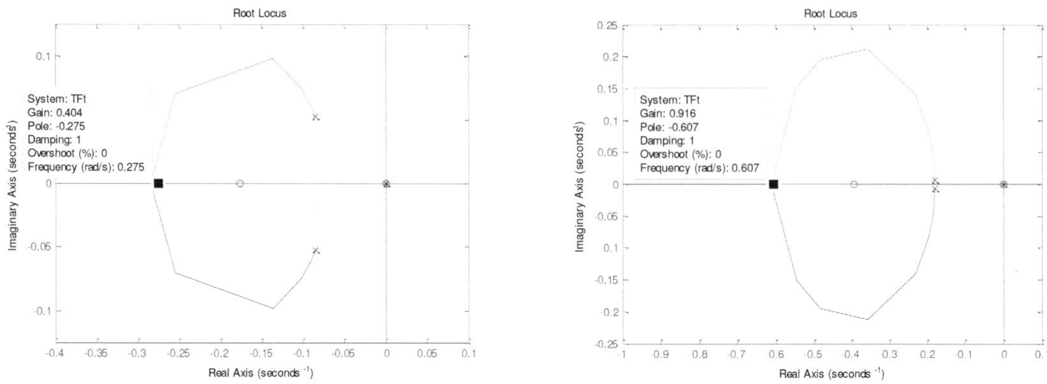

(c) (d)

Figure 4 Root locus plot of pitch attitude hold at trim point (a) 30 m/s, (b) 25 m/s, (c) 20 m/s, and (d) 15 m/s

From root locus on Figure 4, feedback gain chosen for each operating trim point is presented in Table 7.

Table 7 Feedback gain and damping chosen for pitch attitude hold

Velocity (m/s)	Gain	Damping
30	0.214	1
25	0.263	1
20	0.404	1
15	0.916	1

From Figure 5, the response of system with proportional gain did not satisfy the step input, there are steady state error about 10% to 15 % (dashed). To reduce steady state error, integral gain is added (solid). The integral gains chosen for each trim point are 0.04 of proportional gain for 30 m/s and 25 m/s, 0.06 of proportional gain for 20 m/s, and 0.08 of proportional gain for 15 m/s.

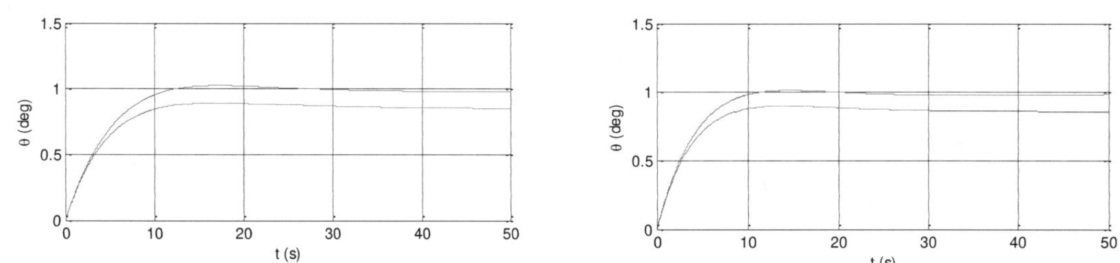

(a) (b)

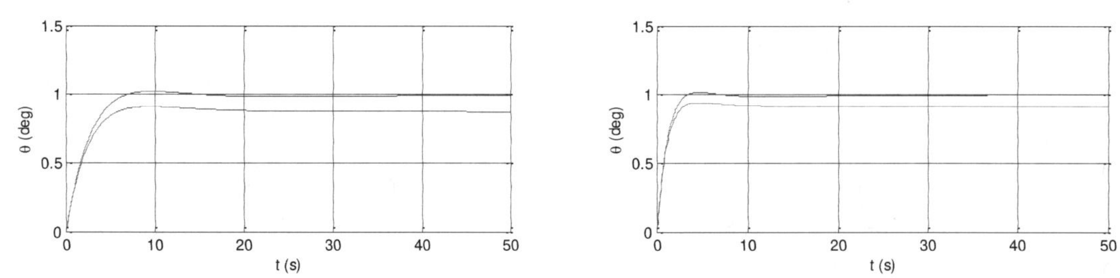

(c) (d)

Figure 5 Step response of pitch attitude hold using P gain (dashed) and PI gain (solid)

Step information for each trim point is presented on Table 8. The lower speed can achieved settling time faster than the higher speed. The settilng time for each trim point did not exceed 10 seconds. Settling time treshold was set to 5%. The overshoot of the lower speed is smaller than the higher speed. The overshoot for each trim point is less than 5 %.

Table 8 Step info of pitch attitude hold for each trim point

Velocity (m/s)	Rise time (s)	Settling time (s)	Overshoot (%)
30	7.5520	9.3564	4.4977
25	6.4381	8.0738	2.8881
20	4.2874	5.3995	2.6497
15	2.0101	2.6115	1.6436

3.2. Altitude Hold

For surveillance mission, altitude hold is very important autopilot. Altitude hold is an autopilot to hold aircraft at certain altitude. To design altitude hold, pitch attitude hold will be used as inner loop. To get better result, the change of altitude must be controlled first. The altitude change (hdot) can be integrated later to get altitude (h). Hdot controller is designed using pitch attitude hold as inner loop. The design of hdot controller is presented on Figure 6.

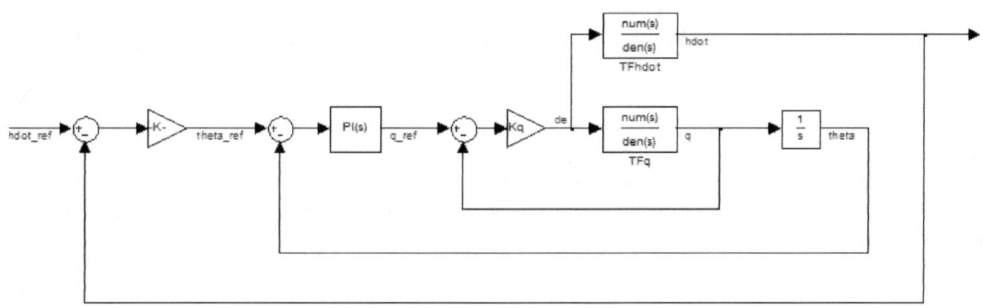

Figure 6 Hdot design with pitch attitude hold as inner loop

Root locus of hdot control for each operating trim point is presented on Figure 7.

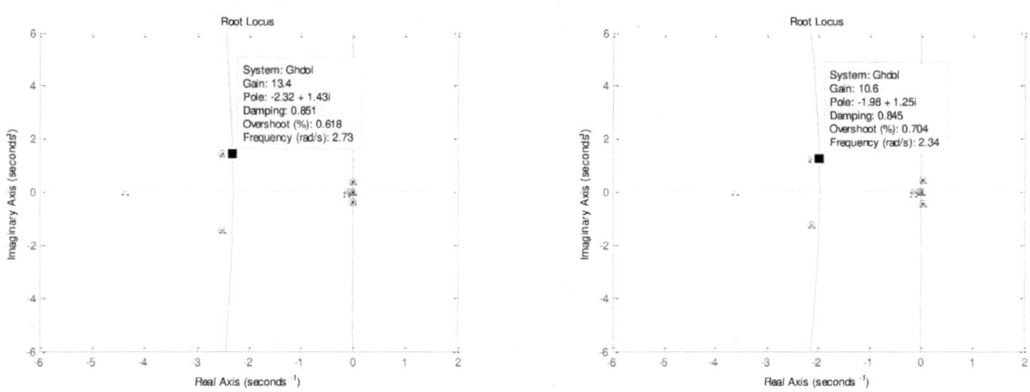

(a) (b)

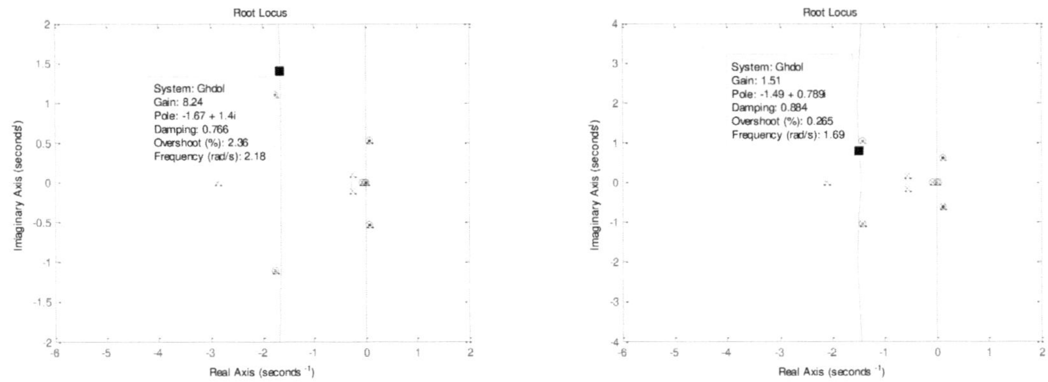

(c) (d)

Figure 7 Root locus of hdot at trim point (a) 30 m/s, (b) 25 m/s, (c) 20 m/s, and (d) 15 m/s

From root locus on Figure 7, feedback gain chosen for each operating trim point is presented in Table 9.

Table 9 Feedback gain and damping chosen for Hdot

Velocity (m/s)	Gain	Damping
30	13.4	0.861
25	10.6	0.845
20	8.24	0.766
15	1.51	0.884

Hdot controls designed obove will be used in altitude hold as inner loop. Hdot was integrated to get altitude (h). Altitude is fedback and compare to reference altitude. The design of altitude hold is presented on Figure 8.

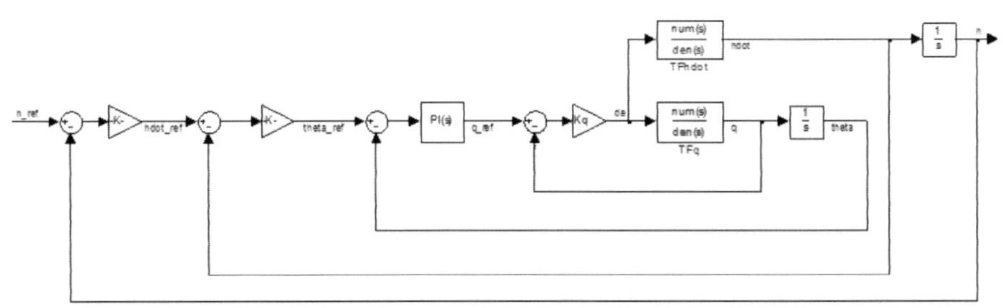

Figure 8 Altitude hold design with hdot control as inner loop

Root locus of altitude hold for each operating trim point is presented on Figure 9.

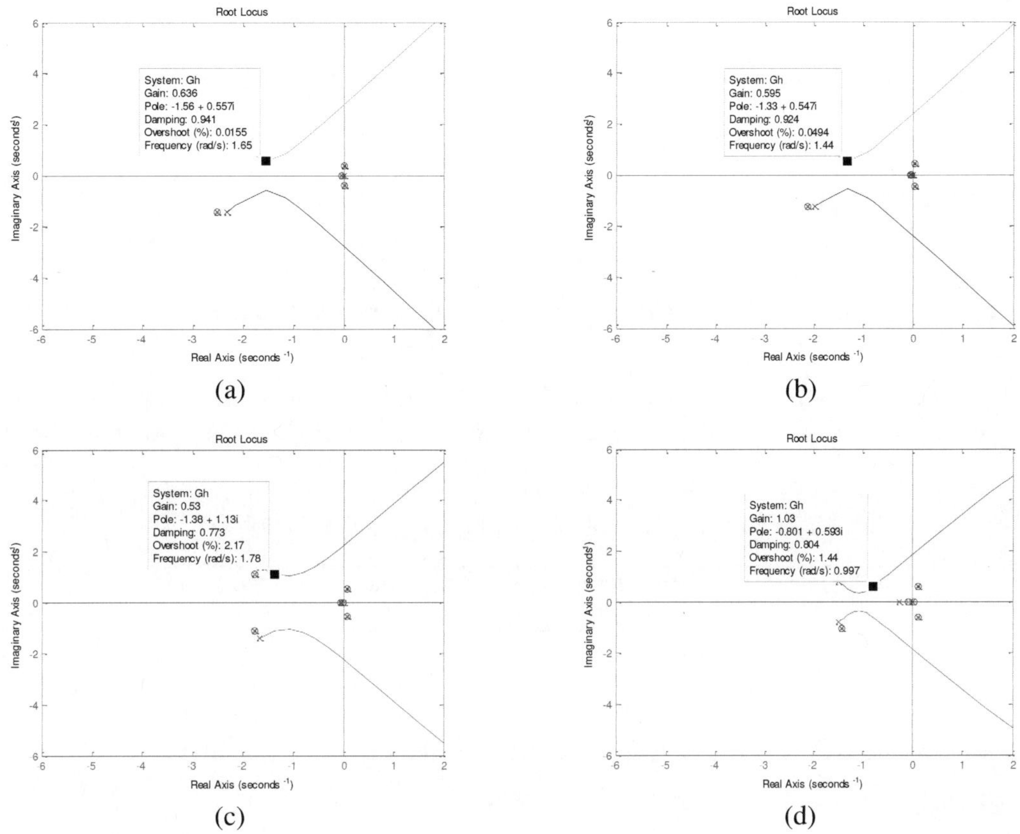

Figure 9 Root locus of altitude hold at trim point (a) 30 m/s, (b) 25 m/s, (c) 20 m/s, and (d) 15 m/s

From root locus on Figure 9, feedback gain chosen for altitude hold at each operating trim point is presented in Table 10.

Table 10 Feedback gain and damping chosen for altitude hold

Velocity (m/s)	Gain	Damping
30	0.636	0.941
25	0.595	0.924
20	0.530	0.773
15	1.030	0.804

Step response of altitude hold for each trim point is presented on Figure 11.

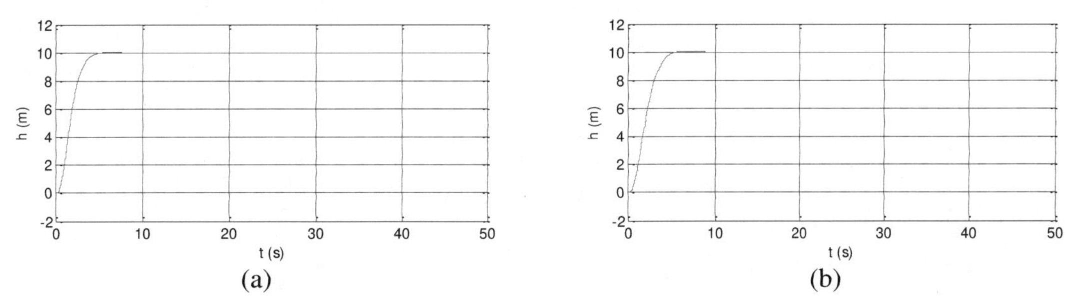

IOP Conf. Series: Journal of Physics: Conf. Series **1005** (2018) 012030 doi:10.1088/1742-6596/1005/1/012030

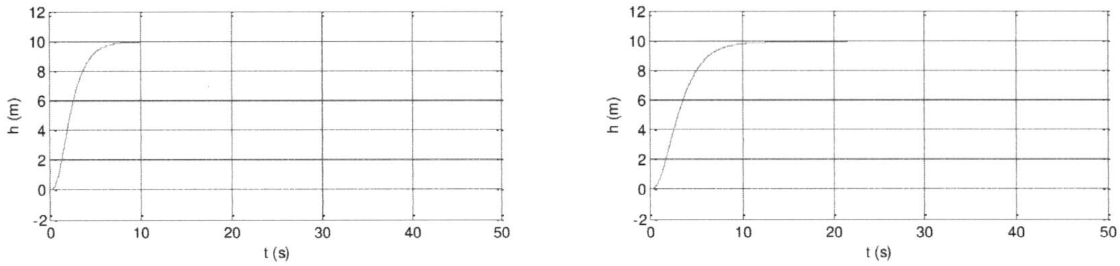

(c) (d)

Figure 10 Step response of altitude hold with speed hold on at trim point point (a) 30 m/s, (b) 25 m/s, (c) 20 m/s, and (d) 15 m/s

From Figure 10, altitude holds designed can achieved 10 meters altitude with settling time less than 8 seconds and overshoot less than 0.5 %. The detail information of altitude hold to step response is presented in Table 11.

Table 11 Step info of altitude hold for each trim point

Velocity (m/s)	Rise time (s)	Settling time (s)	Overshoot (%)
30	2.6327	3.9949	0.2311
25	2.7634	4.0694	0.4777
20	3.7004	5.5781	0.0016
15	5.1836	7.6928	0.1720

3.3. Speed Hold

Speed hold is an autopilot to maintain the speed of aircraft at certain value. It is fedback speed and compare to reference speed. The design of speed hold is presented on Figure 11.

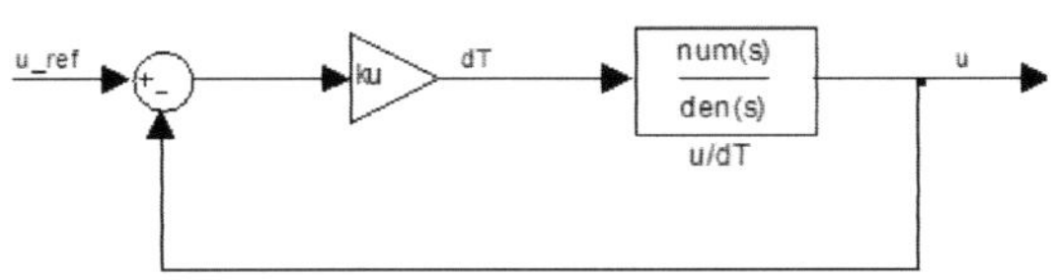

Figure 11 Speed hold design

Root locus of speed hold for each operating trim point is presented on Figure 12.

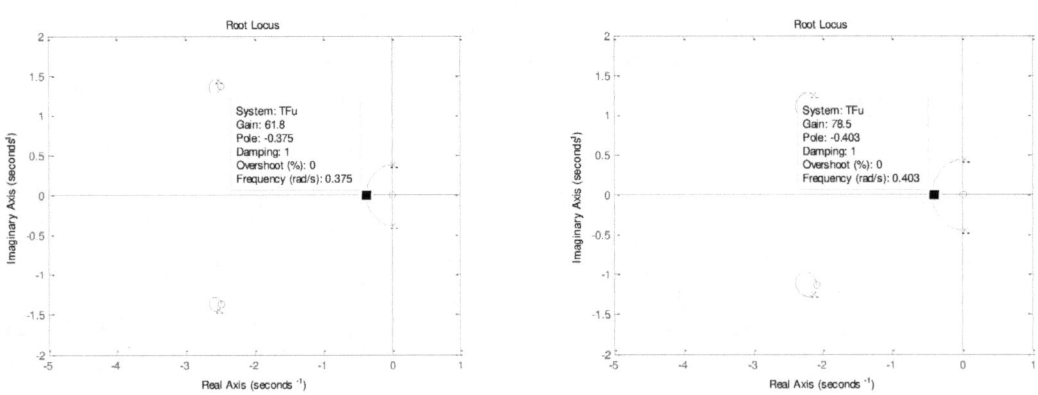

(a) (b)

IOP Conf. Series: Journal of Physics: Conf. Series **1005** (2018) 012030 doi:10.1088/1742-6596/1005/1/012030

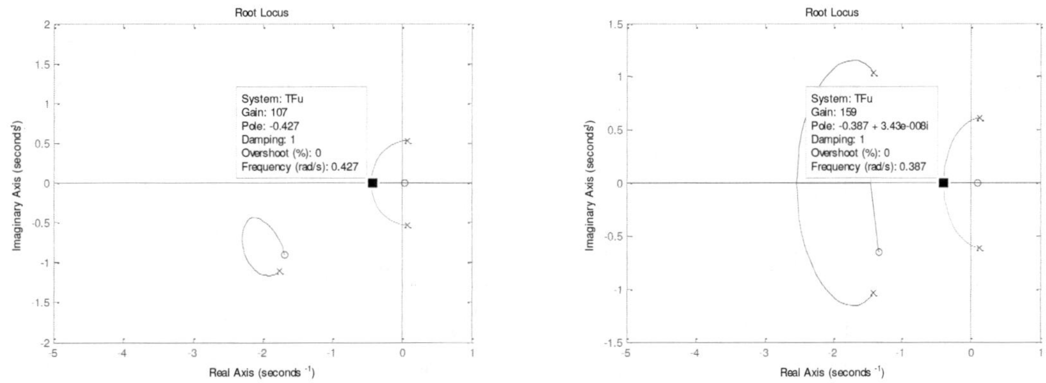

(c) (d)

Figure 12 Root locus of speed hold at trim point (a) 30 m/s, (b) 25 m/s, (c) 20 m/s, and (d) 15 m/s

From root locus on Figure 12, feedback gain chosen for speed hold at each operating trim point is presented in Table 10.

Table 12 Feedback gain and damping chosen for speed hold

Velocity (m/s)	Gain	Damping
30	61.8	1
25	78.5	1
20	107	1
15	159	1

Step response of speed hold for each trim point is presented on Figure 14.

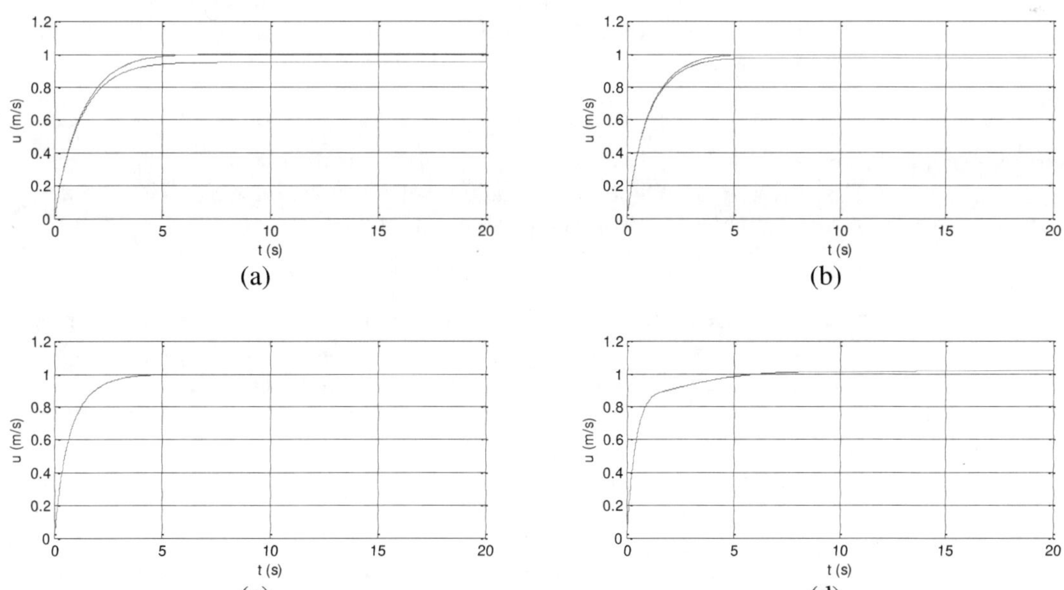

(a) (b)

(c) (d)

Figure 13 Step response for speed hold with altitude hold on at trim point point (a) 30 m/s, (b) 25 m/s, (c) 20 m/s, and (d) 15 m/s

Figure 14 shows step response for speed hold with altitude hold on. Step 1 m/s given to system and the system shows well response but there are steady state error with proportional controller only (dashed), integrator is added to the controller to reduce steady state error (solid). The integral gains chosen are 0.045 of proportional gain for 30 m/s and 0.025 of proportional gain for 25 m/s. Settling time for each trim point did not exceed 5 seconds and overshoot is less than 0.5%. The detail information of speed hold to step response is presented in Table 13.

Table 13 Step info of speed hold for each trim point

Velocity (m/s)	Rise time (s)	Settling time (s)	Overshoot (%)
30	2.6807	3.5639	0.3997
25	2.2430	3.0156	0.1289
20	1.7746	2.4813	0.2401
15	2.3015	4.1161	0

4. Conlusions

Longitudinal automatic control for LSU-05 has been designed at trim point 15 m/s, 20 m/s, 25 m/s, and 30 m/s. Pitch attitude hold autopilot used PI controller shown well results, its settling time did not exceed 10 seconds and overshoot is less than 5%. The higher speed needed more time to steady state than the lower speed. Altitude hold autopilot used P controller shown well results, its settling time did not exceed 8 seconds and overshoot is less than 0.5%. The higher speed needed less time to steady state than the lower speed. Speed hold autopilot used PI controller for speed 25 m/s and 30 m/s, while P controller for 20 m/s and 15 m/s. Its settling time did not exceed 5 seconds and overshoot is less than 0.4%. The controllers designed has met design target, which overshoot/undershoot is less than 5 % and setting time is less than 20 seconds.

The proportional gains chosen for pitch attitude hold are 0.214, 0.263, 0.404, and 0.916 for trim point 15 m/s, 20 m/s, 25 m/s, and 30 m/s respectively. And the integral gains chosen for each trim point are 0.04 of proportional gain for 30 m/s and 25 m/s, 0.06 of proportional gain for 20 m/s, and 0.08 of proportional gain for 15 m/s. The proportional gains for altitude hold are 0.636, 0.595, 0.530, and 1.030 for trim speed 15 m/s, 20 m/s, 25 m/s, and 30 m/s respectively. The proportional gains for speed hold are 61.8, 78.5, 107, and 159 for trim point 15 m/s, 20 m/s, 25 m/s, and 30 m/s respectively. The integral gains chosen are 0.045 of proportional gain for 30 m/s and 0.025 of proportional gain for 25 m/s.

Acknowledgment

This work is a part of Pustekbang-LAPAN research roadmap of Unamanned Aerial Vehicle and joint research with Faculty of Mechanical and Aerospace Engineering, ITB.

References

[1] LSU-05, Pustekbang, LAPAN. URL http://pustekbang.lapan.go.id. Accessed on December 12, 2016

[2] Jenie, S.D. and Budiyono, A. 2006 *Automatic Flight Control System,* ITB, Bandung.

[3] Hogge, Edward F., 2004 *B-737 Linear Autoland Simulink Model*, Lockheed Martin Corporation, Hampton, Virginia.

[4] Eko Budi Purwanto, 2013 *Pemodelan and Simulasi Sistem Kendali Proportional Integral Derivative untuk Kestabilan Dinamika Terbang,* Majalah Sains and Teknologi Dirgantara Vol. 8 No. 2 : pp 48-59.

[5] Sufendi, dkk, 2013 *Design and Implementation of Hardware-In-The-Loop-Simulation for UAV Using PID Control Method*, International Conference on Instrumentation, Communications, Information Technology, and Biomedical Engineering (ICICI-BME), November 7-8, Bandung.

[6] Kada, B. and Ghazzawi, Y., 2011 *Robust PID Controller Design for an UAV Flight Control System*, Proceedings of the World Congress on Engineering and Computer Science Vol II, WCECS, October 19-21, San Francisco, USA.
[7] Ogata, K., 2010 *Modern Control Engineering*, Prentice Hall, New Jersey.
[8] Gouthami, E., and Rani, M.A., 2016 *Modeling of an Adaptive Controller for an Aircraft Roll Control System using PID, Fuzzy-PID and Genetic Algorithm.* IOSR Journal of Electronics and Communication Engineering (IOSR-JECE) Volume 11, Issue 1, Ver.II, pp 15-24.
[9] Nair, M.P. and Harikumar, R., 2015 *Longitudinal Dynamics Control of UAV,* International Conference on Control, Communication & Computing India (ICCC), 19-21 November.
[10] Elsayed, A., 2015 *Design of Longitudinal Motion Controller of a Small Unmanned Aerial Vehicle*, I.J. Intelligent Systems and Applications, 10, pp 37-47.
[11] Mark Drela. AVL. URL http://web.mit.edu/drela/Public/web/avl/. Accessed on February 13, 2017
[12] Hull, D.G. 2007 *Fundamentals of Airplane Flight Mechanics,* Springer, Austin.
[13] Cook, M.V. 1997 *Flight Dynamics Principles,* Elsevier, Oxford.
[14] Nelson, R.C. 1998 *Flight Stability and Automatic Control*, McGraw-Hill, New York.
[15] Allerton, D. 2009 *Principles of Flight Simulation*, John Wiley & Sons Ltd, West Sussex.

IOP Conf. Series: Journal of Physics: Conf. Series **1005** (2018) 012031 doi:10.1088/1742-6596/1005/1/012031

Flight Test of GL-1 Glider Half Scale Prototype

Muhammad Fikri Zulkarnain[1)*], Muhammad Fazlur Rahman[1)], Muhammad Luthfi Imam Nurhakim[1)], Ony Arifianto[1)], Taufiq Mulyanto[1)]

[1)] Aeronautics and Astronautics Engineering
Faculty of Mechanical and Aerospace Engineering
Institut Teknologi Bandung, Indonesia
*fikri0026@students.itb.ac.id

Abstract. GL-1 is a single-seat mid-performance glider, designed to be Indonesian National Glider. The Glider have been developing since 2014. The development produced a half scale prototype called BL-1, which had accomplished static test in 2016, then followed by first flight test at April 20th 2017, and second flight test at May 21st 2017. The purpose of the flight test was to obtain familiarization of the aircraft, aerodynamics characteristics and flow visualization, with data from flight recorded in FDR. The flight test resulted in two flights with total length of 21 minutes. The data from FDR and flight test documents extracted to analyze the characteristics and behavior of the aircraft during flight test. The aerodynamics characteristic was close to analytical results. The control was good; however, the effectiveness of control surface may need to be further analyzed. The result of the flight test will be used as a reference for further improvements and may need further testing.

Keyword: Flight Data, Flight Test, Glider, Sub-scale Prototype

Nomenclature

ρ	Air Density in kg/m^3
ASL	Above Sea Level
CFD	Computational Fluid Dynamics
CL	Coefficient of Lift
FDR	Flight Data Recorder
HTP	Horizontal Tailplane
ISA	International Standard Atmosphere
$m_{aircraft}$	Weight of aircraft in kg
MAC	Mean Aerodynamic Chord
V	Airspeed at flight
VTP	Vertical Tailplane
RoC	Rate of Climb
t	FDR time in minute
W/S	Wing loading

IOP Conf. Series: Journal of Physics: Conf. Series **1005** (2018) 012031 doi:10.1088/1742-6596/1005/1/012031

1. Introduction

GL-1 is a single-seat mid-performance glider, designed to be a national glider to replace the aging fleet of gliders in Indonesia. The development of GL-1 is in collaboration with an aircraft workshop in Pondok Cabe, South Tangerang. The first design of GL-1 had been proposed by Sulc [1] and Pratama [2]. Estimation of GL-1 aerodynamic characteristics and performances had been explored by using analytical method [2], drag polar method based on flight manual data [3][4], CFD simulation [5], and X-Plane simulation [6]. Distribution of aerodynamics forces over GL-1 wings using CFD had also been studied by Amalinadi [7].

A half scale prototype called BL-1 was built in 2016, based on the design by Pratama [2]. It was displayed on Indonesian National Glider Workshop that was held during glider branch of Pekan Olahraga Nasional (PON) 2016, in Subang. BL-1 had conducted a wing static test in 2016 and first flight test at 20th of April 2017. As a result of first flight test, BL-1 had been modified with an addition of outrigger in each wing, a nose landing gear, and an increased surface of rudder and elevator. The incidence angle of HTP had also been fixed. The modification was intended to help the aircraft during takeoff. However, these improvements demand one more flight test to prove the improved design. The purpose of the next flight test was to obtain familiarization of the flying characteristics, and to check control effectiveness and aerodynamic characteristics of the aircraft.

2. BL-1 Specifications

GL-1 half scale prototype (named BL-1) was the test aircraft. The aircraft was controlled using remote control. BL-1 was equipped with detachable 2-piston gas engine with two-blade propeller to be able to conduct both motored and towed flight. A flight controller was installed in BL-1 to store flight test data. The function of the flight controller was to log data (FDR) and send signals to actuator. The data will be recorded directly to the SD card and will be processed after the plane has landed. While the flight controller can only handle low current of servomotor, power supplied to the servomotor will be separated. Though it is capable to fly in several flight modes, only manual mode used in this test to ensure pilot's input will be directly send to the servomotor. Table 1 shows the summary of BL-1 technical data.

Table 1. Summary of BL-1 technical data

Wing Span	7.15 m	VTP Span	0.64 m
Wing Area	3 m^2	VTP Area	0.238 m^2
Wing Aspect Ratio	17	VTP Aspect Ratio	1.72
MAC	0.43 m	Rudder Area	0.083 m^2
MAC location	2.5 cm from Wing Root LE	Airfoil VTP	FX 71-L-150/25
Wing Incidence	3 deg	HTP Span	1.3 m
Wing Twist	-3 deg	HTP Area	0.390 m^2
Airfoil Wing	FX 62-K-153	HTP Aspect Ratio	4.33
Engine Power	10 HP	Elevator Area	0.135 m^2
Propeler Dimension	27x10 inch	Airfoil HTP	FX 71-L-150/25

The Figure 1 below shows the schematics of BL-1 for the second flight test. The centerline defined as reference line for x-axis of the fuselage. The nose and main landing gear were a fixed landing gear, and the tail had a skid. Outrigger was added to each wing tip. The system was located under the canopy area. Engine and propeller placed above the fuselage, with engine boom located at 50% of wing root chord. Nose landing gear and outrigger was intended to help maintain attitude during takeoff. A panel with 60 mm width added to rudder and a panel with 50 mm width added to elevator. VTP and HTP area increased to improve control effectiveness during takeoff.

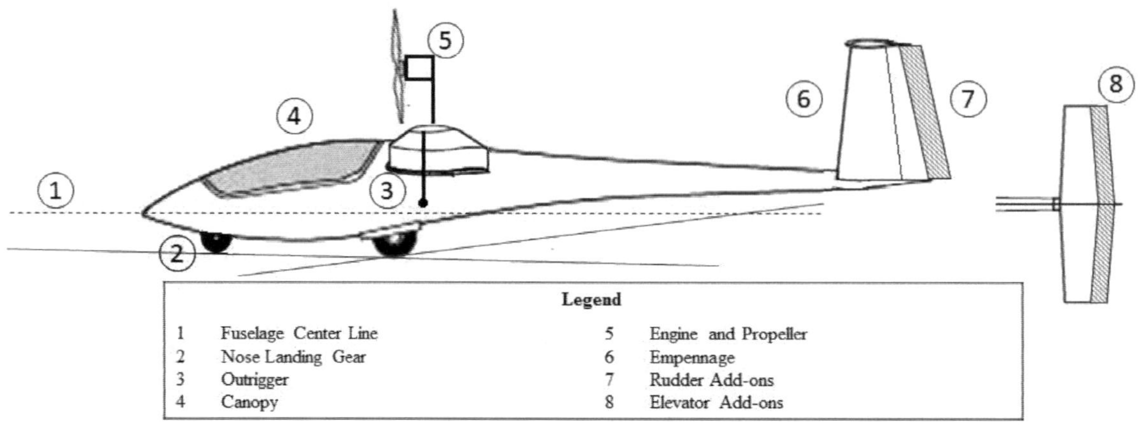

Figure 1. Schematics of BL-1

3. Flight Test Plan

The flight test was located in an airfield in Batujajar, Bandung, at Sunday, 21st of May 2017. The flight test was planned to be conducted in three phases, familiarization with motor, aerodynamics test with towed takeoff (engine off), and flow visualization. The purpose of the flight test was to get familiarization on the flying characteristics, test control effectiveness of the aircraft, and obtain the aerodynamics and stability characteristics of the aircraft [8].

4. Preparation

The preparation was located in two separate place, Pondok Cabe and Batujajar. The preparation in Pondok Cabe was preparation of the airframe and system before the test. At Batujajar, the aircraft would be prepared as follow : geometry measurement and conformity test, system preparation, weight and balance, and engine test [8].

4.1 Preparation in Pondok Cabe

Preparation in Pondok Cabe including the installation nose landing gear, wing outrigger, and the addition of rudder and elevator surface area to increase VTP and HTP surface area. All the control surface was checked and ready, while the flap was fixed because trouble in flap actuators. After that, the aircraft was brought to Batujajar disassembled into four components: fuselage, left wing, right wing, and HTP.

4.2 Assembly

The aircraft was assembled at Batujajar from four components: fuselage, left wing, right wing, and HTP. Because of this, conformity check was needed for the aircraft to be as close to the design as possible.

4.3 Pre-flight Preparation

The airframe and system of the aircraft was checked witho ut any significant trouble.

4.4 Conformity Check

The aircraft needed to be as close to design as possible as confirmation that we could use the test result as an evaluation. The aircraft was checked for the alignment of the components, the incidence of wing and HTP, and the deflection of the control surface.

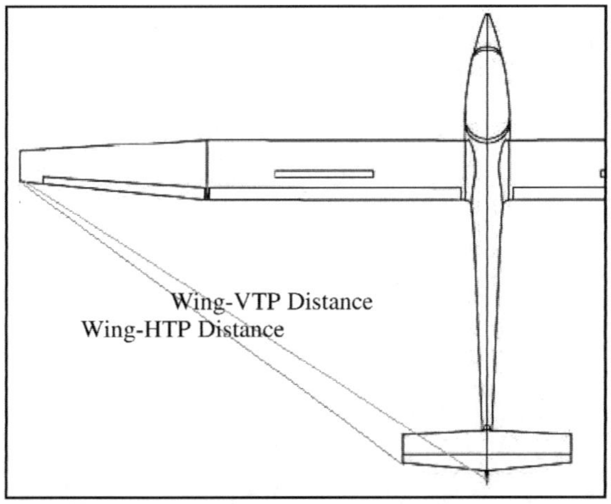

Figure 2. Schematics of Alignment Check

The alignment of the aircraft was checked by measuring the distance between the wing tip-HTP tip and wing tip-VTP tip as shown in Figure 2. The alignment check result was deemed acceptable.The incidence of the wing and HTP was checked at Pondok Cabe. Control surface deflection was also check. The result is shown in table 2.

Table 2. Control surface deflection check result

	Deflections (degrees)	Signal
Aileron	+25/-10	1719/1339
Rudder (on Pondok Cabe)	+20/-20	1830/1311
Elevator	+15/-20	1943/1110

4.5 Weight and Balance

Weight and balance was important to ensure that the aircraft could fly stable and controllable. The plane weigh and balance was measured using three weigh platform located on nose landing gear, left outrigger, and rear skid. The weigh and balance of the aircraft was calculated using the sum of the resultant weight and the sum of moment. Weight and balance was shown on table 3.

Table 3. Weight and Balance result

Weight	C.G. Location (mac)
54 kg	15.5 cm from L.E. mac or 37% MAC

4.6 Engine Check

Engine was checked and ready without any significant trouble.

5. Flight Test Execution

During flight test, the aircraft was able to flew twice in engine configuration, with a total of 21 minutes flight time. Milestones recorded and wind speed data measured using anemometer are shown on table 4 and table 5 consecutively. The second flight trajectory can be shown in Figure 3.

Table 4. Milestones during flight test

First Flight		Second Flight	
Event	**Time GMT+7**	**Event**	**Time GMT+7**
Flight Prep	12:58	Engine Start	15.27

First Flight	
Event	**Time GMT+7**
Engine Start	13.18 – 13.20
Takeoff	13.22
Flypass	13.32
Landing	13.34

Second Flight	
Event	**Time GMT+7**
Takeoff	15.31
Throttle 50%	15.32
Handoff	15.34
Flypass 1	15.38
Flypass 2	15.39
Landing	15.40

Table 5. Wind speed recorded

First Flight		
Windspeed	**Direction**	**Time GMT+7**
3.3 m/s	Northwest	13.17
2.4 m/s	Northwest	13.22
2.4 m/s	Northwest	13.27
1.5 m/s	Northwest	13.32
2.1 m/s	Northwest	13.35

Second Flight		
Windspeed	**Direction**	**Time GMT+7**
0.8 m/s	Northwest	15.25
1.6 m/s	West	15.31
3.2 m/s	West	15.33
2.3 m/s	Northwest	15.35
2.0 m/s	Northwest	15.38
0.6 m/s	Northwest	15.40

Figure 3. Flight trajectory of the second flight.

5.1 First Flight

The first flight was intended for familiarization using engine. The flight took 12 minutes from takeoff to landing. The aircraft had no trouble during takeoff and cruise. During takeoff run, one of the wing outrigger and nose landing gear supported the aircraft. The pilot also said that the control of the aircraft was good. However, the pilot felt that the pitching of the aircraft was too sensitive.

There were several problems during landing for the first flight. The engine was shut down. The rudder effectiveness was considerably decreased. This resulted in a hard landing with quick repairable damage.

Examination after landing had shown that the data of first flight was lost. The low saving rate of micro SD had caused the data lost; thus, all of the data had corrupted. The lost data demanded for a second flight.

For the second flight, the pilot was told to be more careful during landing, and conducting longer approach so the behavior of the aircraft during landing could be observed more. The SD card of the FDR was replaced to ensure the data was saved.

5.2 Second Flight

The second motorized flight was 8 minutes long. Overall performance was almost the same as first flight, with the pitching moment was still too sensitive. Even though the aircraft was maintaining longer approach and considerable thrust, sensitive pitching moment combined with decreased rudder effectiveness resulted in a significant damage during landing. The prototype was unable to continue for towed test and flow visualization test. Fortunately, FDR data was save.

6. Data Processing

As discussed in 5.1 and 5.2, the data from first flight was lost because of a problem in the SD card. Therefore, the data processed are from second flight only. Minute 0 is the time when the FDR is switched on. Data used for analysis is shown on Table 6.

Table 6. Data analyzed

Name	Notes	Data Rate
Time	Time obtained in microseconds, converted to seconds and minutes	Varied
Attitude (pitch, roll, yaw)	Obtained from Magnetometer	1477 Hz
Airspeed	Pitot data error, obtained by subtracting groundspeed with windspeed	295 Hz
Groundspeed	Obtained from GPS	295 Hz
Vertical Speed	Obtained from GPS and calculated from rate of altitude change over time	295 Hz
Altitude	Obtained from GPS	295 Hz
Height	Calculated by altitude - altitude at 0 minutes (ground,)	295 Hz
Gamma	Calculated by finding the angle between horizontal speed (airspeed) and vertical speed	295 Hz
Alpha	Calculated by subtracting pitch angle with gamma angle	No data rate
Elevator, Aileron, Rudder Signal	Obtained from FDR	591 Hz
Throttle Signal	Obtained from FDR	591 Hz

6.1 Major Flight Milestones

Below are tables of flight test milestones from flight test report. The time from flight test report is compared to the data obtained from FDR. The data is shown between 8 to 18 minutes. The time in the relation of times in flight-test report (GMT+7) and from the FDR is shown on Table 7.

Table 7. Milestones of Second Flight

Event	Time GMT+7	Time FDR
Engine Start	15:27	-
Takeoff	15:31	Minutes 8.2 - 8.4
Throttle 50%	15:32	Minutes 8.7
Handoff	15:34	Minutes 10.7
Flypass 1	15:38	Minutes 14.9 - 15.3
Flypass 2	15:39	Minutes 16.2 - 16.5
Landing	15:40	Minutes 17.2 - 17.5

6.2 Takeoff

The behavior and performance of the aircraft during takeoff will be analyzed in this section. The behavior analyzed including the attitude and the effect of addition of the outrigger and nose landing gear. The performance analyzed is the aircraft takeoff run, lift off speed, lift off distance, and lift off angle.

This paragraph will analyze the behavior of BL-1 during takeoff according to Figure 4 to Figure 7. At the start of the takeoff, aircraft is at an attitude with its main landing gear, rear skid, and wing tip gear in ground (5 degrees of pitch angles, circled in red in Figure 5). After the engine started at 8.15 minute, with increasing throttle, the aircraft started to move at 8.2 minutes. After obtaining airspeed of 5 m/s at 8.26 minutes, the aircraft started to pitch down until the nose gear near the ground (pitch angle of -2^0 to 0^0, circled in red in Figure 5) at 8.33 minutes and speed of 15 m/s, this happen without any correction from the pilot (no elevator deflection detected). This condition was maintained until 8.37 minutes with averaged angle of attack of -1^0. The pilot started to deflect the elevator up at 8.345 minutes. After 3 seconds from the nose touch the ground, the nose started pitching up, and at 8.4 minutes, the aircraft started to lift off at 5^0 of pitch and 2^0 of angle of attack (circled in Figure 4 and Figure 6). During takeoff, to adjust the heading of aircraft, the pilot used only rudder.

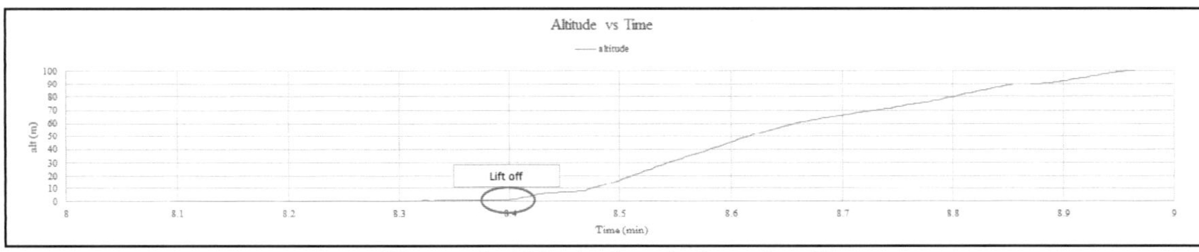

Figure 4. Altitude data during takeoff

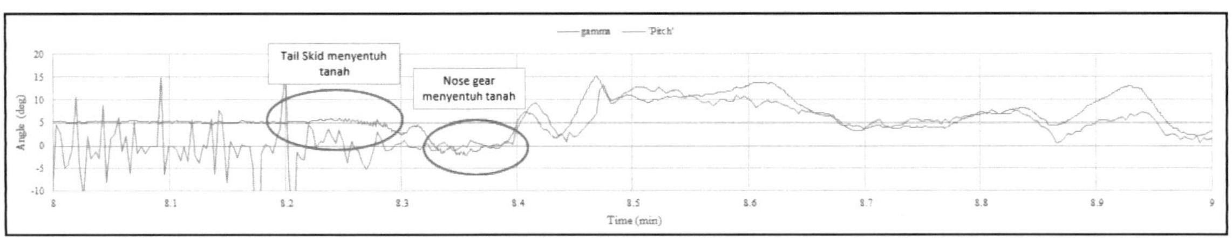

Figure 5. Pitch and gamma during takeoff

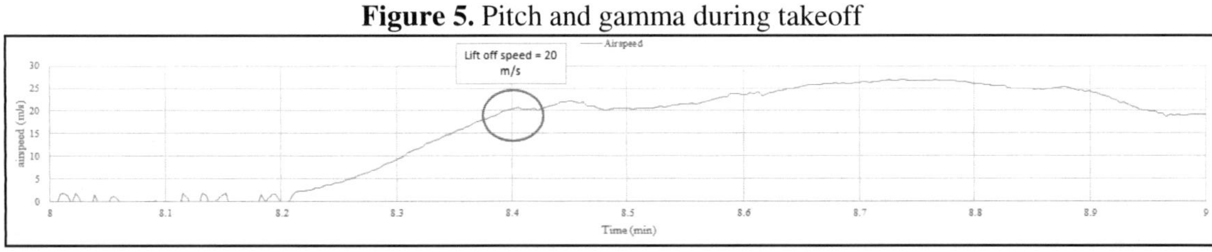

Figure 6. Airspeed data during takeoff

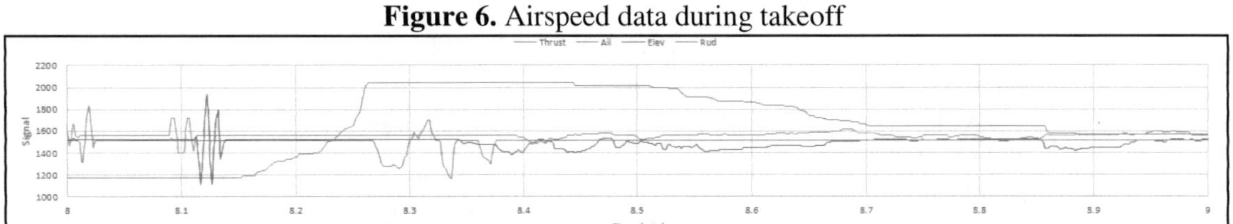

Figure 7. Control signal data during takeoff

Table 8. BL-1 Takeoff Characteristics

Name		Notes
Liftoff distance	114 meters	Ave takeoff speed of 9.5 m/s, for 0.2 minutes.
Lift off airspeed	20 m/s	Height 1.88 m, 8.4 minutes.
Lift off angle of attack	2^0	Height 1.88 m, 8.4 minutes
Takeoff distance	2.0 meters	15 feet, at 8.45 minutes

6.3 Climbing

The behavior and performance of the aircraft during climbing will be analyzed in this section. The performances analyzed including climb rate, climb speed, and climb angle.

According to FDR data, the aircraft started climbing at 8.4 minutes, with average vertical speed of 2.5 m/s and maximum climb rate of 4.2 m/s, until reaching heights of 180 meters at minutes 10.0. The aircraft was climbing with a throttle of 90% at first, and after 9 seconds, the throttle was decreased to 50% for the reminder of the climb. The climb took 1.6 minutes to occur. The data from FDR shown that the aircraft only have roughly 30 seconds total of straight climbing (at 8.7-8.8 and 9.3-9.7 minutes), while the other climb occurred while banking/turning. Table 9 provided the summary climb characteristics obtained from FDR data.

Table 9. BL-1 Climbing Characteristics

Name		Notes
Max RoC	4.2 m/s	90% Throttle, while banking 20-30 degrees at 8.5-8.65 minutes.
RoC	2.5 m/s	50% Throttle, averaged at 8.7-8.8 and 9.3-9.7 minutes.
Climb Angle of Attack	0^0	
Climb Angle	6^0	
Climb Speed	17.8 m/s	

6.4 Cruising

The behavior and performance of the aircraft during cruising analyzed in this section. The performance analyzed is the aircraft cruise speed and cruise angle. Also will be determined the trim used during cruise.

From the FDR data, the aircraft's straight level cruise occurred at 10.1-10.3 minutes, 11.7-11.8 minutes, and 13.0-13.3 minutes. The aircraft performed two fly passes at 14.9-15.2 and 16.20-16.5 minutes. Also, from Table 7, at 15.34 GMT+7 or 10.7-10.9 minutes, the pilot was performing a trimmed and handoff flight for 12 seconds. Throttle used are 50 % during cruise at 10.1-10.3 and 11.7-11.8 minutes, 40% during cruise at 13.0-13.3 minutes, and 30% during fly passes. Handoff flight occurred at 10.7-10.9 minutes. Average speed of 18 m/s, alpha -0.5^0, at 180 m. Noted that during handoff flight, while there is no roll, the aircraft are slowly turning from heading 300 to 330. Table 10 provide the summary of cruise characteristics obtained from FDR data. The trim settings of the aircraft obtained from the remote control was shown in Table 11.

Table 10. BL-1 Cruise Characteristic

Name		Notes
Cruise Speed (50%)	23.4 m/s	Averaged at 10.1-10.3 and 11.7-11.8 minutes
Ave Alpha (50%)	-0.7^0	
Altitude	180 m	
Cruise Speed (40%)	19.7 m/s	Averaged at 13.0-13.3 minutes.
Ave Alpha (40%)	-0.5^0	
Altitude	140 m	
Cruise Speed (30%)	20 m/s	Averaged at 14.9-15.2 and 16.2-16.5 minutes.
Ave Alpha (30%)	-0.5^0	
Altitude	18 m	

Table 11. BL-1 Trim Data

Trim	Trim	Signal	Deflection
Aileron	32 %	1556	10^0
Elevator	56 %	1529	-2^0
Rudder	8 %	1507	-4^0
Throttle	50 %	1520	-

6.5 Landing

The behavior and performance of the aircraft during landing will be analyzed in this section using data from FDR and video recorded during landing. The behavior analyzed including the attitude of the aircraft during landing. The cause of hard landing will also be analyzed.

This paragraph will analyze the behavior of BL-1 during landing according to Figure 8 to Figure 11. The aircraft started preparing for landing after the second fly pass, turning 360 degrees before flying parallel to the runway at 17.15 minutes, at a height of 17 meters above the runway. During approach, the pilot reset the trim of the aircraft to compensate moment induced by decreasing throttle. Once the aircraft is parallel to the runway, the throttle was decreased from 30% to 20%. The pilot was not able to trim the aircraft, indicated by fluctuating pitch angle and elevator deflection in FDR readings. At 17.35 minutes, the aircraft had experienced a rather high angle of attack at 12 degrees, and low speed at only 15 m/s, at a height of only 7 meters from the ground. Maybe, detecting this, the pilot decided to increase the throttle to 100% to nose down the aircraft, while instantly counter by deflecting the elevator down. At this point, the pilot was trimming the aircraft using both engine throttle and elevator deflection. At, 17.38 minutes, the aircraft experienced one high angle of attack again at 12 degrees, low speed of only 16 m/s, at an height of only 4.5 meters from the ground. At 17.40, detecting that the aircraft is pitching down again, the pilot lowered the throttle to 0% and then at 17.42 increasing the throttle to 50% as he sensed the aircraft to pitch up again, while deflecting the elevator up to counter it. At this point, the aircraft have height at 2 meters above the ground, at the speed of 16.8 m/s. At 17.43, the pilot increased the throttle to 100% and deflecting the elevator at maximum negatives, because the aircraft was pitching too down. This is not helping though as in 17.44, the aircraft pitching up very abruptly. After 17.45 minutes, weird things happened. The aircraft started the change heading 180 degrees, experiencing an alpha of maximum 12 degrees at 17.452, at 17.46 the pilot started the deflect the rudder left to counter the spin with no avail, touch the ground at 17.48, and then stopped spinning and moving at 17.5.

IOP Conf. Series: Journal of Physics: Conf. Series **1005** (2018) 012031 doi:10.1088/1742-6596/1005/1/012031

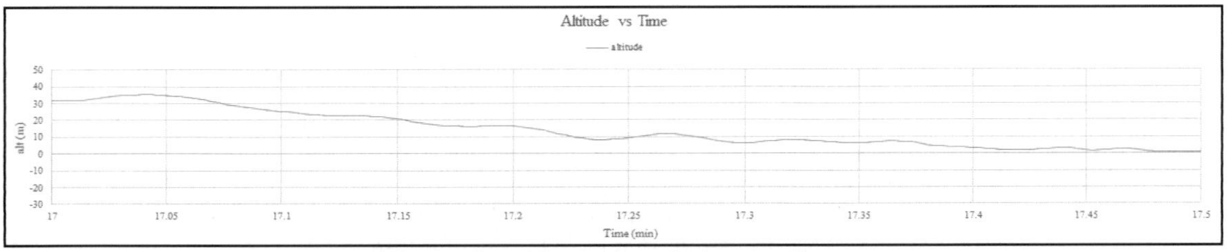

Figure 8. Altitude data during landing.

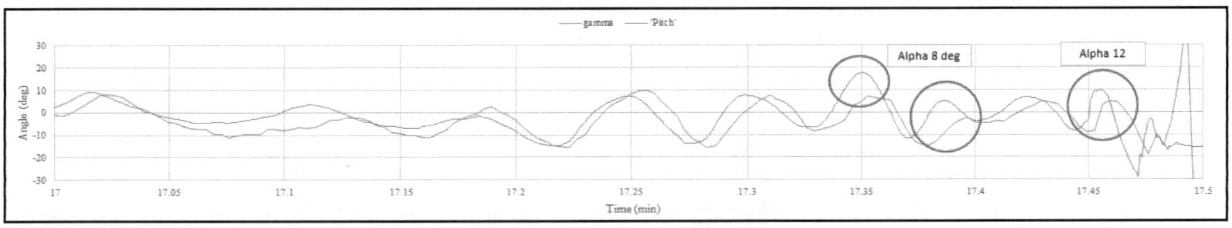

Figure 9. Pitch and gamma during takeoff

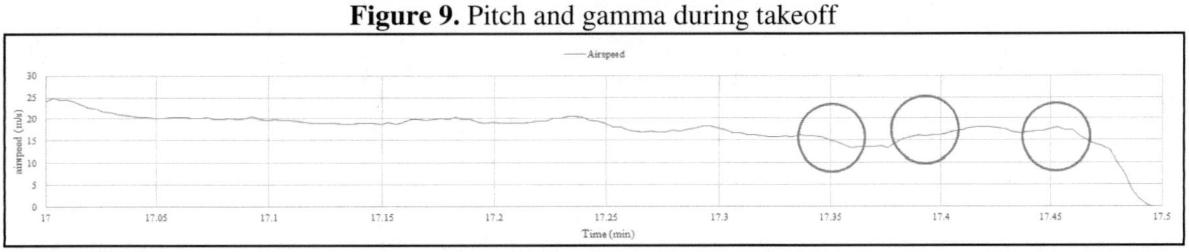

Figure 10. Airspeed data during takeoff

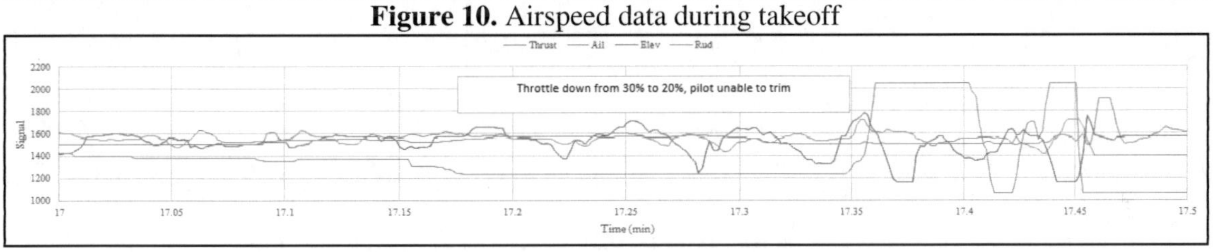

Figure 11. Control signal data during takeoff

6.6 Aerodynamics Characteristics

The aerodynamics characteristics of the aircraft are calculated only at straight and level condition, thus at condition where roll is zero, the heading is constant, with constant speed and altitude. However, the condition was not easily obtainable. Therefore, any condition that acceptable to represent straight and level condition will be used for analysis, as seen in Figure 12. Also, the thrust is assumed to be negligible. As consequences, only the lift characteristics could be analyzed.

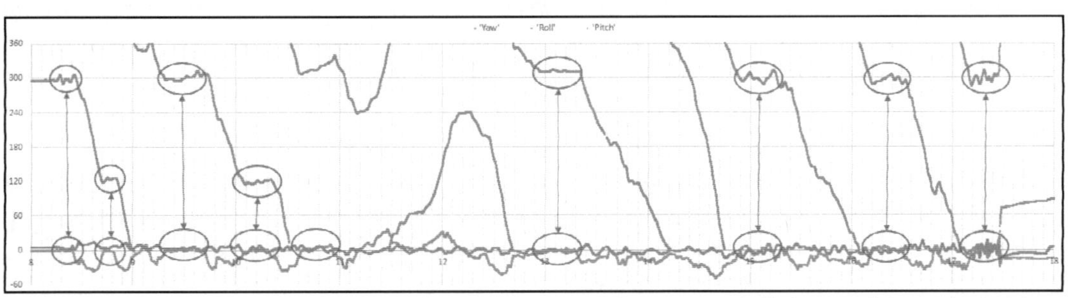

Figure 12. Several data collection at which close to straight and level flight condition (circled in red)

According to Appendix in reference [9], coefficient of lift obtained from the equation:

$$C_L = \frac{W/S}{\frac{1}{2} \times V^2 \times \rho}$$

The density of air is assumed to be equal to density at 800 m ASL ISA+15 (ρ = 1.07 kg/m^3)

The mass of aircraft is decrease during flight due to fuel consumption. However, because there is no data on fuel consumption, the 3 liters of fuel is assumed to be constantly decreasing on the 20 minutes flight until empty. Fuel density assumed 0.8 kg/liters. At the start of the second flight, the fuel is assumed to be used 12 minutes or equal to 1.4 kg, so the weight of the aircraft at the second flight was 52.6 kg. With the rate of 1kg/minutes, the weight of the aircraft during the second flight is calculated as follows:

$$m_{aircraft}(t) = (52.6 - 1 \times (t - 8)/10)\ kg$$

However, the characteristics are averaged from several data to minimize outlier results, as shown in Table 12.

Table 12. Example of averaged result

Time	**Airspeed**	**α**	**CL**	**Time**	**Airspeed**	**α**	**CL**
[minute]	[m/s]	[^{0}]	[-]	[minute]	[m/s]	[^{0}]	[-]
10,165	23,89	0	0,56656	17,34	16,09	11	1,23191
10,185	23,3	-5	0,5956	17,345	15,83	13	0,95452
10,195	23,55	-2	0,58301	17,35	15,07	15	0,84257
10,205	23,82	3	0,56985	17,355	13,83	7	0,83369
10,21	23,12	3	0,60488	17,36	13,19	0	1,83309
Average	23,536	-0,2	0,58398	**Average**	14,802	9,2	1,13915

Figure 13 shows comparison of aerodynamics characteristics obtained from BL-1 flight test, GL-1 DATCOM simulation [2], GL-1 CFD simulation [5], and GL-1 X-Plane simulation [6]. It could be seen that BL-1 lift vs slope coefficients follow the trend as resulted from simulations.

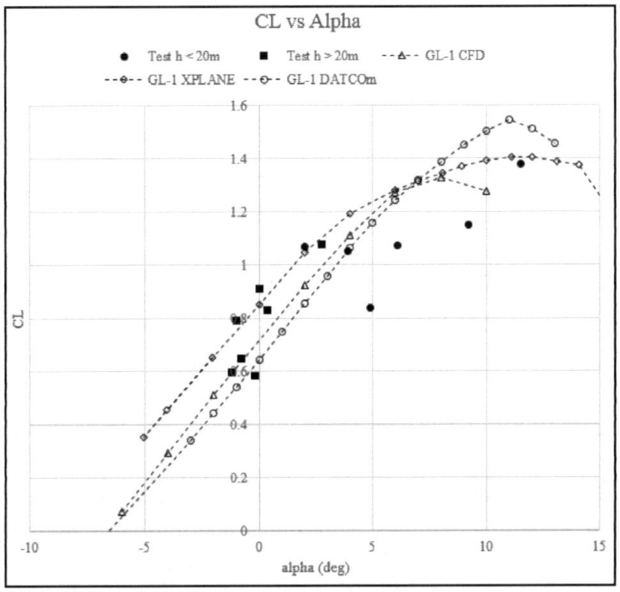

Figure 13. Aerodynamics characteristics of BL-1 obtained in test compared to several methods

7. Conclusion

7.1 Control Effectiveness

As seen in chapter 6, the control effectiveness of BL-1 had mixed results. Based on both FDR data and flight test report, BL-1 was trimmable at 30-50% throttle. However, at 20% and lower throttle, the pilot was unable to find the longitudinal trim. The pilot also stated that during the flight, the elevator deflection was being too sensitive. The aileron was capable of rolling/banking the aircraft throughout the flight, even at very low speed as displayed in figure to figure. The rudder effectiveness may need improvement. At the second takeoff, the FDR data show that rudder could control the aircraft at a low speed of 6 m/s. However, at landing, the rudder was not able to control the aircraft below 16-17 m/s. This could happen because when the rudder was used, the throttle while takeoff was higher than while landing. Higher throttle means that at takeoff, the rudder might have larger induced velocity. In addition, at takeoff, the rudder was deflected much larger than while landing. Those two factors may lead to higher effectiveness of the rudder while takeoff compared to when the aircraft was landing. Also, as it is a scaled down prototype, it may needed a higher velocity for the airfoil to be effective compared to the full-scale counterpart.

7.2 Aerodynamics Characteristics

The lift coefficient calculated from FDR data are very close to the estimation from several methods before, while the closest one being from X-Plane simulations. The aircraft had detected angle of attack as high as 12 degrees, while not stalling. However it seems like the aircraft had too large engine power, resulting in the rate of increase in pitch down moment induced by the engine are too large.

The thrust of the propulsion might have an effect on the aerodynamics characteristics, especially because it had large propulsion. It might lead to better calculation of the aircraft aerodynamics characteristics.

7.3 Design and Future Works

As the result of the second flight test, the design of GL-1 empennage may need to be reconsidered. The prototype may need to be tested without engine, to obtain aircraft characteristics closer to the full scale.

However, the behavior of half-scale and full-scale prototype may differ; thus, the full design may need to be further tested by flight simulations and full-scale prototype.

Acknowledgement
This research was partially supported by P3MI ITB research scheme. We are thankful to our colleagues: Mr. Djobaer Oemar Djaya and team, who provided expertise for helping in the preparation and execution of the flight test.

References

[1] Sulc Jan: *Conceptual Design of East Asia Sailplane: Design Final Report*, Institut Teknologi Bandung, Bandung, 2013.
[2] Pratama, Hendi: *Conceptual Design of Indonesia National Glider GL-01: Master Thesis*, Institut Teknologi Bandung, Bandung, 2015.
[3] Handojo, Vega: *Sailplane Performance Estimation: Student Project Report*, Institut Teknologi Bandung, Bandung, 2013.
[4] Azizah, N. A. D., Nurhakim, M. L. I., Arifianto, O., Mulyanto, T. Glider Uncentered Drag Polar Estimation Based on Flight Manual Data. *Advances in Aerospace Science and Technology in Indonesian Vol.I 2016, 2016.*
[5] Amalinadhi, C., Julistina, R.: GL-1 CFD Simulation Result. *GL-1 Development Archives, 2016.*
[6] Zulkarnain, M.F.: GL-1 X-Plane Simulation Result. *GL-1 Developments Archives, 2017.*
[7] Amalinadhi, C., Julistina, R., Moelyadi, M. A., Mulyanto, T. Comparative Study Between Schrenk and CFD Analysis for Predicting Lift Distribution Along Wing Span of Glider Aircraft. *Advances in Aerospace Science and Technology in Indonesian Vol.I 2016, 2016.*
[8] Zulkarnain, M.F.: BL-1 Test Flight Plan. *GL-1 Developments Archives, 2017.*
[9] Pajno, Vittorio: *Sailplane Design – A Guide for Students and Designers. From Drafting to Flight Test*, IBN Editore, Roma, 2010.

IOP Conf. Series: Journal of Physics: Conf. Series **1005** (2018) 012032 doi:10.1088/1742-6596/1005/1/012032

Wing configuration on Wind Tunnel Testing of an Unmanned Aircraft Vehicle

Yanto Daryanto[1*)], Joko Purwono[2)] and Subagyo [1)]

[1)] National Laboratory for Aerodynamics Aero-elastics and Aero-acoustics, Agency for the Assessment and Application of Technology (BPPT), Indonesia
[2)] Center Of Technology For Defense And Security Industries, Agency for the Assessment and Application of Technology (BPPT), Indonesia

*yanto.daryanto@bppt.go.id

Abstract. Control surface of an Unmanned Aircraft Vehicle (UAV) consists of flap, aileron, spoiler, rudder, and elevator. Every control surface has its own special functionality. Some particular configurations in the flight mission often depend on the wing configuration. Configuration wing within flap deflection for takeoff setting deflection of flap 20° but during landing deflection of flap set on the value 40°. The aim of this research is to get the ultimate CL_{max} for take-off flap deflection setting. It is shown from Wind Tunnel Testing result that the 20° flap deflection gives optimum CL_{max} with moderate drag coefficient. The results of Wind Tunnel Testing representing by graphic plots show good performance as well as the stability of UAV.

1. Introduction

The research development of new aircraft design for both manned and unmanned aircraft using wind tunnel test is a usefull method to get the desired design. This paper discusses the result of wing configuration test of Unmanned Aircraft Vehicle (UAV). It is common that before an UAV's prototype come to manufacture process, it should be known the aerodynamic characteristics, stability and performance provided an estimate aerodynamic characteristics to be done in the design phase simulation. However wind tunnel testing should be done to confirm the result of simulation, especially on the high Angle of Attack (AOA). An aircraft models installed in the wind tunnel which is connected to the external balance by the wing struts was carried out. An aerodynamic forces propelled by wind blows in subsonic regime at a speed of 65 meter per second was measured by the external balance.

A Wind Tunnel Testing (WTT) facility was used to do the test. This facility is parts of National Laboratory for Aerodynamics Aero-elastics and Aero-acoustics (B2TA3), The Agency for the Assessment and Application of Technology (BPPT). The tasks of National Laboratory for Aerodynamics Aero-elastics and Aero-acoustics provide aerodynamic testing services to both the customers from domestic and abroad. By having a 3 x 4 m2 test section, both complete model and component of the aircraft such as two dimensional wing model, half model, tail performance and air intake test can be tested at B2TA3.

The performance of the aircraft on take-off conditions is influenced by several factors, such as aircraft weight, wing design, and control suface configuration. The Control surface of the UAV are generally consist of slat, flap, aileron, elevator, rudder and tail. Among of them it could be combined and use for some missions. Combination of main wing, slat and flap generally which are called by High Lift Devices (HLD) are very useful for to control an UAV or of all type aircraft on generating lift. HLD also will give an enormous influence on how long the distance takeoff and landing of an aircraft. Furthermore the diagram of component aircraft generally can be shown such as in the figure 1[1].

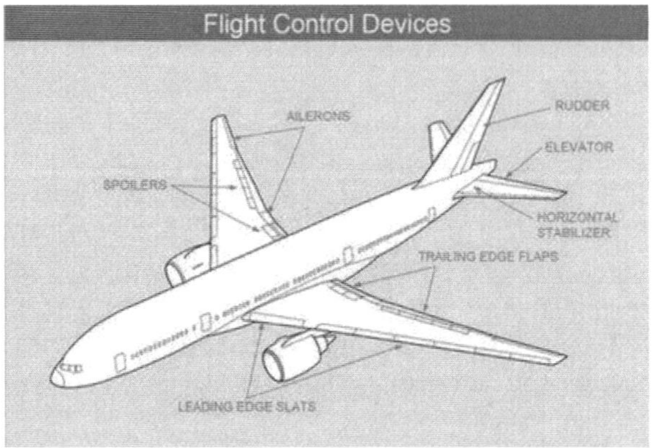

Figure 1. Diagram component of the aircraft

2. Wind Tunnel Testing

Aircraft flight mission are generally contain of take-off, ascending, cruise, decending and landing [2] and those mission were done for both commercial aircraft and UAV. Takeoff and landing are a cycle aspect during flight and to do safely it needs enough lifting force and for landing situation it also needs drag force. Those aerodynamics characteristic could be known by putting an aircraft or UAV model to the wind tunnel testing (WTT). The WTT arrangement and result will be explained including some condition description in the following chapter;

2.1. Model Testing

The model aircraft tested, is a 1: 5 scale test model with 3022 mm wingspan. The sketch of model testing is show in the figure 2.

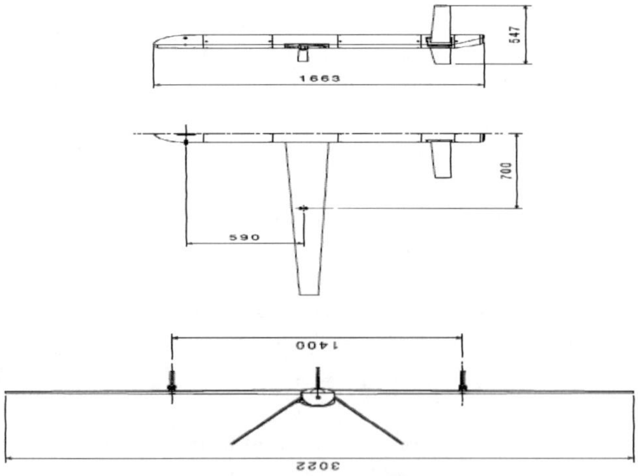

Figure 2. UAV model sketch

Flap of the model lied in the inner control surface. Flap deflection usually setting in the same direction for the right part and the left part to generate more lifting force of an aircraft. Here are the nomenclature for the part of the UAV model for WTT:

W : Wing
B : Body
T : Tail

2.2. Meassurement technic

Methods and techniques of measurement refers to [3],[4][7]. The test model installed using wing support strut that connects to the external balance. The position of the test model is upside down. Wind speed was 65 m / s. For testing the alpha polar, alpha angle is varied from - 12° to 20°, with interval data collection every 1°. As for the angle beta, test model driven from -20° to 20° angle beta, with interval data collection every 1°. Installed test model can be seen in Figure 3.

Figure 3. UAV model installed in the test section

External balance place in the top of test section. The diagram of external balance show in the figure 4. Each external balance component has the capacity and measuring direction show in the table 1.[7].

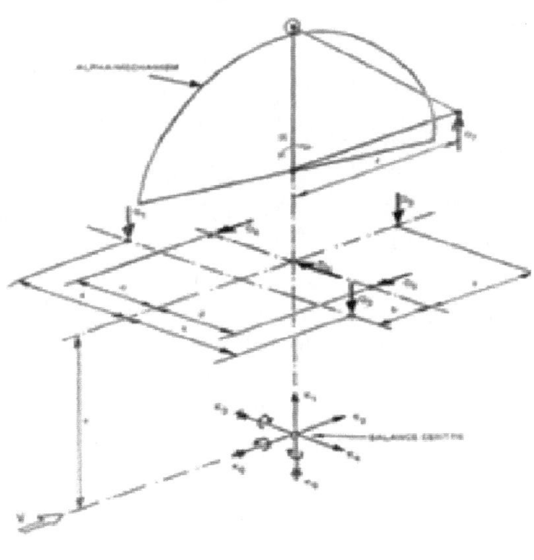

Figure 4. Diagram of external balance

IOP Conf. Series: Journal of Physics: Conf. Series **1005** (2018) 012032 doi:10.1088/1742-6596/1005/1/012032

Table 1. The maximun load each external balance component

No.	Component	Load Capacity	Accuracy(%)
1	*Lift (K1)*	17500 N	0.1 % FS
2	*Drag (K2)*	3500 N	0.1 % FS
3	*Pitching Moment (K3)*	3750 Nm	0.1 % FS
4	*SideForce(K4)*	3500 N	0.1 % FS
5	*Yawing Moment(K5)*	3250 Nm	0.1 % FS
6	*Rolling Moment (K6)*	3000 Nm	0.1 % FS

Furthermore the maximum allowable error calculate by the equation (1)[5]

$$E_i = \left(0.3 + 0.7\frac{P_i}{P_{imax}}\right)0.001fP_{imax} \quad (1)$$

Where *Ei* : The maximum allowable error.
Pi : The force or moment on the *Ki*-th component.
*Pi*max : The maximum capacity of the Ki-th component.
f : A factor, which depend on the force component.

2.3. Forces and Moments Aerodynamics
The measurement characteristic aerodynamic of UAV test model in the wind tunnel measure three forces components i.e. lifts (L = lift), Drag (D = drag), Side Force (SF = side forces) and three components moments are pitching moment (M_P = pitch moment), a rolling moment (M_R = rolling moment), and yawing moment (M_Y = yaw moment) [6],[7]. The six components of forces and moments can be formulated as follows:

Three component forces:

$$L = {}^1/_2\,\rho V^2 S C_L \quad (2)$$

$$D = {}^1/_2\,\rho V^2 S C_D \quad (3)$$

$$S_F = {}^1/_2\,\rho V^2 S C_Y \quad (4)$$

Three component moments:

$$M_P = {}^1/_2\,\rho V^2 S C_M \quad (5)$$

$$M_R = {}^1/_2\,\rho V^2 S C_{Roll} \quad (6)$$

$$M_Y = {}^1/_2\,\rho V^2 S C_{Yaw} \quad (7)$$

Where ρ: air density, V: free stream velocity, S: reference area, C_L: Lift Coefficient C_M: Pitch Moment Coefficient, C_D: Drag Coefficient, C_Y: Side Force Coefficient, C_{Roll}: Rolling Moment Coefficient, C_{Yaw}: Yawing Moment Coefficient.

All of the aerodynamic force and moment measured by load cell corelated with the appropriate position in the external balance. Offset small geometry and strain and friction gives the deviation coefficient of the actual value. Therefore the forces and moments obtained from the measurement results, in software multiplied by the coefficient matrix external balance calibration. The relationship between the aerodynamic forces and moments in units of measurement and raw data matrix shown in the following equation:

$$[K_i] = [a_{ij}] \cdot [R_i] \quad (8)$$

Where: K_i : Forces or moments
a_{ij} : External Balance Calibration Coefficient
R_i : Raw data

Forces and moments of the measurement results are normalized to the dynamic pressure, reference area of wings, and the mean aerodynamic chord. Agreements signs and directions to all the coefficients shown in Figure 5 where CL, CD, CY, C_M, C_{Roll}, C_{Yaw} are the coefficient of lift, drag, side force, moment of pitch, rolling moment, and the yawing moment respectively.

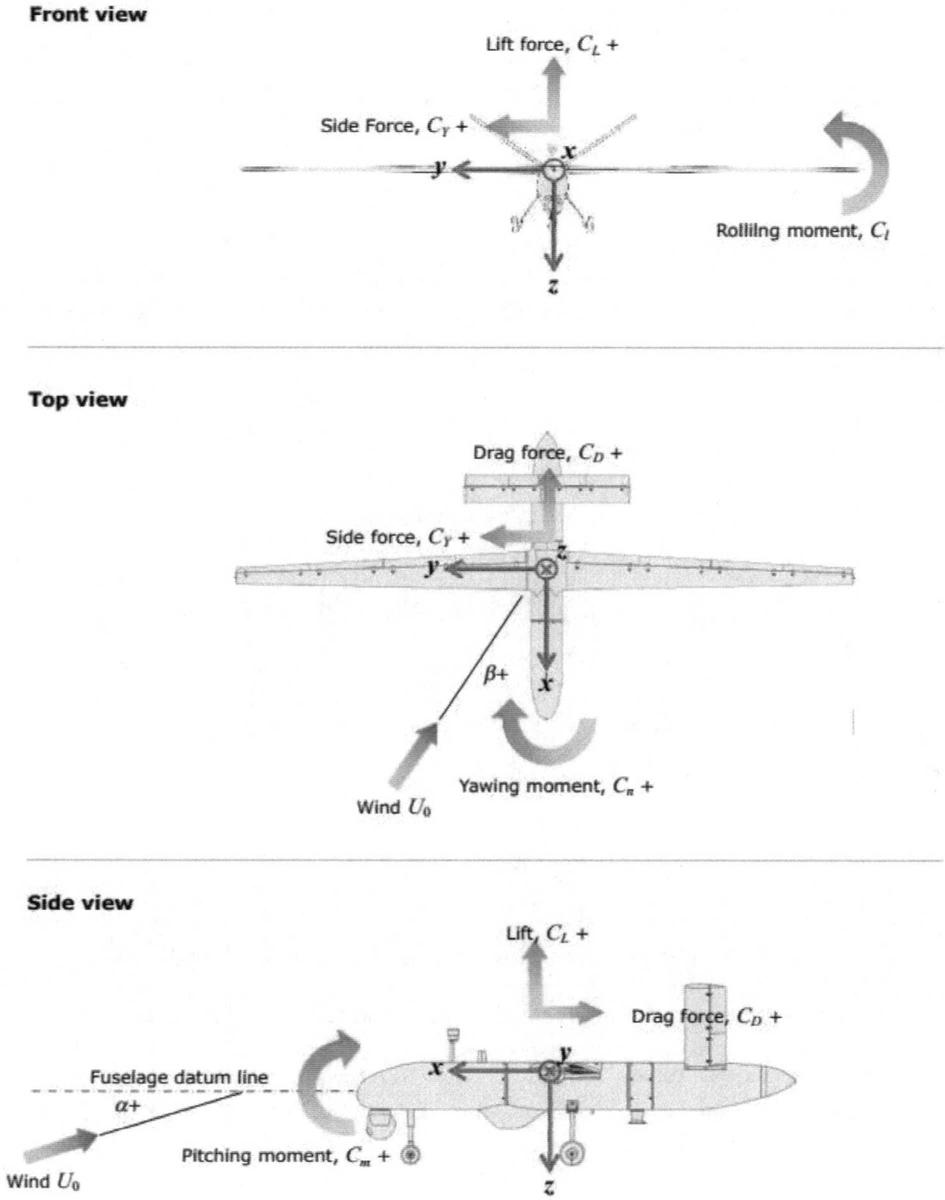

Figure 5. Convention magnitude**,** Sign and direction forces and moments

3. Results and Discussions

During the WTT, some wing configurations combining with main wing flap representing control surface were observed. The flap configurations are written in table 2.

Table 2. Running list Flap investigation

No	RUN	configuration	V(m/s)	AOA	Beta	dFL	dFR	dAL	dAR	dCHL	dCHR
1	1	WBT	65	A	0	0	0	0	0	0	0
2	2	WBT	65	A	0	10	10	0	0	0	0
3	3	WBT	65	A	0	20	20	0	0	0	0
4	4	WBT	65	A	0	30	30	0	0	0	0
5	5	WBT	65	A	0	40	40	0	0	0	0

3.1. Flap effect on the aerodynamics

A series of setting up Flap deflection left (dFL) and right (dFR) are shown in the table 2. The effect of Flap on the wing configurations to lift aerodynamics coeffecient versus Angle of Attack (AOA) show in figure 6.

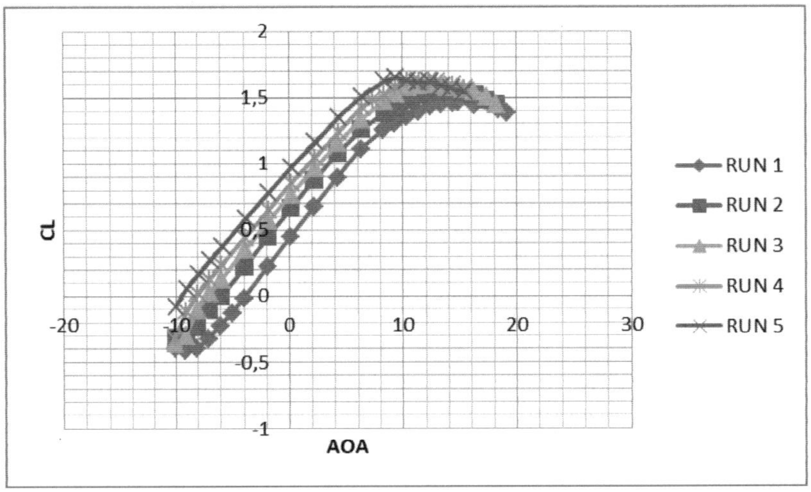

Figure 6. Effect Flap on wing configuration to the Lift coefficient

The Curves in figure 6 shows for different flap setting on wing configuration RUN 1, RUN 2 up to RUN 5 are gradually increase CLmax from 1.4652 up to 1.6538. The maximum lift (CLmax)=1.6538 occurred at RUN 5 which is set at dFL=dFR=40°, the maximum lift for take-off condition (CLmax)=1.6235 occurred at RUN 3 at the flap deflection about 20° and minimum lift (CLmax)=1.4652 occurred at RUN 1.

Figure 7 explained how the drag values change due to the change of flap deflection. Base on the force measurement graphic plot of drag coefficient shown in figure 7, setting Flap on the wing configuraton theoritically will change the frontal surface of UAV. Increasing frontal surface will conseqency change the drag of UAV. The maximum increasing of the drag UAV is about 90 drag account. From view of this result result are reasonable although the deference of flap on the wing configuration have difference purpose. Flap deflection 20° use for takeoff condition but flap setting angle 40° for landing condition. Even flap setting angle have highest CL but also have high value of drag.

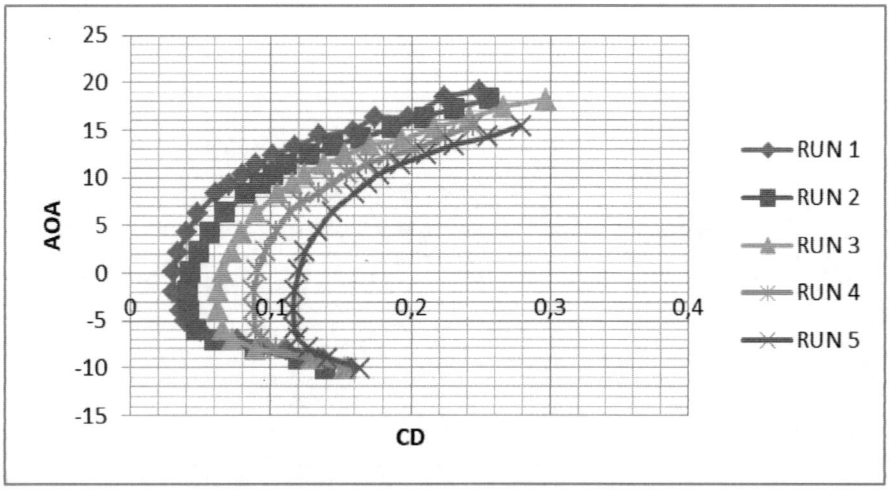

Figure 7. Effect Flap on wing configuration to the drag coefficient

Furthermore the pitching moment coefficient show in the figure 8. Generally the graphich plot of pitching moment coefficient versus AOA have negative slope that is indicate UAV will be stable during flight. However in some point AOA the slope of graphich plot have positive value. Maybe in this case need the dynamics analysis or more information in the flight test.

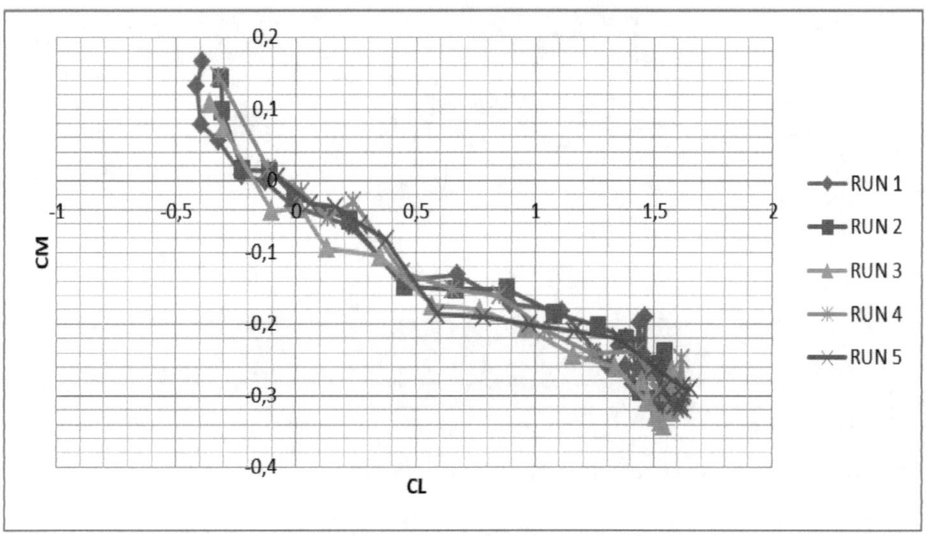

Figure 8. Effect Flap on wing configuration to the pitching moment coefficient

Figure 9 show the graphich plot of side force coefficient CY versus AOA. RUN 1 is the basic condition without setting flap deflection. The value CY of RUN 1 can be seen for zero offset condition. RUN 1 up to RUN 5 have value of CY relatively small varies between -0.0031 up to 0.0091 so we can say side force all of flap set up deflection have CY near zero. Base on this result flap deflection have small effect on the side force so we could not use flap deflection as control surface for directional of UAV.

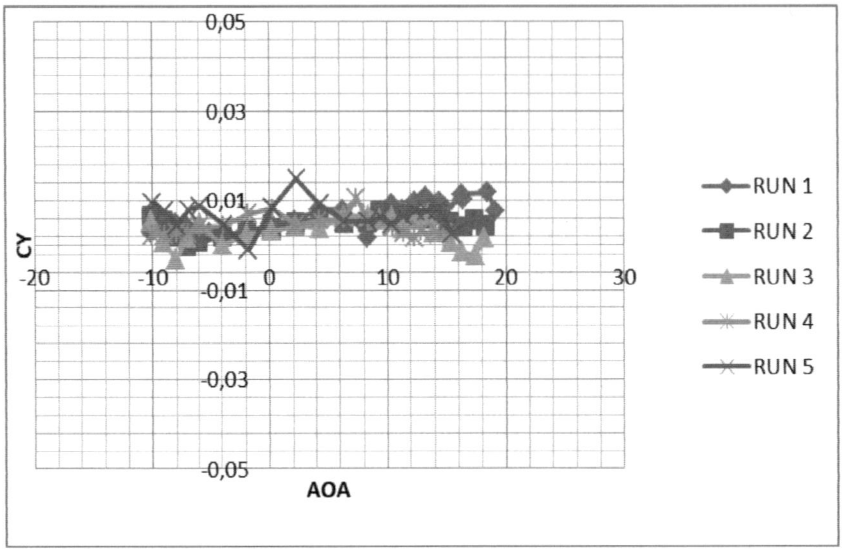

Figure 9. Effect Flap on wing configuration to the side force coefficient

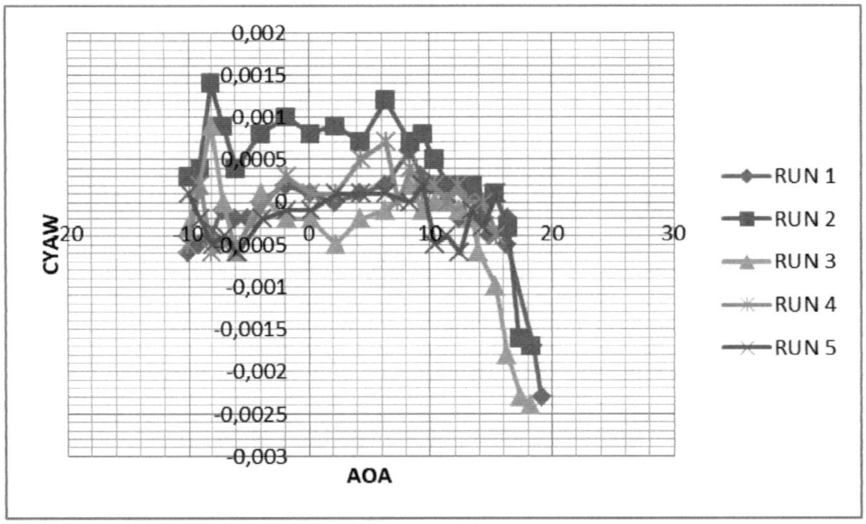

Figure 10. Effect Flap on wing configuration to the yawing moment coefficient

The graphich plot of yawing moment coefficient show in the figure 10. The value yawing moment coefficient as well as side force also very closed to zero. Let us see RUN 1 is the basic condition without setting flap deflection. The value CYAW of RUN 1 can be seen for zero offset condition. Take the difference relative value for RUN 2 up to RUN 5 to RUN 1 we will find the difference value close to zero. These mean flap deflection have the small effect to yawing moment CYAW.

Finally effect of flap setting on the rolling moment coefficient show in the figure 11. Here, RUN 1 is basic condition without setting flap deflection. The value CRoll of RUN 1 can be seen for zero offset condition. Take the difference relative value for RUN 2 up to RUN 5 to RUN 1 we will find the difference value close to zero. These mean flap deflection also have the small effect to rolling moment CROLL. This is show the flap could not use to make UAV rolling to the right or left direction.

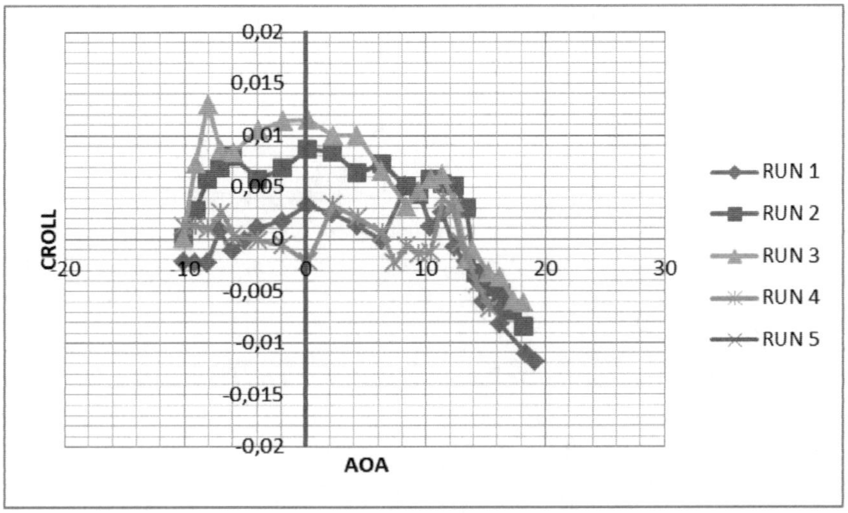

Figure 11. Effect Flap on wing configuration to the rolling moment coefficient

3.2. Performance prediction of UAV

UAV performance base on the characteristic aerodynamics can be predict. Performance of the UAV consist of efficiency, induce drag and endurance. Efficiency of the UAV represent by the ratio of CL/CD. More higher ratio of CL/CD that mean UAV can generate optimal lift with lowest drag. Graphic plot of CL/CD versus AOA show in the figure 12. The maximum value of ratio CL/CD=23 occured at AOA about 6°~7° for RUN 1 with flap deflection setting equal zero. The RUN 4 the bigest flap deflection have lowest value of CL/CD because RUN 4 have highest value of CD. Between the curve CL/CD versus AOA for RUN 1 and RUN 4 there are RUN 2, RUN 3and RUN 5. All of curve have maximum value CL/CD at AOA about 6°~7°.

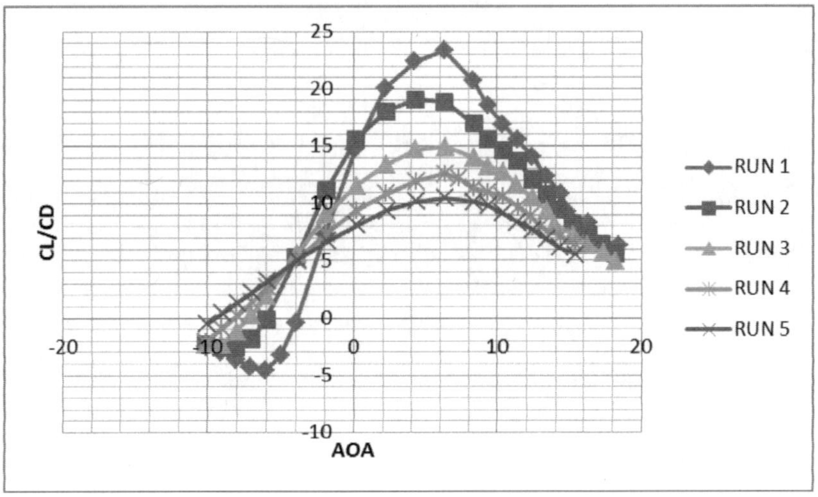

Figure 12. Efficiency of the UAV

Induced drag in term aerodynamics proportional to the value CL^2. Graphic plot of induced drag can be seen in the figure 13. The value of CL^2 versus AOA of all configuration as well as the value of CL are have the maximum value. Figure 13 show the maximum value CL^2 occurred at AOA about 6°~7°.

Induced drag varies depend on the AOA. The increasing of AOA should be inreased the value induced drag.

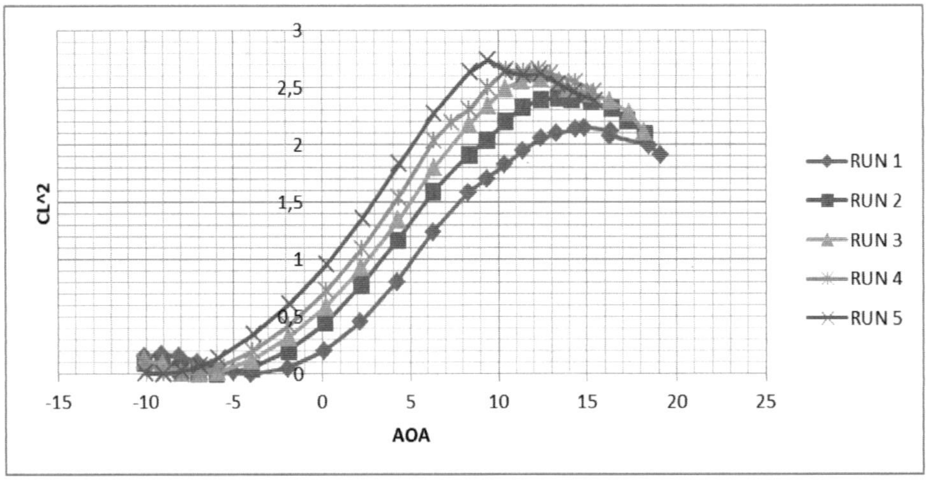

Figure 13. Induced drag proportional to the value CL^2

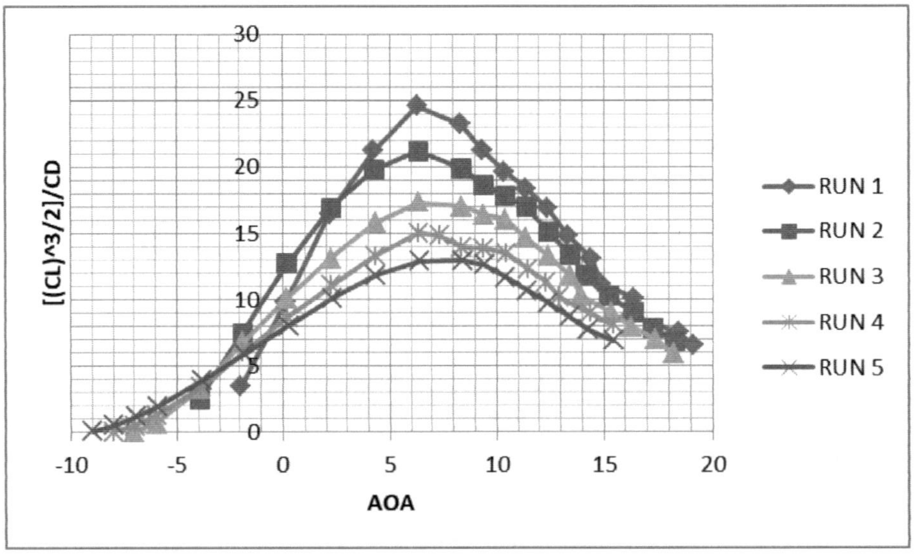

Figure 14. Endurance representation of UAV

Figure 14 show graphic plot endurance representation of UAV. The value of $\{(CL)^{3/2}/\ CD\}$ represent of endurance flight of UAV. All of RUN configuration have maximal value $\{(CL)^{3/2}/\ CD\}$ at the AOA around 6°~7°. It is means UAV will be had longest endurance if UAV flight by that conditions. The highest value of $\{(CL)^{3/2}/\ CD\}$ 25 up to the lowest 13 due to the setting flap deflection from 0° up to 40 ° respectively.

IOP Conf. Series: Journal of Physics: Conf. Series **1005** (2018) 012032 doi:10.1088/1742-6596/1005/1/012032

4. Conclusions

Flight mission of an UAV or aircraft should be supported by control surface devices, such as flap, aileron, rudder, elevator, some equipment and instrumentation. In this study performance and stability of flap settig angle deflection observed by WTT. Base on the meassurement result we can conclude in the folllowing below;

a) Setting flap deflection will contribute lift force of UAV represented by Lift Coefficient.
b) Setting angle deflection of flap 0° have minimal CL but with drag minimal. In this setting good for cruise flight because ratio of CL/CD have maximal value.
c) Setting angle deflection of flap 20° have moderate CL about 1.6235 also with moderate drag. In this setting good for takeoff and cllimb flight.
d) Setting angle deflection of flap 40° have biggest value CL about 1.65 compare with other setting, however this setting have highest value of Drag CD.
e) On the flight mission need high lift and drag force especially at the landing situation so setting angle of flap 40° are the apropriate condition.
f) During operation with flap deflection do not change side force coefficient, Pitching Moment Coefficient, Yawing Moment Coefficient and Rolling Moment Coefficient of UAV significantly.
g) There are Drag Coefficient increasing on the operation of Flap control surface because of frontal surface area of UAV changed.

Acknowledgement

This work is supported by Defense and Security Department on the development drone program. The authors would like to thank director of National Laboratory for Aerodynamics Aero-elastics and Aero-acoustics, Agency for the Assessment and Application of Technology (BPPT) for providing research support facilities.

References

[1] http://ivanhamid.blogspot.co.id/2013/09/jenis-dan-bagian-pesawat-terbang.html
[2] Mohammad H. Sadraey. Aircraft Design: A Systems Engineering Approach, 1st edition, *John Wiley & Sons, 2013.*
[3] Jewel W. Barlow, William H. Rae, Alan Pope. Low speed wind tunnel testing, *John Wiley & Sons, 3nd ed.,1984.*
[4] De Vries, O. Equations for the Data Processing ILST, NLR TR 87122L, *NLR. Amsterdam, 1987.*
[5] NN, ILST External Balance Manual Book, Carl Schenck AG.
[6] Hwankee Cho, Suntae Lee, Cheolheui Han. Experimental study on the aerodynamic characteristics of a fighter-type aircraft model in close formation flight. *Journal of Mechanical Science and Technology, 2014.*
[7] G.Wijiatmoko, Y.Daryanto, “Pengujian Aerodinamika Model Uji Pesawat Udara Nir Awak dengan Empennage berjenis V-Tail.”, Proceeding AMTeQ, 2016.

IOP Conf. Series: Journal of Physics: Conf. Series **1005** (2018) 012033 doi:10.1088/1742-6596/1005/1/012033

Grain Propellant Optimization Using Real Code Genetic Algorithm (RCGA)

Muhammad Farraz Al Farizi[1)], Romie Oktovianus Bura[2)], Soleh Fajar Junjunan [3)] and Bagus H. Jihad [4)]

1,2) Flight Physics Research Group, Faculty of Mechanical and Aerospace Engineering, Institut Teknologi Bandung, Jl. Ganesha 10 Bandung 40132, Indonesia
3,4) Rocket Technology Center, National Institute of Aeronautics and Space (LAPAN), Indonesia

*Muhammad.farraz93@gmail.com

Abstract. Grain propellant design is important in rocket motor design. The total impulse and ISP of the rocket motor is influenced by the grain propellant design. One way to get a grain propellant shape that generates the maximum total impulse value is to use the Real Code Genetic Algorithm (RCGA) method. In this paper RCGA is applied to star grain Rx-450. To find burn area of propellant used analytical method. While the combustion chamber pressures are sought with zero-dimensional equations. The optimization result can reach the desired target and increase the total impulse value by 3.3% from the initial design of Rx-450.

1. Introduction

One of the parameters of the rocket motor is the total impulse. To get the maximum impulse required a good grain propellant design that is grain propellant has a little sliver, high volumetric loading, and high combustion chamber pressure. To get it is required iteration process. The iteration process is iterate the variable independent of propellant (star grain has 7 independent variables, there are r1, r2, ω, η, ξ, N, R_0). Currently, Lapan rocket motor process design still use trial and error method (change the value of 7 independent variables manually), so it takes a long time in the design process and not sure to get optimal results. One method that can be used to speed up the design process is genetic algorithm. Moreover, the use of this method will produce an optimal design. Genetic algorithm (GA) method is widely used because Genetic algorithm (GA) can be applied to functions that have discontinuity (Robust), can capture global optimum and does not require initial design to do optimization process. In addition GA has a very high level of code simplicity so it is very easy to make the code. Raza and Liang (2010) compared Genetic algorithms (GA), Simulated Annealing (SA), and Combined GA and SA for dual thrust propellant optimization. From the results, they found that Genetic algorithm (GA) produces the largest results, while the method of hybrid GA and SA can reduce the computation time. Then Kamran et al (2010) using Genetic algorithm (GA) to get optimal star grain results. Kamran use a drafting method to calculate changes in the area of burning by connecting software optimization with CAD software. For this research the change of the area of combustion is solved by analytic method, the advantage of analytical method to find the area of burning is the process is faster than drafting method, and disadvantage is not flexible (different type of grain, different equation).

Genetic algorithm (GA) is divided into two methode, binary code (Binary Code Genetic Algorithm) and real code (Real Code Genetic Algorithm). Real Code Genetic Algorithm (RCGA) is widely used because it has the advantage of computation time because there is no encoding and decoding process. Besides it, Real Code Genetic Algorithm can be applied to large domain when compared with Binary code Genetic algorithm (BCGA) because the use of domain Large on the Binary code Genetic algorithm will reduce precision if the chromosome length is limited [3]. One problems of this method is difficult to obtain integers during the mutation and crossover process. Fortunatelly, from the previous study [12] was obtained for the case of star grain value of N = 5,6,7 can produce total maximal impulse. So the difficulty of obtaining integers can be resolved by optimizing the value of N at outside optimization.

Rx-450 is onc of Lapan's sonda rocket. This rocket use star grain propellant with a length of 3700 mm. The star propellant is chosen because LAPAN is experienced in making it. In addition, star-shaped propellants can produce large areas of large initial fuel without compromising volumetric efficiency [10]. The large volumetric loading fraction that is owned by star-shaped propellants is 0.75-0.9 and the star propellant has a sliver of 5-10% [1]. Assuming volumetric loading fraction 0.8, sliver 2%, and not optimal throat nozzle affecting 2%, should propellant with existing propellant can produce total impulse equal to 1530 kNs.

$$I_{T\ Optimal} = (254\ s\ \times 0.96 \times 0.92)\ (0.00167 \frac{kg}{cm^3} \times 3.14 \times (21.6\ cm)^2 \times 370\ cm\ \times 0.8 \times 0.98)\ \left(9.81 \frac{m}{s^2}\right)\ 0.98$$

2. Method

2. 1. Optimization Method

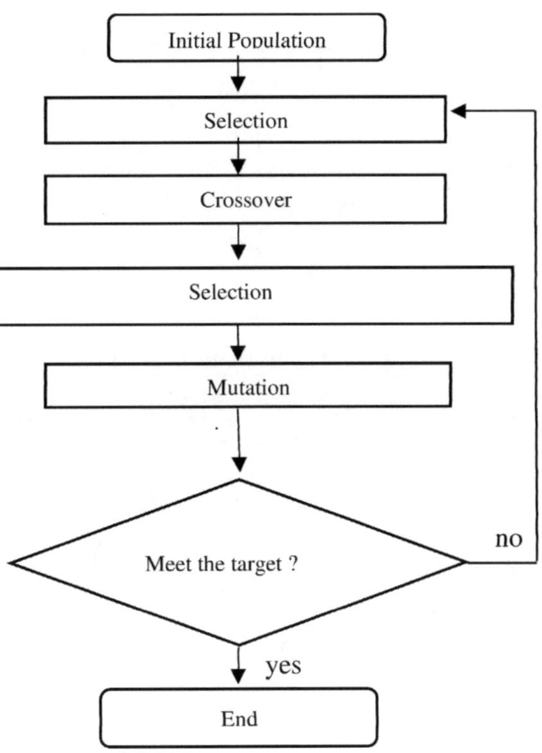

Figure 1 flow chart optimization

Table 1 Parameter Optimization

Max Fitness evaluation	10000
Population	100
Crossover probability	1
Mutation probability	1
Variable	5

The method used optimization is Real code genetic algorithm. But at research the crossover probability and Mutation probability equal one. The best chromosome always do crossover with other chromosome

For the case of grid propellant optimization BLX- α is chosen because this method produces a new population around chromosomes 1 and 2. While the mutation method used is Non-Uniform Mutation. This method is simple and has mutation characteristics occurring at the beginning of a temporary iteration when approaching the maximum limit of individual mutation iteration of mutation results is not large. The following is a crossover process using the BLX- α and Non uniform mutation methods [3].

Crossover

Crossover is an operator to find the optimum value (convergence operator), so crossover probability is set high (crossover probability> 0.95). In this research, crossover probability is not set, but the best population always crossover with other populations, in hopes to get better results from the best population.

Assume $C_1 = (c_1{}^1 \dots c_n{}^1)$ and $C_2 = (c_1{}^2 \dots c_n{}^2)$ is chromosome selected to crossover. An offspring is generated: $H = (h_1, \dots, h_i, \dots, h_n)$. Where h_i is a randomly (uniformly) chosen number of interval $[cmin - I\alpha, cmax + I\alpha]$. With $cmax = max(c_i{}^1, c_i{}^2)$ and $cmin = min(c_i{}^1, c_i{}^2)$ [3].

Mutation

Usually mutation probability is set low because the function of this mutation operator is to keep the population diversity (divergence operator). In this research, there is no mutation probability but mutation always happen at 10 worst population. C'_i is the result of mutation [3]

$$C'_i = \begin{matrix} c_i + \Delta(t, b_i - c_i) if\ \tau = 0 \\ c_i - \Delta(t, c_i - a_i) if\ \tau = 1 \end{matrix}$$

$$c_i \in [a_i, b_i], \tau\ is\ random\ numer\ 0\ or\ 1$$

$$\Delta(t, y) = y\left(1 - r^{\left(1-\frac{t}{gmax}\right)^b}\right),$$

$$t\ is\ generation\ number, gmax\ is\ \max number\ generation$$

2. 2. Grain Burn Back Analysis

The method used for the analysis of burning area of propellant is analytical method. The analytic method has advantages in fast computing time and simple code. One of the things that must be considered when using the analytical method for optimization propellant is the order of phase combustion. Star grain burning can be divided into 4 Phases. This is the analytic equations for star grain for each phase and the order of phase combustion:

Variabel star grain $= (R_0, N, r1, r2, \omega, \eta, \xi)$

$R_p = R_0 - r_1 - \omega$

$\text{Sudut} = 360 - ((180 - 180/N) + \eta + 90) - (180/N)$

$ymax = \left(\left(R_0 - R_p\cos(\xi)\right)^2 + (R_p\sin(\xi))^2 \right)^{0.5} - (r_1 + r_2)$

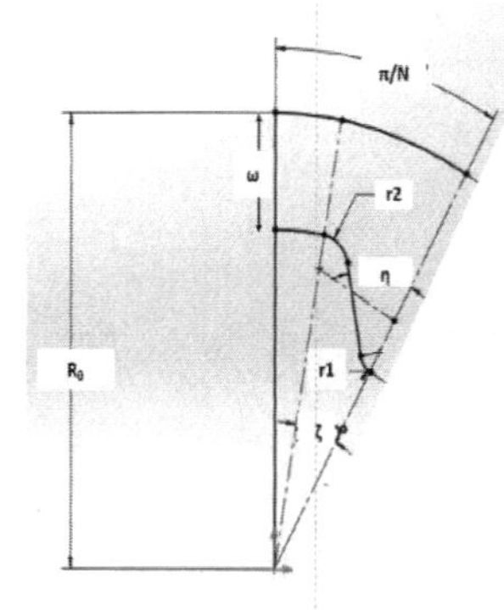

Figure 2. Star Grain Propellant [7]

Phase 1 (when $r_2 > 0$)

Burn Area (A_b)

$$S1 = \left(R_p\frac{\sin(\xi)}{\sin(\eta)}\right) - (y + r_1 + r_2) \times (\cot(\eta))$$

$$S2 = (\Delta y + r_1) \times \left(\left(\frac{\pi}{2}\right) - (\eta) + \xi\right)$$

$$S3 = \left(R_p + y + r_1\right) \times \left(\left(\frac{\pi}{N}\right) - \xi\right)$$

$$\mathbf{S4} = \mathbf{2\pi r_2}\left(\frac{\mathbf{sudut}}{\mathbf{360}}\right)$$

$$A_b = (2N \times (S1 + S2 + S3 + S4)) \times L$$

Port Area (A_p)

$$Ap1 = \left(0.5R_p\sin(\xi) \times \left(R_p\cos(\xi) + R_p\sin(\xi)\tan(\eta)\right)\right)$$

$$Ap2 = (-1) \times 0.5 \times \left(\left(\left(\frac{R_p \times \sin(\xi)}{\sin(\eta)}\right) - (y + r_1 + r_2) \times \cot(\eta)\right)^2\right) \times \tan(\eta)$$

$$AP3 = (0.5)(y + r_1)^2 \times \left(\left(\frac{\pi}{2}\right) - (\eta) + \xi\right)$$

$$AP4 = (0.5)\left(R_p + y + r_1\right)^2 \times \left(\left(\frac{\pi}{N}\right) - \xi\right)$$

$$AP5 = (S1 \times r_2) + (\pi \times (r_2)^2 \times (sudut/360))$$

$$A_p = (2N \times (AP1 + AP2 + AP3 + AP4 + AP5))$$

Furthermore there are two possible processes of burning propellant.

1. **S1 finished, after that S3 finished, after that S2 finished**

Phase 2 (when $r_2 = 0$, $S1 > 0, S3 > 0, S2 > 0$)

Burn Area

$$S1 = \left(R_p\frac{\sin(\xi)}{\sin(\eta)}\right) - (y + r_1) \times (\cot(\eta))$$

$$S2 = (y + r_1) \times \left(\left(\frac{\pi}{2}\right) - (\eta) + \xi\right)$$

$$S3 = \left(R_p + y + r_1\right) \times \left(\left(\frac{\pi}{N}\right) - \xi\right)$$

$$S4 = 0$$

$$A_b = (2N \times (S1 + S2 + S3 + S4)) \times L$$

Port Area

$$Ap1 = \left(0.5R_p\sin(\xi) \times \left(R_p\cos(\xi) + R_p\sin(\xi)\tan(\eta)\right)\right)$$

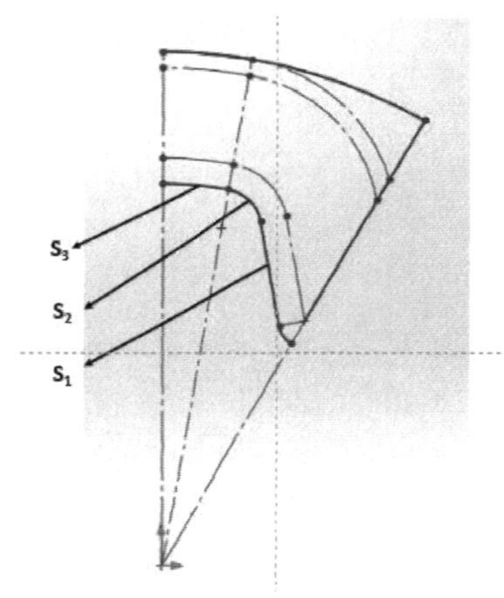

Figure 3. Case 1

$$Ap2 = (-1) \times 0.5 \times \left(\left(\left(\frac{R_p \times \sin(\xi)}{\sin(\eta)}\right) - (y + r_1) \times \cot(\eta)\right)^2\right) \times \tan(\eta)$$

$$AP3 = (0.5)(y + r_1)^2 \times \left(\left(\frac{\pi}{2}\right) - (\eta) + \xi\right)$$

$$AP4 = (0.5)\left(R_p + y + r_1\right)^2 \times \left(\left(\frac{\pi}{N}\right) - \xi\right)$$

$$AP5 = 0$$

$$A_p = (2N \times (AP1 + AP2 + AP3 + AP4 + AP5))$$

Phase 3 (when $r_2 = 0, S1 = 0, (y) < (\omega)\ [S3 > 0], S2 > 0$)

Burn Area

$$S1 = 0$$

$$S2 = (y + r_1) \times \left(\left(\frac{\pi}{2}\right) - (\eta) + \xi - \tan^{-1}\left(\frac{\sqrt{(y + r_1)^2 - \left(R_p \sin(\xi)\right)^2}}{R_p \sin(\xi)}\right)\right)$$

$$S3 = \left(R_p + y + r_1\right) \times \left(\left(\frac{\pi}{N}\right) - \xi\right)$$

$$S4 = 0$$

$$A_b = (2N \times (S1 + S2 + S3 + S4)) \times L$$

Port Area

$$Ap1 = \left(0.5 R_p \sin(\xi) \times \left(R_p \cos(\xi) + \sqrt{(y + r_1)^2 - \left(R_p \sin(\xi)\right)^2}\right)\right)$$

$$Ap2 = 0$$

$$AP3 = (0.5)(y + r_1)^2 \times \left(\left(\frac{\pi}{2}\right) + \xi - \tan^{-1}\left(\frac{\sqrt{(y + r_1)^2 - \left(R_p \sin(\xi)\right)^2}}{R_p \sin(\xi)}\right)\right)$$

$$AP4 = (0.5)\left(R_p + y + r_1\right)^2 \times \left(\left(\frac{\pi}{N}\right) - \xi\right)$$

$$AP5 = 0$$

$$A_p = (2N \times (AP1 + AP2 + AP3 + AP4 + AP5))$$

Phase 4 when $r_2 = 0, S1 = 0, (y) \geq (\omega)\ [S3 = 0], y < ymax$ [S2>0 (sliver)]

$$\beta = \left(\left(\frac{\pi}{2}\right) - (\eta) + \xi\right)$$

$$\gamma = \left(\frac{\sqrt{(y + r_1)^2 - \left(R_p \sin(\xi)\right)^2}}{R_p \sin(\xi)}\right) - \eta$$

$$\varepsilon = \pi - \cos^{-1}\left((-1) \times \left(\frac{R_0{}^2 - R_p{}^2 - (y + r_1)^2}{2R_p(y + r_1)}\right)\right)$$

$$\mu = \cos^{-1}\left(\frac{(y + r_1)}{R_0} \sin(\pi - \varepsilon)\right)$$

IOP Conf. Series: Journal of Physics: Conf. Series **1005** (2018) 012033 doi:10.1088/1742-6596/1005/1/012033

Burn Area

$S1 = 0$

$S2 = (y + r_1)(\beta - \gamma - \varepsilon)$

$S3 = 0$

$S4 = 0$

$A_b = (2N \times (S1 + S2 + S3 + S4)) \times L$

Port Area

$AP1 = (0.5)(R_0)^2 \times ((\frac{\pi}{N}) - \xi + \mu)$

$Ap2 = 0.5\,(y + r_1)^2(\beta - \gamma - \varepsilon)$

$$AP3 = 0.5\left(\left(R_p\sin(\xi)\right)\left(R_p\cos(\xi)\right) + \left(R_p\sin(\xi)\right)\left((y + r_1)^2 - \left(R_p\sin(\xi)\right)^2\right)^{0.5}\right)$$

$$AP4 = -0.5\left(\left(R_p\sin(\mu)\right)\left(R_p\cos(\mu)\right) + \left(R_p\sin(\mu)\right)\left((y + r_1)^2 - \left(R_p\sin(\mu)\right)^2\right)^{0.5}\right)$$

$A_p = (2N \times (AP1 + AP2 + AP3 + AP4))$

2. S3 finished, after that S1 finished, after that S2 finished

Phase 2 (when $r_2 = 0$, $S1 > 0, S3 > 0, S2 > 0$)

Burn Area

$$S1 = \left(R_p\frac{\sin(\xi)}{\sin(\eta)}\right) - (y + r_1) \times (\cot(\eta))$$

$$S2 = (y + r_1) \times \left(\left(\frac{\pi}{2}\right) - (\eta) + \xi\right)$$

$$S3 = \left(R_p + y + r_1\right) \times \left(\left(\frac{\pi}{N}\right) - \xi\right)$$

$S4 = 0$

$A_b = (2N \times (S1 + S2 + S3 + S4)) \times L$

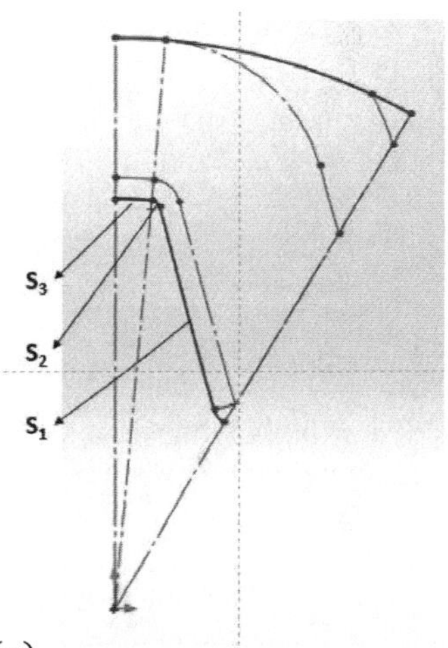

Figure 4. Case 2

Port Area

$$Ap1 = \left(0.5R_p\sin(\xi) \times \left(R_p\cos(\xi) + R_p\sin(\xi)\tan(\eta)\right)\right)$$

$$Ap2 = (-1) \times 0.5 \times \left(\left(\left(\frac{R_p \times \sin(\xi)}{\sin(\eta)}\right) - (y + r_1) \times \cot(\eta)\right)^2\right) \times \tan(\eta)$$

$$AP3 = (0.5)(y + r_1)^2 \times \left(\left(\frac{\pi}{2}\right) - (\eta) + \xi\right)$$

$$AP4 = (0.5)\left(R_p + y + r_1\right)^2 \times \left(\left(\frac{\pi}{N}\right) - \xi\right)$$

$AP5 = 0$

$A_p = (2N \times (AP1 + AP2 + AP3 + AP4 + AP5))$

Phase 3 (when $r_2 = 0, S1 > 0, S3 = 0, S2 > 0$) (Sliver)

$$\beta = \left(\left(\frac{\pi}{2}\right) - (\eta) + \xi\right)$$

$$\gamma = 0$$

$$\varepsilon = \pi - \cos^{-1}\left((-1) \times \left(\frac{R_0{}^2 - R_p{}^2 - (y + r_1)^2}{2R_p(y + r_1)}\right)\right)$$

$$\mu = \cos^{-1}\left(\frac{(y + r_1)}{R_0}\sin(\pi - \varepsilon)\right)$$

Burn Area

$S1 = \left(R_p \frac{\sin(\xi)}{\sin(\eta)}\right) - (y + r_1) \times (\cot(\eta))$

$S2 = (y + r_1)(\beta - 0 - \varepsilon)$

$S3 = 0$

$S4 = 0$

$A_b = (2N \times (S1 + S2 + S3 + S4)) \times L$

Port Area

$Ap1 = 0.5\left(R_p\sin(\xi) \times \left(R_p\cos(\xi) + R_p\sin(\xi)\tan(\eta)\right)\right)$

$Ap2 = (-1) \times 0.5 \times (\left(\left(\frac{R_p \times \sin(\xi)}{\sin(\eta)}\right) \times (y + r_1) \times \cot(\eta) \right)^2) \times \tan(\eta)$

$AP3 = (0.5)(R_0)^2 \times (\left(\frac{\pi}{N}\right) - \xi)$

$AP4 = 0.5 (R_0)^2 \mu - 0.5 \left(R_p\sin(\mu)\right)\left(R_p\cos(\mu)\right) - 0.5 \times \left(R_p\sin(\mu)\right) \times \left((y + r_1)^2 - \left(R_p\sin(\mu)\right)^2\right)^{0.5} + 0.5 (y + r_1)^2(\beta - \varepsilon)$

$AP5 = 0$

$A_p = (2N \times (AP1 + AP2 + AP3 + AP4 + AP5))$

Phase 4 when $r_2 = 0, S1 = 0, (y) \geq (\omega)$ $[S3 = 0], y < ymax$ [S2>0 (sliver)]

Same with case 1

$$\beta = \left(\left(\frac{\pi}{2}\right) - (\eta) + \xi\right)$$

$$\gamma = \left(\frac{\sqrt{(y + r_1)^2 - \left(R_p\sin(\xi)\right)^2}}{R_p\sin(\xi)}\right) - \eta$$

$$\varepsilon = \pi - cos^{-1}\left((-1) \times (\frac{{R_0}^2 - {R_p}^2 - (y + r_1)^2}{2R_p(y + r_1)}) \right)$$

$$\mu = cos^{-1}(\frac{(y + r_1)}{R_0} \sin(\pi - \varepsilon))$$

Burn Area

$S1 = 0$

$S2 = (y + r_1)(\beta - \gamma - \varepsilon)$

$S3 = 0$

$S4 = 0$

$A_b = (2N \times (S1 + S2 + S3 + S4)) \times L$

Port Area

$AP1 = (0.5)(R_0)^2 \times (\left(\frac{\pi}{N}\right) - \xi + \mu)$

$Ap2 = 0.5 (y + r_1)^2(\beta - \gamma - \varepsilon)$

$AP3 = 0.5 \left(\left(R_p\sin(\xi)\right)\left(R_p\cos(\xi)\right) + \left(R_p\sin(\xi)\right)\left((y + r_1)^2 - \left(R_p\sin(\xi)\right)^2\right)^{0.5}\right)$

$AP4 = -0.5 \left(\left(R_p\sin(\mu)\right)\left(R_p\cos(\mu)\right) + \left(R_p\sin(\mu)\right)\left((y + r_1)^2 - \left(R_p\sin(\mu)\right)^2\right)^{0.5}\right)$

$A_p = (2N \times (AP1 + AP2 + AP3 + AP4))$

IOP Conf. Series: Journal of Physics: Conf. Series **1005** (2018) 012033 doi:10.1088/1742-6596/1005/1/012033

2. 3. Fitness Function

The purpose of this optimization is to obtain the maximum total impulse value with the weight limit of propellant should not exceed 815 kg, $\frac{A_p}{A_t} \leq 0.65$, combustion chamber pressure ≤ 70 bar, $\frac{A_b}{A_p} \leq 190$ by optimizing independent variables of grain propellant (**Figure2**) . With the upper and lower limits of the geometry variable are as follows:
Diameter = 432 mm
N=6

40	≤	w	≤ 200
0.5	≤	r_2	≤ 20
0.5	≤	r_1	≤ 15
10°	≤	ζ	≤ 54°
10°	≤	η	≤ 60°

The diameter of propellant and N becomes input optimization so that the variable to be optimized is

$$X = (r1,\ r2,\ \omega,\ \eta,\ \xi)$$

From the optimization objective, mathematically objective function and fitness function can be written

$$F(x) = max(I_T)$$

With constraint:

$$\boldsymbol{Weight = G1(x) \leq 815\ kg}$$

$$\boldsymbol{Chamber\ pressure = G2(x) \leq 70\ bar}$$

$$\boldsymbol{\frac{A_p}{A_t} G3(x) \leq 0.65}$$

$$\boldsymbol{\frac{A_b}{A_p} = G4(x) \leq 190}$$

Where A_p : Port Area, A_b: Burn Area,A_t: Throat Area

Due to Rao's book, the optimization methods are used to find the minimum value, to be similar to the book, the objective function and the fitness function to be

$$F(x) = max(I_T) = min(-I_T)$$

From the optimization goal, the grain propellant optimization is a non-linear optimization with constraint. To simplify the case and code generation, this case will be forced into non-linear optimization without constraint by using the concept of punishment/penalty, so it is expected that when a chromosome selection process that results in value does not meet the constraint it will be eliminated. The penalty/punishment is:

$$If\ (P_c > 70\ bar\ \cup Berat > 815\ kg\ \cup \frac{A_p}{A_t} > 0.65\ \cup\ \frac{A_b}{A_p} > 190\ , I_T = -1 \times I_T$$

The punishment will cause the chromosome that does not meet the constraint has a fitness function value worse than the chromosome that meets the constraint. The value of the total impulse can be searched by solving the following equations

$$F(x) = min(-I_T) = min(-\int_0^{t_b} F\,dt) = min(-\int_0^{t_b} P_c\, A_t\, C_{factual}\, dt)$$

With $P_c\, A_t\, C_{factual}$ from equation:

$$C_{f\,actual} = \lambda\, \eta_f C_{f\,ideal}$$

$$C_{f\,ideal} = \sqrt{\frac{2\gamma^2}{\gamma-1}\left(\frac{2}{\gamma+1}\right)^{\frac{\gamma+1}{\gamma-1}}\left(1-\left(\frac{P_e}{P_c}\right)^{\frac{\gamma-1}{\gamma}}\right.} + \left(\frac{P_e-P_{amb}}{P_c}\right)\varepsilon$$

$$\frac{dP_c}{dt} = \frac{1}{v(t)}\left[RT_c\left(\rho_p A_b(t) a P_c{}^n - \frac{P_c A_t}{C^*}\right) - \frac{P_c dv}{dt}\right]$$

$$aP_c{}^n = r = \frac{dy}{dt}$$

With the value of the variable from the above equation with the value as follows:

$$a = 0.03467\ dan\ n = 0.3180\ (from\ strand\ burner), A_t = 176.25\ cm^2, \lambda = 0.9872\ , \varepsilon = 6.55\ ,$$
$$, \gamma = 1.19\, , P_{amb} = 1\ atm\, , \eta_f = 0.92, c^*{}_{actual} = 1490\frac{m}{s}$$

P_c: *Chamber Pressure,* P_e: Exit Pressure, P_{amb}: Ambient Pressure, C_f : Coefficient thrust, t_b: Burning time, $\lambda\, \eta_f$: Nozzle efficiency, η_c : Combustion efficiency, r: Burn rate, a: Coefficient burn rate (Saint Robert's burn rate law), n: Coefficient burn rate (Saint Robert's burn rate law), c^*: Characteristic velocity, γ: Specific heat ratio, ε: Expansion heat ratio

3. Result

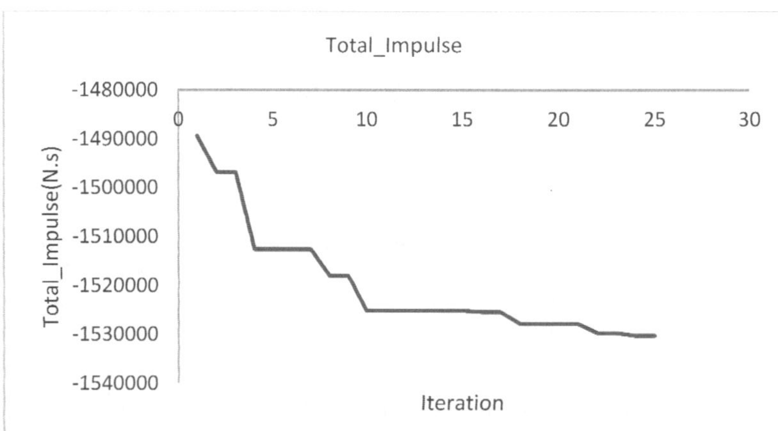

Figure 5 Iteration Process

Figure 5 Show that the iteration process leads to a convergent value. From optimization is obtained graph of thrust as shown at **Figure 6**.

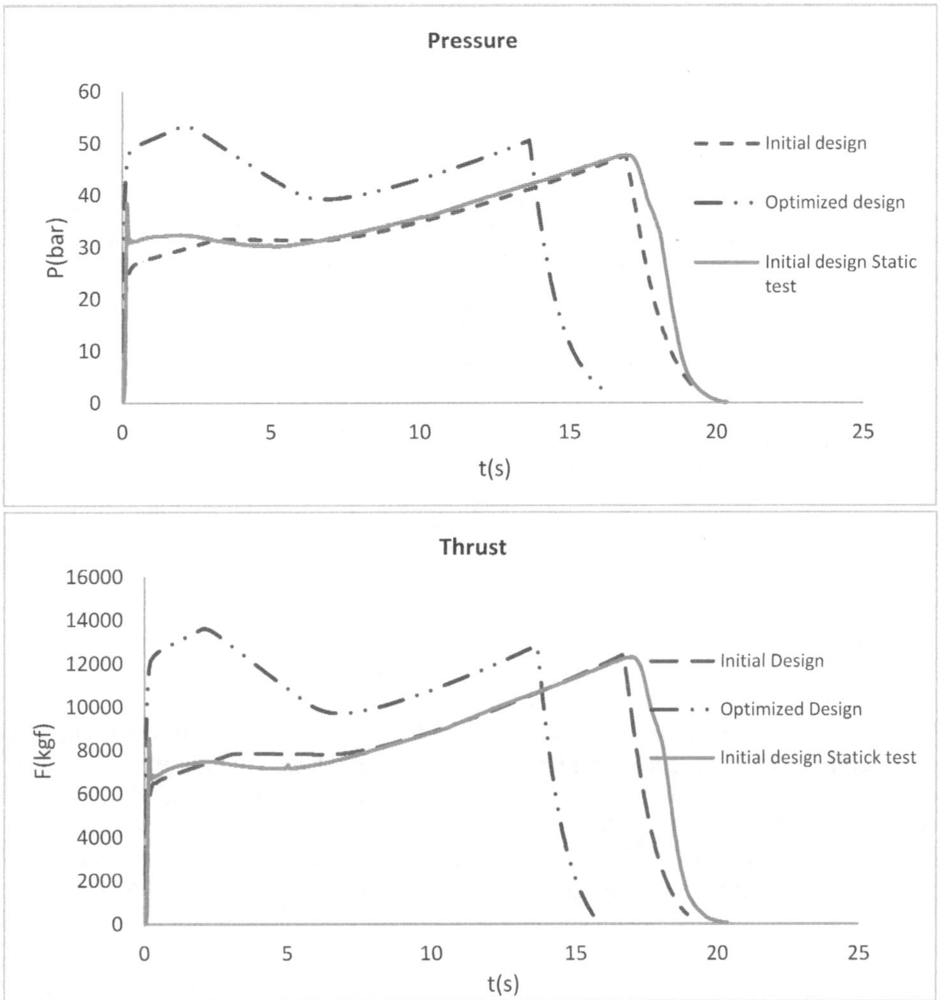

Figure 6 Comparison Initial design and optimized design (Pressure and Thrust)

Table 2 Comparison Initial design and optimized design

	initial	optimization	Increment
weight (kg)	735.2	736.07	0.11 %
Sliver weight(kg)	12.85	11.73	-8.7%
Total Impulse (kNs)	1480	1530	3.3%
I_{sp} (s)	205.2	211.8	3.3%

The optimization process takes about 5 min. Optimization results with the Real Code Genetic Algorithm increase the total impulse by 3.3%, this is because the sliver weight of the optimization result is less than the initial design, the rocket carrying more propellant mass and the combustion chamber optimization is greater than the initial design. Optimized combustion chamber pressures greater than the initial design due to the burn area of optimization results is greater than the initial design.

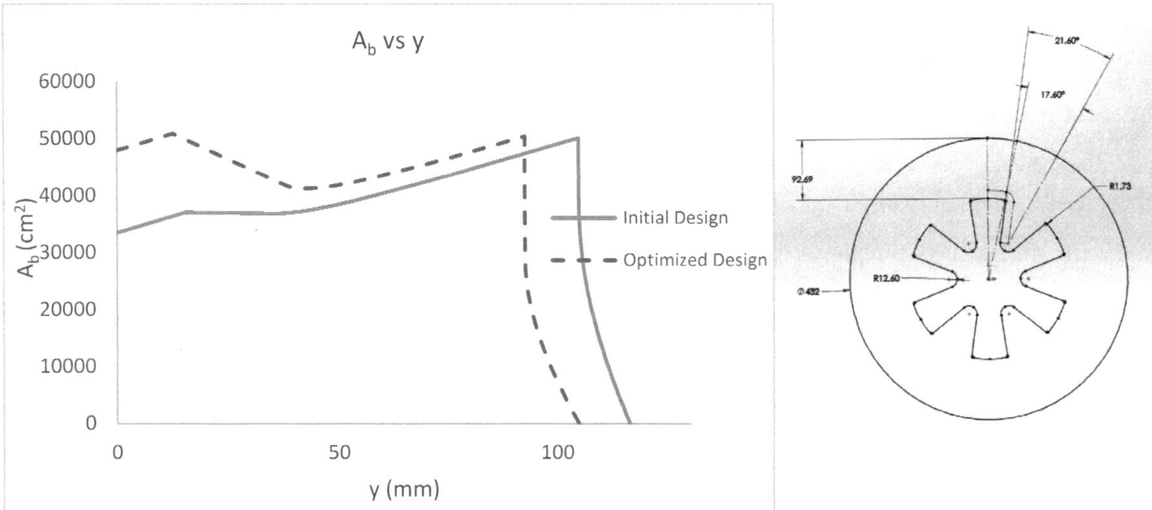

Figure 7 Optimize Design and Change of burn area

4. Conclusion

Real Code Genetic algorithm method can be applied in rocket motor design process. The use of RCGA optimization method can increase the total impulse by 3.3% and takes about 5 min.

Future research will investigate new strategies for improving the speed. To accelerate the process, we must research about influence of GA parameter (population, crossover rate, and mutation rate) on grain propellant optimization case. Another future research, we will change the objective function from max total impulse to profile thrust which are desired (input from user).

5. Acknowledgement
The authors would like to thank director of Rocket Technology Center (Pustekroket) LAPAN for providing research support facilities.

References

[1] Davenas, Alain. Solid rocket propulsion technology. *Newnes, 2012.*

[2] Hartfield, Roy, et al. "A review of analytical methods for solid rocket motor grain analysis." *39th AIAA/ASME/SAE/ASEE Joint Propulsion Conference and Exhibit. 2003.*

[3] Herrera, Francisco, Manuel Lozano, and Jose L. Verdegay. "Tackling real-coded genetic algorithms: Operators and tools for behavioural analysis." *Artificial intelligence review 12.4 (1998): 265-319.*

[4] Junjunan, S. F. Design And Performance Prediction of Solid Rocket Motor For RX450 Ballistic Rocket.

[5] Kamran, A., Guozhu, L., Rafique, A. F., Naz, S., & Zeeshan, Q. "Star grain optimization using genetic algorithm." *Proceedings of the 51st AIAA/ASME/ASCE/AHS/ASC Structures, Structural Dynamics, and Materials Conference, Orlando, FL, 12-15 April, AIAA 2010.* Vol. 3084. 2010.

[6] Design, Solid Propellant Grain, and Internal Ballistics. *"NASA/SP-8076." NASA (March 1972) (1972).*

[7] Rao, Singiresu S., and Singiresu S. Rao. Engineering optimization: theory and practice. *John Wiley & Sons, 2009.*
[8] Raza, Muhammad Aamir, and Wang Liang. "Design and Optimization of 3D Wagon Wheel Grain for Dual Thrust Solid Rocket Motors." *Propellants, Explosives, Pyrotechnics 38.1 (2013): 67-74.*
[9] Stein, S. D. (2007). Benefits of the Star grain Configuration for a Sounding Rocket. *From http://spacegrant.colorado.edu/COSGC_Projects/symposium_archive/2007/papers/S07_14_Star_Grain_Configuration_for_Sounding_Rocket.pdf*
[10] Sutton, George P., and Oscar Biblarz. Rocket propulsion elements. *John Wiley & Sons, 2001.*
[11] Al Farizi, M.F, Bura,R.O., Junjunan,S.F. Jihad,B.H. Pemilihan jumlah –n pada star grain propellant dengan metode trial and error. *SIPTEKGAN.2017*

IOP Conf. Series: Journal of Physics: Conf. Series **1005** (2018) 012034 doi:10.1088/1742-6596/1005/1/012034

Pseudoplasticity of Propellant Slurry with Varied Aluminium Content for Castability Development

A Restasari[1)*], R S Budi[1)] and K Hartaya[1)]

1) Rocket Technology Center, National Institute of Aeronautics and Space (LAPAN), Indonesia

*afni.restasari@lapan.go.id

Abstract. The modification of the percentage of aluminium is necessary to obtain certain specific impulse. But, it affects the pseudoplasticity of propellant in elapsed time that is important in casting. Therefore, this research attempts to investigate the pseudoplasticity of propellant slurry with varied aluminium contents and as time elapsed, the range of percentage of aluminium and time that allows propellant slurry to be well processed. The methods include measuring the viscosity of propellant slurries that contain 6, 8, 10, 12, 14, 16 and 18% of aluminium at varied shear rates until 40 minutes after mixing by using Brookfield viscometer. The graphs of viscosity versus shear rate were made to determine pseudoplasticity index. After that, the graph volume fraction versus pseudoplasticity index were made to be investigated. It is concluded that the more aluminium contents, the slurries with 6 to 12% aluminium contents exhibit more pseudoplastic behaviour, but the slurries with 12 to 16% aluminium exhibit less pseudoplastic. While, slurry of 18% aluminium exhibit high pseudoplasticity. In the correlation with the time, the slurry compositions of 6, 8, 14, 16% aluminium become more pseudoplastic as time elapsed. While, for compositions of 10, 12 and 18% aluminium, the trend becomes contrary. Based on the pseudoplasticity index, propellant slurries that contain 10 and 14% of aluminium are suitable for pressure casting. While for slurries with 6, 8 and 16% of aluminium are also suitable for vacuum casting. All of those suitability are possesed until 40 minutes after mixing. While, the composition of slurries that contain 12 and 18% of aluminium need to be modified to enhanced its castability.

1. Introduction

Composite solid propellant slurry comprises 10 – 15% of liquid contents and 85 – 90% of solid contents. The liquid contents that form binder for propellant, involve polyol such as Hydroxy Terminated Polybutadiene (HTPB) and curing agent such as Toluene diisocyanate (TDI). Beside those, plasticizer such as Dioctyl adipate (DOA) is usually enhanced. While, solid contents of propellant involve almunium (Al) powder as metal fuel, and ammonium perchlorate (AP) as inorganic oxidator. Modification of solid contents that comprises the percentage, shape, size and kind of solid contents, are developed to obtain certain burning rate and specific impulse (Isp). For aluminium, modification of its percentage is usually done. According to Tuzun (2005), the optimum specific impulse can be reached with 18% of aluminium powder in propellant composition [1]. Any modification potentially affects the rheological properties of propellant slurry [2], [3], [4].

Rheology is the study of the deformation and flow of matter. It clarifies how a matter flows in different shear, temperature and time. The flowability of the matter is measured as viscosity (η). In propellant slurry, solid contents are adsorbed on specific areas of binder. Therefore, the viscosity of interstitial liquid (η_0), volume fraction (ϕ) and maximum volume fraction of solid contents (ϕ_m) affect the viscosity of propellant slurry, as shown in equation 1. While, volume fraction of solid contents can be determined by using equation 2 which w is weight fraction and ρ is density [5][6][7]. Concerning those equations, the modification of aluminium percentage with constant percentage of solid contents automatically deducts AP percentage, changes the value of volume fraction as well as maximum volume fraction, affects the viscosity of propellant slurry. The change in viscosity also happen as long as the elapsed time after the addition of curing agent. It is because of polimerization of liquid contents that forms propellant binder. As the viscosity of liquid contents increase, the effect of the percentage of aluminium in the viscosity of slurry propellant also change [8].

$$\eta = \eta_0 \,.\, [1 - (\phi/\phi_m)]^{-2} \qquad (1)$$

$$\phi = \{(w_{Al} : \rho_{Al}) + (w_{AP} : \rho_{AP})\} : \{(w_{Al} : \rho_{Al}) + (w_{AP} : \rho_{AP}) + (1 - w_{Al} - w_{AP})\} \qquad (2)$$

The change in viscosity affects the change in flow behaviour. As a high concentration suspension, propellant slurry is discovered posseses Non-Newtonian behaviour. Its viscosity depends on shear rate conducted on it. Based on previous studies, propellant`s slurry is pseudoplastic or shear thinning. Its viscosity declining in increasing shear rate. This property follows equation 3 and 4. In equation 3, the correlation between shear stress (τ) and shear rate (γ) is affected by pseudoplasticity index (n). This index also affects the corrrelation between viscosity and shear rate in equation 4, as well as index of consistency (K). Index of pseudoplasticity clarifies the degree of pseudoplastic behaviour. Propellant slurry that posses index value of which closer to 1 indicates behaviour that approaching Newtonian. By arranging the graph of η *vs* γ, the value of $n - 1$ can be found as m. By using equation 5, the value of n can be calculated. According to Dombe *et al* (2008), slurry of composite propellants usually have n of 0,6 – 1. [9], [3], [10]

$$\tau = \tau_0 + \eta\, (\gamma)^n \qquad (3)$$

$$\eta = K.\, (\gamma)^{n-1} \qquad (4)$$

$$n = m + 1 \qquad (5)$$

Pseudoplasticity plays important roles in mixing and casting of propellany slurry. In mixing, the viscosity of slurry around an impeller changes with increase of speed. It potentially leads to the formation of a zone of intense motion around the impeller (cavern) with essentially stagnant regions elsewhere (dead zone) [11], [12]. Therefore, pseudoplasticity become the basic reason of determining some mixing parameters such as impeller geometry, numbers of impellers, mixing time, impeller speed, type and maximum power of mixer. The optimation of those parameters leads to a better mixing performance [13], [14], [15], [16], [17] .

In casting process, pseudoplasticity plays important roles in the success of the flow of propellant slurry into the intricate parts of rocket motor. Good flow in casting process is desired to avoid the forming of voids and other defects. It is because those defects can increase the burning surface and lead to abrupt increase in motor pressure and malfunctioning of motor. In order to be suitable with pseudoplastic behaviour of the propellant slurry, various casting technologies such as vacuum casting and pressure casting have been developed. Vacuum casting is reported to be suit with slurry that has pseudoplasticity index of 0,8 – 1. While, pressure casting is suitable for slurry with pseudoplasticity index of 0,6 - 1 [10][18]. Therefore, developing knowledges of pseudoplasticity of propellant slurry with the changes in percentage of aluminium is crucial.

Based on pseudoplasticity index, Muthiah *et al* (1992) reported that propellant slurries with 5 – 10% of aluminium have the best processibility until 6 hours after mixing. This research involves aluminium of 10 – 12 microns. Based on Prangili *et al* (2006), different size of aluminium results different viscosity in propellant slurry, the range that is mentioned by Muthiah *et al* can be different with the different size of aluminium [8], [19].

This research attempts to investigate the pseudoplasticity of propellant slurry with varied aluminium contents and as time elapsed, the range of percentage of aluminium with 30 micron in size and time that allows propellant slurry to be well casted. Those ranges are benefit in determining propellant composition with certain percentage of aluminium in order to reach certain specific impulse.

2. Method

Materials that were used in this research include HTPB, TDI, DOA, AP and Al manufactured by *Dalian Chlorate Ltd.* The compositions by weight included 3% of DOA, 12% of HTPB:TDI 15:1, 85% of solid content. TDI comprises 57,8% 2,6-TDI and 42,2% 2,4-TDI. The solid contents included the variations of aluminium (30 microns) which are 6, 8, 10, 12. 14, 16 and 18 %, and AP. The composition of AP was coarse AP (200 microns) : fine AP (50 microns) is 1:1.

Propellant slurries were made by mixing HTPB, DOA, Al, fine AP and coarse AP for 95 minutes at 60 rpm in horizontal blade mixer. After that, TDI was added into the mixer and mixed well until 20 minutes. The mixing processes were conducted in 46 – 50 ^{0}C. The viscosity of the slurries were measured by using Viscometer Brookfield RVT *spindle* 07 with variation of rotational speed of 0,5; 1; 1,5; 2 rpm. The shear rate was determined by using equation 8. Equation 8 was derived from equation 6. In equation 6, N is rotational speed (rpm), R_c is radius of container (2,25 cm), R_b is radius of spindle (0,16 cm) and x is radius at which shear rate was being calculated (cm). As viscosity of pseudoplastic slurry is very sensitive toward shear rate, so the calculation was adjusted at $x^2 = R_b^2$ and equation 7 was obtained. With the subtitutions of the values of R_b and R_c, equation 8 was obtained [20][21]. The viscosities were measured at 10, 20, 30 and 40 minutes after the mixing process was finished. The graph of shear rate versus viscosity was made for each composition to get the value of m and index of pseudoplasticity based on equation 5. Volume fraction was calculated by using equation 2 with ρ_{AP} is 1,95 g/cm^3 and ρ_{Al} is 2,7 g/cm^3. The graph of time versus pseudoplasticity index and the graph of the value of volume fraction versus pseudoplasticity index were also made

$$\gamma = \{2.(2\pi/60).N.R_c^2.R_b^2\} : \{x^2 (R_c^2 - R_b^2)\} \qquad (6)$$

$$\gamma = \{(4\pi/60).N.R_c^2\} : (R_c^2 - R_b^2) \qquad (7)$$

$$\gamma = N.\ 0{,}21 \qquad (8)$$

3. Results and Discussions

Table 1. Volume Fraction of Compositions

No	% Mass Al	% Mass AP	Volume Fraction
1	6	79	0,740
2	8	77	0,739
3	10	75	0,738
4	12	73	0,736
5	14	71	0,735
6	16	69	0,734
7	18	67	0,732

The increasing of percentages of aluminium deduct AP portion in propellant slurry, so that affect volume fraction and maximum volume fraction of solid contents of propellant slurry as well as its viscosity. Therefore, the volume fractions were calculated by using equation 2 and its results are shown in Table 1. In table 1, as the percentage of aluminium increase, the percentage of AP and volume fraction decrease. It is because the size of aluminium (30 microns) is smaller than fine AP (50 microns) and coarse AP (200 microns).

In propellant slurry, after mixing, as time elapsed, polimerization reaction between hydroxyl groups of HTPB and isocyanate groups of TDI occurs and leads to form binder for solid contents as well as increase the viscosity of propellant slurry [22]. To investigate the effect of increasing viscosity on pseudoplasticity index, viscosity of propellant slurry is measured at 10, 20, 30, 40 minutes after mixing at various shear rates below 1 s^{-1}. The measurements were limited only until 40 minutes because the longest duration of casting process that is desired is 40 minutes. While, the use of shear rates that are below 1 s^{-1} is to represent gravity casting that is usually used [18]. The datas from those measurements are plotted on the graph of shear rate versus viscosity. Those graphs are shown on Figure 1.

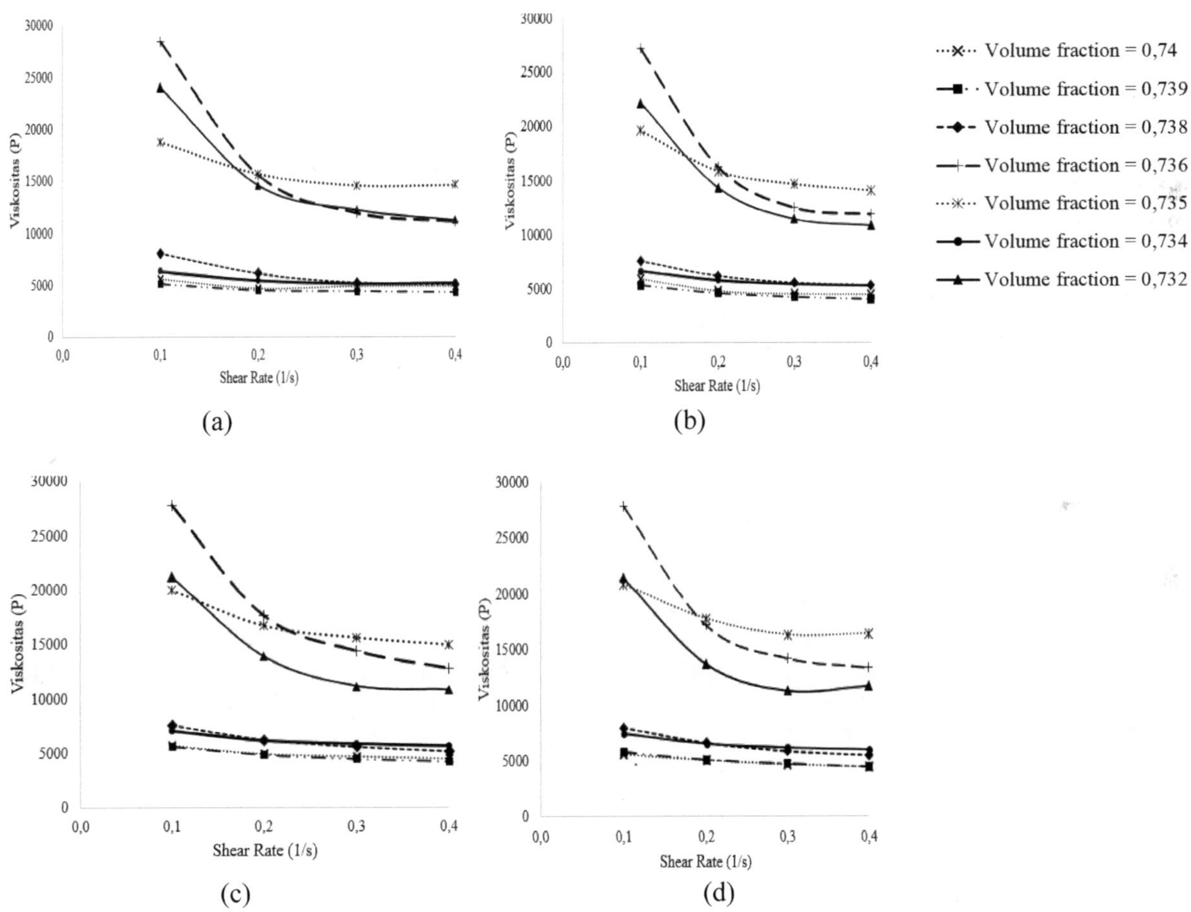

Figure 1. The graph of shear rate versus viscosity of propellant slurry at 10 minutes (a), 20 minutes (b), 30 minutes (c) and 40 minutes (d)

It is shown on figure 1 that the viscosity of all propellant slurry composition decrease as shear rate increase. It exhibites pseudoplastic behaviour. This behaviour is caused by molecular structure of the binder and packing density of the filler. Binder that is used is polyurethane that is the result from the reaction between HTPB and TDI [23]. While, based on the packing density of the filler, various parameters such as particle size, interactions between particles, and interactions between particles and suspending fluid play important roles [24][25]. The percentage of solid contents that are used in this research is 85% of which reported exhibites pseudoplasticity [8]. With the increase in aluminium content, the effects of those mentioned above are vary and to more understanding those effects on pseudoplasticity of slurry, the graph of volume fraction *vs* pseudoplasticity index is made, as shown in figure 2.

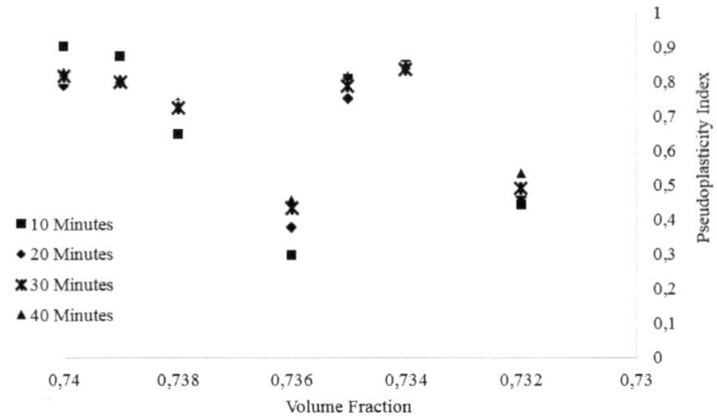

Figure 2. Pseudoplasticity Index of Each Composition.

In figure 2, propellant slurries with volume fraction of 0,736 – 0,74 (6 - 12% Al) shows that with the increase of aluminium content, propellant slurry become more pseudoplastic. In figure 1, as the aluminium content increase, the viscosity of propellant slurry also increase. To explain this trend, the effect of particle size of aluminium become significant. The size of aluminium that is used is 30 microns which almost equals to 0,155 times of the size of largest particle used (coarse AP (200 microns)). That size of aluminium is reported thread throughout the binder effectively [8]. With small content of aluminium (6 and 8%), ability of aluminium flow along with the binder, when the shear conducted on it, is represented by pseudoplasticity index that close to 1 which are 0,8 – 0,9 and low viscosity. As the amount of aluminium increase, until 12%, the interaction among aluminium particles increase, forming flocs. The flocs inhibit the flow of propellant slurry so that the viscosity also increase as shown in figure 1. As the shear rate increase, the flocs break down so that the viscosity become lower and the slurry become more pseudoplastic [26]. It is shown clearly in figure 2 that the pseudoplasticity index decrease until 0,3 – 0,5 (12% of aluminium).

For propellant slurry of which volume fraction of 0,734 – 0,736 (12 – 16%), the trend that is mentioned above become contrary of which with the increase of aluminium content, the viscosity decrease as shown in figure 1 and slurry become less pseudoplastic. It is shown in figure 2 that pseudoplasticity index increase from 0,3 – 0,5 (12% of aluminium) until 0,8 – 0,9 (16% of aluminium). This trend shows the trend in reaching optimum packing network. In ideal packing network, the particles with different size are packed, filling the voids such that voids are minimum so that the viscosity become lower and less psedoplastic [27][25].

Propellant slurry with 18% of aluminium (volume fraction of 0,732) has higher viscosity and more pseudoplastic than propellant slurry with 16% of aluminium. It can be caused by the agglomeration of aluminium. Agglomeration increases the viscosity because the binder entrapped between agglomerated particles so the binder hard to flow. De-agglomeration happens when shear rate increase and it causes the lower viscosity [28].

An agglomerate is formed as a combination of many small flocs. The amount of small flocs as well as viscosity of the binder causes different flow behaviour in elapsed time. The more time elapsed, the more molecular weight of polimeric binder, the more viscous binder, the more viscous the slurry, the larger flocs that are formed, the more pseudoplastic the slurry [29],[22],[26]. This trend is shown significantly from the measurement of 10 minutes to 20 minutes. It can because of the reactivity of 2,4-TDI is higher than 2,6-TDI [23]. As shown in figure 2, this trend happens to the slurry with low aluminium content (6 and 8% aluminium) and the slurry with optimal packing network (14 and 16%

aluminium). While, for 10, 12 and 18% of aluminium, as time elapsed, the slurry become less pseudoplastic. It because, many aluminium flocs that can not be broken down by the shear rates that are used.

Based on the pseudoplasticity index and criterias that reported by Dombe *et al* (2008) [10], propellant slurries that contain 6, 8 and 16% of aluminium are suitable for vacuum and pressure casting because its pseudoplasticity indexes are in range of 0,8 – 1. The slurries that contain 10 and 14% of aluminium are suitable for pressure casting only because its pseudoplasticity indexes are in range of 0,6 – 0,8. While, the composition of slurries that contain 12 and 18% of aluminium need to be modified to be suitable for casting.

4. Conclusions

The pseudoplasticity of propellant slurry with varied aluminium contents and as time elapsed have been investigated. Size of aluminium, packing network, agglomeration and flocculation of aluminium as well as the rising viscosity of the binder become complicated reasons. For compositions of 6 to 12% aluminium contents, the more aluminium contents, the higher the viscosity and the more pseudoplastic the propellant slurry. For compositions of 12 to 16% aluminium contents, the more aluminium contents, the lower the viscosity and the less pseudoplastic the slurry. While, the slurry of 18% aluminium possess high viscosity and pseudoplasticity. In the correlation with the time, the slurry compositions of 6, 8, 14, 16% aluminium become more pseudoplastic as time elapsed. While, for compositions of 10, 12 and 18% aluminium, the trend becomes contrary. Based on the pseudoplasticity index, propellant slurries that contain 6, 8 and 16% of aluminium are suitable for vacuum and pressure casting. The slurries that contain 10 and 14% of aluminium are suitable for pressure casting. All of those suitability are possesed until 40 minutes after mixing. While, the composition of slurries that contain 12 and 18% of aluminium need to be modified to enhanced its castability.

Acknowledgement

The authors would like to thank director of Rocket Technology Centre (Pustekroket) LAPAN for providing research support facilities and to all staffs of Propellant`s Basic Composition Laboratory for helping the research.

References

[1] F. N. Tüzün, "The Effect of Aluminum Content Variation on Burning Rate, Pressure-Propellant Area Ratio Relationship, and Other Properties of Hydroxyl Terminated Polybutadiene Based Composite Propellants," *J. ASTM Int.*, vol. 2, no. 4, pp. 1–15, 2005.

[2] S. K. Manu, "Glycidyl Azide Polymer (GAP) as a High Energy Polymeric Binder for Composite Solid Propellant Applications," Mahatma Gandhi University, 2009.

[3] A. K. Mahanta, M. Goyal, and D. D. Pathak, "Rheokinetic Analysis of Hydroxy Terminated Polybutadiene Based Solid Propellant Slurry," *E-Journal Chem.*, vol. 7, no. 1, pp. 171–179, 2010.

[4] S. Jain, S. Nandagopal, P. P. Singh, K. K. Radhakrishnan, and B. Bhattacharya, "Size and Shape of Ammonium Perchlorate and their Influence on Properties of Composite Propellant," *Def. Sci. J.*, vol. 59, no. 3, pp. 294–299, 2009.

[5] H. A. Barnes, J. F. Hutton, and K. W. F.R.S, "An Introduction to Rheology." Elsevier Science Publishers B.V., Amsterdam, p. 11, 1989.

[6] J. P. Guillemin, L. Brunet, O. Bonnefoy, and G. Thomas, "A flow time model for melt-cast insensitive explosive process," *Propellants, Explos. Pyrotech.*, vol. 32, no. 3, pp. 261–266, 2007.

[7] P. J. Blatz, "On the Microstructure of Composite Propellants." California Institute of Technology, California, 1966.

[8] R. Muthiah, V. N. Krishnamurthy, and B. R. Gupta, "Rheology of HTPB Propellant . 1 . Effect of Solid Loading, Oxidizer Particle Size, and Aluminum Content," *J. Appl. Polym. Sci.*, vol. 44, pp. 2043–2052, 1992.

[9] A. K. Mahanta, I. Dharmsaktu, and P. K. Pattnayak, "Rheological Behaviour of HTPB-Based Composite Propellant: Effect of Temperature and Pot Life on Casting Rate," *Def. Sci. J.*, vol. 57, no. 4, pp. 435–442, 2007.

[10] G. Dombe, M. Jain, P. P. Singh, K. K. Radhakrishnan, and B. Bhattacharya, "Pressure casting of composite propellant," *Indian J. Chem. Technol.*, vol. 15, pp. 420–423, 2008.

[11] S. Nagata, M. Nishikawa, H. Tada, and S. Gotoh, "Power Consumption of Mixing Impellers Pseudoplastic Liquids," *J. Chem. Eng. Japan*, vol. 4, no. 1, pp. 72–76, 1971.

[12] H. Ameur and M. Bouzit, "Mixing In Shear Thinning Fluids," *Brazilian J. Chem. Eng.*, vol. 29, no. 2, pp. 349–358, 2012.

[13] P. Prajapati and F. Ein-Mozaffari, "CFD Investigation of the Mixing of Yield- Pseudoplastic Fluids with Anchor Impellers," *Chem. Eng. Technol*, vol. 32, no. 8, pp. 1211–1218, 2009.

[14] F. Ein-mozaffari and S. R. Upreti, "Investigation of Mixing in Shear Thinning Fluids Using Computational Fluid Dynamics," in *Computational Fluid Dynamics*, H. W. Oh, Ed. Intechopen, 2010, pp. 77–102.

[15] S. M. Shekhar and S. Jayanti, "Mixing of Pseudoplastic Fluids Using Helical Ribbon Impellers," *AIChE J.*, vol. 49, no. 11, pp. 2768–2772, 2003.

[16] V. N. Krishnamurthy, R. Manjari, and R. Muthiah, "Rheology of HTPB Propellant : Mixing Effect Time of Mixing Speed and," *Def. Sci. J.*, vol. 43, no. 2, pp. 167–172, 1993.

[17] R. S. Budi and S. Prangili, "Analisis Kinerja Mixer Kapasitas 425 Liter Berdasarkan Produk Propelan Yang Dihasilkan Menggunakan Parameter Uji Kualitas," in *Energi, Wahana dan Muatan Antariksa*, Jakarta: Massma Publishing, 2009.

[18] A. Davenas, "Composite Propellants," in *Solid Rocket Propulsion Technology*, Oxford: Pergamon Press, 1991, pp. 439–460.

[19] S. Prangili and D. Setyaningsih, "Pengaruh Penggunaan Logam – Logam Aluminium pada Teknis Pembuatan dan Karakteristik Propelan," in *SIPTEKGAN X*, 2006.

[20] Brookfield Engineering Laboratories, *More Solutions for Sticky Problems*. Massachusetts: Brookfield Engineering Laboratories, Inc, 2014.

[21] N. Triantafillopoulos, *Measurement of Fluid Rheology and Interpretation of Rheograms*, 2nd ed. Michigan: Kaltec Scientific, 2000.

[22] R. Gogoi, M. S. Alam, and R. K. Khandal, "Effect of Increasing NCO/OH Molar Ratio on The Physicomechanical and Thermal Properties of Isocyanate Terminated Polyurethane Prepolymer," *Int. J. Basic Appl. Sci.*, vol. 3, no. 2, pp. 118–123, 2014.

[23] M. Szycher, *Polyurethanes*, 2nd ed. Boca Raton: CRC Press, 2013.

[24] A. B. Yu and N. Standish, "A study of the packing of particles with a mixture size distribution," *Powder Technol.*, vol. 76, no. 2, pp. 113–124, 1993.

[25] H. Esiyok and M. E. Candarli, "Theoritical and Experimental Packing Density Study of Hydroxyl Terminated Polybutadiene Ammonium Perchlorate Based Propellant," vol. 13, no. 5, pp. 455–469, 2014.

[26] R. R. Miller, E. Lee, and R. L. Powell, "Rheology of solid propellant dispersions," *J. Rheol. (N. Y.)*, vol. 35, no. 5, pp. 901–920, 1991.

[27] C. Erisken, A. Gocmez, U. Yilmazer, F. Pekel, and S. Ozkar, "Modeling and Rheology of HTPB Based Composite Solid Propellants," *Polym. Compos.*, vol. 19, no. 4, pp. 463–472, 1998.

[28] R. Asthana, A. Kumar, and N. Dahotre, *Materials Science in Manufacturing*. Burlington: Elsevier Inc, 2006.

[29] P. J. T. Mills, J. P. K. Seville, P. C. Knight, and M. J. Adams, "The effect of binder viscosity on particle agglomeration in a low shear mixer/agglomerator," *Powder Technol.*, vol. 113, no. 1–2, pp. 140–147, 2000.

IOP Conf. Series: Journal of Physics: Conf. Series **1005** (2018) 012035 doi:10.1088/1742-6596/1005/1/012035

Combustion Performance of a Staged Hybrid Rocket with Boron addition

D. Lee[1) *] and C. Lee[1) *]

[1)] Department of Aerospace Engineering, Konkuk University, Korea

*cjlee@konkuk.ac.kr

Abstract. In this paper, the effect of boron on overall system specific impulse was investigated. Additionally, a series of combustion tests was carried out to analyze and evaluate the effect of boron addition on O/F variation and radial temperature profiles. To maintain the hybrid rocket engine advantages, upper limit of boron contents in solid fuel was set to be 10 wt%. The results also suggested that, when adding boron to solid fuel, it helped to provide more uniform radial temperature distribution and also to increase specific impulse by 3.2%.

1. Introduction

Staged hybrid rocket engine (HRE) is a newly introduced concept of HRE by the recent study. It can minimize the loss of combustion performance with the oxidizer-to-fuel (O/F) shifting, while maintaining the key advantages of HRE. It uses the hybrid gas generator (GG) as a primary combustor, which utilizes solid fuel and gas/liquid oxidizer to produce fuel-rich effluent, which is then expelled and mixed with additional aft-injected secondary oxidizer in the secondary combustor for thrust generation.

The use of oxygen and nitrogen mixture oxidizer is one of techniques to generate fuel-rich effluent that meets the design requirement of GG. Equivalence ratio had increased significantly with decreasing oxygen (O2) content (by volume) in the mixture oxidizer, resulting in substantial decrease in the effluent temperature. In the results, with the use of hydroxyl-terminated polybutadiene (HTPB) and 28% oxygen content mixture oxidizer, the fuel-rich effluent with a temperature of 1290 K was successfully produced, which was in the range of the design requirement of GG. However, a loss of Isp (specific impulse) was inevitable due to deliberately decreased oxygen content in mixture oxidizer. Though the staged HRE offers many advantages, its Isp should be improved to be at least competitive with conventional chemical rockets to outweigh its advantages.

At the same time, several technical issues can be resolved. One of issues is the low Isp(specific impulse) of overall system. A loss of overall system Isp was inevitable due to deliberately decreased oxygen content in mixture oxidizer [1]. Though the staged HRE offers many advantages such as; minimized performance variation with respect to O/F shifting and engine stop-restart ability, Isp performance of overall system should be improved to be at least competitive with conventional chemical rockets to outweigh its advantages.

Meanwhile, an addition of metal additives to solid fuel is widely used technique to improve the Isp performance as well as fuel density [6-8]. Hence, with use of metal additives in staged HRE, this could bring an enhancement of Isp performance and therefore possibly be competitive with other chemical rockets with respect to Isp performance. However, loading limit of metal additives in solid fuel should be carefully chosen for possible combustion instability. In this regard, Evans [9] examined a detailed regression rate behavior of HTPB and metalized solid fuels in hybrid rockets, and reported that a loading of 13 wt% of metal additives demonstrated a limit of stable combustion. As this is likely to be the first attempt to study the effect of metal additive in hybrid GG, this study has limited metal additives content up to 5 wt%. For a current study, boron(B) was used as a metal additive, as it is reported to have the highest combustion enthalpy per mass and volume [8, 10]. However, according to many literatures related to boron metal additives, its implementation in HRE has been limited due to difficulties associated with ignition and efficient combustion [10-13]. Foelsche [10] studied boron particle ignition and combustion, and reported that boron ignition is hindered by a regenerating protective oxide coating. And, boron particle ignition is highly dependent on the chamber pressure, particle size and combustion temperature [10]. Also, Shin [13] reported a problem associated with a curing reaction during manufacturing process of boron added HTPB solid fuel. In this study, boron bead, which is a mixture of PMMA(PloyMethyl MethAcrylate), AP and boron, was adopted from Shin [13], with an only intention to prevent the curing reaction, not to increase the pressure exponent as intended by the reference. A detailed development process and its compositions of boron bead are described in reference [13].

The key purpose of boron addition is to compensate the Isp loss; therefore, the main objective of this study is to determine if an improvement in Isp is possible with the addition of boron. Additionally, a series of experimental tests was carried out to investigate the combustion characteristics with additives in hybrid GG such as; O/F variation and radial temperature profiles. Also, based on the results, an experimental evaluation of the beneficial effects of each additive were presented.

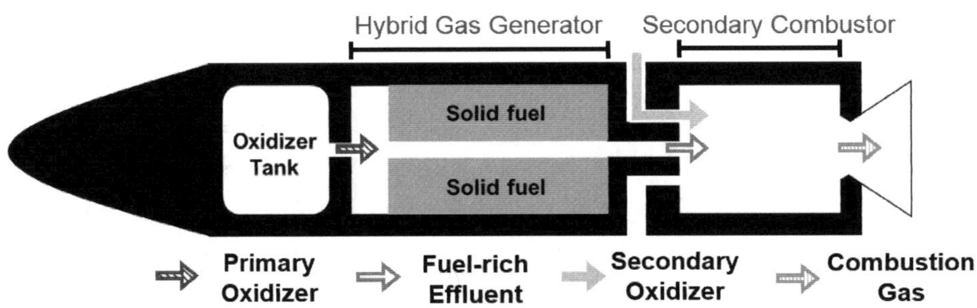

Fig. 1 Schematic of staged HRE

2. Design of combustion test

A series of combustion tests was conducted with a laboratory-scale HRE, as shown in Fig. 2. Solenoid and check valves were used to control oxidizer feeding. The oxidizer mass flow rate for was controlled up to 25 g/s by using the mass flow controller. Nitrogen(N_2) gas was used to purge the combustion by programmable logic controller control. Piezoelectric-type sensors were installed to measure the combustion pressure. A DAQ board and the LabVIEW program were also implemented for the data acquisition process. The dimension of the baseline fuel grain had an outer diameter of 50 mm and an inner diameter of 20 mm. In the baseline configuration, chamber lengths of the main, pre, and post chamber were fixed as 200, 45, and 200 mm, respectively. A water-cooled nozzle was used, in which the throat diameter was 6.5 mm. Ignition was achieved with an A-type model rocket engine, which was first ignited with a small pyrotechnic igniter.

The NASA CEA(Chemical Equilibrium and Application) program was used to calculate the theoretical oxidizer-to-fuel ratio(O/F_{stoi}) for each test case. The actual O/F ratio(O/F_{act}) was calculated with the total mass of injected oxidizer divided by the total burned fuel mass during burning time. Then, the equivalence ratio was calculated with the ratio of O/F_{stoi} to O/F_{act}. The regression rate was determined at each test with the space-time averaged method. For temperature measurements, seven exposed K- and R-type thermocouples were used in all cases, which were located at specific axial and radial locations in the post chamber, as shown in Fig. 2.

To quantify the O/F variation, the initial and final O/F(O/F$_{ini}$ and O/F$_{fin}$) were calculated based on the expressions suggested by Karabeyoglu et al. [14]. Since the O/F variation was highly dependent on the oxidizer flux, G$_{ox}$, and exponent n, a study was focused on O/F behavior with respect to those parameters. In the literature, the classical HRE was reported to have an O/F variation as high as 1.17, meaning the O/F increased by 17% from the initial value [15]. A detailed explanation can be found in reference [1]

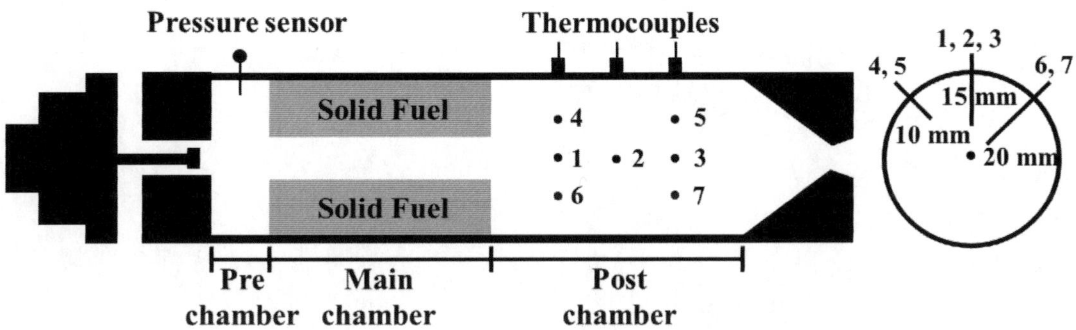

Fig. 2 HRE assembly and location of thermocouples in post-chamber

3. Results and Discussions

Table 1. Summary of test results in GG combustion

Test	Additive, wt%	O/F$_{stoi}$	$\dot{m}_{ox}$, g/sec	O/F$_{act}$	Φ	$T_{Effluent}$, K	O/F variation	a	n
1	None (HTPB)	11.35	15	2.50	4.53	1290	1.04	0.03	0.59
2			20	2.84	3.99	1410	1.05		
3			25	3.09	3.67	1580	1.05		
4	AP10	10.09	10	1.89	5.34	1400	1.02	0.03	0.56
5			15	2.31	4.37	1570	1.02		
6			20	2.56	3.93	1780	1.03		
7	AP10/B5	9.92	11	2.08	4.78	1460	1.03	0.03	0.57
8			16	2.33	4.25	1590	1.03		
9			20	2.71	3.66	1750	1.03		
10	AP10/B10	9.75	10	1.78	5.49	1570	-	0.05	0.44
11			15	2.23	4.37	1800	-		

3.1. Effect of boron addition on Isp performance

As for the GG, it would be necessary to produce fuel-rich effluent with a good radial temperature profile for improving combustion efficiency in a secondary combustor and overall system performance [1]. However, because of the diffusional nature of combustion in a hybrid GG, the radial variation of fuel-rich effluent temperatures seems to be inevitable. Temperature measurements of HTPB fuel is presented in Fig. 3a). The maximum temperature is found at location 7 and the minimum at location 4.

Higher temperatures are measured at the downstream, since unburned fuels are continuously burning as they pass through the post chamber. The radial temperature profile of AP10/B5 is shown Fig. 3b). As shown, AP10/B5 fuel showed an almost half of the temperature variation reported by the pure HTPB fuel case. These results could be mainly due to improved mixing performance with AP and boron additions. Hence, better combustion performance can be expected with boron added solid fuel.

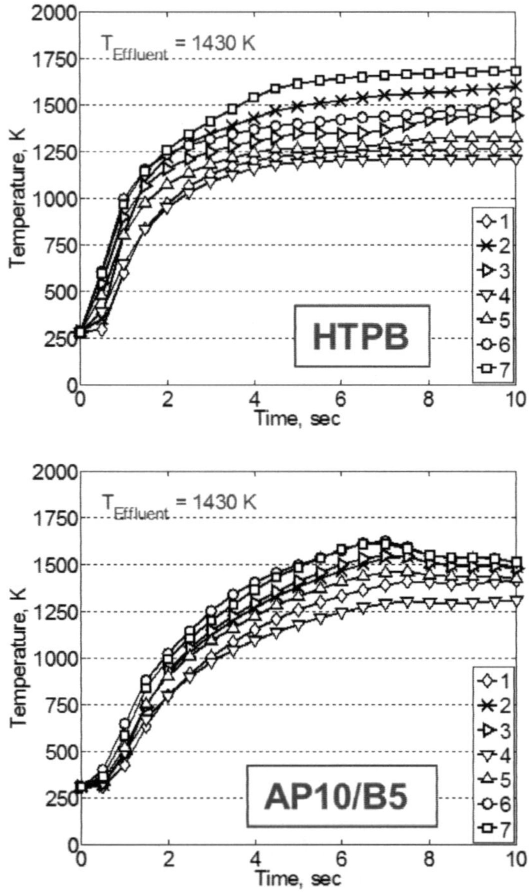

Fig 3. Radial temperature profile of Test a)HTPB and b)AP10/B5

3.2. Effect of boron addition on Isp performance

Vacuum specific impulse is a measure of the efficiency of the overall engine (i.e., staged HRE) performance. Even the addition of boron to solid fuel can produce better radial temperature profiles; the most significant benefit acquired from boron addition is the increase in overall system Isp performance. Thus, the addition of boron can compensate for the loss of Isp due to the use of mixture oxidizer in the GG. Fig. 4 presents the estimations of overall Isp performance of various fuels (i.e., fuel-rich effluent) with respect to the engine O/F ratio by using the NASA CEA code. The estimation was done at the chamber pressure and expansion ratio of 1000 psi (6.89 MPa) and 10, respectively. As for the boron added fuel, peak value of Isp increased about 1.3% compared to that of AP10 fuel and about 3.2% compared to the pure HTPB fuel case. Results also suggested that the peak values of Isp increase up to about 2.8% with AP addition(AP15 fuel) when compared to that of pure HTPB fuel. Note that the current study is done only with fuels of 15 wt% AP and 5 wt% boron additions Thus, the increase in AP weight from 15% and boron weight from 5% can bring about a further increase in peaks of Isp, making up for substantial drops of Isp in the overall system due to the use of mixture oxidizer in the GG. Also, Lee [1] reported that overall system Isp performance could be increased with decreasing effluent temperature. Since current analysis was done at an effluent temperature of around 1430 K, further increment in Isp could be expected with lower effluent temperature that meets the requirement of hybrid GG (below 1300 K). Considering that the HRE has Isp of around ~310 sec for classical [6], and ~320 sec for solid propellant GG [5], it would be possible to make staged HRE to be comparable to other chemical rockets, if higher contents of additives were used as well as optimization in effluent temperature.

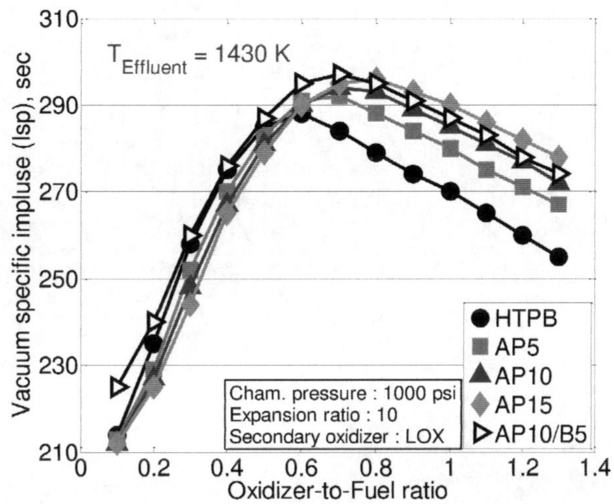

Fig 4. Overall system vacuum I_{sp} performance versus O/F ratio (oxidizer: LOX, fuel: fuel-rich effluent)

3.3. Secondary combustion

Additional tests were carried out to assess the suitability of the effluent and to investigate the sequencing of events in staged combustion. One of the key components of the effluent was the ignition capability in the secondary combustor, and this was the focus of these tests. Lee [1] reported that the effluent temperature was maintained relatively low to allow the use of low-temperature tolerant materials for the hybrid GG. However, it was important for the effluent to maintain a high enough temperature to initiate the combustion in the secondary combustor when additional oxidizer was injected. The entire staged HRE assembly is shown in Fig. 5. In the configuration, chamber lengths of pre, hybrid GG, and secondary were fixed at 45, 400, and 200 mm, respectively.

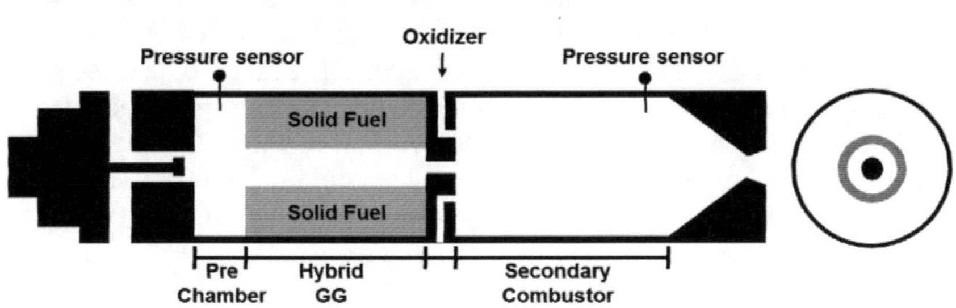

Fig 5. Staged HRE assembly and injector design

Table 2. Sequence of test operation

Action	Timeline, sec
Mixture oxidizer on	0.0

Electric ignitor on/off	1.0
(Only hybrid GG is operating)	(1.0 - 6.0)
GOX oxidizer on	6.0
(Entire staged HRE is operating)	(6.0 – 11.0)
Mixture oxidizer off	11.0
GOX oxidizer off	11.0
N_2 gas on	11.0
N_2 gas off	13.0

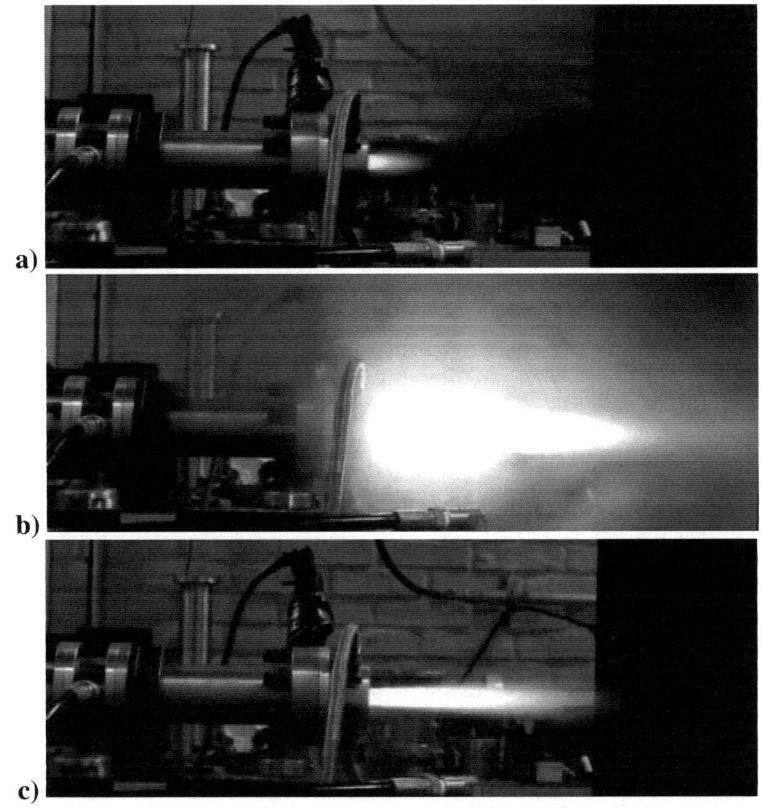

Fig 6. Snapshot of staged combustion at a) 2, b) 7, and c) 10 seconds of test operation

Since the goal was to investigate whether the effluent temperature was high enough to guarantee spontaneous ignition, and to test the sequence of staged operation, tests were conducted with a large nozzle throat area to prevent any possible nozzle blockage. The coaxial-type injector was used, which had inner, outer, and annulus diameters of 4.7, 10.7, and 21.5 mm, respectively. The test 2 condition was used to simulate the effluent coming out from the hybrid GG, and GOX was used as the aft-injected oxidizer. As this test is the first attempt to test the staged combustion of staged HRE, the amount of GOX injected into the secondary combustor was intentionally increased from the stoichiometric value for the fuel-lean combustion in order to decrease the engine performance. Table 2 shows a test sequence done in this test. As shown, only the hybrid GG was operated from 1.0 to 6.0 sec of the period. Then, GOX was injected into the secondary combustor for the next 5 sec. As soon as GOX was additional injected to the secondary combustor, spontaneous ignition was achieved. Fig. 6a), b), and c) shows the snapshot of staged combustion at 2, 7, and 10 sec of test operation, respectively. Based on the test results, an effluent temperature of 1410 K was confirmed to be high enough to initiate spontaneous ignition when additional oxidizer was injected to the secondary combustor.

In addition, combustion efficiency based on characteristic velocity (η_{c*}) was calculated. For the experimental characteristic velocity calculation, the definition given in Eq. (1) was used.

$$c^*_{\exp} = \frac{P_c A_t C_{Dt}}{\dot{m}_f + \dot{m}_{ox}} \tag{1}$$

where, P_c is chamber pressure in Pa, A_t is nozzle throat area in m^2, $\dot{m}_f$ is mass flow rate of effluent in kg/s, $\dot{m}_{ox}$ is mass flow rate of oxidizer in kg/s and C_{Dt} is discharge coefficient at nozzle throat which was assumed to be 1.0.

Then, the combustion efficiency based on characteristic velocity was calculated by using the following definition.

$$\eta_{c*} = \frac{c^*_{\exp}}{c^*_{th}} \tag{2}$$

where, $c^*_{\exp}$ is experimental characteristic velocity in m/s and c^*_{th} is theoretical characteristic velocity in m/s. The theoretical characteristic velocity was calculated using the CEA at the appropriate operating conditions, such as pressure and compositions of fuel and oxidizer. Finally, combustion efficiency based on characteristic velocity of the tested staged combustion was calculated to be 82%.

For the current stage of research, the optimal configurations and flow conditions for the secondary combustion have not yet been determined. Once the optimization was done, the effluent temperature, which was capable of initiating the secondary combustion, could be decreased, since better mixing between the effluent and the oxidizer would give more potential for the effluent to initiate the secondary combustion. Also, higher combustion efficiency (>90%) in secondary combustor could be expected with the optimization of the secondary combustor.

4. Conclusions or Concluding Remarks

This study conducted a series of combustion tests with AP and boron additives in solid fuels to understand the effect of boron addition on the overall system specific impulse. Though the staged HRE offers many advantages such as; minimized performance variation with respect to O/F shifting and engine stop-restart ability, effluent temperature controllability as well as overall system Isp performance should be improved. Therefore, boron as a metal additive was introduced to investigate its effects on GG combustion characteristics as well as on Isp performance. Loading weight of was limited in order to retain the HRE advantages and to prevent possible reported effects, and therefore boron content of 10 wt% were used in this study. Even though the addition of boron did not show much of differences with respect to regression rate and effluent temperature compared to 10 wt% AP solid fuel, the results showed that the addition of boron tended to provide more uniform radial temperature distribution. More importantly, boron addition showed 3.2% increase in Isp performance compared to pure HTPB fuel case, and 1.3% increase compared to 10 wt% AP solid fuel. Considering this study was done with only case of boron content(5 wt%), further enhancement in Isp could be expected with higher boron content. Given that boron addition was done with only one case, the results are not conclusive. Future studies will carry out more combustion tests with higher boron contents to validate the results. In addition, staged combustion tests will be carried out to optimize the secondary combustor and to improve the combustion efficiency.

Acknowledgement

This work was supported by the National Research Foundation(NRF-2015R1D1A1A01058070) in the Republic of Korea.

References

[1] Lee, D., and Lee, C., "Hybrid Gas Generator for a Staged Hybrid Rocket Engine," Journal of Propulsion and Power, 2016, accessed July 4, 2016. doi: http://arc.aiaa.org/doi/abs/10.2514/1.B36036

[2] George, P., Krishnan, S., Varkey, P. M., Ravindran, M., and Ramachandran, L., "Fuel Regression Rate in Hydroxyl-Terminated-Polybutadiene/Gaseous-Oxygen Hybrid Rocket Motors," Journal of Propulsion and Power, Vol. 17, No. 1, 2001, pp. 35–42. doi:10.2514/2.5704

[3] Frederick, R. A., Jr., Whitehead, J. J., Knox, L. R., and Moser, M. D., "Regression Rates Study of Mixed Hybrid Propellants," Journal of Propulsion and Power, Vol. 23, No. 1, 2007, pp. 175–180. doi:10.2514/1.14327

[4] Einav, O., Peretz, A., Hashmonay, B., Birnholz, A., and Sobe, Z., "Development of a Lab Scale System for Hybrid Rocket Motor Testing," 45th AIAA Joint Propulsion Conference, AIAA Paper 2009-4888, August 2010.

[5] Pilon, B., and Louwers, J., "Development of Staged Combustion Aft-Injected Hybrid (SCAIH) Propulsion at Cesaroni Technology Inc," 46th AIAA Joint Propulsion Conference, AIAA Paper 2010-6786, July 2010.

[6] Karabeyoglu, A., Stevens, J., Geyzel, D., Cantwell, B., and Micheletti, D., "High Performance Hybrid Upper Stage Motor," 47th AIAA Joint Propulsion Conference, AIAA 2011-6025, August 2011.

[7] Cantwell, B., Karabeyoglu, A., and Altman, D., "Recent Advances In Hybrid Propulsion," International Journal of Energetic Materials and Chemical Propulsion, Vol. 9, No. 4, 2010, pp. 305-356. doi: 10.1615/IntJEnergeticMaterialsChemProp.v9.i4.20

[8] Young, G., "Metallic Nano-Particles as Fuel Additives in Air-Breathing Combustion," Ph.D. Dissertation, Aerospace Engineering Dept. Univ. of Maryland, College Park, MD, 2007.

[9] Evans, B., Boyer, E., Kuo, K. K., Risha, G., and Chiaverini, M., "Hybrid Rocket Investigations at Penn State University's High Pressure Combustion Laboratory: Overview and Recent Results," 45th AIAA Joint Propulsion Conference, AIAA 2009-5349, August 2009.

[10] Foelsche, R. O., Burton, R. L., and Krier, H., "Boron Particle Ignition and Combustion at 30-150 ATM," Combustion and Flame, Vol. 117, No. 1-2, 1999, pp. 32-58. doi: 10.1016/S0010-2180(98)00080-7

[11] Risha, G. A., Evans, B. J., Boyer, E., Wehrman, R. B., and Kuo, K. K., "Nano-sized Aluminum- and Boron-based Solid-fuel Characterization in a Hybrid Rocket Engine," 39th AIAA Joint Propulsion Conference, AIAA Paper 2003-4593, July 2003.

[12] Thomas, J. C., Petersen, E. L., DeSain, J. D., and Brady, B. B., "Enhancement of Regression Rates in Hybrid Rockets with HTPB Fuel Grains by Metallic Additives," 51st AIAA Joint Propulsion Conference, AIAA Paper 2015-4041, July 2015.

[13] Shin, K. H., Won, J., Tak, H., Choi, S. H., Lee, W., and Lee, C., "A Static Combustion Study on Fuel Rich Propellant for Ducted Rocket Gas Generator," 50th AIAA Joint Propulsion Conference, AIAA Paper 2014-4045, July 2014.

[14] Karabeyoglu, A., Toson, E., and Evans, B., "O/F Shift in Hybrid Rockets," 50th AIAA Joint Propulsion Conference, AIAA 2014-3851, July 2014.

[15] Humble, R. W., Henry, G. H., and Larson, W. J., Space Propulsion Analysis and Design, Space Technology Series, McGraw–Hill, New York, 1995, pp. 365–441.

[16] Zilliac, G., and Karabeyoglu, M. A., "Hybrid Rocket Fuel Regression Rate and Modeling," 42nd AIAA Joint Propulsion Conference, AIAA Paper 2006-4504, July 2006.

Comparative Study of Wing Lift Distribution Analysis for High Altitude Long Endurance (HALE) Unmaned Aerial Vehicle

Faber Y. Silitonga[1)*] and M. Agoes Moelyadi[1)]**

[1)] Department of Aeronautics and Astronautics, Faculty of Mechanical and Aerospace Engineering, Institut Teknologi Bandung, Indonesia.

*faber.yosua@gmail.com **moelyadimochammad@gmail.com

Abstract. The development of High Altitude Long Endurance (HALE) Unmanned Aerial Vehicle (UAV) has been emerged for both civil and military purposes. Its ability of operating in high altitude with long endurance is important in supporting maritime applications.Preliminary analysis of HALE UAV lift distribution of the wing presented to give decisive consideration for its early development. Ensuring that the generated lift is enough to compensate its own weight. Therotical approach using Pradtl's non-linear lifting line theory will be compared with modern numerical approach using Computational Fluid Dynamics (CFD). Results of wing lift distribution calculated from both methods will be compared to study the reliability of it. HALE UAV ITB has high aspect ratio wing and will be analyze at cruise flight condition. The result indicates difference between Non-linear Lifting Line and CFD method.

Keywords: *aerodynamic, lift distribution, HALE UAV.*

Nomenclatures

L	: Lift
L'	: Lift distribution (per chord)
C	: Chord
C_l	: 2D Lift coefficient
C_L	: 3D Lift coefficient
ρ	: Density
P	: Static pressure
V	: Velocity
S	: Wing are
q	: Dynamic pressure
α	: Angle of attack
b	: Span
AR	: Aspect ratio
x,y	: Coordinates

1. Introduction

Indonesia is a very large country that contains of vast amount of islands, not only a huge archipelago, also rich of natural resources the spread from the land throughout the ocean. Large territory which holds thousands of island, challenges to provide modern technology that able to monitor its large area within certain amount of times. Not only being able to monitor large area with long endurance, human resources availability needs to be taken as consideration. HALE UAV ITB (Institut Teknologi Bandung) development is to fulfill the need of reliable autonomous vehicle in order to monitor Indonesian country border and natural resources, that can also be controlled by remote control and telecommand from ground station. ITB bas three years plan to develop HALE UAV that divided into three stages. The first stage produced a 12 m wingspan prototype with four boom configuration with four electric propulsion system for about 20000 ft altitude, as shown in figure 1. First prototype had some problems with side wind effect that cause heavy bending moment, damaging the wing. The second stage, which is currently ongoing project, will have a 16 m wingspan that able to operate at 20000 ft altitude. The second prototype, showed in figure 2, will be using single boom with dual electric motor, which is obviously provides smaller weight than four boom configuration. Larger wingspan means larger aspect ratio, which expected to be producing spanwise lift distribution more similar with rectangular lift distribution approach due to relatively smaller induced drag produced. If the hypothesis is satisfied thus, winglet addition can be ignored that will reducing wing bending moment caused by side wing effect.

Spanwise lift distribution prediction can be used to predict the structural load, which is important in designing the structural initial sizing and configuration of the wing. Prandtl's Non-linear Lifting Line Theory, is choosen rather than Prandtl's Classic Lifting Line Theory due to is ability to predict the non-linear C_l vs alpha curve that usually happened at high angle of attack thus, expected to produce more accurate result with relatively less time consumed, to CFD simulation. On the other hand, CFD requires longer amount of time but calculate more accurate lift distribution. The comparison between both method will test the reliability of Prandtl's Non-linear Lifting Line Theory ability to predict the spanwise lift distribution of HALE UAV ITB.

Figure 1. First generation of HALE UAV-ITB prototype

2. HALE UAV Configurations

2.1. Basic Configurations

HALE UAV-ITB has high wing and single tail-boom configuration with vertical and horizontal tail as stabilizer. The length is 4 m with 16 m wing span. Airfoil that been used for the wing is EMX 07 and for the stabilizers are NACA 0012.

Table 1. HALE UAV-ITB Configuration

MTOW	20 kg
Wingspan	16 m
MAC	0.4 m
Taper Ratio	1
Twist Angle	0 degree
Aspect Ratio	40
S (Wing Area)	6.4 m^2
Altitude	20000 ft

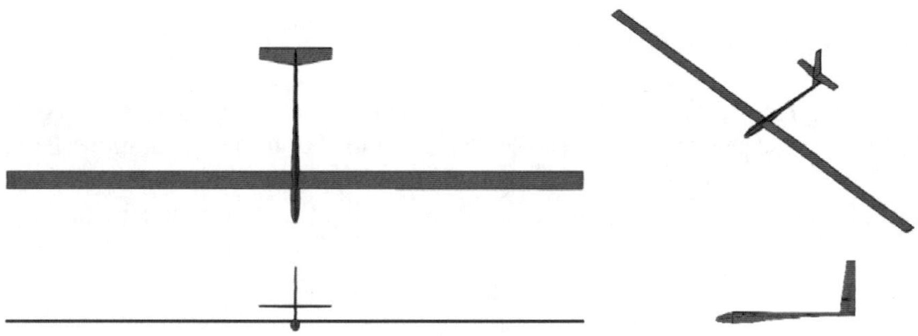

Figure 2. HALE UAV-ITB three view drawing

a.)

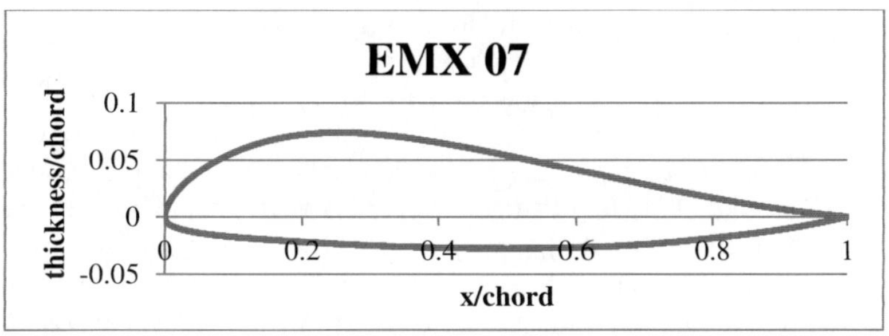

b.)

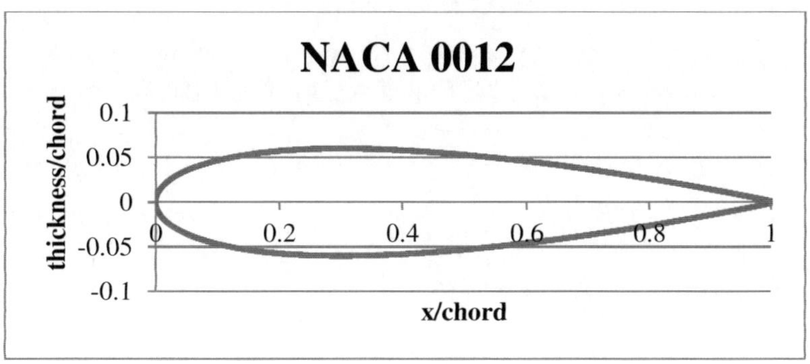

Figure 3. Airfoil: a.) EMX-07 b.) NACA 0012

2.2. Flight Configurations

For surveillance, the designed HALE UAV ITB will fly at cruise stage as follow:

Table 2. HALE UAV ITB Flight Condition

Cruise speed	60 km/hour
Angle of Attack (α)	3 degree
Altitude	20000 feet
Air density	0.73 kg/m^3
Ambient Temperature	250 K
Ambient Pressure	46650 Pa

3. Prandtl Non-linear Lifting Line Theory

The following procedure are Prandtl's Non-linear Lifting Linear Theory method by Anderson[1]. Wing will be divided into a number of k+1 spanwise stations and the initial circulation (Γ) value, at given angle of attack (α), for each station will be assumed as elliptical lift distribution. Then, the induced angle of attack α_i for each section will be calculated by solving the following equation:

$$\alpha_i(y_n) = \frac{1}{4\pi V_\infty}\int_{-b/2}^{b/2} \frac{(d\Gamma/dy)\,dy}{y_n - y} \tag{1}$$

Using effective angle of attack (α_{eff}) at each station can be calculated by using α_i from equation (1) and the wing condition angle of attack, by solving the following equation:

$$\alpha_{eff}(y_n) = \alpha - \alpha_i(y_n) \tag{2}$$

Now, a new Γ distribution is obtained by the latest value of effective angle of attack, and will be compared to the previous value of Γ. If a difference occur between the old and new value of Γ, a new input value of Γ, generated from $\boldsymbol{\Gamma_{input} = \Gamma_{old} + D \cdot (\Gamma_{new} - \Gamma_{old})}$ where D is a damping coefficient[1], to perform iteration until the latest value of Γ agree with its previous value within specified margin of error. After that, Kutta – Joukowski theorem, shown by equation (3), can be performed using the Γ value to generate lift at each section.

$$L'(y_n) = \rho_\infty V_\infty \Gamma(y_n) = \frac{1}{2}\rho_\infty V_\infty{}^2 c_n (c_l)_n \tag{3}$$

Total lift that produced by the wing will be obtained by integrating the value of $L'(y_n)$ spanwise.

4. Computational Fluid Dynamics (CFD) Simulation

Computational Fluid Dynamics (CFD) will be used as the numerical approach to calculate the spanwise lift distribution of the wing. In this simulation ANSYS CFX will be used as the solver of CFD simulations. CFX solver calculation is based on Reynolds Averaged Navier Stokes (RANS) method, which is derived from Direct Navier Stokes (DNS) method. Instead of Large Eddy Simulation (LES) method, RANS method was choosen because it delivers enough accuracy within faster processing rather than LES, even though it has higher accuracy but takes a lot of memory and running time. The following equations are the governing equations used by ANSYS CFX Solver, including continuity (4), momentum (5) and total energy (7):

Continuity eq.

$$\frac{\partial \rho}{\partial t} + \nabla \cdot (\rho V) = 0 \tag{4}$$

Momentum eq.

$$\frac{\partial (\rho V)}{\partial t} + \nabla \cdot (\rho V \times V) = -\nabla p + \nabla \cdot \tau + s_M \tag{5}$$

Stress tensor,τ,
strain rate relation

$$\tau = \mu(\nabla V + (\nabla V)^T - \frac{2}{3}\delta \nabla \cdot V) \tag{6}$$

Total energy eq.

$$\frac{\partial(\rho h_{tot})}{\partial t} - \frac{\partial p}{\partial t} + \nabla \cdot (\rho V h_{tot}) = \nabla \cdot (\lambda \nabla T) + \nabla \cdot (V \cdot \tau) + V \cdot s_M + s_g \tag{7}$$

Total enthalpy, h_{tot}

$$h_{tot} = h + \frac{1}{2}V^2 \tag{8}$$

Shear Stress Transport (SST) method will be used to evaluate the simulation turbulence viscous effect.

4.1. Simulation Model

CFD simulation method is done by the following workflow scheme:

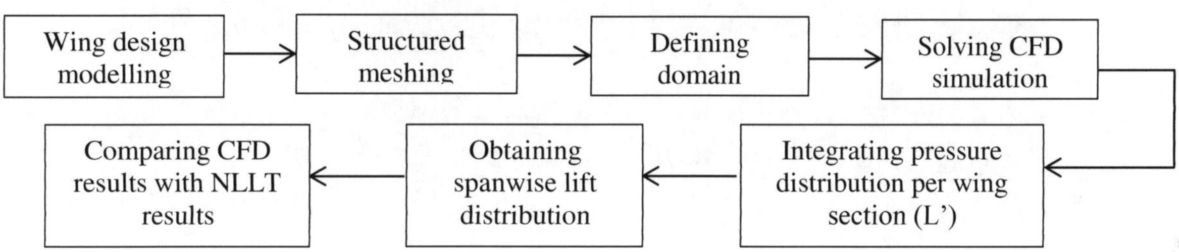

Figure 4. CFD workflow scheme

The simulation will be simplified as a wing-only half spanwise thus, means actual total lift produced is two times the lift calculated by the simulation. The simulation will be using viscid assumption, to compare the lift distribution result of viscid assumption, by CFD, with invisced assumption,by Non-linear Lifting Line Theory. Structured mesh blocking method by ICEM CFD will be used to the wing simulation domain. Computational domain that being used are:

1) INLET boundary condition in front of leading edge
2) SYMMETRY boundary condition for wing root side of the domain
3) WALL boundary condition for the wing itself
4) OPENING boundary condition for the other domain.

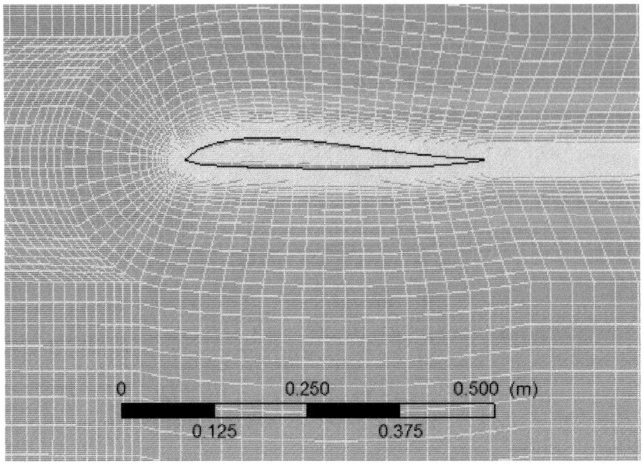

Figure 5. Side view of generated meshes for wing domain

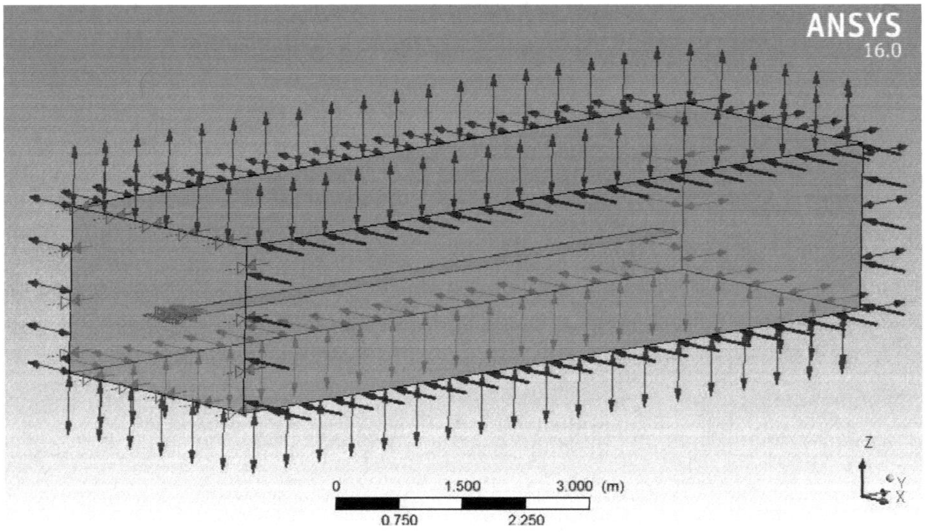

Figure 6. Computational domains for wing

5. Result

Wing will be divided into 10 section for extracting the CFD results (y = 0, 0.8, 1.6, 2.4, 3.2, 4.0, 4.8, 5.6, 6.4, 7.2, 7.95 m) to obtain Lift/chord (L') for each station which will be used to obtain the spanwise lift distribution. Chord wise static pressure distribution for each section will be extracted, then by integrating the pressure distribution for upper and lower chord we will obtain L' for each section.

Here are some notable pressure distribution graph alongside airfoil chord at section y = 0, 7.2, and 7.95 m

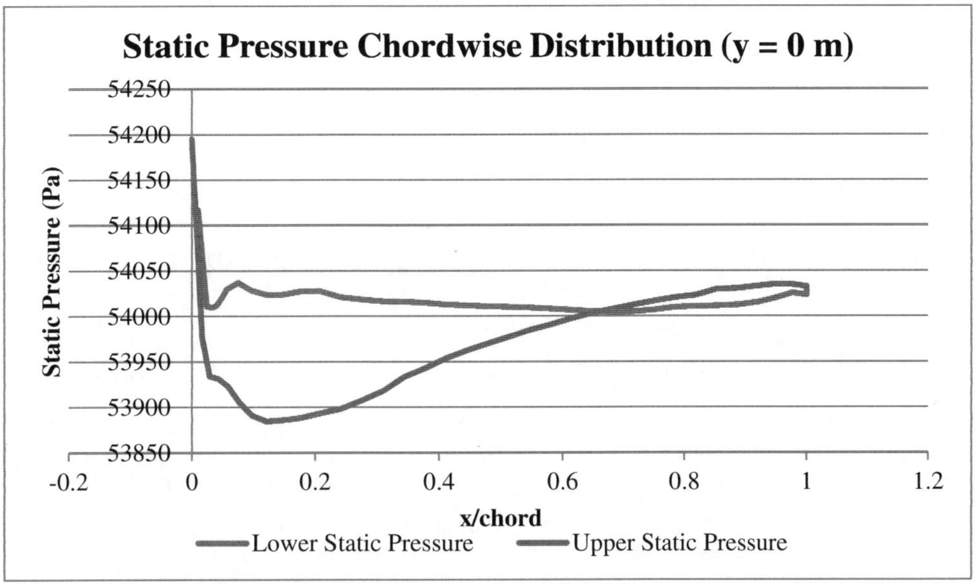

Figure 7. Static pressure distribution chordwise at section y = 0 m

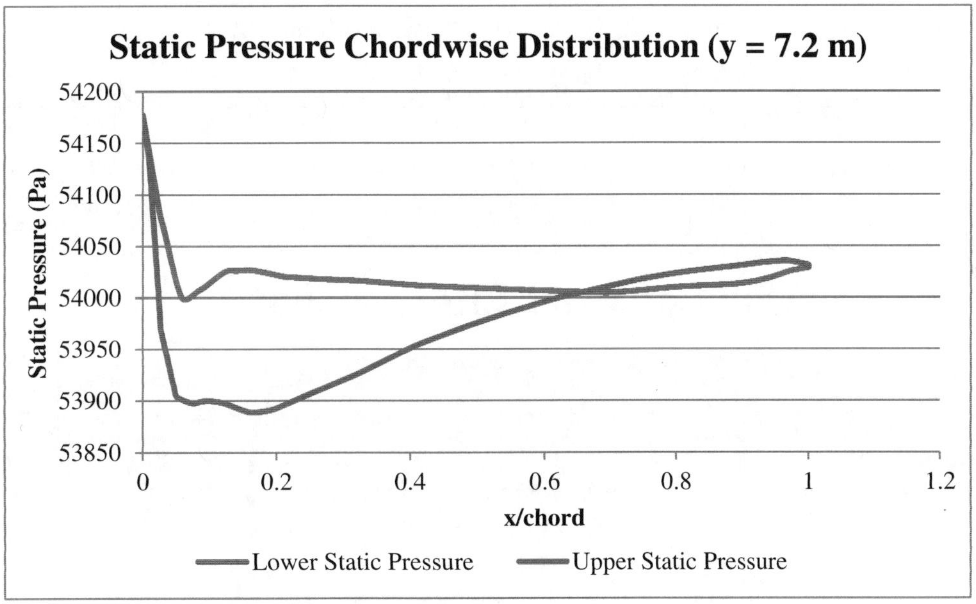

Figure 8. Static pressure distribution chordwise at section y = 7.2 m

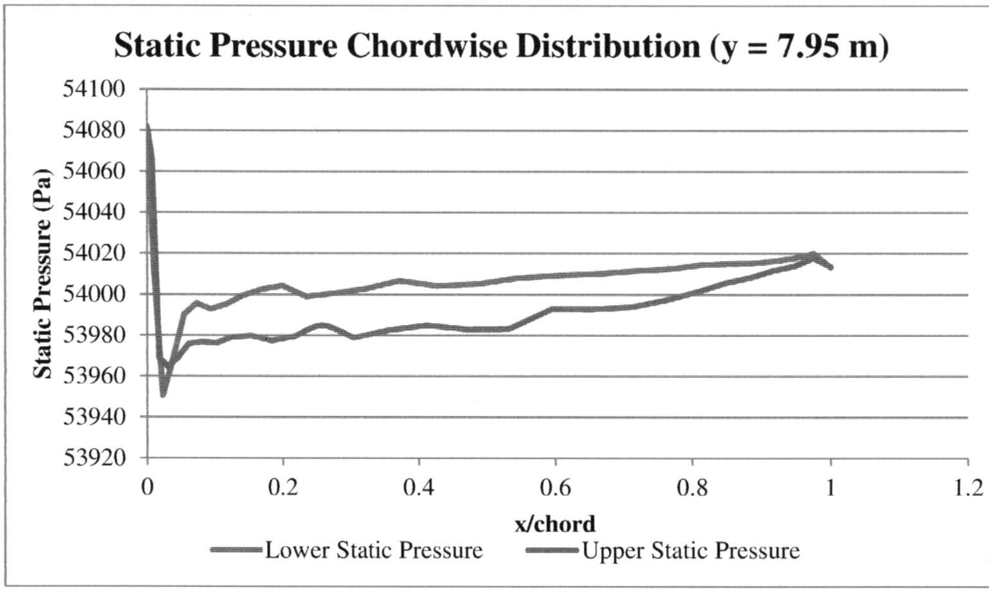

Figure 9. Static pressure distribution chordwise at section y = 7.95 m

By integrating the pressure distribution, lift/chord (L') for each section will be obtained.

Table 3. Lift/chord for each section

Section	1	2	3	4	5	6	7	8	9	10	11
y (m)	0	0.8	1.6	2.4	3.2	4	4.8	5.6	6.4	7.2	7.95
L' (N/m)	18.97	18.90	18.88	18.87	18.84	18.78	18.76	18.76	18.60	17.78	6.26

a.)

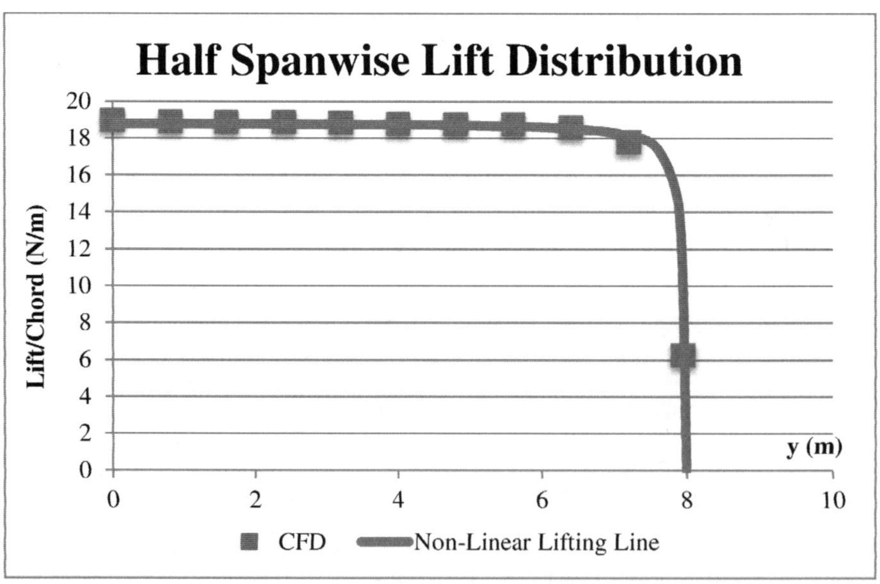

b.)

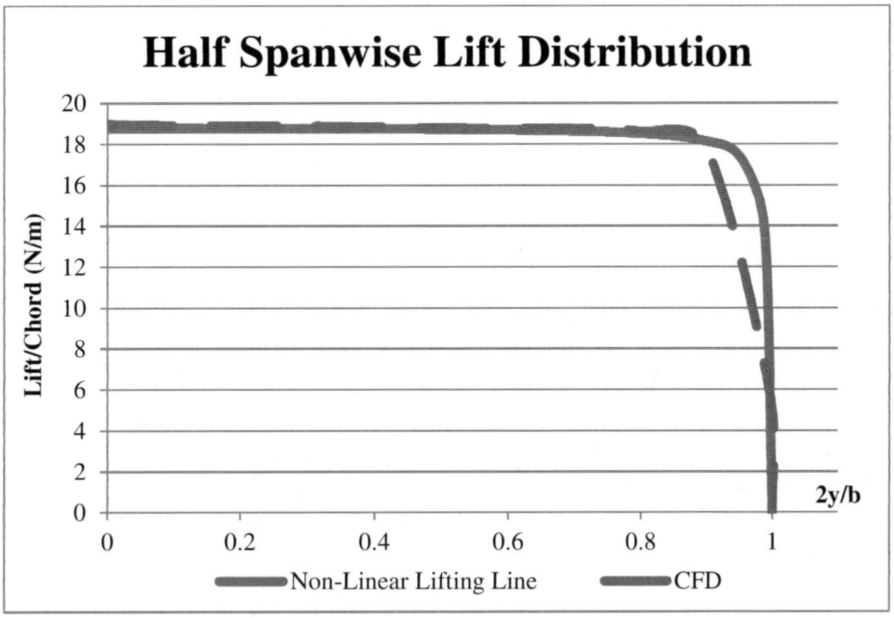

Figure 10. NLLT and CFD lift distribution comparison a.) CFD as data per section b.) CFD as curvature

By integrating the spanwise lift distribution and times it by two we can obtained the total lift performed by whole wing.

Table 4. Total Lift produced and margin of difference by NLLT and CFD method

Total Lift		
CFD Method	**288.38**	N
NLLT Method	**292.59**	N
Margin of Error	**1.46**	%

6. Analysis and Discussion

There is a slight difference of total lift calculated by CFD and NLLT with very little margin of error, around 1.46% , which is still an acceptable value. The distribution of lift spanwisely, according to Figure 10, shows very littly difference distribution besides the wing tip area. Those slight difference are suspected to be caused by mathematically error of the integrating method that been used, which is trapzoidal rule, in order to achive lift/chord for each section. But the trend of distribution is quite the same and still acceptable. It is also due to the simulation that operated in low angle of attack, which is still within the linear region of Cl vs alpha curve thus, lead to more accurate NLLT calculation. Lift distribution differences between both method around wing tip are suspected to be caused by the method of calculating induced drag, which is caused by vortex that created by flow leakage from lower to upper wing. CFD calculation relatively simulate more accurate results towards real condition, due to it's ability to calculate the static pressure distribution then derive it into induced drag effect to lift distribution. In other hand, NLLT calculate the circulation instead of static preasure distribution which caused less accuracy than CFD calculation. But, the difference between both method is still small for this case due to the wing configuration that has high aspect ratio and constant chord size without any twisting.

7. Conclusion

The application of NLLT to predict wing spanwise lift distribution for preliminary analysis is reccomended for HALE UAV, due to it's assured accuracy, small margin of error (1.46%) against CFD calculation results, and also requires relatively easier and shorter amount of time to perform.

Acknowledgements

This research was supported by ITB and Direktorat Riset Teknologi dan Pendidikan Tinggi (RISTEKDIKTI).

References

[1] Anderson, John D. *Fundamentals of Aerodynamics (Fourth Edition)*. McGraw-Hill Education, New York. 2009. (Book)
[2] Anderson, John D. *Computational Fluid Dynamics*. McGraw-Hill Education, New York. 1995. (Book)
[3] Carlos E. S. Cesnik, Patrick J. Senatore, Weihua Su, Ella M. Atkins. *X-HALE: A Very Flexible UAV for Nonlinear Aeroelastic Tests.* The University of Michigan. 2010. (Paper)
[4] NACA 0012, http://airfoiltools.com/airfoil/details?airfoil=n0012-il
[5] EMX-07, http://www.aerodesign.de/profile/profile_sp.htm#emx07
[6] Rapinett, Annabel. *Zephyr: A High Altitude Long Endurance Unmanned Air Vehicle..* Department of Physics University of Surrey.2009 (Paper)

IOP Conf. Series: Journal of Physics: Conf. Series **1005** (2018) 012037 doi:10.1088/1742-6596/1005/1/012037

Obstacle Avoidance for Quadcopter using Ultrasonic Sensor

Muhammad Fazlur Rahman, Rianto Adhy Sasongko

Faculty of Mechanical and Aerospace Engineering, Institut Teknologi Bandung, Indonesia

Email: fazlur.rahman.ae@gmail.com

Abstract. An obstacle avoidance system is being proposed. The system will combine available flight controller with a proposed avoidance method as a proof of concept. Quadcopter as a UAV is integrated with the system which consist of several modes in order to do avoidance. As the previous study, obstacle will be determined using ultrasonic sensor and servo. As result, the quadcopter will move according to its mode and successfully avoid obstacle.

1. Introduction

Application of obstacle avoidance in UAV is highly needed nowadays since it grows rapidly. The avoiding system consist of two main parts which consist of detecting obstacle within a safe operation range and determining a safe flightpath in avoiding the obstacle [3].

Several flight controllers have been used widely but none of them include an obstacle avoidance system. Previous paper discussed the obstacle detection using ultrasonic sensor [1] and this paper discussed about its implementation in quadcopter. The detected obstacle will be used as a data to construct an ellipsoid restricted zone to determine the avoidance path.

Figure 1 illustrate the ellipsoid restricted zone which will be used to generate a contact point as the guidance. It use an optimal path between the heading angle, flight velocity, ellipsoid shape, and clearance [5]. The method is in default used for a fixed wing type of aircraft. Since this research uses quadcopter, some adjustment to the flight manoeuvre and a simpler contact point algorithm is used.

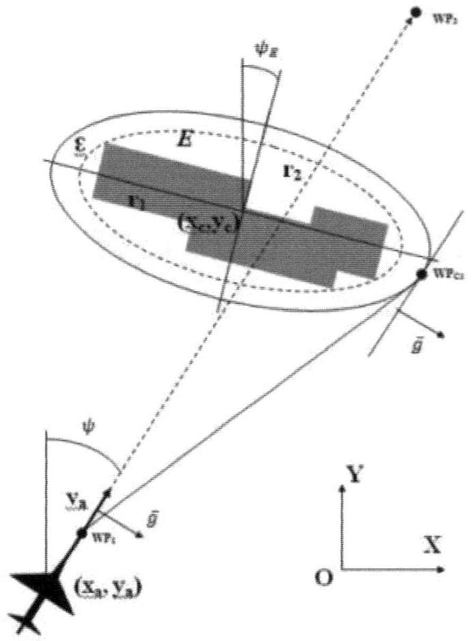

Figure 1. Ellipsoid Restricted Zone [4]

2. Hardware

Designed system should be able to avoid the obstacle using on board processing. Flight between waypoints will be provided by the flight controller and the avoidance will be performed by microcontroller. All the hardware mentioned on previous section combined as shown in Figure 3.

2.1. Arduino

Arduino UNO is a microcontroller using ATMmega328. It provides 14 digital input/output pins (including 6 PWM output pins), 6 analog pins, USB connection, and a power connector [6]. It will be used to receive information from sensor, communicate with ground station, and control another device to perform obstacle avoidance.

2.2. Pixhawk

Pixhawk is an open source autopilot or flight controller developed by Computer Vision and Geometry Lab of ETH Zurich and Autonomous System Lab with collaboration from other researchers. Developed from 3DRobotics and ArduPilot Group, Pixhawk mainly used for hobby, academic, and industrial purposes.

Pixhawk comes with several features such as: 168 MHz Cortex M4F CPU with 256 kb RAM and 2 MB Flash; 3 axis accelerometer, gyroscope, magnetometer, and barometer; Integrated backup, override, and failsafe processor with mixing; and microSD slot, 5 UARTs, CAN, I2C, etc [7].

On this research, quadcopter firmware is used to match the platform. User input comes with a pulse-position modulation (PPM) signal and the output is pulse-width modulation (PWM) signals. Since Arduino as the microcontroller only provides PWM signal output, PWM to PPM converter module is used to match the communication.

2.3. Ultrasonic Sensor

HRLV-MaxSonar-EZ Products is an ultrasonic sensor provided by MaxBotix. The sensor capable of 1 mm resolution with maximum range of 5000 mm. Object closer than 30 cm will be counted as

30 cm. Another specification is the output signal comes in various form such as analog voltage, RS232 or TTL serial, and pulse-width. The sensor operates from 2.5V to 5.5V. With all the specification and its small and light weight criteria, the sensor is suitable for research application yet comes within affordable price.

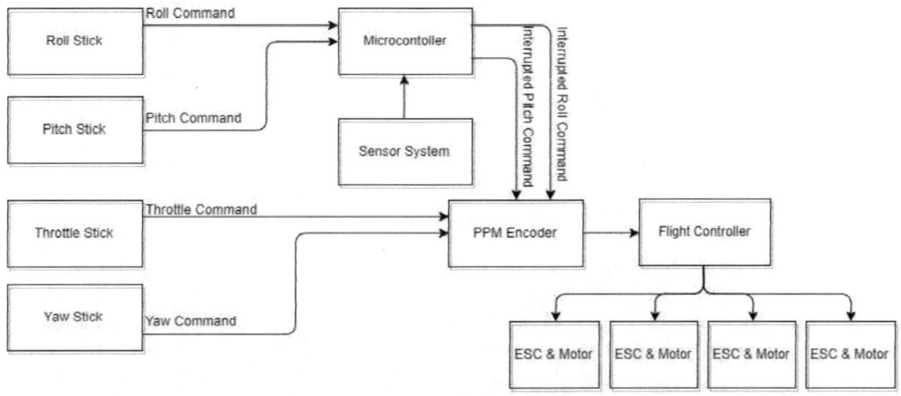

Figure 2. System Architecture

2.4. Servo Gimbal

Servo gimbal used in this research is a 2-axis gimbal, in pan and tilt axis. Pan axis used in obstacle detection by using it to rotate ultrasonic sensor within defined angle. The tilt axis function as a stabilizer so the ultrasonic will always pointed frontward.

2.5. Telemetry

Several data such as distance to obstacle, time stamp, mode, and user input need to be recorded so further analysis can be done. Telemetry is the easiest way to record the data. All those variables inside microcontroller will be sent to laptop via Arduino IDE Serial Monitor. The data stored in excel file and being processed later.

2.6. Airframe

Quadcopter is chosen as the airframe. First consideration is this type of airframe have the ability to hover so lower computing ability of microcontroller is acceptable. Another one is the control requirement is relatively simple as user command directly turns into a non-coupling movement of the quadcopter. All the components have been mounted to the aircraft as shown in figure 3.

Figure 3. Integrated Hardware

3. Microcontroller Algorithm

3.1. Obstacle Avoiding Algorithm

The main mission of this algorithm is to enable the quadcopter fly between predefined waypoints and to avoid obstacle between those waypoint if it exist. There are 5 mode for the algorithm from the quadcopter starts to take-off manually, fly autonomously, and avoiding the obstacle.

Several parameters are made as switch between modes. First parameter is *clear distance*, defined as the minimum distance the quadcopter can fly autonomously. Another one is *safe distance* as the minimum distance the quadcopter can start scanning. On the avoidance mode where quadcopter aimed to fly forward and side at the same time, the distance to obstacle should've decreased. So another parameter is defined that is *clear to avoid distance*, which the quadcopter should not move any closer than this distance when avoiding the obstacle. Illustration of the parameters is shown in Figure 4.

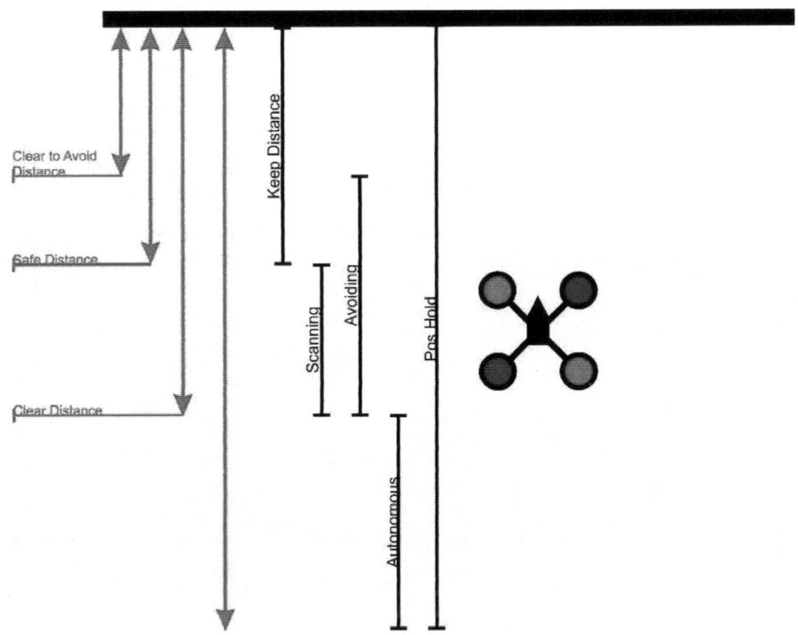

Figure 4. Avoiding Scenario

Position Hold, or generally known as PosHold, is a flight mode which designed to maintain a constant location, heading, and altitude by utilizing data from inertial measurement unit (IMU), GPS module, and compass. This flight mode is preferable rather than others since it provide a natural control sense by directly control vehicle's lean angle based on pilot stick inputs [7].

To control horizontal position on the ground, roll and pitch control stich is utilized with default maximum lean angle of 45 degrees. It means when the stick hit the maximum position on the remote control, the command will be interpreted as attitude reference to keep lean angle at that position. This maximum lean angle can be adjusted with ANGLE_MAX parameter using ground station control [7].

Autonomous mode, known as Auto, is a flight mode which enabled the quadcopter to follow a pre-programmed mission script which is stored in the flight controller by setting waypoints as the navigation command and some other action command [ardupilot.org]. Since this flight mode combine altitude hold (AltHold) and Loiter to control altitude and position, an attempt of those flight mode should be done before utilizing this Auto. As well as the positon hold, Auto flight mode require the same conditions applied before it can take-off.

Keep distance mode is a part of avoidance mode which prevent the quadcopter from hitting the obstacle. This mode is activated whenever the quadcopter distance to obstacle is less than safe distance. Autonomous mode will be switch to position hold mode and microcontroller will send signal to flight controller to fly backwards based on measured distance to the obstacle. The rate of quadcopter fly backwards decreases as the distance to obstacle increases.

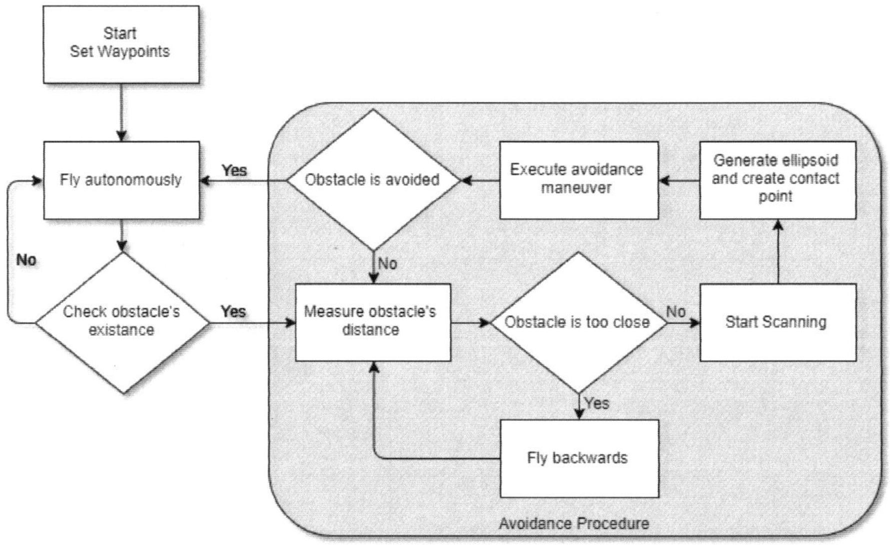

Figure 5. Avoiding Procedure Flowchart

The avoiding process will be start from scanning the obstacle first during scanning mode. Since ultrasonic sensor only capable of measuring distance for a single point, it will be rotated by using servo. Rotating ultrasonic is in limited angle since reflection angle higher than 25 degree will not be detected [2]. These data later will be quantified into an elliptical shape obstacle and a contact point will be resulted as path planning method result.

Figure 5 shows the relation between autonomous flight and the avoidance procedure. Keep distance, scanning, and avoidance mode is represented by fly backwards block, start scanning, and execute avoidance manoeuvre respectively. All these mode is depended on the sensor measurement.

After the contact point is obtained, the quadcopter changes to avoiding mode. In this mode, microcontroller send PWM signal to pitch and roll channel based on the planned path. Quadcopter will change to waypoint following mode after the obstacle is already passed. It happens when the measured distance is more than the clear distance. Since quadcopter needs to approach the obstacle first, it is possible that the quadcopter fly too close to the obstacle. Therefore, the flying mode will switch to keep distance.

3.2. Signal Processing Algorithm

The quadcopter used ultrasonic as a sensor to measure distance and the data will be used to trigger the avoidance mode. During flight, the measurement experience noise which will disturb the flight mode. Therefore, a filter should be applied to reduce those noise.

Filter used in this research is a *moving median*. This filter have a purpose to remove an outlier in the data which can be interpreted as noise. For instance, there are six data to be computed its median and there is one data which is relatively high than the others. This data will be sorted to the end of the window set and never used as data [8].

$$y[n] = median(x[n], x[n-1], x[n-2], \dots, x[n-k]) \quad (1)$$

$y[n]$ = Chosen data
$x[n]$ = Data from measurement

n = Current data
k = Size of filter window

Figure 6 shows the result of these median filter with 11 data window applied to a moving quadcopter. It is noticed that some noise successfully reduced as in 18 seconds in sub-figure (a) and in 12 second in sub-figure (b). This noise can cause the quadcopter directly switched to keep distance mode but fortunately stay at autonomous mode because of the filter.

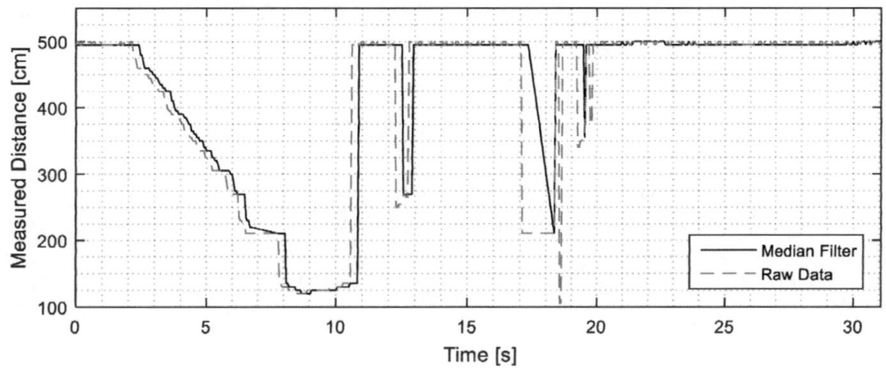

(a)

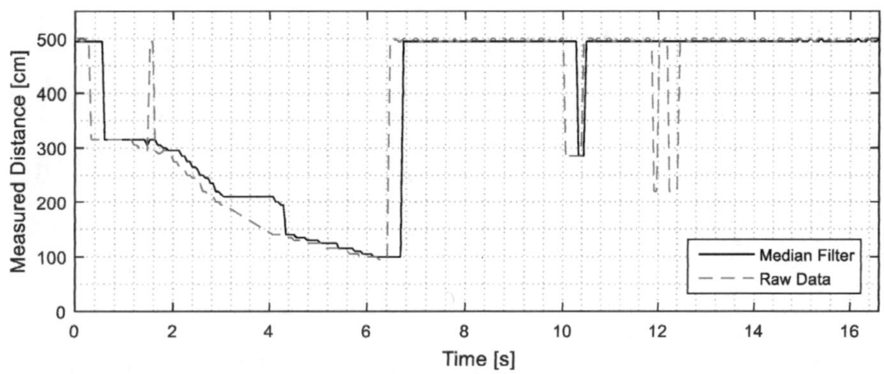

(b)

Figure 6. Signal Processing

4. Results and Discussion

Two avoiding scenarios being tested, which result all flight mode being tested. Firstly, the quadcopter's waypoint navigation speed is at 75 cm/s and the second one is at 60 cm/s. The distance parameters interval is set at constant which is 210 to 140 cm to available for scanning. Gain for the avoidance maneuver is set to be 40 value which will affect the speed during avoiding.

Table 1. Test Parameters

Case No.	Waypoint Navigation Speed [cm/s]	Clear - Save Distance [cm]	PWN Gain for Avoidance [µs]
1	75	210-140	40
2	60	210-140	40

Figure 7 and 8 show roll and pitch command during flight test. From around 3.5 to 5 seconds the signal goes low in a constant value for roll and decreasing value for pitch which shows the quadcopter is moving to left and near the obstacle to avoid it. While 5 to 6 seconds is the keep distance maneuver which is shown by a positive pitch which means the quadcopter is fly backwards.

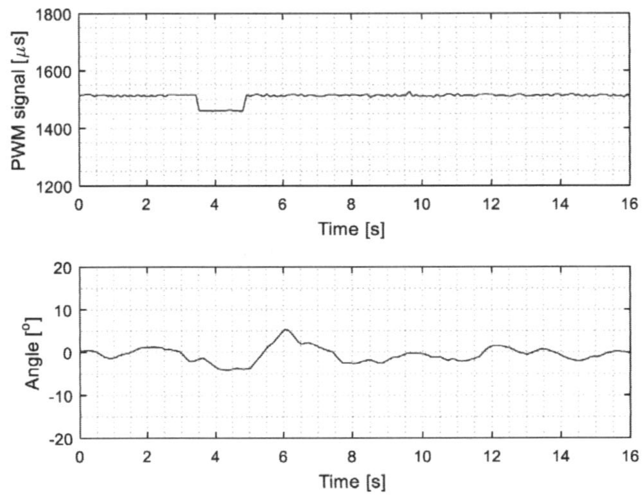

Figure 7. Roll Command and Attitude - Avoidance Case 1

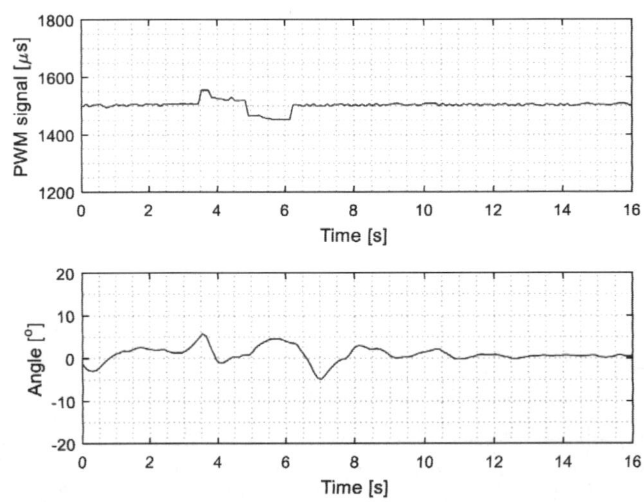

Figure 8. Pitch Command and Attitude - Avoidance Case 1

IOP Conf. Series: Journal of Physics: Conf. Series **1005** (2018) 012037 doi:10.1088/1742-6596/1005/1/012037

Figure 9 shows the measured distance and the flight mode as shown by Arduino. It is shown that the quadcopter slowly reduces it distance and successfully avoid the obstacle when the measurement is raising to about 500 cm. The rest shows a constant reading except for an error which does not change the flight mode.

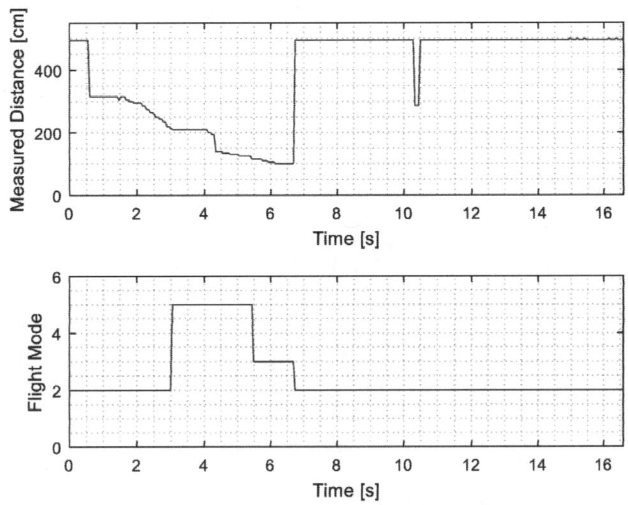

Figure 9. Measured Distance and Flight Mode Based on Arduino – Case 1

Figure 10 shows the elliptical quantification result from the Arduino. The blue point, which shows the contact point is not located in the ellipsoid due to a roundup error. The blue-star is the indicator of measured obstacle in front of it which is acceptable since the quadcopter is in fact do not fly perpendicular to the obstacle when it start to scan. Figure 11 shows the trajectory of quadcopter which successfully follow this guidance but then it fly too close to the obstacle and switch to keep distance mode to avoid crash.

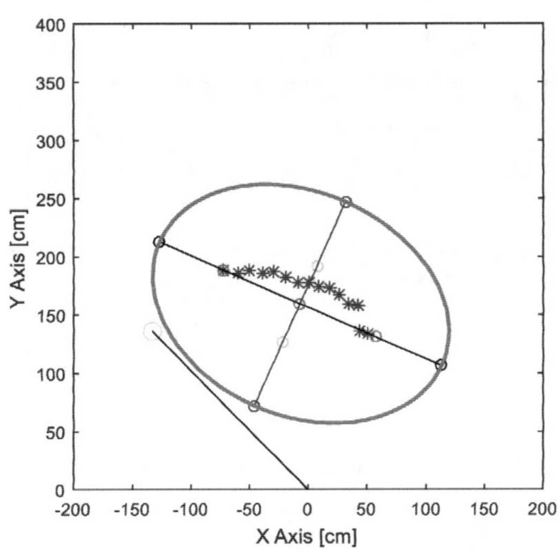

Figure 10. Elliptical Quantification Result – Case 1

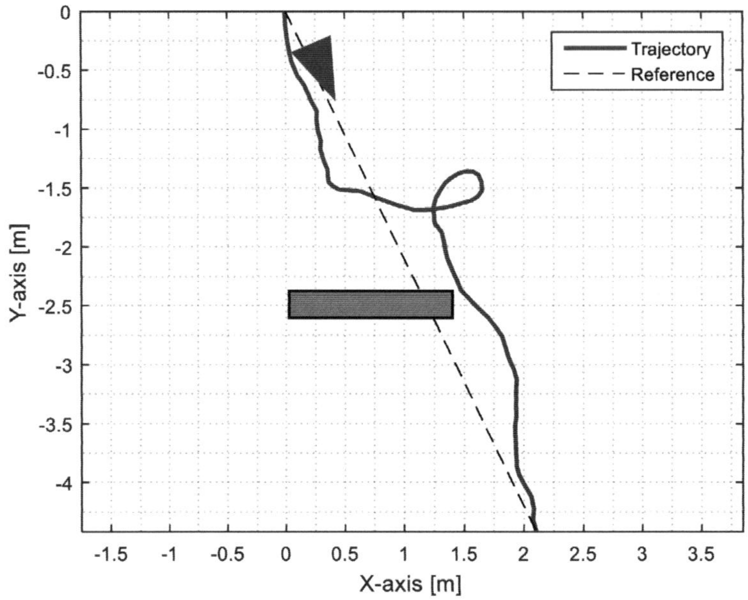

Figure 11. Trajectory – Case 1

Figure 12 and 13 show roll and pitch command during flight test for this second case. From around 5 to 7.5 seconds the roll signal is the same as previous, which can be explained by the same guidance. While the pitch signal is relatively high at first attempt of avoidance and decreased slowly until successfully avoid the obstacle.

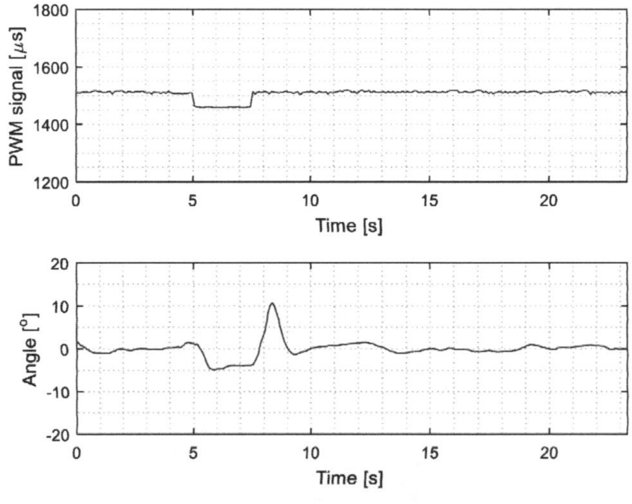

Figure 12. Roll Command and Attitude - Avoidance Case 2

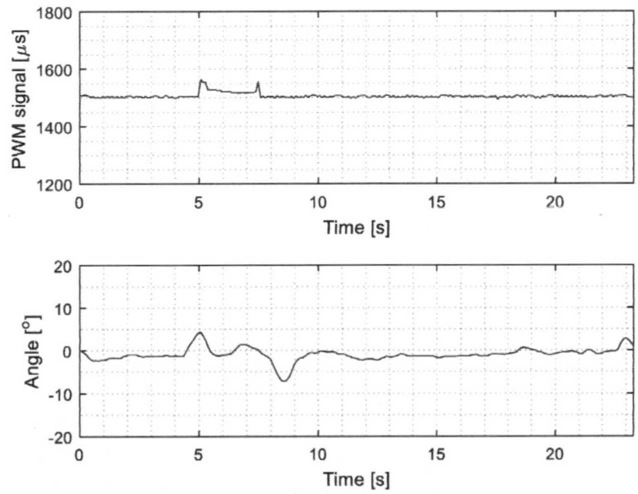

Figure 13. Pitch Command and Attitude - Avoidance Case 2

Figure 14 shows the distance is decreased start from 5 to the next 2.5 seconds. It is the measurement when the quadcopter fly near the obstacle. Next is the sudden raise which shows the obstacle is avoided and the flight mode changed to autonomous mode until it completes the whole mission.

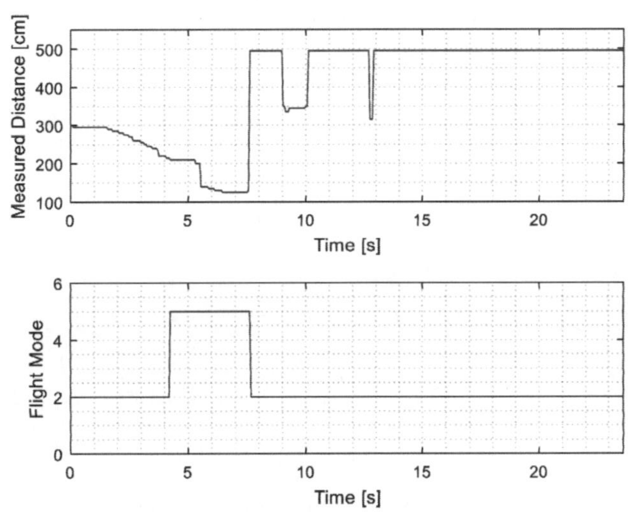

Figure 14. Measured Distance and Flight Mode Based on Arduino – Case 2

Figure 15 shows the elliptical quantification in this second test flight. As the previous case, the quadcopter tend to turn left as it is the most optimal path for this method. The trajectory, as shown in figure 16, shows the quadcopter successfully follow the guidance and keep it from hitting the obstacle before finally turn to autonomous mode again. In the early maneuver of avoidance mode, the quadcopter fly in combination command between roll and pitch and the next maneuver mostly consist only in roll direction since the quadcopter keep it distance to the obstacle.

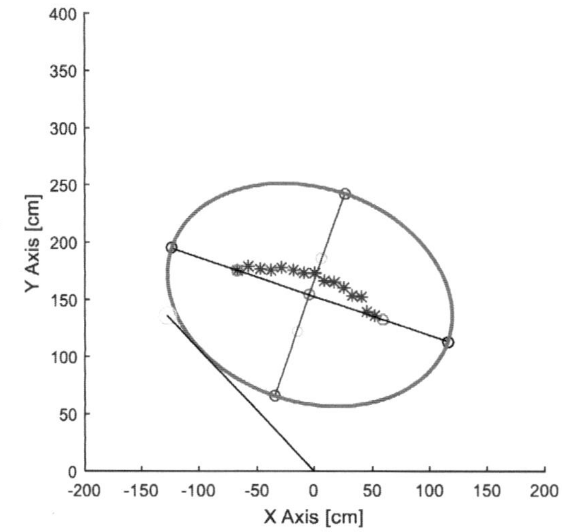

Figure 15. Elliptical Quantification Result – Case 2

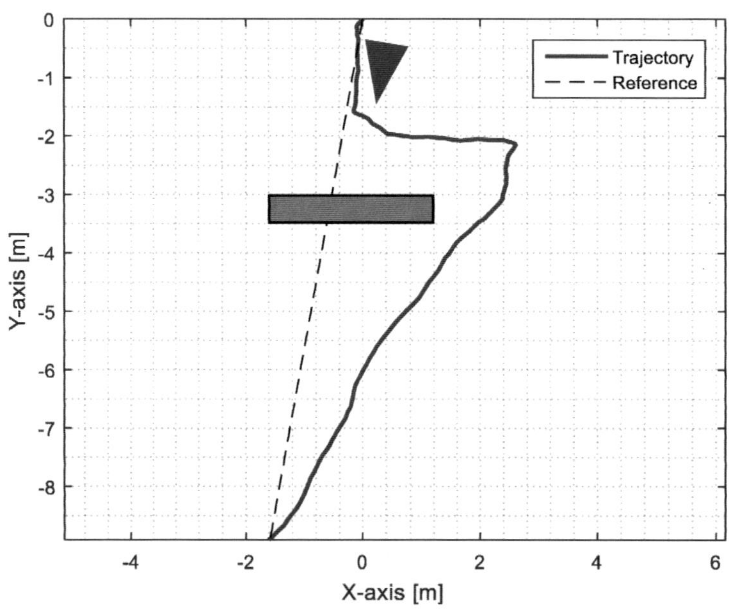

Figure 16. Flight Trajectory – Avoidance Case 2

As a comparison, different waypoint navigation speed lead to different measured distance when the quadcopter enter the scanning mode. A faster velocity means closer obstacle when it start to avoid. This can be an adjustment when the quadcopter fly at a higher speed that the clear distance can be reduced.

5. Conclusion

The implementation of obstacle avoidance for quadcopter using ultrasonic sensor have been done. The result shows that quadcopter successfully change mode to avoiding mode or keep distance mode when

IOP Conf. Series: Journal of Physics: Conf. Series **1005** (2018) 012037 doi:10.1088/1742-6596/1005/1/012037

the obstacle's distance is less than defined distance parameter. The quadcopter then switch to scanning mode and result a set of distance from rotating ultrasonic. When obstacle is not detected, the quadcopter will switch to autonomous mode and continue waypoint following.

All flight command, attitude, and trajectories have proven the quadcopter can avoid the obstacle. An additional evidence of the measured distance also have been shown with its flight mode. Though, a more various flight speed and avoiding speed should be conducted.

References

[1] Rahman, M. F., Wisnu, Sasongko, R.A. Obstacle's Contour Detection using Ultrasonic Sensors. *Advance in Aerospace Science and Technology in Indonesia, Vol. 1*

[2] Borenstein, J. and Koren, Y. Obstacle Avoidance with Ultrasonic Senors, *IEEE J.*, 1998, pp.213-218

[3] Rawikara, S. S. Desain dan Simulasi Sistem Obstacle Avoidance untuk Misi *Waypoint Following* pada Pesawat Udara Nirawak, *Tugas Akhir Program Studi Aeronautika dan Astronautika FTMD ITB*, 2013

[4] Sasongko, R. A., Sembiring, J. Muhammad, H. Design of Obstacle Avoidance Algorithm for Waypoint Following Control System, *Regional Conference in Mechanical and Aerospace Technology,* Manila, Phillipines, 2011

[5] Sasongko, R. A., Rawikara, S. S., Tampubolon, H. J., UAV Obstacle Avoidance Algorithm Based on Ellipsoid Geometry, *Journal of Intellegent and Robotic System*, 2017

[6] http://www.arduino.cc/en/Reference

[7] http://www.ardupilot.org

[8] MATLAB and SIMULINK user guide, Mathworks Inc., 2007

IOP Conf. Series: Journal of Physics: Conf. Series **1005** (2018) 012038 doi:10.1088/1742-6596/1005/1/012038

Solar Cell to Support Perpetual Flight of High Altitude Long Endurance UAV ITB

Muhammad Luqmanul Hakim*, Faber Y. Silitonga, Nurhayyan H. Rosid and Dr. Ing. Mochammad Agoes Moelyadi

Department of Aeronautics and Astronautics, Faculty of Mechanical and Aerospace Engineering, Institut Teknologi Bandung, Indonesia
* luqmanulhakimmuhammad@yahoo.co.id, faber.yosua@gmail.com, nurhayyan.halim.rosid@s.itb.ac.id, moelyadi.mochammad@gmail.com

Abstract. Research on a High Altitude Long Endurance (HALE) Unmanned Aerial Vehicle (UAV) is currently being conducted at Bandung Institute of Technology to reach the flight duration needed and to get the solution of today's challenges, minimizing pollution. Besides the good aerodynamic efficiency needed, energy resource is now becoming important. The energy resource must have a good endurance, easy to get, and of course, less pollution. Discussion in this paper is about the analysis of power needed by HALE UAV while takeoff and cruise flight conditions, and then determine the amount of solar cell and battery needed by the UAV.
Key Words: HALE UAV, solar cell, power consumption

Nomenclatures

T	*Thrust*	P_{SC}	*Power that stored by one solar cell*
D	*Drag*	P_{TSC}	*Power that stored by all solar cell*
W	*Weight*	ρ	*Air density*
L	*Lift*	V	*Air speed*
S	*Wing area*	CL	*Lift coefficient*
A_{SC}	*One solar cell area*	CD	*Drag coefficient*
A_{TSC}	*All solar cell area*	H	*Designated altitude*
γ	*Climbing angle*	X	*Climbing distance*
φ	*Bank roll angle*	t	*Time*
P_a	*Power available*	η_{motor}	*Motor efficiency*
P_r	*Power required*	η_{ESC}	*Electronic Speed Controller efficiency*
P_e	*Excess power*	η_{MPPT}	*MPPT efficiency*
P_{av}	*Avionic power*	η_{SS}	*Supporting system efficiency*
$P_{payload}$	*Payload power*	N_{BTT}	*Amount of batteries*
P_{servos}	*Servos (actuator) power*	N_{series}	*Amount of series batteries setting*
P_{SS}	*Supporting system power*	N_{SC}	*Amount of solar cell*
P_C	*Power for charge*	W_{cell}	*Weight of one cell battery*
P_{CC}	*Power for charge one cell battery*	W_{BTT}	*Wright of all batteries*

1. Introduction

Communication is one of the important things for human life. As the earthlings, a social human must have good communication skills in every way so a human can express whatever they want [1]. Especially in military aspects, communication is the most vital and important to make the command and secret data transmission transferred safely. Either long or short-range communication between earthlings, this communication activity uses satellite to widen the outreach and faster the data transmission [2]. Therefore, it's a good thing if the satellite is owned by the military department itself

IOP Conf. Series: Journal of Physics: Conf. Series **1005** (2018) 012038 doi:10.1088/1742-6596/1005/1/012038

to guarantee the secret communication. However, to make, to launch, and to operate the satellite is very expensive. In addition, if the launching is failed, the satellite couldn't reach the orbit wanted or even broke when on a mission. The cost to fix it will be very expensive because there will be a spacecraft that needed to be launched again to fix it. So, the vehicle to substitute the satellite needed must be cheaper and easy to operate. HALE UAV is the answer for the challenge. The ability to fly high and has high endurance can be used as a repeater that can receive a signal from the earth's surface, amplify it, and transmit it again to the earth's surface.

To obtain a high endurance required a qualified energy source. As we know, most electric motors that used for UAVs use batteries that will eventually run out of energy. Candidate system that can overcome the limitation one of them is a solar cell. Solar cells are able to convert energy from the sun into electrical energy. Because it requires the sun, the weakness of solar cell is can not be used at night.

This paper performs a study to overcome weaknesses in the solar cell so it can produce a high enough endurance in this HALE UAV by finding the exact configuration of solar cell energy sources with batteries in which the solar cell will act as a major energy converter from the sun light into electricity and the batteries as a storage of energy from the solar cell [3]. To find out how much energy the energy source must produce is perform the calculation of the power required in some flying conditions such as take off and cruise.

2. HALE UAV Model

2.1. Model Geometry

HALE UAV use rectangular wing type with fuselage design that often used for a glider. Fuselage model designed aerodynamically to be able to minimize drag. Four brushless motors are planned to be mounted on its wing. Tail configuration uses middle HTP. All electrical components are kept inside the fuselage, except its propulsion system (motor, ESC, Battery) are kept in its own compartment. Here is it the design of the vehicle by design group of HALE UAV ITB using CATIA V5 software.

Figure 1 HALE UAV model

IOP Conf. Series: Journal of Physics: Conf. Series **1005** (2018) 012038 doi:10.1088/1742-6596/1005/1/012038

Table 1. HALE UAV Specification

Wing Airfoil	EMX-07
Wing Chord (m)	0.42
Wing Span (m)	16
Wing Area (m^2)	6.72
MTOW (Kg)	20
Wing Loading (Kg/m^2)	2.976
Material	Fiberglass and carbonfiber composites

2.2. Flight Operation

HALE UAV surveillance mission starts when approaching its cruise condition at desirable altitude and condition, as listed in the following table:

Table 2. HALE UAV flight condition

Cruise Speed (m/s)	11
Altitude (m)	6096
Air density (kg/m^3)	0.652694
Teperature (K)	248.5
Pressure (Pa)	46563.3

HALE UAV mission duration is expected to be twenty (20) hours.

2.3. Power Required

In order to achieve a certain amount of electrical power needed, propulsion power has to determined first. The propulsion power has to meet enough power for aircraft to take off and cruise flighting at maximum duration. Take off phase holds crucial role, due to its high power consuming because, in this stage, aircraft will climb into a certain amount of altitude. HALE UAV designed to use spring as its additional take off mechanism in order to increase its initial acceleration. Because it does not touch the runway, so the take off is only composed of a single phase that is climbing with a spiral trajectory. The vehicle continues to increase the altitude to designated altitude while turning. Actually, turning is produced by aileron with helped by rudder and elevator. So that for turning performance only seen from the lateral dimension. The cruise flight begins when the vehicle has reached the desired altitude of 6096m asl. That phase is the longest phase of the HALE UAV.

2.3.1. Take Off

Climbing phase indicated by the value of γ is bigger than zero ($\gamma > 0$), therefore thrust has to be bigger than drag in order to produce extra power to increase its altitude. The calculation simplified by using assumptions such as aircraft is climbing with constant velocity, bank angle (φ) and climb angle (γ). Motors are mounted parallel equally within aircraft fuselage to produced $\alpha_T = 0$. The angle of attack also to be assumed as the small value [4]. First of all, longitudinal dimension needs to be analyzed to achieved conservation of power equation.

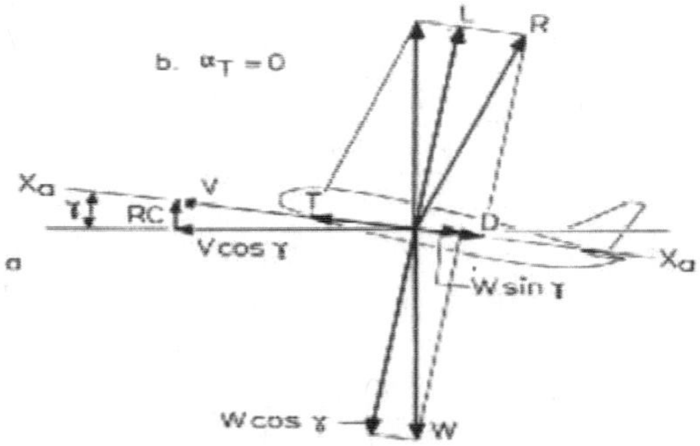

Figure 2. Free body diagram longitudinal of climbing UAV. Reprinted from force equilibrium of climbing flight illustration, Retrieved June 2017 from [4].

Equilibrium equations is performed on the wind coordinate order,

- X-axis:

$$T - D - WSin\gamma = 0 \tag{1}$$

$$Sin\gamma = \frac{T}{W} - \frac{D}{W} \tag{2}$$

- Z-axis:

$$WCos\gamma - L = 0 \tag{3}$$

$$WCos\gamma = L \tag{4}$$

Lateral dimension analysis is performed to see how bank angle affect aerodynamic forces while aircraft in turning flight condition, and its relation with longitudinal dimension,

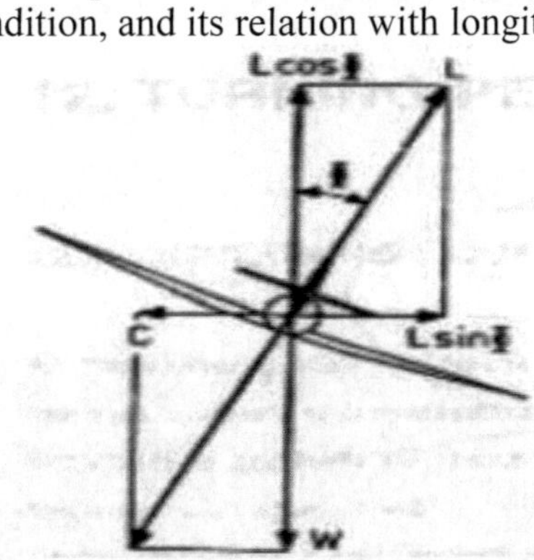

Figure 3. Free body diagram lateral of turning HALE UAV. Reprinted from force equilibrium of turning flight illustration, Retrieved June 2017 from [5].

$LCos\varphi$ equal to L on the longitudinal dimension, so

$$LCos\varphi = WCos\gamma \tag{5}$$

$$L = \frac{WCos\gamma}{Cos\varphi} \tag{6}$$

Back to the longitudinal dimension. With multiplying equation (1) with velocity (V), we get the equilibrium equations of Power as follows,

$$TV - DV - WVSin\gamma = 0 \tag{7}$$

TV=Pa, Power that must be provided to support certain flying condition
DV=Pr, Power that needed to against the drag
WVSinγ=Pe, Excess power that can be used to increase altitude

IOP Conf. Series: Journal of Physics: Conf. Series **1005** (2018) 012038 doi:10.1088/1742-6596/1005/1/012038

$$Pa = Pr + Pe \tag{8}$$

To obtain velocity, substitution equation (4) with the common aerodynamic equation of Lift.

$$\frac{WCos\gamma}{Cos\varphi} = \frac{1}{2}\rho V^2 SCl \tag{9}$$

$$V = \sqrt{\frac{2WCos\gamma}{\rho SCLCos\varphi}} \tag{10}$$

Calculation of Sinγ to obtain Pe reused equation (1),

$$Sin\gamma = \frac{T}{W} - \frac{DCos\gamma}{LCos\varphi} \tag{11}$$

$$Sin\gamma = \frac{T}{W} - \frac{CDCos\gamma}{CLCos\varphi} \tag{12}$$

To obtain the drag magnitude, done the comparison between lift and drag, and also substitution with equation (6),

$$\frac{D}{L} = \frac{CD}{CL} \tag{13}$$

$$D = \frac{CD}{CL}\frac{WCos\gamma}{Cos\varphi} \tag{14}$$

So,

$$Pr = W\sqrt{\frac{2WCD^2Cos^3\gamma}{\rho SCL^3Cos^3\varphi}} \tag{15}$$

From above equation, to obtain CL and CD required drag polar equations,

$$CD = 0.0228 - 0.0013CL^2 + 0.0157CL^3 \tag{16}$$

So, need to know CL magnitude first when take-off to obtain CD magnitude with the equation above.

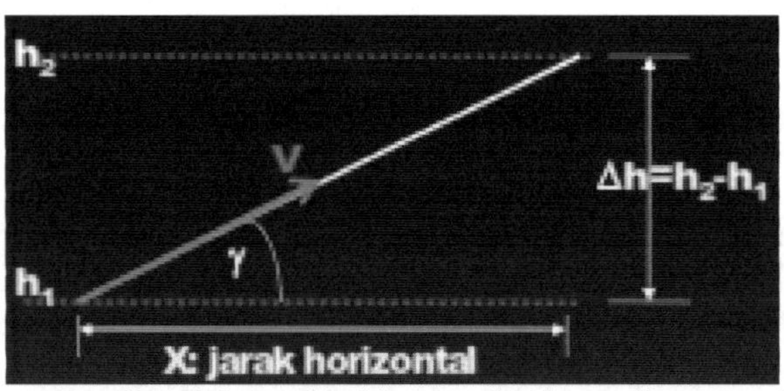

Figure 4. Climbing scheme. Reprinted from climbing angle illustration, Retrieved June 2017 from [4].

$$tan\gamma = \frac{\Delta H}{X} \tag{17}$$

In accordance with the flight operation data, HALE UAV will reach 20000ft (6096m) height directly from the ground (0 m) with a spiral path. The horizontal distance from the path is 13000m. To obtain the time from take-off to the desired height,

$$t = \frac{X}{cos\gamma V} \tag{18}$$

While the bankroll angle limited to 30°, so $\varphi = 0.5236\,rad$.

For more conservative calculation, air density of 6096 m altitude has been used with $\rho = 0.652694\ kg/m^3$ (based on ISA Calculator). So, calculation becomes,

Table 3. Take off calculation

CL (Max)	1.25
CD	0.045
γ	0.4385 rad
φ	0.5236 rad
V (g=9.8m/s^2)	8.65 m/s
t	1659.88 s
Power Available (Pa)	785.43 Watt

2.3.2. Cruise

Cruise is also called flat flight where the angle of attack usually small (wings parallel to local horizontal plane) where the wing parallels with the local horizon. The use of power on this flight phase is very influential on the overall performance of the aircraft because of longest phase. Assumed cruise condition is stationary and analyzed on the wind coordinate order and the angle of attack is so small, so it can be ignored.

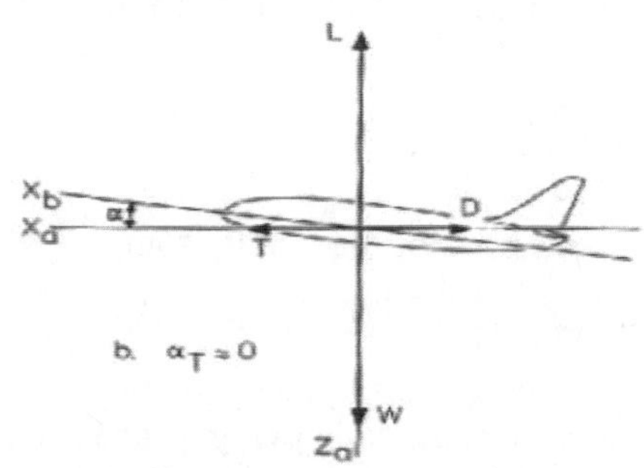

Figure 5. Free body diagram longitudinal of cruising UAV. Reprinted from force equilibrium of cruising flight illustration, Retrieved June 2017 from [6].

Equilibrium equations is performed on the wind coordinate order,

- X-axis:

$$T - D = 0 \tag{19}$$

$$T = D = \frac{1}{2}V^2SCD \tag{20}$$

- Z-axis:

$$W - L = 0 \tag{21}$$

$$L = W = \frac{1}{2}\rho V^2SCL \tag{22}$$

Cruise speed has been determined 10m/s^2, the to obtain CL altitude when cruising from equation (20),

$$CL = \frac{2W}{\rho V^2 S} \tag{23}$$

CD calculation is done by equation (16) with entering CL value when cruise flight. Whereas to obtain power that needed when cruise flight with using equation (19) by multiplying it by V.

$$Pa = TV = DV = \frac{1}{2}V^3SCD \tag{24}$$

IOP Conf. Series: Journal of Physics: Conf. Series **1005** (2018) 012038 doi:10.1088/1742-6596/1005/1/012038

Table 4. Cruise calculation

V_{cruise} (g=9.8m/s^2)	11 m/s
CL	0.739
CD	0.03
Power Available (Pa)	136.1 Watt

2.3.3. Power supporting system

In addition to calculating the power for the propulsion system, it also necessary to calculate the power for its supporting systems such as avionics system, actuator system, and payload that are carried.

$$P_{SS} = \frac{P_{av} + P_{payload} + P_{servos}}{\eta_{SS}} \tag{25}$$

Table 5. Power supporting system

Avionics	1.71 watt
Payload	2 watt
Servo	24.48 watt

Efisiensi is assumed to be 85%, so the total power required for supporting system is 33.16 watt.

3. Component and System Energy Source

3.1. Component

Here are some important components as parts of the energy source system whose the existence is very influential on the power required.

1. Batteries
   The batteries used are Li-ion type because it is easy to recharge and have a large capacity for the weight of only 46 grams. Each of its cell having a capacity of 3350 mAh and voltage 3.6 V. Meanwhile, recharging process of this type of battery requires the power of 6.825 watts per cell [7]. The batteries are assembled into 6S in series in order to get a voltage of 21.6 V.
2. Solar cell
   Solar cell used has a single cell size of 125 mm x 125 mm, efficiency value of 28% per cell, and the power that can be generated is 3.5 W per cell [8].
3. MPPT
   MPPT (Maximum Power Point Tracking) is a device that controls the energy flow from solar cells to both battery and another system installed in the UAV [9]. The MPPT used has an efficiency of 85%.
4. Propulsion System
   The propulsion system consists of brushless motor and Electronic Speed Controller (ESC), each having efficiency of 85%.

3.2. Energy Source

As we know, solar cells have the ability to convert solar energy into electricity at day, while at night solar cells then become useless since there's no sunlight. Therefore, a battery is required as a storage that can be used for night operation. It will recharge its power during the day and can be used for the following night flight. The mission of this HALE UAV ITB will be for 20 hours as shown in the following scheme.

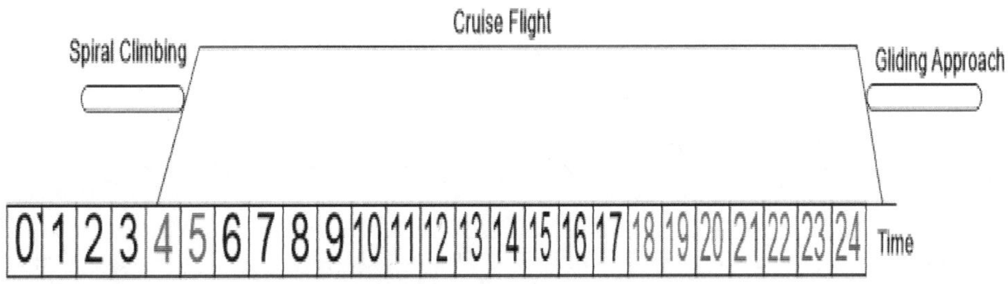

Figure 6. Flight scheme

The UAV take off at around 4:30 am with a full energy source from the battery because of the high current requirement and there is no sunlight yet. The UAV reach the cruise heights in about 27 minutes from take-off (calculation as shown in Table 3). In this phase, the UAV still use the battery as its only energy source, but with a much lower power requirement since it is now in cruise flight phase. When the sun rises up at around 6:00 am, MPPT will open a circuit to deliver energy from the solar cell to the battery charger and to the UAV's propulsion system, disconnecting the circuit of the battery to the propulsion system. When the sunlight disappears at about 6:00 pm, MPPT will break the circuit to the battery charger and propulsion system, then reconnect the battery with the propulsion system. It is expected that at the end of the mission the UAV still has remaining energy left in the battery as an anticipation to any unwanted condition, even though the UAV is designed to landing by gliding (fly without thrust). From the above scheme, the cruise flight will be conducted for 12 hours exposed to the sun and 8 hours fly without exposure to the sun (1 hour is done early after takeoff).

3.2.1. Battery calculation

In order to get how much battery to use, it is necessary to analyze the phase of the night cruise and take off where the UAV fully use power energy from the battery. The power used is the total power that has corrected to the efficiency [10].

$$P = \frac{Pa}{\eta_{Motor}\eta_{ESC}} + P_{SS} \tag{26}$$

Next, the amount of current required is obtained by dividing P with voltage,

$$I = \frac{P}{21.6} \tag{27}$$

Then, to find out how much capacity is required, multiply equation (27) with the duration of the power being used.

$$Cap. = Ixt \tag{28}$$

Once the amount of capacity required is known, then this amount is divided by the capacity available from the battery so that we can know the number of batteries in parallel (currents summation). Furthermore, multiplied it with the number of batteries used in series (voltage addition)

$$N_{BTT} = \frac{Needed\ Cap.}{Battery\ Cap.} xN_{series} \tag{29}$$

$$W_{BTT} = N_{BTT} x W_{cell} \tag{30}$$

Here is the table of the calculation,

IOP Conf. Series: Journal of Physics: Conf. Series **1005** (2018) 012038 doi:10.1088/1742-6596/1005/1/012038

Table 6. Capacity calculation

P_{cruise_7hours}	221.5 Watt
I_{cruise_7hours}	10.25 A
$Cap._{cruise\_7hours}$	71773 mAh
$P_{takeoff}$	1120.26 Watt
$I_{takeoff}$	51.86 A
$Cap._{takeoff}$	23913.51 mAh
$Cap._{cruise\_1hours}$	10253.22 mAh
$Cap._{takeoff+cruise\_1hour}$	34167 mAh

From **Table 6** above, capacity when take-off and one-hour cruise is smaller than a seven-hours cruise. By therefore the required capacity data is used from the capacity during night cruise for 7 hours. For safety reason, the multiplier factor 1.15 is used. So, the required capacity is 82538 mAh. Then equation (29) is used to obtain the required amount of battery.

$$N_{BTT} = \frac{82538}{3350} x6 = 150\ cell$$

So, it takes 150 batteries where each installed 6 pieces of series and 25 pieces of parallel.

$$W_{BTT} = 150x4.6 = 6900gr = 6.9\ Kg$$

3.2.2. *Solar Cell*

To obtain how much solar cell is needed, the analysis is done on a day cruise (12 hours) where the solar cell supplies energy to charge batteries and also to the propulsion system. Power and current calculations use the same equation (26) and equation (27). To determine power that needed to recharge the batteries is as follows,

$$P_C = P_{CC} x N_{BTT} \tag{31}$$

So, the total power required when cruising with solar cell is,

$$P_{TSC} = \frac{P_{charge} + P_{cruise\_12hours}}{\eta_{MPPT}} \tag{32}$$

Amount of solar cell is,

$$N_{SC} = \frac{P_{TSC}}{P_{SC}} \tag{33}$$

While the required area is,

$$A_{TSC} = N_{SC} x A_{SC} \tag{34}$$

After the calculation obtained the following results,

Table 7. Solar cell calculation

P_{charge}	1023.75 Watt
P_{TSC}	1464.96 Watt
N_{SC}	419 pcs
A_{TSC}	6.54 m^2

From **Table 7**, total area required by solar cell system is 6.54 m^2. The area is slightly smaller than the wing area vehicle, so it still possible to be installed in HALE UAV.

4. Results and Discussions

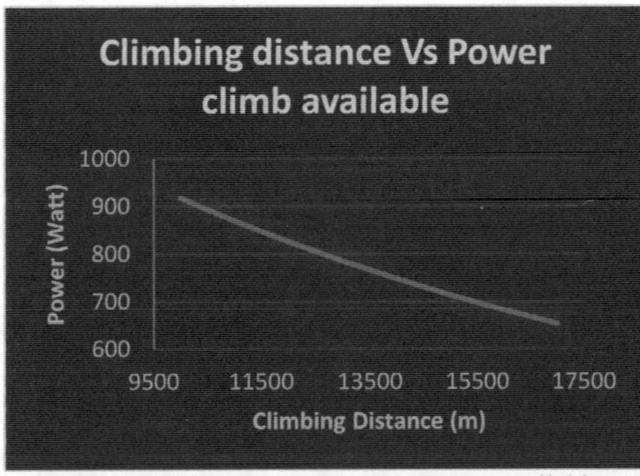

Figure 7. Climbing distance Vs Power available

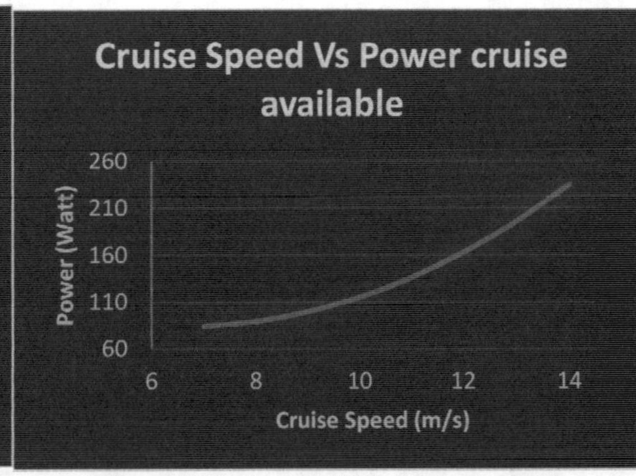

Figure 8. Cruise speed Vs Power available

Figure 7 above shows that climbing distance will affect to power that must be available for take off, but for cruise flight, there is no change (has no effect). Higher climbing distance with same designated altitude will give a lower climbing angle, so the power needed to take off is less but it will make the time for take off to be longer than high climbing angle. **Figure 8** shows that cruise speed will affect to power that must be available for the cruise, but for the take off (climbing) flight there is no change (has no effect). Slower cruise speed will give less power for cruise. Because of cruise flight is the longest flight phase, so the cruise speed must be considered carefully. If we want to accelerate the mission, it must provide great power. But if we want a small power, cruise speed must be small as well but still higher than stall speed.

The planform area of the solar panel has to be as wide as the desired power generation. In order to generate a great amount of power, it is required a large of the solar panel too. However, the large solar panel gives result in considerable weight. Heavier weight will force the UAV to give more power. Therefore, calculation of power required is very important because it can affect the required wing area and how much the fill factor is. Fill factor is the ratio between size of solar panel area compared to wing area. The fill factor of HALE UAV ITB is 97% which means almost of the entire wing is covered with the solar panel. The stiffness of the solar panel should be a consideration since it has to fit the airfoil shape. The elastic solar panel is preferable since it has to follow the shape of the wing surface when being deflected. The 150 used batteries will be stored in a space located in the fuselage. Those batteries are connected directly to MPPT while MPPT also connects to solar panel and propulsion system. MPPT must be able to manage the power supply, when the supply should be from the solar panel and when it should be not.

5. Conclusions

HALE UAV ITB with its configuration that has fill factor 97% (6.54m^2 area of the solar panel) and 150 cell 3350mAH batteries still capable of doing continuous flight until 20 hours at altitude 6096 m when take off at 4.30 am and landing (with gliding descent and approach). The strategy is at the beginning of the flight battery was fully charged and drain for climbing flight (about 27 minutes) and for cruise flight (about 1 hour) until sun light appears and solar cell ready to convert and give the energy for flight. In the day, energy fully stored by the solar cell. In the night, batteries capacity will run out and that is time for landing with gliding flight which is no need external energy at 12 a.m.

For real perpetual flight (24 hours continuous flight), the configuration must be adjusted because it will need more battery capacity and solar panel area which is more weight and it must be equal or under designed MTOW (20kg). For the next, to makes calculation close to reality, solar cell calculation must be considered the position of the vehicle, like latitude and longitude because it will affect to sun light intensity and will affect to energy that produced by the solar cell.

Acknowledgement
This work is supported by Institut Teknologi Bandung and Direktorat Riset Teknologi dan Pendidikan Tinggi Republik Indonesia.

References
[1] Communication: Meaning, Purpose, Importance and Principles, http://www.yourarticlelibrary.com/management/communication/communication-meaning-purpose-importance-and-principles/60291/ .
[2] Military communications, https://en.wikipedia.org/wiki/Military_communications .
[3] Solar Cell, https://en.wikipedia.org/wiki/Solar_cell.
[4] Muhammad, Hari. *Prestasi Terbang Menanjak*. Department of Aerospace and Astronautics, Faculty of Mechanical and Aerospace Engineering, ITB, 2015.
[5] Muhammad, Hari. *Prestasi Terbang Belok*. Department of Aerospace and Astronautics, Faculty of Mechanical and Aerospace Engineering, ITB, 2015.
[6] Muhammad, Hari. *Prestasi Terbang Jelajah*. Major of Aerospace and Astronautics, Faculty of Mechanical and Aerospace Engineering, ITB, 2015.
[7] Panasonic NCR18650B, https://batterybro.com/products/panasonic-b-ncr18650b-3400mah-4-8a.
[8] Sunpower Maxeon Solar Cell, https://aliexpress.com/item/Sunpower-Maxeon-Solar-Cell-21-8-Efficiency-3-34W-Semi-flexible-125x125mm-Mono-1pcs-Solar-Cell/1541942602.html.
[9] Maximum Power Point Tracking, http://www.leonics.com/support/article2_14j/articles2_14j_en.php
[10] Oettershagen, P., Stastny, T., Mantel, T., Melzer, A., Rudin, K., Gohl, P., Agamennoni, G., Alexis, K., Siegwart, R.. *Long-Endurance Sensing and Mapping using a Hand-Launchable Solar-Powered UAV*. In ICRA 2015, Seattle. 2015.

IOP Conf. Series: Journal of Physics: Conf. Series **1005** (2018) 012039 doi:10.1088/1742-6596/1005/1/012039

Self-equilibrated Tapered Three-stage Tensegrity Mast

C L Oh[1)] *, K K Choong[2)], T Nishimura[3)] and S W Lee[1)]

[1)]Faculty of Civil Engineering, Universiti Teknologi MARA,40450 Shah Alam, Selangor, Malaysia
[2)] School of Civil Engineering, Universiti Sains Malaysia, Malaysia
[3)] Department of Architectural Engineering, Institute of Disaster and Environmental Science, Kanazawa Institute of Technology, Ishikawa, Japan

* chailian_o80@yahoo.com

Abstract. Investigation of tensegrity structures for the space application is ongoing owing to the characteristics of being lightweight and flexible. Tensegrity structures consist of struts and cables are self-stressed and stable under gravitational loading. Form-finding is an important process to obtain the configuration of tensegrity structures that are in self-equilibrated state. Form-finding of tensegrity structures involves a complex computational strategy in solving the geometrical and forces of the structures. This paper aims to form-finding for a tapered three-stage tensegrity mast. The form-finding strategy involves the assemblage of the tensegrity mast, establishment of equilibrium equations and determination of one possible set of coefficient beta. Several cases of configurations with various twist angles with range of 20°-40° are investigated. A configuration with 9 struts and 42 cables satisfying the material elastic conditions was successfully found. The scalable self-equilibrated tensegrity mast is recommended for space applications.

1. Introduction

Tensegrity is an amazing system with characteristics of being lightweight, self-stressed, flexible and controllable. The design of tensegrity is different from the conventional ways with continuous transmission of compression. In tensegrity (Figure 1), the tensional network is assembled in order to support the floated compression. Most importantly, the structure can maintain its shape via self-equilibrium without any supports under gravitational loading. There is also a potential to convert local pressure into global deformation and subsequently seek for another balanced configuration in tensegrity.

Over the past decade, extensive investigations on tensegrity structures have been performed. Deployable and transformation capabilities of tensegrity structures have been utilized in active deployable structures (particularly for space engineering) as well as the robotic and automation community in the recent years. A design of a deployable tensegrity hollow rope footbridge with three actuation schemes were explored by Rhode-Barbarigos, et al. [1]. The crawling ability of a six-strut tensegrity robot was affirmed experimentally by Shibata, et al. [2] . Several methods to move the tensegrity robot have been highlighted: adopting shape memory alloy coil or motor driven wire as cables, pneumatic cylinder as struts or employing actuator along the cable connectivity to two specific struts.

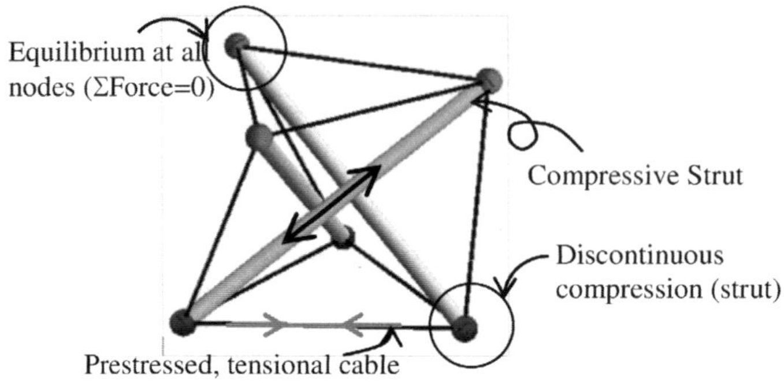

Figure 1. Tensegrity

For the space applications, deployable tensegrity in ring shape as reflectors for small satellites was suggested by Tibert and Pellegrino [3]. The tensegrity reflector has lower construction cost and higher precision in geometrical assemblage. R. Skelton recently proposed tensegrity technology in a design of a robotic system that allows growth of a habitat in space [4]. The design aims to solve several space travel problems specifically the growth, radiation protection, and the gravity problems.

There are many established form-finding strategies on tensegrity structures which are reviewed by Tibert and Pallegrino [5]. Li et al. [6] presented Monte Carlo method whereas Gan et al. [7] proposed a combination of numerical technique and genetic algorithm for form-finding of irregular tensegrity structures. Zhang and Ohsaki [8] used fictitious materials properties in a nonlinear programming approach for form-finding of tensegrity structures. Form-finding is an important process to obtain the configuration of tensegrity structures that are in self-equilibrated state. Form-finding of tensegrity structures involves a complex computation on the geometrical and forces of the structures, and the study generally remains open. This paper aims to form-finding of a tapered three-stage tensegrity mast, with a thought that the technology could be applied as space structures

The remainder of the paper is organized as follows: Section 2 presents the search strategies for form-finding of self-equilibrated configuration of the tapered three-stage tensegrity mast. Section 3 presents form-finding results especially the self-equilibrated configuration and the axial forces of the mast. Lastly, concluding remarks are presented in Section 4.

2. Form-finding Computational Strategy

2.1. Description of Tensegrity Mast

Triangular cell is chosen for the tensegrity mast, owing to its simplicity of application. Figure 2 shows the triangular cell consisting of three struts and nine cables. Three cables connect the lower (as well as upper) end of struts to form lower (as well as upper) triangle surface. The remaining three cables connect the lower and upper end of struts vertically. The triangular cell is prestressed after the upper and lower triangles are rotated by an angle termed twist angle, alpha α and the above-mentioned vertical cables become diagonal cables.

IOP Conf. Series: Journal of Physics: Conf. Series **1005** (2018) 012039 doi:10.1088/1742-6596/1005/1/012039

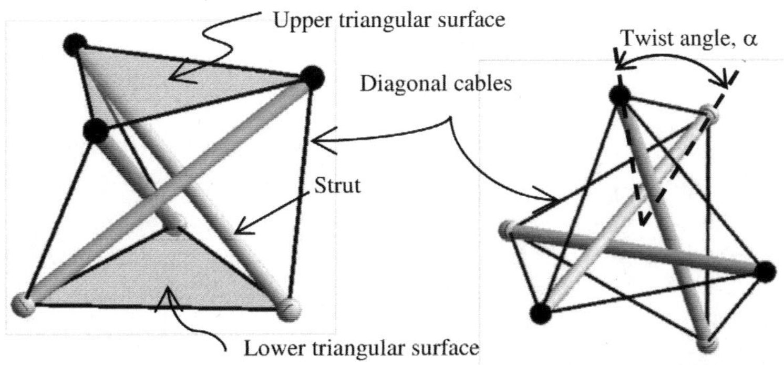

(a) Perspective view (b) Plan view

Figure 2. Triangular cell

Tensegrity can be classified into class 1, 2 and *k* relying on the numbers of strut at a node. A class 1 tensegrity mast is chosen for the study. Specifically the mast is assembled with the principle of discontinuous struts connected solely by cables. The tensegrity mast with three triangular cells are assembled according to the chart as shown in Figure 3. The lower and upper triangular surfaces of ith (i = 1, 2, 3) triangular cell are denoted as L*i*B and L*i*T, respectively. The triangular cell that is stacked up on the current i^{th} triangular cell is denoted as the $(i+1)^{th}$ cell. All the elements are classified into specific groups as shown in Figure 3.

During the stacking of the tensegrity mast, the odd-numbered i^{th} triangular cell is twisted in clockwise direction whereas the triangular cell with even-number is twisted in counterclockwise direction. The horizontal cables forming the upper and lower triangular surfaces (i.e. the interface of two triangular cells) are replaced by six saddle cables which are eventually arranged in hexagonal pattern. The height of N-stage tensegrity mast is not equal to total height of three triangular cells due to the saddle height *h*.

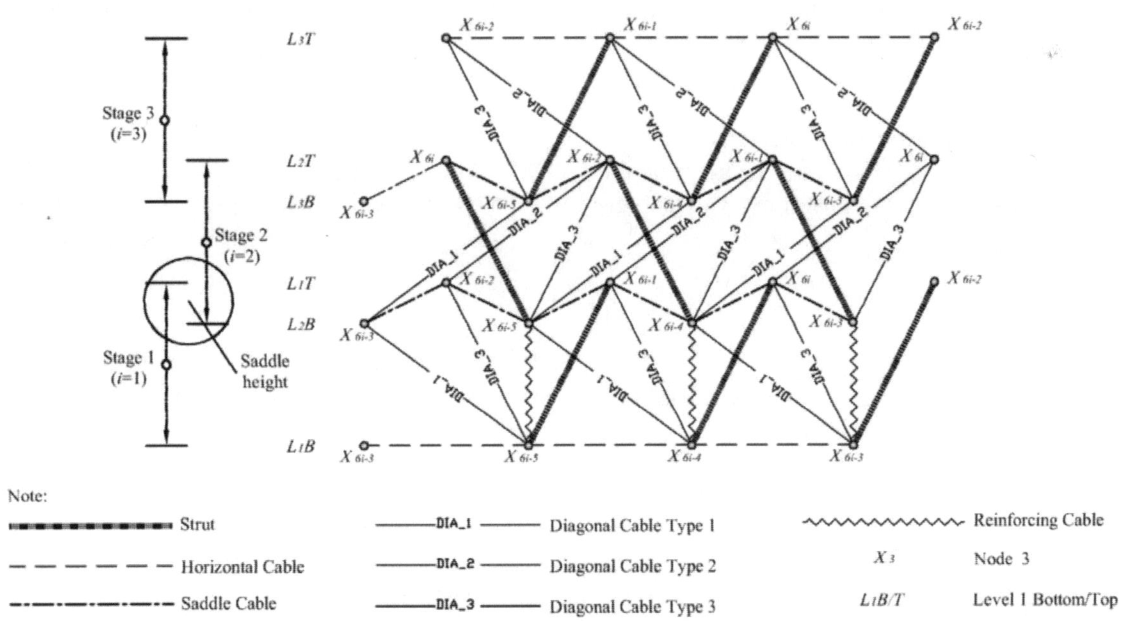

Figure 3. Element connectivity

2.2. Equilibrium Equations

The connections for the struts and cables of the tensegrity mast are assumed to be pinned connected. Hence, the equilibrium equations of a tensegrity mast could be derived based on the basic formulations of pin-jointed structures.

The tensegrity mast in this study consists of m elements, n nodes, n_c constraints and n_u ($3n$-n_c) unconstrained degree of freedoms. Consider an element of the temsegrity mast k connecting node i and j ($i < j$) in a Cartesian coordinate system O-$X\ Y\ Z$. The coordinates of node i and j are denoted as $\boldsymbol{x}_i$ and $\boldsymbol{x}_j$, respectively. The length of element k is evaluated using the following expression:

$$L_T = \sqrt{\left(\boldsymbol{x}_j - \boldsymbol{x}_i\right)^T \left(\boldsymbol{x}_j - \boldsymbol{x}_i\right)} \tag{1}$$

Considering equilibrium of forces at all nodes, the following equation could be obtained:

$$\boldsymbol{F} = \boldsymbol{Bn} \tag{2}$$

where $\boldsymbol{B}$ is a $n_u \times m$ matrix consisting of directional cosines λ , $\boldsymbol{F}$ is a vector of nodal forces with size n_u and $\boldsymbol{n}$ is a vector of axial forces with size m. The directional cosine λ of an element k is expressed:

$$\lambda = \frac{\left(\boldsymbol{x}_j - \boldsymbol{x}_i\right)}{L_T} \tag{3}$$

The Moore-Penrose generalized inverse is used to obtain the generalized inverse of the matrix $\boldsymbol{B}$ with rank deficiency (i.e. rank $< n_u$). For the case of self-equilibrated state without external forces (i.e. $\boldsymbol{f}=\boldsymbol{0}$), by making use of Moore-Penrose generalized inverse, the solution for the axial force vector $\boldsymbol{n}$ for the tensegrity mast is given as:

$$\boldsymbol{n} = [\boldsymbol{I}_m - \boldsymbol{B}^+\boldsymbol{B}]\boldsymbol{\beta} \tag{4}$$

where $\boldsymbol{\beta}$ is a vector of arbitrary coefficient of size m.

An optimization process to find one possible optimal combination of the coefficients β_i is performed with the following optimization problem:

$$\min f(x) = -\beta_1 \times \beta_2 \times .... \times \beta_i \tag{5}$$

where i is the numbers of self-stress mode.

A combination of β_i is used to determine the possible axial force $\boldsymbol{n}$ in an iterative way that satisfy the following inequality constraints which comes from the elastic condition of the material properties.:

$$0 \le n_c \le \sigma_c A_c \tag{6a}$$

$$-\frac{\pi^2 E_s I_s}{l_s^2} \le n_s \le \sigma_s A_s \tag{6b}$$

where n_c, σ_c and A_c are axial forces, yield stress and cross sectional area for cable elements, respectively; n_s, E_s, I_s, L_s, σ_s and A_s are axial forces, Young modulus, moment inertia of section, element current length, yield stress and cross sectional area for strut elements, respectively. In this study, all strut elements have circular cross sectional area.

2.3 Form-Finding Algorithm

The flowchart in Figure 4 shows the form-finding algorithm of the tapered three stage tensegrity mast. The form-finding algorithm begins with the preparation of geometrical input of the three-stage class 1 tensegrity mast with triangular cells. Equation for vector of element axial force (Equation 4) satisfying self-equilibrium condition is established. An initial estimate for coefficients β_i through the solution of the optimization problem as shown in Equation (5). It is noted that, only one possible combination of

IOP Conf. Series: Journal of Physics: Conf. Series **1005** (2018) 012039 doi:10.1088/1742-6596/1005/1/012039

coefficients β_1, β_2 and β_3 which satisfies the inequality constraints as shown in Equations (6a and b) is chosen through an iterative process. If the combination of coefficient β_I does not satisfy Equation (6a and b), a new trial by new nodal coordinates of the tensegrity mast is given and the algorithm is repeated as shown in Figure 4. The form-finding algorithm seek for a possible configuration of three-stage tensegrity mast that is in self-equilibrated state. By using Equation 4, the initial axial forces for the tensegrity mast could be obtained with the chosen combination of the coefficients β_i.

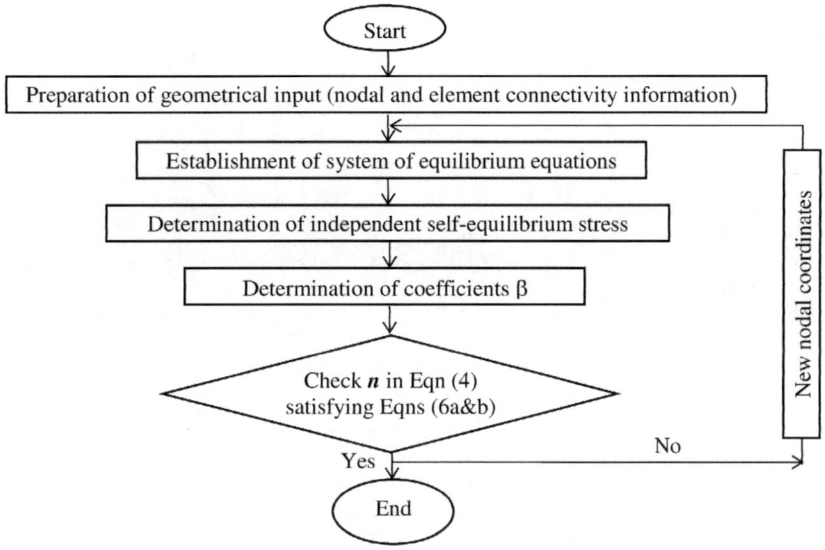

Figure 4. Form-finding algorithm

3. Results and Discussions

This section presents the initial parameters for the form-finding process as well as the form-finding results such as the topology and the initial forces of the self-equilibrated tensegrity mast.

3.1. Geometrical and Material Parameters

The tapered three-stage tensegrity mast with aspect ratio (width/height) of 1:3.3 is modelled with the base width and top width measuring 300 mm and 100 mm, respectively. The triangular cell with nominal height of 450 mm is used. The saddle depth between two triangular cells is 180 mm. The tensegrity mast of 9 struts and 42 cables is assembled. Table 1 shows the material properties for the tensegrity mast. Three self-equilibrium stress modes are obtained.

Table 1. Material properties

Elastic Modulus, E (MPa)	Yield Stress, σ_s (MPa)	Density, γ (kg/mm^3)	Strut Cross sectional Area, A_s (mm^2)	Cable Cross sectional Area, A_c (mm^2)
200 000	250	7.85×10^6	113.10	19.63

3.2. Self-equilibrated Tensegrity Mast

In the form-finding of self-equilibrated tensegrity mast, twist angle of each triangular cell with range of 20°-40°, with 5° increment are investigated. Figure 5 shows the number of slackened cables after the application of various twist angles to the tensegrity mast. Significant increment in the number of slackened cables is observed in the cases where the twist angle is 30° and onwards. Since the case of the tensegrity mast with the zero slackened cables represents the solution to the self-equilibrated configuration, configuration with twist angle 25° is one of the possible tensegrity mast at self-equilibrated state.

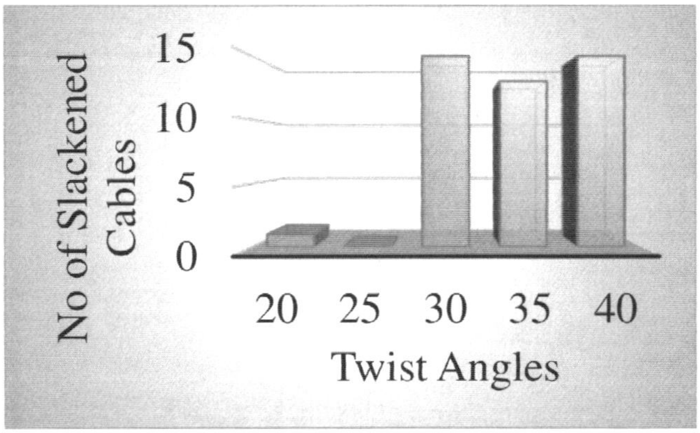

Figure 5. Number of slackened cables

Table 2 and Figure 6 show the nodal coordinates and member connectivity for the initial and final configurations for the tensegrity mast. It is emphasized that the obtained solution for the tensegrity mast represents one possible self-equilibrated configuration, and it could be more to be discovered.

Table 2. Nodal coordinates and member connectivity

Nodal coordinates			Member connectivity					
Node	Initial configuration (Twist angle 0°)	Final configuration (Twist angle 25°)	Members			*i,j*,type of elements (i.e. 1 is strut, 2 is cable)		
1	-150.00,-86.60,0.00	-150.00,-86.60, 0.00	1	18	35	1,6,1	5,9,2	8,11,2
2	150.00,-86.60,0.00	150.00 ,-86.60, 0.00	2	19	36	2,4,1	9,6,2	9,12,2
3	0.00,173.21,0.00	0.00,173.21, 0.00	3	20	37	3,5,1	6,7,2	7,13,2
4	-104.55,-60.36,450.00	-120.26,-10.52,450.00	4	21	38	7,12,1	7,4,2	8,14,2
5	104.55,-60.36,450.00	69.24,-98.89, 450.00	5	22	39	8,10,1	10,14,2	9,15,2
6	0.00,120.72,450.00	51.02,109.41, 450.00	6	23	40	9,11,1	14,11,2	4,10,2
7	-122.73,70.86,270.00	-81.28,116.08, 270.00	7	24	41	13,18,1	11,15,2	5,11,2
8	0.00,-141.71,270.00	-59.89,-128.43,270.00	8	25	42	14,16,1	15,12,2	6,12,2
9	122.73,70.86,270.00	141.17,12.35, 270.00	9	26	43	15,17,1	12,13,2	13,16,2
10	-77.28,44.61,720.00	-77.28,44.61, 720.00	10	27	44	1,2,2	13,10,2	14,17,2
11	0.00,-89.23,720.00	0.00,-89.23, 720.00	11	28	45	2,3,2	1,4,2	15,18,2
12	77.28,44.61,720.00	77.28,44.61, 720.00	12	29	46	3,1,2	2,5,2	10,16,2
13	0.00,110.22,540.00	0.00,110.22,540.00	13	30	47	16,17,2	3,6,2	11,17,2
14	-95.45,-55.11,540.00	-95.45,-55.11,540.00	14	31	48	17,18,2	1,7,2	12,18,2
15	95.45,-55.11,540.00	95.45,-55.11,540.00	15	32	49	18,16,2	2,8,2	1,8,2

IOP Conf. Series: Journal of Physics: Conf. Series **1005** (2018) 012039 doi:10.1088/1742-6596/1005/1/012039

16	0.00,57.74,990.00	24.40,52.33, 990.00	16	33	50	4,8,2	3,9,2	2,9,2
17	-50.00,-28.87,990.00	-57.51,-5.04,990.00	17	34	51	8,5,2	7,10,2	3,7,2
18	50.00,-28.87,990.00	33.12,-47.29, 990.00						

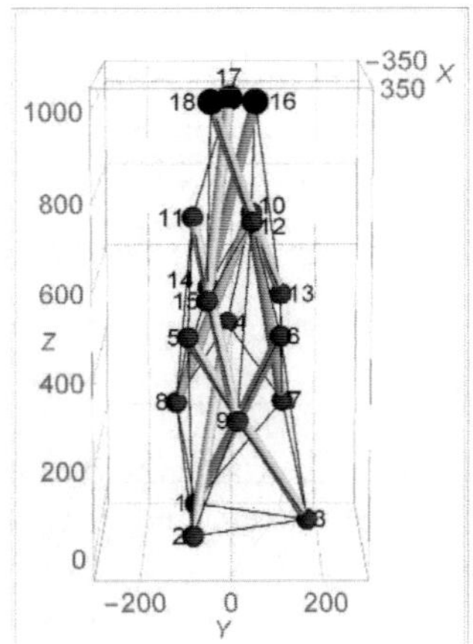

Figure 6. Self-equilibrated tappered three-stage tensegrity mast

3.3. *Initial Axial Forces*

The tensegrity mast was successfully searched when the axial forces of the mast lie between the upper and lower axial force limits (see equation 6a and b). Calculation of axial forces as in Equation 4 depends of the magnitude of the set of coefficients beta β_i (i=1,2,3). Figure 7 shows the initial axial forces (pre-stressed force) of the mast with the determined value for coefficients β are -12146, -5931, -6689. It is noted that all the cables are in tension and struts in compression. No axial forces of the mast exceed the upper axial force limits (i.e. 4610 N and -9050 N for cables and struts, respectively).

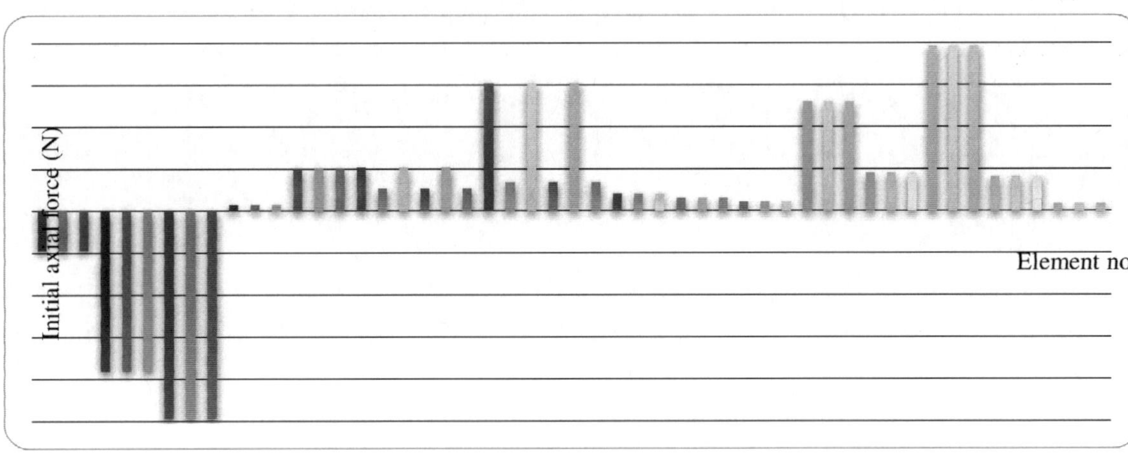

Figure 7. Initial axial force (N)

4. Concluding Remarks

This paper presents the form-finding of a tapered three-stage Class 1 tensegrity mast that is at self-equilibrated state. The effort of form-finding through various trial cases involving different twist angles are presented. The results of slackened cables are used to obtain the solution to the self-equilibrated mast. The determination of the coefficients β is a challenge especially the class 1 masts are to be searched. The axial forces of the searched tensegrity mast lie within the upper and lower axial force limit which satisfy the material elastic conditions. The self-equilibrated tensegrity mast is recommended for the space applications such as deployable structures or tensegrity robots.

Acknowledgement
The authors would like to thank Ministry of Higher Education Malaysia through FRGS grant (FRGS/1/2017/TK01/UITM/02/1) for funding this study. We would like to express our appreciation to University Teknologi MARA Selangor in providing the facilities

References

[1] L. Rhode-Barbarigos, N. B. H. Ali, R. Motro, and I. F. Smith, "Design aspects of a deployable tensegrity-hollow-rope footbridge," *International Journal of Space Structures,* vol. 27, pp. 81-96, 2012.
[2] M. Shibata, F. Saijyo, S. Hirai, and Ieee, "Crawling by Body Deformation of Tensegrity Structure Robots," in *Icra: 2009 Ieee International Conference on Robotics and Automation, Vols 1-7*, ed, 2009, pp. 3617-3622.
[3] A. Tibert and S. Pellegrino, "Deployable tensegrity reflectors for small satellites," *Journal of Spacecraft and Rockets,* vol. 39, pp. 701-709, 2002.
[4] R. Skelton. (May 13, 2016). *Tensegrity Approaches to In-Space Construction of a 1g Growable Habitat*. Available: https://www.nasa.gov/feature/tensegrity-approaches-to-in-space-construction-of-a-1g-growable-habitat
[5] A. Tibert and S. Pellegrino, "Review of form-finding methods for tensegrity structures," *International Journal of Space Structures,* vol. 26, pp. 241-255, 2011.
[6] Y. Li, X.-Q. Feng, Y.-P. Cao, and H. Gao, "A Monte Carlo form-finding method for large scale regular and irregular tensegrity structures," *International Journal of Solids and Structures,* vol. 47, pp. 1888-1898, 2010.
[7] B. S. Gan, J. Zhang, D.-K. Nguyen, and E. Nouchi, "Node-based genetic form-finding of irregular tensegrity structures," *Computers & Structures,* vol. 159, pp. 61-73, 2015.
[8] M. Ohsaki and J. Zhang, "Nonlinear programming approach to form-finding and folding analysis of tensegrity structures using fictitious material properties," *International Journal of Solids and Structures,* vol. 69, pp. 1-10, 2015.

Half Wing N219 Aircraft Model Clean Configuration for Flutter Test On Low Speed Wind Tunnel

Sayuti Syamsuar[1)], Budi Sampurno[2)], Katia Mayang Mahasti[2)], Muchamad Bayu Sakti Pratama[2)], Triyono Widi Sasongko[1)], Nina Kartika[1)], Adityo Suksmono[1)], Mohamad Ivan Aji Saputro[1)], Dimas Bahtera Eskayudha[1)]

1) Pusat Teknologi Sistem dan Prasarana Transportasi, Kedeputian Teknologi Industri Rancang Bangun dan Rekayasa, Badan Pengkajian dan Penerapan Teknologi (BPPT), Puspiptek, Serpong, Indonesia
2) Flutter Specialist, PT Dirgantara Indonesia, Bandung

Email add: sayutisyamsuar@yahoo.com; and sayuti.syamsuar@bppt.go.id

Abstract. Flutter is a rapid self-feeding motion which is caused by the interaction of aerodynamic, structural and inertial forces. Flutter can cause major damage on aircraft structure which can lead to fatal accident in aviation. Several methods have been evolved to avoid the flutter phenomena occur during the flight envelope of aircraft design. On this study, method was developed by Indonesian Aerospace which consist of Finite Element Method (FEM) analysis, Ground Vibration Test (GVT), and Wind Tunnel Flutter Test (WTT). Based on the study, FEM have similar results toward to Wind Tunnel Flutter Test conjunction the clean configuration of N219 aircraft half wing model.

Keywords: fluid flow, particle, aerodynamic forces, moments, flutter speed, NASTRAN.

1. INTRODUCTION

Flutter is a rapid self-feeding motion which is caused by the interaction of aerodynamic, structural and inertial forces. Flutter can cause major damage on aircraft structure which can lead to fatal accident in aviation. The aims of this research are to validate and compare the results of wind tunnel test to Finite Element Method (FEM) and to earn the critical speed on flutter phenomena based on wind tunnel test. This research was conducted to avoid and ensure the flutter phenomena will not occur when flight test of aircraft is held.

In other paper, flutter suppression has been spotlighted because of its destructive nature (Westin et al, 2009). The materials use is suppressed, especially PZT materials. Instead of the passive suppression will increase the mass of aircraft. The result of the test is to measure of flexible wing design which is suppressed on materials choice. Piezoelectric transduction is used by active and passive flutter suppressions for highly flexible wing (Tsushima et al, 2017). By properly place the piezoelectric actuators and energy harvesters, it was possible to stabilize the wing, while extracting a certain amount energy, both of which would contribute to improving the performance of wing and aircraft.

Furthermore, the methods of flutter analysis of aeroelastic system includes modelling uncertainties is more efficient (Lokatt, 2017). Both of structural and aerodynamic uncertainties can have notable effect on the damping of the flutter modes. The method shows that main advantage is the ability to analyse the combine effect of structural and aerodynamic uncertainties.
Scope:
This paper consists of four sections. Section 1 presents scope of work, objectives as introduction. Section 2 briefly describes procedure of the Wind Tunnel Flutter and FEM analysis result. Section 3

presents the result of Wind Tunnel Flutter Test and comparison with the analysis FEM results. Finally, section 4 describes the conclusions of Low Speed Wind Tunnel Flutter Test.

2. TEST METHOD

2.1 FEM Analysis

Based on method which is described in flow chart below on Figure 1 and then updated with GVT, mode shape and its natural frequency final design half wing model that is mentioned on Table 1.

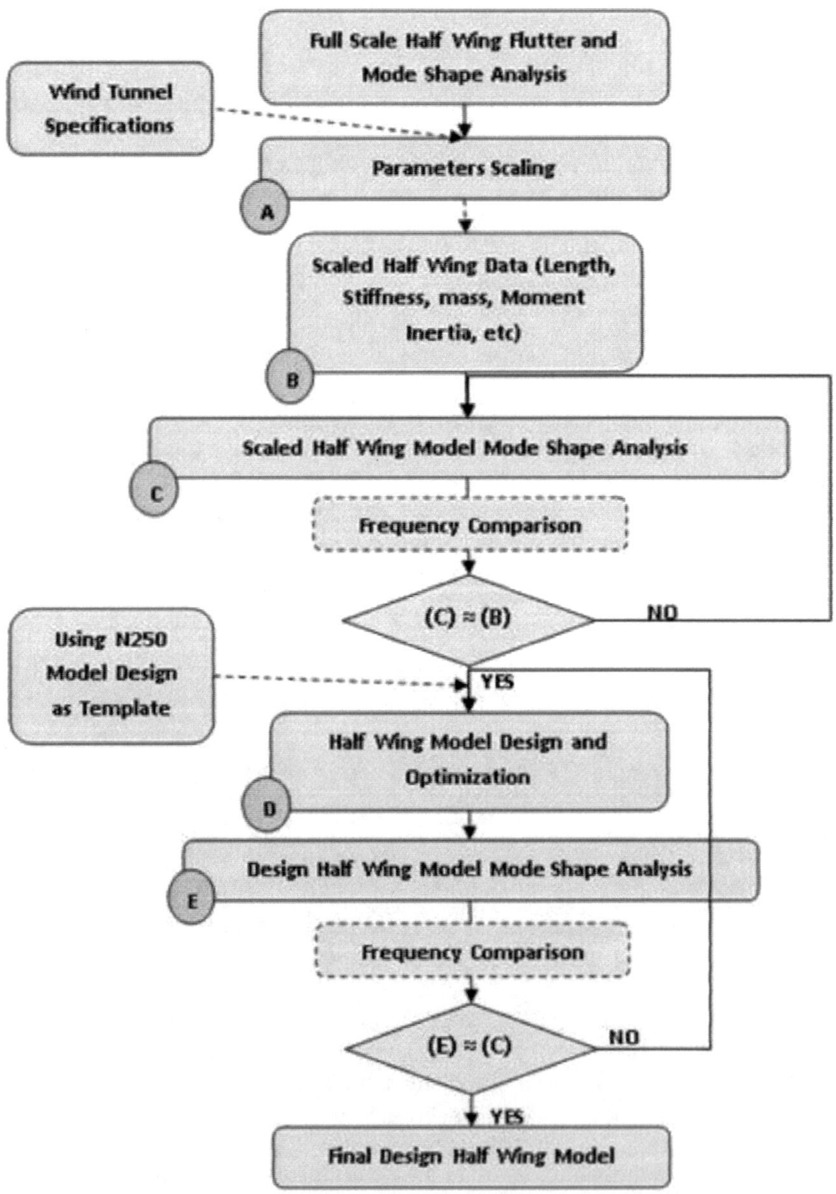

Figure 1 Flowchart of half wing model determination

Table 1 Mode shape and half wing natural frequency

No	Mode Shape	Analysis (Hz)	GVT (Hz)	Analysis Update (Hz)
1	Wing Vertical Bending	3.65	3.20	3.23
2	Wing Inplane Bending	6.90	5.09	5.08
3	Wing Vertical Bending	15.28	13.2	12.46
4	Wing Torsion	24.77	18.20	18.14
5	Wing Inplane Bending	28.16	21.11	20.37
6	Wing Torsion	34.14	28.31	29.09

Then, half wing flutter model will be tested on wind tunnel. The model attached on wind tunnel wall using moderate spring and soft spring. Therefore, spring shall be modeled on FEM and tested on GVT in order to get better accuracy.

2.2. Ground vibration test (GVT)

Ground Vibration Test (GVT) is useful for validating dynamic characteristics of test model from analytical model. The purpose of this GVT is to get structural modes parameters such as natural frequency, mode shapes, and damping. Those modes parameters from GVT result, will be used on flutter analysis and dynamic aeroelastomechanics system responses.

The main instruments that will be used for GVT exciter are accelerometer, and frequency analyzer as showed in Figure 2. The model should have fix supported at the wing root (Fixed on T_x,T_y, T_z, R_x, R_y & R_z). T is torsion which react to external force on model and R is translation force which react to external force on model that both of them is placed on wing root. While x, y, z are the direction based on axis. The excitation method that will be used to produce mechanical vibration on test model is a random excitation ranged up to 50 Hz. Responses on the observation points will be measured using transducer accelerometers. The signal coming from accelerometers and force transducer will amplified using amplifier and continue to be processed on the analyzer for getting its frequency response function. With curve fitting method then the modal parameters of the half wing flutter model can be achieved. The listing of Detail Accelerometer Locations as showed in Table 2.

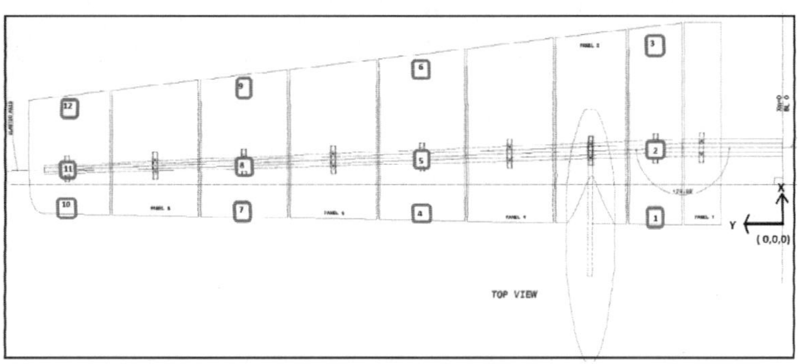

Figure 2. Sketch of Accelerometer Locations

Table 2 Listing of Detail Accelerometer Locations

Accelerator No	Coordinate			DOF
	X	Y	Z	
1	0.001	0.277	0.004	Z
2	0.166	0.277	0.004	Z
3	0.458	0.277	0.004	Z
4	0.014	0.782	0.034	Z
5	0.150	0.782	0.034	Z
6	0.390	0.782	0.034	Z
7	0.023	1.162	0.057	Z
8	0.137	1.162	0.057	Z
9	0.338	1.162	0.057	Z
10	0.032	1.542	0.080	Z
11	0.125	1.542	0.080	Z
12	0.287	1.542	0.080	Z

The Ground vibration test (GVT) result of N219 Half Wing Flutter Model are discussed in ref. [11]. The GVT was performed in one mass configurations that the mass configurations are described in Table 3.

The GVT result will be use for updating the structural dynamic model of the half wing flutter test model design, furthermore the updated structural dynamic model will be use for updating the flutter analysis result. Thus, the flutter speed reference will be use to compare wind tunnel flutter test result with flutter speed from updated flutter analysis result.

Table 3 Mass Distribution Half Wing Model N219-B11 Test Configuration (MC01)

Section	Weight (kg)			Center of Gravity (m)			Moment of Inertia (kg Sq.m)					
	Beam	Segment	Total	X	Y	Z	Ixx	Iyy	Izz	Ixy	Ixz	Iyz
SECTION 0	0.136	0.000	0.136	1.206	0.066	0.621	2.034E-04	1.021E-05	2.079E-04	-1.486E-09	9.071E-10	1.599E-07
SECTION 1	0.086	0.374	0.460	1.213	0.174	0.628	3.852E-04	4.000E-03	4.000E-03	-6.074E-07	-7.217E-07	9.413E-05
SECTION 2	0.118	0.315	0.433	1.210	0.276	0.630	5.909E-04	3.000E-03	3.000E-03	2.225E-05	-1.459E-05	7.408E-05
SECTION 3	0.142	1.181	1.323	1.104	0.416	0.608	3.000E-03	2.400E-02	2.400E-02	2.167E-05	-1.239E-04	-2.000E-03
SECTION 4	0.148	0.350	0.498	1.199	0.589	0.646	2.000E-03	3.000E-03	4.000E-03	6.879E-05	-4.504E-05	6.882E-05
SECTION 5	0.127	0.332	0.459	1.193	0.779	0.656	1.000E-03	2.000E-03	4.000E-03	6.377E-05	-4.185E-05	5.732E-05
SECTION 6	0.108	0.302	0.410	1.187	0.968	0.666	1.000E-03	2.000E-03	3.000E-03	5.732E-05	-3.769E-05	4.533E-05
SECTION 7	0.090	0.262	0.352	1.180	1.158	0.676	1.000E-03	1.000E-03	2.000E-03	4.942E-05	-3.252E-05	3.357E-05
SECTION 8	0.074	0.234	0.308	1.174	1.347	0.686	9.084E-04	1.000E-03	2.000E-03	4.337E-05	-2.859E-05	2.540E-05
SECTION 9	0.047	0.216	0.263	1.168	1.527	0.695	6.238E-04	7.556E-04	1.000E-03	3.097E-05	-5.301E-06	1.879E-05
Total Half Wing Flutter Model N219	**1.076**	**3.567**	**4.643**	**1.168**	**0.653**	**0.641**	**7.930E-01**	**5.300E-02**	**8.330E-01**	**4.900E-02**	**6.000E-03**	**-4.000E-03**

IOP Conf. Series: Journal of Physics: Conf. Series **1005** (2018) 012040 doi:10.1088/1742-6596/1005/1/012040

2.3 Finite Element Model of Half Wing Flutter Model Test N219

According to the boundary condition (support system) of the tested model, the analysis FEM for GVT needs to be modified to represent the support system. These support systems are connected to Beam combination Spring. Figure 3 is show the FEM of the Half Flutter Model for GVT.

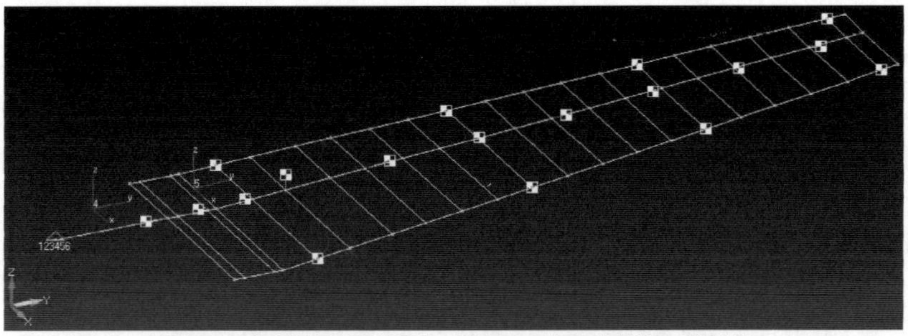

Figure 3 FEM Half Wing Flutter Model Test N219

The analysis FEM that is optimized and updated is the structure FEM described in Ref. 4. The updating is based on the GVT results described in Ref. 7. It must be remembered that frequencies and modes from GVT results actually include some damping of the tested Model.

The natural frequencies and modes of the structural FEM are obtained by solving the real 'undamped' eigenvalue equation (1).

$$\left[[K]-\omega_i^2\,[M]\right]\{\phi\}_i=\{0\} \tag{1}$$

Where:

$[M]$	= generalized mass matrix	size (ndof x ndof)
$[K]$	= generalized stiffness matrix	size (ndof x ndof)
$\{\phi\}_i$	= modal eigenvector	size (ndof x 1)
$[\Phi]=[\{\phi\}_1,\{\phi\}_2,\ldots\{\phi\}_{n\,\mathrm{mode}}]$	= modal matrix	size (ndof x nmodes)
ω_i	= natural frequency	(rad/sec)
ndof	= number of total degrees of freedom	
n mode	= number of modes extracted	

The above equation of motion is solved using MSC/NASTRAN SOL 103 (Normal Mode Analysis).Care must be taken in comparing frequencies between damped GVT results and undamped analysis results of the structural FEM. The actual amount of damping should not significantly affect the matching of frequencies and the correlation of modes.

IOP Conf. Series: Journal of Physics: Conf. Series **1005** (2018) 012040 doi:10.1088/1742-6596/1005/1/012040

A method that is used to establish the degree of correlation between the real eigenvectors from test and from the analysis model is visual examination by plotting and animation of the test and analytical modes. The test modes are related to the analytical modes that correspond.

The visual method evaluates the quality of modes correlation. In this document, the grades of modes correlation are divided into three categories. The first category is '***good***', this category explains similar mode shape between test and analysis FEM. The second category is '***fair***', this one declares the dominant modes are similar. Finally the third category is '***poor***', this one affirms weak or no corresponding between the test and analytical FEM.

In In a simplistic sense the Bayesian parameter estimation procedure has some similarity to the minimization of error by the method of least squares. The former permits a balanced model updating approach by incorporating uncertainties in both test data and in modeling parameters. In this approach each 'piece' of test data has a confidence associated with it. Then the test data as a whole is weighed relative to the initial model as a whole. The Bayesian parameter estimation procedure is an iterative approach wherein as error is minimized. The error is written as:

$$E = wt * \sum WR * (RT - RA)^2 + wp * \sum WP * (PF - PO)^2 \tag{2}$$

Where:

E	= a measure of the error between test data and analysis
RT	= target response quantities (i. e. test data)
RA	= responses from the analysis
WR	= weighting factors (confidences) for response
PF	= parameters for the final model
PO	= parameters of the original model
WP	= weighting factors (confidences) for the parameter
wt	= scalar weighting for the test data as a whole
wp	= scalar weighting for the model parameter as a whole

The above equation is solved iteratively to find PF, the final model parameter that minimize the error.

Model response and test data can be static displacements, resonant frequencies and mode shapes, and element forces, stresses, and strains any quantity that can be measured or computed with MSC/NASTRAN. Model parameters consist of beam areas and inertias card entries. Weighting matrices are usually diagonal, and are scaled such that they are divided by the square of the initial value. The units for all computations are therefore, fractional changes rather than absolute changes to avoid numerical problem caused by a wide variance of units.

The minimization of Equation 2 is performed with the use of MSC/NASTRAN SOL 200. This approach follows closely the paper presented in Ref.9.

The GVT results are considered as the baseline standard data. This means that the GVT data are assumed correct a priori. The relative error is defined as the discrepancy between the corresponding measures frequencies of GVT and the computed frequencies of the structural FEM. The minimum of the relative error in the corresponding frequencies of GVT and structural FEM implies the optimization of the member properties corresponding to the selected design variable in the optimization process. Considering the selection criteria described above, these design variable are the member properties I1, I2, and J of the equivalent beam replacements of the structural FEM. Recall the optimization process to minimize the relative error using MSC/NASTRAN SOL 200.

IOP Conf. Series: Journal of Physics: Conf. Series **1005** (2018) 012040 doi:10.1088/1742-6596/1005/1/012040

2.4. Wind Tunnel Flutter Test

The test model is valid to use on wind tunnel flutter test after confirmed through GVT and validation tests. Likewise the wind tunnel itself must be already fulfilled the wind tunnel safety requirement.
The main objectives of the wind tunnel flutter test are:

- To compare the dynamic and flutter characteristics between the computational/analysis model with the test model.
- To calculate the level of accuracy and representativeness of the test model on dynamic and flutter characteristics of wing of the N219.
- To simulate and understand flutter testing process and methodology.

The wind tunnel model flutter test will be carried out by sweeping the speed which started from the lowest speed up to flutter speed (if possible).
The execution of test will be conducted into two steps sweep speeds are as follows:

- The first step is to obtain preliminary data and as an exercise test procedure (Familiarization) prior to testing at higher and more critical speed (close to flutter speed).The speed sweepsof test will be conducted gradually started from speed 5.0 m/s, 10.0 m/s and 15.0 m/s.
  From the previous analysis document, it shows that the cruising speed on wind tunnel is around 15 m/s and the diving speed is around 21.0 m/s.
- The second step will focus to observe interaction of structural dynamic with aerodynamic force (Flutter phenomenon), and to predict flutter occurrence speed. The second step will started from speed 20.0 m/s, and continue with 5.0 m/s speed increment(situational), if the test is closing to flutter speed, then the speed increment can be reduce to increase the flutter speed prediction accuracy. The test will be continue until the test model structural dynamic behavior show signs of near flutter occurrence. This wind tunnel flutter test is not destructive testing, so the test will be stop if the test model shows sign of destructive possibility (i.e. large displacement).

According to spring value, there are 3 configurations model: Soft spring, Moderate spring, and hard spring. The wing tunnel flutter test only performed on soft spring model and moderate spring model. The Wind Tunnel Flutter Test for N219 half wing ideally has procedure shown on Figure 4. Remark: the block diagram of Flutter Margin Use Zimmerman is not recognized.

IOP Conf. Series: Journal of Physics: Conf. Series **1005** (2018) 012040 doi:10.1088/1742-6596/1005/1/012040

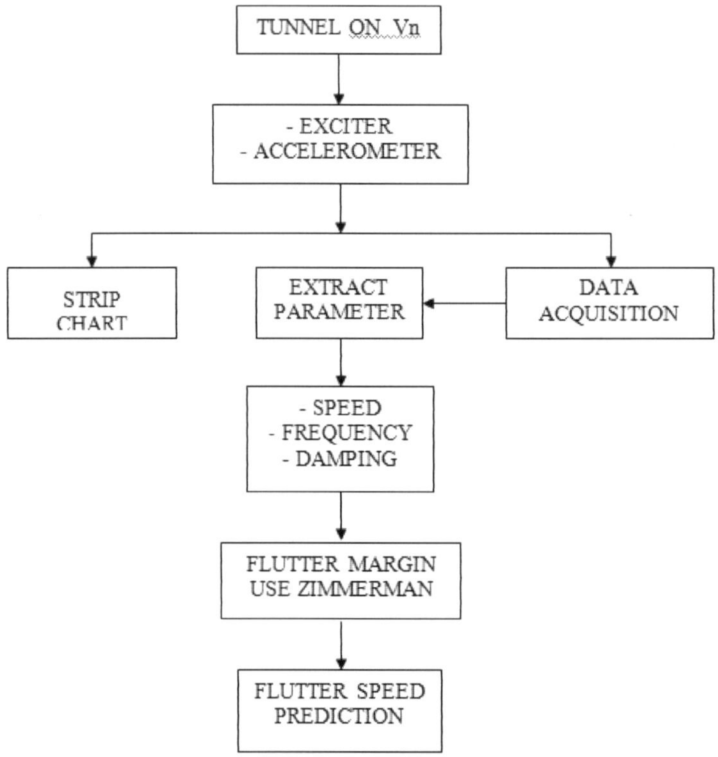

Figure 4 Procedure of Wind Tunnel Flutter Test

The half wing model was placed in wind tunnel then the accelerometer will received response from the exciter. The response will be read by data acquisition system then forwarded to extract parameter software to determine flutter margin. Figure 5 is the model installation in test section of wind tunnel.

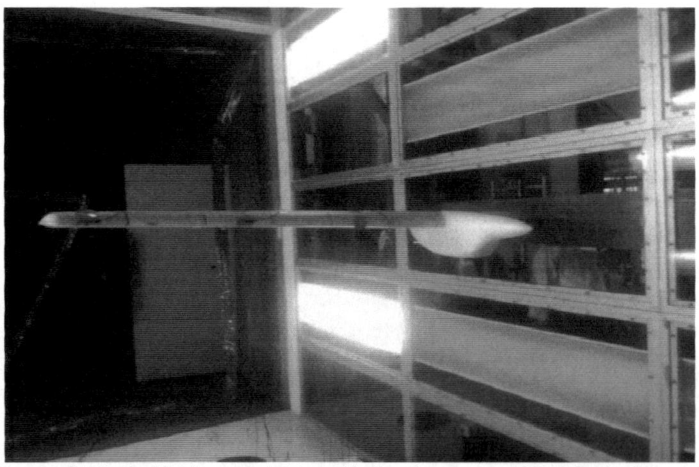

Figure 5 The half wing model installation in test section of wind tunnel.

The flutter margin cannot be resulted online due to delay of data transfer from data acquisition system to the extract parameter, therefore flutter speed from FEM analysis was used as speed limitation. The strip chart also cannot be used due to technical error in the software.

Data acquisition system forwarded data response to be analyzed by extract parameter software. The extract parameter was generated by Matlab coding will result speed, frequency, and damping data by curve fitting process. The extract parameter data for soft spring and moderate spring as showed in Figure 6.

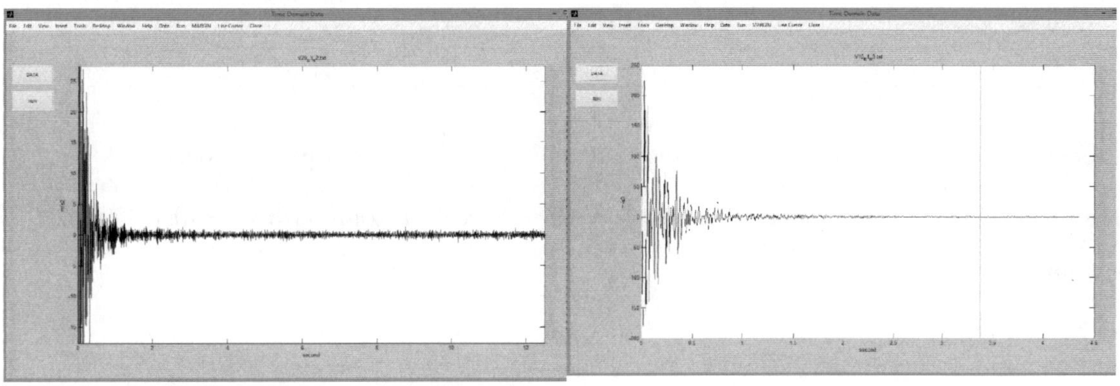

(a)...(b).

Figure 6 Extract parameter data for soft spring (a) and moderate spring (b).

3. RESULTS

Table 4 Comparison Mode Shape & Frequency FEM Test, GVT and FEM Up Date Soft Spring model.

No	Mode Shape	Analysis (Hz)	GVT (Hz)	Analysis Update (Hz)
1	Wing Vertical Bending	3.23	3.11	3.23
2	Wing Inplane Bending	5.08	5.21	4.58
3	Wing Torsion		5.93	6.20
4	Wing Vertical Bending	12.46	12.88	13.15
5	Wing Torsion	18.14		
6	Wing Inplane Bending	20.37	20.03	20.61
7	Wing Torsion		25.91	
8	Wing Bending Torsion	29.09	28.51	29.37

Table 5 Comparison Mode Shape & Frequency FEM Test, GVT and FEM Up Date Moderate Spring model.

No	Mode Shape	Analysis (Hz)	GVT (Hz)	Analysis Update (Hz)
1	Wing Vertical Bending	3.23	3.06	3.19
2	Wing Inplane Bending	5.08	5.21	5.03
3	Wing Torsion		8.72	8.62
4	Wing Vertical Bending	12.46	12.70	13.28
5	Wing Torsion	18.14		
6	Wing Inplane Bending	20.37	20.02	20.61
7	Wing Torsion		26.58	
8	Wing Bending Torsion	29.09	28.60	29.41

There are 86% of the test modes that correspond to the analytical FEM Update modes.

The Figure7 and Figure 8 are the results of the NASTRAN calculation of half wing model of N219 aircraft.

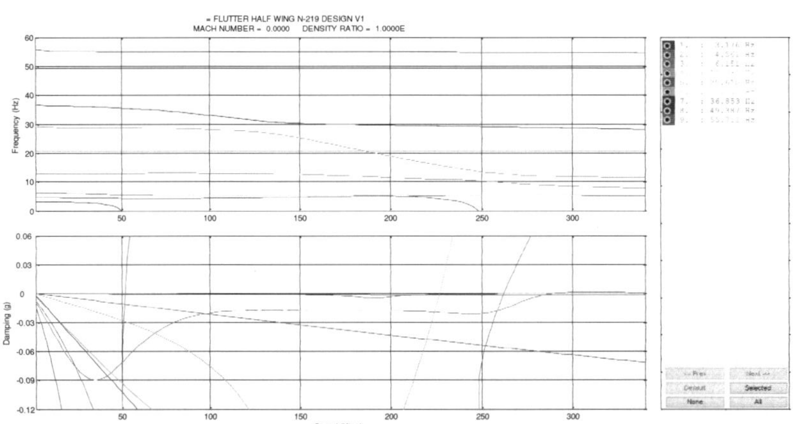

Figure 7 V-g-f Curve of Model Half Wing N219-B11 update Based on GVT BBTA3 Soft Spring Model

Flutter Speed 51 ktas (26.23m/s) at damping 0% and 52 ktas (26.75 m/s) at damping 3%.
Main Flutter Mechanisms *Wing Bending (3.23 Hz) couple with Wing Torsion (6.20 Hz).*

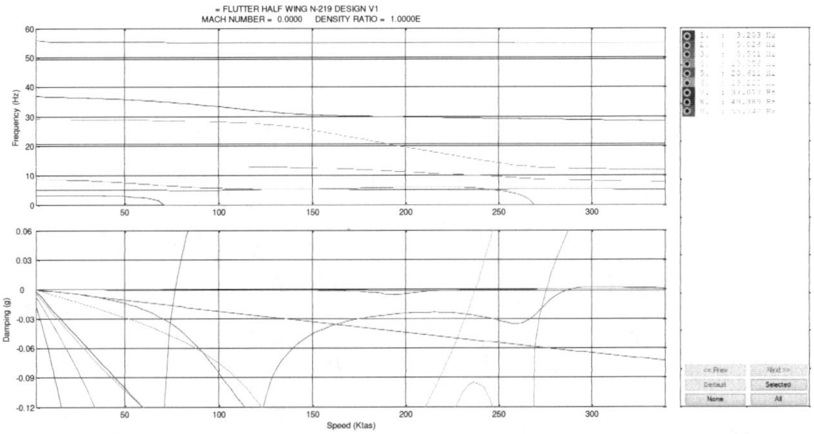

Figure 8 V-g-f Curve of Model Half Wing N219-B11 update Based on GVT BBTA3 Moderate Spring Model

Flutter Speed 76.29 ktas (39.24m/s) at damping 0% and 79 ktas (40.83m/s) at damping 3%.
Main Flutter Mechanisms *Wing Bending (3.26 Hz) couple with Wing Torsion (8.6 Hz).*

Based on data result from extract parameter there was coupling between bending mode and torsion mode on moderate spring model, shown in Figure 9. During a coupling the damping value approached zero on moderate spring model, it can be seen in Figure 10.

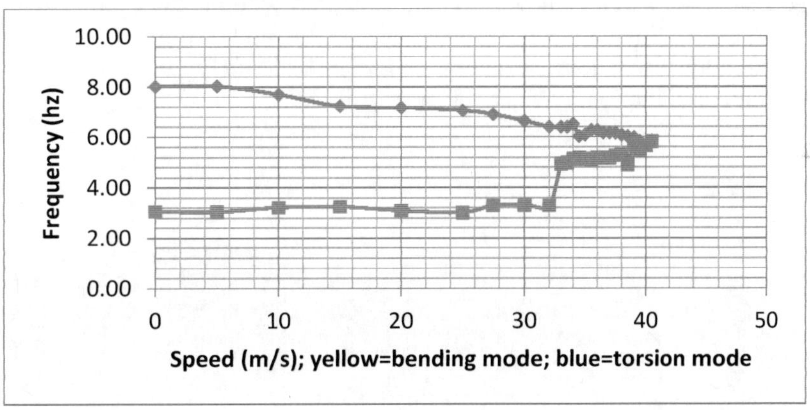

Figure 9 Frequency of bending and torsion mode at moderate spring model

IOP Conf. Series: Journal of Physics: Conf. Series **1005** (2018) 012040 doi:10.1088/1742-6596/1005/1/012040

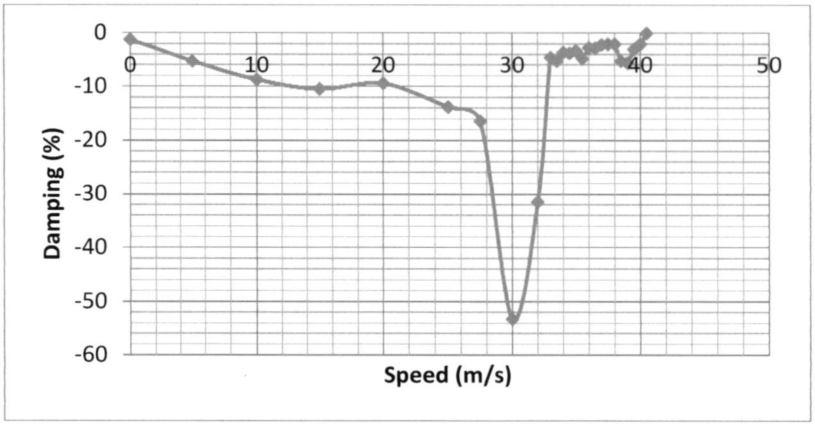

Figure 10 Damping of torsion mode at moderate spring model

The bending frequency increased and torsion mode decreased then coupling occurred at speed 40.5 m/s, frequency 5.8 Hz and damping 0.06 %. According to these data can be concluded that flutter speed happened at 40.5 m/s while FEM analysis result showed flutter speed at 40.83 m/s. Therefore the flutter speed based on FEM analysis and Wind Tunnel Test was nearly the same.

For the soft spring model case, the data that was sent by data acquisition system cannot be processed properly in extract parameter due to spring property. The spring was very responsive therefore other frequency appeared.

When wind speed reached 24 m/s, half wing model vibrated without excitation, it showed that half model wing already unstable. Frequency of bending and torsion mode were difficult to be extracted thus frequency of coupling mode also hard to be determined. Figure 11 and Figure 12 are shows the frequency and damping of bending and torsion mode at soft spring model.

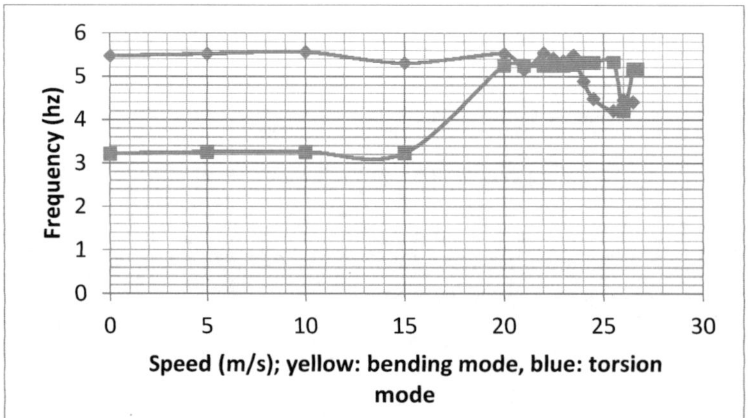

Figure 11 Frequency of bending and torsion mode at soft spring model

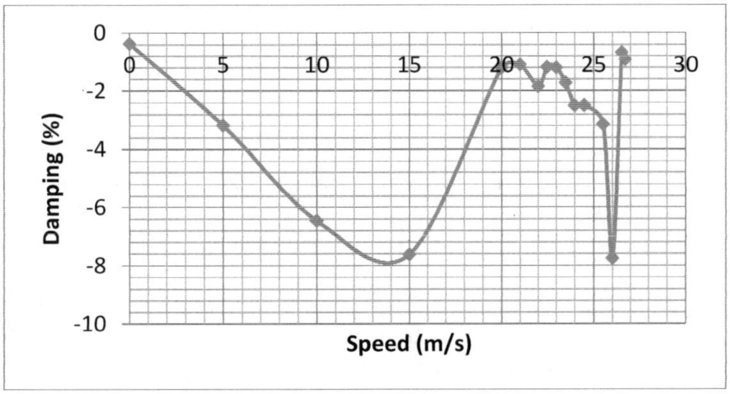

Figure 12 Damping of torsion mode at soft spring model

Based on the V-f curve and V-g curve, there can be seen that there is an existence of coupling mode at speed of 24.0 m/s, yet it does not indicate the phenomenon of flutter. The noise made it difficult to extract the data, therefore this coupling mode was not necessarily caused by the frequency of bending or torsion mode. In order for the data extraction process to be more convenient, it is recommended that the data is filtered beforehand.

4. CONCLUSION

Wind tunnel flutter test represents the real flight condition, therefore its result was used to validate Finite Element Model analysis result. Comparison between FEM analysis and wind tunnel flutter test show at Table 6.

Table 6 Comparison between Analysis FEM and Wind Tunnel Test Result

Case	Flutter Speed (m/s)	
	Analysis	Wind Tunnel Test
Moderate Spring	40.8	40.5
Soft Spring	25.7	24.0

In moderate spring case, flutter speed resulted by FEM analysis and wind tunnel flutter test showed nearly the same result. In soft spring case, FEM analysis resulted flutter speed at 26.75 m/s however in wind tunnel flutter test at speed 24.0 m/s the half wing structure already unstable. According to the comparison between FEM analysis and wind tunnel flutter test, can be concluded that the methodology of FEM analysis has been correctly validated.

References

[1] Blakely, K, *MSC/NASTRAN Basic Dynamic Analysis*, Los Angeles, MacNeal-Schwendler Corporation, 1993.
[2] Blackely, Ken, "*Updating MSC/NASTRAN Models to Match Test Data*", MSC 1991 World User's Conference, Volume II, March 11-15, 1991.
[3] Brenner, M.J., "Aeroservoelastic Modeling and Validation of a ThrustVectoring F/A-18 Aircraft", *NASA Technical Paper 3647*, September 1996.
[4] Cheng, P.Y. and Hirner, T.J., "Aircraft Aeroservoelastic Compensation Using Constrained Optimization", *AIAA Structures, Structural Dynamics, and Materials Conference*, AIAA-1992-2399, April 1992.

[5] Daniel Norrison, Nomad Flutter and Flow Simulation Acceleration for Elastics Wing, School of Mathematical and Geospatial Sciences, College of Science, Engineering and Health, RMIT University, 2009

[6] Gupta, K.K., Brenner, M.J., and Voelker, L.S., "Development of an Integrated Aeroservoelastic Analysis Program and Correlation with Test Data", *NASA Technical Paper 3120*, May 1991.

[7] Lokatt, Mikaela *Aeroelastic flutter analysis considering modeling uncertainties*. Journal of fluids and structures:2017

[8] Moore, G. J. "MSC/NASTRAN, Design Sensitivity and Optimization, User's Guide V68", May 1994.

[9] Pendleton, E., Lee, M., and Wasserman, L., "Application of Active Flexible Wing Technology to the Agile Falcon", *AIAA Journal of Aircraft*, Vol. 29, No. 3, May-June 1992.

[10] Pratt, R.W., Taylor, R., and Caldwell, B.D., "Aeroservoelasticity: Key Issues Affecting the Design of Flight Control Systems", *IEEE Control Conference Publication No. 389,* March 1994

[11] Sayuti Syamsuar, Muhammad Kusni, Adityo Suksmono, Mohamad Ivan Aji Saputro, The Analysis Of Half Wing Flutter Test N219 Aircraft Model In The Low Speed Wind Tunnel, *Warta Ardhia, Volume 42 No. 4, ISSN: 0215-9066*, Kemen-Hub, Jakarta, Desember 2016

[12] Sayuti Syamsuar, Leonardo Gunawan, Martina Widiramdhani, Nina Kartika, The The Design and Experiment of Ground Vibration Test of N219 Aircraft Half Wing Model, *Warta Ardhia, Volume 42 No. 3, ISSN: 0215-9066*, Kemen-Hub, Jakarta, Desember 2016

[13] Schaeffer, H.G, *MSC/NASTRAN Primer—Static and Normal Modes Analysis,* Milford, Wallace Press, 1988.

[14] Strass, H.K., and Stephens, E.W., "An Engineering Method for Determination of Aeroelastic Effects upon the Rolling Effectiveness of Ailerons on Swept Wings", *NACA Research Memorandum L53H14,* November 1953 22 of 23 American Institute

[15] Tang, D and E. H Dowell, Experimental Aeroelastic Models Design and Wind Tunnel Testing for Correlation with New Theory, *Aerospace*, 2016

[16] Tsushima, Natsuki and Su, Weihua. *Flutter suppression for highly flexible wings using passive and active piezoelectric effects*. Aerospace science and technology:2017.

[17] Westin, Michelle F et all. *Aeroservoelastic modeling of a flexible wing for wing tunnel flutter test*. Proceedings of COBEM 2009

[18] Wilson, J.R., "*Active Aeroelastic Wing: a New/Old Twist on Flight*", Aerospace America, September 2002.

The use of new facility by means internal balance with sting support for wide range Angle of Attack aircraft

Subagyo [1)] *, Yanto Daryanto [1)] and Novan Risnawan[1)]

[1)] National Laboratory for Aerodynamics Aero-elastics and Aero-acoustics (BBTA3), Agency for the Assessment and Application of Technology (BPPT), Indonesia

*subagyo@bppt.go.id

Abstract. The development of facilities for the testing of wide range angle of attack aircraft in the wind tunnel at subsonic regime has done and implemented. Development required to meet the test at an angle of attack from -20 ° to 40 °. Testing the wide range angle of attack aircraft with a wide variation of the angle of attack become important needs. This can be done simply by using the sting support-equipped by internal balance to measure the forces and moments component aerodynamics. The results of development and use on the wide range angle of attack aircraft testing are aerodynamics characteristics in the form of the coefficient three components forces and the three components of the moment. A series of test aircraft was successfully carried out and the results are shown in the form of graphs of characteristic of aerodynamics at wind speed 70 m/s.

1. Introduction

Aircraft testing in the wind tunnel is necessary in the design cycle of an aircraft and came up with the design for a prototype. On the condition of take-off and landing, aircraft to fly at low speed conditions associated with the testing regime should to do either subsonic commercial aircraft or wide range angle of attack aircraft. In this condition is very crucial situation.

In terms of wide range angle of attack aircraft testing, the angle of attack (AOA) is varied so that it requires a mechanism that does not like the testing of commercial aircraft. AOA varies from -20 ° to 40 ° requires facilities specially prepared support along with the measuring instrument aerodynamic forces and moments. This condition can be done by using a sting support equipped with measuring aerodynamic force and moment inside in the model i.e. internal balance.

The Wind Tunnel Test (WTT) facility show in figure 1 is belonging to National Laboratory for Aerodynamics Aero-elastics and Aero-acoustics, Agency for the Assessment and Application of Technology (BPPT). The tasks of National Laboratory for Aerodynamics Aero-elastics and Aero-acoustics provide aerodynamic testing services to both the customers from domestic and abroad.Test objects that can be done in National Laboratory for Aerodynamics Aero-elastics and Aero-acoustics not only complete model of the aircraft, but also allows for the test model which is a component of the aircraft, such as two dimensional wing model, half model, tail performance and air intake test.

2. Background Theory and Measurement Method

When an aircraft is flying, then there are aerodynamic forces and moments acting on it due to the interaction between the airflow with aircraft. These events can simulate in WTT of aircraft models. The aircraft model installed in the test section and then give varies of AOA at the constant freestream airflow. Forces and moments that arise measured by a measuring tool called the internal balance. Internal balance is placed in a model test and connected to the Sting Support Mechanism which can move at the position alpha(AOA) or beta (β) angle is varied. In the wind tunnel, the measurement of aerodynamic forces and moments can do in many ways such as using a pressure measurement method on the surface, the measurement by external balance, or internal balance.

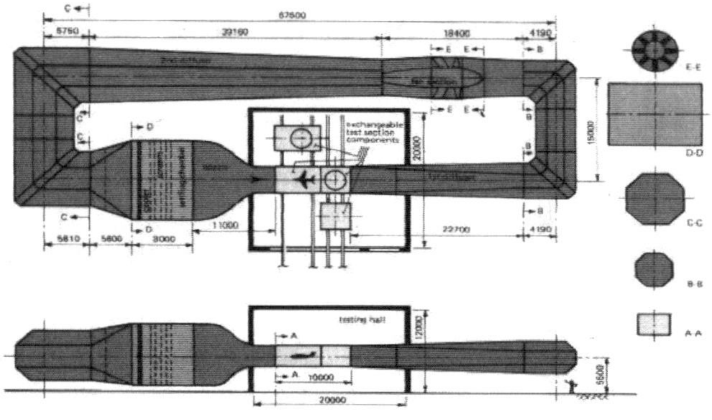

Figure 1. Indonesia Low Speed Wind Tunnel (ILST) facility

Measurement technich by using external balance there are any limitation on the angle of attack (AOA) mechanism. The limitation of AOA mechanism can not adjust AOA varies between -20° up to 40°. In this opportunity National Laboratory for Aerodynamics Aero-elastics and Aero-acoustics, Agency for the Assessment and Application of Technology (BPPT) develope equipment to conduct wide range angle of attack aircraft test. In order to satisfy such mechanism sting support is one apropriate equipment

AOA mechanism incorporated with internal balance. Comparation between sting support and other support show in the figure 2 [1].

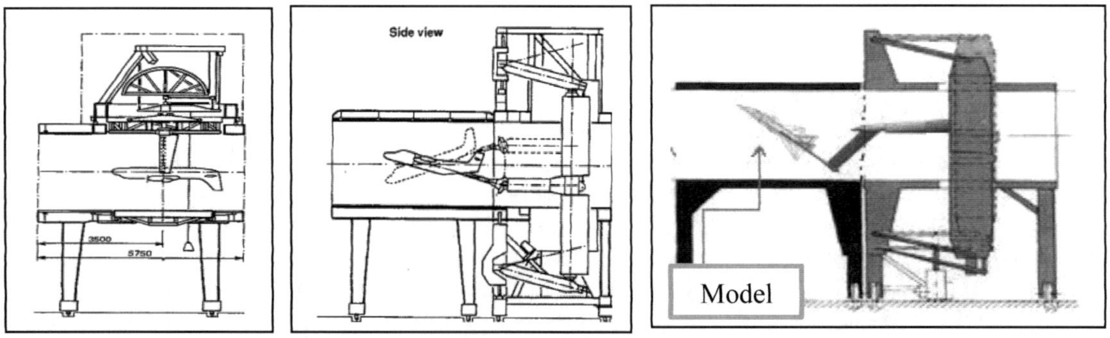

(a) External Balance with centrat strut

(b) Internal Balance with Sting Support

(c) Developed Internal Balance with Sting Support for wide range angle of attack WTT

Figure 2. Support system for WTT

2.1. Develope of wide range angle of attack aircraft test facility

Almost 30 years experience conduct WTT of several civil aircraft i.e. CN-235, N-250, N-2130 also N-219 by existing support system. Recently Indonesian goverment have policy to build aircraft with wide range angle of attack. In order to fullfill WTT wide range angle of attack aircraft , National

Laboratory for Aerodynamics Aero-elastics and Aero-acoustics(BBTA3), Agency for the Assessment and Application of Technology (BPPT) develope test facility for wide range angle of attack aircraft, see figure 2.(c). New Sting support system develope and make complementation to the existing equipment. New sting support equipped with internal balance to measure three component aerodynamics forces and three component aerodynamics moment. As our goal to meet varies AOA, the two stings was designed and produced have a lot a common, they treated as parts of a one integrated sting. The selected concept is, therefore, to have two rear sting parts of similar design (Figure.3), one with the 15° bend and another with the 55° bend, and also to have a common front part of the sting, to be used with either of the bent rear parts. Some parts to be interchangeable between the 15° and 55° sting assemblies. The bent stings will each consist of two cylindrical pods that will be connected to sword of the existing mechanism. The rear ends of the pods have a cylindrical extensions with dimensions compatible with the existing sting.

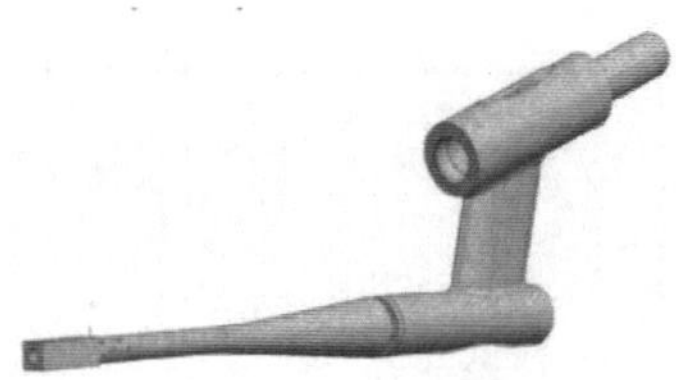

(a) The 3-D Drawing of Sting Bent 15°

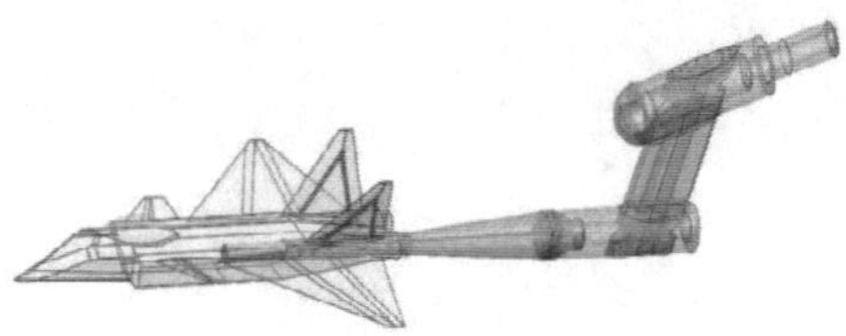

(b) 3-D Design of Sting Bent 15° with a Model Attached

(c) The 3-D Drawing of Sting Bent 55°

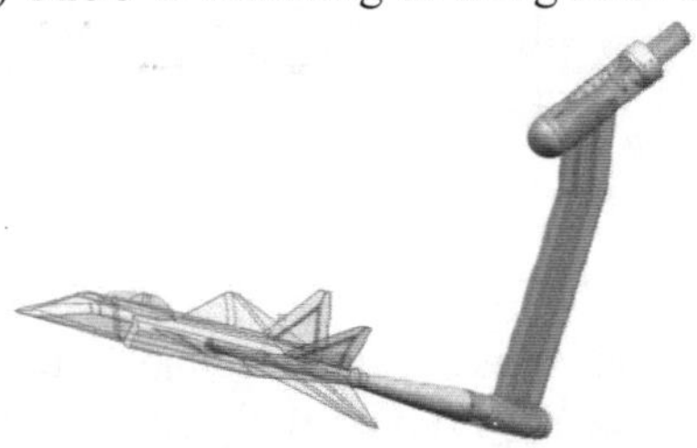

(d) 3-D Design of Bent Sting 55° with a Model Attached at the Sting

Figure 3. Design concept of sting support system

2.2. Wind Tunnel Test

Characteristic aerodynamics of aircraft can be obtained by WTT. For that porpose the aircraft model install in the test section supported by sting. Three component aerodynamics forces and three component aerodynamics moment measure by internal balance. Specification of the internal balance can be seen in table 1 and the physics of internal balance shown in figure 4[2],[6].

Table 1. Internal Balance Specification

No.	Component	Load Capacity	Accuracy (%)
1	*Axial force(Fx)*	3000 N	0.1 % FS
2	*Side force (Fy)*	3200 N	0.1 % FS
3	*Normal Force (Fz)*	8500 N	0.1 % FS
4	*Rolling Moment (Mx)*	400 Nm	0.1 % FS
5	*Pitching Moment (My)*	720 Nm	0.1 % FS
6	*Yawing Moment (Mz)*	540 Nm	0.1 % FS

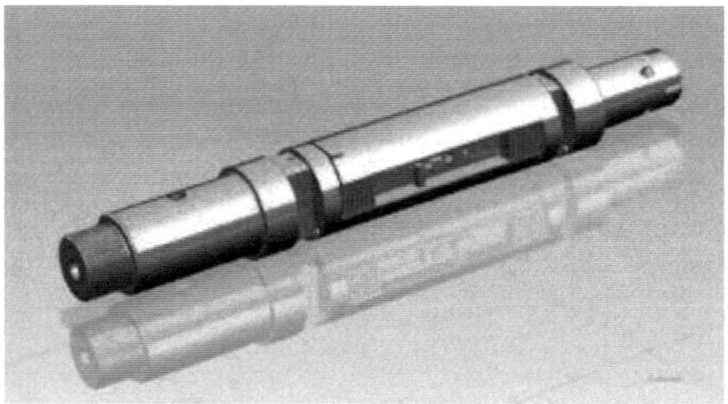

Figure 4. Internal Balance for WTT

Typical testing wide range angle of attack aircraft, AOA have wide range from -20 ° up to 40 °. Here some important run-down test list of wide range angle of attack aircraft show in table 2. Where A1 is variation of AOA from -10 ° up to 40 °.

Table 2. WTT table of wide range angle of attack aircraft

CONFIG	TRIP	VO	ALPHA	BETA	LEF	TEFL	TEFR	HTL	HTR	RUDL	RUDR	LG	GB	SB
WBH1V1	ON	70	A1	0	30	0	0	-30	-30	0	0	-	-	-
WBH1V1	ON	70	A1	0	30	0	0	30	30	0	0	-	-	-
WBH1V1	ON	70	A1	0	30	0	0	0	0	0	0	-	-	-

2.3. Aerodynamics Characteristic

The measurement characteristic aerodynamic of wide range angle of attack aircraft test model in the wind tunnel measure three forces components i.e. lifts (L = lift), Drag (D = drag), Side Force (SF = side forces) and three moments components are pitching moment (M_P = pitch moment), a rolling moment (M_R = rolling moment), and yawing moment (M_Y = yaw moment) [3]. The six components of forces and moments can be formulated as follows:

Three component forces:

IOP Conf. Series: Journal of Physics: Conf. Series **1005** (2018) 012041 doi:10.1088/1742-6596/1005/1/012041

$$L = {}^{1}/_{2}\,\rho V^2 S C_L \quad (1)$$
$$D = {}^{1}/_{2}\,\rho V^2 S C_D \quad (2)$$
$$S_F = {}^{1}/_{2}\,\rho V^2 S C_Y \quad (3)$$

Three component moments:

$$M_P = {}^{1}/_{2}\,\rho V^2 S C_M \quad (4)$$
$$M_R = {}^{1}/_{2}\,\rho V^2 S C_{Roll} \quad (5)$$
$$M_Y = {}^{1}/_{2}\,\rho V^2 S C_{Yaw} \quad (7)$$

Where ρ: air density, V: free stream velocity, S: reference area, C_L: Lift Coefficient C_M: Pitch Moment Coefficient, C_D: Drag Coefficient, C_Y: Side Force Coefficient, C_{Roll}: Rolling Moment Coefficient, C_{Yaw}: Yawing Moment Coefficient.

Methods of measuring forces and moments wide range angle of attack aircraft in WTT using a measuring instrument types Internal Balance, where the model is supported by Sting Support Mechanism, allowing the model to move aircraft maneuvering alpha or beta angle. The working principle of force measurement and torque by using the Internal-Balance; this is to measure physical quantities and is converted into an electrical signal, and then converted back into physical quantities in the form of aerodynamic forces on the computer Data Acquisition and Reduction System Software or DARS [4],[5].

2.4. Model Installation

The model is one to tenth scaled down to the actual size of wide range angle of attack aircraft. It is design and install mounted to three-component force & three-component moment internal balance supported by an aft mounted sting support mechanism in test section see figure 5. The model configuration consists of wing-body-tail, with all-movable horizontal tail as differential stabilizer, and four sets of vertical tails

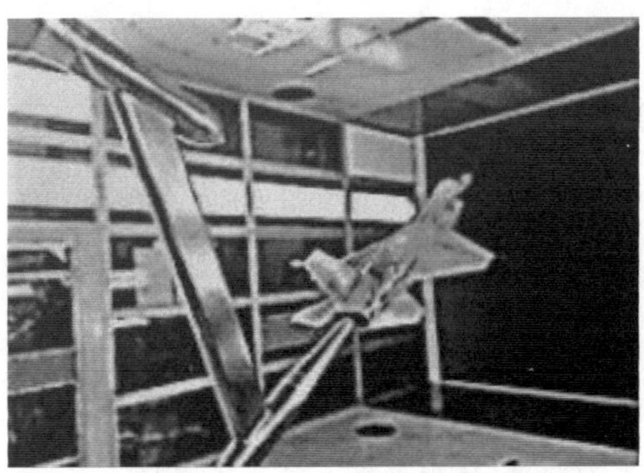

Figure 5. Model of wide range angle of attack aircraft installed in test section

3. Results and Discussions

Developed facility wide range angle of attack aircraft test is implemented in the WTT. Aerodynamic testing data of wide range angle of attack aircraft model in ILST by using internal balance presented in the form of forces and moments Coefficient.

3.1. Lift Force

Lift coefficient wide range angle of attack in the WTT, for configuration LEF = 30° deflection angle of horizontal Ruder zero (HT-R / L = 0°) is greater than CL compared with horizontal tail condition HT-R / L = -30° and HT- R / L = 30° are shown in Figure 6.

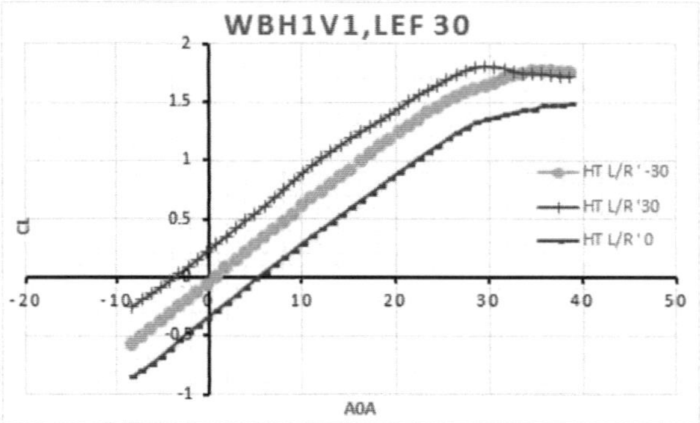

Figure 6. Lift coefficient versus AOA

3.2. Drag force

The drag coefficient CD versus the angle of attack for configuration LEF = 30° is shown in Figure 7. The drag coefficient CD with setting HT-R / L = 0° have lowest drag coefficient at AOA about 3° and followed by drag coefficient CD with setting HT-R / L = 30° at AOA about 8°. Finally highest drag coefficient CD with setting HT-R / L = -30° at AOA about 0°.

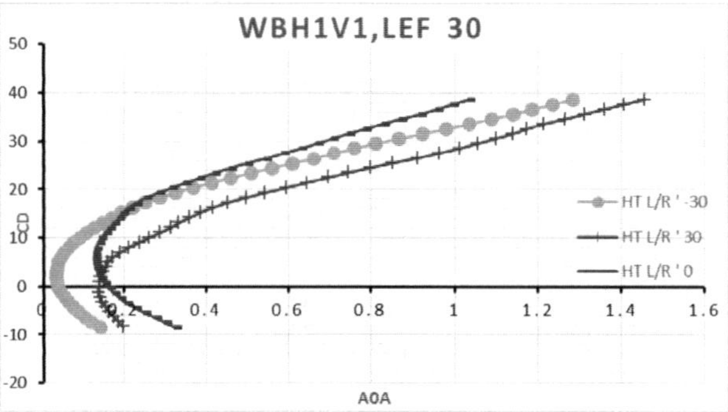

Figure 7. Drag coefficient versus AOA

3.3. Side force

The curve in Figure 8 shows the configuration with setting LEF=30°; HT-R / L = 0° whose CY value use as zero offset. The configuration with setting LEF=30°; HT-R / L = 30°, the value CY is greather than setting with HT-R / L = 0° and also greather than HT-R / L = -30°.

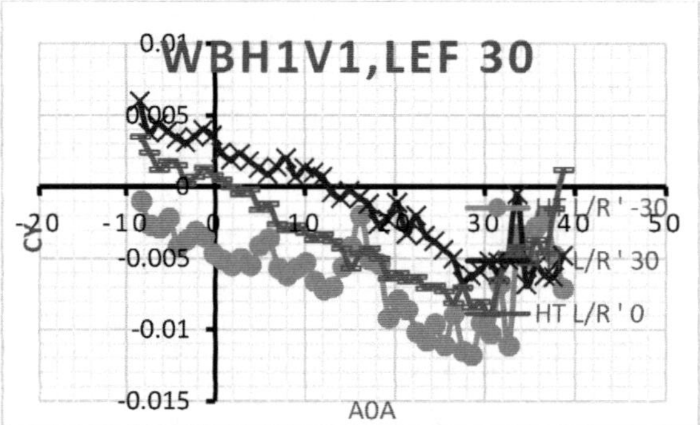

Figure 8. Side force coefficient versus AOA

3.4. Yaw Moment

Directional stability criteria for aircraft is dCYAW/dβ greather than 0. Directional stability investigated as a function of the angle of attack [6] and the results are shown in Figure 9.

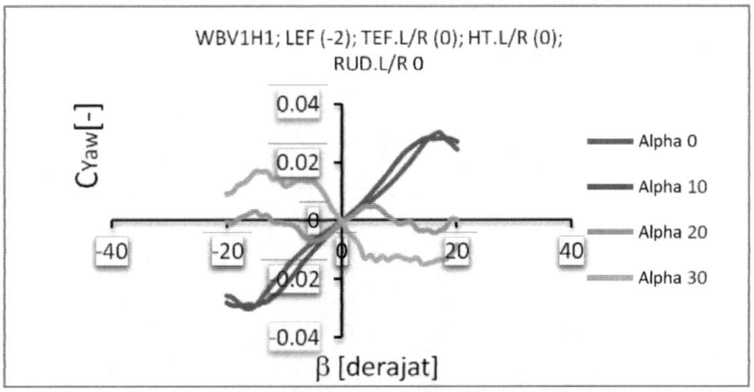

Figure 9. Yaw moment coefficient versus β (side angle)

The CYAW curve in Figure 9 indicated that the gradient CYAW versus a side angle beta changes its value from the positive toward negative at the AOA greater than or equal to 20° so it can be concluded that the aircraft began to experience instability directional at the attack angle greater than or equal to 20°.

3.5. Pitch Moment

Longitudinal static stability of the aircraft with variation of LEF configuration but without deflection of control surface is shown in the figure 10. The wide range angle of attack aircraft is designed to be unstable static due to the high maneuverability, in contrast to conventional aircraft criteria that require stable static. As a result of unstable static characteristics of the wide range angle of attack aircraft fully rely on automatic control system that is controlled by a computer.

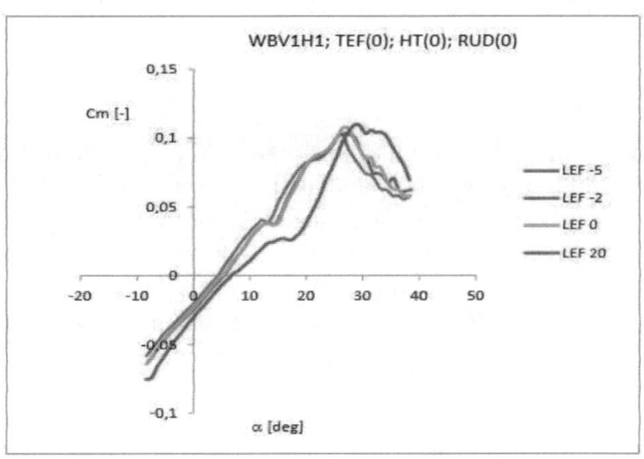

Figure 10. Pitching moment coefficient versus AOA

3.6. Rolling Moment

The characteristic of lateral stability for the AOAvariation is shown in Figure 11 and Figure 12.

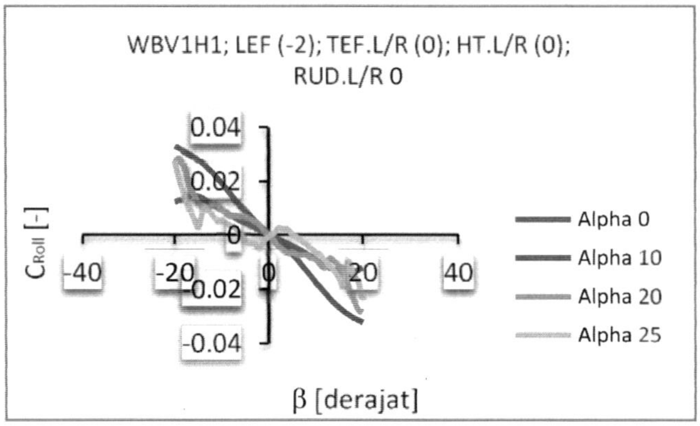

Figure 11. Roll moment coefficient versus β (side angle)

From the image curve in figure 11 and figure 12 shows that the aircraft is stable in the lateral direction varied to the measured angle of attack in this test.

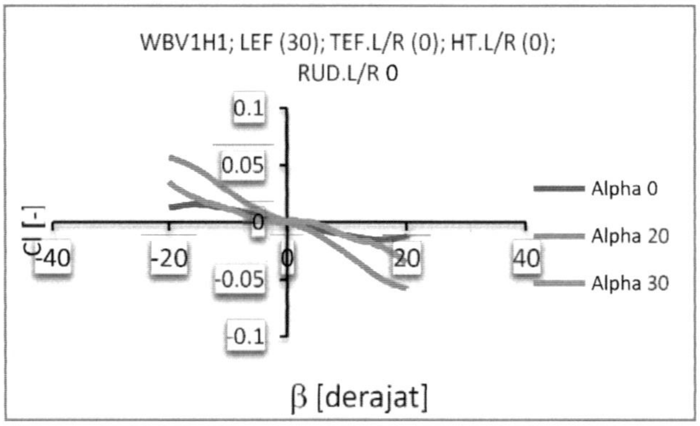

Figure 12. Roll moment coefficient versus β (side angle)

4. Conclusions

Measurement of forces and moments of a wide range angle of attack aircraft model in the wind tunnel with a monoblock type of internal balance with resistance strain gauge showed good results and accurate.Values lift coefficient (CL) is influenced by the angle of attack, look for the AOA up to 30° CL is increasing, while AOA have value above 30° angle showed the CL value decline. This is caused by the occurrence of flow separation around upper wing of the model. While the value of the coefficient of drag (CD) is influenced by frontal area of aircraft model to the direction of the wind.

Static longitudinal stability expressed by the derivative moment coefficient to the AOA is negative. Characteristics longitudinal static stability of wide range angle of attack aircraft are not stable, so the wide range angle of attack aircraft relying solely on automatic control system that is controlled by the computer. Curve of rolling moment coefficient versus a side angle with the variation of AOA, for attack angle configuration 0°, 10°, 30° and with LEF=30° are investigated. The result showed aircraft still stable laterally. While the directional stability observed as a function of the AOA, in the attack angle greather than 20 °, the gradient of the beta CRoll changed from positive to negative, so the aircraft began to experience instability directional.

The last but not least the functional of the new Internal-Balance with sting support mechanism can measure and simulate the force and moment in extreme flying conditions such as on wide range angle of attack aircraft with great accuracy.

IOP Conf. Series: Journal of Physics: Conf. Series **1005** (2018) 012041 doi:10.1088/1742-6596/1005/1/012041

Acknowledgement
This work is supported by Defense and Security Department on the development wide range angle of attack aircraft program. The authors would like to thank director of National Laboratory for Aerodynamics Aero-elastics and Aero-acoustics, Agency for the Assessment and Application of Technology (BPPT) for providing research support facilities.

References

[1] Jaarsma F. General Design Aspects of Low Speed Wind Tunnels. *NLR TECHNICAL PUBLICATION, TP 96555 U, 1996.*
[2] https://en.wikipedia.org/wiki/Sting_(fixture)
[3] Hwankee Cho, Suntae Lee, Cheolheui Han. Experimental study on the aerodynamic characteristics of a fighter-type aircraft model in close formation flight. *Journal of Mechanical Science and Technology, 2014.*
[4] De Vries, O. Equations for the Data Processing ILST, *NLR TR 87122L, NLR. Amsterdam. 1987.*
[5] Jewel W. Barlow, William H. Rae, Alan Pope. Low speed wind tunnel testing, *John Wiley & Sons, 3nd ed.,1984.*
[6] AIAA: Calibration and Use of Internal Strain Gauge Balances with Application to Wind Tunnel Testing, *AIAA R-091-2003 , Reston, 2003.*

Short Term Single Station GNSS TEC Prediction Using Radial Basis Function Neural Network

Buldan Muslim[1,*)], Asnawi Husin[1)], Joni Efendy[2)]

[1)] Space Science Center, National Institute of Aeronautics and Space (LAPAN), Indonesia
2) National Agency of Geospatial Information, Indonesia

*buldan.muslim@lapan.go.id

Abstract. TEC prediction models for 24 hours ahead have been developed from JOG2 GPS TEC data during 2016. Eleven month of TEC data were used as a training model of the radial basis function neural network (RBFNN) and 1 month of last data (December 2016) is used for the RBFNN model testing. The RBFNN inputs are the previous 24 hour TEC data and the minimum of Dst index during the previous 24 hours. Outputs of the model are 24 ahead TEC prediction. Comparison of model prediction show that the RBFNN model is able to predict the next 24 hours TEC is more accurate than the TEC GIM model.

1. Introduction

The ionosphere is the upper atmosphere layer consisting of charged particles in amounts that can affect radio wave propagation. It is well known that the ionosphere is complex system, various physical and energy transport processes takes place in the ionosphere. The space weather, associated with the solar flares, coronal mass ejection and geomagnetic storms are the phenomena effect to the ionosphere. Magnetic storms are measured by geomagnetic scales; the most widely used ones are the Disturbance Storm Time (Dst). During magnetic storm, the solar wind energy deposited into the magnetosphere and will be dissipated into the ionosphere and eventually followed by disturbances in the ionosphere. The disturbed ionosphere is manifested as a large increase or depletion of electron density and Total Electron Content (TEC) from their normal level. Many author has intensive research to description and prediction of the features of the ionospheric related to geomagnetic activities and its processes [1],[2],[3],[4].

Ones of the effect of the ionosphere on radio wave propagation used by GNSS is the ionospheric time delay of pseudo range code (positive delay) or time acceleration of phase carrier wave (negative delay) of the GPS signal propagation from the satellite to the receiver. The satellite distance from the receiver estimated from the time of GNSS signal propagation at the light speed has an error that proportional to the TEC. Such distance measurement error can cause GNSS position error. Thus for precision positioning, ionospheric TEC needs to be estimated for the determination of ionosphere correction on the GNSS signal. The reduction of the ionospheric effect on GNSS positition can be done with various methods depending on which GPS device used. For dual frequency GPS, ionospheric TEC can be calculated from GPS observation data through ionospheric-free linear combination. A single frequency GPS can reduce ionosphere errors by differential GPS techniques (DGPS) requiring at least 2 receivers. Especially for single frequency GPS and unavailable other GPS data for DGPS implementation, ionospheric TEC maps are absolutely necessary for PPP positioning.

For real-time GNSS positioning and operational guidance the GNSS survey, the ionospheric TEC information on the day of the survey and a few hours to come are required. Short term TEC prediction models (1 to 3 days ahead prediction) are available online such Global Ionospheric Maps (GIM). But the GIM TEC prediction data is derived from IGS stations where in Indonesia region there are only 2 stations used such BAKO and JOG2. While the territory of Indonesia is largely unrepresented with two GPS stations for ionospheric TEC observation. Whereas in Indonesia there have been GNSS data from more than 100 stations, but its not used for modeling and predictions of daily TEC. Therefore, daily TEC predictions from GNSS stations in Indonesia are important to obtain more accurate predictions of TEC over Indonesia.

Recently, several short-term TEC prediction models have been developed. Multi regression analysis has been used for prediction of foF2 and M (3000) F2 [5]. A short-term ionospheric forecasting empirical regional model (IFERM) can be used for predicting limited ionospheric parameters for the next 3 hours and specifically above Europe [6]. Radial Basis Function Neural Network (RBFNN) has been used for 30 minutes ahead TEC predictions by [7] that yielded predictive TEC accuracy of about 90% for middle latitudes and 75% for low latitudes. This paper describes the results of the next 24-hour low latitude ionospheric TEC prediction model using an RBFNN from GNSS data which in this early stage was limited to only single GNSS station, JOG2 during 2016.

2. Radial Basis Function Neural Network

The radial basis function network consists of 3 layers: input layer, hidden layer, and output layer. As shown in Figure 1, the input layer accepts a set of data

$$x = [p1, p2, p3, \ldots, pd]^T \quad (1)$$

A set of data, x, is sent by the input layer to the hidden layer. Each node in the hidden layer contains an activation function

$$h_n(x) = h(\|x - c_n\|, \sigma_n) \quad (2)$$

Where n = 1, 2, 3, ..., N is the number of the node, $\|\cdot\|$ is the norm of the vector, $h_n(k)$ is the function of the radial basis localized around cn and with the level of the locality is parameterized by σn. All activations are weighted and sent to the output layer. Each node output represents a unique task and has a special weight that connects the hidden layer to the output layer. The weighted activation of all hidden layer nodes is summed for each target (output) associated. With the notation $W_k = [w_{ok}, w_{1k}, w_{2k}, \ldots, w_{Nk}]^T$, which connects the hidden layer nodes to the output at n to k then output to k as the input response x can be written as:

$$f_i(x) = \sum_{j=1}^{N} w_{i,j} h_j(x) \quad (3)$$

i = 1, 2, 3, … k, indicte the outputs number ,
j = 1, 2, 3, … N, indicate the numbers of activation functions,
The weigths number is N x K.

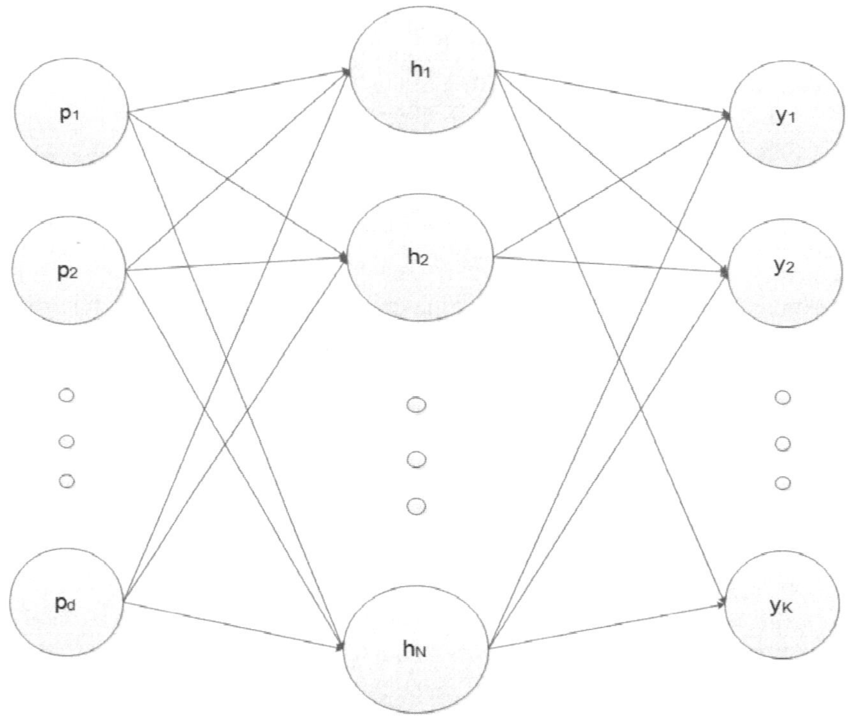

Figure 1. Radial Basis Function Neural Network Architecture

3. Data and Methodology

The TEC data used in this model is obtained from two sources: TEC derived from rinex data of GNNS station in Yogyakarta (JOG2) and TEC from Global Ionospheric Maps (GIM) model released by Center for Orbit Determination in Europe (CODE).

Derivation of TEC from JOG2 GPS data is done by using GOPI software [8]. An output of GOPI included the average TEC of all GPS satellite observations at elevations above 20 degrees. This GOPI GPS TEC data is used as a reference to TEC data which is assumed to be close to the actual TEC and as a model error measure calculated from its deviation from GOPI GPS TEC data.

GIM TEC data released by CODE are used as a comparison of RBFNN TEC prediction results, hence RBFNN model performance can be known. TEC CODE has a spatial resolution of 2.5o of latitude and 5o of longitude. Early the CODE GIM TEC has a time resolution of 2 hours [9]. But since the last few years, TEC CODE has a resolution of 1 hour. Therefore GOPI TEC data having the resolution every 30 seconds needs to be adjusted by calculating the average TEC every hour from 00:00 to 23:00.

For model completion, the calculation of model weights, equation (3) is written in matrix form [10].

$$A\hat{W} = H\hat{Y} \tag{4}$$

With the design matrix H as follows:

$$H = \begin{bmatrix} h_1(p_1) & h_2(p_1) & \cdots & h_m(p_1) \\ h_1(p_2) & h_2(p_2) & \cdots & h_m(p_2) \\ \vdots & \vdots & \ddots & \vdots \\ h_1(p_d) & h_2(p_2) & \cdots & h_m(p_d) \end{bmatrix} \tag{5}$$

IOP Conf. Series: Journal of Physics: Conf. Series **1005** (2018) 012042 doi:10.1088/1742-6596/1005/1/012042

And A^{-1}, as the variant matrix i.e

$$A^{-1} = \left(H^T H + \Lambda\right)^{-1} \tag{6}$$

Where Λ contains a zero-valued component except for the regulatory parameters located on its diagonal and $\hat{Y} = \left[\hat{y}_1 \; \hat{y}_2 \; \hat{y}_3 \cdots \hat{y}_p\right]$ is the learning output vector in this context the TEC data. As the input (x) is also the TEC data but for the previous 24 hours data from the output TEC data.

The solution of (4) is what is called a normal equation

$$\hat{W} = A^{-1} H^T \hat{Y} \tag{7}$$

To solve equation (7), available toolbox Matlab is used and method of determining JFBR model is described in Figure

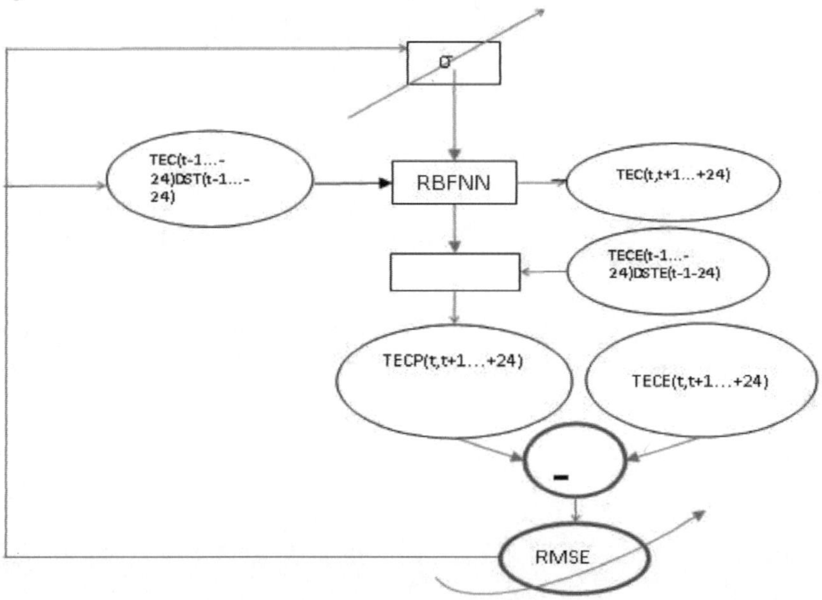

Figure 2. Flow chart of RBFNN training for optimizing the locallity parameter using test data

With a set of 24 hours of previous TEC learning data and the previous 24 hours output during 2016 from January to November, we created the TEC model using RBFNN with the locality parameter value σ, which determined the width of the activation function in the form of gaussian function and the σ is the locality parameter.

$$h(X) = e^{-\frac{z^3}{2\sigma^2}} \tag{8}$$

The RBFNN training model with the σ value is used for TEC prediction with TEC data of December 2016. The prediction model error from the test data is measured by root mean square error (RMSE). The next procedure, change the value σ for producing the different RMSE values for each σ. From a series of σ , series of RMSE is obtained. The selected model is the RBFNN model using σ which produces the smallest RMSE or determined at the beginning of the modeling.

4. Results and Discussions

The trend of the RBFNN model errors to the change of the locality parameter value σ is shown in Figure 3 which shows that for value of σ from 10000 to 200000 the standard deviation value falls sharply. And at σ = 948.000, the error of the RBFNN is about 3.8075 TECU.

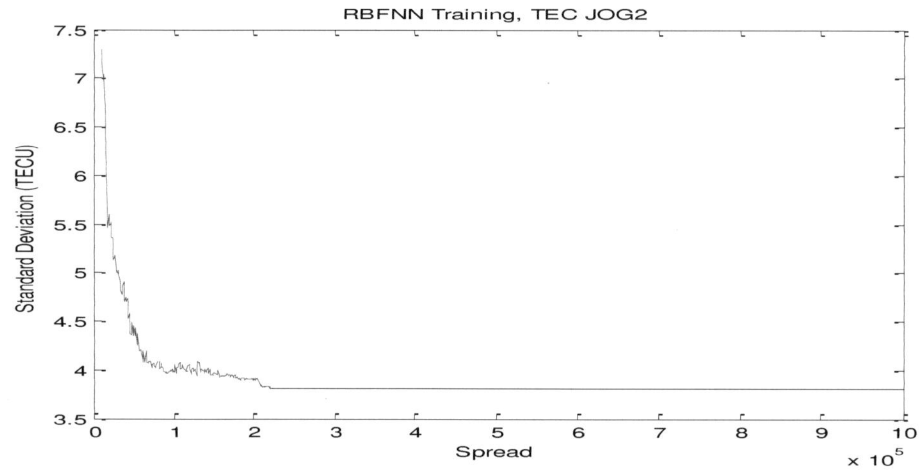

Figure 3. The locality parameter value of σ = 948.000, the RBFNN model has a minimal error with the RMSE is about 3.8075 TECU.

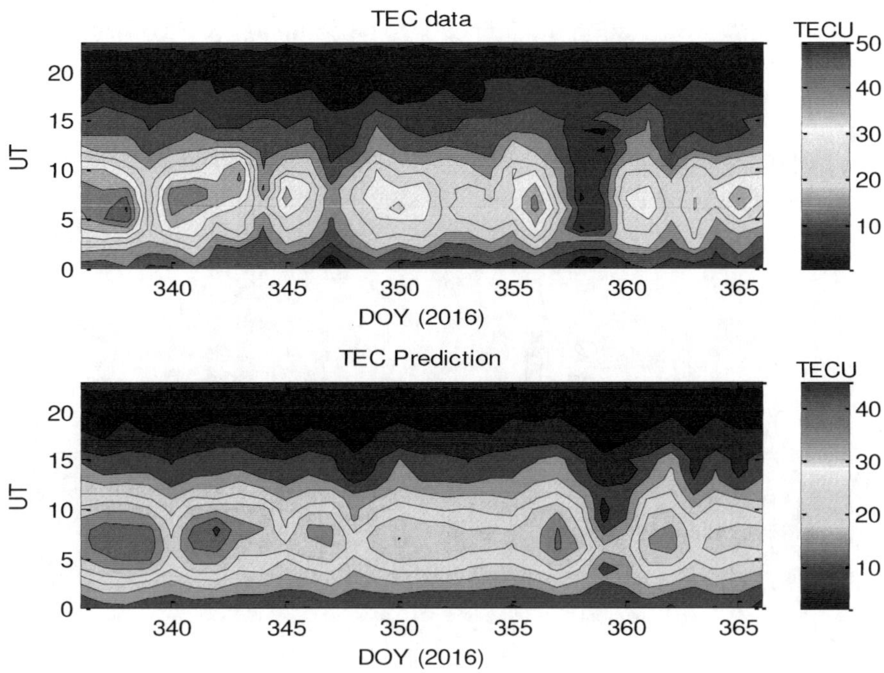

Figure 4. Comparison of TEC prediction patterns with TEC observations from JOG2 GPS station.

It can be seen in Figure 4 that the TEC pattern in the contour is almost similar but there is a slight shift in maximum time such as on day 356 (TEC data) while the prediction model is 357 days, and the maximum value of TEC data at 345 but TEC prediction on 346th day.The overall error of the JSTFBR model of the learning data (used for modeling) is shown in Figure 5, wherein the errors vary from about 2 to 12 TECU, and there appears to be a seasonal variation of model error with a maximum value around March - April and September

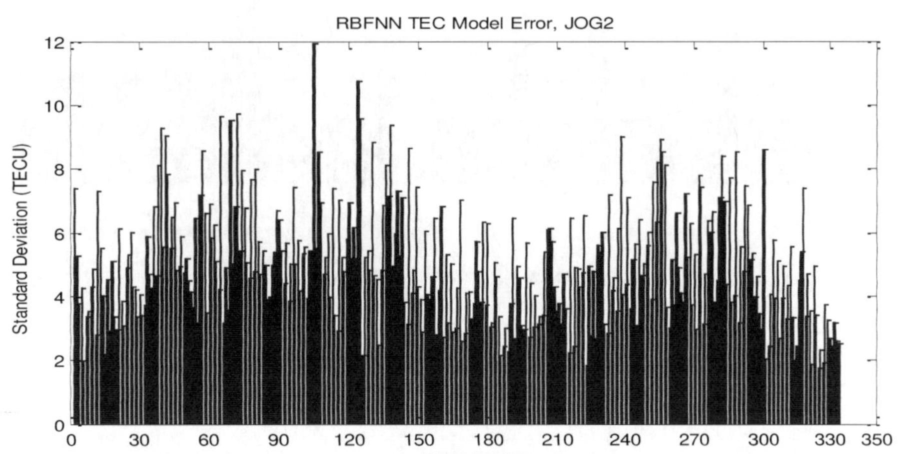

Figure 5. Daily variation of the errors of RBFNN model (Januari - Novemver 2016)

IOP Conf. Series: Journal of Physics: Conf. Series **1005** (2018) 012042 doi:10.1088/1742-6596/1005/1/012042

But by simply comparing the model with TEC data, the RBFNN model performance can not be known, then the RBFNN prediction model needs to be compared with other TEC model data. The comparison of the RBFNN and GIM model to the TEC data is shown in Figure 6 and 7. Figure 6 shows the GIM RMSE ranging from 2 to 14.5 TECU, whereas in Figure 7 it is shown that the TEC RBFNN prediction error ranges from 1.5 to 10 TECU, RMSE of both models are 3.75 and 5.42 TECU respectively.

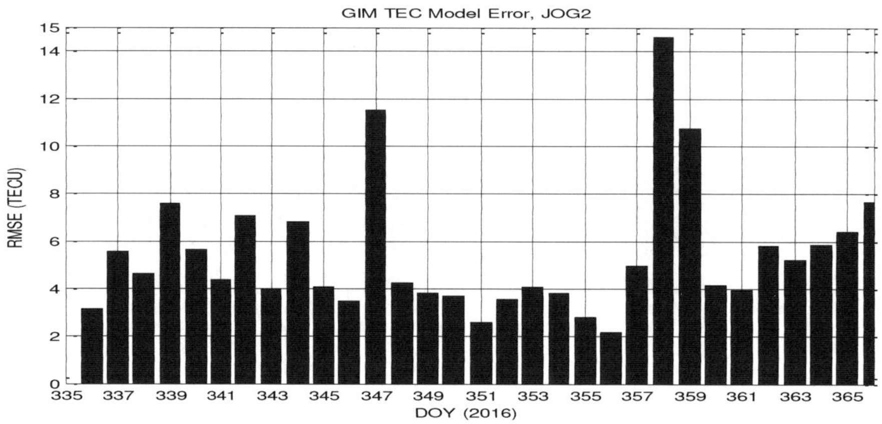

Figure 6. RMSE of GIM TEC model from GOPI TEC value

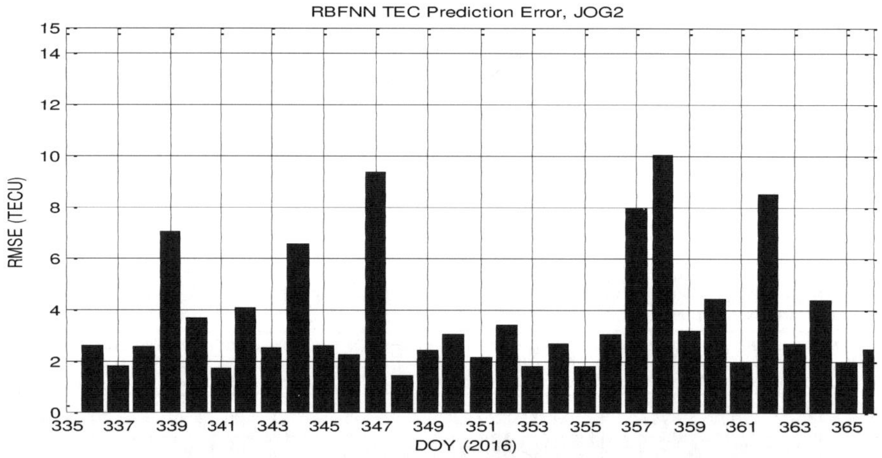

Figure 7. RMSE of RBFNN TEC model from GOPI TEC value

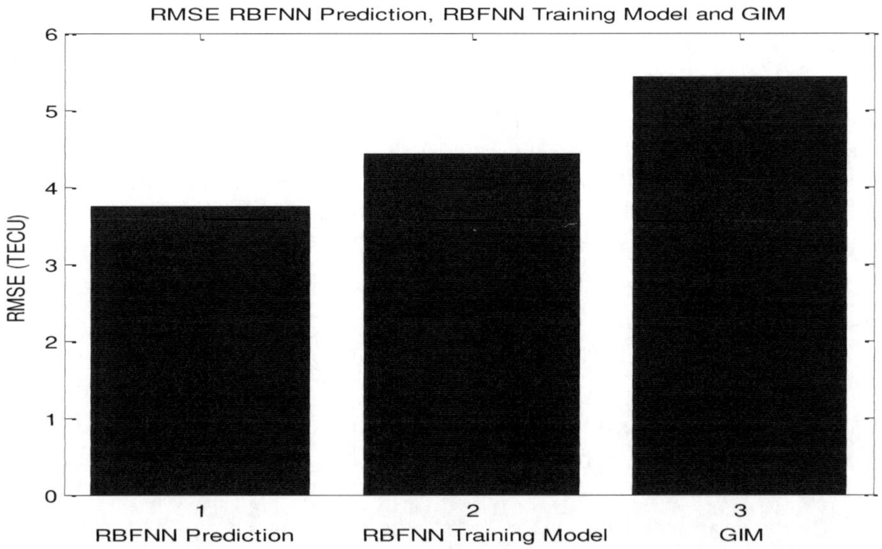

Figure 8. RMSE comparison of RBFNN prediction model, RBFNN training model, and GIM TEC measured from GOPI TEC of JOG2 station.

The average RMSE predicted RBFNN, Model RBFNN, and GIM data TEC data can be seen in Figure 8, which shows that the predicted TEC of RBFNN has a smaller error than TEC GIM and TEC RBFNN model for learning from January - November 2016.

5. Conclusions

The RBFNN TEC model with inputs of the previous 24 hours TEC and the minimum of TEC of one day earlier and with the upcoming 24-hour TEC as outputs has been developed based on the JOG2 GPS TEC data from January - November 2016, and TEC model test data during December 2016. A comparison of RBFNN TEC prediction in December 2016 in which the TEC data are not used in the learning process indicates that the RBFNN TEC has better accuracy than the GIM TEC.

Acknowledgement
The authors would like to thank Head of Space Science Center of LAPAN for providing research support facilities and to Indonesian National Agency of Geospatial Informastion for providing GNSS data, and to International GNSS services for GNSS data sharing.

References

[1] Kutiev I and Muhtarov P. Modeling of mid-latitude F region response to geomagnetic activity *J. Geophys. Res. 106 (A8) 15501–15509. 2001*

[2] Fejer B G. Low latitude ionospheric electrodynamics *J. Atm. Terr. Phys. 64 1401– 1408. 2002*

[3] Liu L et al. The low latitude ionospheric effects of the April 2000 magnetic storm near 1200E. *Earth Planet Space 56 607– 612. 2004*

[4] Lin C H et al. Large-scale variations of the low-latitude ionosphere during the October–November 2003 superstorm: Observational results *J. Geophys. Res. 110 (A09S28) doi:10.1029/2004JA010900. 2005*

[5] Dabas, R. S., K. Sharma, Rupesh M. Das, N. K. Sethi, K. G. M. Pillai, A. K. Mishra, Ionospheric modeling for short- and long-term predictions of F region parameters over Indian zone, *Journal of Geophysical Research, Volume 113, Issue A3, DOI: 10.1029/2007JA012539, 2008*

[6] Pietrella, M., A short-term ionospheric forecasting empirical regional model (IFERM) to predict the critical frequency of the F2 layer during moderate, disturbed, and very disturbed geomagnetic conditions over the European area, *Ann. Geophys., 30, 343-355, doi:10.5194/angeo-30-343-2012, 2012.*

[7] Huang, Z., and H. Yuan , Ionospheric single-station TEC short-term forecast using RBF neural network, *Radio Sci., 49, 283–292, 2014 doi:10.1002/2013RS005247.*

[8] Seemala, Rinex GPS-TEC program Version 2.9.2, http://seemala.blogspot.co.id, akses 3 January 2017.;

[9] Schaer, S., Mapping and Predicting the Earth's Ionosphere Using the Global Positioning System, Ph.D. *Thesis, Astronomical Institute, University of Berne, 1999.*

[10] Orr, M.J.L., Introduction to Radial Basis Function Networks, http://www.cc.gatech.edu/~isbell/tutorials/rbf-intro.pdf, akses 2 Juni 2017

IOP Conf. Series: Journal of Physics: Conf. Series **1005** (2018) 012043 doi:10.1088/1742-6596/1005/1/012043

Analytical Approach to the Fuel Optimal Impulsive Transfer Problem Using Primer Vector Method

E Fitrianingsih[1) *], R Armellin[1)]

[1)] Surrey Space Centre, University of Surrey, United Kingdom
*e.fitrianingsih@surrey.ac.uk

Abstract. One of the objectives of mission design is selecting an optimum orbital transfer which often translated as a transfer which requires minimum propellant consumption. In order to assure the selected trajectory meets the requirement, the optimality of transfer should first be analyzed either by directly calculating the ΔV of the candidate trajectories and select the one that gives a minimum value or by evaluating the trajectory according to certain criteria of optimality. The second method is performed by analyzing the profile of the modulus of the thrust direction vector which is known as primer vector. Both methods come with their own advantages and disadvantages. However, it is possible to use the primer vector method to verify if the result from the direct method is truly optimal or if the ΔV can be reduced further by implementing correction maneuver to the reference trajectory. In addition to its capability to evaluate the transfer optimality without the need to calculate the transfer ΔV, primer vector also enables us to identify the time and position to apply correction maneuver in order to optimize a non-optimum transfer. This paper will present the analytical approach to the fuel optimal impulsive transfer using primer vector method. The validity of the method is confirmed by comparing the result to those from the numerical method. The investigation of the optimality of direct transfer is used to give an example of the application of the method. The case under study is the prograde elliptic transfers from Earth to Mars. The study enables us to identify the optimality of all the possible transfers.

1. Introduction

In an interplanetary mission, one of the mission analysis objectives is designing a trajectory with minimum propellant to allow the spacecraft brings more useful mass [1]. The interplanetary flight requires the spacecraft to leave the orbit of one celestial body to enter the orbit of another planet or performing rendezvous with another planet or object in space.

During the preliminary design phase, the designed trajectory can be modeled as a simple conic arc. Alternatively, it is possible to design a more complex transfer involving other celestial bodies passed by the spacecraft during the transfer, commonly known as gravity assist trajectory. Although the latter is much more complicated than a conic arc, it is widely used by a mission designer due to its advantage of significantly reducing the amount of required propellant for the transfer.

Gravity assist is not the only way to reduce the required propellant for the interplanetary transfer. Mission designers often include corrective maneuver(s) to further reduce the amount of required propellant which is known as Deep Space Maneuver (DSM). Such trajectory, which combines gravity assist and deep space maneuver, is referred to as Multiple Gravity Assist - Deep Space Maneuver (MGA-DSM) trajectory [2].

One example of a recent MGA-DSM application is Juno, the interplanetary space mission to Jupiter under NASA's New Horizon program. Juno, which arrived at Jupiter's orbit in July 2016, has

IOP Conf. Series: Journal of Physics: Conf. Series **1005** (2018) 012043 doi:10.1088/1742-6596/1005/1/012043

successfully performed two DSMs along its way in addition to the earth gravity assists. It initially launched into a heliocentric orbit to obtain a short flight time to Jupiter [3]. On its way to Jupiter, Juno performed two DSMs in 2012, approximately one year after launch [4]. The two DSMs were performed to correct its orbit so that it entered the Earth's sphere of influence (SOI) and obtained `free' energy from the Earth to travel to Jupiter. The second maneuver was performed on February 2016, correcting its path to prepare a rendezvous with Jupiter. Figure 1.1 below illustrates Juno's mission profile from its launch in 2011 until its arrival at Jupiter orbit in July 2016. Table 1.1 lists the application of DSM on some of the most recent missions.

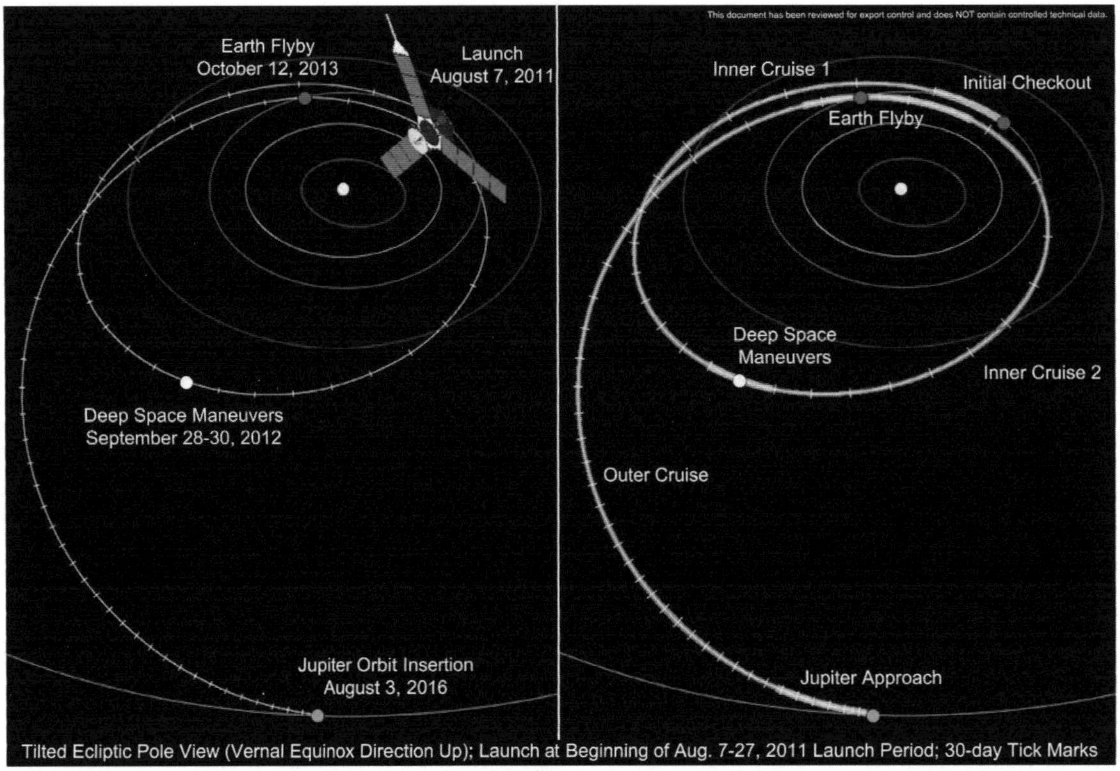

Figure 1. Juno interplanetary mission profile

Table 1. DSM application on interplanetary missions [5], [6], [7]

Mission	Date of performing DSM	V (m/s)	Correction	Target
Exomars	28 July 2016	334.176	DSM-1	Mars
	11 August 2016	17.7	DSM-2	
Juno	28 September 2012	345	DSM-1	Jupiter
	3 August 2016	385	DSM-2	
Messenger	12 December 2005	315.6	DSM-1	Mercury
	17 October 2007	227.4	DSM-2	
	17 March 2008	72.2	DSM-3	
	6 December 2008	222.1, 24.7	DSM-4 (two segments)	
	29 November 2009	117.75	DSM-5	

IOP Conf. Series: Journal of Physics: Conf. Series **1005** (2018) 012043 doi:10.1088/1742-6596/1005/1/012043

An optimal MGA-DSM transfer is solved using modern mathematical tools to obtain the global solution such as genetic algorithm [8] and evolutionary algorithm [9]. However, this approach suffers from the dimensionality of the search space. On the other hand, one can use an analytical approach to solve for an optimum trajectory by using the calculus of variation [10]. One of the most established work in trajectory optimization using an analytical method based on Calculus of Variation is Primer Vector Theory. The theory which was introduced by Lawden in the 1960s [11], providing a method to determine a transfer optimality using a set of Necessary Conditions (NC).

There are two categories of trajectory optimization methods; direct and indirect methods. Trajectory optimization using primer vector is classified as an indirect method because the optimum trajectory is not defined by calculating the cost, instead, it determines the optimality using the NC [12]. The NC allows a mission planner to analyze whether a transfer is optimum or non-optimum by evaluating the primer vector profile along the transfer trajectory and see if it satisfy the NC. Transfer trajectory refers to the trajectory of the spacecraft from orbit around the departure planet to the orbit of around the destination.

A non-optimum trajectory can be altered by modifying the orbital velocity vector as such to satisfy the NC of optimality [13]. The modification of the velocity vector is achieved through the implementation of DSM (corrective maneuver). A linear method to optimize a transfer based on the gradient of the cost functional was first proposed by Jezewski, which estimates the position and time to apply the DSM [14].

The propagation of primer vector profile requires solving the primer vector equation which is a second order differential equation. The primer profile is commonly obtained using a propagation of state transition matrix to solve the primer vector equation. This approach relies on the numerical integration of the STM and the primer vector hence the computation time becomes a disadvantage. The paper by Iorfida [15] presented a novel approach to solve the primer vector equation analytically, giving a focus on the out of plane component (normal to orbital plane). This paper presents the study based on Iorfida's work to solve the in plane component of the primer vector.

2. Method

The primer vector method provides a tool to check the optimality of a transfer by using a set of necessary conditions for transfer optimality. The necessary conditions depend on the types of the transfer and for the impulsive transfer case, there are four necessary conditions for optimality [10]:

1. The primer vector and its derivative are continuous everywhere.
2. The magnitude of the primer vector is always below 1, aside from the instants where the impulse occurs, where $||\mathbf{p}|| = 1$.
3. At the impulse times, the primer vector is a unit vector in the direction of the thrust, $\mathbf{p} = \frac{\Delta \mathbf{V}}{||\Delta \mathbf{V}||}$
4. As a consequence of the above conditions, the first derivative, $\frac{dp}{dt} = \dot{p} = \dot{\mathbf{p}}^{\boldsymbol{T}}\mathbf{p} = 0$ at an intermediate impulse.

In order to evaluate a transfer with regards to the necessary conditions above, the primer profile along the transfer needs to be propagated. The propagation requires solving the primer vector equation in the form:

$$\ddot{\mathbf{p}}(t) = \mathbf{G}(\mathbf{r})\mathbf{p}. \tag{1}$$

$\mathbf{p}$ in equation 1 is the primer vector which belongs to the 3-dimensional space while $\mathbf{G}$ is the gravity gradient matrix. The gravity gradient matrix in the inverse gravitational form is given by

$$\mathbf{G}(\mathbf{r}) = -\frac{\mu}{r^3}\begin{bmatrix} 1 & 0 & 0 \\ 0 & 1 & 0 \\ 0 & 0 & 1 \end{bmatrix} + \frac{3\mu}{r^5}\begin{bmatrix} x^2 & xy & xz \\ xy & y^2 & yz \\ xz & yz & z^2 \end{bmatrix}. \tag{2}$$

A new approach proposed by Iorfida and Palmer to solve the primer vector equation without the use of state transition matrix [15]. The approach used a coordinate transformation from Cartesian coordinate (x, y, z) to local Polar coordinate (r, θ, h). The Polar coordinate system is a local coordinate, which means the coordinate system change with time. The local Polar coordinate is defined as in figure 2.

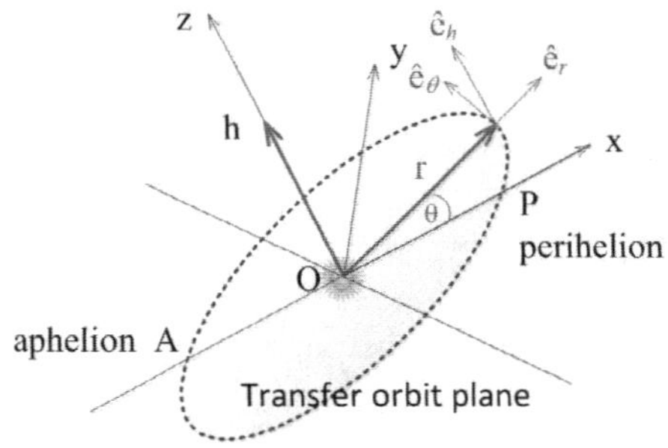

Figure 2. Polar coordinate system definition

This reference coordinate change leads to the decoupling of the in plane (r and θ) and out of plane (h) components of the primer derivative as expressed in equation 3 below:

$$\dot{\mathbf{p}} = \left(\dot{p}_r - \frac{h}{r^2} p_\theta\right) \hat{\mathbf{e}}_r + \left(\dot{p}_\theta + \frac{h}{r^2} p_r\right) \hat{\mathbf{e}}_\theta + \dot{p}_h \hat{\mathbf{e}}_h, \tag{3}$$

hence, the in plane and the out of plane components can be analyzed separately. In equation 3, p_r, p_θ, p_h are the radial, circumferential and normal components of the primer vector (see figure 2). For an in plane transfer case where the initial and final orbit lies in the same orbital plane, the primer vector is obtained by solving the in plane part of equation 3:

$$\dot{\mathbf{p}} = \left(\dot{p}_r - \frac{h}{r^2} p_\theta\right) \hat{\mathbf{e}}_r + \left(\dot{p}_\theta + \frac{h}{r^2} p_r\right) \hat{\mathbf{e}}_\theta. \tag{4}$$

The analytical solution of equation 4 is derived by introducing the relative energy (H_R) and relative angular momentum (L_R) into the equation. (More explanation about the concept of relative energy and relative angular momentum can be found in the paper by Imre and Palmer [16]). As a result, fully analytical solutions to the in plane primer vector component are obtained in the forms:

$$p_r = A_1 \bar{x} + A_3 \left(\frac{1}{x} - 1\right) - A_4 \left[x - \frac{3}{2}(1 - e^2)^{-3/2} M \bar{x}\right] \tag{5}$$

and

$$p_\theta = \frac{A_1}{x} - A_2 x - A_3 \bar{x}(x + 1) + \frac{3}{2} A_4 \left[\frac{1}{x} M (1 - e^2)^{-3/2}\right] \tag{6}$$

where

$$x = r/l, \tag{7}$$

$$\bar{x} = lr/h. \tag{8}$$

M is the mean anomaly of the transfer orbit:

IOP Conf. Series: Journal of Physics: Conf. Series **1005** (2018) 012043 doi:10.1088/1742-6596/1005/1/012043

$$M = {h(1-e^2)^{3/2}t}\big/{l^2} \tag{9}$$

r is the orbital radius, l is the parameter semi latus rectum, e is the eccentricity and h is the true anomaly of the transfer orbit. In equations 5 and 6 above, there are four constants, A_1, A_2, A_3, A_4 which are solved from the boundary conditions of the transfer. In a fixed transfer case (fixed initial and final positions) with fixed transfer time, the boundary values; $p_{r_0}, p_{\theta_0}, p_{r_f}, p_{\theta_f}$ are known.

3. Results and Discussions

In order to validate the analytical solution of the in-plane primer vector components expressed in equations 5 and 6, the propagation results are compared to the result from numerical integration using state transition matrix. The numerical solution of the primer vector is defined in the Cartesian coordinates (x, y, z) with x and y axes lie in the orbital plane and z axis normal to the orbital plane as in figure 3.

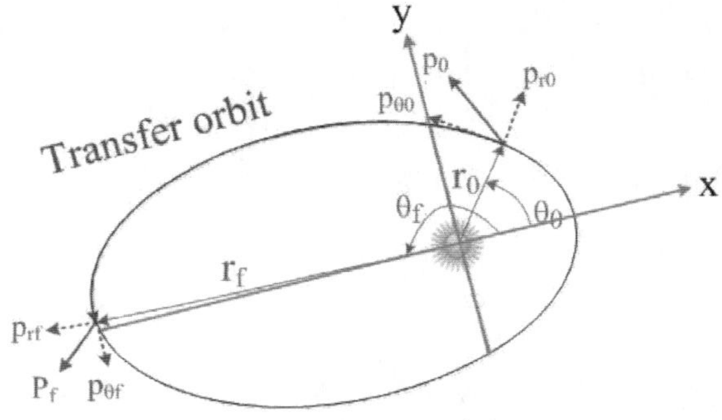

Figure 3. Transfer orbit in Cartesian coordinate frame

In the Cartesian frame, the origin lies in the main body (Sun). The x-axis is in the direction of eccentricity vector. The angular momentum direction is the direction normal to the orbital plane which is defined as the z-axis. The y-axis follows the right-hand rule. The two-points boundary value problem is represented by specifying the initial and final true anomalies of the transfer. For a fixed transfer case, where the eccentricity and semi major axis are known, the radius and velocity vectors at the departure and arrival for a planar case, can be calculated using these formulas [17]:

$$\begin{aligned} \mathbf{r} &= r\cos\theta\,\hat{\mathbf{i}}_x + r\sin\theta\,\hat{\mathbf{i}}_y \\ \mathbf{v} &= -{\mu}/{h}\sin\theta\,\hat{\mathbf{i}}_x + {\mu}/{h}(e+\cos\theta)\,\hat{\mathbf{i}}_y\,. \end{aligned} \tag{10}$$

The polar components of the velocity vector are calculated by multiplying **v** with the rotation matrix from Cartesian to Polar coordinate.

The 3rd axis in Cartesian coordinate (z-axis) is parallel to that of the Polar coordinate (h-axis) so there is no transformation required even if the case is 3-dimensional. Since the true anomaly at departure and arrival are known, the transfer time is solved using Kepler's equation. The algorithm to solve the primer vector using numerical propagation is presented in the flowchart shown in figure 4.

The boundary value in the 4th step of the algorithm in figure 4 came from Lawden's 2nd necessary condition, which states that the magnitude of primer vector at the time of impulse has to be equal to one. This is also used to verify the result from the final step where the modulus of the primer vector, p_f at $t = t_f$ from propagating the primer vector using the initial condition has to be equal to one. It is important to note that for the initial and final conditions, any values of p_r and p_θ can be chosen, as long as $\|\,\mathbf{p}\,\|$ is equal to one. Since the transfer trajectory is fixed, different values of the boundary conditions determine the departure and arrival orbits.

IOP Conf. Series: Journal of Physics: Conf. Series **1005** (2018) 012043 doi:10.1088/1742-6596/1005/1/012043

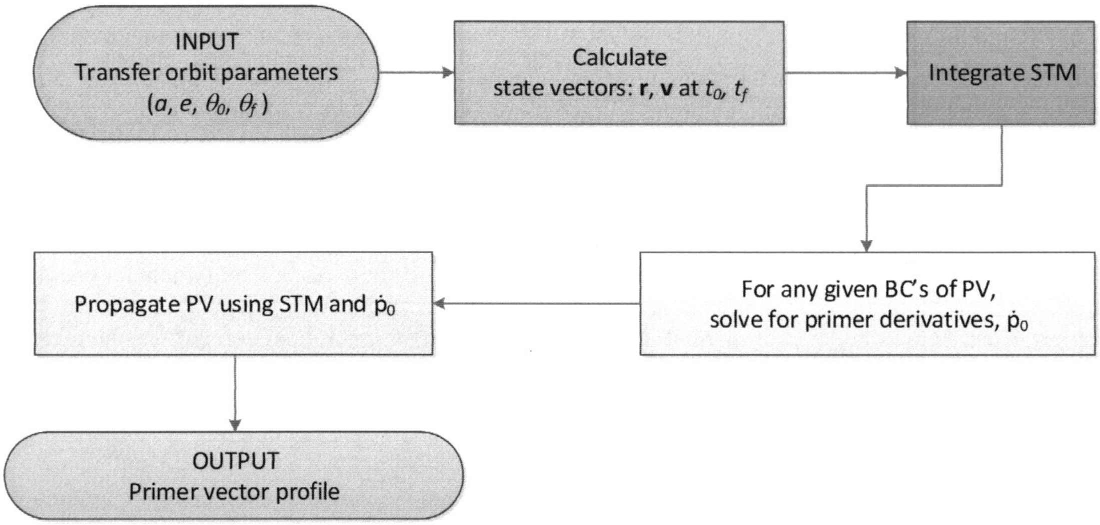

Figure 4. Algorithm for primer vector numerical propagation

The algorithm to propagate the in-plane primer vector profile using the analytical solutions are presented in figure 5.

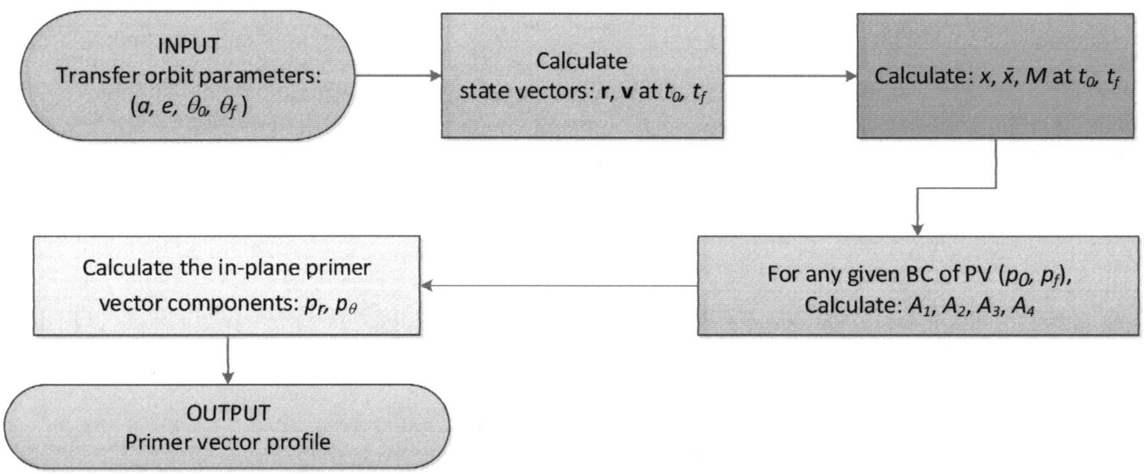

Figure 5. Algorithm for primer vector analytical propagation

The graphs in figure 6 are generated from the analytical and numerical approaches. In the examples presented in those graphs, the chosen transfer trajectory represents Earth to Mars transfer which has these following parameters: semi-major axis = 1.8875 x 10^8 km, eccentricity = 0.6, θ_0 = 20 deg and θ_f = 310 deg. The primer vector boundary conditions in Polar coordinate are given by

$$\mathbf{p}_0 = \begin{bmatrix} p_{r0} \\ p_{\theta 0} \end{bmatrix} = \begin{bmatrix} 0.8 \\ 0.6 \end{bmatrix}; \ \mathbf{p}_f = \begin{bmatrix} p_{rf} \\ p_{\theta f} \end{bmatrix} = \begin{bmatrix} 0.6 \\ 0.8 \end{bmatrix}.$$

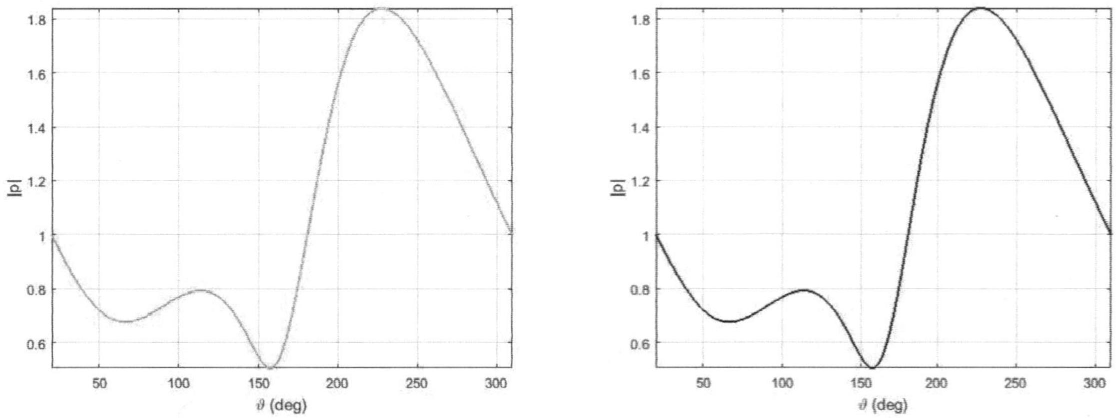

Figure 6. Primer profile for e = 0.6 from analytical (left) and numerical (right) solutions

As can be seen from figure 6, the primer vector from both approaches looks similar. The integration method is validated since both results show that the primer vector magnitudes are equal to one at initial and final points which are the boundary conditions applied to the case under analysis. The results are validated further by comparing the primer profiles for each component (radial and tangential) as in figure 7 which also show similar profiles.

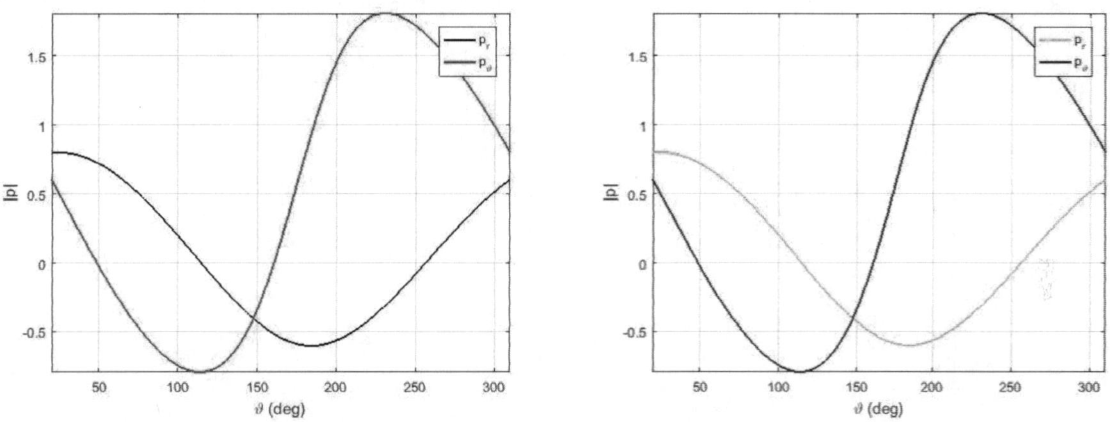

Figure 7. Comparison of primer components from analytical and numerical solutions

The numerical integration of the state transition matrix and the propagation of the primer vectors were done in Matlab using an ODE45 function with an error tolerance of 10^{-12}. The ODE45 function in Matlab is an implementation of the 4th order Runge Kutta algorithm which is essentially Taylor expansion of the 4th order. It can be seen in figure 8, the absolute error and the relative errors between the numerical and the analytical solution are in the order of 10^{-11}, which is closed to the predefined error tolerance.

IOP Conf. Series: Journal of Physics: Conf. Series **1005** (2018) 012043 doi:10.1088/1742-6596/1005/1/012043

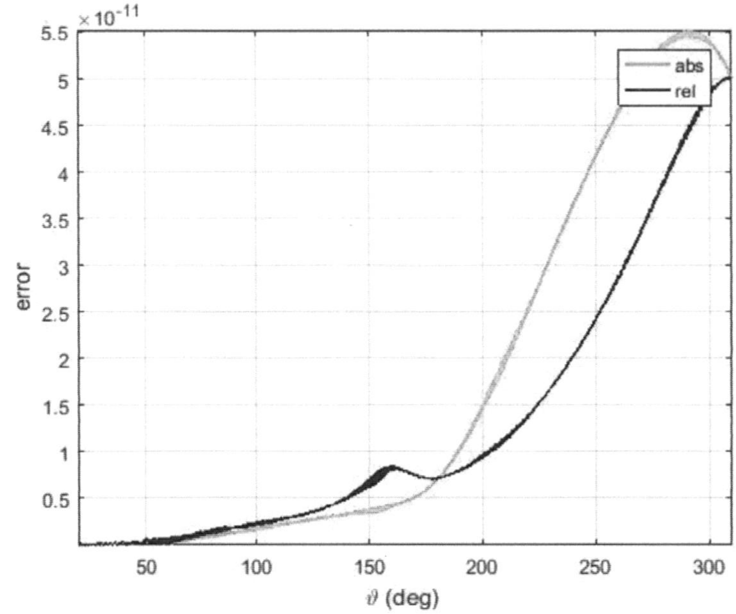

Figure 8. The error between the analytical and the numerical solutions

For further verification, a graph is generated for the same transfer trajectory but with different final conditions of primer vector. Various final conditions of the primer vector are represented by various β angles between the radial components, p_r and the primer vector, **p** at the end of transfer as defined in figure 9.

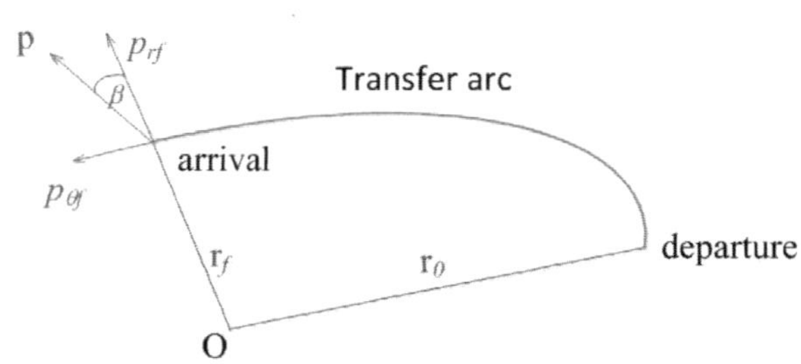

Figure 9. Definition of β angle in the transfer orbit geometry

In figure 10, the primer vector boundary conditions at the end of transfer, p_{r_f} and p_{θ_f} were varied using different β angles. The transfer orbit parameters are kept fixed as well as the primer vector initial conditions.

Another case similar to the test case above is studied, with the difference in the eccentricity value of the transfer orbit which is changed into 0.1. The result is shown in figure 11. As expected, both results show that the errors are in the order of 10^{12}.

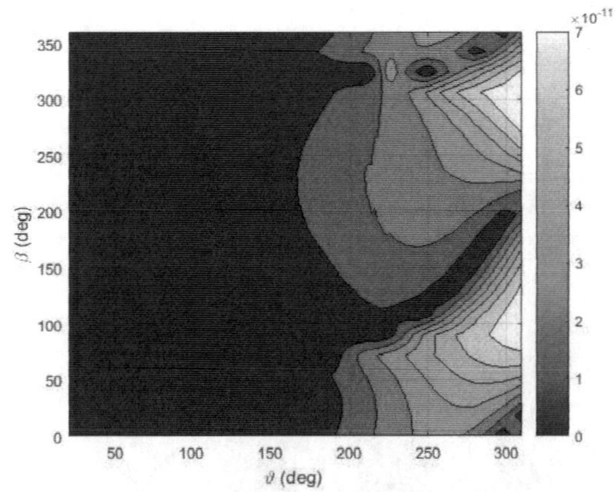

Figure 10. Absolute error for various β angle, e = 0.6

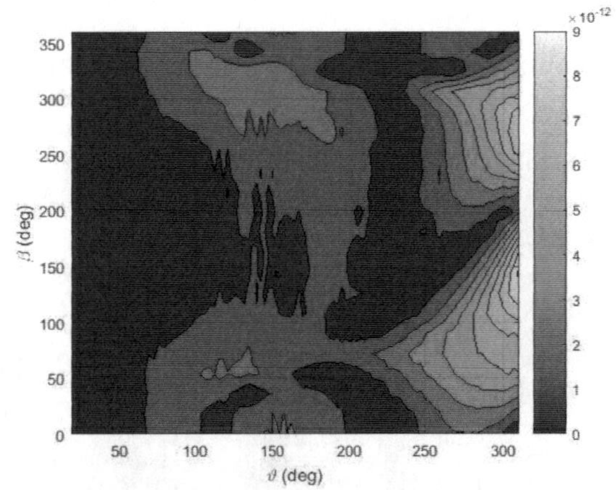

Figure 11. Absolute error for various β angle, e = 0.1

As mentioned before, the primer vector is used to identify the optimality of a transfer so that the mission designer can decide if it is necessary to include the corrective maneuver to reduce the fuel. The next example presents the application of primer vector method to investigate the optimality of direct transfer possibilities from Earth to Mars.

Figure 12 shows the typical ΔV for various options of departure date (T_E) and transfer duration (T_{EM}). It can be seen from the graph that the fuel minimum transfers lie in the darkest blue region which also means that the optimal transfer exist in this region. The lowest ΔV is approximately 5.7 km/s.

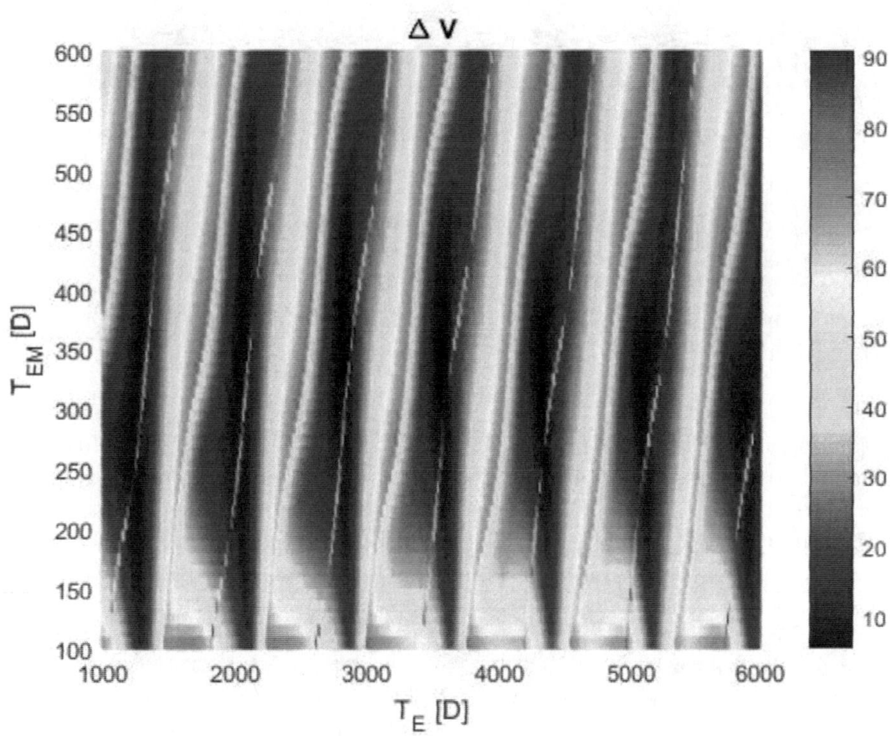

Figure 12. ΔV for Earth – Mars direct transfer (adapted from [2])

IOP Conf. Series: Journal of Physics: Conf. Series **1005** (2018) 012043 doi:10.1088/1742-6596/1005/1/012043

Figure 13 presents the optimality map as a result of the primer vector simulation of the respective transfers. The blue region represents optimum transfers (the region where the modulus of p always less or equal to one) while the non-optimum transfers are represented by the yellow region (the region where the modulus of the primer vector reach a value bigger than one). It means that if the selected transfer lies in the yellow region in figure 13, the required ΔV can be minimized further while if the transfer lies within the blue region it is not necessary to apply a corrective maneuver.

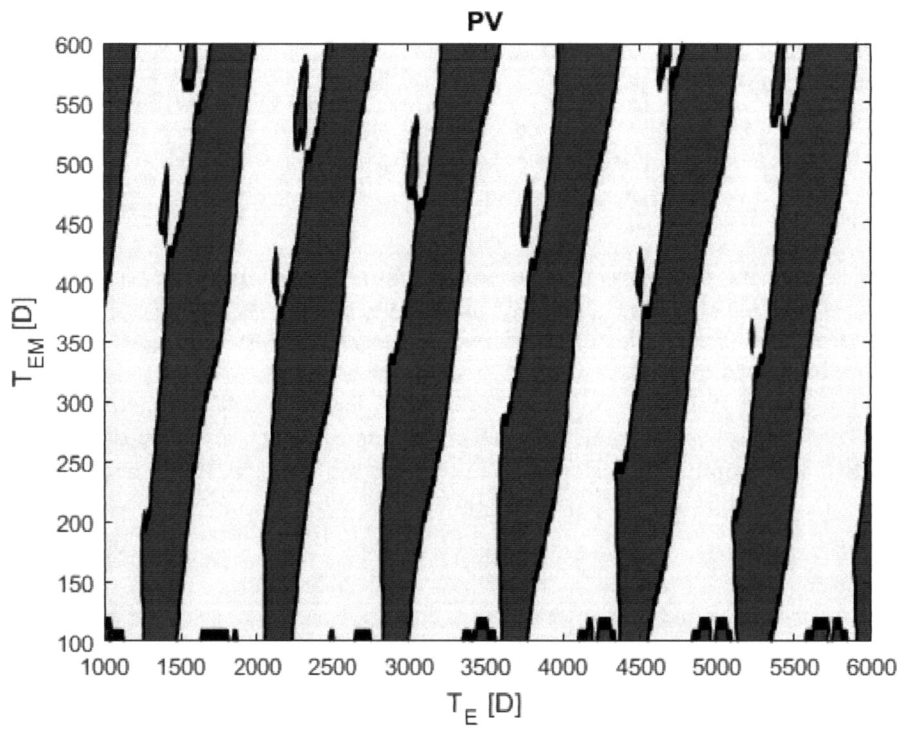

Figure 13. Earth-Mars transfer optimality analysis

In case the selected transfer is not optimum, corrective maneuver can be included to minimize the fuel and the primer vector can provide a good initial guess to apply this maneuver, which is the time and position where its modulus reach its maximum value above one.

4. Conclusions

The analytical solution to the in-plane primer vector equations has been derived using an orbital relative energy and relative angular momentum. The validation of the solution is performed by comparing the results to the numerical solution. The errors between the two approaches are within the expected range of around 10^{-12}. It has been verified that the source of error is the numerical integration process due to the fact that the order of error is roughly equal to the error tolerance specified for the ODE45.

The primer vector can be used to improve the optimization result of the direct method by evaluating if the candidate transfer can be optimize further by applying DSM. The study shows that the analytical solutions of the primer vector can identify the optimum region of the transfer. Having the analytical solution of the primer vector means that it can be derived further to obtain its first derivative. The first derivative is important for identifying the position and time to apply a correction maneuver without having to propagate the whole primer profile of the transfer.

IOP Conf. Series: Journal of Physics: Conf. Series **1005** (2018) 012043 doi:10.1088/1742-6596/1005/1/012043

Acknowledgement
The research is sponsored by Riset-Pro Program, Kemenristek-Dikti (the Ministry of Research and Technology – Higher Education of Indonesia). The authors wish to express the deepest gratitude to the late Prof. Phil Palmer for his guidance. The authors would like to thank the director of Surrey Space Center and the Astrodynamics Research Group for their supports. The authors would also like to thank the Indonesian National Institute of Aeronautics and Space (LAPAN) for their supports.

References
[1] Kemble, S., Interplanetary Mission Analysis and Design. *Springer-Praxis Publishing Ltd, Chicester UK, 2016.*
[2] Zazzera, F. B. et al. Global Trajectory Optimization: can we prune the solution space when considering Deep Space Maneuvers? *Technical report, ESA, 2007.*
[3] Juno mission and trajectory design. *http://spaceflight101.com/juno/juno-mission-trajectory-design/.*
[4] New frontiers program, mission juno. *https://discoverynewfrontiers.nasa.gov/missions/missions_juno.cfml.*
[5] Critical deep-space maneuver targets messenger for its first mercury encounter. *http://www.spaceref.com.*
[6] Exomars-2016 completes deep space maneuvers. *http:/www.russianspaceweb.com/exomars2016-mission-dsm.html.*
[7] Deep space manoeuvre part 2 – the little burn. *http://blogs.esa.int*
[8] Gad, A. H., Space Trajectories Optimization Using Variable-Chromosome-Length Genetic Algorithms. *PhD thesis, Michigan Technological University, 2011.*
[9] Vasile, M., Pascale, P. D, Preliminary design of multiple gravity-assist trajectories. *Journal of Spacecraft and Rockets (AIAA), Vol. 43(No. 4):pp. 794-805, 2006.*
[10] Prussing, J. E., Primer Vector Theory and Applications. *Cambridge University Press, 2010.*
[11] Lawden, D. F., Optimal trajectory for space navigation, *Butterworths Mathematical Text, University of Wisconsin – Madison, 1963.*
[12] Conway, B. A., Spacecraft Trajectory Optimization. *Cambridge University Press, 2010.*
[13] Handelsman, M. and Lion, P. M., Primer vector on fixed-time impulsive trajectories. *AIAA Journal, Vol. 6, No. 1: pp 127-132, 2007.*
[14] Jezewski, D. J., Primer vector theory and applications. *Technical report, Lyndon B. Johnson Space Center, NASA, 1975.*
[15] Palmer, P. L., Iorfida, E., Roberts, M., Geometric approach to the perpendicular thrust case for trajectory optimization. *Journal of Guidance, Control, and Dynamics, Vol. 39(No. 5): pp. 1059-1068, 2016.*
[16] Palmer, P. L. and Imre, E., Relative motion between satellites on neighbouring keplerian orbits. *Journal of Guidance, Control, and Dynamics, Vol. 30(No. 2):pp. 521-528, 2007.*
[17] R. H. Battin. An Introduction to the Mathematics and Methods of Astrodynamics. *American Institute of Aeronautics and Astronautics, Inc. Reston, Virginia, 1999.*

IOP Conf. Series: Journal of Physics: Conf. Series **1005** (2018) 012044 doi:10.1088/1742-6596/1005/1/012044

Preliminary Design of Nano Satellite for Regional Navigation System

L Fathurrohim[1)], R E Poetro[1)*], B Kurniadi[1)], P A Fadillah[1)] and M Iqbal[1)]

1) Faculty of Mechanical and Aerospace Engineering, Institut Teknologi Bandung, Indonesia

*ridanto@ae.itb.ac.id

Abstract. A Low cost Regional Navigation Satellite System employing constellation of nano satellites has been proposed for Indonesian coverage. The constellation of Low Earth Orbit nano satellites off course will not be able to give better position fixed to the GPS. However, the design of navigation system has much lower in cost compare to the current navigation system. This paper tells about preliminary design of the proposed regional navigation satellite system. The results of our satellite design has 3 kg on its weight, 10 W on power requirement at the peak condition, and 2.7 years of lifetime. Payload communication of the satellite will use UHF and TT&C communication will use VHF. Total area of solar panel will be 0.11 m^2.

1. Introduction

Navigation satellites already guide planes, ships, spacecraft and other devices for a wide range of services. The first operational of satellite navigation system was fielded by the U.S Navy in 1964, and was named the Navy Navigation Satellite System [1]. The system, better known as Transit, was used by the U.S submarine fleet to update a ship's position and reset the inertial navigation system. Transit saw a limited civil use in the maritime industry and geodesy starting in 1967, and it was decommissioned in 1996. Unlike Transit, Global Positioning System (GPS) is based on an idea that is both very simple and quite ancient: one's position, e.g., coordinates (x,y,z) can be determined given distances to objects whose positions are known. What is fancy in GPS is the realization of this idea with the technology of the late twentieth century in a global navigation system with capability to provide estimates of position, velocity, and time to an unlimited number of users instantaneously, continuously, and inexpensively [2].

The GPS receiver takes the information from 4 satellites to tri-angulate the user's exact location and fixed the time. Nowadays GPS is widely applied on airplanes, helicopters, naval warships, ground vehicles as well as the military troops. Global Positioning Systems (GPS) have become one of the most important technologies to the U.S. military since its development in the mid-1900 [3]. As well as U.S military, Indonesian military also start using GPS technology to support and improve its military capabilities. Satellite navigation is necessary applied to military missions for navigation purposes in enemy territories, and are especially important in absence of light in night missions. Indonesia, with wide area of territory, depends highly on GPS technology. The popularity of GPS underlined the strategic importance of navigation satellites to military services. In a quest for technology independence, Indonesia should have their own navigation system.

The CubeSat specification was developed at Stanford University and the California Polytechnic State University, San Luis Obispo in 1999 to facilitate the participation of science and engineering students in academic satellite development programs [4]. Evolution of the original picosatellite specification of

IOP Conf. Series: Journal of Physics: Conf. Series **1005** (2018) 012044 doi:10.1088/1742-6596/1005/1/012044

a 10x10x10cm cube with maximum mass of 1 kg (a 1U CubeSat) has since resulted in the standardization of larger 2U and 3U nanosatellites. Additionally, an increasing interest in the use of standardized CubeSat components has become more affordable.

In this paper, we introduced a preliminary design of navigation system using nanosatellites. The potential benefits of nano-satellites lie in significantly lowering mission costs and reducing the time to completion for individual spacecraft. The code name of the satellites is "Nanas", which stands for "Nano Navigation Satellite".

2. Mission Requirement

2.1. Navigation Satellites Comparison

Table 1 shows data of navigation satellite, compiled from references. All of global navigation satellites are located at Medium Earth Orbit (MEO) which has the altitude of 20,000 km and operated at L band spectrum frequency. However, the current navigation satellite has a lifetime for more than 10 years. Therefore, the size of the satellite never be in the microsatellite class. Since the satellites are located in high altitude orbit, they must provide a small amount of error in determining the user position.

Table 1. Current navigation systems

Navigation System	Owner	Number of Satellite	Type	Coverage	Satellite Frequency	Satellite Weight	Accuracy (military)	Source(s)
GPS	United States	32	MEO	Global	L band	~1500 kg	0.02 m	[16] [17]
QZSS	Japan	3	GSO	Regional	L band	~4100 kg	0.02 m	[18] [19]
GLONASS	Russia	24	MEO	Global	L band	~1000 kg	0.02 m	[22]
BeiDou	China	21	GEO MEO	Global	L band	~4600 kg	0.1 m	[21]
Galileo	European Union	30	MEO	Global	L band	~700 kg	0.02 m	[20]
IRNSS	India	7	GEO GSO	Regional	L band S band	~1500 kg	0.1 m	[23]

2.2. Space Segment Constraint

The size of the satellite is another important parameter in the determining the cost in a navigation system. However, Nanas satellite should give an alternative of independent navigation system with the low cost of manufacture. Therefore, the satellite size must be in the class of nanosatellite. Nanosatellite has a weight in less than 10 kg (see Table 2) [5]. In this study, the design of navigation system is restricted with the total cost of less than $4,000,000 and has at least 2 years last for its mission.

Table 2. Typical small satellite mass classification [5]

Classification	**Mass Range [kg]**
Femtosatellite	0.001 – 0.1
Picosatellite	0.1 - 1
Nanosatellite	1 – 10
Microsatellite	10 – 100
Minisatellite	100 - 1000

2.3. User Segment Objectives

In this study, we specify the requirement of Nanas navigation system. As it will become a regional navigation for Indonesia, Nanas navigation must provide a coverage in all Indonesia territory. However, the accuracy of Nanas navigation system should be less than 10 meters in determining the user position and produce bit error rate in less than 10^{-5}. Another important thing, time access for user, should be always available in all the time of operation. The design requirements of Nanas navigation system are shown in the Table 3.

Table 3. Design requirements

Time access	Continuous access
Accuracy	10 m
Bit Error Rate	10^{-5}
Lifetime	2 years

3. Constellation Design

3.1. Satellite Constellation

Preliminary design of navigation system start with study of satellites constellation. Satellites constellation was designed well in order to cover all Indonesia region with a tolerable error. The altitude of the satellites are restricted at LEO orbit which has altitude 500 – 1500 km. However, the accuracy of navigation system depends on the value of dilution of precision (DOP), which is the idea that position error that results from measurement errors depends on the user/foghorn relative geometry [6]. As th Therefore, the DOP value decreases as the satellites separate slightly between each other. Based on DOP requirement, Nanas satellites should be located in the northern part and shouthern part of Indonesia region once they passed above. As the altitude of the satellite goes down, the number of satellites should increase in order to have more access from a particular user. The altitude of the satellite was set at 1500 km, as we want to minimize the number of satellites required to produce a good repetition in the user access.

However, Nanas satellite parameters orbit were designed based on fact that Indonesia region is located around equatorial. Therefore, satellites inclination is set at 20 degree in order to give an access to the border region of Indonesia. We set the total number of satellites at 32 within Nanas constellation, which are divided into 4 groups by 4 different orbit planes (see Table 4 and Figure 1) [7]. The difference between each plane was set by RAAN value and each member in a group was separated by the value of true anomaly. Nanas navigation system has a mission to provide a regional navigation system around

IOP Conf. Series: Journal of Physics: Conf. Series **1005** (2018) 012044 doi:10.1088/1742-6596/1005/1/012044

Indonesia territory. According the mission, the Nanas satellites will be activated once they are in the Indonesia region autonomously.

Table 4. Satellite's orbit parameters

Parameter	Satellite Group 1	Satellite Group 2	Satellite Group 3	Satellite Group 4
Orbit's Shape	Circular	Circular	Circular	Circular
Altitude [km]	1500	1500	1500	1500
Inclination [deg]	20	20	20	20
RAAN	0	90	180	270
Number of Satellite	8	8	8	8
Separation Method	True Anomaly separated by 45^0	True Anomaly separated by 45^0	True Anomaly separated by 45^0	True Anomaly separated by 45^0

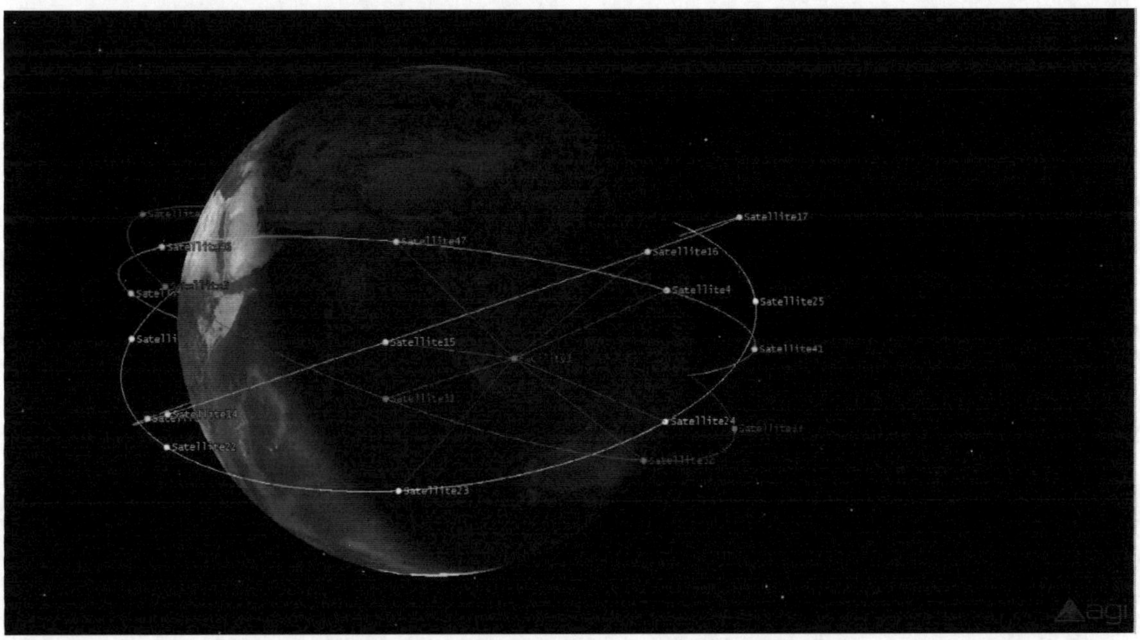

Figure 1. Nanas constellation illustration [24]

3.2. Launcher Selections

The cost of small satellite development is driven down through use of simplified and standardized commercially available bus, the impact of launch cost becomes highly significant compared to the total mission budget [5]. The current opportunities for launch of small satellites to LEO, see Table 5 and Table 6, are distributed between dedicated launch by a small vehicle provider, launch as part of a rideshare agreement or cluster launch, or by piggyback where the satellite is classed as secondary payload and utilizes excess capacity on a scheduled launch. In each case, a compromise between the cost of the launch, date of launch, and access to the desired orbit is required.

IOP Conf. Series: Journal of Physics: Conf. Series **1005** (2018) 012044 doi:10.1088/1742-6596/1005/1/012044

Table 5. Current LEO launch vehicle [8]

Vehicle	Origin	Manufacturer	Reference Payload [kg]	Cost [USD]
Proton-M	Russia	Khrunichev	21,600	$93M
Delta IV	United States	United Launch Alliance	22,950	$300M
Falcon 9	United States	Space X	16,625	$68M
Ariane 5E	Europe	EADS Astrium	21,000	$220M
Atlas V	United States	United Launch Alliance	17,100	$222M
PSLV	India	ISRO	3,800	$16M
Rokot	Russia	Khrunichev	1,950	$14M

Table 6. Available rideshare and piggyback launch price [9]

Vehicle	Form/Mass	Cost [USD]
Athena IIc	3U	$300k
Space X Falcon 9	3U	$200k-$325k
Antares	180 kg	$4.95M
PSLV	300 kg	$6.95M

The clear advantage of dedicated launch however, is that the destination orbit of the payload can be selected to best fit the mission and the date of launch can be chosen to coincide with the payload development and mission operation schedule. More value of dedicated launch however is the flexibility of the spacecraft bus and subsystems, which in turn has an effect on the cost of development and manufacture of the spacecraft itself. For nanosatellite and picosatellite classes, in particular those designed and built by educational instituttions, the cost of dedicated launch is usually can not be afforded by the system budget. In order to reduce the cost of launch, Nanas satellite is designed to rideshare or piggyback launch.

4. Subsystems and System Sizing

4.1. VHF UHF Transceiver

The VHF receiver (see Figure 2) is a single frequency crystal-controlled device, operating at a selected frequency in the 140-150 MHz band. The uplink modulation scheme is audio frequency shift keying (FSK) which ensure that the receiver is compatible to our existing Ground Station uplink facilities. The nominal uplink data rate is 1200 bit/s, which ensures that the signal bandwidth remains within an amateur radio communications.

However, The UHF transmitter is designed to operate between 420-450 MHz. It delivers around 500 mW of radio-frequency (RF) power. The downlink modulation scheme is nominally binary-phase-shift-keying (BPSK) with G3RUH scrambling, but Gaussian-minimum-shift-keying (GMSK) with G3RUH scrambling are also supported. The nominal data rate on NANAS was fixed at 1200 bps. Total cost for communication handling is $3,960.

IOP Conf. Series: Journal of Physics: Conf. Series **1005** (2018) 012044 doi:10.1088/1742-6596/1005/1/012044

Figure 2. ISIS VHF uplink/UHF downlink Full Duplex Transceiver [10]

4.2. Power System

The spacecraft was designed to generate power through four deployable solar panels, each populated with 20% efficient GaAs cells (see Figure 3) producing, nominally 0.5A per panel at 12V. The solar panel includes an integrated temperature sensor. Because of the mechanical configuration of the satellite, the total orbit average power available is approximately 8.4 Watt. However, the minimum required bus power is only 800 mW (which represents only the receiver and power system being on), thus, under nominal power conditions these systems, the attitude and orbit control system (AOCS) cab be continuous use. Each solar panel has its own independent battery charge regulator (BCR). The BCRs charge a single 8.1 Whr NiMH battery to give a nominal 6V. The total mass of the battery pack is 375g, which gives a relatively high energy density of 21.6 Whr/kg. Total cost for power system is $4,000.

Figure 3. GaAs cells [11]

4.3. Attitude and Orbit Control System

The attitude and Orbit Control System (AOCS) is designed to allow 3-axis momentum-biased operation, with autonomous orbit manoeuver. Attitude knowledge is provided via a compact CubeSense (see Figure 4 left). It provides an accurate sun and nadir pointing, also compatible with CubeSat standard. Processing of nadir and sun pointing can be done onboard, has dual dual FPGA/SRAM system for redundancy. The sensors required 100 mW for its power consumption and it has total mass 80g. The attitude stabilization and control is provided by a compact MAI-400 Reaction Wheel (see Figure 4 right). It has specification of momentum storage nominally around 9.35 mNms at 10,000 rpm. It delivers maximum torque 0.635 mNm. Total cost for AOCS sub-system is $2,600.

IOP Conf. Series: Journal of Physics: Conf. Series **1005** (2018) 012044 doi:10.1088/1742-6596/1005/1/012044

Figure 4. Compact CubeSense (left) [12] and MAI-400 Reaction Wheel (right) [13]

4.4. On Board Computer
The On-Board Computer (OBC) uses a $00 MHz ARM9 processor in a compact ISIS On-Board Computer (see Figure 5). It nominally operates at 3.3V and has total mass 94g including FM daughter board. Data storage is provided using redundant SD Card storage 2x2GB with failsafe FAT journaling file system. The On-Board Computer enhances the spacecraft's capabilities, providing automatic control of the spacecraft. Total cost for OBC sub-system is $6,200.

Figure 5. ISIS on board computer [14]

4.5. Mechanics
The Nanas satellite primary structure consists mainly of the standard aerospace aluminium-alloy 3U module boxes (see Figure 6) were used throughout the spacecraft. It will greatly simplify design, reduces manufacture time, and maximizes the ability to deal with design changes in the future works. A simple deployable solar panel was not designed yet for NANAS satellite to allow it to be deployed into the proper position. The total cost for this sub-system is $3100/satellite.

IOP Conf. Series: Journal of Physics: Conf. Series **1005** (2018) 012044 doi:10.1088/1742-6596/1005/1/012044

Figure 6. 3U CubeSat structure [15]

5. Conclusion and Recommendation

The new Indonesian navigation system has been designed at the preliminary phase that cost $3.6M, has met the design requirements. The Nanas navigation system was designed to last for 2.7 years with a continuous access in Indonesia region. Figure 7 shows an illustration of Nanas satellites as well as Table 7 shows their system parameters.

Figure 7. Nanas satellite illustration

Table 7. Nanas satellite specifications

Satellite size	15 cm x 15 cm x 20 cm
Satellite weight	3 kg
Payload frequency	UHF
TT&C frequency	VHF uplink UHF downlink
Solar array	0.11 m^2 deployable GaAs cells
Battery	1350 mAh NiMH
Max. power consumption	8.4 Watt
Lifetime	2.7 years

The data can be used as a preliminary data to the next study in the detail design. The further calculation about determining error position of this navigation system should be done. Finally, this result can be optimized further to meet a better accuracy results, to meet the military requirements.

Acknowledgement
This work is supported by Program Riset dan Inovasi KK ITB 2016. Additionally, the authors would like to thank FMAE ITB for providing research support facilities.

References

[1] Getting, Ivan Alexander. "Perspective/navigation-the global positioning system." *IEEE spectrum* 30.12 (1993): 36-38.
[2] Misra, P., and P. Enge. "Special issue on global positioning system." *Proceedings of the IEEE* 87.1 (1999): 3-15.
[3] Eissfeller, B. E. R. N. D., et al. "Performance of GPS, GLONASS and Galileo." *Photogrammetric Week*. Vol. 7. 2007.
[4] Sumantyo, Woellert, Kirk, et al. "Cubesats: Cost-effective science and technology platforms for emerging and developing nations." *Advances in Space Research* 47.4 (2011): 663-684.
[5] Bouwmeester, J. and Guo, J., 2010. Survey of worldwide pico-and nanosatellite missions, distributions and subsystem technology. *Acta Astronautica*, *67*(7), pp.854-862.
[6] Kaplan, Elliott, and Christopher Hegarty. *Understanding GPS: principles and applications*. Artech house, 1996: 262-266
[7] Ribbah, M.Z. (2016). Perancangan Sistem Navigasi Regional Berbasis Nanosatelit "INNSAV" (Indonesian Nanosatellite Navigation System), Bachelor Thesis. Institut Teknologi Bandung.
[8] https://en.wikipedia.org/w/index.php?title=Comparison_of_orbital_launch_systems&oldid=590182048
[9] Crisp, Nicholas, Katharine Smith, and Peter Hollingsworth. "Small satellite launch to LEO: a review of current and future launch systems." *Transactions of the Japan Society for Aeronautical and Space Sciences, Aerospace Technology Japan* 12.ists29 (2014): Tf_39-Tf_47
[10] https://www.isispace.nl/product/isis-uhf-downlink-vhf-uplink-full-duplex-transceiver/
[11] https://www.isispace.nl/product/custom-solar-panels/
[12] https://www.cubesatshop.com/product/cubesense/
[13] https://www.cubesatshop.com/product/mai-400-reaction-wheel/
[14] https://www.isispace.nl/product/on-board-computer/
[15] https://www.cubesatshop.com/product/3-unit-cubesat-structure/
[16] https://en.wikipedia.org/wiki/Global_Positioning_System
[17] https://en.wikipedia.org/wiki/GPS_Block_IIIA
[18] http://www.navipedia.net/index.php/QZSS
[19] https://directory.eoportal.org/web/eoportal/satellite-missions/q/qzss
[20] https://en.wikipedia.org/wiki/Galileo_(satellite_navigation)
[21] https://en.wikipedia.org/wiki/BeiDou_Navigation_Satellite_System
[22] https://en.wikipedia.org/wiki/GLONASS
[23] https://en.wikipedia.org/wiki/Indian_Regional_Navigation_Satellite_System
[24] http://www.agi.com/products/stk/modules/default.aspx/id/stk-pro

IOP Conf. Series: Journal of Physics: Conf. Series **1005** (2018) 012045 doi:10.1088/1742-6596/1005/1/012045

Data Processing of LAPAN-A3 Thermal Imager

R Hartono[1)*], P R Hakim[1)] and AH Syafrudin[1)]

[1)] Satellite Technology Centre, National Institute of Aeronautics and Space (LAPAN), Bogor 16310, Indonesia

* rommyhartono@ymail.com

Abstract. As an experimental microsatellite, LAPAN-A3/IPB satellite has an experimental thermal imager, which is called as micro-bolometer, to observe earth surface temperature for horizon observation. The imager data is transmitted from satellite to ground station by S-band video analog signal transmission, and then processed by ground station to become sequence of 8-bit enhanced and contrasted images. Data processing of LAPAN-A3/IPB thermal imager is more difficult than visual digital camera, especially for mosaic and classification purpose. This research aims to describe simple mosaic and classification process of LAPAN-A3/IPB thermal imager based on several videos data produced by the imager. The results show that stitching using Adobe Photoshop produces excellent result but can only process small area, while manual approach using ImageJ software can produce a good result but need a lot of works and time consuming. The mosaic process using image cross-correlation by Matlab offers alternative solution, which can process significantly bigger area in significantly shorter time processing. However, the quality produced is not as good as mosaic images of the other two methods. The simple classifying process that has been done shows that the thermal image can classify three distinct objects, i.e.: clouds, sea, and land surface. However, the algorithm fail to classify any other object which might be caused by distortions in the images. All of these results can be used as reference for development of thermal imager in LAPAN-A4 satellite.

1. Introduction

LAPAN-A3/IPB is a 3rd microsatellite developed by LAPAN. The assembly, integration, and testing of LAPAN-A3/IPB is done completely by satellite technology center of LAPAN. LAPAN-A3 micro-satellite was launched by Indian PSVL as piggy back on 22th of June 2016[1)]. LAPAN-A3 satellite project aims to support the national food sustainability program. Therefore the payloads requirement are defined by the Bogor Agricultural Institute (IPB). IPB in an academic institution that is specialized in agricultural science and technology, which is used as national academic reference for the Indonesian food sustainability program. Based on such cooperation, the satellite is also named as LAPAN-IPB satellite. The main payload of the satellite is a 4-bands multispectral imaging line-imager (using 4 Kodak 8023 sensor with spectral filter on 450-520 nm, 520-600 nm, 630-690 nm, and 760-900 nm) with 1000 Mm Nikon Lens, to produce spatial resolution of 18 m, swath width of 110 km. The satellite also carry digital camera similar to LAPAN-A2, so that it may provide image with 3.5 m resolution. Another payload is a low resolution video camera[2)], with has CCD video camera with effective picture element of 752 x 582. 100 mm lens is mounted on the camera so that it can produce image with 35 km swath. The last payload of satellite LAPAN A3 is thermal imager which is called micro-bolometer, which the hardware is provided by FLIR, and has resolution of 640x512 with 60mm lens and swath width of 92 KM.

Payload of micro-bolometer/thermal imager has specifications of spectral bands 8-12 μm, operating voltage range 4-6 Vdc, temperature range between -40^0 until $+80^0$ celcius, and video output using PAL standard video format. Micro bolometer actually has 14 bits image data, but actually the data that goes from satellite to ground station is only 8 bits (0-255), which the data processing from 14 bits to 8 bits is done automatically in the camera[3)]. For the processing of transmitting data camera micro bolometer from satellite to ground station uses S-band transmission system. The system uses the frequency of 2200 Mhz and FM video modulation, the transmission has 5 W output, and uses helical antenna. Meanwhile, ground station which receives video data from LAPAN-A3/IPB satellite uses antenna orbital with 3.0m diameter antenna, computer for controller antenna, FM analog video receiver and PAL Video record or display, and an additional computer to convert analog video into digital image. Figure 1-1 shows micro-bolometer camera on payload block of LAPAN-A3/IPB satellite, where micro-bolometer is place on the top of digital camera with additional brackets to disprove.

Figure1-1. Payload block and micro-bolometer camera of LAPAN-A3 satellite

Processing of video data into an image takes several steps. First, video data of thermal imager will be downloaded at highest elevation in Rancabungur-Bogor ground station, most of downloading video data is at elevation of 45^0 degree or more. Next process is, person who was in charge at acquisition of satellite LAPAN A3/IPB data will prepare everything such as update of TLE antenna orbital, controlling antenna software, turn on video receiver, DVD player and anything else, so that video thermal imager is ready to be recorded at the ground station. After that, next process is image acquisition, where the data video is converted to image sequence, blended, stitched into a mosaic image, and the final classify the image.

This research aims to describe a simple mosaicing and classifying process of LAPAN-A3/IPB thermal imager based on several videos that have been produced by the imager. Three mosaicing methods are used in this research, i.e.: manual approach using ImageJ software, auto stitching method provided by Adobe Photoshop, and also image cross-correlation method developed by using Matlab software. The performance of these three approaches will be compared in terms of image quality produced, time processing needed as well as flexibility of the methods. For classification purpose, simple algorithm is used based on digital number of the image, i.e.: lower digital number means lower temperature and vice versa. This research uses video data captured from LAPAN-A3/IPB thermal imager. Most of observation area is restricted to neighborhood of West Java for real-time acquisitions, while other Indonesia region can also be observed using recorded acquisition mode.

2. Methodology

The goal of this research is to mosaic images of LAPAN-A3/IPB thermal imager and then to classify the objects in the images. First step that is done is pre-processing video data received into a sequence of grayscale images. This is done using Avidemux by determining the wanted image frame rate from original video data of 28 frames per second. Second step is to mosaic the resulted image sequences into single mosaiced image. For illustration, usually thermal imager of LAPAN-A3/IPB satellite will observe the target area for averaged two minutes, which means that there will be more than 500 images in case of 4 frames per second is used, that needs to be mosaiced. There are three methods that will be used for this mosaicing process, i.e.: manual approach using ImageJ software, stitching method provided by Adobe Photoshop and also image cross-correlation method developed in Matlab software. These three methods are run independently, thus the result can be compared each other to determine which method is most suit to be used on actual LAPAN-A3/IPB thermal imager data processing. Third step in this research is to reference the mosaiced image using referenced map, and then classify the image produced into few distinct object such as cloud, sea and land surface. This done by using simple classification based on image digital-number and edge detection. The overall research flowchart and methodology can be seen in Figure 2-1.

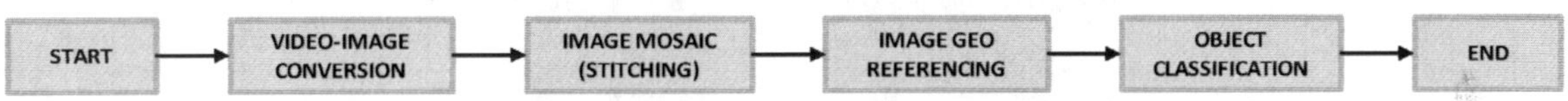

Figure 2-1. General research methodology used in this research

Figure 2-2 shows flowchart of image cross-correlation procedure used for mosaicing process which is developed by using Matlab software. The algorithm consists of four stages, which are image edge detection[4,5,6)], image cross-correlation[7,8)], image resampling and image blending process[9,10)]. All of these processes are pretty simple in nature thus the whole algorithm can be executed in reasonable processing time.

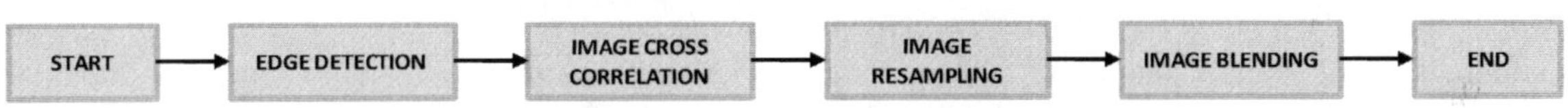

Figure 2-2. Flowchart of image cross-correlation method for image mosaicing

3. Results and Discussion

This section discusses the result of this research which describes data processing of LAPAN-A3/IPB thermal imager, which consists of video-to-image data conversion by using Avidemux software, image mosaicing by using three different approaches, as well as objects classification of the mosaiced image.

3.1. Video-to-Image Conversion using Avidemux

Avi-Demux is an open source software used for video and image editing, as converter from video to an image. The video of micro-bolometer that have been downloaded by S-band antenna then recorded by video recorder, and the resulting recorded video are opened by Avi-demux software. After that the video will be cut into pieces every four frame per second, so that if the thermal imager video have been recorded about 4 minutes then the resulting image is about 960 images. The next process is to organize the image into a whole image or image stitching process.

3.2. Image Mosaicing (Stitching)

Three different methods will be compared, which are manual approach by using ImageJ software, image auto stitching method by using Adobe Photoshop software and image cross-correlation method developed in Matlab software. Each method has its own advantages and disadvantages which will be compared in terms of mosaiced image quality, size of image area that can be processed and time processing needed by each method.

3.2.1. Manual Approach using ImageJ software

ImageJ is a public domain, Java-based image processing program developed at the National Institutes of Health. ImageJ also has a Java plugin and custom recordable macros. Custom acquisition, analysis and processing of plugins can be developed using the built-in ImageJ editor and Java compiler. User-written plugins allow to solve many image processing and analysis problems, from cell imaging to three-dimensional images for radiological image processing, to a comparison of multi-imaging systems. ImageJ plugin architecture and built-in development environment have made it a popular platform for teaching image processing. Example of manually mosaiced LAPAN-A3/IPB thermal imager images using ImageJ software by determining several pairs of GCP of two consecutive images in image sequence. Major results image geometry and radiometry produced is good but it needs long processing time. Figure3-1 below shows some manual image processing done by using ImageJ software.

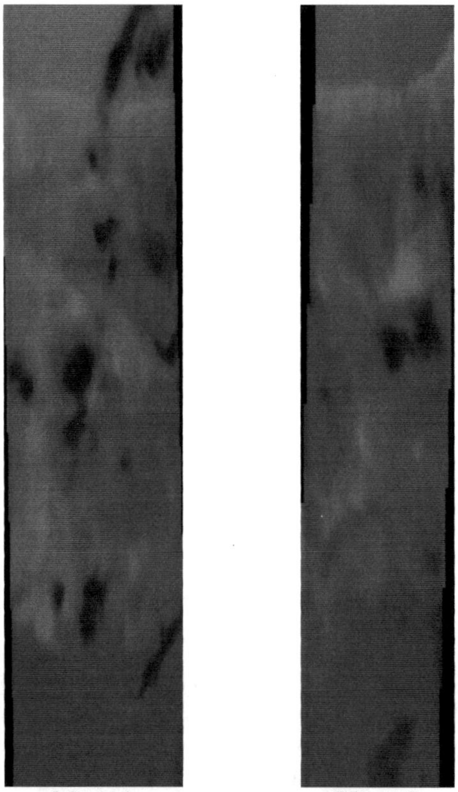

Jakarta-1
2548 images

Jakarta-3
3900 images

Figure 3-1. Image mosaic using ImageJ

3.2.2. Image Stitching using Adobe Photoshop

Image stitching using Adobe Photoshop to combine more than 500 images into a single complete image is very easy and simple. The step is to select automate, then photomerge with auto layout and select the image to be blend, and finally mosaic of thermal camera images is produced. It seems that it requires a good picture quality from downloaded thermal camera in ground station. Major result shows that this method ideally will produces good mosaiced image, but it has limitations, which are in terms

IOP Conf. Series: Journal of Physics: Conf. Series **1005** (2018) 012045 doi:10.1088/1742-6596/1005/1/012045

of long processing time and inabillity to process all of the images at once. Figure 3-2 shows example of image mosaic results using Adobe Photoshop software.

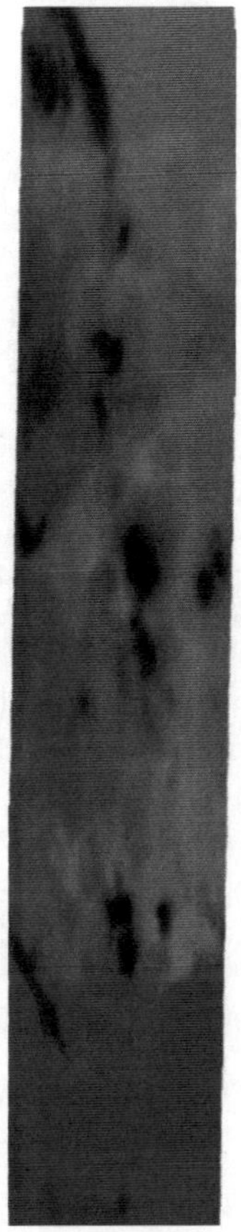

Jakarta-1
2548 images

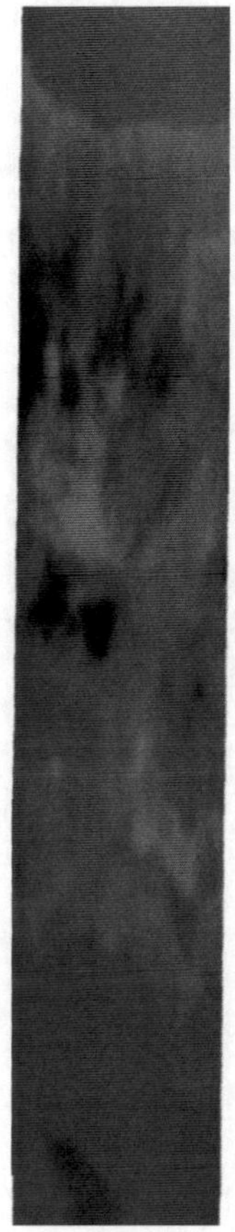

Jakarta-3
3900 images

Figure 3-2. Image mosaic using Adobe Photoshop

3.2.3. Image Cross-Correlation using Matlab

To avoid long processing time and provide the ability to process all of the images at once, image mosaic process is done by using image cross-correlation method implemented in Matlab software. Figure 3-3 shows the results of several mosaiced images of LAPAN-A3/IPB thermal imager video data. It can be seen that although the algorithm can handle all the images sampled from the video, the quality produced is not as good as the other methods. However, the time needed to process even the longest video captured is not more than five minutes, much faster compare to the other methods. There are two disadvantages of using this method. First only simple blending algorithm is performed in this method which causes intersection area of two consecutive images will slightly discontinues. Second, the algorithm often fails to find actual image angle, which correspond to satellite attitude of yaw angle. This incorrect image angle in turn will affect direct geo-referencing process which will produces incorrect latitude-longitude information.

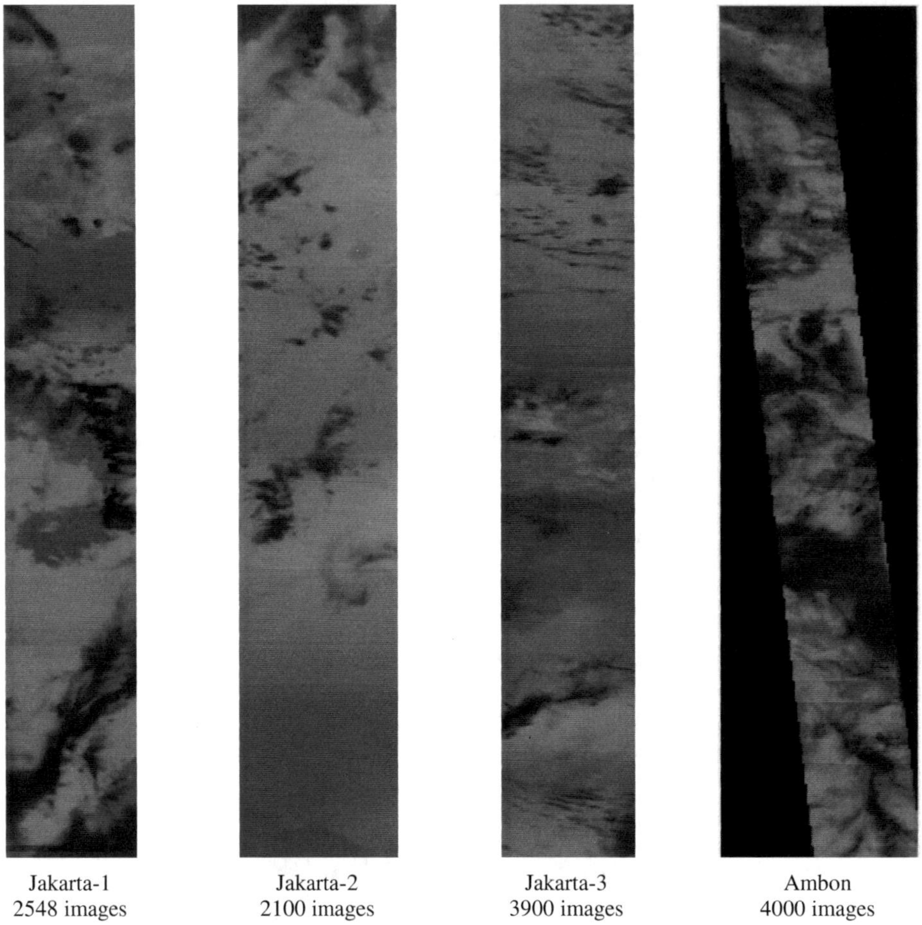

Figure 3-3. Image mosaic using image cross-correlation method

3.2.4. Comparison of Image Mosaicing Methods

Based on several image mosaicing that have been done, it can be concluded that each method has its own advantages and disadvantages. Manual approach using ImageJ software for example, could produce very accurate result in terms geometry and color blending. However, the process needs a lot of hard works to find control-point pairs of each successive images, and considering that the a video data consists of at least 1000 images, the full image-sequence processing can take hours to complete. Meanwhile, mosaicing with Adobe Photoshop software offers good geometry accuracy and image blending result but sometimes it cannot produce accurate mosaicing geometry, meaning that the image sequences fail to be stitched, usually for long sequence of images. The processing time needed is also a burden for this method as 500 image processing could take one hour to complete. The other method by using Matlab software to do image-mosaicing offers another advantage compare the other two methods, mainly in terms of processing time. For 1000 image sequence processed, it took only around five minutes to complete, although the quality both in geometry and image blending is not optimal. Therefore, this method is suitable for quick-look application which needs fast processing algorithm. Table 3.1 shows complete comparison between the tree image-mosaicing methods that have been employed to process thermal imager data of LAPAN-A3 satellite.

IOP Conf. Series: Journal of Physics: Conf. Series **1005** (2018) 012045 doi:10.1088/1742-6596/1005/1/012045

Table 3.1 Comparison between image-mosaicing methods

Parameter	ImageJ	Photoshop	Matlab
Mosaicing-geometry	Very good, if done carefully	Very good in general, but sometimes fail to stich the images	Generally poor, but sometimes can produces fairly good result
Image-blending	Good, as it uses good algorithm	Very good, as it uses proven algorithm	Fairly good, as it uses simple algorithm
Processing time	Very long, need a lot point matching	Fairly long, especially for big data	Very fast, even for long video data
User work-rate	A lot of effort is needed to match all successive pairs of images	No effort needed by the user, just wait and see	No effort needed by the user, although the algorithm have to be developed by satellite data owner
Suited application	For accurate result for small area of images	For casual user who wants fairly good result without big effort and time constraint	For quick-look application which needs fast processing and allows medium result

3.3. Object Classification

After converting LAPAN-A3 thermal imager video data into sequence of images and mosaicing the images into one mosaiced image, the final step is to classify objects that appear on the image. To provide valid image reference, image produced by thermal imager is compared to image produced by LAPAN-A3 analog video camera. The analog video camera image processing is similar to thermal imager one, but the mosaicing process is easier since the camera has visible spectrum color and therefore the images produced are easier to manipulate. Figure 3-4 shows comparison between mosaiced imaged produced by thermal imager and analog video camera using kappa camera of LAPAN-A3/IPB satellite.

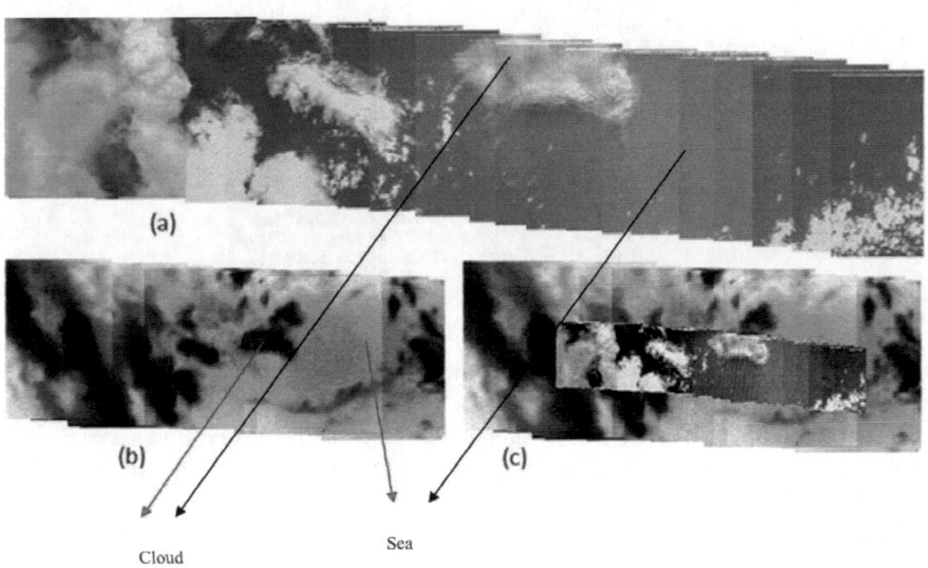

Figure 3-4. Mosaiced images of LAPAN-A3 analog video camera (a), thermal imager camera (b), and comparison between the two mosaiced images (c)

IOP Conf. Series: Journal of Physics: Conf. Series **1005** (2018) 012045 doi:10.1088/1742-6596/1005/1/012045

It can be seen from analog video camera image that sea is represented by blue color as usual while cloud is represented by mostly white and some of yellow. In the other hand in thermal image, cloud is represented by black color since the temperature is low while the sea is represented by white color since normally sea temperature is higher compare to cloud temperature. These two mosaiced images can be directly compared because field of view (FoV) of analog video camera is inside the FoV of thermal imager. Although the comparison is not trivial, but with careful matching it can be seen that the shape of many clouds that appear in both images are similar. These fact shows that LAPAN-A3 thermal imager can differentiate the cloud from the sea. However, the object detection capability of the camera should be more than detecting between cloud and sea. Figure 3-5 below shows that the thermal imager also capable of differentiating land surface from both sea and cloud.

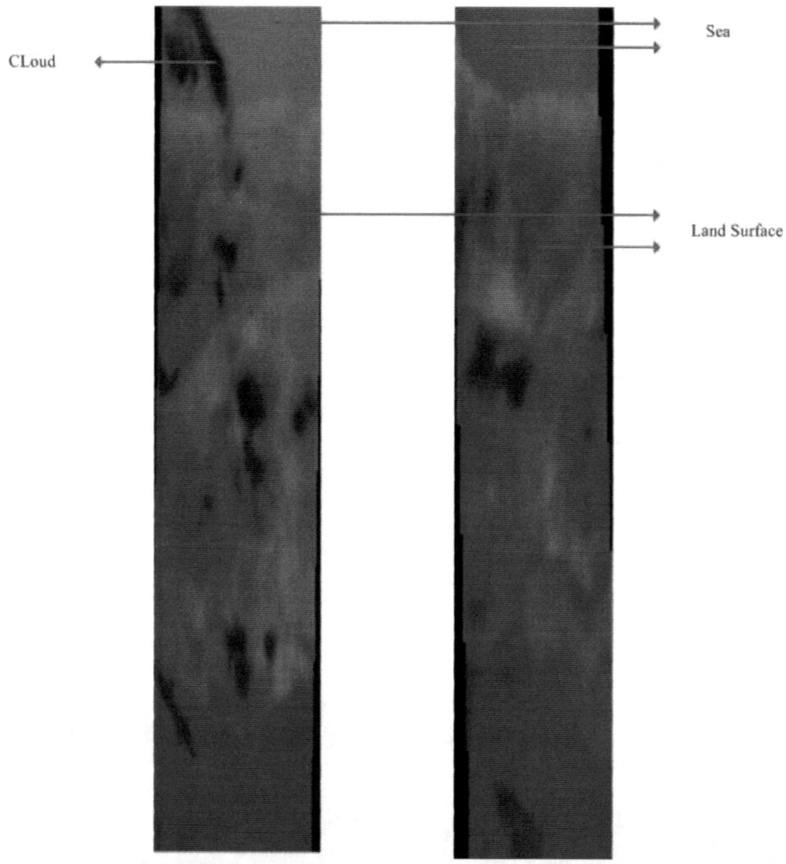

Figure 3-5. Capability of LAPAN-A3 thermal imager to classify land surface, sea and cloud

However, although LAPAN-A3 thermal imager can differentiate between land surface, sea and cloud as shown in Figure 3-5, it seems that with current image quality produced, the imager cannot classify any other objects which appear on the image. Natural object such as river or mountain, or human-made object such as building, farm or road, cannot be seen on the images. This happen because the thermal imager has 143 meter resolution, therefore the imager could not detect small object. Beside, the imager works based on surface temperature, so any object on the surface with relatively similar temperature will looks like the same. This is why the objects with significant temperature differences, such as land surface, sea and cloud, and perhaps if there is fire on the forest, can be identified by LAPAN-A3 thermal imager.

4. Conclusions

This research discusses about mosaic and classification process of LAPAN-A3/IPB thermal imager. Results show that image mosaic by using Adobe Photoshop produces good result, both in geometry and radiometry point of view, but it needs long time processing time and can only process small image sequence. Manual approach using ImageJ software can produce a very good result but needs a lot of works and time-consuming. Meanwhile, mosaic process using image cross-correlation method can process all of the images at once in significantly shorter time processing, but the quality produced is not as good as other methods. Direct geo-referencing has also been done but the quality is only moderate because the process does not use any metadata at all such as satellite orbit and satellite attitude. Image classification that has been done also shows that simple algorithm can only classify three objects in the images, which are clouds, sea, and land surface, and fail to classify any other object in the images.

Acknowledgments

The authors would like to thank Director of LAPAN Satellite Technology Center and Chief Engineer of LAPAN-IPB satellite, for their support and assistance so that this work can be well completed.

References

[1] W. Hasbi, and Suhermanto, 2013. Development of LAPAN-A3/IPB Satellite an Experimental Remote Sensing Microsatellite, *Proceedings of 34th ACRS*, Bali, Indonesia.

[2] A.H. Syafrudin, W. Hasbi, and A. Rahman, 2013. Camera Payload Systems for LAPAN-A2 Experimental Microsatellite, *Proceedings of 34th ACRS*, Bali, Indonesia.

[3] FLIR Commercial Systems, 2010. *Tau 640 Slow Video Camera: User's Manual*, Document Number: TAU-0640-00-10, Version 100.

[4] M.H. Shokhan, 2014. An Efficient Approach for Improving Canny Edge Detection Algorithm, *International Journal of Advances in Engineering and Technologies*, Vol. 7, Issue 1, pp. 59-65.

[5] P. Tiwari, 2015. Edge Detection Algorithm-A Review, *International Journal of Computer Science and Information Technology Research*, Vol. 3, Issue 4, pp. 9-12.

[6] F.A. Pellegrino, W. Vanzella, and V. Torre, 2004. Edge Detection Revisited, *IEEE Transactions on Systems, Man, and Cybernetics*, Vol. 34, No. 3, pp. 1500-15018.

[7] J.N. Sarvaiya, S. Patnaik, and S. Bombaywala, 2009. Image Registration by Template Matching Using Normalized Cross-Correlation, *International Conference on Advances in Computing, Control, and Telecommunication Technologies*, IEEE, pp. 819-822.

[8] N. Jayanthi, and S. Indu, Comparison of Image Matching Techniques, *International Journal of Latest Trends in Engineering and Technology*, Vol. 7, Issue 3, pp. 396-401.

[9] A. Levin, A. Zomet, S. Peleg, and Y. Weiss, 2004. Seamless Image Stitching in the Gradient Domain, *8th European Conference on Computer Vision*, Prague.

[10] W. Wang and K.N. Michael, 2012. A Variational Method for Multiple-Image Blending, *IEEE Transactions on Image Processing*, Vol. 21 No. 4, pp. 1809-1822.

IOP Conf. Series: Journal of Physics: Conf. Series **1005** (2018) 012046 doi:10.1088/1742-6596/1005/1/012046

Analysis of type II and type III solar radio bursts

J.V Wijesekera[1], K.P.S.C Jayaratne[2], J. Adassuriya[3]

Department of Physics, Faculty of Natural Sciences, The Open University of Sri Lanka, Nawala, Nugegoda

Abstract. Solar radio burst is an arrangement of a frequency space that variation with time. Most of radio burst can be identified in low frequency range such as below 200 MHz and depending on frequencies. Solar radio bursts were the first phenomenon identified in the field of radio astronomy field. Solar radio frequency range is from 70 MHz to 2.2 GHz. Most of the radio burst can be identified in a low frequency range such as below 200 MHz. Properties of low-frequency radio were analyzed this research. There are two types of solar radio bursts were analyzed, named as type II and type III radio bursts. Exponential decay type could be seen in type II, and a linear could be indicated in type III solar radio bursts. The results of the drift rate graphs show the values of each chosen solar radio burst. High drift rate values can be seen in type III solar flares whereas low to medium drift rate values can be seen in type II solar flares.

In the second part of the research the Newkirk model electron density model was used to estimate the drift velocities of the solar radio bursts. Although the special origin of the solar radio burst is not known clearly we assumed. The chosen solar radio bursts were originated within the solar radius of 0.9 - 1.3 range from the photosphere. We used power low in the form of $(x) = A \times 10^{-bx}$ were that the electron density related to the height of the solar atmosphere. The calculation of the plasma velocity of each solar radio burst was done using the electron density model and drift rates. Therefore velocity of chosen type II solar radio bursts indicates low velocities. The values are 233.2499 Km s^{-1}, 815.9522 Km s^{-1} and 369.5425 Km s^{-1}. Velocity of chosen type III solar radio bursts were 1443.058 Km s^{-1} and 1205.05 $Km\ s^{-1}$.

1 INTRODUCTION

Solar activity is one of the most significant events take place in the climate of the solar system which is unpredictable. Solar activities include solar flares, coronal mass ejections, high-speed solar wind, and solar energetic particles[1][2]. All solar activities are driven by the changes occur in the solar magnetic field. Solar flare is an intense burst of electromagnetic wave coming from the sun due to release of magnetic energy associated with sunspots.

Solar radio burst is an arrangement of a frequency space that variation with time [3]. Most of radio burst can be identified in low frequency range such as below 200 MHz and depending on frequencies. Solar radio bursts (SRB) were the first phenomenon identified in the field of radio astronomy field.

IOP Conf. Series: Journal of Physics: Conf. Series **1005** (2018) 012046 doi:10.1088/1742-6596/1005/1/012046

Solar radio frequency range is from 70 MHz to 2.2 GHz[3][4]. Most of the radio burst can be identified in low frequency range below 200 MHz Depending on the frequencies radio burst are classified into five categories as mentioned below[1][2][5].

- Type 1 burst (noise storm)
- Type II burst (slow drift rate)
- Type III burst (fast drift rate)
- Type IV burst
- Type V burst

Type II and type III solar radio bursts were detected by e-CALLISTO network system, a system operates in 67 locations around the world. In this study using the data gathered by CALLISTO network system, type II and type III solar radio bursts were analyzed to obtain useful information

2 METHODOLOGY

In this research, few SRB events were selected covering type II and III radio bursts using e-Callisto system which is used to calculate drift rates, electron density and plasma velocity of solar radio bursts to make a comparison between type II and type III.

2.1 Analysis of drift rates of solar radio bursts

Type II is a slow drift from high to low frequency in dynamic radio spectra burst, but type III has a fast drift rate. For us to distinguish them, the frequency drift rate of a solar burst is a power law in the frequency of emission. Drift rate (df/dt) is a displacement of the peak in frequency per unit time. It can be determined by taking start time to end time and start frequency to end frequency of the solar burst type II and type III[6]. The equation of drift rate of solar radio burst of type II and III was defined in equation 01[7][8][9]

$$drift\ rate;\ \frac{df}{dt} = \frac{(f_e - f_s)}{(t_e - t_s)}\ \left(\frac{MHz}{s}\right) ------ \quad (01)$$

f is a frequency of starting time, f is a frequency of ending time, t_e is an end time and t_S is a start time. MATLAB software could be used for analysis of drift rate graphs and selecting few solar radio stations which belong to e-callisto network system were selected.

Few SRB flares from each type were taken for getting better comparison between type II and type III. Drift rate graphs of solar flares can be plotted by using frequency (which corresponding to highest intensity) with time. This method is used to get the best drift rate graph of solar radio burst flares. Each type of solar radio flares was extracted by using MATLAB software.

Several noises could be realized in these images. These noises can be reduced by isolating the solar radio burst part, using the crop tool. This method was successful to reduce noise in certain levels, but not for all. Using cropped image, maximum intensity frequency corresponding to time can be calculated. Each maximum frequency point and related time were obtained. Subsequently feat those points, maximum frequency vs time graph was plotted to obtain the drift rate graph. The above method could be used for each solar flares which were selected. However a certain amount of noise was still present in the graph. That noise frequency needs to be eliminated to get the best distribution graph. The drift rate could not be retrieved from the shape of distribution under the noise frequency. That might possibly affect our final comparison. Noise frequency was eliminated by deleting the repeating same patterns of frequencies in this research. This method was repeated to all other selected images. After recognizing the noise area of the images help to eliminate the noise frequency. The frequency time graph from each image could be obtained by plotting those points. Comparing these graphs it is possible to identify, what is type II and what is type III. Model equations for each solar radio burst type could be initiated by using frequency vs time graph. The drift rate vs frequency graph was plotted after the frequency vs time graph. Thus to find a drift rate of each solar radio bursts, should differentiate each model equation separately. After obtaining those derivatives we could see a variation of drift rates for each solar radio burst. There by finding the mean value of these drift rates, one could obtain an average value for the particular solar radio burst.

2.2 Analysis of electron density and height of solar radio bursts

The major issue of this research was to find the electron density on chromosphere in the sun. The several solar chromosphere models can be used to find electron density. Newkirk model is one of the best models that can be used. Newkirk model could be used reversely to verdict each electron density and height of solar radio bursts. The frequency corresponds to electron densities should be found for calculating a height of each solar radio burst. The electron densities of each radio burst could be initiated by using equation 02. The main frequency of plasma radiation F is proportional to electron density $\sqrt{N((r)}$ [3][10][11][12].

$$F_{elec} = 8.973 \times 10^{-3}\sqrt{Ne(r)}\ MHz------(02)$$

Where F_{elec} is a plasma frequency in MHz and $\sqrt{Ne(r)}$ is the electron density in cm^{-3}.

A wide range of frequency was selected to get better distributions and it reduced error factor. The electron density vs plasma frequency graphs could be plotted by collaring the electron densities related to the plasma frequency. This research only got the highest intensity frequencies as mentioned earlier and by using same frequency value could use to find related electron densities and heights as well
The height distribution of electron density could be found by using Newkirk model as well. Newkirk model equation is shown in equation 03[3][12][13].

$$Ne = N \times 10^{4.32Ro/R} -------(03)$$

Where $N_O = 4.2 \times 10^4\ cm^{-3}$. - Concentration, Ro - solar radius, R - distance from solar center to source of type burst.

The distribution graph of electron density vs height could be plotted after getting height values corresponding to electron densities. The above step was repeated for other chosen solar radio bursts.

2.3 Analysis of plasma velocity of solar radio bursts

Calculation of electron densities and ejection heights were used to calculate its plasma velocities. For this calculation following parameters need to be used as well.

I. Plasma frequencies of each solar radio bursts
II. Change of Electron density with respect to height
III. Drift rate value for each solar radio bursts.

Frequency vs time graphs as mentioned in above paragraphs was used to calculate the plasma frequency of each solar radio burst. Calculation of the change of electron density with respect to height was the next step and the electron density vs height graphs was used for it. The each value of the change of electron density with respect to height should be calculated by differentiating above mentioned graph.

By using the above mentioned parameters (plasma frequency, electron density, drift rate, change of electron density with respect to height) in the equation 04, could successfully give the plasma velocities of the solar radio bursts[2][14].

$$v = \frac{(2\frac{df}{dt}N)}{f\frac{dN}{dr}} - - - - - (04)$$

Where

$v - plasma\ velocity\ (km\ s^{-1}), \frac{df}{dt} - drift\ rate(\ (MHz\ s^{-1}),$

$N - electron\ density\ (cm^{-3}),\ f - plasma\ frequency\ (MHz)$

$\frac{dN}{dr} - change\ of\ electron\ density\ with\ respect\ to\ height\ (cm^{-3}{R_o}^{-1})$

3 RESULTS AND DISCUSSION

There were five major solar radio bursts images selected to in this research and they are shown in figure1, 2, 3 and 4.

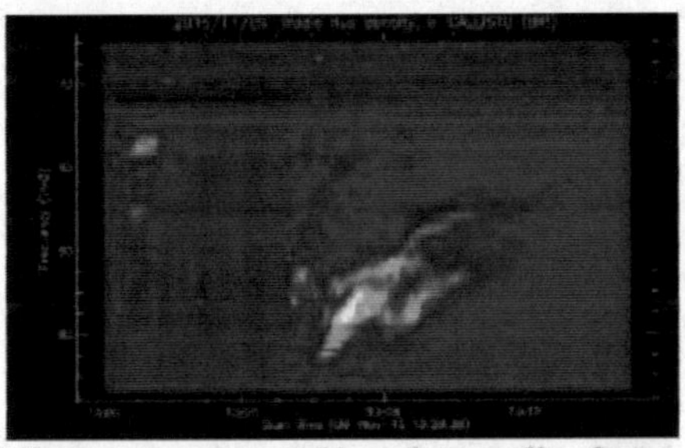

Figure 1- Type II solar radio burst extracted from BIR station 09-11-2015 time period between 13.04UT- 13.12U

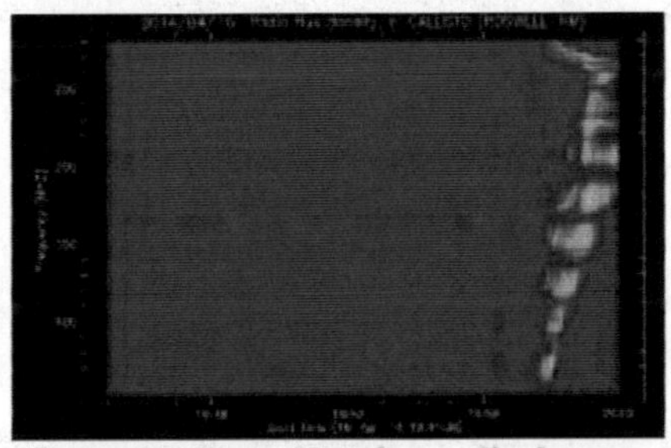

Figure 2- Type II solar radio burst extracted from ROSWELL station 16-04-2014 time period between 19.56UT- 20.00UT

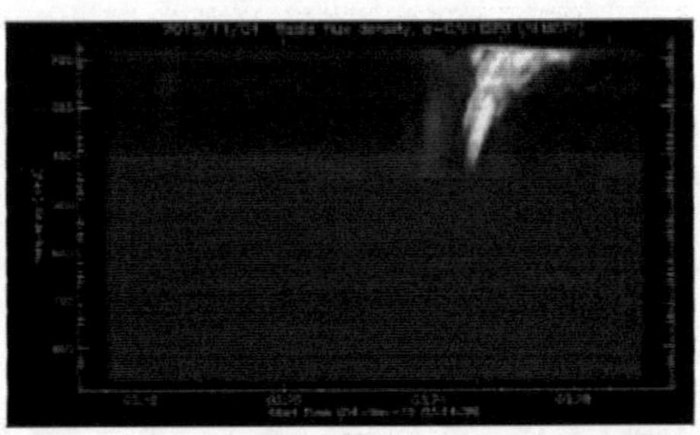

Figure 3- Type II solar radio burst extracted from ALMATY station 04-11-2015 time period between 03.24UT- 03.28UT

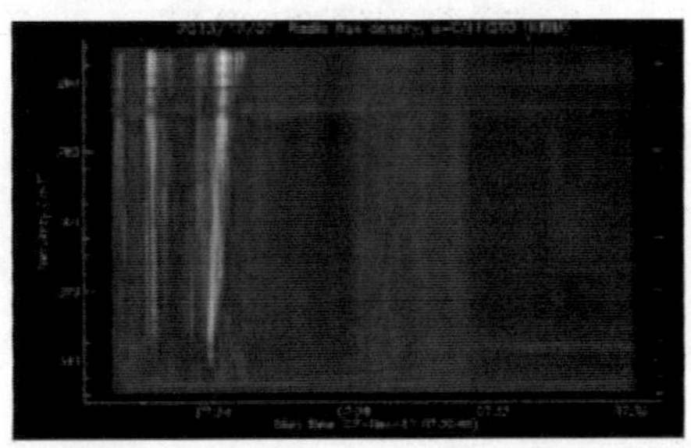

Figure 4- Type III solar radio burst extracted from KRIM station 07-12-2013 time period between 07.24UT- 07.25UT

Drift rates and frequency vs time graphs were obtained in the first part of the research. Variations of drift rates between type II and type III solar radio bursts can be observed from these plots. A difference between type II and type III was also observed by the distribution of solar radio bursts which were shown in the frequency vs time graphs of figure 5 and 6. The research thoroughly and successfully analyzed the five images of solar radio bursts. Using those analyzed images variations of type II and type III could be retrieved. Details of frequency vs time graphs are shown in Table 1. According to Table 1,

IOP Conf. Series: Journal of Physics: Conf. Series **1005** (2018) 012046 doi:10.1088/1742-6596/1005/1/012046

we could see that the type II solar radio burst indicates some exponential curve fitting model and type III solar radio bursts indicates straight line curve fitting model. Following these data we could find the drift rates of each solar radio burst types. Drift rate graphs tables are shown in Table 2.Table 2 shows the drift rate values obtained from graphs of each chosen solar radio burst. High drift rates values can be seen in type III solar flares whereas low to medium drift rate values can be seen in type II solar flares. Fast drift rate values indicate the fast distribution type and this may range from few seconds to minutes. Slow drift rate values indicate the slow distribution type and this may takes quite a few minutes. Table 2 shows the main difference between type II and type III solar radio bursts.

In the second part of this research, the electron density of solar radio bursts, ejection height and plasma frequency were calculated. Using Newkirk model, we could get the above mentioned parameters. According to the results of plasma frequencies and electron densities some distributions look like their original image forms. However, their height values depict different values. That meaning to say that each solar radio burst ejects different heights [3][12]. Hence we could see that the chosen solar radio bursts split into sub types. According to Hanslmeier and A., & Messerotti , chosen solar radio burst types belong to type II Bb, type II FS and type III B [3]. These sub types do not provide a perfect difference between type II and type III because chosen solar radio burst does not shows high variations. All the chosen solar radio bursts were located in solar radius of 0.9 - 1.3 range, but main point was that they always obey the Newkirk model. The research results demonstrated that the electron density vs height graph always follows power equations in the form of $f(x) = A \times 10^{-bx}$, as shown in figure 7.

IOP Conf. Series: Journal of Physics: Conf. Series **1005** (2018) 012046 doi:10.1088/1742-6596/1005/1/012046

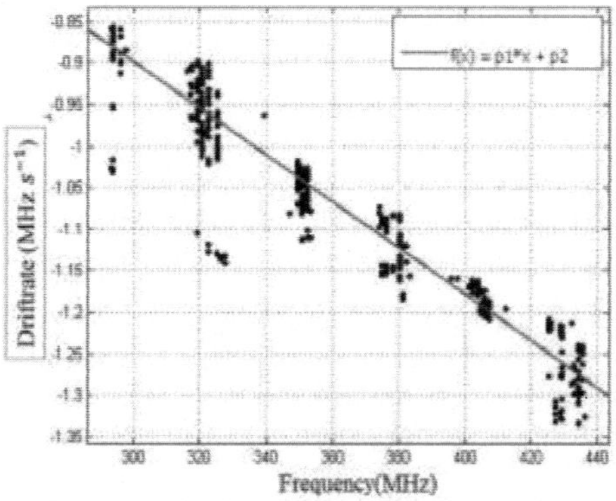

Figure 5- Type II solar drift rate vs frequency graph exctracted from flare ROSWELL-NM 16-04-2014

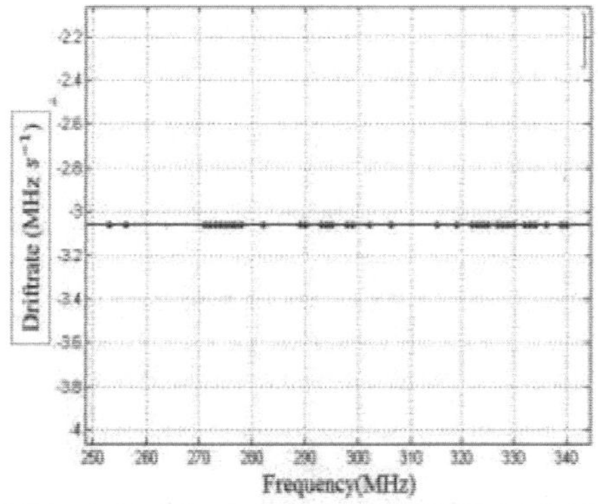

Figure 6- Type III solar drift rate vs frequency graph exctracted from flare KRIM 07-12-2013

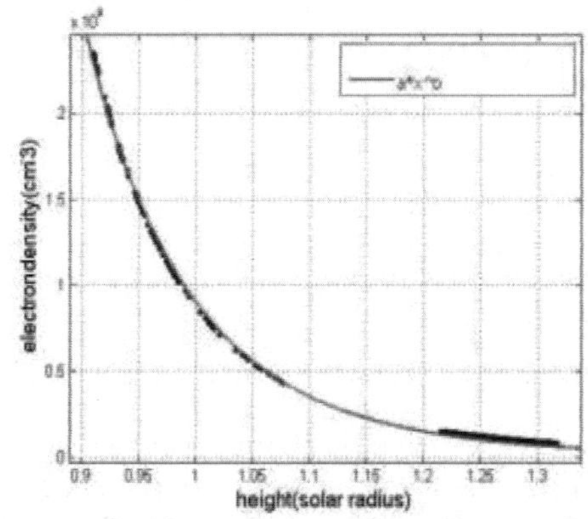

Figure 7- Type III and II solar radio burst electron density vs height graph

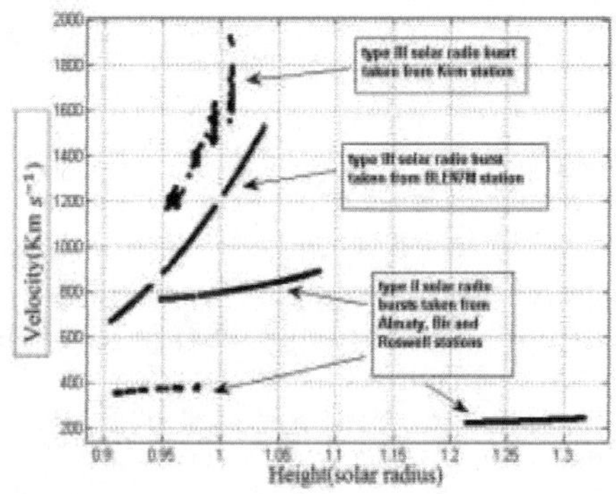

Figure 8- Type II and type III solar radio bursts velocity vs height graphs taken from chosen station

Final part of this research is the calculation of plasma velocity of each solar radio burst. Accordingly we could see low and medium velocities for type II solar radio bursts and a high velocity for the type III solar radio burst. This conclusion is in agreement with the figure 8. When Coronal Mass Ejection (CME) was associated with type II solar radio bursts it helps to improve the velocity. In this research mean velocities of 233.24 Km s^{-1}, 815.95 Km s^{-1} and 369.54 Km s^{-1} for type II solar radio bursts and 1443.05 Km s^{-1}, 1205.058 Km s^{-1} for type III solar radio burst were obtained. According to G. A. Newkirk, type II solar radio bursts could get the mean velocities less than 500 Km s^{-1} and if the CME was associated, they might get fast mean velocity values.

Table 1

Solar radio burst type	Curve fitting model
TYPE II	Exponential curve fitting $f(x) = A \times e^{-bx}$
TYPE III	Straight line curve fitting $f(x) = -mx + c$

Table 2

Solar radio burst type	Drift rate value($\boldsymbol{MHz\ s^{-1}}$)
ALMATY TYPE II 2015-11-04	-1.458
BIR TYPE II 09-11-2015	-0.098
ROSWELL-NM TYPE II 16-04-2014	-1.079
BLEN7M TYPE III 10-06-2014	-2.375
KRIM TYPE III 07-12-2013	-3.062

4 CONCLUSION.

The properties between Type II and type III of solar radio burst was accrued. Drift rate values of type II, solar radio bursts were less than type III. However, the drift rate value of type III solar radio burst could be seen in type II roughly and rarely. Type II and type III solar radio bursts always obey the Newkirk model and different between of electron density vs height graphs could not be indicated. Additionally, subtypes of type II and type III was identified, thus the variation of height was only indicated. Plasma velocity of type III was higher than type II. However, one of a type III solar radio bursts was originated from type region. Properties of type II and type III could be seen in that type III solar radio burst.

In this research, we have calculated the drift rate of type II and type III solar radio bursts more accurately. The drift rate was obtained by differentiating each point on the model fitted line of frequency vs time graph. Applying the parameters of the chosen solar flares to the Equation 1, drift rates were calculated. The calculated drift rates are approximately equal to the values previously obtained. This was done to verify the validity of and accuracy of results by this method.

5 REFERENCE

[1] National Aeronautics and Space Administration, "Solar Storm and Space Weather - Frequently Asked Questions," 2015. [Online]. Available: http://www.nasa.gov/mission_pages/sunearth/spaceweather/index.html.

[2] R. Behlke, "Solar radio bursts and low frequency radio emissions from space," SE-755 91 Uppsala, Sweden, 2001.

[3] M. . Hanslmeier, A., & Messerotti, *Motions in the solar atmosphere*, vol. 239, no. 9. SPRINGER-SCIENCE+BUSINESS MEDIA, B.Y., 1999.

[4] IPS AUSTRALIA, "Solar Radio Burst Classifications," pp. 1–4, 2009.

[5] Z. S. Hamidi, U. Ferwani, S. Ungku, and N. N. M. Shariff, "Theoretical Review of Solar Radio Burst III (SRBT III) Associated With of Solar Flare Phenomena," vol. 3, no. 2, pp. 20–23, 2013.

[6] C. Monstein, "Catalog of dynamic electromagnetic spectra," *Physics, Astron. Electron. Work Bench*, pp. 1–16, 2015.

IOP Conf. Series: Journal of Physics: Conf. Series **1005** (2018) 012046 doi:10.1088/1742-6596/1005/1/012046

[7] S. White, "Solar radio bursts and space weather," *Asian J. Phys*, vol. 16, pp. 189–207, 2007.
[8] H. A. S. Reid and H. Ratcliffe, "A review of solar type III radio bursts," *Res. Astron. Astrophys.*, vol. 14, no. 7, pp. 773–804, 2014.
[9] M. I. Aguilar-Rodriguez, E., Gopalswamy, N., MacDowall, R., Yashiro, S., & Kaiser, "A Study of the Drift Rate of Type II Radio Bursts at Different Wavelength," *Proc. Sol. Wind 11 / SOHO 16, "Connecting Sun Heliosphere" Conf. (ESA SP-592)*, p. 65.1-65.4, 2005.
[10] N. H. Zainol, Z. S. Hamidi, N. N. M. Shariff, S. Arifin, and C. Monstein, "Investigation of Drift Rate of Solar Radio Burst Type II due to Coronal Mass Ejections Phenomenon," *Int. Lett. Chem. Phys. Astron.*, vol. 48, no. i, pp. 146–154, 2015.
[11] C. Monstein, "E-Callisto solar spectrometer." [Online]. Available: http://www.e-callisto.org/.
[12] G. A. Newkirk, "Structure of the solar corona," in *Structure of the solar corona*, vol. 53, no. 9, 1967, pp. 1689–1699.
[13] M. O. Ali, Z. S. Hamidi, N. N. M. Shariff, and C. Monstein, "An Analysis of the Electron Density and Drift Rate of Solar Burst Type III During 13 th of May 2015," *World Sci. News*, vol. 31, no. May 2015, pp. 1–11, 2016.
[14] H. Ratcliffe, E. P. Kontar, and H. A. S. Reid, "Large-scale simulations of solar type III radio bursts: flux density, drift rate, duration, and bandwidth," *Astron Astrophys.*, vol. 572, p. A111, 201

Madden-Julian Oscillation (MJO) Signal over Kototabang, West Sumatera Based on the Mini Automatic Weather Station (MAWS) Data Analysis Using the Wavelet Technique

E Hermawan[1)] *

[1)] Center of Atmospheric Science and Technology, National Institute of Aeronautics and Space (LAPAN), Indonesia

*eddy.hermawan@lapan.go.id and eddy_lapan@yahoo.com

Abstract. This study is mainly concerned an application of Mini Automatic Weather Station (MAWS) at Kototabang, West Sumatera nearby the location of an Equatorial Atmosphere Radar (EAR) side. We are interest to use this data to investigate the propagation of the Madden-Julian Oscillation (MJO). We examined of daily MAWS data for 3 years observations started from January 2001 to Mei 2004. By applying wavelet analysis, we found the MJO at Kototabang have 32 days oscillations as shown in Fig.1 below. In this study, we concentrate just for local mechanis only. We will show in this paper that at the phase of the MJO with a dipole structure to the convection anomalies, there is enhanced tropical convection over the eastern Indian Ocean and reduced convection over the western Pacific. Over the equatorial western Indian Ocean, the equatorial Rossby wave response to the west of the enhanced convection includes a region of anomalous surface divergence associated with the anomalous surface westerlies and pressure ridge. This tends to suppress ascent in the boundary layer and shuts off the deep convection, eventually leading to a convective anomaly of the opposite sign. Over the Indonesian sector, the equatorial Kelvin wave response to the east of the enhanced convection includes a region of anomalous surface convergence into the anomalous equatorial surface easterlies and pressure trough, which will tend to favour convection in this region. The Indonesian sector is also influenced by an equatorial Rossby wave response (of opposite sign) to the west of the reduced convection over the western Pacific, which also has a region of anomalous surface convergence associated with its anomalous equatorial surface easterlies and pressure trough. Hence, convective anomalies of either sign tend to erode themselves from the west and initiate a convective anomaly of opposite sign via their equatorial Rossby wave response, and expand to the east via their equatorial Kelvin wave response.

Keywords: MAWS, MJO, rainfall and Kototabang

1. Introduction

In the tropics weather is not as predictable as in mid-latitudes. That is because in mid-latitudes the weather variables (clouds, precipitation, wind, temperature, and pressure) are largely governed by the upper-tropospheric Rossby waves, which interact with surface weather in a process called baroclinic instability. In the tropics there is no such dominant instability or wave motion, and therefore the weather is less predictable for the 1-10 day period. Until recently it was believed that tropical weather variations on time scales less than a year were essentially random.

In 1971 Roland Madden and Paul Julian[1)] stumbled upon a 40-50 day oscillation when analysing zonal wind anomalies in the tropical Pacific. They used ten years of pressure records at Canton (at 2.8° S in the Pacific) and upper level winds at Singapore. The oscillation of surface and upper-level winds was remarkably clear in Singapore. Until the early 1980's little attention was paid to this oscillation, which became known as the *Madden and Julian Oscillation* (MJO), and some scientists questioned its global significance. Since the 1982-83 El Niño event, low-frequency variations in the tropics, both on intra-annual (less than a year) and inter-annual (more than a year) timescales, have received much more attention, and the number of MJO-related publications grew rapidly.

The MJO, also referred to as the 30-60 day or 40-50 day oscillation, turns out to be the main intra-annual fluctuation that explains weather variations in the tropics. The MJO affects the entire tropical troposphere but is most evident in the Indian and western Pacific Oceans. The MJO involves variations in wind, sea surface temperature (SST), cloudiness, and rainfall. Because most tropical rainfall is convective, and convective cloud tops are very cold (emitting little longwave radiation), the MJO is most obvious in the variation of outgoing longwave radiation (OLR), as measured by an infrared sensor on a satellite.

Figure 1 below from Elleman (1997)[2)] shows how the OLR anomalies in the eastern hemisphere propagate to the east at around 5 m/s. The OLR signal in the western hemisphere is weaker, and the recurrence interval for the eastward propagating OLR anomalies in the eastern hemisphere is about 30 to 60 days. How exactly the anomaly propagates from the dateline to Africa (i.e. through the western hemisphere) is not well understood. It appears that near the dateline a weak Kelvin wave propagates eastward and poleward at a speed exceeding 10 m/s.

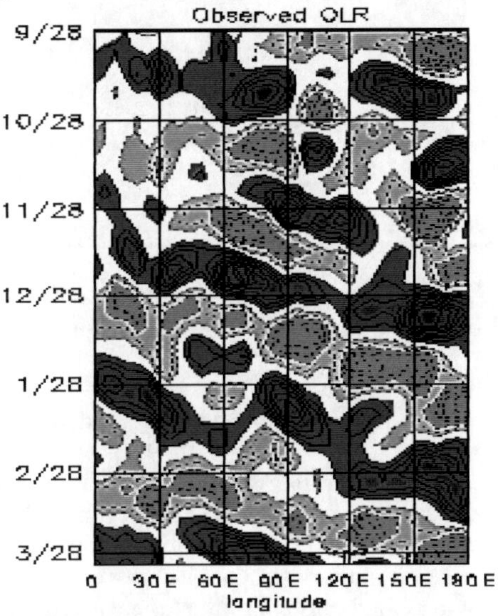

Figure 1: Departures from normal outgoing longwave radiation around the globe between 5S and 5N during 6 months (10/'91 through 3/'92). The contour interval is 5 Wm^{-2}. Areas in blue have a negative anomaly exceeding 5 Wm^{-2}, and areas in red have positive anomalies over 5 Wm^{-2}. The data have been 'filtered' to remove high-frequency (<30 day) variations.

Associated with the propagation of convective anomalies, the MJO involves variations in the global circulation. The MJO affects the intensity and break periods of the Asian and Australian monsoons and interacts with El Niño. Wet spells in the Australian monsoon occur about 40 days apart. Fairly weak correlations with the midlatitude rainfall patterns and jet stream characteristics have also been found[3)]. It has long been known that dry spells as well as wet spells occur in the north Australian monsoon season[4)]. The wet spells usually last longer than the dry spells, which last from a few days to two weeks. Wet spells in northern Australia are associated with westerly winds at the surface, and easterly wind anomalies aloft. Many wet spells seem to originate as cold air outbreaks over Southeast Asia, where it is winter.

A synoptic analysis of the evolution leading up to a wet spell was performed for a large number of cases during 1980-1992[5)]. The surges are initially evident over one of three regions: the South China Sea, the northwest Pacific (near Japan) and Indochina. Five days later there is an increase of northerly

IOP Conf. Series: Journal of Physics: Conf. Series **1005** (2018) 012047 doi:10.1088/1742-6596/1005/1/012047

surface wind over Indonesia. The onset of a wet westerly monsoon wind over the Top End of Australia then occurs another 2-5 days later, usually leading to high rainfalls there.

With the center of suppressed convection, clear skies associated with a stronger-than-normal trade wind inversion allow more shortwave radiation to reach the ocean surface as shown in Figure 2, causing a slight SST increase as the wave travels eastward[6)]. The Trade winds too are stronger than normal, explaining enhanced evaporation from the sea surface.

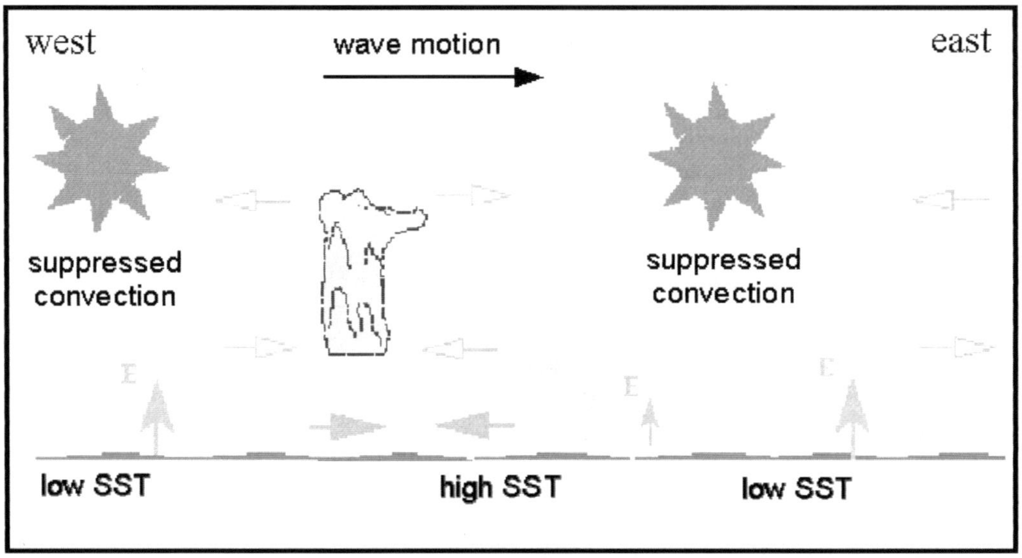

Figure 2: Schematic of the MJO. The cross section represents the equatorial belt around the globe, or just the eastern hemisphere. E stands for evaporation, SW for net shortwave radiation absorbed by the ocean. The converging bold green arrows indicate the location of strongest moisture convergence. The hollow green arrows show the anomalous circulation associated with the MJO. The areas of enhanced convection are indicated by the yellow schematic thunderstorm. (adapted from Elleman 1997)

Easterly winds (and the evaporation rate) weaken near the western edge of the suppressed convection region, and this leads to low-level moisture convergence. This triggers deep convection, leading to the other half of the OLR oscillation, i.e. the region of enhanced convection. This region is comprised of one or more super cloud clusters (SCCs) that move eastward along with the MJ wave. Within the SCCs, westward-moving cloud clusters form at the eastern edge of the SCC and die at the western edge. These smaller clusters have a lifetime of one to two days. In turn, the individual mesoscale convective systems within these smaller clusters typically move eastward, usually by discrete propagation, and have a lifetime of 6-12 hours. The SCCs travel eastward at 5-10 m/s, not as a long-lived storm complex, but rather as a moving wave or oscillation, i.e. the MJO. The MJO has a wavenumber of 1-2, that is at any time there are one or two areas around the equator with enhanced convection, and one or two with suppressed convection.

Equatorially trapped waves (Kelvin and Rossby waves) that explain the evolution of an El Niño event are also the driving mechanism for the MJO. These waves occur in the entire troposphere from 30□ N to 30□ S, mainly in the eastern hemisphere. Surface air flows away from the suppressed convection in both zonal directions towards enhanced convection regions. In the upper troposphere, anomalous easterlies exit the west side of the enhanced convection (Figure 2). The strong westerlies from the east side of the enhanced convection flow into the region of suppressed convection as shown in Figure 3 below[7)].

IOP Conf. Series: Journal of Physics: Conf. Series **1005** (2018) 012047 doi:10.1088/1742-6596/1005/1/012047

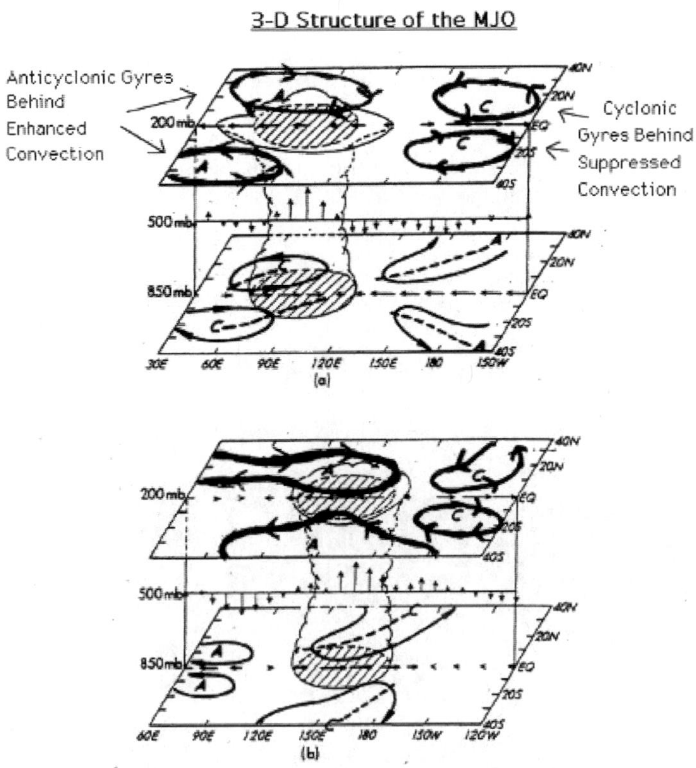

Figure 3 : The three dimension structure of the MJO (*after* Rui and Wang, 1990)

When suppressed convection is strong from the Indian Ocean to the middle Pacific Ocean, anomalous cyclonic gyres at 200 mb trail the region of suppressed convection. Similarly, anticyclonic gyres at 200 mb trail the enhanced convection region once it becomes strong in the Indian and western Pacific Oceans. Gyres in the opposite sense are produced at surface, but they are much weaker than the ones at the tropopause. The zonal circulation and horizontal gyres are important processes by which the MJO shuffles mass around the tropics.

The explanation above is simplistic, in that it idealises the oscillation, as it isolates it from other variations. As mentioned before, the speed and size are variable, and the MJO mainly affects rainfall patterns in Indonesia and surrounding areas. Not all of the elements of the MJO -- convection, zonal wind, moisture convergence, and SST anomalies -- are always visible[8)]. It is only when the 30-60 day oscillations are extracted from a series of MJO events that the idealised picture of the MJO emerges. Consecutive oscillations have varying amplitudes, periods, and wavelengths.

The MJO exhibits the mixed Kelvin-Rossby wave structure over the eastern hemisphere, but over the western hemisphere, it only shows a Kelvin wave structure. It moves through the eastern hemisphere at around 5 m/s and through the western hemisphere at a higher speed (at least 10 m/s). The oscillation is stronger in the northern hemisphere winter. It is also in this season that the negative OLR anomalies are most likely to propagate along the equator from the Indian Ocean to the central Pacific Ocean. In the northern hemisphere summer, many of the anomalies veer away from the tropics before they make it to the central Pacific[9)]. Notwithstanding its complexity and dependence on convection, the essence of the MJO (its periodicity, structure and zonal asymmetry) can be simulated in a GCM[10)].

2. Method

Based on the basic concept of MJO above, we used the daily of MAWS (Mini Automatic Weather Station) data installed by Shimane University, Japan and located near by the EAR (Equatorial Atmosphere Radar) at position on 0.2°S ; 100.32°E and 865 m from MSL (mean sea level) in period of

Januari 2001 to May 2004. We used the *wavelet transforms* analysis taken from Matlab software to obtain the global wavelet spectrum. Wavelet transforms are a mathematical means for performing signal analysis when signal frequency varies over time. For certain classes of signals and images, wavelet analysis provides more precise information about signal data than other signal analysis techniques.

3. Results and Discussions

The result of *wavelet analysis* can be shown in Figure 4 below. We need to compare this result with another study which have done by Renggono *et. all* (2001)[11].

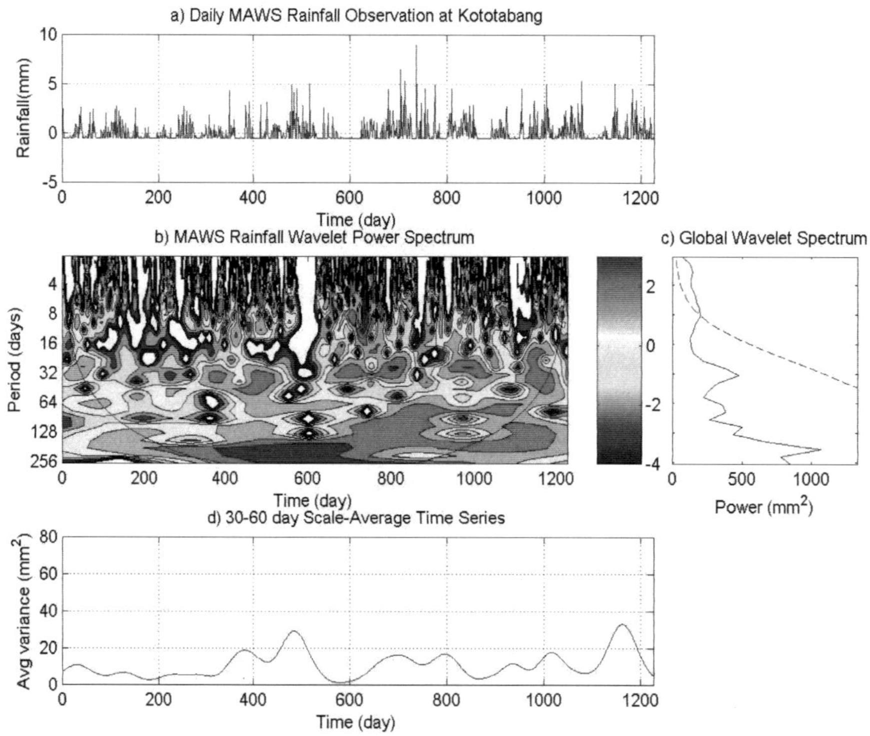

Figure 4: The daily MAWS Rainfall Observation at Kototabang

The upper panel of Figure 4 shows the time series of daily MAWS data, while the middle and upper panel show the wavelet power spectrum and average variance of MAWS data analysis, respectively. If we look at carefully the middle panel of this figure, we can see that the most predominant peak of MAWS data is not MJO oscillation, it is close to 256 days oscillation. We suspect it is related well to the Monsoon phenomena which has oscillation in between 6 month to 1 year oscillation. We call this phenomena as Annual Oscillation (AO).

Although, it not as peak dominant oscillation, but we found another peak oscillation near by 32 days oscillation. We strongly suspect that it is close to the MJO phenomena, since some researchers before determined this phenomena is located in between 30 to 60 days oscillations. This is an important result, since MJO has already suspected have a strong correlation with Dipole Mode (DM) oscillation which occurred over the Tropical Indian Ocean and Western Pacific including Indonesia and have 3 to 6 months oscillation period. We tried to explain this phenomena by showing this Figure 5 below.

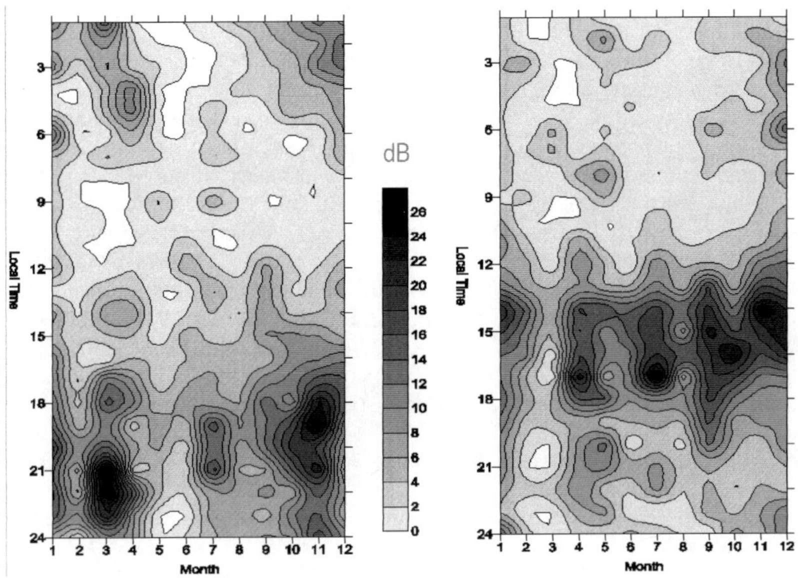

Figure 5 : The differences between stratiform and convective cloud over Kototabang observed with the Boundary Layer Radar (BLR) (after Renggono, *et. all,* 2001)

The left and right panel of Figure 5 shows the comparison between stratiform and convective cloud over Kototabang using the BLR data, respectively. We can see the differences between stratiform and convective cloud clearly. Stratiform cloud has two peak dominant oscillations, in March and November, although it has another peak in July, but still it is not clear. While, convective cloud has more rapid oscillation, close to every 2-3 months oscillation, in April, July, September/October and December. We suspect it mostly effected by the Dipole Mode phenomena.

4. Summary

We have already presented here an application of MAWS (Mini Automatic Weather Station) data in determining a fine structure of MJO (Madden-Julian Oscillation) near by EAR (Equatorial Atmosphere Radar) site at Kototabang, West Sumatera, Indonesia. By using *wavelet* analysis taken from Matlab software, we sucsessfully found the MJO oscillation at Kototabang area that is 32 days oscillation. Theoreritically, the MJO oscilation is between 30-60 days oscillation and most dominant appeared near by the Western part of Sumatera Island.

5. Future Works

We need to continue this study to improve the significant correlation between MJO and Dipole Mode oscillation, since both phenomena are suspected appeared near the cost of Sumatera Island. We need to know also the propagation wave of both oscillation, especially from startosphere to troposphere still near by EAR site. For this reason, we strongly recommand to use the EAR data, since this instrument can detect these oscillations from 2 km to 22 km above mean sea leave (MSL) and spatial designed to observed an atmospheric parameter with good time and spatial high resolution.

References:

[1] Madden, R. A., and P. R. Julian, “Detection of a 40-50 day oscillation in the zonal wind in the tropical Pacific”, *J. Atmos. Sci.*, **28**, 702-708 (1971).

[2] Elleman, R., “Predicting the Madden and Julian Oscillation Using a Statistical Model” (unpublished), (1997).

[3] Madden, R. A., and P. R. Julian, "Observations of the 40-50 day tropical oscillation: a review", *Mon. Wea. Rev.*, **122**, 814-837 (1994).
[4] McBride, J., "The Australian summer monsoon. In: Reviews of monsoon meteorology" (Ed. Krishnamurti), 203-231 (1988).
[5] Suppiah, R. and X. Wu, "Surges, cross-equatorial flows and their links with the Australian summer monsoon circulation and rainfall". *Aust. Meteor. Mag.* (1998).
[6] Jones, C., D.E. Waliser and C. Gautier, "The influence of the Madden-Julian Oscillation on ocean surface heat fluxes and sea-surface temperatures", *J. Climate,* **11**, 1057-72 (1998).
[7] Rui, H., and B. Wang, "Development characteristics and dynamic structure of tropical intraseasonal convection anomalies", *J. Atmos. Sci.*, **47**, 357-379 (1990).
[8] Jones, C. and B.C. Weare, "The role of low-level moisture convergence and ocean latent-heat fluxes in the Madden and Julian Oscillation", *J. Climate,* **9**, 3086-104 (1996).
[9] Elleman, R., "Predicting the Madden and Julian Oscillation Using a Statistical Model" (1997). (unpublished).
[10] Hayashi, Y. and D.G. Golder, "Tropical 40-50 and 25-30 day oscillations", *J. Atmos. Sci.,* **50**, 464-94 (1993).
[11] Renggono, F, H. Hashiguchi, S. Fukao, M.D. Yamanaka, S.Y. Ogino, N. Okamoto, F. Murata, B.P. Sitorus, M. Kudsy, M. Kartasasmita, and G. Ibrahim, "Precipitating Clouds Observed by 1.3-GHz Boundary Layer Radars in Equatorial Indonesia", Annales Geophysicae. **19**, 889-897 (2001).

IOP Conf. Series: Journal of Physics: Conf. Series **1005** (2018) 012048 doi:10.1088/1742-6596/1005/1/012048

Development and Testing of a USM High Altitude Balloon

A S Mohamed Thaheer[1)] *, N A Ismail[1)], S H Md. Yusoff[1)], and M A Nasirudin[1)]

[1)] School of Aerospace Engineering, Universiti Sains Malaysia, Malaysia

*aenorilmi@usm.my

Abstract. This paper discusses on tests conducted on the component and subsystem level during development of the USM High Altitude Balloon (HAB). The tests conducted by selecting initial components then tested individually based on several case studies such as reliability test, camera viewing, power consumption, thermal capability, and parachute performance. Then, the component is integrated into sub-system level for integration and functionality test. The preliminary result is utilized to tune the components and sub-systems and trial launch is conducted where the sample images are recorded and atmospheric data successfully collected.

1. Introduction

High-altitude balloon (HAB) is an unmanned balloon, typically filled with light gases such as helium. It is capable of reaching approximately 50 km above the sea level which has seized the interest of the public in recent years with more HAB projects being released into the stratosphere due to the advancement of low cost communication technology and easy accessibility to civilian global positioning system (GPS). In addition, the payload carried by HAB is also retrievable, which gives the opportunity for services and maintenance and helps in reducing space debris. A typical HAB a weather balloon used to acquire atmospheric data throughout the flight course [1].

The objective of USM HAB is to collect atmospheric data and capturing images of the earth horizon. The secondary objective is to instill the interest of spacecraft design knowledge among undergraduate's students. The detail concept of operations and mission timeline is explained in [2].

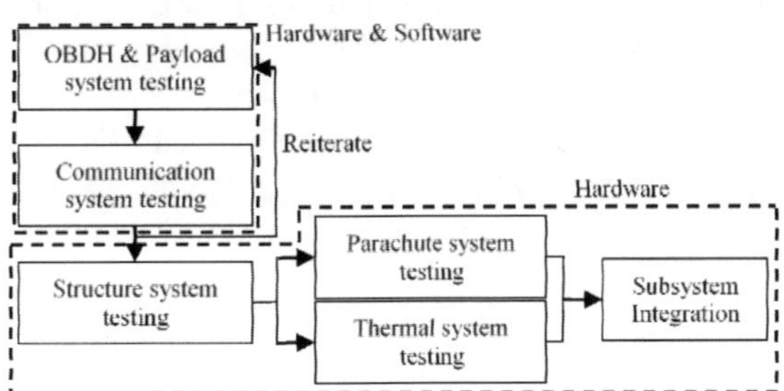

Figure 1. Subsystem's verification flowchart.

System integration verification is vital to achieve the mission objectives. Six tests are conducted including one integration test, to ensure reliability, capability, performance, and the efficiency of the components and subsystems in various conditions. The risks associated with these

IOP Conf. Series: Journal of Physics: Conf. Series **1005** (2018) 012048 doi:10.1088/1742-6596/1005/1/012048

subsystems have been identified and the management plan for the risks have been developed. Each subsystem test drives the next subsystem test as seen in Figure 1.

2. System Description

The main system of HAB is regularly divided into five subsystems, which are balloon and navigation, communication, on-board data handling (OBDH) and payload, structure, and thermal subsystem. The structure uses polystyrene material in shape of a cube (30 cm x 30 cm x 30 cm) which is lightweight, waterproof, and able to float. Preliminary design suggested that the components are placed in stacking level where power system is placed at the bottom level, OBDH and payload are placed in middle level, and finally, the communications are placed on top level of the stack. The heat packs are placed at the bottom of each stack plate for thermal control. Temperature and humidity sensor as well as the antenna are positioned outside of the box in order to function better. The payload is then connected with a parachute and tied with the balloon line shown in Figure 2.

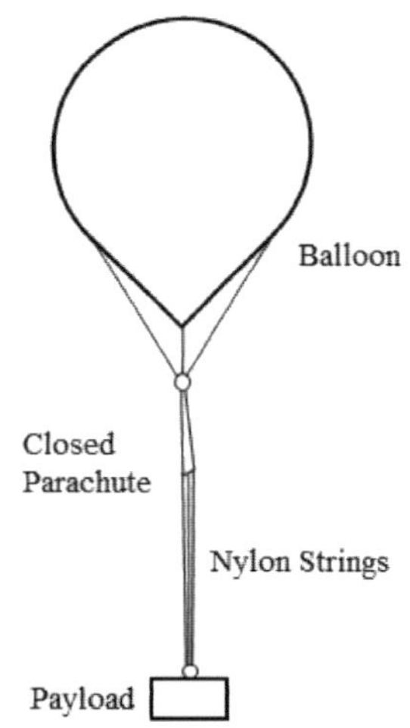

Figure 2. HAB configuration

3. Components Test

The critical subsystems are the OBDH and payload, and communication subsystem, although other subsystems are of equal importance depending on the mission's objective. In order to avoid mission failure, it is of utmost importance that these two subsystems are thoroughly tested and verified before launching. This would ensure that the microcontroller programming is compatible with the sensors as well as the communication system. After it finishes, the communication subsystem test follows, which is done by testing the transmitter, receivers, cameras, and the GPS system. As sensors are prone to noise, which leads to error in reading, both OBDH and payload, and the communication subsystem are run through a small integration test to check its susceptibility. This to ascertain whether the communication system interferes with the sensors' reading, and if it does, the effect shall be minimal.

Next, the structure test is carried out to check the design flaws and the ability of selected material to withstand the landing impact and the capability to float on water for a long time. Following that, a thermal test is performed to determine component functionality at optimum temperature and at low temperature condition.

In additional, the parachute is tested by integrating the final design and respective mass. During this process, the descent rate and the stability of the structure are observed. The test result become the input for redesigning parachute if necessary. Finally, an integration test is conducted on the interface and mounting between the balloon, the payload, and the parachute are done before the launching.

IOP Conf. Series: Journal of Physics: Conf. Series **1005** (2018) 012048 doi:10.1088/1742-6596/1005/1/012048

3.1 On-Board Data Handling (OBDH) Components & Payload

Figure 3 shows the schematic diagram of the sensor connecting to the Arduino microcontroller with power supply.

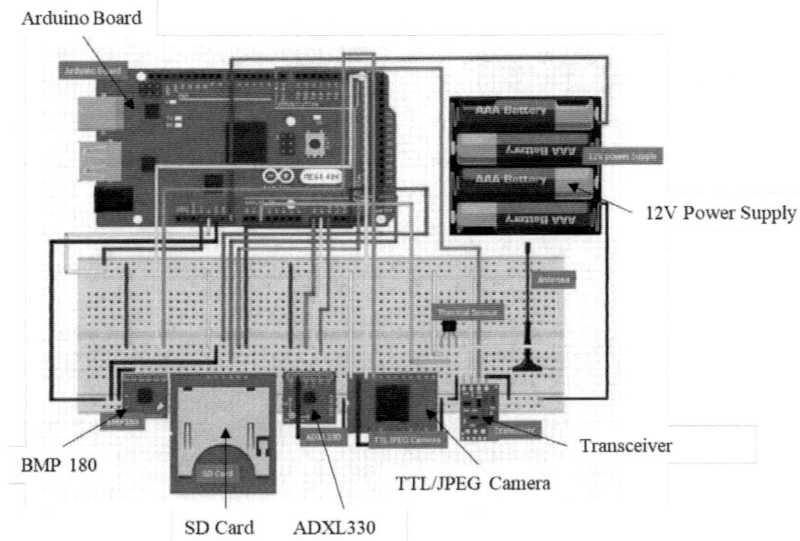

Figure 3. Breadboard wiring and schematic diagram of LM35, SD Card shield, BMP180, ADXL330, camera into Arduino Mega 2560.

The sensors and components sensitivity is tested in different environment as tabulated in Table 1. This test is to ensure that the sensor can read negative value in order to show a negative temperature. This due to the HAB will experience different temperature range during its flight [3]. The camera plays an important role in the project as the objectives of the project is to be able to capture images of the horizon. Thus, three tests are developed to evaluate the capability of the payload.

Table 1. Sensor sensitivity analysis

Components	Study Case
BMP180 + LM35	Different temperature and pressure environment
SD Card Shield	Multiple sensor logging data at the same time
ADXL330	Different direction and elevation

3.1.1 Camera Viewing Angle

The camera is first positioned with no angle shown in Figure 4 (a). This is to give a general idea on what is going to be in the camera frame. Due to the mission objectives, the camera angle is tilted to 35° depicted in Figure 4 (b). This is to fully capture the image of the earth surfaces as well as the horizon.

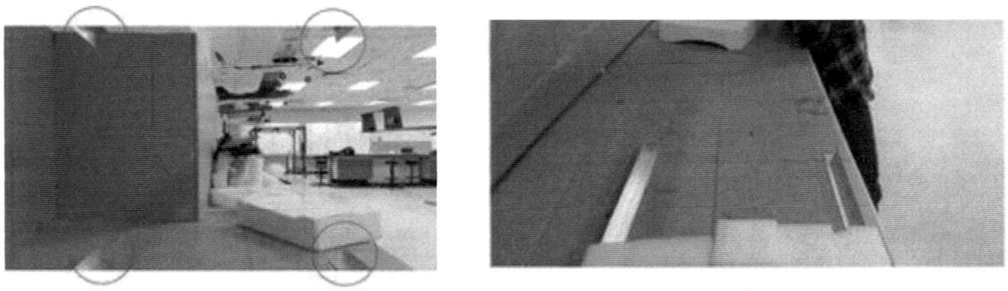

(a) (b)

Figure 4. (a) Camera 1st location testing with no angle. (b) New camera position with angle of 35°.

3.1.2 Camera Battery Endurance

The test is to evaluate the time taken of the battery to fully discharged during recording and how much power it required. The duration of test started from the camera has been switched on and video recording is started until it fully discharged. The total recorded footage time is assumed to be the total discharged time. The battery has a capacity of 900 mA·h and the duration of recorded footage is 143 minutes and 3 seconds. It has an operating voltage of 3.7 V. From this, the power consumption from the camera can be calculated using Equation (1).

$$Power = \frac{Capacity \times V}{3600} \tag{1}$$

Where, $Capacity$ is the battery capacity in mA·h and V is the operating voltage of the camera. The power consumption of the camera calculated is 0.925 mW and power subsystem will design the power management based on the value calculated.

3.1.3 Camera Storage Capacity

In order to determine the amount of storage used for the camera, early estimation can be done by analyzing the size of each of footage saved. Since the camera in-built settings that automatically split the footage for every 30 minutes recording, it takes up 2.11 GB of memory. Thus, the memory size required per second can be calculated using Equation (2).

$$MBps = \frac{Size_{footage}}{t} \tag{2}$$

Here, $Size_{footage}$ refers to the size of footage file in MB and t is the recording time in seconds. The calculated memory per unit time is 1.167 MBps, therefore, a 4-hours of a HAB flight requires approximately 16 GB of memory. However, the memory size required depends on the color intensity of footage [4]. Thus, the file size for each footage is varied. Hence, a 32 GB memory is suitable for the project.

3.2 Communication

In keeping track of HAB, there are two ways to be able to do so as shown on Figure 5.

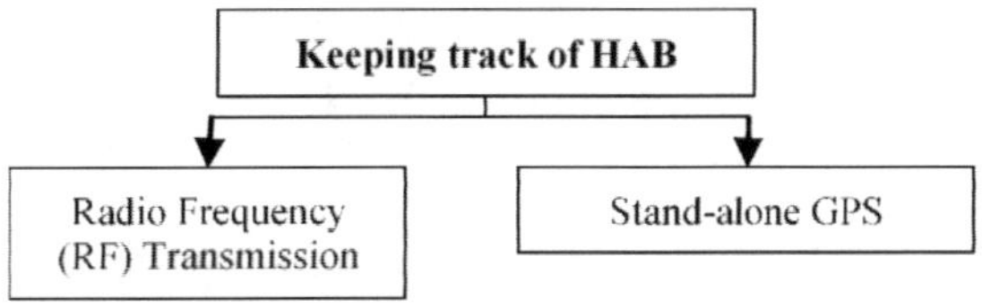

Figure 5. Tracking Method of HAB.

3.2.1 Radio Frequency (RF) Transmission

Communication system is using the ham radio network coverage, in which offers an extremely wide horizontal range of coverage and helps the detection of balloon during ascent and descent. As shown in Figure 6 (b), a custom-made Arduino shield called the Trackuino is used to transmit data from HAB. It consists of a transmitter, GPS receiver, and audio beeper. The Trackuino transmits the flight data in the form of Automatic Packet Reporting System (APRS) packet data. These APRS packet data can be readily received by radios, and relayed around by digipeaters before the data are passed to the internet via an IGate. The data is sent to a website called aprs.fi, where the balloon's live flight path can be viewed.

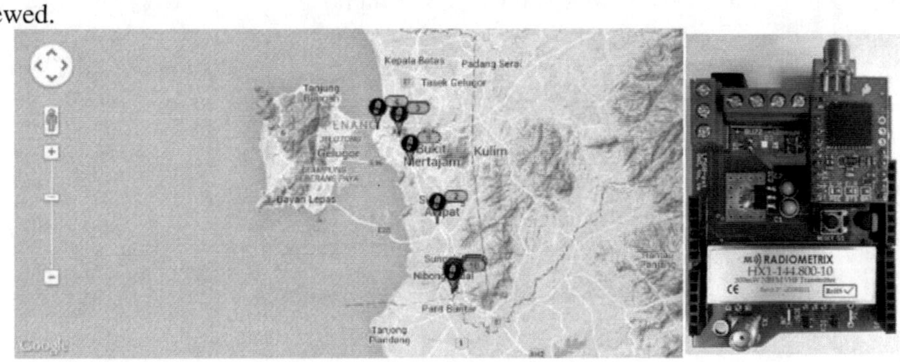

(a) (b)

Figure 6. (a) Tracking of SPOT GPS. (b). Trackuino with Radiometrix HX1 VHF Narrow Band FM 300mW Transmitter & GPS Module.

3.3 Power

Power subsystem test is required to make sure the power supplied by the power source according to the estimated power requirements. The entire communications system was turned on and the actual operating voltage and current consumption was measured shown in Figure 7. The required power consumption and battery capacity are determine using Equation (3).

$$P = IV \tag{3}$$

(a) (b) (c)

Figure 7. (a) Operating voltage measured. (b) Current measured during idle. (c) Current measured during transmission.

IOP Conf. Series: Journal of Physics: Conf. Series **1005** (2018) 012048 doi:10.1088/1742-6596/1005/1/012048

Based on the results from Figure 7, the Trackuino power consumption is tabulated in Table 2. Thus, the total power consumption is listed in Table 3 is 1.89 W. Therefore, suitable power source is needed by considering every components power rating.

Table 2. Trackuino power consumption

Description	Value
Operating Voltage, V_o	5.11 V
Idle Current, I_o	0.15 A
Transmit Current, $I_{transmit}$	0.29 A
Idle Power, P_o	0.77 W
Transmit Power, $P_{transmit}$	1.48 W

Table 3. Power consumption

Subsystem	Power (W)
OBDH	0.405
Payload	0.000
Communication	1.480
Structure	0.000
Thermal	0.000
Balloon & Navigation	0.000
TOTAL	1.885

3.4 Structure

3.4.1 Drop Test

A drop test is conducted to observe the stability of the payload due to high probability of the payload landing on the sea depicted in Figure 8. If the payload toppled over during landing, the sea water may enter inside the payload bus and damaging the components. The test is performed by placing a 2 kg of weight inside the payload bus and fully attached with the parachute system. It is then drop from a height of 25 meter. The payload falls accordingly and it lands perfectly upright.

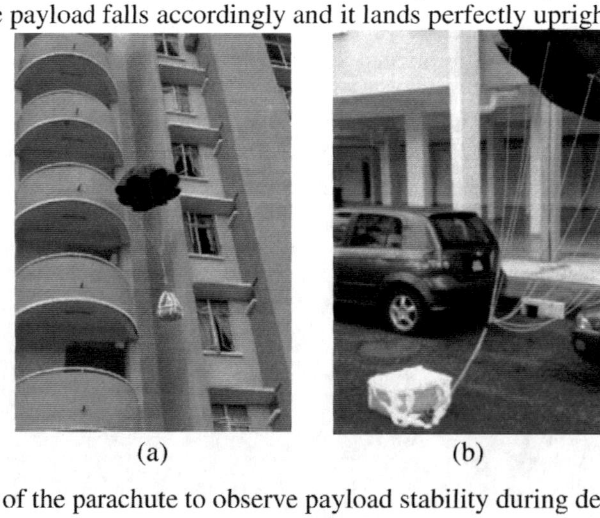

(a) (b)

Figure 8. (a) Opening of the parachute to observe payload stability during descent. (b) Landing of the payload.

IOP Conf. Series: Journal of Physics: Conf. Series **1005** (2018) 012048 doi:10.1088/1742-6596/1005/1/012048

3.4.2 Impact Test

Impact test is done to make sure that the main structure can resist the impact during landing when the parachute system is malfunctioned. This test is also to understand the critical part of the structure that will likely to fail first upon landing. The main structure is filled with 2 kg of weight and has been drop at height of 7 meter. The drop focuses on the edge of the structure that probably have the highest stress concentration. As the structure drops, the motion is captured with a high-speed camera. As shown in Figure 9 (a), deformation occurs at the edge of the structure and the stress propagates to the top of the of structure producing cracks as shown in Figure 9 (b). The cracks produced are caused by the compression and tension of the structure under load.

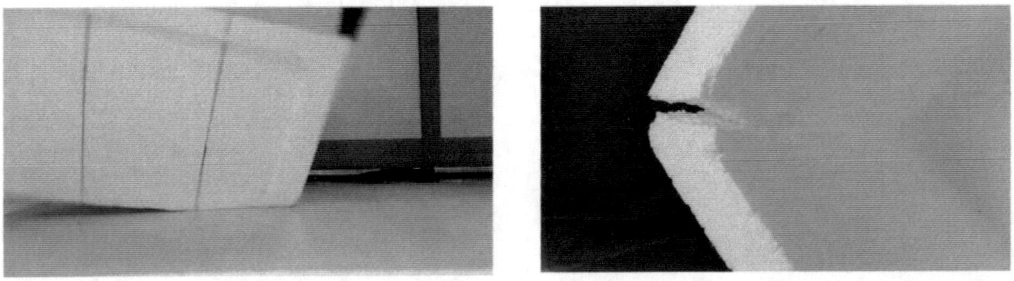

(a) (b)

Figure 9. (a) Impact testing whereby the box is drop on the edge and deformation occurs. (b) Close up view of the cracks.

3.4.3 Buoyancy Test

The retrieval period may take a long time and the payload will be exposed to the waive at the sea. Thus, the payload bus must be able to float on the sea with capable of withstanding force of the wave. The test is carried out by placing a weight of 2 kg inside the payload bus. The main structure is then wrapped with a plastic to make it water tight and attached with a long rope for retrieval process and tested to be fall on the water as in Figure 10 (a) and (b). The payload is then retrieved and has been visually inspect for any water leaking inside the structure.

(a) (b)

Figure 10. (a) Buoyancy test preparation. (b) Impact result when it hits the water.

3.5 Thermal

Thermal tests are done on two separate case which are with -20°C freezer and -80°C freezer. BMP180 sensor along with digital thermometer and thermocouple sensor are used. The heat source for this test are heat packs. The sensors and heat packs are placed as shown in Figure 11.

IOP Conf. Series: Journal of Physics: Conf. Series **1005** (2018) 012048 doi:10.1088/1742-6596/1005/1/012048

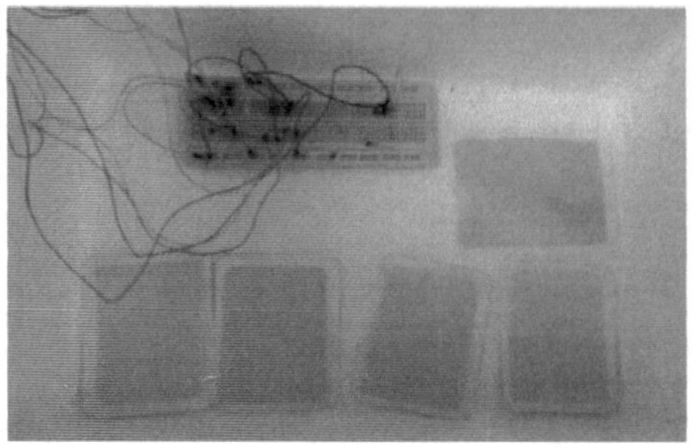

Figure 11. Experimental setup of BMP 180 sensor and heat pack positioning

Figure 12 shows that the heat pack can generate heat as the temperature inside the payload is higher than the surrounding temperature.

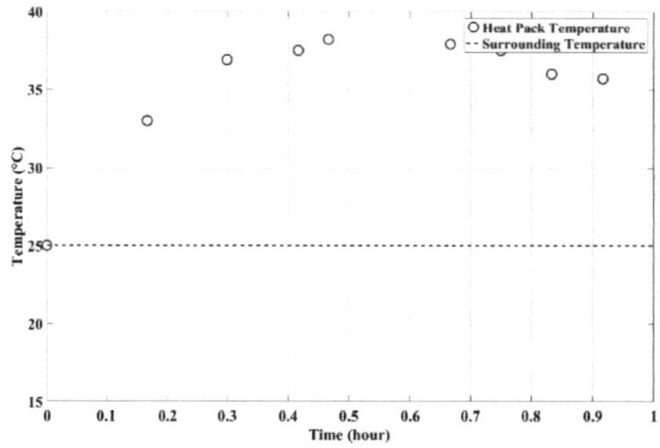

Figure 12. Heat pack temperature test comparing with surrounding temperature.

This initial experiment lead to the next experiment with different thermal condition inside the structure as listed in Table 4. The Arduino is set up and it is calibrated shown in Figure 13 (a) based on block diagram drawn in Figure 13 (b).

Table 4. Thermal environment analysis

Condition	Thermal Components	Analysis Time
Case 1	No heat packs & thermal reflector	1 hour 30 minutes
Case 2	5 heat packs	
Case 3	5 heat packs & thermal reflector	

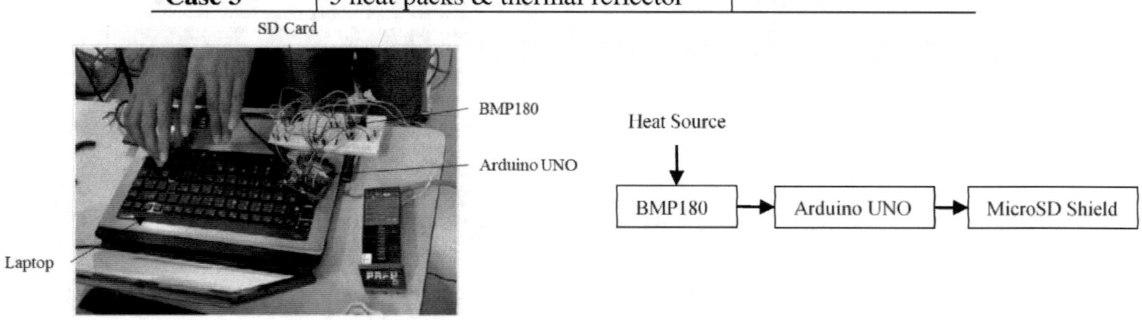

(a) (b)

Figure 13. (a) Component setup and sensor calibration for -80 °C freezer experiment. (b) Thermal test block diagram.

The result for each condition are shown in Figure 14 (a). All condition started at average initial temperature of 24°C. As the time approaches minute of 20, there is an approximately 100% decrease from the initial temperature and this is due to the payload bus incapable to withstand the cold temperature. On top of that, the heat packs for case 1 and 2 are not fully released the heat. Since, heat packs are passive thermal components, it needed some time in order to release the heat. However, as time goes on to minute of 60, there is a slight increase in temperature as the heat pack started to give out its maximum heat and stabilizes the internal temperature.

It can be observed that, a heat packs along with thermal reflector are better in containing the heat inside the payload bus as it records higher temperature compare to only with heat packs. As for case 1, the internal temperature drops and it only depends on the payload bus material to withstand the cold temperature. From this test, it can be concluded that the heat packs with thermal reflector are the best in handling cold thermal environment.

The test continues using -80°C freezer with an objective to simulate the actual thermal environment in the stratosphere. This also give us the chance to test the BMP180 and Arduino UNO operating condition as both components are not able to work under -40 °C based on its product datasheet. The outer payload structure is sprayed with ethanol solution to make sure there is no contamination in the freezer. It is assumed that the ethanol solution does not have any impact to the temperature change in the payload structure.

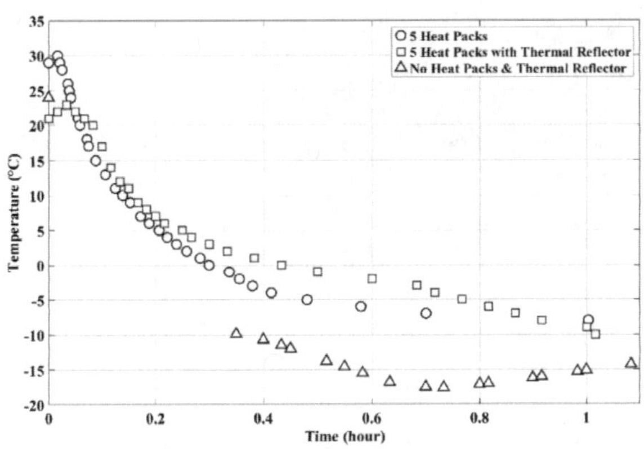

(a)

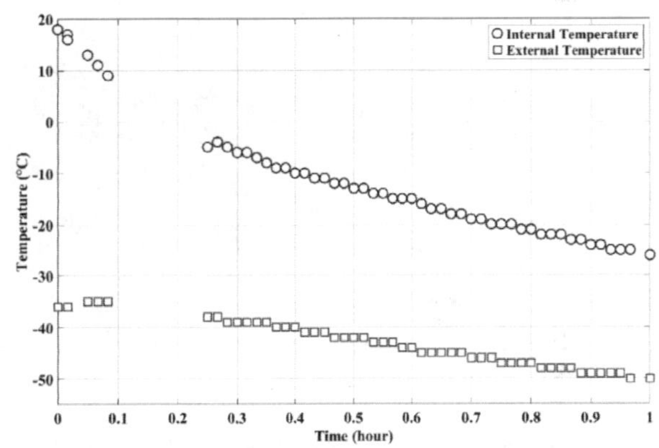

(b)

Figure 14. (a) Comparison of thermal components case study. (b) Internal and external temperature test result using -80°C freezer.

As shown in Figure 14 (b), the internal and external temperature drops linearly. Nevertheless, the heat packs and thermal reflector can withstand the -80 °C cold temperature and able to maintain above -40 °C which is above the operating temperature for BMP180 and Arduino UNO.

3.6 Balloon & Navigation

The parachute is tested by dropping a dummy payload with a mass of 0.5 kg from height of 22.4 m. Two types of parachute are developed with two different design. The parachute is made up from plastic and nylon fabric material. Each one of the parachute has a hole at its center namely Apex Vent. The function of Apex Vent is decrease the drag by enabling the air to flow through the parachute. Without the Apex Vent, the air will accumulate at the parachute center and this increases drag [5]. The diameter of the Apex Vent is set at 0.1 m. However, selecting suitable material also gives a significant impact on the performance of the parachute. This can be seen in Table 5.

Table 5. Descent time & velocity for different types of parachutes

Type of Parachutes	Descent Time (s)	Descent Velocity (ms^{-1})
Plastic Parachute (without AV)	14.0	1.60
Plastic Parachute (with AV)	11.6	1.93
Nylon Fabric Parachute (without AV)	8.0	2.80
Nylon Fabric Parachute (with AV)	7.0	3.20

Based on the testing result, nylon fabric parachute seen in Figure 15 (b) has a higher velocity compare to both plastic parachutes shown in Figure 15 (a). This is due to the weight of the parachute. Heavier parachute resulting a higher descent rate. The effect of Apex Vent can be seen in the table as with the same material, for instance, the nylon fabric parachute, shows a different descent time and descent rate results. Nevertheless, this is still an estimation whereas it does not take into account effect of cross-wind. During flight, the bus will experience different type of aerodynamic forces such as wind which are random. Thus, the performance of the parachute may vary at different flight conditions.

(a) (b)

Figure 15. (a) Plastic parachute. (b) Nylon Fabric parachute. Both without the Apex Vent.

3.7 Final Design

After finishing the component tests, the final design configuration of the HAB can now be finalized. The block diagram of the system is as shown in Figure 16.

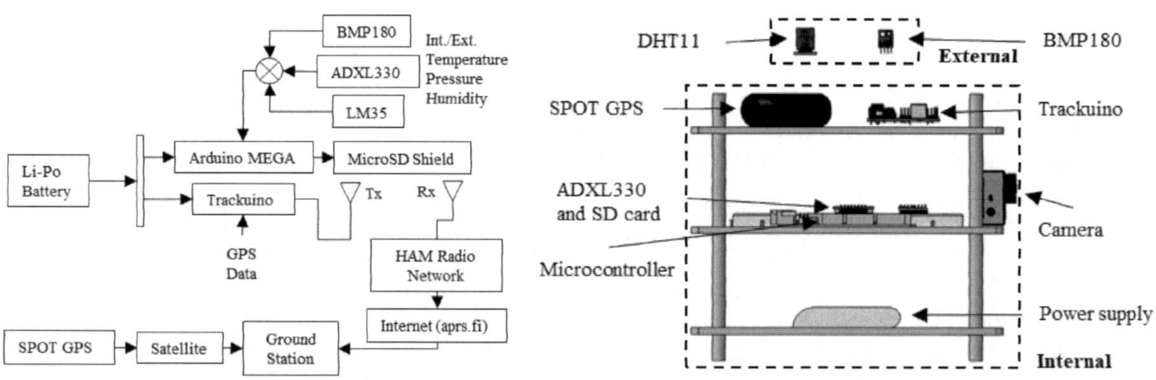

(a) (b)

Figure 16. (a) HAB payload components integration block diagram. (b) HAB subsystem front view.

4. Flight Test

A trial launch flight test was conducted to test the system integration. The system comprises of all of the subsystems except Trackuino due to the unavailability to obtain the operating license. However, the only way to track the balloon is by using SPOT Messenger GPS. This is risky due to the capability of the SPOT Messenger that can only be used up until 6 km and the GPS might lose connection in case the balloon floats away from the launch site. Nevertheless, the flight test is done successfully as the payload are able to retrieve. The data is then compared with the International Standard Atmosphere (ISA) data and from the Malaysia Meteorological Department (MET Malaysia).

4.1 Atmospheric data

Four atmospheric data are recorded during the flight test which are altitude, temperature, pressure, and humidity. Based on Figure 17 (a), the highest altitude recorded is 24.2 km. It takes approximately 39 minutes to achieve its maximum altitude. The total flight time is 1 hour and 21 minutes, though, retrieval time took the total mission time to nearly 3 hours. Based on the same figure, the ascent rate and descent rate can be calculated which are 9.2 ms^{-1} and 10.4 ms^{-1} respectively.

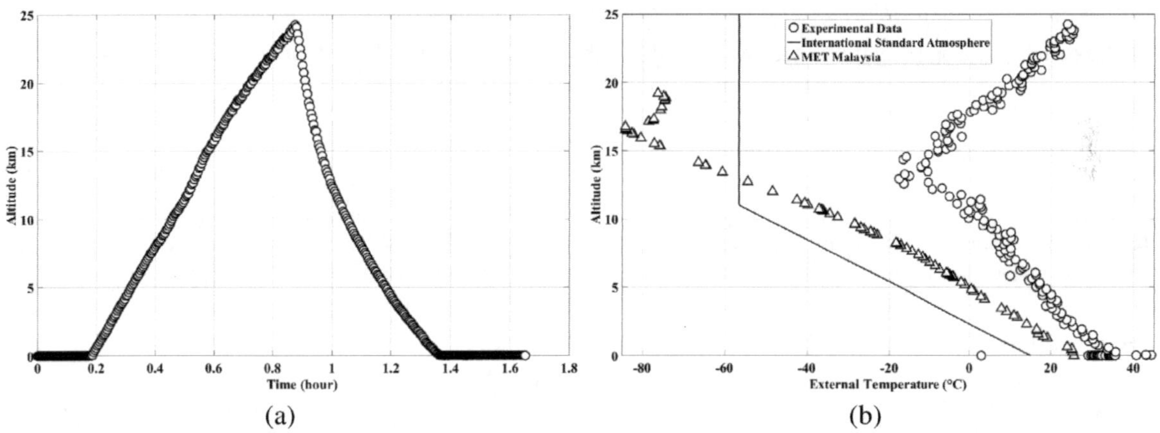

Figure 17. (a) Recorded altitude during trial launch. (b) Comparison of external temperature of payload bus with ISA data and MET Malaysia at different altitude.

Figure 17 (b) shows the temperature variation at different altitude. The ISA data are based on formula that are developed in 1920 by A. Toussaint [6]. The formula provides a simple and reasonable representation of the mean annual temperature results in United States yet it is not completed. The Standard Atmosphere was introduced as a reference atmosphere for the aviation industry and was never meant to be used to predict the actual barometric pressure at a particular location and was in fact developed by people who clearly recognized that there would be substantial local variations caused by latitude and seasons of the year [7]. In [8], it is clearly shown that different month and different places produces different temperatures variations. Plus, temperature change based on solar activity as there is high correlation between solar activity with the earth's surface temperature [9].

It can be seen that in Figure 17 (b), the experimental data recorded by the BMP180 sensor is higher compared too other two data. This is due to the placement of the BMP180 sensor that might be too close to the transmitter and the heat contained in the payload affecting the result. Yet, the data is quite promising as the pattern is the same as the MET Malaysia data.

Observing the data shown in Figure 18 (a), the pressure recorded by the sensor is fairly accurate with the ISA data as well as with MET Malaysia data. The pressure decreases exponentially with recorded pressure of $0.158x10^4$ N/m^2 at the maximum altitude.

Figure 18 (b) shows the variation of humidity results compared with data from MET Malaysia. The data from MET Malaysia are scattered but generally, the pattern is the same as experimental data where the humidity reduces as the altitude increases. This is because of the decreasing temperature resulting in colder air. The sensor works by measuring the temperature of the air as warmer air will have higher humidity content.

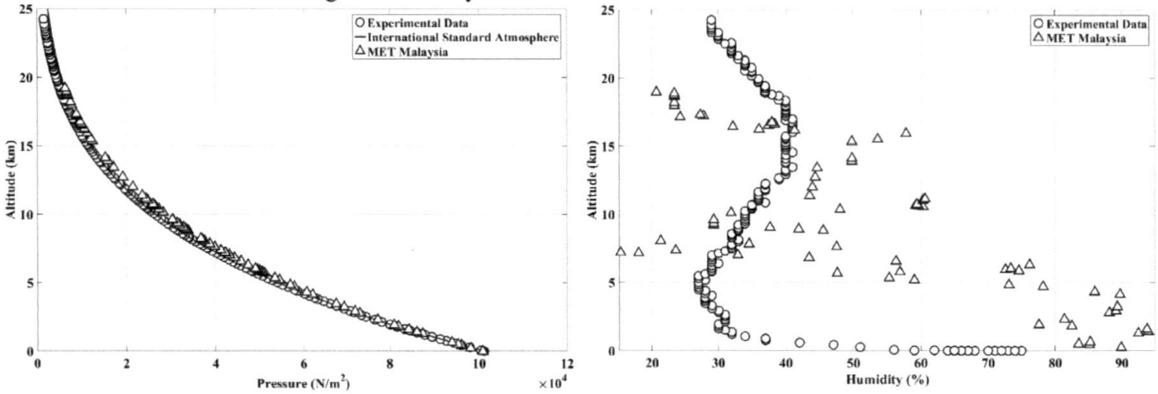

(a) (b)

Figure 18. (a) Comparison of pressure from experimental data with ISA data at different altitude. (b) Variation of humidity data at different altitudes.

4.2 Pre-launch Trajectory

Before the launching of HAB, the balloon trajectory is done to estimate the landing position of HAB. The prediction is done using two web-based predictors that is Cambridge University Spaceflight (CUSF) Landing Prediction by the University of Cambridge [10] and the Balloon Trajectory Forecasts by University of Wyoming [11]. Unfortunately, during the pre-launch day, there is unavoidable circumstances and delayed the balloon released by 2 hours. Hence, flight path prediction is done again on new flight time and it is shown in Figure 19.

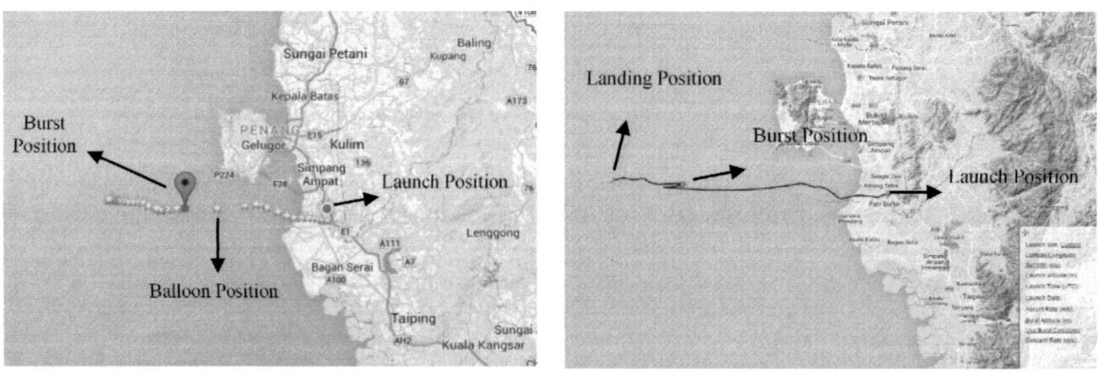

(a) (b)

Figure 19. Path tracking prediction result. (a) Wyoming's predictor. (b) CUSF landing predictor.

Based on Figure 19 (a), the prediction shows the launch position (blue point), burst position (red tag), and balloon position (star symbol). It is predicted that the flight time is 2 hours with a landing coordinate of 5°17'31"N, 99°28'14"E.

By using the CUSF landing predictor, the total flight time is approximately 2 hours as well compared to Balloon Trajectory Forecasts by Wyoming University observed in Figure 19 (b). The predicted landing position is at 5°17'31.3"N, 99°28'14.0"E which is identical to the Wyoming's result. Figure 20 (a) shows the position recorded using SPOT Messenger. Notice that, between position-feed 8 and position-feed 9, there is a huge gap. There is where the commercial communication limit which is at the altitude of 6 km. However, SPOT Messenger manage to keep alive all the time up until 24.2 km and send signal back to the ground station when it reaches 6 km during descending. The landing coordinates registered by the SPOT Messenger is 5°17'31"N, 99°28'14"E.

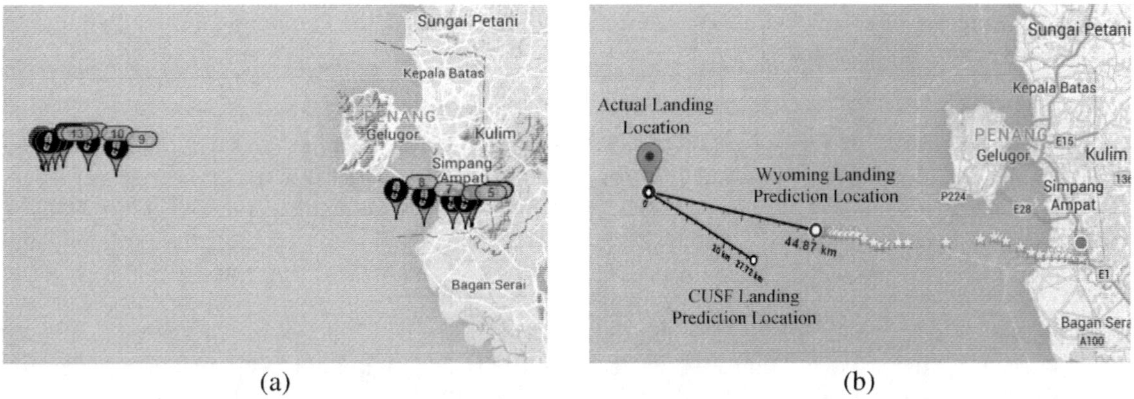

Figure 20. (a) Balloon trajectory recorded using SPOT Messenger during trial launch. (b) Comparison of predicted results and actual trajectory for trial launch.

According to Figure 20 (b), the flight path for all three tracking results are fairly similar. This shows that both predictors are good enough to give an early estimation for the balloon trajectory but still falls short from the actual data as there is a possibility of unknown assumptions are used during the building of the software. In addition, actual flight is affected by the actual wind behavior that is unpredictable.

4.3 Image Results
Several images are taken during the flight including the curvature of the earth and the image of Penang island during landing are shown in Figure 21.

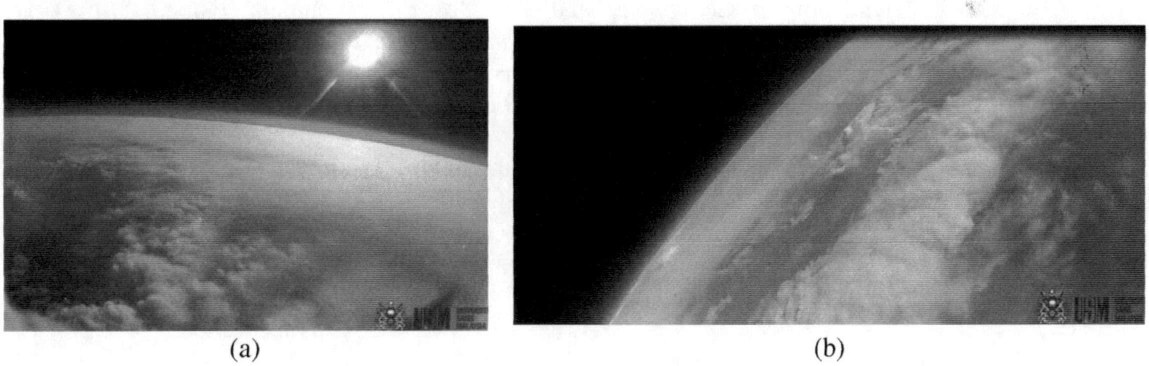

Figure 21. Sample images of the earth curvature taken by the camera.

5. Conclusion
In this paper, the development and testing of USM High Altitude Balloon is investigated. The project is aimed to be able to collect atmospheric data as well as capturing images of the earth horizon whilst instill the interest in spacecraft design knowledge among undergraduate's students.

The project is done by undergoing component testing to test each component reliability and it is followed by integration testing. Each testing has its own case study. Next, the HAB is trial launched to get a general idea on launching a high-altitude balloon. This also gives a chance to test the components in actual condition. The result from the trial launch is promising as data recorded is fairly identical to

IOP Conf. Series: Journal of Physics: Conf. Series **1005** (2018) 012048 doi:10.1088/1742-6596/1005/1/012048

the actual data. However, during actual launch, the payload could not be retrieved because of misinterpreting the data from SPOT Messenger and the Trackuino and weather problem on the sea that holds off the retrieve mission

As for future works, there will be an implementation of autonomous guidance system that will help in returning the payload back to the launch site and upgrading the communication subsystem to have better tracking of balloon.

Acknowledgment

The authors would like to express their deepest gratitude to Astronautical Association of Malaysia (AAM), Kelab Peminat Radio Amateur & Rekreasi Kerian Perak (REAKSI), Malaysia Maritime, Malaysian Meteorological Department (MET Malaysia), Department of Civil Aviation (DCA), Department of Survey and Mapping Malaysia (JUPEM), School of Chemical Engineering, USM, and School of Aerospace Engineering, USM for the support and encouragement in completing the project.

References

[1] Voss, P., Advances in Controlled Meteorological (CMET) Balloon Systems, *AIAA Balloon Systems Conference, vol. 53, no. May, pp. 1689–1699, 2009.*

[2] Mohamed Thaheer, A. S. and N. A. Ismail, Mission Design and Analysis of USM High-Altitude Balloon, *Journal of Mechanical Engineering UiTM, 2017.*

[3] Anderson Jr., J. D., The Standard Atmosphere, in *Introduction to Flight, 4th ed., Boston: WCB/McGraw-Hill, 2000, p. 766.*

[4] Parsons, J. J. and D. Oja, Colour and Compression, in *Computer Concepts: Illustrated Introductory, 9th ed., Cengage Learning, 2012, p. 296.*

[5] Dadvand, M., D. Abdolrahman, and A. Hassanzadeh, A Numerical and Experimental Study of the Aerodynamics and Stability of a Horizontal Parachute, *ISRN Aerospace Engineering, vol. 2013, p. 8, 2013.*

[6] J.D. Anderson Jr., Aircraft performance and design. *1999.*

[7] West, J. B., Prediction of barometric pressures at high altitudes with the use of model atmospheres, *American Physiological Society, pp. 1850–1854, 1996.*

[8] Jursa, A. S., Handbook of Geophysics and The Space Environment, *1st ed. Air Force Geophysics Laboratory, Air Force Systems Command, U.S. Air Force, 1985.*

[9] ZHAO, X. and X. FENG, Periodicities of solar activity and the surface temperature variation of the Earth and their correlations, *Chinese Science Bulletin (Chinese Version), vol. 59, no. 14, p. 1284, 2014.*

[10] Sowman, J., A. Greig, D. Richman, R. Wareham, R. Anderson, F. Noble, and E. Moore, Cambridge University Spaceflight (CUSF) Landing Prediction, *University of Cambridge, 2013.* [Online]. Available: http://predict.habhub.org.

[11] Oolman, L., Balloon Trajectory Forecasts, *University of Wyoming.* [Online]. Available: http://weather.uwyo.edu/upperair/balloon_traj.html.

IOP Conf. Series: Journal of Physics: Conf. Series **1005** (2018) 012049 doi:10.1088/1742-6596/1005/1/012049

Reentry Capsule for Sample Return from Asteroids in the Planetary Exploration Missions

Yoshifumi Inatani

Institute of Space and Astronautical Science (ISAS),
Japan Aerospace Exploration Agency(JAXA),
Sagamihara, Japan inatani@isas.jaxa.jp

Abstract. For carrying sample from the bodies of interplanetary space, a wide range of knowledge of reentry technology is needed. HAYABUSA(MUSES-C) was an asteroid explorer returned to the earth after the 7 years of voyage, and its capsule reenters into the Earth's atmosphere, which was a good example of reentry technology implemented to the flight vehicle. It performed a safe reentry flight and recovery. For the design of the capsule, many considerations were made due to its higher entry velocity and higher aerodynamic heating than those of normal reentry from the low earth orbit. Taking into account the required functions throughout the orbital flight, reentry flight, and descent/recovery phase, the capsule was deigned, tested, manufactured and flight demonstrated finally. The paper presents the concept of the design and qualification approach of the small space capsule of the asteroid sample and return mission. And presented are how the reentry flight was performed and a brief overview of the post flight analysis primarily for these design validation purposes and for the better understanding of the flight results.

INTRODUCTION

MUSES-C spacecraft was launched by M-V-5 launch vehicle in May 9. 2003, toward the asteroid 1998SL36. After the launch, the name "Hayabusa" was given to the spacecraft. In September 2005, it approached to the target body, observation through the proximity operations, and several attempts to touchdown and to try to collect sample. Finally it returned to the earth, and performed its reentry flight and recovery in June 13, 2010.

The major objective of the mission is to demonstrate technical issues necessary to conduct an approach and rendezvous to one of the near earth asteroids[1] using a high performance ion-engine system during its cruising flight to and from the asteroid. In toching down to the asteroid, the surface material will be taken by a sampleing mechanism and the sample is to be carried by the capsule that directly reenter into the earth's atmosphere along the hyperbolic trajectory with super-orbital speed from the asteroid as presented in Fig.1.

IOP Conf. Series: Journal of Physics: Conf. Series **1005** (2018) 012049 doi:10.1088/1742-6596/1005/1/012049

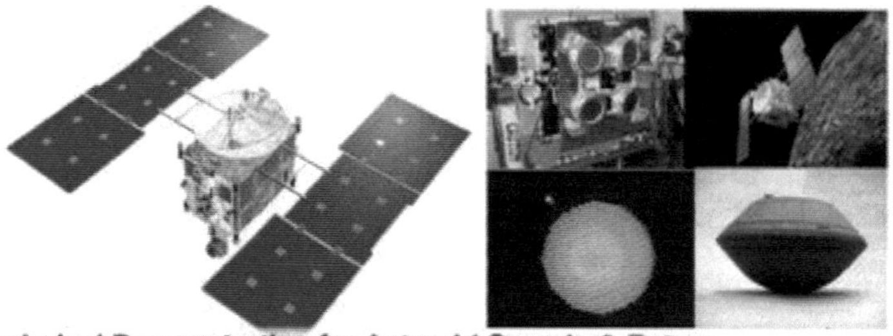

Figure. 1 Asteroid Explorer Hayabusa and its Technical Challenges

The returning capsule's conceptual design work was initiated in early 90th based on the related technical activities and background both for the aerothermodynamics, thermal protection materials and its responses under heating, descent and recovery subsystems in the several flight lessons at ISAS in the past. The major technical issues are due to hypervelocity atmospheric reentry flight in which the entry velocity is as high as 12 km/sec. Unlike other reentry vehicle, its flight speed is the fastest as ever experienced for the entry into the earth's atmosphere. It will also impose higher heating rate to the thermal protection system. For these reasons, the estimation of the reentry flight environment and the thermal protection material and structure are essential to the design of the capsule. Due to the limitation of the size and weight allocated to the capsule, it has to be accommodated by the various functional parts inside the extremely small capsule, such as a sample canister, parachute and its jettisoning mechanisms, beacon for the localization, power supply subsystem, thermal control, and so on. These various new requirements imposed several technical challenges in the design. An overview of the design of the capsule at the completion of the development is summarized and issued as a collection of technical papers in Ref. 2.

In the flight of Hayabusa spacecraft to and from the targeted asteroid that was named "Itokawa", many unexpected things happened. It performed the proximity operations, touch-down and trials to collect samples from the surface of the asteroid in 2005 to 06. During these operations, many troublesome affairs happened but finally it recovered the operation on the way back to the Earth, it was suffered by the damages and, as a result, serious limitations for the spacecraft operation in the propulsion subsystems, attitude control subsystems, power/communication subsystems and so on. However it finally returned to the earth in June 2010 after the additional revolution of the Helio-centric orbit. Then the reentry capsule finally took an important part of the mission at its very end.

The capsule came back to the Earth and the final reentry flight of the capsule was conducted in June 13. 2010 at the Woomela prohibited area (WPA) Australia. In the final approach of the spacecraft to the reentry atmospheric flight interface to the landing area, a series of careful trajectory maneuver operations was conducted from safety assurance point of view, and receive the final permission to be issued by the Australian authority. The view of reentry from the ground is shown in Fig.2. After the entry a beacon signal was received at the right time as planned by the final parameter setting made use of the measurement of the deceleration by aerodynamic drag in accordance with the increase of the dynamic pressure.

IOP Conf. Series: Journal of Physics: Conf. Series **1005** (2018) 012049 doi:10.1088/1742-6596/1005/1/012049

The capsule position during descent by parachute was localized, then the capsule was found after 30 minutes after the estimated landing, at the place less than one kilometer from the planned landing position. As presented in the previous chapters, the separated parachute cap and foreside heat shield were also found and recovered. After the inspection of the sample canister inside the capsule, dusts from the asteroid were identified and we obtained the extraterrestrial material for the first time from one of the small asteroids. The investigation of the asteroid surface dust is still underway and it will help better understand the evolution of our solar system. As a result Hayabusa mission and the capsule completed its mission.

Here in the present paper, an overview of the aerodynamic / aerothermodynamic design of the capsule, the flight results and ongoing post flight analysis are presented in the following Chapters

Figure 2. Hayabusa returned at Woomela in south Australia in June 13, 2010

OVERVIEW OF THE DESIGN OF THE RETURNING CAPSULE

Remarks in Aerodynamics of Hypervelocity Atmospheric Entry Flight

In the design of the reentry capsule for the present purpose, many new design considerations should have been made in aerodynamics, aerothermodynamics, thermal protection system and its responses at very high temperature, flight mechanics, and so on. Because of the entry speed into the earth atmosphere is 12km/sec, also because the resources allocated to the capsule in terms of its weight and envelope were very limited as well. In Table 1, important issues to be thought serious at the stage of starting the design work when the program was approved.

IOP Conf. Series: Journal of Physics: Conf. Series **1005** (2018) 012049 doi:10.1088/1742-6596/1005/1/012049

Table 1. Aerothrmodynamic Issues to be Thought for the Sample Return Capsule Design

High Enthalpy Flow Characterization in Super-Orbital Speed (12km/s) Entry to the Earth's atmosphere
· Highly Nonequilibirium Flow · Thermal Responses of Ablative Materials under High Heating · Strong Interaction between Shock Layer Flow & Ablation Products · Radiation Heating · Injection Induced Boundary Layer Transition · Leeward Flow Phenomena and Back Face Heating Environment
Aerodynamic Characteristics
· Static Instability in Rarefied Regime · Transonic Dynamic Instability
New Parachute Deployment System, Verification & its Accommodation

There would be many causes of uncertainties in prediction, and to reduce these uncertainty was one of the ways in the preliminary aerodynamic design such as geometry trade-off phase, relation among the entry trajectory condition, nose radius, aerodynamic stability and so on as shown in Fig.3. Also considered was to take into account for maximizing the inner envelope for the instrumentation such as parachute, sample canister and avionics. For resolving these issues, following measures were taken on each design issue. Most of these considerations are covered and summarized in Ref. 2.

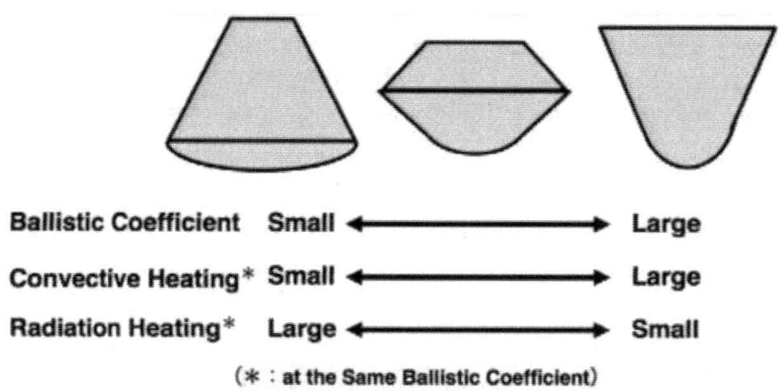

Figure 3 Reentry Capsule geometry and Flight & Heating Characteristics

Radiation Heating

The heating by radiation from highly excited states of the high temperature gas in the shock layer is indispensablel on this mission. For that reason, a numerical radiation estimation code; SPRADIAN(structured package of radiation analysis) is newly developed and used for the radiation analysis by the aid of the flow field solver. The results were compared with the existing and simpler modeling technique such as NEQAIR.

Interaction Between Outer Flow and Ablation Products

A massive ablation product out of the body surface will affect the shock layer flow greatly. For the charring ablator such as a carbon-phenolic material which is often used for the thermal protection in relatively high heating rate, the reaction at the surface itself is divided into oxidation dominant regime below 3000K of the surface temperature and sublimation dominant regime for the higher temperature than 3000K. In the latter regime, the surface recession rate increases rapidly with the temperature, a massive flow of ablation product must be taken into consideration in addition to the pyrolysis gassing. Then, a fully thermal and chemical nonequilibrium shock layer flow analysis was made under the condition that there is a carbonaceous ablation product with concerned surface temperature regime.

Base Flow of the Capsule

Generally the backward surface heating is far less than that of the foreside, however a careful estimate is needed because of the complexity of the flow and the light-weight design requirement and so on. Since the wake flow phenomenology behind the capsule is so much complicated, it sometimes has difficulties in estimating the heating environment there. The result quantified that the radiation enhancement by ablation products both for the oxidation and sublimation for the present flight environment of the capsule will not be negligible. We had to be careful about the design heat flux setting because of many functioning parts in the surface of the leeward.

Boundary Layer Transition under Massive Ablation Products

A possibility of injection-induced boundary layer transition has been pointed out. Since the nature of the transition, even a flow phenomenology is not well understood and some qualitative explanations are being proposed. Ablation products will play as an injectant through the surface will augment the disturbance in the boundary layer, which may result in the earlier transition than that occurs without such an injection of ablation products. A model of the injection-induced turbulence is incorporated to the foreside flow analysis.

Ablative Thermal Protection and its Characterization of Thermal Responses

The selection and evaluation of the heat shield material are ones of the most important issues in the design of the capsule. The present mission offers a considerably high heating load up to $15MW/m^2$ for about 30 seconds, then a charring ablator is to be a unique solution. A carbon-phenolic material is chosen and there are many possibilities in fabric layout and manufacturing processes. Starting with the preliminary evaluation and screening for choosing the fabric-resin arrangement by arc-heating test, basically a cloth-layered carbon-phenolic ablator was selected. In the evaluation process, a careful inspection was made to characterize both the thermo-chemical reactions and the mechanical durability of the ablator such as pyrolysis, surface recession by both oxidation and sublimation, spallation, delamination, and so on. Besides these heating test-based evaluation, a numerical simulation model to simulate the thermal responses was made and refined for the present application.

The flow enthalpy of heating facility such as arc-heaters is far less than that in the real flight, and the numerical-simulation-based correlating studies is necessary to bridge from the test results to that at the flight condition. On this context, modeling of the thermal responses of the ablator is essential to predict the ablator behavior in the flight environment. At the surface of the heated material, many effects are taken into account, such as air mixture of incoming flow, composition of pyrolysis gas, and products of the surface reaction by oxidation and sublimation. The angle of fabric layers against the surface is one of the important parameters taking into

IOP Conf. Series: Journal of Physics: Conf. Series **1005** (2018) 012049 doi:10.1088/1742-6596/1005/1/012049

account the in-depth thermal conductivity and avoiding delamination due to out-gassing. The refinement of these parameters was conducted primarily by heating-test basis.

Aerodynamics and Flight Mechanics of the Capsule

Since the reentry capsule is flown by totally passive aerodynamic stability, the flight characteristics are to be carefully investigated both statically and dynamically, throughout the entire reentry flight and descending flight followed by the parachute deployment. Due to the nature of the aerodynamics of the disk-shaped body used here, it inherently has a dynamic instability in pitching motions in the transonic flight regime. In addition, it was pointed out that the static pitching instability may occur in the rarefied gas regime[3].

This trend would be critical to the attitude motion at initial entry. The static stability is primarily examined by a conventional wind tunnel test or by a simple estimation technique, however, the dynamic stability is difficult to know generally. Fig.4 illustrates the methodology to establish the aerodynamic database of MUSES-C reentry capsule. Above-mentioned important characteristics were needed to careful and intense considerations in relation to the reentry flight mechanics and following parachute opening.

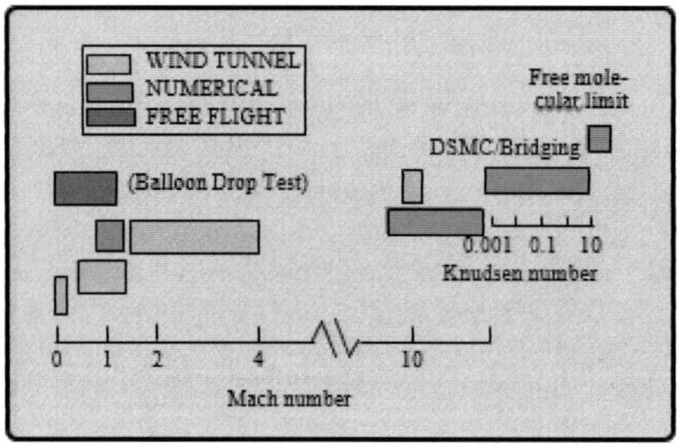

Figure 4. Methodology for Constructing Aerodynamic Database

Descent Subsystem and Subsonic Parachuting

The descent sub-system of the capsule has several special features. These are to cope with the difficulty in an arrangement of the descent subsystems inside the small capsule, such as light-weight cross parachute with radar reflector, torus-shaped packing, drogue concept, newly developed parachute ejection devices, and so on. Due to the pitching motion instability from low supersonic to subsonic speed, supersonic parachuting could have been a solution. The characterization of this instability showed and resulted no tumbling may occur by this instability. Therefore, a subsonic parachuting was chosen. In addition, for the sake of simplicity, we employed the design in which jettisoned parachute cover plays a roll of drogue chute. The sequence of parachute deployment and foreside heat shield separation is illustrated in Fig.5. We employed that the foreside heat shield is to jettison, in order to avoid heating from the hot heat shield to the sample container inside the instrument module. Making use of the ballooning flight test opportunities at ISAS, two drop-tests for supersonic descending flight of the capsule were conducted for the demonstration of both transonic attitude instability and parachute opening.

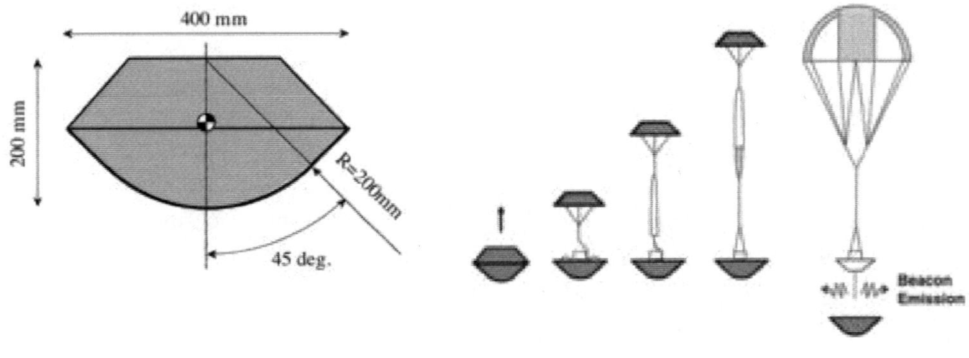

Figure 5. Resulted Capsule size, Geometry and Descent Subsystem

HAYABUSA REENTRY FLIGHT AND RESULTS

Reentry Flight and Recovery Operation Overview

The spacecraft came back to the Earth and the final reentry flight of the capsule was performed in June 13. 2010, at the Woomela Prohibited Area (WPA) Australia. In the final approach of the spacecraft to the reentry atmospheric flight interface to the landing area, a series of careful trajectory maneuver operations was conducted from ground-safety-assurance point of view, and the final permission was issued by the Australian authority.

Figure.6 shows the final trajectory foot-print and 4-sigma error ellipse of capsule's landing position which was estimated in the planning phase before the final trajectory correction maneuver. This estimation was made based on the GRAM-99[4] atmosphere model and the initial condition, whose position and velocity of capsule at 200km in altitude, estimated according to the orbital determination before the final trajectory correction maneuver. Thus, its initial condition used in this estimation includes the error of the orbital determination and the trajectory correction maneuver. The altitude of the parachute deployment was set to 10 km in the planning phase. The solid ellipse in Fig.6 shows the 4-sigma dispersion ellipse of landing position of the capsule. Its ellipse was expected to extend to 150km in the east-west direction and 100km in the north-south direction due to the uncertainty of the orbital determination and the wind profile. The dash dot line shows the WPA boundary. We had to define the target landing position to set the 4-sgima error ellipse of the landing position into WPA.

The capsule passed 200km altitude in 13:51(UT) and the luminescence of HAYABUSA mother ship and the capsule was observed from the ground stations. In 5 minutes after the reentry, the parachute was deployment at about 5km altitude and the beacon signal was transmitted as scheduled. All of the station acquired the beacon signal. The headquarters of the recovery team analyzed the information from the stations and localized result in real time using the position analysis system. The capsule was landed near the predicted landing position at 14:09(UT). The final prediction was made with initial condition estimated from orbital determination data which was delivered at 0:00(UT) 13. June, before reentry operation and modified atmospheric model which takes account of National Center Environmental Prediction's (NCEP) [5] atmospheric forecast data which delivered 0:00(UT) in 13. June.

In the final decision for parameter setting, the altitude of the parachute deployment is 5 km by setting on-board G-trigger and delay timer. According to the final prediction, the capsule reenters from the west-northwest direction and changes the descending course to northerly direction due to the jet stream. After the parachute deployment at 5km altitude, capsule with a parachute moves to the west direction at about 3km due to the eastward wind near the ground surface. The discrepancy between the actual recovery position and the final predicated landing position was less than 1km as shown in Fig.7. The reentry and descending trajectory prediction contributed to the recovery operation greatly. The jettisoned heat shields and parachute cover were found and recovered also very near to the predicted positions.

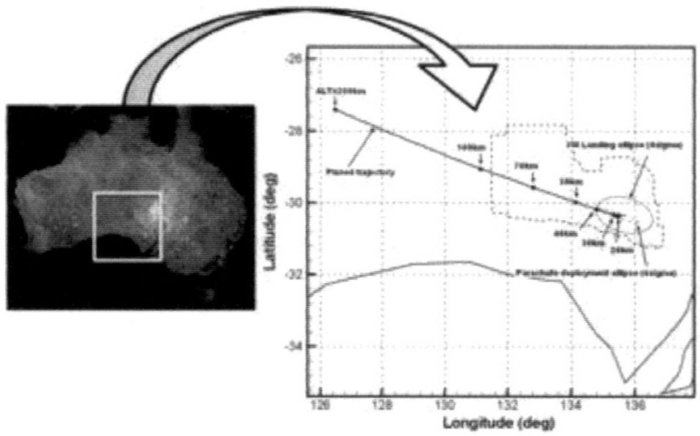

Figure 6. Nominal reentry trajectory and 4-sigma error ellipse of the landing position

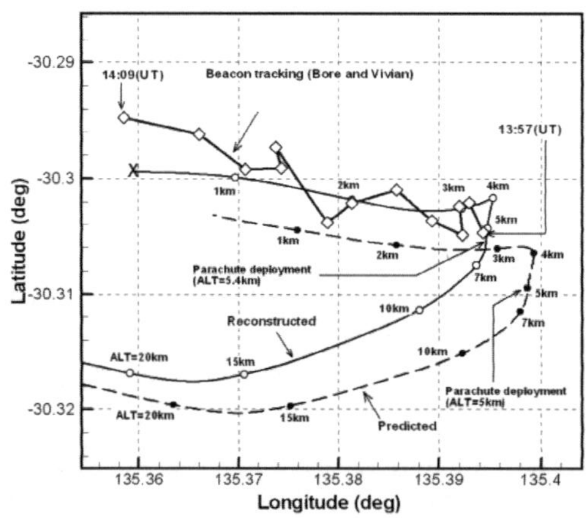

Figure 7. Planned and resulted trajectory and important locations (Actual recovered position of the instrument module is designated by **X**)

Reconstruction of Reentry Trajectory

Reentry trajectory was reconstructed made use of most reliable data acquired in the reentry operation for a post flight analysis. The most reliable initial condition, which is a set of time, position and velocity of capsule at altitude 200km, is listed in Table 2. That is based on the final orbital determination carried out at the position nearest to the Earth. Table 2 also shows the dispersion (1-sigma) in addition to nominal value. The most reliable atmosphere model including the wind profile was made to take in consideration most recent NCEP forecast data which was delivered 12:00(UT) in 13. June. Figure 8 is wind profile comparison between the nominal data and dispersion of GRAM99 statistical database and NCEP forecast data. In the present reconstruction, the head shield jettison was made at the altitude of 5.5 km determined by onboard 5G trigger and 244 second delay timer. The parachute deployed in 1 second after the head shield jettison.

Table 2. Initial condition of reentry trajectory reconstruction

	Final orbital determination data		Shift value for reconstruction
	Nominal	Dispersion( )	
Time	13:51:12.16(UT)	0.091 (sec)	--
Altitude	200043 (m)	13(m)	--
Latitude	-27.37608 (deg)	0.00269(deg)	+0.00029deg(= 32m)
Longitude	126.42890 (deg)	0.00693(deg)	+0.0045 deg(=434m)
Velocity	11651.87 (m/s)	0.0163(m/s)	--
Azimuth	112.8867 (deg)	0.003454(deg)	--
Path angle	-12.68093 (deg)	0.004513(deg)	--

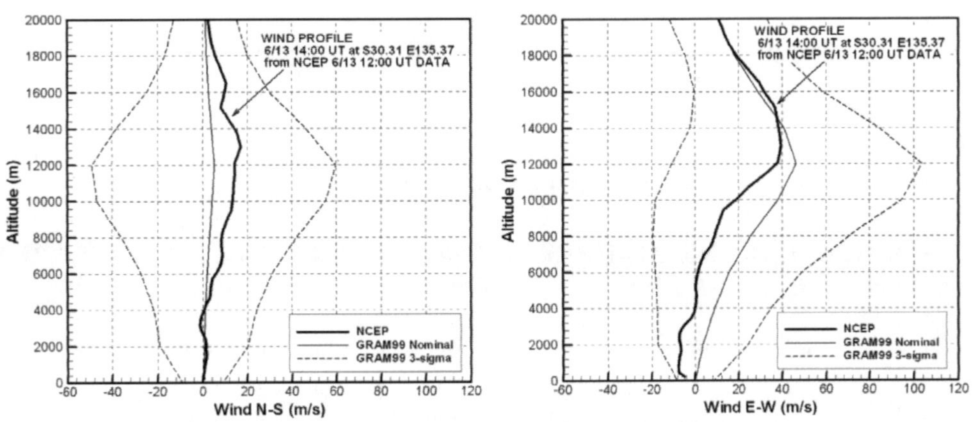

Figure 8. Wind profile comparison with GRAM-99 and NCEP forecast data delivered 12:00(UT) in 13. June 2010

For the reconstruction of the reentry trajectory, the descent trajectory of the capsule was calculated according to the above conditions at first. Next, the initial position is shifted from the final orbital determination data to match the calculated landing point with the actual recover position. The shifted value of initial condition from final orbital determination is 0.00029 deg in latitude and 0.0045 deg in longitude. These values are within 1-sigma dispersion. The reentry trajectory reconstructed from shifted initial condition is expected to be reasonable reentry trajectory.

IOP Conf. Series: Journal of Physics: Conf. Series **1005** (2018) 012049 doi:10.1088/1742-6596/1005/1/012049

The trajectory reconstructed by above mentioned procedure is shown in Fig. 9. For showing the agreement between the predicted, real-time-estimated and resulted/reconstructed trajectory, these three trajectories are presented in Fig.7. The real-time-estimated trajectory is that by using the data taken by two ground-tracking stations located at "Vivian" and "Cattle Bore". The agreement of the landing point of sample container for example is less than1km, which was far better than the preflight prediction taken into account various causes of the uncertainty.

The parachute deployment position of reconstructed trajectory was almost identical to the first determined capsule position by two stations. After deploying parachute, the capsule (now it is a sample container suspended by parachute as shown in Fig. 5 in the previous section) moved from east to west according to reconstructed trajectory. That movement also agrees with the tracking results. Then it was confirmed that the reconstructed data is valid. Figure 6 shows relation between altitude, velocity and downrange and acceleration and stagnation heat flux time history. The stagnation heat flux is estimated by the Tauber's equation[6].

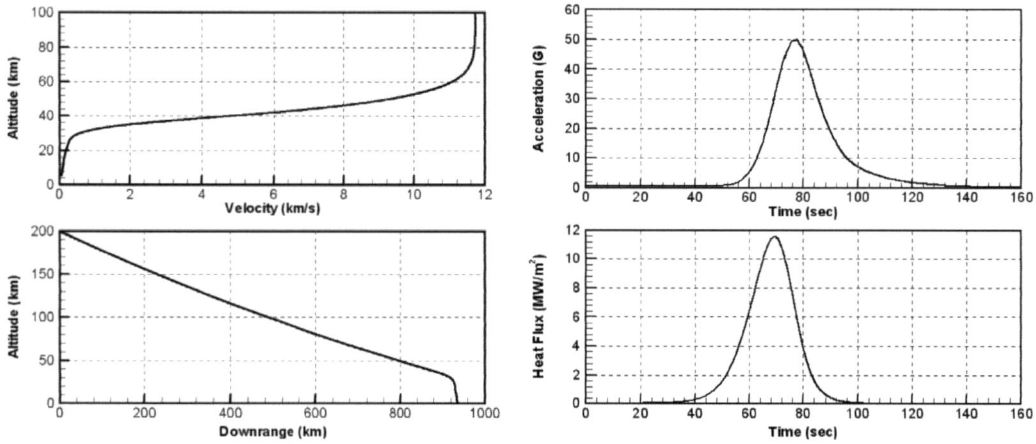

Figure 9. Reconstructed trajectory about relation between altitude, velocity and downrange and acceleration, stagnation heat flux time history

INSPECTION OF THE RECOVERED CAPSULE AND IN-FLIGHT SURFACE TEMPERATURE MEAUREMENT

The recovered parts of the reentry capsule were looking all very soundly worked as presented in Fig. 1o. No serious damages were observed, and no hazardous and anomalous things were observed both in the heat shields, parachute and sample container. The surface of the foreside heat shield looks very smooth, and in the rear side heat shield, there remained the surface coating for the in-orbit thermal control, which may tell smaller heat input than expected.

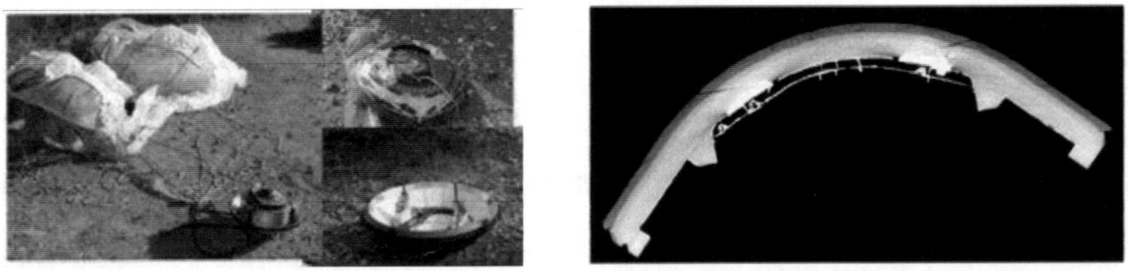

Figure 10. Recovered parts of the Capsule and X-ray Cross section image of the foreside heat shield

By inspection, the surface of the foreside region was very smooth and no hazardous damage and / or delaminations were observed, which means preflight evaluation of the surface recession and out-gassing of pyrolysis gas were properly made. Also no major change in shape and color of internal surface of the heat shield and exterior surface of the instrument container, which tells us the validity of the thermal design of the capsule system during reentry flight were properly made. Then the parachute and payload container was kept in the proper temperature, then, as a result, the samples in the sample canister was estimated and kept in good condition. In addition, in the special parts such as the gaps and junctions of neighboring heat shield, holes for venting, cable penetration, and so on, flight result shows that all looks very soundly worked out.

A non-intrusive technique such as X-ray CT-scanning has been applied as much as possible, in order not for unnecessary destruction of the recovered items. The linear absorption coefficient (LAC) of the X-ray is strong function of the density and the element composition of the material. Though the element well-correlated to the ablator density in advance by using arc heated test piece with a given density. Although there are many causes of difficulties in conducting non-destructive investigation of the internal properties, the correlation between the heating test pieces and recovered items were made. By adjusting the beam power and the band width, we could successfully obtained the LAC distribution map in the regions of char layer, pyrolysis layer, and virgin layer over the heat shield. This result is used for the correlation between thermal response analysis as presented in the following sections. The right hand side of the Fig. 7 shows an example of raw X ray image of the fore-body heat shield.

On the surface of the rear side heat shield, there remained the thermal control film adhered to the surface of the heat shield for the in-orbit thermal-control purposes. It may mean that the rear-surface heating was smaller than expected. Many of studies and discussions were made in the design phase on these phenomena in rear-side. Considerably large heat flux (10% of that at stagnation point) was assumed as previously described. The recovered rear-side heat shield shows us that the maximum temperature will be about 400 °C from the decomposition of the polyimide film. As a result, the design heat flux has much margin and that is the case. As the flow phenomenology in the wake region is so much complicated, we have to be more careful about the next estimation and design on this context.

In addition to these inspections, the surface radiation measurement during reentry was carried out. In order to study the radiation and resulted heating environment and in order to know the surface temperature of the capsule, the radiation observation was made both from the ground and the airborne measurement system [7][8]. As a result, the radiation temperature is fixed

IOP Conf. Series: Journal of Physics: Conf. Series **1005** (2018) 012049 doi:10.1088/1742-6596/1005/1/012049

by taking account of optical conversion efficiencies. The temperature profile based on the blackbody radiation from the capsule surface has been estimated based on this measurement and the results will be presented in the following sections.

POST FLIGHT ANALYSIS OF THE THERMAL RESPONSES OF THE HEATSHIELD

In order to make better understanding of the flight results, the thermal response of ablative TPS of the capsule is calculated along the most probable reentry trajectory by using the integrated analysis method [9]. In the present study, several physical models hitherto proposed to predict the ablation phenomena are tested for the purpose of comparison. The calculated results are compared with the flight data obtained at the present time.

In the present study, the thermal response of ablative TPS of the capsule is calculated by using the several physical models related to the ablation phenomena. Calculation conditions used in this study are summarized in Table 3. Four cases are considered for the purpose of comparison[10-14]: In case 1, we consider the conventional surface reaction model that was used to calculate the heating environments over the Stardust entry capsule. The effect of nitridation reaction is considered in case 2. In case 3, the effect of the enhanced probability value due to surface roughness[13] is considered both for atomic oxygen and atomic nitrogen. The effect of turbulence due to the pyrolysis gas[14] is included in case 4.

Table 3. Calculation Conditions

Case	Surface reaction probability			Boundary layer	Notes
	Oxidation	Nitridation	Sublimation		
1	A	N/A	E	Laminar	Conventional surface reaction set
2	A	C	E	Laminar	Case 1 + nitridation
3	B	D	E	Laminar	Fully rough surface for O & N
4	A	N/A	E	Turbulent	Case 1 + injection-induced turbulence

Reaction	Surface condition	Reaction probability model	Symbol	References
Oxidation $C(s) + O \rightarrow CO$	smooth	$\alpha_{oxi} = 0.63\exp(-1160/T)$	A	Park, *AIAA J.* 14(11) 1976
	rough	$\alpha_{oxi} = 1.0$	B	Park, textbook , 1989
Nitridation $C(s) + N \rightarrow CN$	smooth	$\alpha_{nit} = 8.441\times10^{-3}\exp(-2322/T)$	C	Suzuki, *JTHT* 24(1), 2010
	rough	$\alpha_{nit} = 1.0$	D	Park, textbook, 1989
Sublimation $3C(s) \rightarrow C_3$		Hertz-Knudsen-Langmuir equation	E	Blottner, NASA SP-252, 1970

Figure 11 shows the temporal variations of surface temperature at the stagnation point. As shown in this figure, the temperature begins to increase at 40s after the passage of altitude of 200 km. A maximum temperature value of about 3,200 K is then obtained at 70 s of flight time for the cases 1, 2 and 4. After that time, the surface temperature decreases due to the radiative cooling effect. As shown in this figure, there are no significant differences in surface temperature between the cases 1 and 2. The effect of nitridation reaction on surface temperature is believed to be small because the rate of nitridation given by Symbol C is smaller than the rate of oxidation given by Symbol A by a factor of 100.

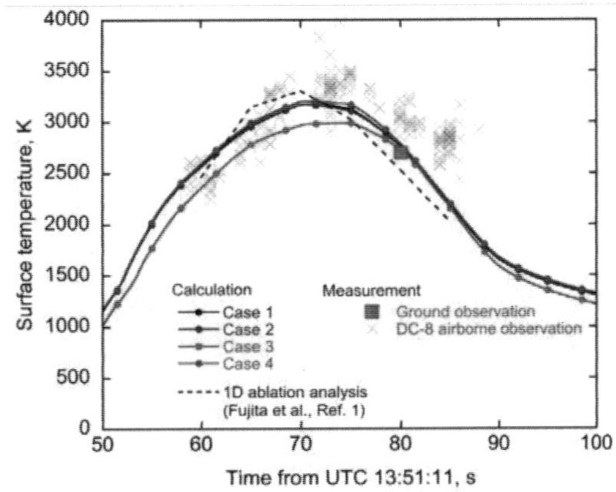

Figure 11. Comparison of surface temperature variations at the stagnation point between the calculation and the measurements

Figure 11 also shows the surface temperature data estimated from the black body spectrum obtained by both the ground observation[7] and the airborne observation presented in the previous section[8]. As shown in this figure, the temperature value estimated from the airborne observation

data is about 2,500K at 60s of flight time, and the maximum value of 3,300K is obtained at 70s. After that, the temperature decreases down to about 2,700K at 85s, although the temperature values are characterized by very strong scattering after the peak heating point. On the other hand, the surface temperature is estimated to be about 2,700K at 80s from the ground observation result. By comparing these estimated surface temperatures with the present calculated results, it is found that the calculation for the cases 1, 2 and 4 duplicates well the temperature data obtained by the airborne observation, although the calculated peak temperatures slightly underestimate the measurement at 70s of flight time.

Based on the X-ray image of the capsule as presented in the previous sections, the char depth and the thickness of remained virgin layer of the ablative TPS are measured, and the results are also shown in Figs. 12(a) and 12(b). By comparing the flight results and the calculated results, it is found that the density distributions calculated for the cases 1 and 2 reproduce the flight data moderately both at the stagnation point and the downstream region. Especially in the downstream region, the calculated thickness of char layer overestimates the flight data when we consider the effect of turbulence due to the pyrolysis gas, while it underestimates the flight data when the effect of surface roughness of ablative TPS is taken into account.

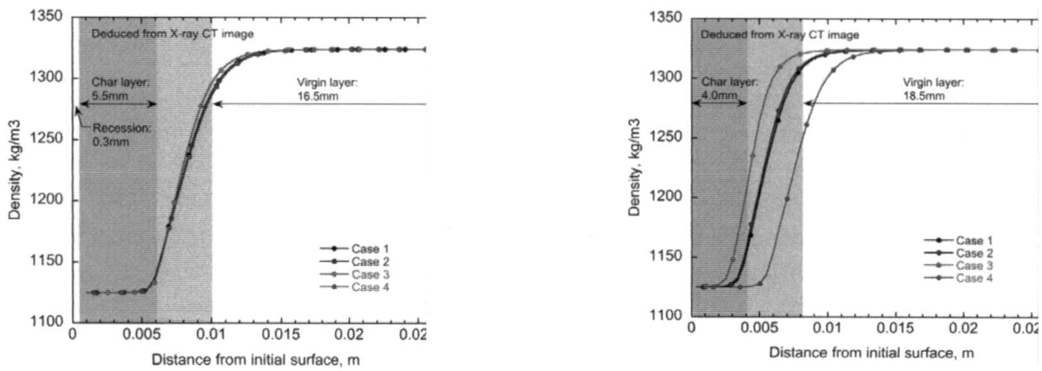

(a) Stagnation point (b) Downstream region (s=0.2m)

Figure 12. Comparison of char depth and thickness of remained virgin layer between the calculations and the flight data

CONCLUDING REMARKS

A flight opportunity of an asteroid sample return mission offered us many new study topics in high speed aero- and aerothermodynamics, thermal protection systems, descent subsystems and in the design of the whole capsule system. For these technical issues, an approach to being ready for the flight and the flight results are summarized in the present paper. One of the major achievements is that the capsule performed the reentry flight from the asteroid, and recovered totally as planned. As a result, it would be understood that the design and verification processes were satisfactorily made in terms of the high enthalpy flow characteristics around the capsule with emphasis on the radiation heating, interaction between outer-flow and ablation-product, and so on. A characterization of the ablative heat shield durable to the high heating condition was also one of the major and new design issues, and it is also satisfactorily performed. The post flight analysis is still going on in many aspects, however the present paper highlighted the relation between the design and flight results. On this context, issues such as how much design margin we should have for the heat shield, was one of the key design issues. Also observed were the conservative design for the rear side heat shield. These topics will need further studies for more accuracy and a lighter vehicle construction. The flight result of landing point dispersion was very small, which means that the aerodynamic characteristics, entry initial conditions, atmosphere modeling and so on were sufficiently accurate and appropriate. From aerodynamic and flight mechanical point of view, however, the flight results tells less, because only few onboard measurements for the motion analysis were done because of the limited resources due to its small capsule size. No serious discrepancy is foreseen between the prediction and the results on this flight mechanics such as attitude motion of the capsule, but we have to know that the design credibility has not yet been perfectly proved. These issues will be passed to the next ventures. Based on the success of the Hayabusa mission's final reentry and recovery, many follow-on missions toward other asteroids and bodies in the planetary space. The excersises and lessons learned in this Hayabusa capsule studies and flight demonstration contributed greatly toward these next ventures.

IOP Conf. Series: Journal of Physics: Conf. Series **1005** (2018) 012049 doi:10.1088/1742-6596/1005/1/012049

REFERENCES

[1]"Asteroid Sample Return Mission Analysis," reported by Asteroid Mission Working Group, the Institute
of Space and Astronautical Science, March 1994.

[2] Inatani, Y. *ed.*, "Aerodynamics, Thermophysics, Thermal Protection, Flight System Analysis and
Design of Asteroid Sample Return Capsule," ISAS Report SP-17, Institute of Space and Astronautical
Science, March, 2003

[3] Fujita, K., Inatani, Y., and Hiraki, K.,"Attitude Stability of Blunt-Body Capsules in Hypersonic Rarefied Regime, Journal of Spacecraft and Rockets, Vol. 41, No.6, pp.925-931, 2004

[4] C.G Justus, D.L. Johnson: The NASA/MSFC Global Reference Atmoshperic Model – 1999 Version
(GRAM-99), NASA/TM-1999-209630, May 1999.

[5] NCEP (National Centers for Enviromental Prediction) HP, http://www.ncep.noaa.gov/

[6] Tauber, M.E., Bowles, J.V., "Use of Atmospheric Braking During Mars Missions" J.Spacecraft Vol.27,
No.5, pp.514-521, 1990

[7] Fujita, K. *et al.*, "Numerical Reconstruction of HAYABUSA Sample Return Capsule Flight
Environments," AIAA 2011-3477.

[8] Yamada, T. *et al.*, "Reentry of Hayabusa Sample Return Capsule and Post-Flight Analysis of the
Recovered Heatshield," AIAA 2011-3322.

[9] Suzuki, T. *et al.*, "Post-Flight TPS Analysis of Hayabusa Reentry Capsule," AIAA 2011-3759.

[10] Park, C., "Effect of Atomic Oxygen in Graphite Ablation," *AIAA Journal*, **14**(11), 1976, pp. 1640-1642.

[11] Suzuki, T. *et al.* "Experimental Study of Graphite Ablation in Nitrogen Flow," *Journal of
Thermophysics and Heat Transfer*, **22**(3), 2008, pp. 382-389.

[12] Blottner, F.G., "Prediction of Electron Density in the Boundary Layer on Entry Vehicles with Ablation,"
NASA SP-252, 1970, pp. 219-240.

[13] Park, C., *Nonequilibrium Hypersonic Aerothermodynamics*, Wiley, New York, 1989, pp. 350-353.

[14] Park, C., "Injection-Induced Turbulence in Stagnation Point Boundary Layer," *AIAA Journal*, **22**(2),
1984, pp. 219-225.

IOP Conf. Series: Journal of Physics: Conf. Series **1005** (2018) 012050 doi:10.1088/1742-6596/1005/1/012050

Longitudinal Static Stability and wake visualization of high altitude long endurance aircraft developed in Bandung institute of technology

Irsyad Lukman E.[1)*], and M. Agoes Moelyadi [1)]
[1)] Faculty of Mechanical and Aerospace Engineering, Institut Teknologi Bandung, Indonesia.

*irsyaded@s.itb.ac.id

Abstract. A High Altitude Long Endurance (HALE) Unamanned Aerial Vehicle (UAV) is currently being researched in Bandung Institute of Technology. The HALE is designed to be a pseudo-sattelite for information and communication purpose in Indonesia. This paper would present the longitudinal static stability of the aircraft that was analysed using DATCOM as well as simulation of the wing using ANSYS CFX. Result shows that the aircraft has acceptable stability and the wake from the wing at climbing condition cannot be ignored, however it does not affect the horizontal tail.

Nomenclature

ρ	: Fluid Density	S_M	: Momentum Source Components
t	: Time	S_E	: Energy Source Components
U	: Velocity Vector	λ	: Thermal Conductivity
p	: Pressure	h_{tot}	: Total Enthalpy
$\boldsymbol{\tau}$	: Stress Tensor	T	: Temperature

1. Introduction

In recent decades, the use of Unmanned Aerial Vehicles (UAV) in military applications has increased. The potential to substitute satellite to High Altitude Long Endurance (HALE) as a relay for communications is one such applications[1]. ITB is currently being supported by Direktorat Riset Teknologi dan Pendidikan Tinggi (RISTEKDIKTI) in doing research and development of this type of aircraft. The aircraft that was analyzed in this paper is the second prototype that was made. The first prototype had four propulsion systems and was made mainly of balsa wood. However, after the structural failure of the first prototype, the second one was made mainly by carbon fiber composite.

IOP Conf. Series: Journal of Physics: Conf. Series **1005** (2018) 012050 doi:10.1088/1742-6596/1005/1/012050

Figure 1 First Prototype of ITB HALE UAV

In addition to that, the second prototype seeks to reduce the number of propulsion system to save weight and improve stability. The design also reduces the number of tail section as can be seen from Figure 1 and Figure 2 in order to reduce drag while increasing wingspan to generate more lift. The other improvement is that this second prototype will be using solar panels to generate power during the day. However, this paper only discuss about the wake analysis from the wing, and longitudinal static stability of the aircraft.

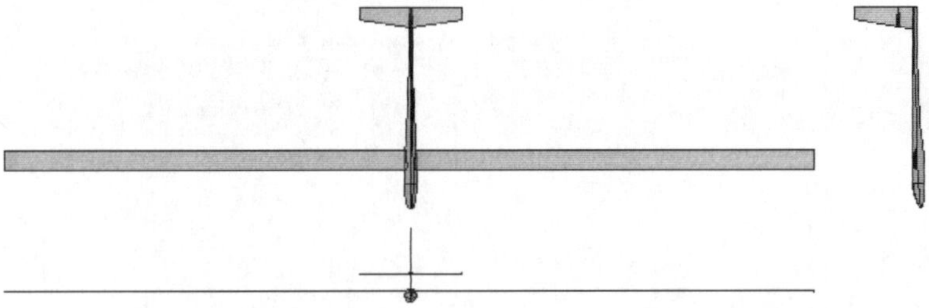

Figure 2 Second Prototype of ITB HALE UAV Design

Table 1 Second Prototype Specifications

MTOW	[kg]	20
Wingspan	[m]	16
MAC	[m]	0.4
Aspect Ratio		40

Even though this aircraft uses natural laminar airfoil EMX 07, it was expected that the aircraft would need a high angle of attack during climbing and this would provide the motivation to check the validity of inviscid assumptions. When the flow over the airfoil is dominated by viscous effect, then the boundary layer at the top would likely separate and form a large wake downstream [2]. This wake may affect the airflow near the tail [3]. Generally, the inviscid assumptions hold for high Reynolds number hence to simplify the problem, the CFD simulation was made at atmospheric condition at operating altitude where the density is lower. Since Reynolds number is related to density, this simplification would give a conservative result.

Lastly, longitudinal static stability will be analyzed using the software DATCOM on the final configuration of the aircraft. The static stability parameters and its derivative are taken from the output and the value will be crosschecked from theory to determine whether the aircraft is stable or not.

IOP Conf. Series: Journal of Physics: Conf. Series **1005** (2018) 012050 doi:10.1088/1742-6596/1005/1/012050

2. Methodology

2.1 Aircraft Overview

The aircraft cruising speed is 16 m/s with wing geometry as shown in Figure 3 (all dimensions in mm). It will operate at 5 km above sea level and the atmospheric properties would refer to International Standard Atmosphere[4]. The simulation would be done by varying the angle of attack to 0, 5, and 10 degrees.

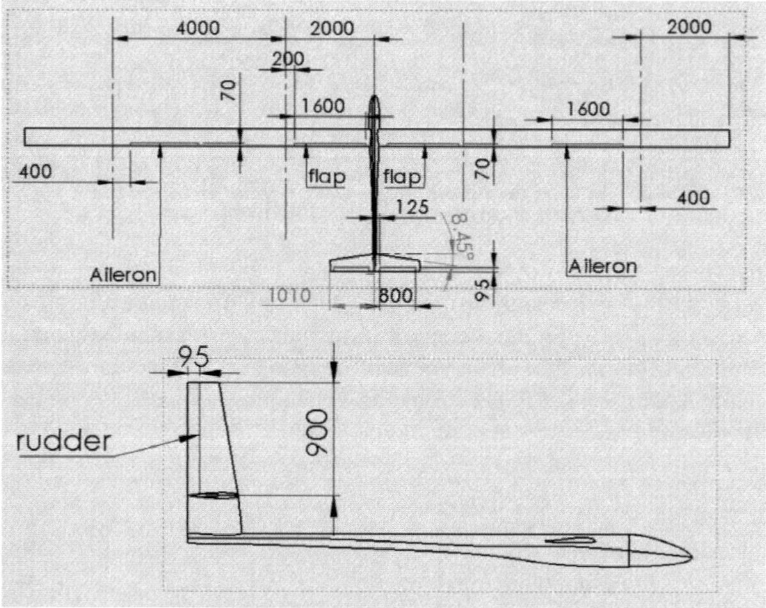

Figure 3 Second Prototype Dimensions

2.2 ANSYS CFX Solver

The equations that was used in ANSYS CFX solver is the unsteady Navier-Stokes equations in conservation form[5]. Below is presented the continuity equation, momentum equation, and total energy equation that was used by ANSYS CFX.

$$\frac{\partial \rho}{\partial t} + \nabla \cdot (\rho \boldsymbol{U}) = 0$$

$$\frac{\partial (\rho \boldsymbol{U})}{\partial t} + \nabla \cdot (\rho \boldsymbol{U} \times \boldsymbol{U}) = -\nabla p + \nabla \cdot \boldsymbol{\tau} + \boldsymbol{S_M}$$

$$\frac{\partial (\rho h_{\boldsymbol{tot}})}{\partial t} - \frac{\partial p}{\partial t} + \nabla \cdot (\rho \boldsymbol{U} h_{tot}) = \nabla \cdot (\lambda \nabla T) + \nabla \cdot (\boldsymbol{U} \cdot \boldsymbol{\tau}) + \boldsymbol{U} \cdot \boldsymbol{S_M} + \boldsymbol{S_E}$$

In order to solve this equation a fluid domain that was discrete must be created. The discretization process was called meshing and for this simulation, a structured mesh was created. The resulting mesh can be seen in the Figure below. Next, boundary and initial condition must be defined to initiate the simulation. In this simulation the wing is defined as a no slip wall where the velocity at the wall is zero, and the fluid is assumed to be incompressible.

IOP Conf. Series: Journal of Physics: Conf. Series **1005** (2018) 012050 doi:10.1088/1742-6596/1005/1/012050

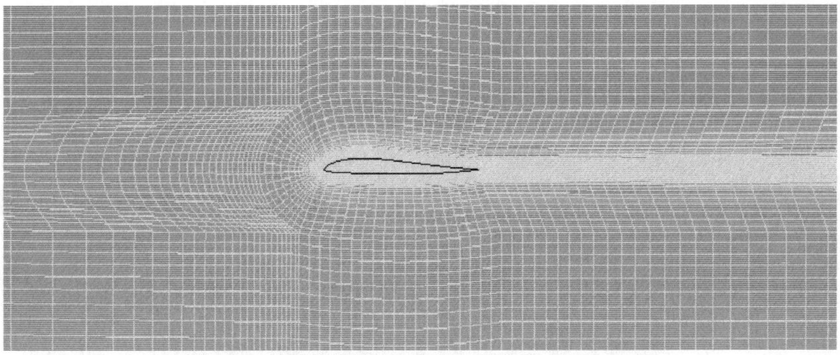

Figure 4 Side View of The Meshed Domain

2.3 Stability Parameters and DATCOM

An aircraft is said to be stable if the response of the aircraft due to small disturbance is to return to its original state [6]. To quantify this condition some aerodynamic parameters must be considered. These parameters were derived from an equilibrium equation where the aircraft was in trim condition [7].

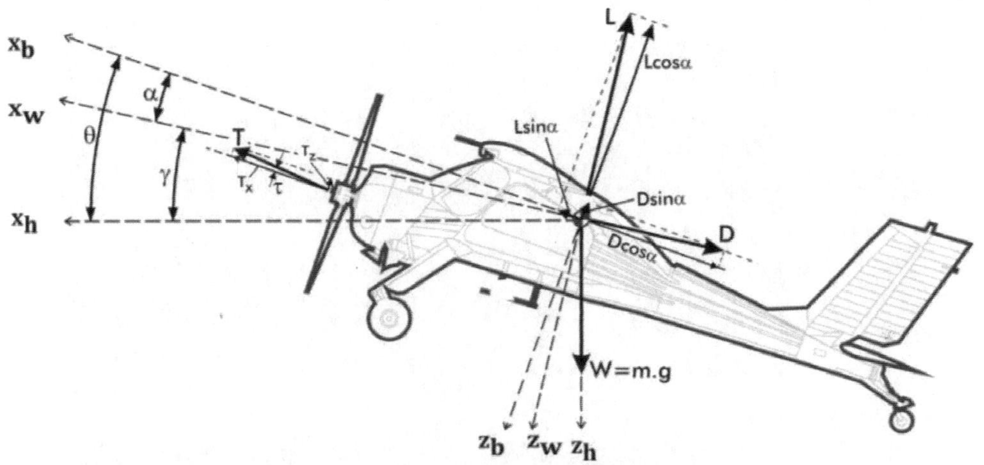

Figure 5 (Source: Reference 7)

Referring to the Figure above, the equilibrium equation is

$$T\cos(\alpha+\tau) - D - mg\sin\gamma = 0$$
$$-T\sin(\alpha+\tau) - L + mg\cos\gamma = 0$$
$$M_A + M_T = 0$$

where M_A denotes the moment due to aerodynamic forces and M_T the moment due to thrust.

Using some assumptions and non-dimensional conversion (details can be read from the references) it can be derived that if there are changes in the value of the variables in the equation, the aircraft would return to its equilibrium state if the conditions below is satisfied.

a. The change of pitching moment due to elevator deflection $C_{m_{\delta e}} < 0$
b. The change of pitching moment due to angle of attack $C_{m_\alpha} < 0$

The software that was used to find out the value of these parameters is DATCOM. DATCOM uses a semi-empirical approach by interpolating results from known parameters in a specific aircraft. The figure below shows the geometry that was generated to be analyzed in DATCOM.

IOP Conf. Series: Journal of Physics: Conf. Series **1005** (2018) 012050 doi:10.1088/1742-6596/1005/1/012050

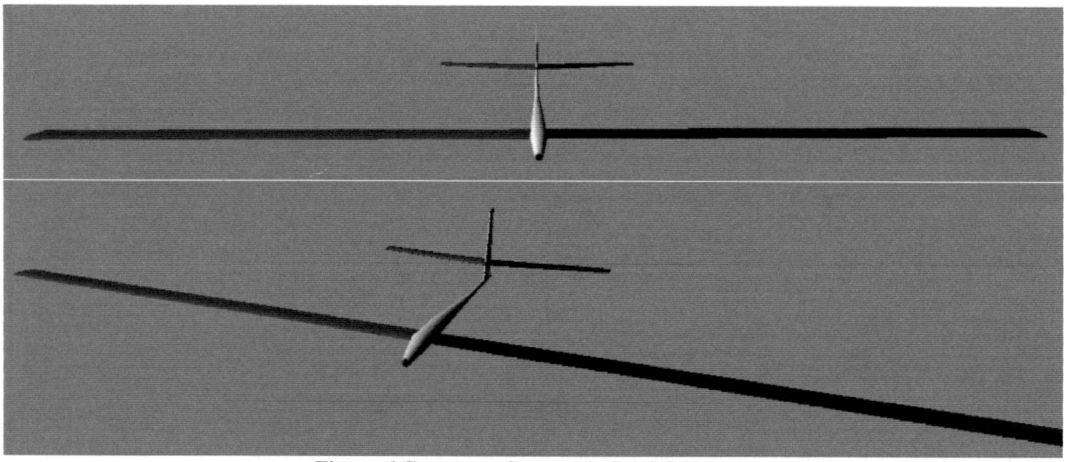

Figure 6 Geometry Created in DATCOM

3. Results and Discussions

3.1. Wake Visualization

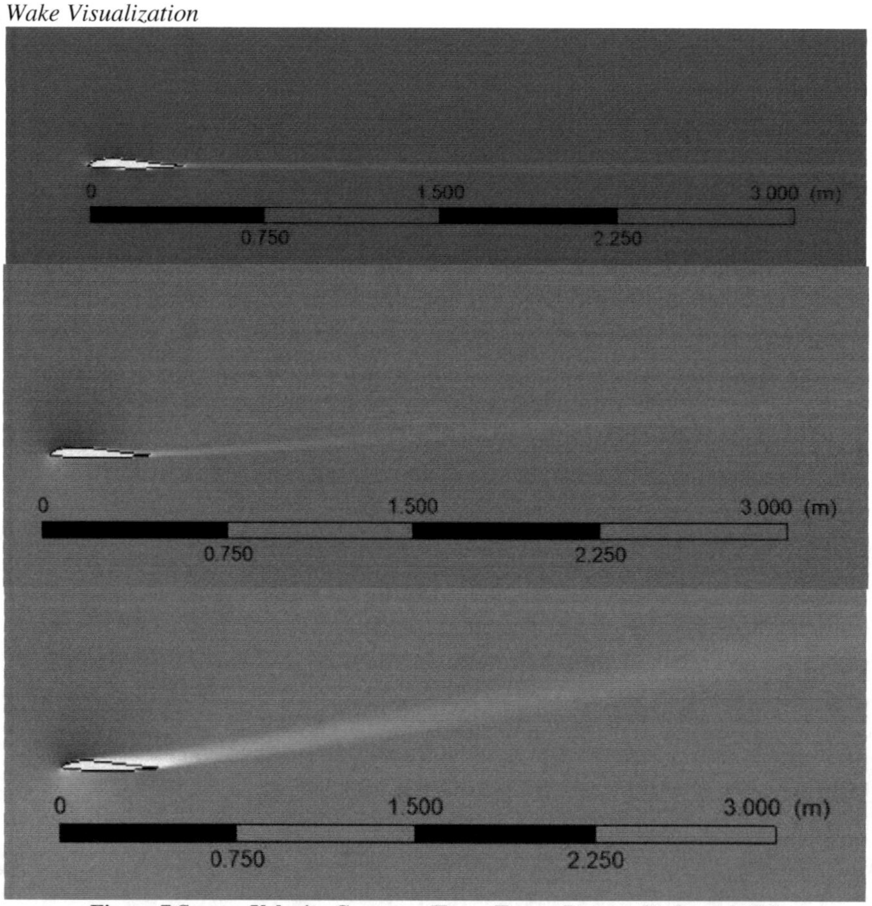

Figure 7 Stream Velocity Contour. (From Top to Bottom 0°, 5°, and 10°)

IOP Conf. Series: Journal of Physics: Conf. Series **1005** (2018) 012050 doi:10.1088/1742-6596/1005/1/012050

Figure 7 above shows the wake from the wing at different angle of attack. It can be seen that at 5° angle of attack the boundary layer thickens (shown by the white color at the upper surface of the wing in Figure 8), and at 10° angle of attack separation had already occurs at the thickest point of the airfoil. This shows that the wake produced when the aircraft was climbing (where the angle of attack is around 5°) can alter the flow downstream of the wing. However, the final configuration has horizontal tail placement that was not affected by the wake from 5° angle of attack.

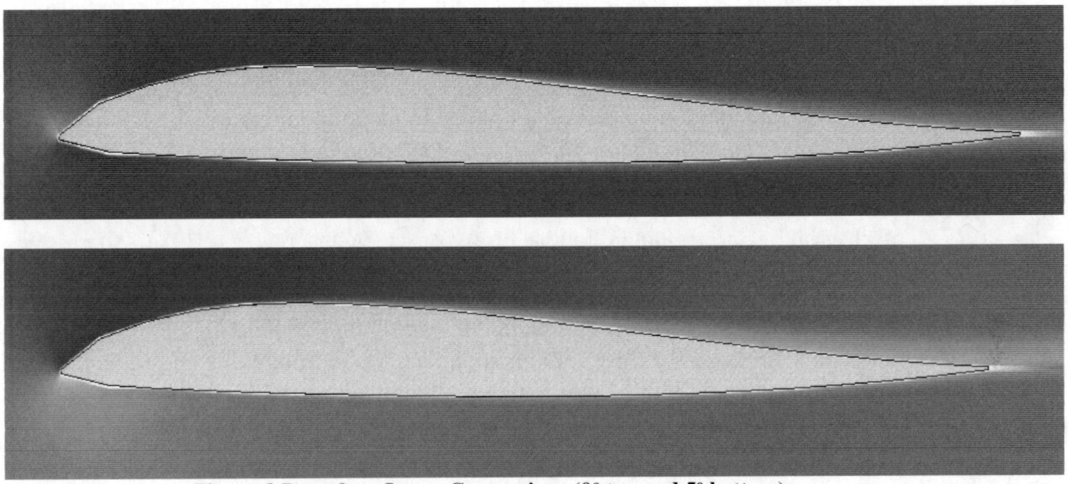

Figure 8 Boundary Layer Comparison (0° top and 5° bottom)

3.2. Stability Results

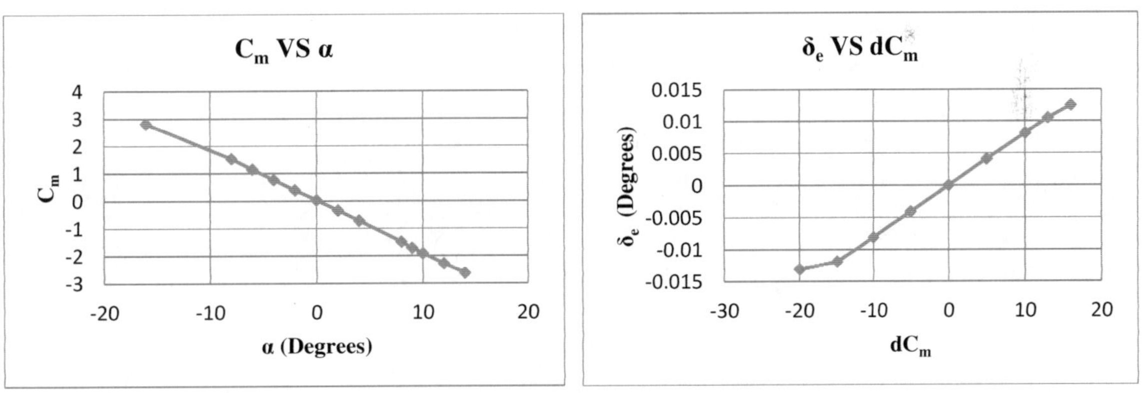

Figure 9 Stability Parameters Curve

The Figure above shows the curve that is taken from DATCOM output. The left figure show that C_{m_α} of the HALE is negative, which means if the angle of attack of the aircraft is increased, it will have a tendency to pitch down and vice versa. This is an indication that the aircraft is stable.

The right figure shows the output from DATCOM, which is the increment of pitching moment coefficient due to elevator deflection. Hence, to obtain the $C_{m_{\delta e}}$ parameter, an integration of the curve

IOP Conf. Series: Journal of Physics: Conf. Series **1005** (2018) 012050 doi:10.1088/1742-6596/1005/1/012050

must be carried out[8]. By using trapezoidal method, it is obtained that the $C_{m_{\delta e}} = -0.05$, which also indicates that the aircraft is stable.

4. Concluding remarks

The current design of the HALE prototype has good longitudinal static stability and the wake from the wing has been simulated to prove that it will not hinder the performance of the horizontal tail. However, flight test must be conducted to ascertain the result of analysis.

Acknowledgement
This research was supported by ITB and Direktorat Riset Teknologi dan Pendidikan Tinggi (RISTEKDIKTI).

References

[1] Romeo, G., G Frulla, and E Cestino. *Design of a High-Altitude Long-Endurance Solar-Powered Unmanned Air Vehicle for Multi-Payload and Operations*. Proceedings of the Institution of Mechanical Engineers, Part G: Journal of Aerospace Engineering, Vol 221 (2007) : 199-216.

[2] Anderson, John D. *Fundamentals of Aerodynamics*. McGraw-Hill Education, 2009.

[3] Jones, Robert T., and Leo F. Fehlner. *Transient Effects of The Wing Wake on The Horizontal Tail*. Washington: Technical Notes National Advisor Committee for Aeronautics, No 771, 1940.

[4] Anderson, John D., Jr.: *Introduction to Flight*, McGraw-Hill Book Company, Boston, 2008.

[5] ANSYS CFX Solver Theory Guide, 2008.

[6] McCormick, B.W. *Aerodynamics, Aeronautics, and Flight Mechanics*. New York: John Wiley & Sons, 1979.

[7] Muhammad, Hari and Yazdi Ibrahim Jenie . *Diktat Kuliah: Dinamika Terbang*. Bandung: Institut Teknologi Bandung, 2014.

[8] The USAF Stability and Control DATCOM Volume 1, User's Manual. Missouri: McDonnel Douglas Astronautics Company, 1999.

Institute of Physics
Dirac House, Temple Back
Bristol BS1 6BE UK

CURRAN ASSOCIATES INC.
proceedings.com

ISSN: 1742-6588
ISBN 978-1-5108-6286-9

9 781510 862869